Lecture Notes in Computer Science   1310
Edited by G. Goos, J. Hartmanis and J. van Leeuwen

Advisory Board: W. Brauer   D. Gries   J. Stoer

Springer
*Berlin
Heidelberg
New York
Barcelona
Budapest
Hong Kong
London
Milan
Paris
Santa Clara
Singapore
Tokyo*

Alberto Del Bimbo (Ed.)

Image Analysis and Processing

9th International Conference, ICIAP '97
Florence, Italy, September 17-19, 1997
Proceedings, Volume I

Springer

Series Editors

Gerhard Goos, Karlsruhe University, Germany
Juris Hartmanis, Cornell University, NY, USA
Jan van Leeuwen, Utrecht University, The Netherlands

Volume Editor

Alberto Del Bimbo
Università di Firenze, Dipartimento di Sistemi e Informatica
Via di Santa Marta, 3, I-50139 Firenze, Italy
E-mail: delbimbo@aguirre.ing.unifi.it

Cataloging-in-Publication data applied for

Die Deutsche Bibliothek - CIP-Einheitsaufnahme

Image analysis and processing : 9th international conference ;
proceedings / ICIAP '97, Florence, Italy, September 17 - 19, 1997.
Albert DelBimbo (ed.). - Berlin ; Heidelberg ; New York ; Barcelona
; Budapest ; Hong Kong ; London ; Milan ; Paris ; Santa Clara ;
Singapore ; Tokyo : Springer
   Literaturangaben
Vol. 1 (1997)
(Lecture notes in computer science ; Vol. 1310)
ISBN 3-540-63507-6

CR Subject Classification (1991): I.4, I.5, I.3.3, I.3.5, I.3.7, I.2.10

ISSN 0302-9743
ISBN 3-540-63507-6 Springer-Verlag Berlin Heidelberg New York

This work is subject to copyright. All rights are reserved, whether the whole or part of the material is
concerned, specifically the rights of translation, reprinting, re-use of illustrations, recitation, broadcasting,
reproduction on microfilms or in any other way, and storage in data banks. Duplication of this publication
or parts thereof is permitted only under the provisions of the German Copyright Law of September 9, 1965,
in its current version, and permission for use must always be obtained from Springer-Verlag. Violations are
liable for prosecution under the German Copyright Law.

© Springer-Verlag Berlin Heidelberg 1997
Printed in Germany

Typesetting: Camera-ready by author
SPIN 10551833   06/3142 – 5 4 3 2 1 0   Printed on acid-free paper

Message from the General Chair

This volume collects proceedings of ICIAP'97, September 17-19, 1997, Florence, Italy. ICIAP'97 is the ninth meeting of the International Conference on Image Analysis and Processing, organized biennially by the Italian Chapter of the International Association for Pattern Recognition (IAPR). Following the successful 1995 meeting in Sanremo, ICIAP'97 is held in the magnificent city of Florence, one of the most beautiful and famous cities in the world, renown for its artistic and cultural heritage. The 1997 ICIAP conference is one of the largest ever, with over 200 participants coming from almost every part of the world. This confirms the success of this initiative of the IAPR Italian Chapter, as well as the very good work carried out by the organizers of the previous ICIAP meetings.

We received a very large submission of 304 papers from 40 different countries, confirming the intense and ever growing activity in imaging technology research and development, worldwide. Papers covered basic research topics in image analysis, pattern recognition and computer vision, as well as applications of these technologies to real problems. Basic topics addressed included image enhancement, image segmentation, image compression, motion analysis, object recognition, image understanding, and special hardware architectures and systems. Applications were in the fields of biomedicine, character recognition, safety and surveillance, object identification and inspection, and quality control in manufacturing, among others. Growing and emerging research and application topics, such as image and video databases, vision-assisted man-machine interaction, and color image processing were also strongly represented.

The reviewing process resulted in the selection of 173 papers. Only papers that received high ranks by all the reviewers were accepted for presentation at ICIAP'97. Oral presentations were limited to 42, organized in 12 sessions. Four poster sessions included 131 papers. In setting the conference program, we favored large poster sessions to encourage interactivity between researchers and promote exchanges and the establishment of new links. We invited four distinguished speakers, Dr. Dragutin Petkovic, from IBM Almaden Research Center, Prof. Jake Aggarwal, from Texas University at Austin, Prof. Linda Shapiro, from Washington University in Seattle, and Prof. Ramesh Jain, from the University of California at San Diego, to predict the state of imaging technologies in 2000 and suggest research perspectives and trends for the near future. For the first time, ICIAP'97 hosts a special session devoted to successful ongoing or recently completed projects in image analysis and processing and computer vision, developed under European Community programs. A total of 9 poster presentations were accepted. Dr. Kostas Glinos, EU officer from DG III in Brussels, was invited to provide a view of forthcoming EU programs and opportunities in these fields for research and industry communities. This session was prepared in cooperation with APRE, the Florence Agency for European Research Development, and will hopefully stimulate interaction and technology transfer between research and industrial communities.

I would like to thank IAPR Italian Chapter for allowing us to organize this conference in Florence and IAPR for its sponsorship. Moreover, I gratefully thank Provincia di Firenze, and particularly its Vice-President Riccardo Conti, for their financial backing and sponsorship of this initiative and wise sensitivity in understanding our effort in this task. Thanks are also due to CESVIT SpA, Florence, and its President Sergio Bertini and General Manager Silvestro Mitolo; to Bassilichi Sviluppo SpA and its president Luca Bassilichi; to OTE SpA, Florence, and its President Carlo Lastrucci; to Logitron SpA, Florence, and its President Andrea Ripasarti; to SESA SpA, Empoli and its President Paolo Castellacci; as well as to the University of Florence and its Dean Prof. Paolo Blasi, and to the Italian National Council of Research, who all generously supported this event. I also thank Claudia Bianconi, local coordinator of the APRE, who greatly helped us in organizing the special session on EU projects together with the European Community.

An excellent program committee and their colleagues did great work in carefully reviewing an unexpectedly large number of papers, thus easing the task of selecting the best contributions. Their work is sincerely acknowledged. Special thanks go to Carlo Colombo and Pietro Pala, who made a fundamental, voluntary, contribution to this conference, helping in managing, working on, and resolving those many problems that a large event like this presents. All the student volunteers of the Visual Information Processing Laboratory at the University of Florence are also gratefully acknowledged. Finally, I thank Consulta Umbria Srl and its administrative staff, especially Simona Sarti and Giuseppina Meniconi, who assisted us in the organization of the conference and helped us in too many situations to be remembered here.

I wish to all delegates a very successful conference and hope that many new links will be established, and long lasting friendships will be set and reinforced.

Florence, July 1997   Alberto Del Bimbo

General Chair

Alberto Del Bimbo University of Florence, I

Program Chairs

Vito Cappellini University of Florence, I
Alberto Del Bimbo University of Florence, I

Program Committee

Carlo Arcelli CNR Arco Felice Naples, I
Carlo Braccini University of Genoa, I
Michael Brady University of Oxford, UK
Virginio Cantoni University of Pavia, I
Roberto Cipolla University of Cambridge, UK
Luigi P. Cordella University of Naples, I
James L. Crowley INPG Grenoble, F
Leila De Floriani University of Genoa, I
Ernst Dickmanns Universität Bundeswehr München, D
Vito Di Gesù University of Palermo, I
Marco Ferretti University of Pavia, I
Herbert Freeman Rutgers University, USA
Giovanni Garibotto ELSAG BAILEY, Genoa, I
Marco Gori University of Siena, I
Concettina Guerra University of Padoa, I
Sebastiano Impedovo University of Bari, I
Anil K. Jain Michigan State University, USA
Xiaoyi Jiang University of Bern, CH
Josef Kittler University of Surrey, UK
Walter Kropatsch Technical University of Vienna, A
Stefano Levialdi University of Roma, I
Piero Mussio University of Brescia, I
Dragutin Petkovic IBM Almaden, USA
Matti Pietikäinen University of Oulu, SF
Vito Roberto University of Udine, I
Masao Sakauchi University of Tokyo, J
Alberto Sanfeliu Universitat Politecnica de Catalunya, E
Gabriella Sanniti di Baja CNR Arco Felice Naples, I
Jorge L.C. Sanz IBM Argentina, ARG
Linda G. Shapiro University of Washington, USA
Arnold W.M. Smeulders University of Amsterdam, NL
Renato Stefanelli Politecnico di Milano, I
Anastasios N. Venetsanopoulos University of Toronto, CAN
Gianni Vernazza University of Cagliari, I
Juan José Villanueva Universidad Autonoma de Barcelona, E
Sergio Vitulano University of Cagliari, I
Hezy Yeshurun Tel Aviv University, IL
Bertrand Zavidovique Université Paris XI, F

Local Organizing Committee

Luciano Alparone	University of Florence, I
Stefano Baronti	IROE-CNR Florence, I
Carlo Colombo	University of Brescia, I
Jacopo M. Corridoni	University of Florence, I
Alberto Del Bimbo	University of Florence, I
Marco Lusini	University of Florence, I
Pietro Pala	University of Florence, I
Enrico Vicario	University of Florence, I

Sponsored by:

IAPR – Italian Association for Pattern Recognition
DSI – Dipartimento Sistemi e Informatica, Università degli Studi di Firenze

Supported by:

Università degli Studi di Firenze
CNR – Consiglio Nazionale delle Ricerche
Provincia di Firenze
CESVIT SpA – Firenze
APRE – Firenze
Bassilichi Sviluppo SpA – Firenze
Logitron SpA – Firenze
OTE SpA – Firenze
SESA SpA – Empoli

Table of Contents – Volume I

Keynote Address

Challenges and Opportunities for Pattern Recognition and Computer Vision Research in Year 2000 and Beyond 1
D. Petkovic

Session 1: Segmentation

Multiscale Gradient Magnitude Watershed Segmentation 6
O.F. Olsen, M. Nielsen

Segmentation of Multispectral Images of Works of Art through Principal Component Analysis 14
S. Baronti, A. Casini, F. Lotti, S. Porcinai

Session 2: Image Analysis & Pattern Recognition

Multiscale Edge Detection via Normal Changes 22
C.-J. Sze, H.-Y.M. Liao, H.-L. Hung, K.-C. Fan, J.-W. Hsieh

Extending Adjacency to Fuzzy Sets for Coping with Imprecise Image Objects 30
I. Bloch, H. Maître

Adaptive Selection of Image Classifiers 38
G. Giacinto, F. Roli

Classification Reliability and Its Use in Multi-classifier Systems 46
L.P. Cordella, P. Foggia, C. Sansone, F. Tortorella, M. Vento

Poster Session A: Color & Texture, Enhancement, Image Analysis & Pattern Recognition, Segmentation

Color Linear Model 54
C.-Y. Kim, Y.-S. Seo, I.-S. Kweon

A Computational Approach to Color Illusions 62
D. Marini, A. Rizzi

Improved Textured Images Segmentation Using an Energy Functional 70
A. Grau, J. Saludes

Contribution to the Colour Segmentation by Means of an Algorithm Which Reduces the CCDs Saturation Problems 79
J. Regincós Isern, J. Batlle Grabulosa

Pyramid-Based Multi-sensor Image Data Fusion with Enhancement of Textural Features 87
B. Aiazzi, L. Alparone, S. Baronti, V. Cappellini, R. Carlà, L. Mortelli

Texture Analysis Using Pairwise Interaction Maps 95
D. Chetverikov

Estimation of the Color Image Gradient with Perceptual Attributes 103
P. Pujas, M.-J. Aldon

Contour Line Extraction from Color Images of Scanned Maps 111
M. Lalonde, Y. Li

Subjective Analysis of Edge Detectors in Color Image Processing 119
P. Androutsos, D. Androutsos, K.N. Plataniotis, A.N. Venetsanopoulos

Similarity Measures for Binary and Grey Level Markov Random Field Textures 127
A. Çarkacioğlu, F.T. Yarman-Vural

A Simple and Effective Edge Detector 134
C. Cafforio, E. Di Sciascio, C. Guaragnella, G. Piscitelli

Improvements to Image Magnification 142
A. Biancardi, L. Lombardi, V. Pacaccio

Refining Surface Curvature with Relaxation Labeling 150
R.C. Wilson, E.R. Hancock

Dynamic Scale-Space Theories 158
A.H. Salden

Reconstructing Digital Sets from X-Rays 166
E. Barcucci, A. Del Lungo, M. Nivat, R. Pinzani, A. Zurli

Pattern Recognition from Compressed Labelled Trees of Fuzzy Regions .. 174
L. Wendling, J. Desachy, A. Paries

Optimality Analysis of Edge Detection Algorithms for Range Images 182
X. Jiang

Analysis Situs and Image Processing 190
F. Sloboda, B. Zat'ko

Defining Cost Functions and Profitability Measures for Digraphs Associated with Raster Dems 198
P. Matsakis, J. Gadiou, J. Desachy

Using Proximity and Spatial Homogeneity in Neighbourhood-Based Classifiers 206
J.S. Sánchez, F. Pla, F.J. Ferri

Image Segmentation by Means of Fuzzy Entropy Measure 214
C. Di Ruberto, M. Nappi, S. Vitulano

Efficient Region Segmentation through "Creep-and-Merge" 223
A. Basman, J. Lasenby, R. Cipolla

An Automatic Transformation from Bimodal to Pseudo-Binary Images ... 231
J.M. Iñesta, P.J. Sanz, Á.P. del Pobil

A New Deformable Model for 3D Image Segmentation 239
Z. Zhang, M. Braun, P. Abbott

Evolutionary Image Segmentation 247
P. Zingaretti, A. Carbonaro, P. Puliti

Discontinuity Adaptive MRF Model for Synthetic Aperture Radar Image
Analysis 255
P.C. Smits, S.G. Dellepiane, G. Vernazza

Region Growing Euclidean Distance Transforms 263
O. Cuisenaire

COP: A New Method for Extracting Edges and Corners 271
S.C. Bae, I.S. Kweon

An Integrated Approach for Segmentation and Representation of Range
Images 279
O.R.P. Bellon, C.L. Tozzi

Session 3: Segmentation & Coding

Two-Dimensional Fractal Segmentation of Natural Images 287
V. Anh, J. Maeda, T. Ishizaka, Y. Suzuki, Q. Tieng

Fast Segmentation of Range Images 295
M. Haindl, P. Žid

Image Compression Based on Centipede Model 303
B. Kurt, M. Gökmen, A.K. Jain

Session 4: Color & Texture

Unsupervised Texture Segmentation Using Feature Distributions 311
T. Ojala, M. Pietikäinen

Color Based Object Recognition 319
T. Gevers, A.W.M. Smeulders

Color Texture Classification by Wavelet Energy Correlation Signatures ... 327
G. Van de Wouwer, S. Livens, P. Scheunders, D. Van Dyck

Cross-Media Color Matching Using Neural Networks 335
E. Boldrin, R. Schettini

Keynote Address

Object Recognition and Performance Bounds 343
J.K. Aggarwal, S. Shah

Session 5: Shapes & Surfaces

Relating Image Warping to 3D Geometrical Deformations 361
A.L. Yuille, M. Ferraro, T. Zhang

Using Top-Down and Bottom-Up Analysis for a Multiscale Skeleton
Hierarchy 369
G. Borgefors, G. Ramella, G. Sanniti di Baja

A New Algorithm for 3D Profilometry Based on Phase Measurement 377
L. Di Stefano, F. Boland

Keynote Address

Surface Modeling and Display from Range and Color Data 385
*K. Pulli, M. Cohen, T. Duchamp, H. Hoppe, J. McDonald, L. Shapiro,
W. Stuetzle*

Session 6: Matching & Recognition

An Improved Active Shape Model: Handling Occlusion and Outliers 398
N. Duta, M. Sonka

Perspective Matching Using the EM Algorithm 406
A.D.J. Cross, E.R. Hancock

Identifying Human Face Profiles with Semi-Local Integral Invariants 414
J. Sato, R. Cipolla

Poster Session B: Active Vision, Motion, Shape, Stereo

Adaptive Fovea Structures for Space-Variant Sensors 422
P. Camacho, F. Arrebola, F. Sandoval

Structural Characterization of Image Processing Operators 430
P. Bottoni, L. Cinque, S. Levialdi, P. Mussio, B. Nebbia

Easy Calibration of Pan/Tilt Camera Heads and Online Computation of the
Epipolar Cerrespondences 438
S. Spiess, M. Li

Integration of Spatio-Temporal Information for Motion Detection by Means of
Fuzzy Reasoning 446
M. Barni, F. Bartolini, V. Cappellini, F. Lambardi

Adaptive Motion Estimation and Video Vector Quantization Based on
Spatiotemporal Non-linearities of Human Perception 454
J. Malo, F. Ferri, J. Albert, J.M. Artigas

Integral Based Approach for Determining Motion Vector Fields 462
A. Nomura

A Practical Algorithm for Structure and Motion Recovery from Long Sequence
of Images 470
M. Trajković, M. Hedley

Object Pose by Affine Iterations 478
F. Dornaika, C. Garcia

Robust Motion Estimation Using Chrominance Information in Color Image
Sequences 486
J. Magarey, A. Kokaram, N. Kingsbury

Temporal Prediction of Video Sequences Using an Image Warping Technique
Based on Color Segmentation 494
N. Herodotou, A.N. Venetsanopoulos

Motion and Intensity-Based Segmentation and Its Application to Traffic
Monitoring 502
J. Badenas, M. Bober, F. Pla

A Geometrically Deformable Contour Model 510
A. Raji, E. Petit, J. Lemoine, S. Djeziri

Non-visible Deformations 519
J.-D. Durou, L. Mascarilla, D. Piau

Two-Step Parameter-Free Elastic Image Registration with Prescribed Point
Displacements 527
W. Peckar, C. Schnörr, K. Rohr, H.S. Stiehl

Learning for Feature Selection and Shape Detection 535
R. Cucchiara, M. Piccardi, M. Bariani, P. Mello

Experiments on the Decomposition of Arbitrarily Shaped Binary Morphological
Structuring Elements 543
G. Anelli, A. Broggi, G. Destri

Bézier Modelling of Cracks 551
A. Varley, P. Rayner

An Adaptive Deformable Template for Mouth Boundary Modeling 559
A.R. Mirhosseini, K.-M. Lam, H. Yan

A Two-Stage Framework for Polygon Retrieval Using Minimum Circular Error
Bound 567
L.H. Tung, I. King

Topology and Shape Preserving Parallel Thinning for 3D Digital Images –
A New Approach 575
P.K. Saha, D.D. Majumder

Convergence of Model Based Shape from Shading 582
M.S. Lew, M. Chaudron, N. Huijsmans, A. She, T.S. Huang

Quantitative Assessment of Two Skeletonization Algorithms Adapted to Rectangular Grids 588
M. Ciuc, D. Coquin, P. Bolon

An Algorithm for the Global Solution of the Shape-from-Shading Model .. 596
M. Falcone, M. Sagona

A Statistical Classification Method for Hierarchical Irregular Objects 604
M.Peura

Multi-level Dynamic Programming for Axial Motion Stereo Line
Matching 612
R.K.K. Yip

Analysis of Grey-Level Features for Line Segment Stereo Matching 620
O. Schreer, I. Hartmann, R. Adams

3D Object Positioning from Monocular Image Brightnesses 628
T. Shioyama, H.Y. Wu, W.B. Jiang, S. Terauchi

Camera Calibration Based on 3D-Point-Grid 636
X.-F. Zhang, A. Luo, W. Tao, H. Burkhardt

A Geometric Modeling Tool for Stereo-Matching and Reconstruction of a Model
of 3D-Scene 644
L. Sommellier, E. Tosan, D. Vandorpe

Session 7: Motion & Stereo

Estimating Translation/Deformation Motion through Phase Correlation .. 653
F. Pla, M. Bober

Robust Fitting of 3D CAD Models to Video Streams 661
C. Meilhac, C. Nastar

Experiments with a New Area-Based Stereo Algorithm 669
A. Fusiello, V. Roberto, E. Trucco

Adaptive Stereo Matching in Correlation Scale-Space 677
C. Menard, W.G. Kropatsch

Hierarchical Depth Mapping from Multiple Cameras 685
J.-I. Park, S. Inoue

Session 8: Recognition

Fast Computation of Error-Correcting Graph Isomorphisms Based on Model Precompilation 693
B.T. Messmer, H. Bunke

Function-Described Graphs Applied to 3D Object Representation 701
F. Serratosa, A. Sanfeliu

Cooperative Vision in a Multi-Agent Architecture 709
N. Oswald, P. Levi

Author Index 717

Table of Contents – Volume II

Keynote Address

Content-Centric Computing in Visual Systems 1
R. Jain

Session 9: Image Databases

Color Image Retrieval Fitted to "Classical" Querying 14
J. Martinez, S. Guillaume

Quality Measures for Interactive Image Retrieval with a Performance Evaluation of Two 3 × 3 Texel-based Methods 22
D.P. Huijsmans, M.S. Lew, D. Denteneer

Holographic Image Representations: The Fourier Transform Method 30
A.M. Bruckstein, R.J. Holt, A.N. Netravali

Image Databases Are Not Databases with Images 38
S. Santini, R. Jain

Poster Session C: Compression, Hardware & Software, Image Databases, Neural Networks, Object Recognition & Reconstruction

Customizing MPEG Video Compression Algorithms to Specific Application Domains: The Case of Highway Monitoring 46
N. Zingirian, P. Baglietto, M. Maresca, M. Migliardi

A New Lossless Image Compression Algorithm Based on Arithmetic Coding 54
B. Carpentieri

Analysis of a Two Step MPEG Video System 62
L. Teixeira

Dedicated Hardware Processors for a Real-Time Image Data Pre-processing Implemented in FPGA Structure 69
K. Wiatr

Wavelet Transform Architectures: A System Level Review 77
M. Ferretti, D. Rizzo

Lossless Compression of Pre-press Images Using a Novel Colour Decorrelation Technique 85
S. Van Assche, W. Philips, I. Lemahieu

Real Time Hardware Architecture for Visual Robot Navigation 93
F. Marino, E. Stella, N. Veneziani, A. Distante

Speeding Up Fractal Encoding of Images Using a Block Indexing
Technique 101
R. Distasi, M. Nappi, S. Vitulano

Adding Associative Meshes to the PACCO I.P. Environment 109
A. Biancardi, A. Mérigot

Smoothing of MPEG Multi-program Video Coding for Packet Networks .. 117
L. Teixeira, T. Andrade

Audio-visual Processing for Scene Change Detection 124
C. Saraceno, R. Leonardi

Weighted Walkthroughs in Retrieval by Content of Pictorial Data 132
E. Vicario, W.X. He

A New Approach to Computation of Curvature Scale Space Image for Shape
Similarity Retrieval 140
F. Mokhtarian, S. Abbasi, J. Kittler

Optimal Keys for Image Database Indexing 148
M.S. Lew, D.P. Huijsmans, D. Denteneer

The Terminological Image Retrieval Model 156
C. Meghini, F. Sebastiani, U. Straccia

Novel Block Truncation Coding of Image Sequences for Limited-Color
Display 164
S.-C. Pei, C.-M. Cheng

Image Registration with Shape Mixtures 172
S. Moss, E.R. Hancock

Image Retrieval by Color Regions 180
A. Del Bimbo, M. Mugnaini, P. Pala, F. Turco, L. Verzucoli

Interactive Model-Based Matching Retrieval 188
L. Cinque, S. Levialdi, A. Malizia, R. Mancini

Where Are the Ball and Players? Soccer Game Analysis with Color Based
Tracking and Image Mosaick 196
Y. Seo, S. Choi, H. Kim, K.-S. Hong

Histogram Families for Color-Based Retrieval in Image Databases 204
C. Colombo, A. Rizzi, I. Genovesi

Image Retrieval by Multidimensional Elastic Matching 212
P. Pala, S. Santini

Optimization Methods in Multilayer Classifier Networks for Automatic Control
of Lamellibranch Larva Growth 220
G.G. Vass, M. Daoudi, F. Ghorbel

Neural Networks for Region Detection 228
G. Cucurachi, G. Tascini, F. Piazza

Static and Dynamic Attractors of Auto-associative Neural Networks 238
D.O. Gorodnichy, A.M. Reznik

A Brain-Like Approach to Multistage Hierarchical Image Processing 246
L.I. Timchenko, Y.F. Kutaev, M.A. Grudin, S.V. Cepornyuk, D.M. Harvey, A.A. Gertsiy

Contextual Edge Detection Using a Recurrent Neural Network 254
A.J. Pinho, L.B. Almeida

A Divide-and-Conquer Strategy in Recovering Shape of Book Surface from Shading 262
S.I. Cho, H. Saito, S. Ozawa

Reconstruction of 3D Shape and Texture by Active Rangefinding 270
Y. Sato, T. Ishikawa, M. Otsuki

Wavelets for Multiresolution Shape Recognition 276
M.G. Albanesi, L. Lombardi

Invariant Object Representation and Recognition Using Lie Algebra of Perceptual Vector Fields 284
J. Chao, A. Karasudani, K. Minowa

A Fast Approach for Determining of Visibility of 3D Object's Surfaces ... 292
N.M. Sirakov

Direct Aspect-Based 3D Object Recognition 300
M. Pontil, A. Verri

Visualizing Parametric Surfaces at Variable Resolution 308
L. De Floriani, P. Magillo, E. Puppo

Learning Visual Ideals 316
M. Burge, W. Burger

Session 10: Recognition & Reconstruction

Adaptive Non-cartesian Networks for Vision 324
J.R. Serra, J.B. Subirana

Adaptive Logic Networks for Facial Feature Detection 332
D.O. Gorodnichy, W.W. Armstrong, X. Li

An Appearance-Based Approach to Gesture-Recognition 340
J. Martin, J.L. Crowley

Exponential Vector Field Tomography 348
K. Stråhlén

Poster Session D: Biomedical Applications, Detection, Control & Surveillance, Inspection, Optical Character Recognition

Detection of Rib Shadows in Digital Chest Radiographs 356
S. Sarkar, S. Chaudhuri

A Markov Random Field Model for Bony Tissue Classification 364
J.M. Pardo, D. Cabello, J. Heras

A New Methodology to Automatically Segment Biomedical Images 372
A. Garrido, N. Pérez De La Blanca, M. García-Silvente

Histogram-Based Image Registration for Digital Subtraction Angiography 380
T.M. Buzug, J. Weese, C. Lorenz, W. Beil

A Method for Segmentation of CT Head Images 388
S. Lončarić, D. Kovačević

Automatic Recognition of Spicules in Mammograms 396
H. Jiang, W. Tiu, S. Yamamoto, S.-i. Iisaku

Specialized Environment for Medical Radiological Image Visualization 404
V. Di Lecce, A. Guerriero

Interactive Segmentation of 3D Ultrasound Using Deformable Solid Models and Active Contours 412
C.R. Dance, M.H. Syn, R.W. Prager, J.P.M. Gosling, L.H. Berman, K.J. Dalton

Computer Aided Diagnosis System for Lung Cancer Based on Helical CT Images 420
Y. Kawata, K. Kanazawa, S. Toshioka, N. Niki, H. Satoh, H. Ohmatsu, K. Eguchi, N. Moriyama

A Generalized Geometry and Intensity Based Partial Volume Correction for Magnetic Resonance Images 428
F. Bello, A.C.F. Colchester, S.A. Röll

A Regularization Method for Unfolding the Measured Data of Different X-Ray Spectrometers in Compton Scattering Tomography 436
C. Bonifazzi, G. Maino, A. Tartari

Fast Tissue Segmentation Based on a 4D Feature Map: Preliminary Results 445
S. Vinitski, T. Iwanaga, C. Gonzalez, D. Andrews, R. Knobler, J. Mack

Texture Features in the Classification of Melanocytic Lesions 453
J. Kontinen, J. Röning, R.M. MacKie

Segmentation of Sputum Color Image for Lung Cancer Diagnosis Based on Neural Networks 461
S. Rachid, N. Niki, H. Nishitani, S. Nakamura, S. Mori

Fast Face Detection via Morphology-Based Pre-processing 469
C.-C. Han, H.-Y.M. Liao, G.-J. Yu, L.-H. Chen

Generalization of Shifted Fovea Multiresolution Geometries Applied to Object Detection 477
F. Arrebola, P. Camacho, F. Sandoval

A Long Term Change Detection Method for Surveillance Applications 485
C.S. Regazzoni, A. Teschioni, E. Stringa

Automatic Pedestrian Recognition Using Real-time Motion Analysis 493
P. Vannoorenberghe, C. Motamed, J.-M. Blosseville, J.-G. Postaire

Person Identification System Based on a Trapezoid Pyramid Architecture of a Grey-Level Image 501
M. Kosugi, K. Yamashita

Autonomous Plant Inspection and Anomaly Detection 509
M. Gregori, L. Lombardi, M. Savini, A. Scianna

Nobel Chile Jalapeño Sorting Using Structured Laser and Neural Networks Classifiers 517
F. Hahn, R. Mota

Developement of Image Processing Technique for Detection of the Rescue Target in the Marine Casualty 524
T. Sumimoto, K. Kuramoto, S. Okada, H. Miyauchi, M. Imade, H. Yamamoto, T. Kunishi

Bimodal Histogram Transformation Based on Maximum Likelihood Parameter Estimates in Univariate Gaussian Mixtures 532
N. Schultz, J.M. Carstensen

A Robust Structural Fingerprint Restoration 544
M.H. Ghassemian Yazdi

A System for the Automatic and Real Time Recognition of V.L.P.'s (Vehicle Licence Plate) 552
X.F. Hermida, F.M. Rodríguez, J.L.F. Lijò, F.P. Sande, M.P. Iglesias

Spatial Correlation Features for SAR Images in a Small Sample Size Context 560
R. Vaccaro, S. Dellepiane

Combination of Active Sensing and Sensor Fusion for Collision Avoidance in Mobile Robots 568
T.C.H. Heng, Y. Kuno, Y. Shirai

Underwater Vegetation Detection in High Frequency Sonar Images: A Preliminary Approach 576
R. Bozzano, A. Siccardi

Leather Inspection through Singularities Detection Using Wavelet
Transforms 584
A. Branca, M.G. Abbate, F.P. Lovergine, G. Attolico, A. Distante

Zoning Design for Handwritten Numeral Recognition 592
G. Dimauro, S. Impedovo, G. Pirlo, A. Salzo

Improving the Use of Contours and Skeletons for Off-Line Cursive Script
Segmentation 600
A. Chianese, M. De Santo, A. Picariello

Optical Character Recognition Without Segmentation 608
M.A. Özdil, F.T. Yarman-Vural, N. Arica

Automatic Recognition of Printed Arabic Text Using Neural Network
Classifier 616
A. Amin, M. Kavianifar

A Novel Pair-Wise Recognition Scheme for Handwritten Characters in the
Framework of a Multi-expert Configuration 624
A.F.R. Rahman, M.C. Fairhurst

A General and Flexible Deskewing Method Based on Generalized
Projections 632
E. Del Ninno, G. Nicchiotti, E. Ottaviani

Logical Structure Analysis by Typographic Characteristics Extraction 639
L. Duffy, F. Lebourgeois, H. Emptoz

Combining High-Level Features with Sequential Local Features for On-Line
Handwriting Recognition 647
J. Hu, A.S. Rosenthal, M.K. Brown

Handwritten Chinese Character Recognition Using Displacement Extraction
Based on Directional Features 655
Y. Mizukami, K. Koga

Session 11: Biomedical Applications

Optical Image Acquisition, Analysis and Processing for Biomedical
Applications 663
D.L. Farkas, B.T. Ballou, C. Du, G.W. Fisher, C. Lau, R.M. Levenson

Segmentation of Ultrasound Image Data by Two Dimensional Autoregressive
Modelling 672
P. Abbott, M. Braun

Comparison and Application of Selected Statistical Shape Models in Medical
Imaging 680
A. Neumann, C. Lorenz

Two Motion Detection Algorithms for Projection-Reconstruction Magnetic
Resonance Imaging: Theory and Experimental Verification 688
R. Van de Walle, I. Lemahieu, E. Achten

Session 12: Miscellaneous Applications

Image Analysis and Synthesis Using Physics-Based Modeling for Pearl Quality Evaluation System 697
N. Nagata, T. Dobashi, Y. Manabe, T. Usami, S. Inokuchi

Computer Vision and Image Processing in Postal Automation 705
G. Garibotto, C. Scagliola

Adaptive Pen User Interface with Supervised Competitive Learning 713
T.D. Kimura

Special Session on European Projects

Joint Detection, Interpolation, Motion and Parameter Estimation for Image Sequences with Missing Data 719
A.C. Kokaram, S.J. Godsill

The VIRSBS Project: Visual Intelligent Recognition for Secure Banking Services 727
M. Tistarelli, E. Grosso, J. Bigun, C. Sacerdoti, J. Santos-Victor, D. Vernon

The ESPRIT LTR Research Project: "Nonlinear Model-Based Analysis and Description of Images for Multimedia Applications (NOBLESSE)" 735
V. Pahor, G. Ramponi, R. Castagno

Analysis and Segmentation of Remote Sensing Images for Land-Cover Mapping 743
P.C. Smits, S.B. Serpico

Integration of Optical and Acoustical Imaging Sensors for Underwater Applications 749
G.G. Pieroni, G.L. Foresti, V. Murino

Project CROMATICA 757
L. Khoudour, J.P. Deparis, J.L. Bruyelle, F. Cabestaing, D. Aubert, S. Bouchafa, S.A. Velastin, M.A. Vincencio-Silva, M. Wherett

The COMPARES Project: COnnectionist Methods for Preprocessing and Analysis of REmote Sensing Data 765
J. Austin, G. Giacinto, I. Kanellopoulos, K. Lees, F. Roli, G. Vernazza, G. Wilkinson

Reconstruction of Severely Degraded Image Sequences 773
A.C. Kokaram

The CRASH Project: Defect Detection and Classification in Ferrite Cores 781
M. Mari, C. Dambra, D. Chetverikov, J. Verestoy, A. Jozwik, M. Nieniewski, L. Chmielewski, M. Sklodowski, W. Cudny, M. Lugg

Author Index 789

Challenges and Opportunities for PR&CV Research in Year 2000 and Beyond

Dr. Dragutin Petkovic

IBM Almaden Research Center
San Jose CA USA

1 Introduction

In this short position paper we attempt to point out to Computer Vision, Pattern Recognition and Artificial Intelligence (AI) communities some important and challenging opportunities for future research, development, and applications related to visual information. The emphasis is on both "challenging and important", since we believe both need to be addressed in any successful R&D effort. This paper is intended for researchers, directors of R&D programs and young researchers planning their research programs. Paper is based on authors experience ranging from biomedical to industrial computer vision, multimedia information retrieval and user interfaces, both as a researcher and as a product manager. In addition, this paper is based on authors extensive communication with IBM colleagues, with academic colleagues, especially at MIT Media Lab (Prof. Roz Picard and Sandy Pentland), and with "customers" i.e. users of such technologies in manufacturing and media industry. At the end, we suggest some readings and Web sites simply as examples of interesting work known to the author, and certainly not as an exclusive list.

We all know that majority of human communication is based on visual information. However, there were two basic impediments so far to make machines an effective partner in this information processing: a) our basic knowledge how this process is done; and b) lack of adequate technology for processing, storage, capture and transmission. Last few years the technology situation improved rapidly. We have increasingly powerful processors, recently getting specialized instruction sets for signal/image processing (i.e. Intel Pentium with MMX). Such processors progressed so rapidly that making special purpose and expensive vision chips is not cost effective any more. On the storage side, we envision laptops having 10 GBytes disk space soon, with 128 Mbyte RAM memory. Tape/optical/disk libraries can manage Terabytes of data economically. Recent DVD standard for video CDRoms promises full featured movies (2 hours of digital video) on one CDRom. Networks are also improving dramatically. Another big change on the hardware and technology front was of course Internet. It finally made it clear to all commercial companies that they can reach all sorts of users, thus significantly speeding up the conversion of their main business data (including images and video) into digital form. Promise of Internet is one of the biggest changing factors where our community will have to play - it will drive both software, hardware and algorithms and the way computers are used. In addition, desktop machines increasingly are becoming equipped with sensors like cameras, scanners and video cameras. Digital cameras are beginning to proliferate making it feasible to create a lot if images and video in digital (even MPEG compressed) form.

The above changes will enable radically new environment where visual data will be easily and economically available in all areas of human endeavor, and therefore will also be sought after and used by wide variety of the users. This is the key driving factor which our community will benefit from - without it our work would be largely relegated to narrow areas. Unless political situation in the word changes, the key driving factor for R&D will be commercial, not government funding. Therefore, our community has to look much more closely into these commercial applications in the future.

While the progress on technology was immense, our progress in basic understanding of image analysis and intelligence is still very moderate. While we have to continue to push for basic understanding of these processes, we also have to seize the opportunity new technology offers and look into what our community can contribute both in the near future (2-3 years) and long term (3-10 years). After all, Web years are equivalent to 3 months nowadays!

So far, majority of our focus was the "ultimate goal": full analysis and recognition from images, nothing less than completely replacing humans. This "noble" goal remains elusive and actually might be ill posed, maybe unsolvable in our times, but also not always necessary for quite effective progress in real applications. Discovering how to do full semantic analysis of images and video may take years, but why not try to go step by step? Instead of trying to recognize people in video, can we at least reliably find out are there people in the video and what basic actions they are doing? Instead of describing set of objects in the images automatically, can we combine some simpler pattern recognition with any other available information with the images like URL and text data pointing to it? Instead of trying to solve 100% or AI problem, why not let machines do what they are best (count, measure, analyze some narrow domain) and combine them with what people do best (meta knowledge, top control)? We are still far from famous Hal computer from "2001 - Space Odyssey "(4).

Here are some significant opportunities that should be attacked by our community. They are both challenging (require significant scientific and technological advances) and important (their solutions would bring real economic benefits). This list is of course not inclusive, but in authors opinion offers a good starting point. It includes content based retrieval; future video compression standards; and vision enhanced user interfaces.

2 Content Based Retrieval

This area already started and is attracting a number of researchers, conferences and companies (1-3). The idea is to automatically index images, video and audio using computer vision, pattern recognition and artificial intelligence, and thus augment (in some cases replace) tedious, expensive and inconsistent human indexers. In many cases, it is the only way to index large number of data elements (i.e. Web crawlers). While progress has been made, much more needs to be done. Here are some ideas:

- Automatically extract some simple semantics such as: is this black&white or color image, are there people or faces in the image, how many, indoor/outdoor, buildings or horizons etc.

- Segment videos into background and objects, compute some basic patterns (color, motion, texture, shape) from these objects, also compute some basic semantic attributes
(people vs. buildings etc.).
.
- Use more of machine learning to help index the images - let user annotate a number of them,
then let the system finish the job by for example trying several possible models.

- Use audio information for basic video segmentation (noise, pause, explosions, talking), spot some apriory defined words and index them, enable query by audio content ("get me tunes like this...").

- Make intelligent Web crawlers that can combine clues from ALL available sources (URLs, accompanied text, image/video/audio content baser retrieval) for automated indexing

- Enable very fast indexing (e.g. K-NN in very high multidimensional feature space) and integrate it with traditional data management models like relational databases and Web crawlers.

3 Video Compression

Currently two new MPEG standards are in the works: MPEG4 (5) and MPEG7. MPEG4 refers to very low bit rate compression involving both real and synthetic images/video and will require basic video segmentation - the idea is to segment objects from background and encode them separately. Emerging MPEG7 attempts to standardize content based retrieval of video. Both of these areas require significant image and video analysis of the type our community is doing. Challenge will be to develop a set of basic algorithms that are reliable and robust, and can also be computed economically and work on all sorts of data. Similarly, for very low bit rate video conferencing extraction of participants' parameters like facial expressions can significantly increase compression ratio. Techniques for computer vision are in heart of such techniques.

4 Vision Enhanced User Interfaces

So far, user interfaces basically did not "close the loop" between the machine and the user. Computer has no knowledge of the state of the user (his/her gaze, facial position and expression etc.). The basic user input is still the keyboard and mouse. Why not gaze, gesture, body language - this is after all how we naturally communicate. Being able to understand this and add one more channel to user input has a potential to significantly improve the whole human - computer interaction, maybe first in some niche areas (disabled users, medical), but then many others. Current desktops are or can easily be equipped with video cameras and microphones and speech recognition is becoming a reality. Why not combine ALL of these inputs into one user interface system controlled by some intelligent central process? Success in this area will require computer vision and AI technology, but also significant human - computer interaction expertise, something our community has to incorporate in its research methodology. Doing gesture analysis for its own sake, without understanding how people could use it, will not be enough.

Once we can "sense" more information about the user, we can go even further, not only improve the basic control of computer input. "Affective computing" is another concept being discussed recently and is very promising (7). The basic idea is to enable computers to sense and use users' emotions (anger, joy, frustration etc.). Why would one want to do this? In many applications we want computers to be impartial and exact (like controlling a power plant). But in many applications like learning, we would like to make computers "softer" and more "human-like", allowing them the adjust their operation based among other things, on user emotion. This is after all characteristic of all good teachers, performers, speakers and coaches. Sensing user's frustration computer could change the training pace, maybe play some soothing music? It has been demonstrated that some basic emotions could be derived from videos of user's face (6). Once we perfect this process we can attempt to make much more human-like computer systems that could significantly improve training, learning, information manipulation/creating etc.

In summary, our community has important and challenging opportunities in the future, but we have to seize them and work effectively to solve them. technology, especially Internet in all its forms, works strongly in our favor. In order to make progress, we have to apply more strict research methodologies. For example, there is no excuse any more not to test our algorithms on very large data sets, they are available and there is compute power to do these tests. Another change for our community comes from the need to address a number of commercial applications, rather than government ones. For anybody who thinks they are less challenging, we encourage to try them and to make an average person be able to use such systems - the challenges are many. Finally, user perspective has to be given much more weigh, since majority of systems will be used by people who are often non-computer experts.

Given all this, we have every reason to be optimistic and look forward to exciting, productive and fun R&D careers!

5 References, Suggested Readings and Web Sites

1. H. Zhang, P. Agrain, D. Petkovic: "Introduction to Special Issue on Representation and Retrieval of Visual Media in Multimedia Systems", Kluwer Academic Publishers, Vol. 4, No.1, January 1997

2. Interesting Web sites related to content based retrieval: webseer.cs.uchicago.edu; www.media.mit.edu; www.ctr.columbia.edu/webseek/; www.informedia.cs.cmu.edu; wwwqbic.almaden.ibm.com, www.virage.com

3. T. P. Minka, R. W. Picard: "Interactive Learning Using a Society of Models", Proceedings CVPR, San Francisco, 1996, pp 447-452.

4. D. Stork ed.: "Hal's Legacy - - 2001's Computer as Dream and Reality", MIT Press, 1997

5. Web site related to MPEG4: www.cselt.stat.it/ufv/leonardo/mpeg/cfp/snhccfp.htm

6. I. Essa, S. Pentland: " A Vision System for Observing and Extracting Facial Action Parameters", Proceedings CVPR, Seattle, WA, 1994, pp. 76-83

7. R. W. Picard: "Affective Computing", MIT Media Laboratory, Perceptual Computing TR, N. 321, MIT, 1995

Multi-Scale Gradient Magnitude Watershed Segmentation

Ole Fogh Olsen[1] and Mads Nielsen[2]

[1] DIKU, University of Copenhagen, Universitetsparken 1, DK-2100, Copenhagen E, Denmark, fogh@diku.dk
[2] 3D-Lab, School of Dentistry, University of Copenhagen, malte@lab3d.odont.ku.dk

Abstract. A partitioning of an nD image is defined as the watersheds of some locally computable inhomogeneity measure. Dependent on the scale of the inhomogeneity measure a coarse or fine partitioning is defined. By analysis of the structural changes (catastrophes) in the measure introduced when scale is increased, a multi-scale linking of segments can be defined. This paper describes the multi-scale linking based on recent results of the deep structure of the squared gradient field[1]. An interactive semi-automatic segmentation tool, and results on synthetic and real 3D medical images are presented.

1 Introduction

The goal of an image segmentation is a description of the shape of some image structure of predefined semantics. However, addressing the shape of an object is not simple since the shape is not intrinsically defined[2]; *shape is defined through an interpretation of measurements.* This introduces the measurements apparatus and its intrinsic resolution as an important part of a shape definition. This is well-known from the definition of coast-lines.

In this paper, we use a Gaussian probe as a linear measurement apparatus (i.e. Gaussian convolution) and thereby introduce the Gaussian scale-space formalism [3, 4]. We base the shape definition on the local scale-space n-jet. In general, the definition of shapes cannot be based solely on local information; global information may constrain local decisions. Following this line, segmentations have been defined as the minimum of a energy functional [5, 6]. This is computationally expensive and difficult to tune to prior information unless extensive statistical material is available [7, 8]. Also split and merge techniques[9] have been introduced. However, this strategy is captured more elegantly in multi-scale linking approaches [4, 10, 11, 12, 13].

Locally we compute a measure of dissimilarity of the image, at a certain scale. The watersheds of this measure defines the segmentation. Watershed cannot be identified locally, i.e. they capture global properties of the image.

A segment boundary is defined as a watershed of a dissimilarity measure in turn defined using a certain width (scale) of the Gaussian aperture function. When varying the scale parameter, the watersheds deform continuously until a transition point where a watershed appears/disappears. Analysis of such transitions in the multi-scale structure has been carried out for a number of local image functionals (i.e. feature detectors) which may be used as dissimilarity measure. Damon established the catastrophe theory for diffused images [14] and also analysed ridge measures [15], Lindeberg analysed blob detectors [16], and Rieger analysed edge and corner detectors [17]. We analysed the gradient magnitude[1].

Multi-scale watershed segmentation has been carried out based on the intensity and ridges: Gauch[12] used the image intensity function directly as local measure of homogeneity. Eberly [13] defined a homogeneity measure based upon local "ridgeness". Griffin [18] used the image intensity or the image intensity gradient and based the segmentation on a multi-grid method. We use the recent results on the multi-scale structure of the gradient magnitude [1] to establish the multi-scale linking. The watersheds in the gradient magnitude intuitively partition the image where the gradient is large.

An object is defined through a root segment and its linking to a localization scale. To interactively select roots and scales, an interactive tool (serving same task as Pizer et al.'s [11]) has been constructed. Since the multi-scale structure can be pre-computed and hashed, interaction is fast.

The following section defines the scale space and the local dissimilarity measures. Then, watersheds, catchment basins, segments and multi-scale linking are defined in Section 3. Section 4 describes the interactive segmentation tool. Section 5 presents experimental results. Finally, in Section 6, we summarise.

2   Scale-space and local dissimilarity measures

Definition 1 Scale-space. The scale-space $L(\cdot, t)$ is generated from an image $I(\cdot) \equiv L(\cdot, 0)$ by Gaussian blurring $L(x,t) \equiv \int I(x')g(x-x',t)dx'$ where $g(\cdot, t)$ is a Gaussian and $t = \sigma^2/2$ it's spread.

Derivatives of the scale-space can be obtained robustly by differentiation of the Gaussian prior to convolution.

For images where segments are assumed to have homogeneous intensity we use the gradient magnitude $|\nabla L|^2 = L_x^2 + L_y^2 + L_z^2$ as dissimilarity measure. In images where only the texture differs from segment to segment a local texture measure is used: the local frequency contents of an image can be measured with a Fourier transform under a Gaussian window function:

$$\hat{L}(x,k,t) = \int I(x')g(x-x',t)e^{-ikx'}dx'$$

where k is the wave-vector and the total filter is an oriented Gabor function. When spatially differentiating this the local phase shift is taken into account so that $\tilde{\partial}_x \hat{L}(x,k,t) \equiv (\partial_x - ik \cdot e_x)\hat{L}(x,k,t)$ is assumed to be small in regions of same texture. e_x denotes a unit vector in the x direction. We define a local dissimilarity measure for texture segmentation (K is a subset of frequencies chosen to discriminate textures) as

$$m(\cdot, t) = \sum_{k \in K, x_i \in \{x,y,z\}} |\tilde{\partial}_{x_i} \hat{L}(\cdot, k, t)|^2 \quad . \tag{1}$$

3 Segments and linking

This section defines segments based on watersheds of an arbitrary local dissimilarity measure. The notion of *watersheds* and *catchment basins* arises when a function is viewed as a topographic relief with height identified with the image intensity. The watersheds are the boundaries between areas, the so-called catchment basins, which drain to one local minimum.

Definition 2 Catchment basin. A catchment basin of a local minimum is the inner points of the closure of the union of all steepest descent lines ending in the minimum.

Definition 3 Watersheds. The watersheds are the boundaries of the catchment basins.

A property especially interesting for segmentation is the fact that *watersheds form closed hyper-surfaces for Morse functions*. Hence, the watersheds of a function give a full partitioning of the multi-dimensional image domain; there is no need for closing or connecting edges to get a partition. This partition has a very flexible topology. As an example in 2D, any number of segments may generically meet in a point. On the contrary, a partition based upon zero-crossings of a feature detector will generically only exhibit 2- and 4-junctions.

Each catchment basin contains exactly one local minimum, the seed of the basin. Instead of directly analysing the multi-scale structure of the watersheds, we can analyse the dual : the local minima. This makes analysis feasible in terms of catastrophes[19]. We suggest:

Definition 4 Segment. A segment is the catchment basin for a local minimum of a dissimilarity measure.

Often the image structure is probed on a much finer scale than the scale of the structures of interest, giving rise to over-segmentation. A common solution [20] is to "flood" the image. Maes et al.[21] post-processed the segmentation by merging neighbouring regions using a MDL Principle. Griffin et al. [18] simplified the image stepwise by treating districts (bounded by maximum gradient paths) as one point and recalculating the slopelines. We suggest to detect objects at coarse scale and localise them at finer scale. In order to do this, the structures must be linked across scale.

Single scale watershed segmentation on the gradient is well known [22, 23]. The singularities of the gradient magnitude and with them the seeds of segments occur in the critical points of the image but also in the points where the second order structure of the image vanishes in one direction. These points evolve when scale is changed, and at certain catastrophe points in scale-space, they interact: appear or annihilates.

The only generic events in scale-space of the gradient magnitude image is a fold catastrophe and a cusp catastrophe involving a minimum[1]. The duality between segments and the minima of the gradient magnitude suggests the linking

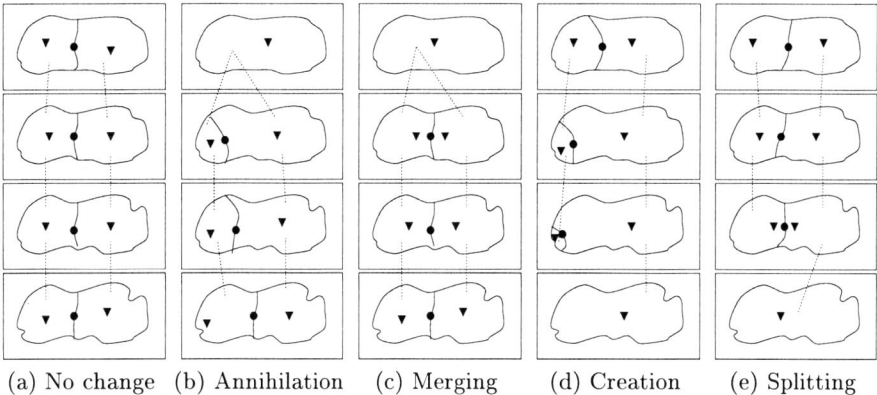

Fig. 1. Multi-scale linking of generic events in watersheds of the gradient magnitude. The events (annihilation, merging, creation, splitting) are named after the interaction between the saddle and the minimum (or minima). Minima and saddles are symbolised with triangles and circles, respectively. A line from a segment to a segment indicates the linking.

scheme. A cusp is the interaction between three singularities, in the present case two minima and a saddle. The two minima and the saddle either meet and become one minimum or the reverse event. A fold is the interaction between two singularities, in the present case one minimum and a saddle. The two singularities meet and annihilate or the reverse creation event.

Figure 1 illustrates the idea in 2D with scale increasing upwards. In the cases of annihilation (b) and merging (c) two minima and a saddle are reduced to one minimum, corresponding to the disappearing of a border between the two segments. The cases of splitting (d) and creation (e) are the reverse events where the emerging saddle corresponds to the appearing of a border between the segments (dual to the two minima). Hence, the linking is in all cases given by the saddle connecting the involved minima.

The implementation of the linking exploits the fact that image structure changes smoothly with scale, therefore a spatial maximum correlation between segments at neighbouring scales can be used as linking criterion. Lindeberg [16] used a similar idea for blob linking.

4 The interactive segmentation interface

A user-interface has been constructed for accessing the multi-scale segment structure (Figure 2). Raising the detection scale gives generally fewer segments and vice versa. Raising the localisation scale results in more smooth boundaries and vice versa. The user gets interactive 3D feedback on the selections limited in speed mainly by the computers rendering capabilities.

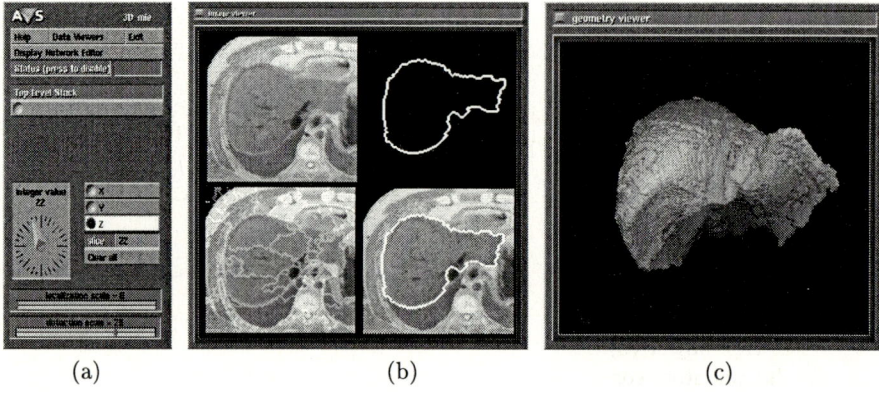

(a) (b) (c)

Fig. 2. User interface windows. In window (a), the localization and detection scale is selected as well as a slice in one of the three Cartesian directions. This gives a partition of the domain. Window (b) displays the image slice (top left), the partition superimposed on the image slice (bottom left), the union of the selected segments (top right) and the selected segments superimposed on the image slice (bottom right). The object is defined by selecting/deselecting volume segments in one of images in window (b). The third window (c) continuously renders the union of the selected volume segments.

5 Results and verification

This section presents results on three types of images using the gradient magnitude squared as the dissimilarity measure. The images are a software simulated liver phantom (Figure 3), a CT head scan of a patient with abnormal growth (Subfigures 5.c,d) and digital photos (red channel) from the visible human project (Subfigures 5.a,b). The tasks are to segment the phantom, jaw muscles and the liver, respectively. Furthermore results of texture segmentation on a toy image is shown in Figure 6.

Figure 3 (a) displays a rendering of the true phantom. Different levels of noise has been added: respectively 0, 25, 50 and 75 percent of the voxels have been modified with Gaussian additive noise with zero mean and standard deviation of 80% of the phantom to background contrast. We shall refer to the different noise level as the 0, 25, 50 and 75 percent case. Segmentation has been performed using an appropriate high detection scale in order to define the object as one single segment, which has automatically been tracked to a lower localisation scale (see Figure 3).

Segmentation has been performed using an appropriate high detection scale in order to define the object as one single segment, which has automatically been tracked to a lower localisation scale (see figure 3).

The errors in localisation of the boundary are due to two different sources. The noise pixels influences the multi-scale linking so that a noisy boundary is created at low scales. By increasing the localization scale, a smooth boundary can be constructed. However, at this scale, the Gaussian blurring has deformed the

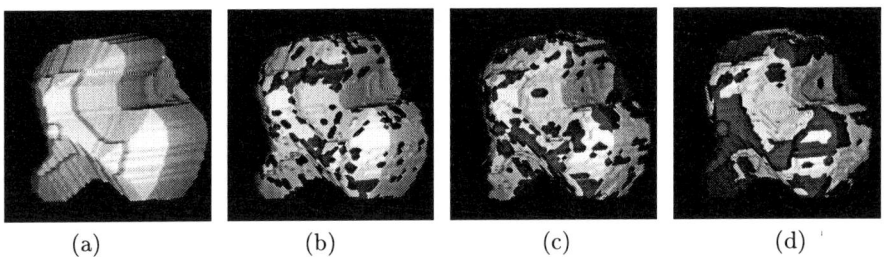

Fig. 3. The true object is presented in (a) as a bright surface rendering. In (b), (c) and (d) is a bright surface rendering of the segmentation for noise level 25, 50 and 75, resp. . The true object (a) is for comparison superimposed as the dark surface in (b), (c), (d). The phantom consists of 57708 voxels in a 64^3 volume.

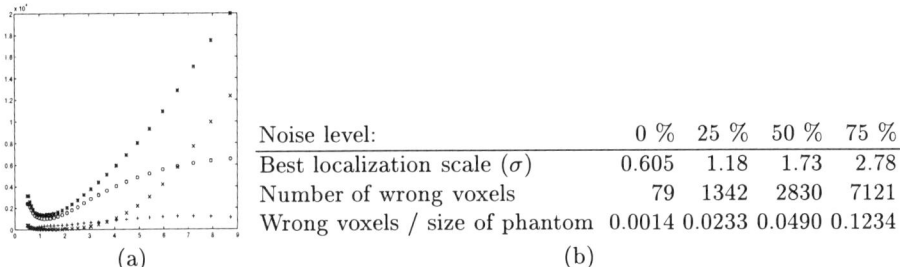

Noise level:	0 %	25 %	50 %	75 %
Best localization scale (σ)	0.605	1.18	1.73	2.78
Number of wrong voxels	79	1342	2830	7121
Wrong voxels / size of phantom	0.0014	0.0233	0.0490	0.1234

(a)   (b)

Fig. 4. Number of erroneous voxels as a function of localization scale for noise levels 25. Bold crosses indicate total number of voxels, circles indicate missing voxels on surface, crosses indicate missing interior voxels, and pluses indicate additional voxels. The qualitative shape of the curves are similar for the other noise levels. The statistics of phantom segmentation is summarised in (b)

object deterministically so that sharp corners (convex or concave) are rounded. An optimal scale may be established from prior information on noise level, object size, etc. This in done empirically in Figure 4. In Figure 3 (d) the phantom is generally exposed in convex patches while the segmentation is exposed in concave patches due to the deterministic shape distortion at higher scales.

6 Summary

A framework for multi-scale segmentation has been presented. The partition by the watersheds of the gradient magnitude has been analysed and implemented in the case of Gaussian scale space. The multi-scale linking has been defined on the basis of results from catastrophe theory.

The selection mechanism of interactively picking objects at an appropriate scale and combining the result at localization scale provides a fast way of doing semi–automatic segmentation.

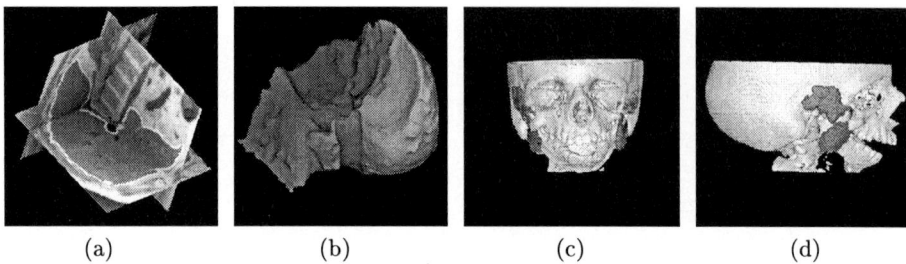

(a) (b) (c) (d)

Fig. 5. Liver segmentation (a) and (b) from a cube of size 128x128x112 voxels each 1.758 mm³. In (a), the liver boundary is superimposed on three orthogonal slices of the subject cube. The same liver segment is visualised in (b) as a surface rendering. The view from the spine (b) clearly reveals the imprint from other internal organs. The segmentation is difficult for mainly two reasons: The high similarity between liver tissue and the neighbouring muscle tissue; and the inhomogeneity of the liver tissue itself. **Jaw muscles.** The segmentation was verified by Professor Sven Kreiborg. The subject is a 256x256x64 cube of size 1x1x2 mm³. The muscular structures are located next to bone (high value), skin (low value) and salivary glands (approximately same value) which makes the task difficult for standard techniques. A fine detection scale must be used because the muscles are flat structures, that is fine scale structure in one direction. The coarse structures of a muscle was selected with a few ($<$ 5) mouse clicks using a coarse detection scale ($\sigma \approx 3.06$ pixels), and the segmentation was then refined with a few ($<$ 10) clicks using a finer detection scale ($\sigma \approx 0.805$ pixels). Localisation scale is 0.5 pixels.

Fig. 6. Multi-scale texture segmentation based on local frequency differences defined by Gabor functions. Two distinctive textures with Gaussian noise added is segmented using the dissimilarity measure defined in Eq. 1. These are preliminary results serving only as indication of the generality of the multi-scale watershed segmentation approach.

The definition of segments can be changed by using another measure of dissimilarity instead of the gradient magnitude. This is possible within the same general framework although different structural changes might occur generically for other measures and diffusion schemes.

Acknowledgements We thank S. Kreiborg, P. Larsen, and A. B. Dobrzeniecki, 3D-Lab, School of dentistry, University of Copenhagen, for providing the CT data and phantom image and supervising the interactive segmentation.

References

1. O.F.Olsen and M.Nielsen, "Generic events for the gradient squared with application to multi-scale segmentation." SS97, Utrecth, 1997.
2. J. J. Koenderink, *Solid Shape*. Cambridge, Mass.: MIT Press, 1990.
3. A. P. Witkin, "Scale space filtering," in *Proc. International Joint Conference on Artificial Intelligence*, (Karlsruhe, Germany), pp. 1019–1023, 1983.
4. J. J. Koenderink, "The structure of images," *Biol. Cybern.*, vol. 50, pp. 363–370, 1984.
5. D. Mumford and J. Shah, "Boundary detection by minimizing functionals," in *Proc. IEEE Conf. on CVPR*, (San Francisco), 1985.
6. Y. G. Leclerc, "Constructing simple stable descriptions for image partitioning," *IJCV*, vol. 3, pp. 73–102, 1989.
7. U. Grenander, Y. Chow, and D. Keenan, *Hands. A Pattern Theoretic Study of Biological Shapes*. Springer Verlag, 1991.
8. T. Cootes, J. Taylor, D. Cooper, and J. Graham, "Active shape models–their training and application," *CVIU*, vol. 61, pp. 38–59, January 1995.
9. R. C. Gonzales and R. E. Woods, *Digital Image Processing*. Addison Wesley, 1993.
10. K. L. Vincken, C. N. de Graaf, A. S. E. Koster, M. A. Viergever, F. J. R. Appelman, and G. R. Timmens, "Multiresolution segmentation of 3D images by the hyperstack," in *VBC*, pp. 115–122, Los Alamitos, CA: IEEE CS Press, 1990.
11. L. M. Lifshitz and S. M. Pizer, "A multiresolution hierarchical approach to image segmentation based on intensity extrema," *IEEE Trans. Pattern Analysis and Machine Intelligence*, vol. 12, no. 6, pp. 529–541, 1990.
12. J. M. Gauch, W. R. Oliver, and S. M. Pizer, "Multiresolution shape descriptions and their applications in medical imaging," in *IPMI 10*, 1988.
13. D. Eberly and S. M. Pizer, "Ridge flow models for image segmentation," Tech. Rep. TR93-056, University of North Carolina, Dept. of Computer Science, 1993.
14. J. Damon, "Local morse theory for solutions to the heat equation and gaussian blurring," *Journal of Differential Equations*, vol. 115, January 1995.
15. J. N. Damon, "Properties of ridges and cores for twodimensional images." Unpub.
16. T. Lindeberg, *Scale-Space Theory in Computer Vision*. Kluwer Academic Publishers, 1994. ISBN 0-7923-9418-6.
17. J. H. Rieger, "Generic evolutions of edges on families of diffused greyvalue surfaces," *JMIV*, vol. 5, pp. 207–217, 1995.
18. L. D. Griffin, A. C. F. Colchester, and G. P. Robinson, "Scale and segmentation of grey-level images using maximum gradient paths," *Image and Vision Computing*, vol. 10, pp. 389–402, July/August 1992.
19. R. Gilmore, *Catastrophe Theory for Scientists and Engineers*. Dover, 1981. ISBN 0-486-67539-4.
20. L. Najman and M. Schmitt, "Watershed of a continuos function," *Signal Processing*, vol. 38, pp. 99–112, July 1994.
21. F. Maes, D. Vandermeulen, P. Suetens, and G. Marchal, "Computer–aided interactive object delineation using an intelligent paintbrush technique," in *CVRMed95* (N. Ayache, ed.), pp. 77–83, Springer–Verlag, 1995. Lecture Notes 905.
22. F. Meyer, "Topographic distance and watershed lines," *Signal Processing*, vol. 38, pp. 99–112, July 1994.
23. L. D. Griffin, *Descriptions of Image Structure*. PhD thesis, Uni. of London, 1995.

Segmentation of Multispectral Images of Works of Art Through Principal Component Analysis[1]

Stefano Baronti, Andrea Casini, Franco Lotti and Simone Porcinai

"Nello Carrara" Research Institute on Electromagnetic Waves IROE - CNR,
64, Via Panciatichi, 50127 Florence, Italy

Abstract

Investigation of materials constituting painted layers of works of art (panels, canvas, frescoes) can be profitably done by means of non-destructive optical techniques based on the analysis of reflectance spectra in the visible and near infrared regions.

Accurate and high spectral resolution measurements can be obtained by means of fiber optics spectrophotometers, but only in small spot areas. Image spectroscopy systems can give instead a complete spectral information on the whole examined surface in a great number of bands, allowing direct visual interpretation. The analysis of such amount of data is not trivial. A possible approach is to decorrelate the data and concentrate the significant information in few images, by using principal component analysis (PCA).

In this work segmentation is investigated in order to partition the imaged scene into regions of spectral similarity to facilitate successive analysis. The results on a test tempera panel and on a predella painted in the XVI century show the effectiveness of the proposed approach, also revealing details undetectable by conventional techniques.

1 Introduction

Optical reflectance spectroscopy is an important non invasive technique in the field of preservation and restoration of paintings for the identification of the constituting materials (typically: preparation layers, pigments, varnishes, degradation products, etc.). Objective data can be gathered to monitor the conservation status of paintings.

After ten years of experience in this field, we may state that *fiber-optic reflectance spectroscopy* (FORS) in the visible and short wavelength infrared (400-2000 nm) is almost always effective when a suitable archive of spectra is available as a reference [1, 2]. To this purpose, an exhaustive archive of visible and near-infrared reference spectra of the pigments used in the middle ages and renaissance florentine painting has been set up [4], on the basis of measurements on both pure pigment powders and test panels painted in accordance with ancient techniques.

Among non-invasive spectroscopic techniques, FORS is also the most functional, as it can be operated with transportable equipment and works with fiber-optic probes which can be easily handled to collect data from small spots on the painted surface. However, such a collection of local 1-D spectrograms (spectral resolution is ~ 1 nm), though rich in information to a chemist, cannot be readily correlated with the scene. Such a limitation makes 1-D reflectance spectroscopy less directly useful to restorers and art historians.

A technique which promises to be more efficient than FORS is its two-

---

[1] Work partially supported by the CNR Project *Beni Culturali* and by the EC project *ERA*: Environmental Research for Art Conservation, EV5V CT94 0548.

dimensional extension: *imaging spectroscopy* (IS) or spectral imaging, which supplies very readable data by retaining the aspect of the objects. Imaging spectroscopes in the VIS-NIR regions are not yet available as handy transportable instruments of reasonable cost. For this reason we developed a transportable image spectroscopy system based on a Vidicon camera (400-2000 nm) and a set of several narrow-band interferential filters supervised by a computerized program that manages the whole acquisition process [3, 7].

The large amount of collected data (tenths of multispectral images of 754×572 pixels in the VIS-NIR range) has suggested to perform multivariate statistical analysis, such as principal component analysis (PCA), both to compress the meaningful information into few *components* and to segment the scene, mapping the areas with similar pigment composition. PCA has been already used in a variety of chemical applications and mapping of remotely sensed multispectral data [5-6].

In the first part of this work we report the results of the application of PCA to reflectance IS data acquired from a test tempera panel, after validation with data acquired using a high resolution spectrophotometer.

In the second part of the work an application is described to an ancient oil painted panel of XVI century: Luca Signorelli's Holy Trinity predella. Results showed that multivariate data analysis can help in the recognition of the materials, e.g. by allowing scene segmentation into areas with similar spectral behavior.

2 Multispectral Image Acquisition

The imaging setup used for the collection of the multiwavelength images has been assembled around a PbO-PbS Vidicon camera (*Hamamatsu* C2400-03). Spectral selection has been realized by means of narrow-band optical filters, lodged on 8-filter wheels, which are sequentially mounted in front of the camera lens [7].

A set of 29 narrow-band interferential filters has been used with full width at half maximum of ~10 nm and 0.01% blocking factor out of band (from X-ray to far infrared). Illumination is supplied by two projectors symmetrically placed at the camera sides with ~45° beam incidence angles. The illuminance in the visible is ~12000 lux. In the UV the irradiance is very low (< 0.4 W/m^2), while in the IR (up to ~2.5 µm) it is 67 W/m^2. In order to normalize the amplitude of the collected images, five SPECTRALON® reflectance standards are included in the scene: two of the standards (99% and 3% reflectance) are used for the normalization according to a linear model of the camera response (after gamma correction). The other three standards (75%, 50%, 10%) are used to control the quality of the measurement.

3 Principal Component Analysis

PCA is a method which gives a compressed description of the data variance by means of a small number of uncorrelated variables. In our case, each of the n collected spectra (objects) may be represented by a point in a v-dimensional space, where v is the number of wavelengths (variables). PCA involves orthogonal rotation of the original axes, each representing an original variable, into new axes. The new axes, called *Principal Components*, are the eigenvectors of the data variance-covariance matrix [8, 9]. PCA allows the reduction of the parameter space dimensionality, as the number of new axes a (variables) needed to describe most of the sample data variance is usually much less than v.

4 Analysis of the Test Panel

The test panel is a wooden tablet prepared with a gypsum layer and painted with four pigments using a tempera medium. Formulas and dominant wavelengths of the four pigments are reported on the Table I. Three rectangular stripes, for each pigment, were painted using the pure pigment and two mixtures with 10% w/w and 5% w/w carbon black, thus creating 12 stripes of the same size differentiated either for their dominant wavelength (λ_D), due to the nature of the pigment, or for their brightness, due to the carbon black percentage. Moreover, some carbon black lines had been drawn on the gypsum layer, in order to simulate the presence of underdrawings. The purity of the pigments was checked by means of X-ray diffraction; the average dimension of the carbon black particles, measured by laser scattering granulometry is about 10 µm.

PIGMENT	FORMULA	λ_D (nm)
cinnabar	HgS	619
malachite	$Cu_2CO_3(OH)_2$	510
yellow ochre	$FeO(OH)$	583
chromium oxide	Cr_2O_3	560

Table I: Formula and dominant wavelength (λ_D) of the four pigments of the test tablet.

Table II reports the contributions of the first five PCs to the overall variance as a result of the application of PCA to 29 images of the tablet almost equally distributed in the VIS-NIR region (420-1550 nm). It appears that the first five PCs account nearly entirely for the total variance (99.62 %) of the data set.

PCs	Var. %	Cumul. Var %
1	86.83	86.83
2	6.66	93.49
3	4.23	97.72
4	1.2	98.92
5	0.7	99.62

Table II: Variance and cumulative variance for the first five principal components from 29 VIS-NIR images of the test tablet.

The redistribution of variance has an interesting theoretical interpretation: if the variance is considered as a measure of the information content, then a packing of information has occurred because a lower number of variables contains a large percentage of the total data variance. In our case, most of the information is expected to be contained in the five largest components.

Figure 1 reports the first four PC images of the tablet. The gray scale value of the pixels was obtained by scaling from 0 to 255 their coordinates in the PC domain.

It can be noted that in the PC1 image the gray values of the areas corresponding to pure pigments are generally higher than those corresponding to mixtures with carbon black. However, the clustering in the PC domain of points belonging to the twelve differently painted zones on the tablet can be better appreciated by considering the 2-D score plots. These are the projections of the position vectors (in the PC space) of the objects (pixel spectra) into a 2-dimensional subspace. The score plot which offers the clearest clustering of the 12 painted zones is the projection on the PC1/PC3 plane, as shown in Figure 2, where the pixel densities are represented by their level lines. It is possible to appreciate not only the different pigments but also the zones with the same pigment and different percentages of carbon black.

The study of the trend of the eigenvectors versus wavelength can be helpful in understanding the common features of the groups and why an object (a spectrum of a pixel in our case) is located in a specific position on the score plot. The weight plot reported in Figure 3 shows how the different wavelengths influence the PC1 and PC3 eigenvectors. PC1 eigenvector receives a contribution from the near-infrared bands (83.8%) heavier than from the visible ones (16.2%); the disposition of the weights on the same plot shows that all the near-infrared bands have a positive and almost equal weight on the PC1. Approximately PC1 lies in the subspace defined by the NIR bands; furthermore, as these bands give PC1 almost the same contribution, PC1 is oriented mainly along the diagonal of this subspace.

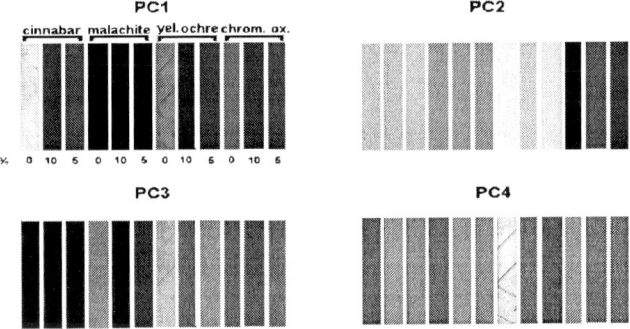

Figure 1 - PC1, PC2, PC3 and PC4 images of the test tablet from a sequence of 29 VIS-NIR images. Carbon black percentages are indicated.

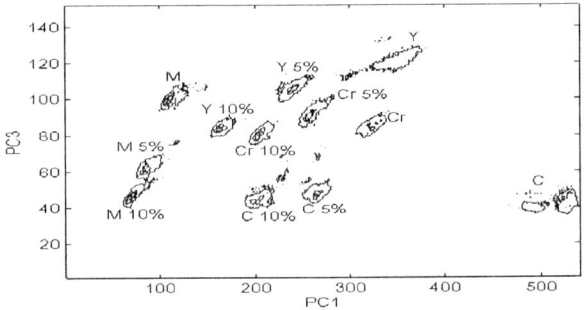

Figure 2 - Score plot PC3/PC1 from PCA on 29 VIS-NIR images of the test tablet. C: cinnabar; Y: yellow ochre; M: malachite; Cr: chromium oxide. The fractions carbon black/pigment (w/w) are reported as percentages.

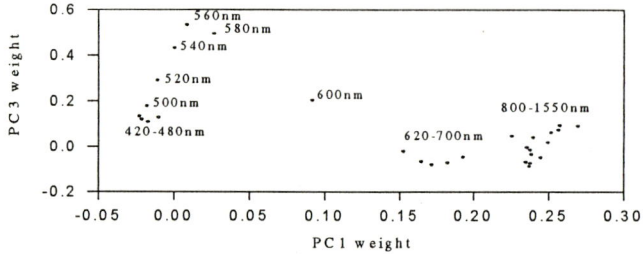

Figure 3 - Weight plot PC3/PC1 from PCA on 29 VIS-NIR images of the test tablet.

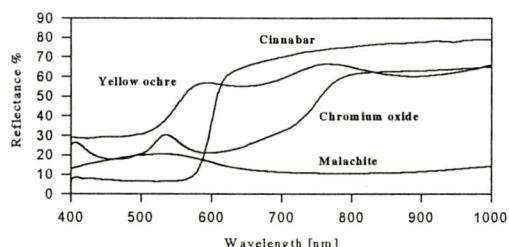

Figure 4 - Spectra of the four pure pigments of the test tablet taken with FORS.

In other words, the most important contribution to variance is given by the spread of the NIR mean reflectances. Therefore, as carbon black reflectance has a constant high absorption through the NIR region, its percentage determines the mean NIR reflectances of the mixtures and the separation of each pigment group into clusters along the PC1 axis (Figure 2). The visible bands in the range 420-600 nm have a strong influence (92.7%) on the third principal component. In particular, the bands at 560 nm and 580 nm give the highest contribution for this test tablet. Among the four pigments, yellow ochre and cinnabar receive the greatest discrimination just by the wavelengths in the range 560-580 nm, (Figure 4); indeed, these pigments are placed at the extrema of the PC3 range in the PC1/PC3 score plot (Figure 2).

It is interesting to note how underdrawings, which are usually made of carbon black, are clearly revealed on PC1 and PC4 images (Figure 1), especially under yellow ochre, which is characterized by a low absorption in the NIR region. Underdrawings become visible in the NIR images because the radiation in this wavelength range penetrates more deeply into the painted layers. It is partially absorbed by the carbon black, which has almost constant low reflectance, and partly reflected by the gypsum background layer, which has high reflectance (but for few well defined characteristic absorption bands).

5. Analysis of a XVI Century Painted Panel

The results obtained on the test tablet furnished some guidelines for the application to works of art. The above explained methodology has been profitably applied to the study of a scene of the *Holy Trinity* predella (on exhibit at the Uffizi Gallery), which was painted by Luca Signorelli in the early years of XVI century. It is an oil painted panel of 32×204 cm, divided into three scenes of Jesus's passion. A sequence of 29 multispectral images of *Flagellation* scene was acquired by our image spectroscopy system in the visible and near-infrared region. In Figure 5 we report the plot of the first eigenvector (PC1) after application of PCA to this set of images. It can be seen that the main contribution to the total variance of the data set is given by the NIR bands, which have similar weight, showing that the variance accounted from them is mainly due to the variations of the average reflectances

The high penetration of NIR radiation suggests to evaluate the spectral behaviour of the pixels in this region, to evidence information from deeper layers. To this purpose, and to avoid that higher variations in the visible range could mask those in the NIR, we limited the PCA to a subset of 13 NIR images (800-1550 nm).

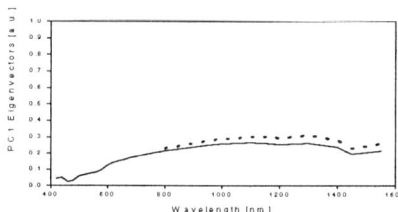

Figure 5 - PC1 Eigenvectors from PCA on 29 VIS-NIR images(solid) and on 13 NIR images (dotted) of the predella.

In Figure 6 the first three PC images are reported which account for more than 99% of the variance in the NIR data set. Again, the variation of the mean reflected light gives the most important contribution to data variance (91.3% accounted by PC1). In fact, the most important eigenvector (PC1) is now uniformly weighted across all the spectral bands (Figure 5). In general we are not interested in this variance which gives poor information concerning chemical properties of the materials and does not help in segmentation. To eliminate this variance, we normalized the reflectance values of each multispectral image $I(x,y,\lambda)$ to the image at 800 nm $I(x,y,\lambda=800)$ before performing PCA:

$$I_{norm}(x,y,\lambda) = I(x,y,\lambda) - I(x,y,\lambda=800)$$

This normalization has the effect of bringing to zero the first element of the eigenvectors associated with each component. Adopting this normalization, PC1 generally accounts for the variance of the baseline slope of the reflectance spectra.

The variance, computed by applying PCA to the normalized data (NIR range) resulted 87% for PC1, 9.42% for PC2 and 0.91 for PC3.

The normalized eigenvector $PC1_{norm}$ is highly correlated with the non-normalized eigenvector PC2 (Fig.7) and the same happens for the $PC2_{norm}$ and PC3.

Figure 6 - Detail of Luca Signorelli's Holy Trinity predella. Clockwise, from top left: grey level representation, PC1 image, PC2 image, PC3 image.

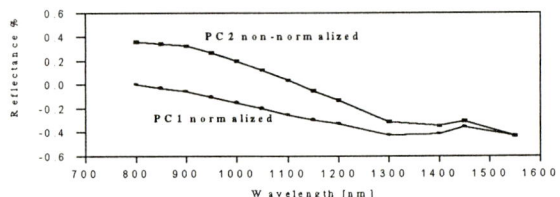

Figure 7 - PC1 and PC2 eigenvectors from PCA on 13 NIR images of the predella.

The correlation coefficient r between the normalized PCi_{norm} and non-normalized $PC(i+1)$ eigenvectors decreases with i because the variance subtracted adopting this normalization is not only due to the energy offset in the reflectance spectra of pixels.

$r(PC1_{norm}, PC2) = 0.994$; $\quad r(PC2_{norm}, PC3) = 0.67$.

The PCs, now accounting for the variance due to finer spectral features of the pixel spectra, become more important in the PC model and allow to evidence zones characterized by different chemical composition of the materials.

Visual inspection of the first two PC images (Figures 8) reveals that the images obtained from the normalized variance-covariance matrix have better contrast and appear to be more useful for visual segmentation purposes. In particular, by means of the PC1 and PC2, it is possible to distinguish zones which appear to be painted with the same pigment. For example in the PC1 image (Figure 8 A), a pilaster behind a whipper appears to be split into two distinct parts. The left part of the pilaster had been likely painted as correction of the original profile, and covers a portion of the dark wall. In fact, the subsequent observation of the reflectance spectra in the NIR region shows lower values of reflectance in the left part than in the right one. Moreover, in the PC2 image (Figure 8 B), the values of Jesus Christ's skin are quite different from the other bodies present in the scene, indicating that the artist used different pigment compositions, not evident on a visual examination.

6 Conclusions

Image Spectroscopy in the visible and near-infrared regions, in conjunction with PCA can be profitably used for segmenting painted zones characterized by similar spectral behavior and therefore to map areas according to their chemical composition or physical properties. The application of multivariate image analysis to the image sequence of the test tablet showed that it is possible to obtain a complete segmentation of zones homogeneously painted.

A B

Figure 8 - Normalized image of the predella from PCA on 13 NIR images. A PC1; B: PC2.

Non-invasivity and objectivity of the instrumentation are the main features that make this procedure particularly suitable for investigating paintings both to monitor their status of conservation and to guide restoration activity. From the results obtained on the Holy Trinity predella by Luca Signorelli, it appears that the use of image spectroscopy together with Principal Component Analysis is a powerful tool in investigating the pigment distribution on paintings.

Segmentation maps can be easily done, grouping areas with similar spectral behaviour, which help both in the interpretation of the painting and in the selection of minimal but meaningful micro-sampling, when destructive analysis is required.

Acknowledgments

Thanks are given to Mr. L. Stefani (IROE) for his technical support during the measurements and to Dr. M. Picollo for helping in the FORS measurements.

References

1. M. Bacci, *Sensors and Actuators B* **29**, 190 (1995).
2. A. Orlando, M. Picollo, B. Radicati, S. Baronti and A. Casini, *Appl. Spectrosc.* **49**, 459, (1995)
3. M. Bacci, S. Baronti, A. Casini, F. Lotti, M. Picollo and O. Casazza, *Material Issues in Art and Archaeology III* **267**, 265-282 (1992).
4. R. Linari, M. Picollo and B. Radicati: IROE Technical report TR/POE/92.7, July, 1992
5. P.J. Ready and P.A. Wintz, *IEEE Trans. Commun.* **21**, 1123-1130 (1973)
6. M.E. Kargacin and B.R. Kowalski, *Anal. Chem.* **5**, 2300, (1986)
7. A. Casini, F. Lotti, M. Picollo, L. Stefani and G. Troup: *Proc*, Atti Fondaz. G. Ronchi, LI, n.1-2, pp.289-303 (1996).
8. K.V. Mardia and J.T. Kent and J.M. Bibby: *Multivariate Analysis*. Birnbaum and Lukacs London, 1979.
9. H. Martens and T. Næs: *Multivariate Calibration* (Wiley and Sons, New York, 1989).

Multiscale Edge Detection via Normal Changes*

Chwen-Jye Sze[1], Hong-Yaun Mark Liao[2], Hai-Lung Hung[3], Kuo-Chin Fan[1], and Jun-Wei Hsieh[4]

[1] Institute of Computer Science and Information Engineering, National Central University, Chung-Li, Taiwan
[2] Institute of Information Science, Academia Sinica, Taiwan
[3] Department of Electrical Engineering and Computer Science, Northwestern University, IL 60208, U.S.A.
[4] Computer and Communication Research Labs, Institute of Technology Research Industry, Taiwan

Abstract. A new edge detection technique based on detection of normal changes is proposed. Most of the existing range image-based edge detection algorithms base their detection criterion on depth or curvature changes. However, the depth change-based approach does not have keen sensitivity in detecting roof (or crease) edges, and the curvature change-based approach suffers from a complicated and tedious principal curvature derivation process. Using normal changes as a detecting criterion, on the other hand, the existence of an edge can be easily detected, even when the change across a boundary is slight. Experimental results using both synthetic and real images demonstrate that the proposed method can efficiently detect both step and roof edges.

1 Introduction

In this paper, we propose a new edge detection technique based on detection of normal changes. The normal value is an important characteristic in differential geometry[10]. We find that by detecting normal changes, both step edges and roof edges can be easily identified. The whole detection procedure is divided into two stages. In the first stage, the normal of every point in a range image is decided. Since all the data points in a range image are discrete, the partial derivatives which are required to derive the normal value cannot be directly computed. For comparison purposes, we propose use of quadratic surface fitting[11], the orthogonal wavelet-based approach[1,12], and the non-orthogonal wavelet-based approach[2], respectively, to approximate the original object surfaces and to then calculate the normal value of every discrete point on the surfaces. After the normal values of all the surface points are determined, the non-orthogonal wavelet transform (dyadic wavelet transform) proposed by Mallat *et al.*[2,13] is applied to detect those points which have significant normal changes as edge

---

* This work is supported by the National Science Council of Taiwan under grant no. NSC86-2745-E-001-004.

points. From the experimental results, we find that the non-orthogonal wavelet-based approach can best approximate the original surfaces from a discrete data set. Further, we also find that edge detection based on normal changes is a more promising alternative than other methods that base their detection criterion on depth or curvature changes. The proposed edge detector can detect both step edges and roof edges without introducing any edge models or heuristics.

2 Range edge detection via normal change

In this section, we shall explain why normal change can be used as a cue for range edge detection. Some properties of a 3D surface from the differential geometry viewpoint which are useful for edge detection will be addressed in Section 2.1. Then, a detailed explanation of why normal change is a better choice for edge detection will be given in Section 2.2.

2.1 Some properties in differential geometry useful for edge detection

In this section, some basic properties of a 3D surface will be addressed from the differential geometry viewpoint. These properties are useful for solving the edge detection problem on range images.

Let S be a differentiable surface and $p(u,v)$ be a point on S with coordinate (u,v). If p_u and p_v are the partial derivatives of $p(u,v)$ with respect to u and v, respectively, then we can say that p_u, p_v form the basis of a tangent plane, $T(p)$, of $p(u,v)$ (Figure 1). The normal of $T(p)$ can be defined as

$$N(u,v) = \frac{p_u \times p_v}{\|p_u \times p_v\|}. \tag{1}$$

Here, the norm of any $N(u,v)$ is always equal to 1, and all $N(u,v)$'s lie on a

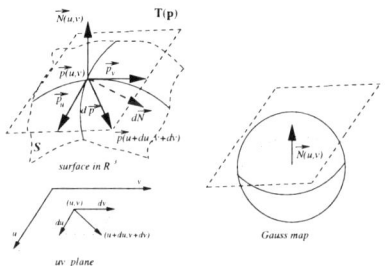

Fig. 1. Relations between a surface point, its corresponding parameter coordinate and the Gauss map.

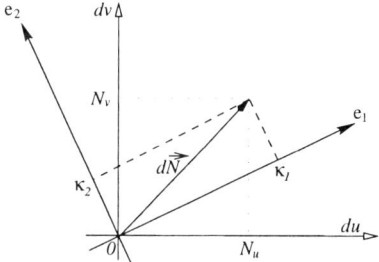

Fig. 2. Relationship between dN and principal curvatures.

unit sphere in R^3. The mapping, $\boldsymbol{N} : \mathcal{S} \to R^3$, is called the Gauss mapping, $G(R^3)$[10,14].

Let $\boldsymbol{N}(u,v)$ be differentiable; the mapping, $d\boldsymbol{N}(u,v)$, is from $G(R^3)$ to a tangent plane, $G(R^3)$, at $\boldsymbol{N}(u,v)$. Since the tangent plane of $\boldsymbol{N}(u,v)$ is equal to that of $\boldsymbol{p}(u,v)$, $d\boldsymbol{N}(u,v)$ is also on the $T(p)$ plane, as shown in Figure 1. Thus, both $d\boldsymbol{p}$ and $d\boldsymbol{N}$ can be represented by the linear combination of du and dv as follows [10,7]:

$$d\boldsymbol{p} = \boldsymbol{p}_u du + \boldsymbol{p}_v dv, \qquad (2)$$

and

$$d\boldsymbol{N} = \boldsymbol{N}_u du + \boldsymbol{N}_v dv. \qquad (3)$$

Here, $p(u,v)$ is the gray level (intensity image) or depth (range image) at position (u,v). Therefore, $d\boldsymbol{p}$ physically means the intensity or depth change with respect to $\boldsymbol{p}(u,v)$'s neighbors. As for $\boldsymbol{N}_u$ and $\boldsymbol{N}_v$, they are mathematically defined as $\boldsymbol{N}_u = \frac{\partial \boldsymbol{N}}{\partial u}$ and $\boldsymbol{N}_v = \frac{\partial \boldsymbol{N}}{\partial v}$. That is, they physically mean the normal change along the u and v directions, respectively.

2.2 Why normal change is better for edge detection

Comparing the normal change-based approach and the curvature-based approach [4,6], the computational complexity of the former is much less than that of the latter. As shown in Figure 2, { du, dv } and { e_1, e_2 } are two independent orthonormal bases on the tangent plane of $\boldsymbol{p}$. Based on differential geometry, the two principal curvatures, κ_1 and κ_2, can be derived by projecting $d\boldsymbol{N}$ on a specific basis. Let { e_1, e_2 } be the basis with e_1 and e_2 corresponding, respectively, to the directions of the two principal curvatures, κ_1 and κ_2[10]. From Figure 2, it is obvious that $d\boldsymbol{N}$ can be represented by a linear combination based on either { du, dv } or { e_1, e_2 }. One thing to be noted is that no matter how $d\boldsymbol{N}$ is represented, its magnitude is independent of the basis selected. The only thing that will vary with respect to the basis change is the orientation of $d\boldsymbol{N}$. Therefore, using the normal change directly to locate edge positions does not require that the values of two principal curvatures be explicitly determined. This fact explains why $d\boldsymbol{N}$ can replace the two principal curvatures in edge detection and why using $d\boldsymbol{N}$ is much more efficient than using κ_1 and κ_2.

3 Calculating normals from discrete surface points

In the previous section, we have discussed how the change of normals at every point on a surface (Equation (3)) can be used to detect edges. The continuous domain normal value derivation process is summarized as follows. Let $\boldsymbol{p}(u,v) = (u, v, f(u,v))$ be a point located on a surface. The partial derivatives of $\boldsymbol{p}(u,v)$ in the u and v directions are

$$\boldsymbol{p}_u = \frac{\partial \boldsymbol{p}(u,v)}{\partial u} = (1\ 0\ f_u(u,v)), \qquad (4)$$

and
$$\boldsymbol{p}_v = \frac{\partial \boldsymbol{p}(u,v)}{\partial v} = (0\ 1\ f_v(u,v)). \tag{5}$$

Plugging these two values into Equation (1), the value of $\boldsymbol{N}(u,v)$ can be rewritten as

$$\boldsymbol{N}(u,v) = (\frac{-f_u}{\sqrt{1+f_u^2+f_v^2}}\ \frac{-f_v}{\sqrt{1+f_u^2+f_v^2}}\ \frac{1}{\sqrt{1+f_u^2+f_v^2}}) \tag{6}$$
$$= (n_1(u,v)\ n_2(u,v)\ n_3(u,v)), \tag{7}$$

where $f_u = \frac{\partial f(u,v)}{\partial u}$ and $f_v = \frac{\partial f(u,v)}{\partial v}$, $n_1(u,v) = \frac{-f_u}{\sqrt{1+f_u^2+f_v^2}}$, $n_2(u,v) = \frac{-f_v}{\sqrt{1+f_u^2+f_v^2}}$ and $n_3(u,v) = \frac{1}{\sqrt{1+f_u^2+f_v^2}}$. The ranges of n_1, n_2 and n_3 are all bounded by (-1,1).

In real implementation, since all the data points acquired in a range image are discrete by nature, the above calculation does not apply. Therefore, we have to find an appropriate method to deal with this problem. In the implemenation stage, we split the edge detection procedure into two steps. In the first step, the normal at every point on a surface should be determined. This step involves calculation of partial derivatives on a set of discrete data points. Then, in the second step, a detector is required to accurately detect the points where significant variations of normals are encountered. In order to calculate the normals on a set of discrete surface points, some existing methods [2,11,15,16] can be applied. For comparison purposes, we choose quadratic surface fitting[7,11], the orthogonal wavelet-based approach[17], and the non-orthogonal wavelet-based approach[2] to calculate the normal value of every point on a surface.

4 Detecting edges based on normal changes

In [2,18,13], Mallat and his students developed some pioneering works for multiscale edge detection based on gray level changes. Here, we shall review part of their work which will be useful in our work. Define two wavelet functions, $\psi^1(x,y)$ and $\psi^2(x,y)$ [2,18], where

$$\psi^1(x,y) = \frac{\partial \theta(x,y)}{\partial x}, \tag{8}$$

and

$$\psi^2(x,y) = \frac{\partial \theta(x,y)}{\partial y}. \tag{9}$$

$\theta(x,y)$ is a smoothing function whose integration over the full domain is equal to 1 and converges to 0 at infinity. These two functions have to satisfy the following conditions:

$$\int_{-\infty}^{\infty}\int_{-\infty}^{\infty} \psi^1(x,y)dxdy = 0 \tag{10}$$

and

$$\int_{-\infty}^{\infty}\int_{-\infty}^{\infty} \psi^2(x,y)dxdy = 0. \tag{11}$$

Let $f(x,y) \in \mathbf{L}^2(\mathbf{R})$. The so-called dyadic wavelet transform [2,18] of $f(x,y)$ at scale 2^j along the x and y directions can be represented, respectively, by

$$W_{2^j}^1 f(x,y) = f * \psi_{2^j}^1(x,y), \qquad (12)$$

and

$$W_{2^j}^2 f(x,y) = f * \psi_{2^j}^2(x,y), \qquad (13)$$

where $\psi_{2^j}^1(x,y) = \frac{1}{2^{2j}}\psi^1(\frac{x}{2^j}, \frac{y}{2^j})$ and $\psi_{2^j}^2(x,y) = \frac{1}{2^{2j}}\psi^2(\frac{x}{2^j}, \frac{y}{2^j})$. In what follows, we shall use the above mentioned dyadic wavelet transform to detect significant normal changes as edge points.

From Equation(3), it is obvious that the vector of the normal change, $d\boldsymbol{N}(u,v)$, can be represented by the linear combination of the two bases on the du-dv plane, i.e., $\boldsymbol{N}_u du + \boldsymbol{N}_v dv$. Also, their associated weights are the gradients of $\boldsymbol{N}$ along the du and dv directions, respectively. Since the dyadic wavelet transform proposed by Mallat [2,13] can be used to calculate the magnitudes of these gradients, we can apply their method directly to calculate $d\boldsymbol{N}(u,v)$. According to the formulation reported in [2,13], the vector dyadic wavelet transform of $\boldsymbol{N}(u,v)$ at scale 2^j can be defined as follows:

$$\mathcal{W}_j \boldsymbol{N}(u,v) = W_{2^j}^1 \boldsymbol{N}(u,v) du + W_{2^j}^2 \boldsymbol{N}(u,v) dv, \qquad (14)$$

where

$$W_{2^j}^1 \boldsymbol{N}(u,v) = (W_{2^j}^1 n_1(u,v),\ W_{2^j}^1 n_2(u,v),\ W_{2^j}^1 n_3(u,v)\ ), \qquad (15)$$

and

$$W_{2^j}^2 \boldsymbol{N}(u,v) = (W_{2^j}^2 n_1(u,v),\ W_{2^j}^2 n_2(u,v),\ W_{2^j}^2 n_3(u,v)\ ). \qquad (16)$$

Since the $\mathcal{W}_j \boldsymbol{N}(u,v)$ vector also lies on the du-dv plane, the magnitude and argument of $\mathcal{W}_j \boldsymbol{N}(u,v)$ can be directly computed. Referring to [2], the norms of $W_{2^j}^1 \boldsymbol{N}(u,v)$ and $W_{2^j}^2 \boldsymbol{N}(u,v)$ should be defined, respectively, as follows:

$$\|W_{2^j}^1 \boldsymbol{N}(u,v)\| = \sqrt{[W_{2^j}^1 n_1(u,v)]^2 + [W_{2^j}^1 n_2(u,v)]^2 + [W_{2^j}^1 n_3(u,v)]^2}, \qquad (17)$$

and

$$\|W_{2^j}^2 \boldsymbol{N}(u,v)\| = \sqrt{[W_{2^j}^2 n_1(u,v)]^2 + [W_{2^j}^2 n_2(u,v)]^2 + [W_{2^j}^2 n_3(u,v)]^2}. \qquad (18)$$

The magnitude of $\mathcal{W}_j \boldsymbol{N}(u,v)$ at scale 2^j can, thus, be computed as follows:

$$M_{2^j} \boldsymbol{N}(u,v) = \sqrt{[\|W_{2^j}^1 \boldsymbol{N}(u,v)\|]^2 + [\|W_{2^j}^2 \boldsymbol{N}(u,v)\|]^2}. \qquad (19)$$

Furthermore, the angle of $\mathcal{W}_j \boldsymbol{N}(u,v)$ with respect to du direction is

$$A_{2^j} \boldsymbol{N}(u,v) = argument(\|W_{2^j}^1 \boldsymbol{N}(u,v)\| + i\|W_{2^j}^2 \boldsymbol{N}(u,v)\|). \qquad (20)$$

From the above calculations, every point in a range image will obtain two values. One is the magnitude of its normal change with respect to its neighbors, and the other is the direction tendency of this point. Like other multiscale edge detection methods[2,18,19], the edge points can be determined by locating those local extrema whose normal changes exceed a preset threshold.

5 Experimental results

In the experiments, a number of synthetic and real range images were adopted as test images to corroborate the effectiveness of the proposed method. Figure 3 showed a synthetic image : agpart. The size of the agpart image was 240×240. We also used two real images with different sizes. In the first stage of the experiment, the normal of every point in a range image had to be decided. In order to make a comparison, we used three different methods, *i.e.*, quadradic surface fitting, and orthogonal and non-orthogonal wavelet-based approaches, to calculate the normal values. For this part we used the synthetic images, "agpart", as the test image. The experimental results of this part are shown in Figure 4. Figures 5(a)-(c) show, respectively, the multiscale edges (2^1, 2^2, and 2^3) detected from three differently approximated surfaces. Among them, It is apparent that, when the quadratic surface fitting or the orthogonal wavelet-based approach was adopted to estimate the normal values, the detected edges contained some spurious results or the original edges delocalized from their original position. On the other hand, the edges detected from the normals as estimated by the non-orthogonal wavelet-based approach were the best results. One thing to be noted is that the first stage of the proposed approach, *i.e.*, the normal determination step, is crucial because a "poor" estimation of normal values may have an irrecoverable effect on the edge detection stage. A poor estimation method may smooth out the original image and, thus, delocalize edges from their correct locations. After comparing the empirical results, the non-orthogonal wavelet-based approximation was chosen because it produced the best results out of the three methods. Figures 6 and 7 show the results of a sequence of experiments based on real range images. All of these results were obtained by applying the non-orthogonal wavelet-based approach to estimate the normal values. From the results, we can find that most of the crease (roof) edges were detected correctly.

6 Conclusion and discussion

In this paper, we have proposed a new edge detection technique based on detection of normal changes. We have found that, by detecting normal changes, both step edges and roof edges can be easily identified. Therefore, the new technique has been proven to be a more promising method than other methods that base their detection criterion on depth or curvature changes.

References

1. J. W. Hsieh, M. T. Ko, H. Y. Mark Liao, and K. C. Fan, "A new wavelet-based edge detector via constrained optimization", *Image and Vision Computing*, 1997, to appear.
2. S. Mallat and S. Zhong, "Characterization of signal from multi-scale edges", *IEEE Transactions on Pattern Analysis and Machine Intelligence*, vol. 14, no. 7, pp. 710–732, July 1992.

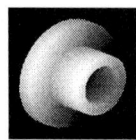

Fig. 3. A synthetic range image: agpart.

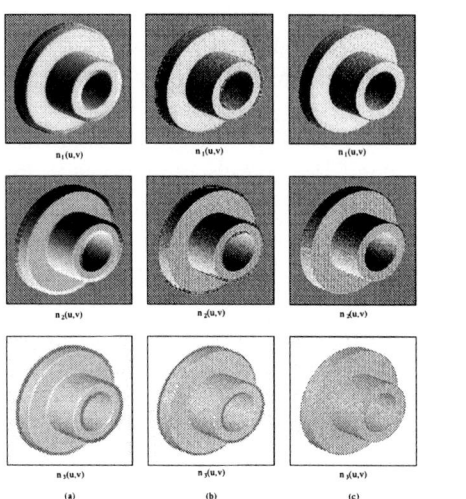

Fig. 4. Three component of normals (n_1, n_2, n_3) estimated by using (a) quadratic surface fitting (b) the orthogonal wavelet-based approach with vanishing moment N=3, and (c) the non-orthogonal wavelet-based approach.

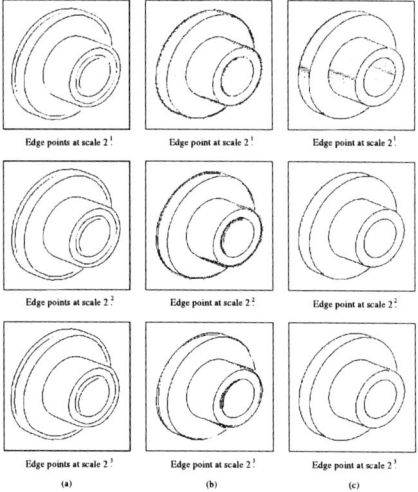

Fig. 5. Detected edge points at different scales using (a) quadratic surface fitting (b) the orthogonal wavelet-based approach with vanishing moment N=3, and (c) the non-orthogonal wavelet-based approach.

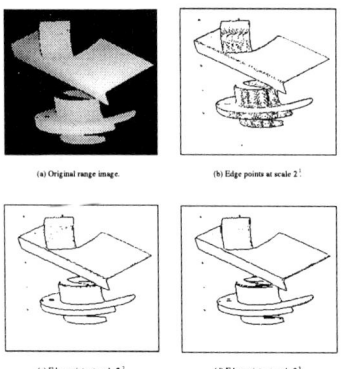

Fig. 6. Detected edge points at different scales from real image - "occl".

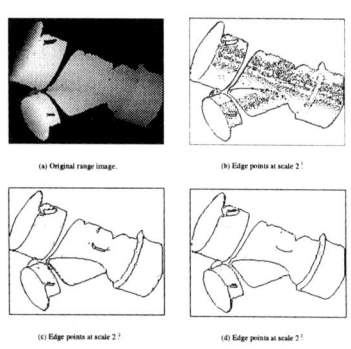

Fig. 7. Detected edge points at different scales from real image - "bigwye".

3. S. Ghosal and R. Mehrotra, "Segmentation of range images: an orthogonal moment-based integrated approach", *IEEE Transactions on Robotics and Automation*, vol. 61, no. 2, pp. 171–189, Aug. 1993.
4. O. Monga and S. Benayoun, "Using partial derivatives of 3D images to extract typical surface features", *Computer Vision and Image Understanding*, vol. 61, no. 2, pp. 171–189, Mar. 1995.
5. O. Monga, R. Deriche, G. Malandain, and J. P. Cocquerez, "Recusive filtering and edge tracking: two primary tools for 3D edge detection", *Image and Vision Computing*, vol. 9, no. 4, pp. 203–214, Aug. 1991.
6. T. J. Fan, G. Medioni, and R. Nevatia, "Segmented descriptions of 3-D surfaces", *IEEE Transactions on Robotics and Automation*, vol. 3, no. 6, pp. 527–538, Dec. 1987.
7. P. J. Besl, *Surface in range image understanding*, Spring-Verlag, New York, 1988.
8. H. M. Cung, P. Cohen, and P. Boulanger, "Multiscale edge detection and classification in range images", in *IEEE International Conference on Robotics and Automation*, 1990, pp. 2038–2044.
9. M. Djebali, M. Melkemi, and D. Vandorpe, "3D range image segmentation based on Deriche's optimun filters", in *Proceedings of IEEE International Conference on Image Processing*, 1994, pp. 503–507.
10. M. P. Do. Carmo, *Differential Geometry of Curves and Surfaces*, Prentice-Hall, Englewood Cliffs, NJ, 1976.
11. P. J. Besl and R. C. Jain, "Invariant surface characteristics for three-dimensional object recognition in range images", *CVGIP*, vol. 33, pp. 33–80, 1986.
12. I. Daubechies, "The wavelet transform, time-frequency localization and signal analysis", *IEEE Trans. Information Theory*, vol. 36, no. 5, pp. 961–1005, Sept. 1990.
13. S. Mallat, "Multiresolution approximation and wavelet orthonormal bases of $l(r^2)$", *Transactions on Amer. Math.*, vol. 15, no. 3, pp. 69–87, Sept. 1989.
14. J. McCleary, *Geometry from a Differentialable Viewpoint*, Cambridge university, Cambridge, 1994.
15. I. Daubechies, "Orthonormal bases of compactly supported wavelets", *Communications Pure Applied Mathematics*, vol. 41, pp. 909–996, 1988.
16. X. Zhou R. O. Wells, "Wavelet interpolation and approximate solutions of elliptic partial differential equations", Tech. Rep., Computational Mathematics Laboratory, Rice University, 1993.
17. J. W. Hsieh, H. Y. Mark Liao, M. T. Ko, and K. C. Fan, "Wavelet-based shape from shading", *Graphical Models and Image Processing*, vol. 57, no. 4, pp. 343–362, July 1995.
18. S. Mallat and W. L. Hwang, "Singularity detection and processing with wavelets", *IEEE Transactions on Information Theory*, vol. 38, no. 2, pp. 617–643, March 1992.
19. J. W. Hsieh, H. Y. Mark Liao, M. T. Ko, K. C. Fan, and Y. P. Hung, "Image registration using an edge-based approach", *Computer Vision and Image Understanding*, 1997, to appear.

Extending Adjacency to Fuzzy Sets for Coping with Imprecise Image Objects

Isabelle Bloch, Henri Maître

École Nationale Supérieure des Télécommunications
Département Images - CNRS URA 820
46 rue Barrault, 75013 Paris, France
Tel: +33 1 45 81 75 85, Fax: +33 1 45 81 37 94, E-mail: bloch@ima.enst.fr

1 Introduction

Adjacency has a large interest in image processing and pattern recognition, since it qualifies an important relationship between image objects or regions. A crisp definition of adjacency between crisp objects often leads to a low robustness in case of noise or segmentation errors. Let us consider for instance a problem of model-based pattern recognition where spatial relationships are an important part of the recognition process. If two model objects are adjacent, we expect the corresponding image objects to be adjacent too, otherwise they will be difficult to recognize. However, if classical crisp adjacency is used, the fact that two objects are adjacent or not may depend on one point only.

In order to include possible errors or imprecision in the processing and in the recognition, we use the framework of fuzzy sets that already proved to be useful for image processing under imprecision. Two ways can be considered for representing imprecision. In the first one, the satisfaction of the adjacency property between two objects is considered to be a matter of degree. The second one consists in introducing imprecision in the objects themselves, and to deal with fuzzy objects, i.e. with objects considered as fuzzy sets on the image space. For instance, spatial imprecision due to the limited quality of image information can be represented in an adequate way by considering fuzzy objects. Then obviously adjacency is also a matter of degree. In both cases, a need exists to give proper definitions for fuzzy adjacency between image regions. Here the second way only will be considered (see [3] for an approach along the first track).

Unfortunately, only a few works address the problem of fuzzy adjacency in the literature. Fuzzy topology was introduced in [7]. In this paper, Rosenfeld defines a fuzzy connectivity between points but without reference to fuzzy neighborhood, or to fuzzy adjacency. Similar approaches can be found in [8], [11], where degrees of connectivity in a fuzzy set are also introduced, but neither the connectivity nor the adjacency between two fuzzy sets are defined. Rosenfeld and Klette [9] define a degree of adjacency between two crisp or fuzzy sets, using a geometrical approach based on the notion of "visibility" of a set from another one. However, this definition is not symmetrical, and probably not easy to extend to higher dimensions. We propose in this paper a completely different approach. The closest to our work is probably the one decribed in [4], where a degree of adjacency between two fuzzy sets is defined by extending binary definitions of contours, frontiers, and neighborhood. Again, the proposed definition of [4] is not symmetrical, and presents some other drawbacks.

In this paper we propose several definitions for degree of adjacency coping with spatial imprecision in image processing. Basic definitions for classical notions of adjacency are given in Section 2, in the discrete domain. Then we account for imprecision in the representation of objects, consequently defined as spatial fuzzy sets, and we define a degree of adjacency between two fuzzy sets. This is detailed in Section 3, using fuzzy extensions of the notions of neighborhood and of boundary of a set. We finally show that fuzzy mathematical morphology provides a consistent framework for expressing the obtained definitions.

2 Crisp adjacency

Here, we restrict ourselves to the discrete case, and use discrete topology as derived from discrete connectivity for defining adjacency between two image regions X and Y, subsets of the discrete space. Consider an n-dimensional discrete space (typically $\mathbb{Z}^n$), and any discrete connectivity defined on this space, denoted c-connectivity (for instance, for $n = 3$, we may consider 6-, 18- or 26-connectivity on a cubic grid). Since we would like to distinguish between connectedness and adjacency relationships, we use the following definition of crisp adjacency, where we denote by $n_c(x, y)$ the Boolean variable stating that x and y are neighbors in the sense of the discrete c-connectivity:

Definition 1 *For any two subsets X and Y in $\mathbb{Z}^n$, X and Y are adjacent according to the c-connectivity if: $X \cap Y = \emptyset$ and $\exists x \in X, \exists y \in Y : n_c(x, y)$.*

We consider for the discrete boundary of a set X its interior boundary defined as: $\partial X = X - E(X, B_c)$ where $E(X, B_c)$ denotes the morphological erosion of X by the structuring element B_c of size 1 defined according to the chosen discrete connectivity [10]. Using this definition, the discrete adjacency can be related to the boundary in the following way:

Property 1 *A consequence of definition 1 is that, if X and Y are adjacent, then any $x \in X$ and $y \in Y$ that satisfy $n_c(x, y)$ belong to the boundary of X and Y respectively.*

Therefore the fuzzy extension of definition 1 can be obtained either by considering only the constraint on the neighborhood, or by considering also the constraint on the boundary, as will be seen in Section 3.2.

Property 2 *Definition 1 can also be expressed equivalently in terms of morphological dilation, as: $X \cap Y = \emptyset$ and $D(X, B_c) \cap Y \neq \emptyset$, $D(Y, B_c) \cap X \neq \emptyset$, where $D(X, B_c)$ denotes the dilation of X by the structuring element B_c.*

This property provides a third way to extend the definition to fuzzy sets, either directly from fuzzy dilation, or by means of distance computation, which is closely related to dilation.

3 Extending adjacency to fuzzy objects

In the rest of this text we consider fuzzy objects (i.e. fuzzy sets defined on the considered space by means of their membership function) and define fuzzy adjacency between such objects. In the discrete case, a fuzzy object is simply defined by its membership function, defined on $\mathbb{Z}^n$ and taking values in $[0, 1]$.

3.1 Methods

This Section presents shortly the possible principles that can be used for extending adjacency to fuzzy sets and the requirements posed to this extension. We consider the general problem of extending a relationship R_B between two binary objects to its fuzzy equivalent R (fuzzy relationship between two fuzzy objects). Instantiations of the described methods to the case of adjacency are provided next.

By using the α-cuts One way to define crisp sets from a fuzzy set consists in taking the α-cuts of this set. Therefore a first class of methods relies on the application of the relationship R_B to each α-cut. This gives rise to two different "fuzzification" methods in the literature.

The first one consists in "stacking" the results obtained with binary operations on the α-cuts: let us denote by μ and ν the membership functions of two fuzzy objects defined on the considered space and taking values in [0,1], the fuzzy equivalent R of R_B is then defined as (see e.g. [5]): $R(\mu,\nu) = \int_0^1 R_B(\mu_\alpha, \nu_\alpha) d\alpha$, or similarly by a double integration (other fuzzification equations exist, but will not be examined here). Examples of this approach concern for instance connectivity [8], fuzzy mathematical morphology [2], distances [5], [1], etc.

The second method is the extension principle [12], which leads, in the general case, to a fuzzy number: $\forall n \in \mathcal{V}(R_B), R(\mu,\nu)(n) = \sup_{R_B(\mu_\alpha, \nu_\alpha)=n} \alpha$, where $\mathcal{V}(R_B)$ denotes the image of R_B, i.e. the set of values taken by R_B. If the relationship to be extended only takes binary values (0/1, or true/false), then the extension principle reduces to: $R(\mu,\nu) = \sup_{R_B(\mu_\alpha, \nu_\alpha)=1} \alpha$. This is typically the case for binary adjacency between binary sets.

By formal translation of equations A second class of methods consists in translating binary equations into their fuzzy equivalent: intersection is replaced by a t-norm, union by a t-conorm, sets by membership functions, etc. This has been used e.g. for defining fuzzy morphology [2]. This translation is straightforward if the binary relationship can be expressed in set theoretical and logical terms. It is obtained in a natural way from the definitions given in Section 2. Moreover, this remark endows methods based on mathematical morphology with a particular interest, since mathematical morphology is mainly based on set theory.

Since methods based on α-cuts appeared to have poorer properties than those based on a formal translation (see [3]), we restrict to this second kind of approach in the following.

Minimal properties required for a fuzzy adjacency The properties we require for fuzzy adjacency are the following: symmetry, consistency with binary definitions, decreasingness with respect to the distance between both sets. A last property, often desirable although not mandatory, is invariance with respect to geometrical transformations.

3.2 Formal translation of binary adjacency equations into fuzzy adjacency

In this Section, we make use of the principle shortly described in Section 3.1 in order to define a degree of adjacency between two fuzzy sets μ and ν. Since binary definitions always involve constraints on the intersection of the two sets and a notion of neighborhood, we first define fuzzy equivalents of these concepts. Then, we extend definition 1, using only neighborhood relationships, and then, we add boundary constraints, as introduced in property 1. We also consider fuzzy adjacency derived from fuzzy dilation and from fuzzy distance.

Degree of intersection between two fuzzy sets The degree of intersection between two fuzzy sets is obtained by translating the set equation $X \cap Y \neq \emptyset$ into fuzzy terms. This equation is equivalent to $\exists x \in \mathbb{Z}^n$, $x \in X \cap Y$. The simplest fuzzy translation provides: $\mu_{int}(\mu, \nu) = \sup_x t[\mu(x), \nu(x)]$, where t is any t-norm [6]. The supremum is taken over the whole space. A degree of empty intersection (or of disjunctness) is then derived as: $\mu_{\neg int}(\mu, \nu) = c[\mu_{int}(\mu, \nu)]$, where c is a fuzzy complementation (for instance defined as $\forall a \in [0, 1], c(a) = 1 - a$). This form is not always adequate for image processing purposes since it does not incorporate any spatial information. Degrees of intersection and of non-intersection can therefore be reformulated in order to better reflect the spatial overlapping by considering the fuzzy hypervolume of the intersection. This may also be interpreted as a translation process, in the sense that we have: $X \cap Y = \emptyset \Leftrightarrow V_n(X \cap Y) = 0$. For defining the hypervolume of a fuzzy set, we simply use the classical fuzzy cardinality. This provides for a fuzzy set μ (with a bounded support) in the discrete case: $V_n(\mu) = \sum_{x \in \mathbb{Z}^n} \mu(x)$. From the hypervolume of $t(\mu, \nu)$, a degree of intersection in $[0, 1]$ is derived. It should be equal to 0 if μ and ν have completely disjoint supports, be high if one set is included in the other, and increasing with respect to the hypervolume of the intersection. The following definition satisfies these requirements:

Definition 2 *The degree of intersection between two fuzzy sets μ and ν, depending on the hypervolume of their intersection, is defined by:*
$\mu_{int}(\mu, \nu) = \frac{V_n[t(\mu,\nu)]}{\min[V_n(\mu), V_n(\nu)]}$. *Here again a degree of non-intersection can be derived from this expression using fuzzy complementation.*

Property 3 *The intersection degrees defined by maximum of intersection and definition 2 are both consistent with the binary definition and invariant with respect to geometrical transformations.*

In the following definitions of fuzzy adjacency, we may use either expressions derived from the height of the intersection or expressions involving the fuzzy hypervolume of the intersection. We will see that this leads to different adjacency degrees in overlapping situations.

Fuzzy neighborhood In this Section, we define a degree of neighborhood n_{xy} between two points x and y in $\mathbb{Z}^n$ endowed with a discrete connectivity. Let us first consider binary definitions of n_{xy}: we set $n_{xy} = 1$ if x and y are neighbors in the sense of the considered discrete connectivity, and $n_{xy} = 0$ otherwise (i.e. $n_{xy} = n_c(x, y)$). With this definition, the consistency with the binary case is guaranteed. In the fuzzy case, n_{xy} can be defined as a decreasing function of the distance between x and y, as proposed in [4] and in [3].

Using neighborhood constraints We propose to fuzzify definition 1 by combining a degree of empty intersection ($X \cap Y = \emptyset$) with a degree of existence of neighbors ($\exists x \in X, \exists y \in Y, n_c(x, y)$) using a t-norm t (expressing the simultaneous satisfaction of both conditions). For the first part, a degree of non-intersection can be used, derived either from the height of the intersection of from its fuzzy hypervolume, as suggested in Section 3.2. For the second part, existence is translated by means of a supremum (taken over the whole space), leading to: $\sup_x \sup_y t[\mu(x), \nu(y), n_{xy}]$, where n_{xy} represents the degree to which x and y are neighbors[1]. It can be either crisp or fuzzy (as defined in Section 3.2). Finally, we obtain the following definition for fuzzy adjacency.

Definition 3 *The degree of adjacency between μ and ν involving only neighborhood constraints is defined as:* $\mu_{adj}(\mu, \nu) = t[\mu_{\neg int}(\mu, \nu), \sup_x \sup_y t[\mu(x), \nu(y), n_{xy}]]$.

Property 4 *The degree of adjacency obtained with this definition is symmetrical, consistent with the discrete binary definition (i.e. in the case where μ and ν are crisp and $n_{xy} = n_c(x, y)$), and decreasing with respect to the distance between the two fuzzy sets. It is invariant with respect to geometrical transformations (for scaling, only if n_{xy} is itself invariant).*

Figure 1 illustrates the results obtained with definition 3 with the t-norm minimum and both definitions of degree of intersection. Using the maximum of the intersection we obtain $\mu_{adj}(\mu, \nu) = 0.36$ and $\mu_{adj}(\mu, \nu') = 0.35$, which are very similar values. On the contrary, using the fuzzy hypervolume, definition 3 accounts for the differences in intersection and provides $\mu_{adj}(\mu, \nu) = 0.67$ and $\mu_{adj}(\mu, \nu') = 0.34$, which are different and better fit the intuition.

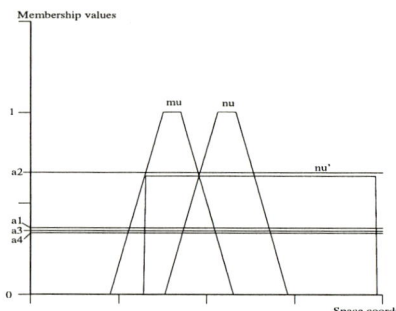

Fig. 1. Illustration of definition 3 when using different definitions for the degree of intersection. Using the maximum of intersection we obtain $\mu_{adj}(\mu, \nu) = a_1 (= 0.36)$ and $\mu_{adj}(\mu, \nu') = a_3 (= 0.35)$, and using the fuzzy hypervolume $\mu_{adj}(\mu, \nu) = a_2 (= 0.67)$ and $\mu_{adj}(\mu, \nu') = a_4 (= 0.34)$.

Let us consider now a 2D example. Figure 2 shows a slice of a magnetic resonance (MR) image of the human brain, where several structures have been segmented and serve as a model (or atlas), and a slice (at approximately the same level) of another MR image where the same structures have to be recognized. A rough fuzzy segmentation of this second image is also shown. The adjacency degrees between some of the obtained fuzzy objects are given in Table 1. They are obtained using definition 3 with the maximum of intersection as intersection degree and the t-norm minimum. 4-connectivity was used. The results are in agreement with what can be expected from the model. In this case, crisp adjacency would provide completely different results in the model and in the image, preventing its use for recognition. This suggests that fuzzy adjacency degree can indeed be used for pattern recognition purposes, of course combined with other spatial relationships.

Adding boundary constraints Another way to extend fuzzy adjacency from definition 1 consists in introducing a constraint on the boundary of the considered

---

[1] In such expressions $t(a, b, c)$ stands for $t[t(a, b), c]$. This notation is adopted for sake of simplicity and justified since any t-norm is commutative and associative.

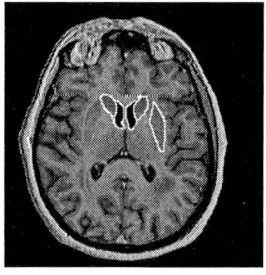 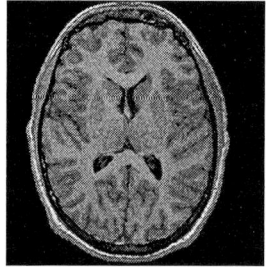 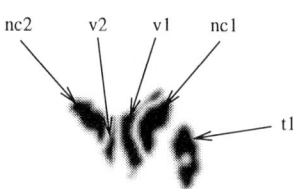

Fig. 2. MR image of a brain with a few segmented structures (left). MR image of another brain (middle). Right: 5 fuzzy objects resulting from a rough fuzzy segmentation of the middle image (membership values rank between 0 and 1, from white to black, the maximum membership value is displayed at each point) and labels (used in Table 1).

Fuzzy object 1	Fuzzy object 2	degree of adjacency	adjacency in the model (crisp)
v1	v2	0.368	1
v1	nc1	0.463	1
v1	t1	0.000	0
v1	nc2	0.035	0
v2	nc2	0.427	1
nc1	t1	0.035	0

Table 1. Results obtained using definition 3 with the maximum of intersection as intersection degree, the t-norm minimum, and 4-connectivity. Labels of structures are given in Figure 2. High degrees are obtained between structures where adjacency is expected, while very low degrees are obtained in the opposite case.

sets, as given by property 1, i.e. the neighbor points involved in definition 1 are on the boundary of the sets.

A similar work has already be done in [4], but although the approach is very attractive, the proposed definitions suffer from several drawbacks, with respect to the requirements we imposed in this paper (not symmetrical, not consistent with the binary case, etc., see [3] for more details).

We propose a new definition that overcomes these drawbacks and better matches our requirements. Our approach consists in defining only the fuzzy boundaries of the fuzzy sets, which are then combined with neighborhood relationship.

Definition 4 *The fuzzy boundary of a fuzzy set μ is defined by the membership function b_μ as: $\forall x \in \mathbb{Z}^n, b_\mu(x) = t[\mu(x), \sup_{z \in \mathbb{Z}^n} t[c(\mu)(z), n_{xz}]]$.*

Property 5 *In the binary case (μ and n_{xz} binary), this definition is consistent with the classical definition of the boundary of a crisp set X (set of points of X that have a neighbor in X^C). It is also invariant with respect to geometrical transformations (for scaling, only if n_{xy} is itself invariant).*

The translation of definition 1 along with the property on boundary leads now to the following definition:

Definition 5 *The degree of adjacency between μ and ν involving neighborhood and boundary constraints is defined by:*
$\mu_{adj}(\mu, \nu) = t[\mu_{\neg int}(\mu, \nu), \sup_x \sup_y t[b_\mu(x), b_\nu(y), n_{xy}]]$, *where the supremum is taken over $\mathbb{Z}^n$.*

This definition is illustrated in Figure 3.

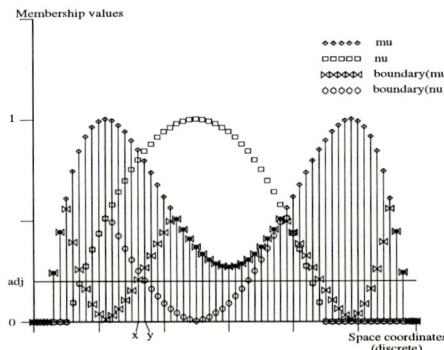

Fig. 3. Fuzzy boundary b_μ and b_ν of two fuzzy sets μ and ν. Fuzzy adjacency is then equal to adj according to definition 5, which is less than a_2 provided by definition 3. The neighbor points for which the adjacency value is attained are again x and y.

Again, as in the previous Subsection, if we assume that the considered fuzzy sets constitute a fuzzy partition, we can ignore the first term corresponding to the degree of empty intersection.

Property 6 *The degree of adjacency in this definition is symmetrical, consistent with the binary definition if μ, ν and n_{xy} are binary, and decreases if the distance between μ and ν increases. It is invariant with respect to geometrical transformations (for scaling, only if n_{xy} is itself invariant).*

Property 7 *This definition is equivalent to definition 3 in the binary case. In the fuzzy case, it is more severe, i.e. leads to a lower degree of adjacency.*

Using fuzzy morphological operators In a morphological context, it may also be interesting to define adjacency from fuzzy dilation, by translating property 2 into fuzzy terms. A direct translation of this property leads to the following definition.

Definition 6 *The degree of adjacency between μ and ν involving fuzzy dilation is defined as:* $\mu_{adj}(\mu,\nu) = t[\mu_{\neg int}(\mu,\nu), \mu_{int}[D(\mu, B_c),\nu], \mu_{int}[D(\nu, B_c),\mu]]$.

This definition represents a conjunctive combination of a degree of non-intersection between μ and ν and a degree of intersection between one fuzzy set and the dilation of the other. B_c can be taken as the elementary structuring element related to the considered connectivity, or as a fuzzy structuring element, representing for instance spatial imprecision (i.e. the possibility distribution of the location of each point).

Property 8 *This definition is symmetrical, consistent with the binary definition if μ, ν and B_c are binary, and decreases if the distance between μ and ν increases.*

Fuzzy dilation can also serve for defining the fuzzy boundary of a fuzzy set, as follows.

Definition 7 *The fuzzy boundary b_μ of a fuzzy set μ is defined from fuzzy dilation as:* $b_\mu(x) = t[\mu(x), D(c(\mu), B_c)(x)]$.

Property 9 *This definition is equivalent to definition 4 if the structuring element is consistent with the choice of the fuzzy neighborhood (typically if we take the elementary structuring element defined from the discrete connectivity used in a binary definition of n_{xy}).*

Property 10 *The degree of adjacency obtained from this boundary definition (with any structuring element) is still symmetrical, consistent with the binary case and decreasing when the distance between both fuzzy sets increases.*

Finally, since in the discrete binary case the equation using dilation means that the minimum (nearest point) distance between X and Y is equal to 1, we can also exploit this fact in the fuzzy case, by using the fuzzy minimum distance, defined from fuzzy dilation as in [1]. We do not go into further details for this approach, since it leads to similar definitions, sharing the same properties as the previous ones.

4 Conclusion

The aim of this research was to investigate notions of fuzzy adjacency that could serve for model-based pattern recognition in image processing under imprecision. We proposed several definitions for extending adjacency to fuzzy objects, that show good properties with respect to binary definitions and to the requirements we imposed, and that provide for a consistent representation and management of imprecision, which is directly represented in the considered objects.

Acknowledgment: This work has been partially supported by the "Conférence des Grandes Écoles" (we would like to thank in particular Prof. Michel Perrin) and has been initiated while the first author was visiting at BISC group, Computer Science Department, University of California at Berkeley (Profs. Zadeh and Anvari).

References

1. I. Bloch. Distances in Fuzzy Sets for Image Processing derived from Fuzzy Mathematical Morphology. In *Information Processing and Management of Uncertainty in Knowledge-Based Systems*, volume III, pages 1307–1312, Granada, Spain, July 1996.
2. I. Bloch and H. Maître. Fuzzy Mathematical Morphologies: A Comparative Study. *Pattern Recognition*, 28(9):1341–1387, 1995.
3. I. Bloch, H. Maître, and M. Anvari. Fuzzy Adjacency between Image Objects. *International Journal of Uncertainty, Fuzziness and Knowledge-Based Systems*, 5(6), 1997.
4. C. Demko and E. H. Zahzah. Image Understanding using Fuzzy Isomorphism of Fuzzy Structures. In *IEEE Int. Conf. on Fuzzy Systems*, pages 1665–1672, Yokohama, Japan, March 1995.
5. D. Dubois and M.-C. Jaulent. A General Approach to Parameter Evaluation in Fuzzy Digital Pictures. *Pattern Recognition Letters*, 6:251–259, 1987.
6. D. Dubois and H. Prade. Combination of Information in the Framework of Possibility Theory. In M. Al Abidi et al., editor, *Data Fusion in Robotics and Machine Intelligence*. Academic Press, 1992.
7. A. Rosenfeld. Fuzzy Digital Topology. *Information and Control*, 40:76–87, 1979.
8. A. Rosenfeld. The Fuzzy Geometry of Image Subsets. *Pattern Recognition Letters*, 2:311–317, 1984.
9. A. Rosenfeld and R. Klette. Degree of Adjacency or Surroundness. *Pattern Recognition*, 18(2):169–177, 1985.
10. J. Serra. *Image Analysis and Mathematical Morphology*. Academic Press, London, 1982.
11. J. K. Udupa and S. Samarasekera. Fuzzy Connectedness and Object Definition: Theory, Algorithms, and Applications in Image Segmentation. *Graphical Models and Image Processing*, 58(3):246–261, 1996.
12. L. A. Zadeh. The Concept of a Linguistic Variable and its Application to Approximate Reasoning. *Information Sciences*, 8:199–249, 1975.

Adaptive Selection of Image Classifiers

Giorgio Giacinto and Fabio Roli

Dept. of Electrical and Electronic Eng., University of Cagliari, Italy
Piazza D'Armi, 09123, Cagliari, Italy - Phone: +39-70-6755874 Fax: +39-70-6755900
e-mail {giacinto,roli}@diee.unica.it

Abstract. Recently, the concept of "Multiple Classifier Systems" was proposed as a new approach to the development of high performance image classification systems. Multiple Classifier Systems can be used to improve classification accuracy by combining the outputs of classifiers making "uncorrelated" errors. Unfortunately, in real image recognition problems, it may be very difficult to design an ensemble of classifiers that satisfies this assumption. In this paper, we propose a different approach based on the concept of "adaptive selection" of multiple classifiers in order to select the most appropriate classifier for each input pattern. We point out that adaptive selection does not require the assumption of uncorrelated errors, thus simplifying the choice of classifiers forming a Multiple Classifier System. Reported results on the classification of remote-sensing images show that adaptive selection can be used to obtain substantial improvements in classification accuracy.

1 Introduction

Recently, in the field of character recognition, the concept of Multiple Classifier Systems (MCSs) was proposed as an approach to develop a high performance recognition system [1,2]. In particular, it has been pointed out that by combining the outputs of an MCS it is easy to exploit complementary characteristics of classification algorithms based on different methodologies and/or using different input features [1]. The potentialities of these recognition systems have been reported also in the remote sensing field [3–5]. Several combination functions have been proposed based on voting rules, statistical theory, Dempster-Shafer evidence theory, belief functions, and many other "integration schemes" [1,2,6–8]. Despite the promising results reported in the literature, performances of MCS greatly depend on the assumption that classifiers exhibit a sufficiently large "uncorrelation" in their classification errors [1,9,10].

In this paper, a different approach to the exploitation of the potential advantages of MCSs is proposed. This approach is based on the concept of "adaptive selection" of multiple classifiers aimed at selecting the most appropriate classifier for each input pattern. This concept is not completely new in the field of pattern recognition. Recently, Srihari et al. pointed out the potentialities of "dynamic classifier selection" [7]. In the neural networks field, Jacobs and Jordan proposed a multiple neural network system that allows for a particular kind of dynamic selection based on the concept of "adaptive mixtures of local experts" [11]. In this paper, we first point out that adaptive selection does not require the assumption of uncorrelated errors (Section 2). Afterwards, a selection algorithm is described (Section 3). In Section 4, a method based on data clustering is proposed to design selection-based multiple classifier systems. In Section 5, experimental results on the classification of remote-sensing images are reported. Conclusions are drawn in Section 6.

2 Multiple Classifier Systems: Selection vs. Combination

Some researchers clearly showed that combination mechanisms can increase classification accuracy only if the assumption of independent classification errors is satisfied. Hansen and Salamon showed that a combination mechanism based on a simple majority decision rule can provide very good performances if classifiers are "independent" [8]. Tumer and Gosh pointed out that classification accuracy increases obtained by combining depend on error correlation more than on the particular combination mechanism adopted [9]. On the other hand, experimental results showed that, in real pattern recognition applications, may be very difficult to design and train independent classifiers, even if based on different methodologies [2,3]. Consequently, very recently, some researchers proposed combination mechanisms aimed at avoiding the independence assumption [4,12]. Methods that identify and remove classifiers that are excessively correlated have also been proposed [10,13].

It is quite easy to see that if we could design an "optimal classifier selector" that always selects the most appropriate classifier for each test pattern, then there would be no longer any need for an ensemble of independent classifiers. For each test pattern, it would be sufficient to have just one classifier that correctly classifies it. Unfortunately, it is just as easy to see that the above optimal classifier selector is more difficult to "design" than the combination mechanisms adopted in the present MCSs. The design of an adaptive classifier selector requires the definition of "selecting conditions" that focuses on choosing the most appropriate classifier for each input pattern. On the other hand, the combination can be implemented more simply, but requires the "selection" of independent classifiers. Therefore, selection mechanisms can greatly simplify that part of the MCS design related to a choice of classifiers. Their drawbacks are mainly related to designing complexity and computational load. The reverse is true for combination mechanisms.

3 The Proposed Algorithm for Adaptive Classifier Selection

The proposed algorithm is based on the definition of a "selecting condition" which makes it possible to select, for each test pattern, the classifier that has more chances to make a correct classification on that pattern. This selecting condition is based on the estimate of *classifier local accuracies* in a "neighbourhood" of the input pattern **X** (neighbourhood(**X**)), defined with respect to a "validation set", i.e., a set of data whose classification is known but that is different from the data set used to train classifiers. The neighbourhood could be also defined with respect to the training set, but it may be very difficult to provide good estimates of classification local accuracies, since, mainly due to the so-called "overfitting problem", classifiers exhibit good performances.

Let us assume that our MCS is formed by K classifiers C_j, $j = 1\ldots K$ and each classifier focuses on solving a pattern recognition problem with M data classes ω_i, $i = 1\ldots M$. For each test pattern **X** the estimate of classifier local accuracies in a "neighbourhood" of the input pattern **X** can be computed with the following formula:

$$\hat{p}(correct_j/\mathbf{X}, neighbourhood(\mathbf{X})) = \frac{N_j}{N} \quad (1)$$

where N is the number of validation patterns forming the neighbourhood(**X**) and N_j is the number of validation patterns that were correctly classified by the classifier C_j. At present, the appropriate dimension of the neighbourhood is decided by experiments or by using heuristic rules.

The ratio computed in the above equation is assumed to be equal to the probability that classifier C_j correctly classifies the test pattern **X**. The rationale of this assumption is a sort of "stationarity" of classification accuracy in a small "partition" of the data set, i.e., all the patterns belonging to the neighbourhood have the same probability of being correctly classified by a given classifier. The general validity of this assumption is very difficult to prove. It strictly depends on the available data set and on the size of the neighbourhood. However, according to our experiments, it seems to apply for most cases and, in particular, it is reasonable for our purposes, since it allows us to compare the classifiers "locally" in order to select the most appropriate for the test pattern.

A "soft" version of equation (1) can be defined as follows:

$$\hat{p}(correct_j/\mathbf{X}, neighbourhood(\mathbf{X})) = \frac{\sum_{i=1}^{N} p_j(\omega_k/\mathbf{X}_i \in \omega_k) W_i}{\sum_{i=1}^{N} W_i} \quad (2)$$

where:

- ω_k ($k = 1\ldots M$) is the correct data class for the neighbourhood pattern $\mathbf{X}_i$;
- $p_j(\omega_k/\mathbf{X}_i)$ is an estimate of the posterior probability provided by classifier C_j. This term constitutes a measure of classifier accuracy on the validation pattern $\mathbf{X}_i$ and, with respect to the "hard" selecting condition defined by equation (1), allows uncertainties related to validation data classifications to be managed more efficiently;
- $W_i = 1/d_i$, where d_i is the Euclidean distance of validation pattern $\mathbf{X}_i$ from the test pattern **X**. This term takes into account the uncertainty due to the heuristic neighbourhood-size definition.

It is easy to see that equations (1) and (2) have a value equal to 1 when the classifier C_j perfectly classifies all the neighbourhood patterns.

The following adaptive classifiers selection algorithm was defined on the basis of the selection conditions described above. Equations (1) or (2) can be used to implement a "hard" or a "soft" selecting condition, respectively.

Adaptive Classifiers Selection Algorithm
Input parameters: classifier confusion matrices on the validation set and
size of the neighbourhood
Begin
For each test pattern **X**:
Do
    $\forall C_j(j = 1 \ldots K)$:
    Begin
    Do
        STEP 1: Compute $\hat{p}(correct_j/\mathbf{X}, neighbourhood(\mathbf{X}))$
        STEP 2: If $\hat{p}(correct_j/\mathbf{X}, neighbourhood(\mathbf{X})) < 0.5$ **Then**
            Reject classifier C_j
    End
    STEP 3: Identify the classifier C_m exhibiting the maximum value of
        $\hat{p}(correct_j/\mathbf{X}, neighbourhood(\mathbf{X})), j = 1 \ldots K', K' \leq K$
    STEP 4: For each classifier $C_l, l = 1 \ldots K'$, compute the following difference
$d_l = [\hat{p}(correct_m/\mathbf{X}, neighbourhood(\mathbf{X})) - \hat{p}(correct_l/\mathbf{X}, neighbourhood(\mathbf{X}))]$
    STEP 5: If $\forall l, l = 1 \ldots K', l \neq m, d_l > Threshold$ **Then**
        Select Classifier C_m **Else**
        Randomly Select one of classifiers for which $d_l < Threshold$
End

Steps 1 and 2 focus on selecting K' classifiers ($K' \leq K$) by removing classifiers that have a probability of less than 0.5 to correctly classify the test pattern **X**. The differences computed at Step 4 are used to compute a sort of "confidence" for the selection. If all the differences are higher than an a-priori fixed threshold (e.g., 0.1), then there is reasonable confidence that classifier C_m is the most appropriate for the test pattern. On the other hand, a random selection is carried out between C_m and the classifiers that exhibit values of the selecting condition close to the value exhibited by C_m. In fact, it is not reasonable to directly select the classifier C_m if there are other classifiers exhibiting similar values of the selecting condition.

4 A Method for Designing MCSs

The basic concepts of this method are the subdivision of the training set into "partitions" and the assignment of each partition to a "specialised classifier". Each specialised classifier is dedicated to correctly classify a partition of the data set and it is consequently trained only on that partition. It is easy to see that the operation mechanism of an MCS based on the above specialised classifiers should be an adaptive selection mechanism.

Let us assume that the training data set Ω is defined by the union of M mutually exclusive data classes ω_k:

$$\Omega = \bigcup_{k=1}^{M} \omega_k \qquad (3)$$

Analogously, each data class ω_k can be defined by the union of M_k data clusters $\omega_{k,m}$, by using one of the many clustering algorithms proposed in the literature [14]:

$$\omega_k = \bigcup_{m=1}^{M_k} \omega_{k,m} \qquad (4)$$

After clustering, the number M_k of clusters is generally different for each data class. Let us assume that ω_i is the data class with the maximum number of clusters M_i. Our goal is to create "partitions" of the data set that correspond to different classification tasks with the same number of M data classes as the initial task, and to assure that the union of these partitions "covers" the data set Ω. To this end, for each data class ω_k, $k \neq i$, "cloned" clusters $\omega_{k,*}$ are generated by a random choice of "natural" clusters in order to obtain a number of clusters equal to M_i for all classes:

$$\omega_k = \omega_{k,1} \cup \omega_{k,2} \cup \cdots \cup \omega_{k,M_k} \cup \omega_{k,*} \cup \cdots \cup \omega_{k,*}, k = 1 \ldots M, k \neq i \qquad (5)$$

As an example, if the class ω_1 has two clusters and M_i is equal to four, then two new clusters for the class ω_1 are generated by randomly choosing among the two natural clusters $\omega_{1,1}$ and $\omega_{1,2}$.

Afterwards M_i partitions of the data set P_z, $z = 1 \ldots M_i$ are defined as follows:

$$P_z = \bigcup_{j=1}^{M} \omega_{j,z} \qquad (6)$$

and a specific classifier C_z is trained on each partition. In most cases, the above partitions are not mutually exclusive. Therefore, the resulting data set covering is redundant.

5 Experimental Results

5.1 Data Set Description

The data set used for our experiments consists of a set of multisensor remote-sensing images related to an agricultural area near the village of Feltwell (UK) [15]. The images were acquired by an ATM sensor with eleven bands and a SAR with twelve channels, both installed on an airplane. For our experiments each pixel was characterised by a fifteen-element "feature vector", using six bands of the ATM and nine channels of the SAR. We selected 10944 pixels belonging to five agricultural classes (i.e., sugar beets, stubble, bare soil, potatoes, carrots) and subdivided them into a training set (5124 pixels), a validation set (582 pixels), and a test set (5238 pixels). We used a very small validation set to simulate real cases where validation data are difficult to obtain.

5.2 Results and Comparisons

Several experiments have been carried out to validate the proposed methods [16]. In the following, for the sake of brevity, we report two main experiments (here called Experiments A and B) that show the main advantages provided by the proposed methods.

Experiment A: We designed an MCS consisting of four classifiers: three multilayer perceptrons (MLPs) neural networks with different architectures to make them as "independent" as possible, and one k-nearest neighbour (k-nn) classifier (we used $k = 21$). With regard to the parameters of our selection algorithm, we used a neighbourhood containing twenty validation patterns and the selecting condition was based on equation (2). Table 1 shows classification accuracies on the test set provided by our selection-based MCS compared to those of individual classifiers. The selection-based MCS substantially improves the classification accuracy without increasing the rejection rate. Table 2 shows the comparison between performances of our selection-based MCS and those of MCSs based on two of the most commonly used combination mechanisms proposed in the literature [1], i.e., the "majority rule" and the "Bayesian average". Both of these methods require the assumption of "independent errors". Results show that selection-based MCSs allows one to improve accuracies provided by MCSs based on combination mechanisms. These results agree with our analysis of correlation among errors made by individual classifiers [15,16]. We also compared the *selection performances* provided by our selection mechanism with the "reference" performances provided by a sort of "oracle" that always predicts the best classifier for each test pattern [16]. The proposed selection algorithm was able to make the correct decision on the most appropriate classifier for 97.22% of the test set.

Classification Algorithm	% Accuracy	% Rejection
Selection-based MCS	93.10	1.83
Neural Network 1 MLP 15-30-15-5	87.30	1.66
Neural Network 2 MLP 15-7-7-5	85.36	1.13
Neural Network 3 MLP 15-15-5	90.71	2.83
k-nn Classifier	90.70	1.89

Table 1. Classification accuracies on the test set provided by our selection-based MCS compared to those of individual classifiers

	% Accuracy	% Rejection
Selection-based MCS	93.10	1.83
Majority-based MCS	90.38	3.37
Average-based MCS	89.48	not available

Table 2. Comparison between performances of our selection-based MCS and those of MCSs based on the combination mechanisms

Experiment B: This experiment focused on evaluating performances of MCSs designed according to the method described in Section 4. For this purpose, a clustering algorithm based on a "hierarchical clustering technique" was performed on training data [14]. Different numbers of cluster were found for the five data classes contained in the selected data set (Class 1: 2 clusters, Class 2: 4 clusters, Class 3: 7 clusters, Class 4: 5 clusters, Class 5: 2 clusters). According to the proposed method, the training set was subdivided into seven partitions and a MCS based on seven classifiers was designed. In particular, we used seven k-nearest neighbour classifiers. With regard to the parameters of our selection algorithm, we used a neighbourhood containing six validation patterns and the selecting condition was based on equation (2). Table 3 shows classification performances of MCS designed according to our method. With respect to the performances of an "optimal" selector, the *selection accuracy* obtained by using this method is 95.73%. It is worth noting that these results cannot be directly compared to those contained in the previous Tables, since different classifiers form the related MCSs.

6 Conclusions

In this paper, we proposed a novel approach to the exploitation of potential advantages of MCSs based on the concept of "adaptive selection". We described an "adaptive classifiers selection algorithm" and reported experimental results related to the classification of remote-sensing images. We showed that the proposed selection-based MCS performs better than classical MCSs based on combination mechanisms. In particular, we showed that our selection algorithm provides performances that are reasonably close to those of an optimal selector. Finally, we proposed a systematic method to design MCSs based on classifiers selection and reported the satisfactory classification accuracies provided by MCSs designed according to this method. To the best of our knowledge, no other adaptive classifiers selection algorithm has been presented in the pattern recognition literature.

	% Accuracy	% Rejection
Selection-based MCS	96.30	3.09

Table 3. Performances of MCS designed according to the method in Section 4

In the field of neural networks, only Jordan's work can be regarded as an implementation of the concept of dynamic classifiers selection, since his "mixture of local experts" is adaptive [11]).

References

1. L.Xu, A.Krzyzak, and C.Y.Suen, "Methods for combining multiple classifiers and their applications to handwriting recognition", IEEE Trans. on Systems, Man, and Cyb., Vol. 22, No. 3, May/June 1992, pp. 418-435
2. R.Battiti, and A.M.Colla, "Democracy in neural nets: voting schemes for classification", Neural Networks, Vol. 7, No. 4, 1994, pp. 691-707
3. F.Roli, G.Giacinto, and G.Vernazza, "Comparison and combination of statistical and neural network algorithms for remote-sensing image classification", Neurocomputation in Remote Sensing Data Analysis, Advances in Spatial Science Series, Springer Verlag Ed. (in press, 1997)
4. G.Giacinto, and F.Roli, "Ensembles of Neural Networks for Soft Classification of Remote Sensing Images", Proc. of the European Symposium on Intelligent Techniques, March 20-21, 1997, Bari, Italy, pp.166-170
5. I.Kanellopoulos et al., "Integration of neural network and statistical image classification for land cover mapping", Proc. IGARSS 93, Tokio, 18-21 August 1993, pp. 511-513
6. Y.S.Huang, and C.Y.Suen, "A method of combining multiple experts for the recognition of unconstrained handwritten numerals", IEEE Trans. on Pattern Analysis and Machine Intelligence, Vol.17, No.1, January 1995, pp.90-94
7. N.Srihari et al., "Decision combination in multiple classifier systems", IEEE Trans. on Pattern Analysis and Machine Intelligence, Vol.16, No.1, Jan. 1994, pp. 66-75
8. L.K.Hansen, and P.Salamon, "Neural network ensembles", IEEE Trans. on Pattern Analysis and Machine Intelligence, Vol. 12, No. 10, October 1990, pp. 993-1001
9. K.Tumer and J.Gosh, "Error correlation and error reduction in ensemble classifiers", Tech. Report, Dept. of ECE, University of Texas, July 11, 1996
10. D.Partridge, W.B.Yates, "Engineering multiversion neural-net systems", Neural Computation, 8, 1996, pp. 869-893
11. R.Jacobs, M.Jordan, S.Nowlan, and G.Hinton, "Adaptive mixtures of local experts", Neural Computation, 3, 1991, pp. 79-87
12. C.Y.Suen et al., "The combination of multiple classifiers by a neural network approach", Int. Journal of Pattern Recognition and Artificial Intelligence, Vol. 9, no.3, 1995, pp.579-597
13. D.Opitz, and J.Shavlik, "Generating accurate and diverse members of a neural-network ensemble", Advances in Neural Information Processing Systems 8, MIT Press, 1996
14. R.O.Duda, P.E.Hart, "Pattern Classification and Scene Analysis", Wiley & Sons, Inc., 1973
15. S.B.Serpico, L.Bruzzone and F.Roli, "An experimental comparison of neural and statistical non-parametric algorithms for supervised classification of remote-sensing images", Pattern Recognition Letters, Vol 17, No. 13, November 1996, pp. 1331-1341
16. F.Roli and G.Giacinto, "Adaptive selection in multiple classifier systems", Tech. Rep., MCS-4-96, University of Cagliari, Italy, 1996

Classification Reliability and Its Use in Multi-classifier Systems

L. P. Cordella, P. Foggia, C. Sansone, F. Tortorella, M. Vento

Dipartimento di Informatica e Sistemistica
Via Claudio 21, I-80125 Napoli, Italy
E-mail: {cordel, foggia, carlosan, tortorel, vento}@nadis.dis.unina.it
WWW: http://amalfi.dis.unina.it

Abstract

In the last years, great attention has been devoted to multiple classifier systems. The implementation of such a system implies the definition of a rule (combining rule) for determining the most likely class, on the basis of the class attributed by each single expert. The availability of a criterion to evaluate the credibility of the decision taken by a classifier can be profitable in order to implement the combining rule. We propose a method that, after defining the reliability of a classification on the basis of information directly derived from the output of the classifier, uses this information in the context of a combining rule. The results obtained by combining four handwritten character recognizers on the basis of classification reliability are compared with those obtained by using three different combining criteria. Tests have been performed using a standard handwritten character database.

1 Introduction

In many Pattern Recognition applications, achieving acceptable recognition rates is conditioned by the large pattern variability, whose distribution cannot be simply modeled. This affects the results at each stage of the recognition system so that, once this has been designed, its performance cannot be improved over a certain bound, despite the efforts in refining either the classification or the description method.
Employing a multiple classifier system can be very useful for tackling such situation [1,2]: in fact, the consensus of a set of experts may compensate for the weakness of the single expert, while each single expert preserves its own strength. The implementation of a multiple classifier system implies the definition of a combining rule for determining the most likely class a sample should be attributed to, on the basis of the class to which it is attributed by each single expert [2]. The availability of a criterion to evaluate the credibility of the decision taken by a classifier can be very profitable in order to implement the combining rule. However, the definition of a parameter measuring such credibility is quite critical: an improper definition may result in giving high credibility to experts whose classification is unreliable, or viceversa, low credibility to really reliable experts.

Most of the combining rules proposed in the literature implement a "weighted voting", in the sense that the vote (i.e. the attribution to a class) expressed by an expert is weighted by the reliability estimated on the basis of the class chosen [1]: the input sample is then assigned to the class for which the sum of the weighted votes is the highest. In this way, however, all the samples attributed to the same class are assigned the same reliability and thus this value could not reflect the actual reliability of the single classification act.

To get out of this problem, we propose to associate a reliability measure to each classification performed by a given expert and to use this value to weight its vote in a multi-expert system. The operative definition of the parameter allowing to recognize situations which can give rise to unreliable classifications and enabling to quantify classification reliability will depend on the considered classifier architecture.

To evaluate the effectiveness of the reliability parameter several multi-expert systems, each obtained by combining various experts according to different combining rules, have been employed. The experts are handwritten character recognizers made of different pairs descriptor-classifier which will be described in the following.

Tests have been carried out using the digits of the NIST Database 19 [3]. Note that the experts have not been selected because they perform particularly well on the considered database. Aim of this paper is only that of quantitatively evaluating the differences of performance obtainable when using different definitions for the weights attributed to the votes of the experts in a multi-expert system. In the next Sections the adopted experts are briefly described and the definition of the classification reliability parameter is introduced together with the used combining rules. Finally, the experimental results are presented.

2 The Adopted Experts

In the field of handwritten character recognition character descriptions can be based on measurements directly performed on the character bit map (not structural descriptions) or given in terms of component parts, coming from a decomposition of the character, and relations among them (structural descriptions). Hybrid techniques that combine the two approaches are also possible.

We have considered each of the three types of description schemes. The not structural description uses as features the pixels of the character image obtained after a suitable filtering and scaling process leading to a small size "gray level picture" whose pixel values are computed by averaging the original image and are normalized so as to fall within the interval [0,1]. The obtained 8x8 matrices of numbers, encoded as vectors of length 64, are the descriptors used by the classifiers. Fig. 1 shows some characters before and after the scaling process.

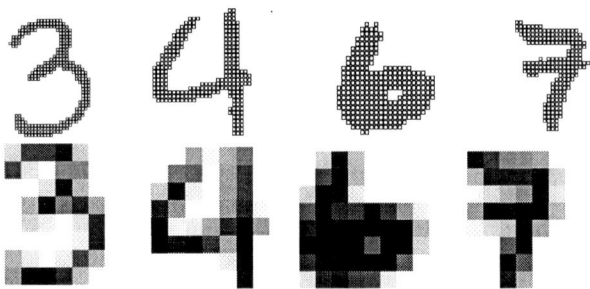

Fig. 1. Some characters of the NIST database and the results of the scaling process (8x8 gray level matrices).

In order to obtain the adopted structural description, characters are thinned (Fig. 2a,b) and then further processed for correcting the shape distortions introduced by thinning. After this correction a character is represented by a set of polygonal lines (Fig. 2c) which are then approximated with pieces of circular arcs (Fig. 2d). The structural description can be conveniently put in the form of an Attributed Relational Graph (ARG) (Fig. 2e), whose node attributes specify span, relative size and orientation of the corresponding arc, while branch attributes specify topologic relations between arc pair projections on the coordinate axes. More details can be found in [4].

The description we have called hybrid combines the structural and not structural approaches. After approximating a character with a set of circular arcs (see Fig. 2d), geometrical moments of this set are computed [5]. Geometric moments up to the 7th order are considered. Moments of zero and first order have been used to make the remaining moments invariant with respect to scale and translation. A character is thus described by a 33 element vector.

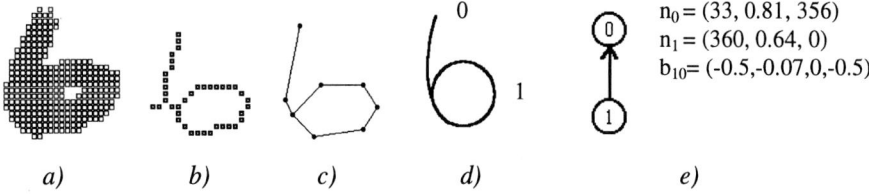

a)   b)   c)   d)   e)

Fig. 2. An example illustrating the structural description process: *a)* Character bit map, *b)* character skeleton, *c)* polygonal approximation of the skeleton, *d)* decomposition in terms of circular arcs, *e)* the corresponding Attributed Relational Graph; attribute values of nodes and branches are listed (n_i represents the i-th node, while b_{ij} denotes the branch connecting the i-th and j-th nodes).

Three classifiers were implemented by two different kinds of neural networks: the Multi-Layer Perceptron (MLP) [6] and the Learning Vector Quantization network (LVQ) [7]; the fourth classifier is of the Nearest Neighbor type (NN).

The above classifiers have been combined with the three previously outlined descriptions, giving rise to the four experts illustrated in the following. The acronyms used to denote the experts specify the classifier and the associated description type. Only handwritten digits were considered for carrying out the test, thus the number of classes is ten for every classifier.

The *MLP-NS Expert:* the MLP-NS expert combines the MLP classifier with the Not Structural description. Therefore the input layer of the classifier is made of 64 neurons each one associated to a pixel of the scaled image. The chosen network architecture has a single hidden layer of 30 neurons and an output layer of 10 neurons corresponding to the ten digits. The learning algorithm is the standard Back-Propagation one, with a constant learning rate equal to 0.5. The sigmoidal activation function was chosen for all the neurons.

The *LVQ-NS Expert:* the LVQ-NS has an input layer composed by 64 neurons and a number of Kohonen neurons fixed to 7 for all classes. The net was trained with a supervised version of the FSCL algorithm to overcome the neuron under-utilization problem [8]. The learning rate was initially set equal to 0.5 and then decreased according to the rules illustrated in [8].

The *MLP-H Expert:* in this expert, the classifier works with the Hybrid description. Thus the input layer of the classifier is made of 33 neurons. All the remaining network parameters are the same used in the MLP-NS expert.

The *NN-S Expert:* it uses the structural description associated to a NN statistical classifier. The distance between two characters is measured by means of a metric defined in the ARG space [9]. In order to reduce the computational effort, the experiments performed with this expert were carried out using only a subset (about 25%) of the training set used in the other cases.

3 Defining the Reliability Parameters

The low reliability of a classification is generally due to one of the following situations: a) the considered sample is significantly different from those present in the training set, i.e., its representative point is located in a region of the feature space far from those occupied by the samples of the training set and associated to the various classes; b) the point which represents the considered sample in the feature space lies where the regions pertaining to two or more classes overlap, i.e., where training set samples belonging to more than one class are present. It may be convenient to distinguish between classifications which are unreliable because a sample is of type a) or b). To this end, let us define two reliability parameters, say ψ_a and ψ_b, whose values vary in the interval [0,1] and quantify the reliability of a classification from the two different points of view. Values near to 1 will characterize very reliable classifications, while low parameter values will be associated with classifications unreliable because the considered sample is of type a) or b). A parameter ψ providing a comprehensive measure of the reliability of a classification can result from the combination of the values of ψ_a and ψ_b. We have chosen the form $\psi = \min\{\psi_a, \psi_b\}$. This is certainly a conservative choice because it implies that, for a classification to be unreliable, just one reliability parameter needs to take a low value, regardless of the value assumed by the other one. By definition, the ideal reliability parameter should assume a value equal to 1 for all the correctly classified samples and a value equal to 0 for all the misclassified samples. However, this will almost never happen in real cases. The operative definition of ψ requires the classifier to provide an output consisting of a vector the values of whose elements make it possible to establish the class a sample belongs to.

As regards the MLP classifier, it can be shown [10] that an effective definition of the reliability parameter ψ_a can be $\psi_a = O_{win}$ where O_{win} is the output of the winner neuron, while a suitable reliability parameter for the case b) is $\psi_b = O_{win} - O_{2win}$. In conclusion, the classification reliability of the MLP classifier can be measured by:

$$\psi = \min\{\psi_a, \psi_b\} = \min\{O_{win}, O_{win} - O_{2win}\} = O_{win} - O_{2win} = \psi_b$$

For an LVQ classifier, the values of the elements of the output vector give the distances of an input sample X from each of the prototypes W_i, i=1,...,M, with M generally greater than the number N of classes. Therefore, the winner neuron is the one having the minimum output value $O_{win} = \min_i \{O_i\} = \min_i \{d(W_i, X)\}$. With this assumption, a convenient form for the first parameter can be $\psi_a = \max\{1 - O_{win}/O_{max}, 0\}$ where O_{max} is the highest value of O_{win} among those relative to all the samples of the training set. For the case b), we have adopted the definition $\psi_b = 1 - O_{win}/O_{2win}$, where O_{2win} is the value of the output neuron having the second lowest distance from the input sample. The classification reliability for the LVQ classifier is thus given by:

$$\psi = \min\{\psi_a, \psi_b\} = \min\left\{\max\left\{1 - \frac{O_{win}}{O_{max}}, 0\right\}, 1 - \frac{O_{win}}{O_{2win}}\right\} \quad (1)$$

The same considerations hold for the NN classifier. In fact it assigns the input sample X to the class including the reference graph having the smallest distance from X. The only differences are that the value of O_{max} has to be computed on a set different from the reference set and O_{2win} is the distance between X and the reference graph having the second smallest distance from X, among all those belonging to a class different from that of O_{win}. Therefore, the classification reliability for the NN classifier is again given by equation (1).

4 Combining Criteria

Most of the combining criteria proposed in the framework of the multi-expert approach use the confidence degree assigned by an expert to each classification it performs. To evaluate the confidence degree of the vote given by the k-th expert, the most common choice [1] is the classification confusion matrix E^k whose generic element e_{ij}^k represents the number of times the k-th expert assigns to the j-th class a sample belonging to the i-th class, divided by the total number of samples belonging to the i-th class.

To investigate the influence of the set used to compute the confusion matrix on the obtained performance, the values of the elements of E^k were computed once using the training set and then using a set of data different from both the training set and the test set (see next Section).

The following combining criteria were used:
1) *Majority Voting (MV)*: each expert votes for one class and the estimated (i.e., the actually assigned) class is the one voted by the majority. If more classes obtain the same number of votes, the values e_{ii}^k are used for tie breaking, i.e. the vote of each expert is weighted by the number representing the reliability of that expert when it assigns a sample to the class it is voting for.
2) *Bayesian Combination (BC)*: the estimated class is the one which maximizes the a posteriori probability. The probability that a sample belongs to the i-th class when the k-th expert assigns it to the j-th class is assumed to be $e_{ij}^k / \sum_{i=1}^{n} e_{ij}^k$.

3) *Dempster-Shafer evidential reasoning* (*DS*): this criterion is based on the Dempster-Shafer theory [11]. According to it, we define for each expert, the "belief" in every possible subset A of the set $\Theta = \{A_1, A_2, ..., A_m\}$, where A_i is a proposition representing the fact that a sample is assigned to the i-th class by the considered expert. The belief bel(.) is calculated from a function, called basic probability assignment, which is denoted *m*(.), by using the equation

$$\text{bel}(A) = \sum_{B \subseteq A} m(B) \qquad (2)$$

where B is any subset of A. Obviously, we have bel(A_i) = $m(A_i)$ and bel(Θ)=1. In our case, when the k-th expert votes for the i-th class, we consider $m(A_i) = e_{ii}^k$ and $m(\Theta) = 1 - e_{ii}^k$. The values *m*(A) supplied by each expert are combined via the Dempster rule, and the values bel(A_i) are calculated using the equation (2). The estimated class is the one that maximizes the value of bel(A_i).

The criteria based on the confidence degree were compared with a fourth criterion based on the reliability parameter:

4) *Majority Voting Using the Reliability Parameters* (*RV*): this criterion differs from the first one only for the values used for tie breaking: in this case, in fact, the reliability parameters defined in the previous section are used.

5 Experimental Results and Conclusions

All the tests were performed using the NIST database 19 [3], which contains 8 sets of images extracted from 3699 Handwriting Sample Forms and digitized at 300 dpi. In particular, we used the sets hsf_3 and hsf_4. Only digits were considered.

The set hsf_3 was split in two sets: a training set (TRS), composed of 34,644 samples, used for training the MLP-NS, MLP-H and LVQ-NS experts, and a so called training-test set (TTS) made of 29,252 samples. As already mentioned, a subset of TRS (8000 samples) was assumed as reference set for the NN-S expert. TTS was used both to compute the confusion matrices and to establish the number of cycles for stopping the learning phase of the MLP-NS, MLP-H and LVQ-NS experts, in order to avoid the overtraining phenomenon [8].

The set hsf_4, made of 58,646 samples, was adopted as test set (TS).

The MLP-NS and MLP-H experts have been trained performing 5000 learning cycles, while for the LVQ-NS expert 2000 learning cycles were performed. The recognition rates obtained by each single expert on the considered sets are reported in Tab. 1.

Expert	TRS	TTS	TS
MLP-NS	97.96	96.52	88.53
LVQ-NS	98.80	96.66	85.90
MLP-H	95.56	94.59	85.63
NN-S	--	90.98	84.11

Tab. 1. Recognition rates obtained by the single experts on TRS, TTS and TS.

Eleven different multi-experts have been considered: 1 of them combines 4 experts, 4 of them combine 3 experts, and each of the remaining 6 combines 2 experts. The multi-experts have been designed so as to test all the significant combinations of experts, each obtained by pairing a classifier with a descriptor. The considered combinations of experts and the experimental results obtained with them are summarized in Tab. 2.

Let us remind that the recognition systems considered have been selected not because they have an outstanding performance on the used database, but in order to perform the test on systems adopting different description and classification paradigms. However the recognition rates obtained are neither low, considering the quality of the characters in the data base. The use of the reliability parameter allows to improve the recognition rates for all the considered multi-expert systems. The improvement is more significant when the number of experts is equal to two: in this case, in fact, the need of tie breaking is more frequent than in presence of three or more experts. The recognition improvement achieved by the multi-experts when using the combining rule RV, although limited to a few percent, should be considered relevant, since it depends only on the fact that the reliability parameter has been used while the combining rules and the experts have been fixed.

Multi-Expert				Combining Rule			
MLP-NS	MLP-H	LVQ-NS	NN-S	RV	BC	DS	MV
X	X	X	X	92.19	91.96	91.68	91.62
X	X	X		90.56	90.35	90.23	90.21
X	X		X	91.73	91.70	90.79	90.79
X		X	X	91.68	91.30	90.78	90.77
	X	X	X	91.30	91.21	90.50	90.50
X	X			89.49	87.91	87.79	87.78
X		X		89.49	87.19	87.23	87.23
X			X	90.74	89.41	88.11	88.11
	X	X		88.15	87.19	86.31	86.31
	X		X	88.24	87.83	86.53	86.52
		X	X	88.40	87.02	86.09	86.08

Tab. 2. Recognition rates obtained by each multi-expert as a function of the combining rule. Values in parentheses refer to the case in which the confusion matrices are computed on the training set instead of the training-test set.

This is still more true, if it is considered that the values shown in Tab. 2 are the result of the average over all the classes, while the improvement due to the reliability parameter is not uniformly distributed among classes. It has been verified that there is a smaller improvement for classes whose samples do not exhibit a large shape variability (e.g., the class of the 1s) while the improvement is significantly higher than the average for classes whose samples are quite different from each other, like the class of the 8s.

As regards the set of data used to evaluate the confusion matrix it has been noted that the use of a confusion matrix computed on TRS makes the recognition performance worst for almost all the multi-expert systems and particularly for the combinations including an LVQ-NS expert. This is due to the fact that for this expert the difference between the recognition rates on TRS and TS is greater than for the other experts. The use of the reliability parameter, whose value does not depend on the choice of a specific set of data, allows to overcome the problem, existing when computing the confusion matrix, of selecting a set of data different from the training set, but adequately representative of the real world.

Finally it is worth noting that to determine the performance of a multi-expert system another important factor is the variety of the component experts. In fact, all the combination with the NN-S expert achieve the best performance; this confirms that the selection of experts as much as possible complementary as regards both description and classification methods can significantly improve the performance of a multi-expert system.

References

[1] L. Xu, A. Krzyzak, C.Y. Suen, "Method of Combining Multiple Classifiers and Their Application to Handwritten Numeral Recognition", *IEEE Trans. on Systems, Man and Cybernetics*, vol. 22, no. 3, pp. 418-435, January 1992.

[2] S.-B. Cho, J.H. Kim, "Combining Multiple Neural Networks by Fuzzy Integral for Robust Classification", *IEEE Trans. on Systems, Man and Cybernetics*, vol. 25, no. 2, pp. 380-384, February 1995.

[3] P.J. Grother, NIST Special Database 19, Technical Report, National Institute of Standards and Technology, 1995

[4] L.P. Cordella, C. De Stefano, M. Vento, "A Neural Network Classifier for OCR using Structural Descriptions", *Machine Vision and Applications*, no. 8, pp. 336-342, 1995.

[5] P. Foggia, C. Sansone, F. Tortorella, and M. Vento, "Character Recognition by Geometrical Moments on Structural Decompositions", *Proc. 4th Int. Conf. on Document Analysis and Recognition*, to appear.

[6] D.E. Rumelhart, J.L. Mc Clelland, *Parallel Distributed Processing - Explorations in the Microstructure of Cognition*, Vol.1: Foundations. MIT Press, Cambridge, Mass, 1986.

[7] T. Kohonen, "The Self-Organizing Map," *Proc. of the IEEE*, vol. 78, no. 9, pp. 1464-1480, September 1990.

[8] R. Hecht-Nielsen, *Neurocomputing*, Addison-Wesley, Reading (MA), 1990.

[9] C. De Stefano, P. Foggia, F. Tortorella, M. Vento, "A Distance Measure for Structural Descriptions using Circle Arcs as Primitives" in *Proc. 13th Int. Conf. on Pattern Recogn.*, IEEE Comp. Soc. Press, Vol. II, pp. 290-294, 1996.

[10] L.P. Cordella, C. De Stefano, F. Tortorella, M. Vento, "A Method for Improving Classification Reliability of Multilayer Perceptrons", *IEEE Trans. on Neural Networks*, vol. 6, no. 5, pp. 1140-1147, September 1995.

[11] J. Gordon, E.H. Shortliffe, "The Dempster-Shafer Theory of Evidence", in B.G. Buchanan, E.H. Shortliffe (Eds.), *Rule-Based Expert Systems*, Addison-Wesley, pp. 272-292, 1984.

Color Linear Model

Chang-Yeong Kim*, Yang-Seok Seo*, In-So Kweon°
* Signal Processing Lab. Samsung Advanced Institute of Technology, P.O. Box 111, Suwon, Korea 440-600 Phone) 0331-280-9217 fax) 0331-280-9207 Cykim@saitgw.sait.samsung.co.kr
° Dept. of Electrical Engineering, Korea Advanced Institute of Science and Technology, 207-43, CheongRyangRiDong, DongDaeMoonGu. Seoul,Korea,130-012
Phone) 02-985-3415 fax)02-968-1638 Kweon@design.kaist.ac.kr

Abstract

In this paper, procedures for creating an effective linear model to represent surface spectra are presented. The model is derived by considering spectral data and the human visual characteristic that depends on wave lengths. Two human visual weighting functions (HVWF) are derived from human visual characteristic. The basis functions of the linear model for the surface reflectance are selected by minimizing least square error in approximating the spectral data weighted by the HVWF. The linear model is shown to perform better than conventional linear models for color constancy, the surface identification related to object recognition, and the characterization of a scanner and a camera.

1 Introduction

Recent computational models of color vision [1-7] demonstrate that it is possible to achieve color constancy over some limited range of illuminant and surface. The algorithm provided by Cohen [1] initially was attempted to find an efficient, low dimensional linear representation of surface reflectance (SR) and spectral power distribution (SPD) of illuminants. Efficient spectral representation is useful in many applications such as color reproduction system [4, 5], rendering in computer graphics, color constancy, and surface identification related to object recognition.

Linear model has been developed in two main directions. First, relationship between the error of reconstructed reflectance and the dimension of basis functions has been observed in the linear model [2,3,7]. Second is to analyze not only the error of reconstructed reflectance but also color difference on color space for real applications [6,10,11] A linear model designed to minimize the error between original SR and reproduced one is inappropriate in efficient color representation because it is not designed to minimize color differences. Moreover, it is important to remember that basis functions should be derived independent of illuminants because color constancy is get "an invariant" under the varying illuminants. Therefore, we try to consider and minimize the human visual characteristic (HVC) independent of illuminant to derive the basis function. In [10], the effectiveness of basis functions considering HVC was proved. In this paper, the CIE 1931 XYZ standard observer and CIE1976 CIEL*a*b* as HVC are investigated.

2 Linear Model

Linear models [1-8] are used to approximate SR with a small number of descriptors in wavelength domain. The basis functions $R_j(\lambda)$ for SR are chosen to minimize the mean square error of reflectance $R(\lambda)$:

$$\underset{R_j}{MIN} \sum_{S \in sample} \int \left[R_s(\lambda) - \sum_{j=1}^{d} \sigma_j(\lambda) R_j(\lambda) \right]^2 d\lambda \quad (1)$$

where *sample* and *d* represent the number of samples and basis functions respectively and σ_j is the weight of each basis function. If the *d*-dimensional basis functions are chosen to approximate the SR $R(\lambda)$, the reflectance function is represented as:

$$R(\lambda) \approx \sum_{j=1}^{d} \sigma_j R_j(\lambda) \equiv R'(\lambda). \qquad (2)$$

The basis functions R_j that minimize the quantity in equation (1) are derived by as singular value decomposition (SVD) of the matrix [9] whose columns are composed of the SR of samples [1,5,6]. Let us consider SR as 31 by 1 matrix with 10 nm sampling wavelengths from 400 nm to 700 nm. If mapping of the original reflectance matrix, *R*, onto the reconstructed spectral reflectance matrix *R'* is expressed by the projection matrix:

$$R' = R_j W = R_j(R_j^+ R) = (R_j R_j^+)R = PR \qquad (3)$$

where *W* is 3 by 1 weight matrix, R_j^+ is the psudoinverse of the basis vectors and *P* is the projection matrix with entries of 31 by 31. The mean square error, E_{ref}, between the original and an approximated spectra become:

$$E_{ref} = \frac{1}{31}\sum_{j=1}^{31} (|R(\lambda_j) - R'(\lambda_j)|^2) = \|R - PR\|^2 \qquad (4)$$

where *31* is the number of sampling points in spectrum and $\|\ \|$ represents the norm of vector. The linear model for illuminant has the same representation as for reflectance except that basis vectors derived from a set of illuminants used. The linear approximation $E(\lambda)$ for illuminant corresponding to equation (2) is

$$E(\lambda) \approx \sum_{j=1}^{d} E_j(\lambda)\varepsilon_j$$

, where $E_j(\lambda)$ are basis functions for illuminant set.

3 Color Linear Model using Human Visual Perception

It is important that the magnitude of spectra errors is not always coincident with the amount of color difference perceived by human eye, such as CIEXYZ or CIELAB:

$$\text{if } (\Delta E_{ref}^a < \Delta E_{ref}^b), \text{ then } (\Delta E^{*a}_{ab} > \Delta E^{*b}_{ab}) \text{ or } (\Delta E^{*a}_{ab} < \Delta E^{*b}_{ab}). \qquad (5)$$

Basically, this inequivalence is due to minimization strategy, equation (1), in which the HVC is not considered. Human visual responses are relatively insensitive at the ends of visible spectrum, while they show different sensitivity characteristics in around the center of visible spectrum. The color constancy, perceived by human visual system, for an object discounts illumination effect not for reflectance but for trichromatic visual response. Wandell et al. derived the basis vectors of color data set by n-mode analysis to minimize the error in the predicted scanner response [6]. Trusell has discussed the related topic that uses principal component analysis (PCA) in color reproduction [5]. The target of this section is to derive the basis functions considering the human visual perception for the surface reflectance set.

3.1 The Proposed Minimization Equation

To derive effective basis functions, the error is minimized for reflectance data based on human characteristic rather than SR itself:

$$\min_{sample} \sum \int \left[H(\lambda) - H'(\lambda)\right]^2 d\lambda = \min_{sample} \sum \int \left[H(\lambda) - H_j(\lambda) \cdot \sigma_{hj}(\lambda)\right]^2 d\lambda \quad (6)$$

where $H(\lambda)$ represent the product of reflectance $R(\lambda)$ and HVC, and $H'(\lambda)$ is an approximated reflectance in human visual domain and σ_{hj} is weight coefficient and $H_j(\lambda)$ are basis functions. The basis functions H_j for the SR are chosen to minimize the mean square error of SR data $H(\lambda)$. If reflectance replaces the continuous function of wave length with sampled functions expressed as vectors, then the human visual error as in equation (4):

$$E_{HV} = \|H - P_H H\| \quad (7)$$

where P_H is a projection matrix and $\| \ \|$ represents the norm of vector.

3.2 Human Visual Weighting Functions (HVWF)

Two examples of HVWF described here are found from the three spectral components of color vector such as CIEXYZ or CIELAB. All spectral data are composed of 31 sample points ranging from 400 to 700 nm in 10-nm steps in our model. Human visual sensitivities are C_i, i=1-3, whose entries compose of three column matrix, $C=[\ C_1,\ C_2,\ C_3]$. Three spectral human visual responses X_{ij} of the surface reflectance R under the illuminant E at each sample wave length can be computed by matrix product $X=C^T ER$, where R and E are 31 by 31 diagonal matrices. Hence, each column matrix of X is expressed as:

$$X_j = R(\lambda_j)\ E(\lambda_j)[\ C_1(\lambda_j),\ C_2(\lambda_j),\ C_3(\lambda_j)]^T \quad (8)$$

where j=1-31. Suppose $T(\lambda)$ as CIE 1931 Standard Observer [8], then human visual response at sample wave lengths becomes in terms of CIE *XYZ* values as:

$$X_j = R(\lambda_j)\ E(\lambda_j)\ K\ T(\lambda_j), \quad (9)$$

where $K = 100 / \sum_{j=1}^{31}\left[E(\lambda_j) \cdot \overline{y}(\lambda_j) \cdot \Delta\lambda\right]$, and $T(\lambda_j)$ is a vector form of trichromatic observer, $[\overline{x}(\lambda_j), \overline{y}(\lambda_j), \overline{z}(\lambda_j)]^T$. Suppose X_j' the recovered human visual response of the reflectance $R(\lambda_j)$, then the norm of the difference of human response vector:

$$\|X_j - X_j'\| = E(\lambda_j)\ K\ \{R(\lambda_j) - R'(\lambda_j)\}\ [\ T_1(\lambda_j)^2 + T_2(\lambda_j)^2 + T_3(\lambda_j)^2]^{1/2} \quad (10)$$

If we substitute the norm $\|X_j - X_j'\|$ instead of $(R(\lambda_j)-R'(\lambda_j))$ in equations (1) and (7), then the basis functions can be chosen to minimize the following quantity:

$$\sum_{sample} \sum_j \left[E(\lambda_j) \cdot K \cdot \|T(\lambda_j)\| \cdot \{R(\lambda_j) - R'(\lambda_j)\}\right]^2. \quad (11)$$

Using equation (11), we are ready to define the HVWF at sample wave lengths as:

$$H_V(\lambda_j) = E(\lambda_j)\ K\ [\ C_1(\lambda_j)^2 + C_2(\lambda_j)^2 + C_3(\lambda_j)^2]^{1/2} \quad (12)$$

Equation (12) displays the HVWF of a general form considering the illuminant and HVC. Equation (11) can be rewritten by:

$$\sum_{sample} \sum_j \left[H_V(\lambda_j)\ R(\lambda_j) - H'(\lambda_j)\right]^2 \quad (13)$$

Besides the above HVWF, it is possible to adopt a human visual sensitivity that is a linear space of the reflectance such as NTSC RGB or ATD space [8]. Suppose an illuminant with uniformed unit energy spectrum, then the HVWF that is independent of illuminant is expressed by:

$$H_V(\lambda_j) = [C_1(\lambda_j)^2 + C_2(\lambda_j)^2 + C_3(\lambda_j)^2]^{1/2}, j=1,...31. \quad (14)$$

The above $H_V(\lambda_j)$ is depicted in Figure 1 and is refereed to color matching function (CMF) in the following sections. CMF represents characteristics that the middle range of visible spectrum has a high sensitivity like human being's but both ends of visible spectrum have relatively low sensitivity. Besides the linear spaces, non-linear space to reflectance like LAB (CIE 1976 L*a*b*) space [8] can be considered as the HVC. The XYZ values X_j of equation (9) can be transformed into LAB domain:

$$L(\lambda_j) = 116 \{R(\lambda_j) E(\lambda_j) K T_2(\lambda_j)/Yn\}^{1/3} - 16 \quad (15)$$

$$A(\lambda_j) = 500 [\{R(\lambda_j) E(\lambda_j) K T_1(\lambda_j)/Xn\}^{1/3} - \{R(\lambda_j) E(\lambda_j) K T_2(\lambda_j)/Yn\}^{1/3}] \quad (16)$$

$$B(\lambda_j) = 200 [\{R(\lambda_j) E(\lambda_j) K T_2(\lambda_j)/Yn\}^{1/3} - \{R(\lambda_j) E(\lambda_j) K T_3(\lambda_j)/Zn\}^{1/3}] \quad (17)$$

where Xn, Yn, and Zn represent CIE X, Y, and Z values of a given illuminant. Let us X_j a column matrix $[L(\lambda_j), A(\lambda_j), B(\lambda_j)]^T$. Suppose an uniformed illuminant, $E(\lambda_j)=[1]$, and human visual response $l(\lambda_j)$, $a(\lambda_j)$, and $b(\lambda_j)$ for a surface with unit reflectance, then the norm of human visual difference vector between original and the estimated one become:

$$\|X_j - X'_j\| = \|T(\lambda)\| \{R(\lambda_j)^{1/3} - R'(\lambda_j)^{1/3}\}, \quad (18)$$

where $\|T(\lambda)\| = \{l(\lambda_j)^2 + a(\lambda_j)^2 + b(\lambda_j)^2\}^{1/2}$. The quantity corresponding to equation (11) is then represented by:

$$\sum_{sample} \sum_j \{\|T(\lambda)\|\{R(\lambda_j)^{1/3} - R'(\lambda_j)^{1/3}\}\}^2 . \quad (19)$$

There are two obstacles to apply HVC with a non-linear color space in a linear model:
1) It is impossible to separate SR $R(\lambda_j)$ from the minimization equation (19), hence human visual weighting function cannot be derived;
2) It is hard to understand LAB values at sample wave lengths because LAB values are defined from the sum of XYZ values along to wave length. Moreover, LAB space is non-linear to reflectance space. Hence, there is difficult to apply LAB values as the HVC in the linear model directly, even though the color space describes the HVC quite well. Hence, assuming a recovered human visual response $H'(\lambda_j)$, then HVWF can be derived by minimizing a different quantity from equation(19):

$$\sum_{sample} \sum_j [\|T(\lambda)\|^{1/p} \cdot R(\lambda_j) - H'(\lambda_j)]^2 \quad (20)$$

where $\|T(\lambda)\|^{1/p}$ is p-norms defined by $\left(|T_1|^p + |T_2|^p + |T_3|^p\right)^{1/p}, p \geq 1$.

Here, we try to define a quantity similar to equation (19) that uses the p-norms of LAB values and extracts p by minimizing the total error for variables. HVWF at sample wave lengths in the above equation is defined by:

$$HVWF(\lambda_j) = [\{L(\lambda_j)\}^p + \{A(\lambda_j)\}^p + \{B(\lambda_j)\}^p]^{1/p}. \quad (21)$$

To get more information about *LABp* the original spectra are reconstructed and the errors are estimated for the dimension of basis vector from three to seven. The errors of the reconstructed spectra decrease as the number of basis increases from three to seven while the maximum errors in XYZ space increase. This is due to minimize not

XYZ error but the error of weighted spectra of $LAB_{P=2}$ for deriving the linear model. The basis vector of parameter p=2 gives the best reconstruction for spectra, XYZ, and LAB. Figure 1 shows three kinds of HVWFs that are uniform, CMF, and LAB the case of p=2, where the LAB has three maximum points that is different from CMF.

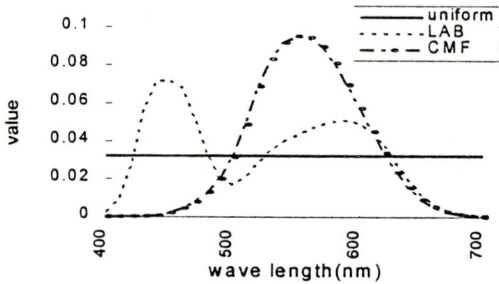

Fig. 1 Example of the HVWFs with uniform sensitivity, CMF, and LAB.

3.3 Basis Functions Using HVWFs

Reflectance set weighted with the HVWF produces weighted spectra set S_h:
$$S_h = H_v R \qquad (22)$$
where entries of diagonal matrix H_v are values of HVWF and R is 31 by 1. For deriving new basis vectors for the spectra weighted by HVWF S_h, SVD performed to minimize the error in the human visual domain:
$$C_h = (1/p) \cdot S_h S_h^+ = U_h D_h V_h^+, \qquad (23)$$
where C_h is a correlation matrix of the spectra weighted by HVWF and D_h is a diagonal matrix. The first d columns of U_h are orthonormal basis vectors for the linear model B_h. The 690 reflectance set is used for obtaining these vectors. Figure 2 represent the first four basis vectors for SR weighted by $LAB_{p=2}$. The feature of the LAB basis vector is different from the conventional and CMF cases [10]. The relation between a matrix of original SR S and a reconstructed one S' becomes as:
$$S' = B_h W_h = B_h (B_h^+ S) = (B_h B_h^+) S = P_h S. \qquad (24)$$
Let a matrix for mapping the original spectra S onto the reconstructed spectra S' be the projection matrix P_h, via the basis vectors. The spectral reflectance estimated from the reconstructed spectra S' can be obtained by:
$$R'(\lambda) = \sum_{j=1}^{d} B_{hj}(\lambda) W_{hj} / HVWF(\lambda). \qquad (25)$$

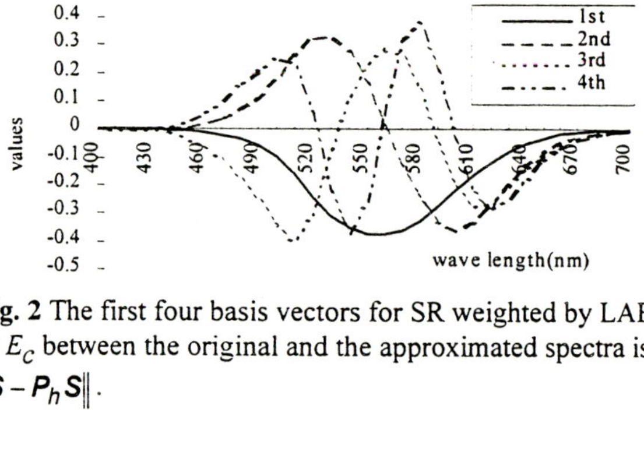

Fig. 2 The first four basis vectors for SR weighted by LAB

The mean error E_c between the original and the approximated spectra is defined as:
$$E_h = \| S - P_h S \|. \qquad (26)$$

The HVWF in equation (24) is found at each term of S' and S:
$$diag.(H_V)R' = P_h \, diag.(H_V)R. \quad (27)$$
Equation (27) can be rewritten by removing H_V in each term:
$$R' = P_h R. \quad (28)$$

4 Experiments and Discussion

A reflectance data set of 690 materials was selected for our experiments. The 690 SR data set consists of the following materials: 1) 400 color chips picked from Munsell color book measured using a PhotoResearch-703 spectro-radiometer; 2) 120 paint chips which selected from the Solid Selection of DuPont Color Sample, 170 natural and man-made objects, including rocks, plants and vegetation, human skin and hair, and fabrics of which complete data are available in ftp.eos.ncsu.edu [7]. In our experiment, we select the data set measured at 10nm interval in 400-700nm range. The averaged errors of reconstructed SR is estimated by the square root of equation (4) and the color differences on XYZ and LAB color spaces of the reconstructed SR can be estimated under the illuminant E, respectively [8].

Table 1 The errors for 3~7 Dimensional Reconstruction in the case of the CMF

Basis Dimen-sion	ERROR						
	LAB		XYZ		Ref.		
	avg.	max.	avg.	max.	avg.	max.	1st E.V.(%)
3	2.931745	28.57438	0.914159	6.413741	0.027008	0.12995	99.11
4	1.665761	14.59375	0.398795	3.818641	0.01715	0.106929	99.62
5	1.070622	14.59240	0.272975	2.981871	0.01314	0.076783	99.84
6	0.554749	6.638414	0.213997	2.654337	0.011001	0.074365	99.91
7	0.491418	6.068511	0.138272	1.923089	0.008812	0.065608	99.96

Table 1 summarizes the error distributions of three cases for the CMF. The errors are estimated under the five CIE standard illuminants, D_{50}, D_{55}, D_{65}, D_{75}, and A for 690 SR using three-seven dimensional basis. The range of XYZ error distribution is under 0.92 and the LAB errors show below 2.93. Table 2 represents the details of the errors for 3-7-Dimensional reconstruction in LAB case. Table 1 and Table 2 show that the reconstuctitve ability of the LAB case is better than the CMF case.

Table 2. The errors for 3~7 Dimensional Reconstruction in the case of $LAB_{p=2}$

Basis Diemsion	ERROR						
	LAB		XYZ		Ref.		
	avg.	max.	avg.	max.	avg.	max.	1st E.V.(%)
3	1.6989	15.0783	0.5555	3.3592	0.0258	0.1248	98.96
4	1.2340	18.4376	0.4621	2.5695	0.0171	0.1034	99.60
5	0.9171	13.7532	0.2967	2.9619	0.0123	0.0731	99.83
6	0.5913	6.8157	0.2379	2.9637	0.0108	0.0731	99.90
7	0.5075	5.4517	0.1716	1.9063	0.0086	0.0652	99.95

Table3 shows mean and maximum errors by three dimensional basis for three kinds of HVWFs. There is no serious difference among mean reflectance errors but maximum error has minimum value at uniformed HVWF. In the case of the $LAB_{p=2}$, the average

LAB error of the reconstructed SR data under five different standard illuminants was 1.569018 that shows the best reproduction in our experiments. It is clear that the LAB error of uniformed HVWF is about 5.7, so the basis vectors of conventional type is difficult to be used for the purpose of special applications which need color discrimination between the original spectra and the reconstructed spectra. We can make a conclusion from table 1, table2, and table3 that the basis vectors derived from the LAB reconstruct the most accurate colors in our experiments.

Table 3 Errors estimated by the linear models using 3-D basis for 690 data set

HVWFs	ERROR					
	LAB		XYZ		Ref.	
	avg.	max.	avg.	max.	avg.	max.
Uniform	5.750791	52.085278	1.569018	11.654885	0.021617	0.101884
CMF	2.931745	28.574381	0.914159	6.413741	0.027008	0.12995
LAB	1.698997	15.078359	0.555544	3.359257	0.025865	0.1248

ILLUMINATION EFFECT IN THE LINEAR MODEL:

The color differences of the LAB unit of the 690 spectrum reconstructed by each linear model for five standard illuminants are given in table 4 in the case of three dimensional basis vectors. The $CMF(D_{65})$ is the CMF estimated by equation (10) under the illuminant D_{65} and $LAB(D_{65})$ is the case of the LAB estimated in D_{65}. CIE D_{65} standard illuminant is assumed as a representative of illuminant. Table 4 summarizes the effect of the representative illuminant in linear models for the case of three dimensional basis. We can observe that the illumination effect on $CMF(D_{65})$ is very serious while the representative of illumination of $LAB(D_{65})$ display the worse effect than the uniformed illuminant of LAB. The main reason for the difference of illumination effect is, we think, that the illumination in equation (21) is non-linear to LAB. The mean LAB color difference of $CMF(D_{65})$ reduced to about 70 % of the that of the CMF case except for the illuminant "A" with about 111%, which is due to the characteristic of representative illuminant. This gives us another important conclusion again that, basically, basis vectors in the linear model have to be derived independent of illumination characteristic. But carefully selected representative illuminant for the special purpose can increase the accuracy of representation of the linear model for SR data. For example, if we consider the linear model under the various day lights, it is easy to guess that the selection of proper representative illuminant will support to increase the accuracy. We can also know from the experiments for varying illuminant that the proposed linear model can be applied in reconstruction of color signals that are the product of spectral reflectance and illuminant with the same accuracy.

Table 4 Illuminant effect in linear models in the case of three dimensional basis

Illuminant	HVWFs				
	uniformed	CMF	$CMF(D_{65})$ (% of CMF)	LAB	$LAB(D_{65})$:
D_{50}	5.633101	2.63924	1.89409(71.59%)	1.36240	1.38171
D_{55}	6.00571	2.91406	2.09597(71.72%)	1.42634	1.44550
D_{65}	6.35477	3.28648	2.42434(73.63%)	1.57878	1.59586
D_{75}	6.61387	3.59474	2.71411(75.48%)	1.75749	1.76890
A	4.14648	2.22418	2.48798(111.5%)	2.36995	2.59246

5 Conclusions

In this paper, we have proposed color linear model, which explicitly takes into account human visual perception. Two kinds of HVC are derived from CIE color matching functions and LAB representation of the color matching functions. Those have been used for the reconstruction of spectral data sets of natural objects and the errors of reconstruction have been analyzed in terms of reflectance, XYZ, and LAB values. Through extensive experiments using 690 samples we observed that the proposed color linear models are superior to conventional linear model for the color representation of SR. The carefully selected representative illuminant increases the accuracy of the linear model for SR and the linear model can be applied in reconstruction of color signals with the high accuracy from the observation of the illumination effect.

References

1. J. Cohen, Dependency of the spectral reflectance curves of Munsell Color chips, Psychnomic Sci. 1, 367-370 (1964).
2. J. Parkkinen, J. Hallikainen, and T, Jaaskelainen, Characteristic spectra of Munsell colors, J. Opt. Soc. Am. A, 6, 318-322 (1989).
3. L. T. Maloney, Evaluation of linear models of surface spectral reflectance with small numbers of parameters, Color Res. & Appl. 14, 325-334 (1986).
4. L. T. Maloney and B. A. Wandell, Color constancy: A method for recovering surface spectral reflectance, J. Opt. Soc. Am. A, 3, 29-33 (1986).
5. M. J. Vrhel and H. J. Trussell, Color Correction using principal components, Color Res. & Appl., 17, 328-338 (1992).
6. D. Marimont and B. A. Wandell, Linear models of surface and illuminant spectra, J. Opt. Soc. Am. A, Vol9, No.11, Nov., 1905-1913 (1992).
7. M. J. Vrhel, R. Gershon, and L. S. Iwan, Measurement and Analysis of Object reflectance Spectra, Color Res. & Appl. 19, 4-9 (1994).
8. G. Wyszecki and W. S. Stiles, Color Science 2nd Ed., John Wiley & Sons (1982).
9. W.H. Press, S.A. Teukolsky, W.T. Vetterling and B.P. Flannery, Numerical Recipes in C-The Art of Scientific Computing, Cambridge Univ. Press (1992).
10. S. D. Lee, C. Y. Kim, and Y. S. Seo, Linear Model of Surface and Scanner Characterization Method, in IS&T/SPIE's Symposium on Electronic Imaging : Device Independent Color Imaging II, Feb., San Jose, California, 84-93 (1995).
11. T. Jaakelainen, J. Parkkinen, and S. Toyooka, Vector-subspace Model for color representation, J. Opt. Soc. Am. A, Vol. 7, No. 4, April,725-730 (1990).

A Computational Approach to Color Illusions

Daniele Marini* and Alessandro Rizzi**

* Dipartimento di Scienze della Informazione
Università degli Studi di Milano
Via Comelico 39 - 20135 Milano, Italy
marini@eidomatica.dsi.unimi.it

** Dipartimento di Elettronica per l'Automazione
Università degli Studi di Brescia
Via Branze 38 - 25123 Brescia, Italy
rizzi@bsing.ing.unibs.it

Abstract. Tri-stimulus theory of color perception is not able to justify effectively some well known perception phenomena as color illusions and color constancy. Retinex theory, by Land and McCann, grounds color perception on a color space based on three lightness computed as relative reflectance along multiple exploration paths of the perceived scene. This paper considers in a new light Retinex theory, as a theory which tries to justify not only color constancy but also illusions arising from simultaneous contrast configurations. An improvement to Retinex computational model is presented in the paper, which selects Retinex computation paths by approximating a brownian path. The algorithm has been tested not only on traditional Mondrian patches, but also on natural pictures and photographs and on typical color illusion patches. The examples demonstrate the ability of the model to emulate human color perception behavior.

1   Introduction

Some common visual experiences show that human color perception cannot be completely explained by recurring to tri-stimulus theory, which dates back to von Helmholtz and Young. Let us consider a picture taken with a photo camera in an interior illuminated by tungsten lamps: a dominant red-orange modifies the original colors. The same happens when we take some shot using a TV camcorder, without "balancing the white". If, on the contrary, we observe an interior scene in different illumination condition, e.g. illuminated by a tungsten lamp or by the sunlight from a window, we do not perceive such deep differences.

This problem is known as *color constancy*, and the human ability to compensate varying light conditions cannot be explained by recurring to a color theory based on a physical model of light-matter interaction and on the tri-stimulus theory. Indeed such theories can explain how the spectral composition of light reflected by a given and known surface can change when the spectral composition of the illumination changes.

Many physiological experiments and analysis have shown that human visual system is grounded on the stimulation of cones by three fundamental ranges of frequencies, so that all experts agree in the assumption that cones have three different pigments that allow them to react differently to these three light frequency ranges. What tristimulus theory is not able to explain is how such triple stimulus is elaborated by the human visual system, so that color perception, in large part, is invariant to varying illumination conditions.

Color constancy is usually considered as the ability to perceive the same color in varying viewing conditions. Brainard & Wandell formalize the problem of color constancy, for a visual configuration typical of the first Land's experiments. Let a surface composed of different matte materials (reflecting by Lambert law) be given; an observer looks at the surface, he has three classes of photoreceptors. The spectral power distribution of the ambient light is $E(\lambda_n)$, where λ_n is given at discrete wavelengths values n=1,...,N. Let $S^x(\lambda_n)$ be the reflectance of the surface at a point x. The light arriving at the observer's eye is the *color signal*, and it is given by:

$$C^x(\lambda_n) = E(\lambda_n) S^x(\lambda_n)$$

The observer has three arrays of photoreceptors that spatially sample the color signal; their response is computed from the color signal and the spectral sensitivity of the photoreceptor's pigment in the k-th class $R_k(\lambda_n)$ (k is three as the retinal cone pigments):

$$\rho_k^x = \Sigma_{n=1..N}\,(C^x(\lambda_n)\,R_k(\lambda_n)) = \Sigma_{n=1..N}\,E(\lambda_n) S^x(\lambda_n) R_k(\lambda_n)$$

The equation can be written in matrix form:

$$\rho^x = \Lambda_E\,\sigma^x \qquad (1)$$

where Λ_E is a 3xN matrix and each entry is $E(\lambda_n)\,R_k(\lambda_n)$, i.e. it depends only from the spectral power distribution of the ambient light and on the receptor spectral sensitivity. Λ_E is also called *lighting matrix*.

ρ^x is a vector of 3 components, representing the response of the three receptor classes, at location x.

σ^x is the surface spectral reflectance at point x, an N-vector.

The problem of color constancy can now be stated: if the receptor response vector is given, and also given are $E(\lambda_n)$ and $R_k(\lambda_n)$ (therefore if the lighting matrix is fully known), to determine the vector σ^x in any point x of the surface it is necessary to invert the linear equation (1). The system has a unique solution if the lighting matrix is square, i.e. if only three wavelength samples are taken, otherwise it has infinite solutions.

More complex is the problem when the ambient light spectral power distribution is not known and it must be estimated from the given parameters, or when a light source spectral distribution is changing in time. Brainard and Wandell discuss under what conditions the ambient light coefficient can be estimated so that the problem can be reduced to the solution of a linear system. In any case more information is required at this aim; in practical applications, like color balancing in video cameras, a reference surface is taken, to which all other colors are normalized.

Based on the above discussion the problem arises of understanding how the human visual system can reconstruct the missing information of light source spectral composition. What we assume, following hypothesis first proposed by E. Land, is that the color perception is the result of a complex comparison process among different visual areas and colored patches. The input to this process is the stimulus produced by cones and rods in the retina, and the process is completed in the higher cortical visual area of the brain.

2 Retinex color theory

Retinex theory, due to Edwin Land, assumes that color perception depends strictly on the neural structure of human visual system; being not clear if the retina or the cortex plays the central role, Land coined the term "Retinex" derived from retina and cortex.

The fundamental observation that drove Edwin Land in Retinex theory development is: " .. the eye, in determining color, never perceives the extra red [produced by a tungsten lamp] because [the color] does not depend on the flux of radiant energy reaching it." The first Land's experiment was based on taking two black and white slides with a red and green filter, and projecting them through the same filters: what appears, counter-intuitively, is an image having all the original colors!

Following other experiments, Land defined a quantity, named *lightness*, which is associated to every object of a scene, and which does not change as illumination conditions and object's location (closer or farther from other objects with different lightness) change. Lightness is perceived by human visual system independently from the light flux that impinges the eye. With a second experiment, Land and McCann verified that a full color sensation of color gamut can be obtained producing a light stimulus to rods and long wavelength (red) sensitive cones: color sensation is again non depending from spectral fluxes reflected by the single objects, but from a comparison of lightnesses. A third experiment demonstrated that color sensation produced by the so called Mondrian, is still independent from reflected spectral flux, or more properly from the product of spectral reflectivity by incident light energy produced by a light source, which is the property of color constancy.

Land and McCann defined a measurable physical quantity that can be correlated to the lightness: it is the ratio between the integrated radiance from a Mondrian area and the integrated radiance from a white paper measured under white illumination, with three filters that perform as cone spectral sensitivity. The color sensation is produced by a comparison of the three resulting values, even if, as Land says, no corresponding neural regions can be exactly identified in human visual system. The three numbers characterize a measure of lightness, allowing to propose a color geometric space, that can be used to measure and compare colors and color scales. Moreover the experiments show that the color sensation is not due only to the spectral characteristics of the light impinging the eye, because in that case the color of the same patch illuminated by two different light sources should be perceived of two different colors. We recall that this effect can indeed be perceived in "void condition", i.e. when the illuminated patch is isolated and the observer has no other stimulus from the surround. The Retinex theory assumes that human vision is based on three retinal-cortical system, each processing independently the low, middle and high frequency of the visible spectrum.

Each system forms a separate image of the world; the images are not mixed but compared and each system discovers independently the reflectance of the various region of the image, independently also from variation of the light source spectrum.

Land & McCann discuss how the relative reflectance computation can be simulated by an electronic equipment and by a neural net. They arrive at the conclusion that a couple of excitatory and inhibitory neurons in a chain can account for the reset mechanism required by the search for the lightest area (i.e. the white).

The theory assumes that the three Retinex systems receive their stimulus by the retinal cones, that have an absorption non uniform in their wavelength range sensitivity. So what is effectively fed into the computation cycle is the integral of the spectral product of absorbency of the cone pigments times the irradiance times the reflectance of the patch.

3 A computational implementation of Retinex theory

The relative reflectance of a colored patch is the mean value of relative reflectance computed along a number N of random paths to that patch:

$$R^\Lambda(i) = \frac{\sum_{j=1}^{N} R^\Lambda(i,j)}{N}$$

where Λ is the low, middle or high wavelength range and:

$$R^\Lambda(i,j) = \sum_k \delta \log \frac{I_{K+1}}{I_k}$$

is the reflectance of a colored patch i relative to a patch j, and:

$$\delta \log \frac{I_{K+1}}{I_k} = \begin{cases} \log \frac{I_{K+1}}{I_k} & if \left|\log \frac{I_{K+1}}{I_k}\right| > threshold \\ 0 & if \left|\log \frac{I_{K+1}}{I_k}\right| < threshold \end{cases}$$

I_k is the intensity in location k of the image, and should be computed, for a given wavelength range Λ, as:

$$I_k = \int_{\lambda \in (400,700)} a^\Lambda(\lambda) L(\lambda) \rho(\lambda) d\lambda$$

where: $a^\Lambda(\lambda)$ is the absorbency of the cone pigment for the range Λ, $L(\lambda)$ is the spectral irradiance on the patch and $\rho(\lambda)$ is its spectral reflectivity.

These computations are to be executed for the three fundamental wavelengths range Λ, corresponding to low, middle and high bands.

The above model depends on the randomness and number of paths that are chosen for the computation of the relative reflectance, moreover it depends on the value of the threshold that makes less relevant low reflectance ratios, that correspond to smooth change in color due to non uniform illumination. The basic Land algorithm, moreover, has a reset mechanism: if during a path computation a lighter area is found the cumulated relative reflectance is forced to 1, making the average computation to restart form this area. The effect of the reset mechanism is to consider the lightest area of an image as the reference value of the color white.

A critical problem in the above algorithm is the choice of the random path. An analysis of this problem has allowed us to propose a new solution, based on the Brownian motion. This assumption derives from results inresearch on human cortical visual system, where the distribution of the receptive fields of area V4 (which is considered the most responsive to color signal, see: Zeki) shows a Brownian path aspect.

3.1 The *Brownian-path* algorithm

To implement the Brownian motion to approximate a path in the Retinex Algorithm we have adopted the random mid point displacement technique (see: Saupe). The number of recursions to generate the brownian path is a power of 2. During the path traversal the algorithm scan converts the pixels computing the relative reflectance for each pixel along the edge:

foreach chromatic channel compute sequentially each pixel as follow:
   generate N random points in the image and
   foreach random point:
      generate a path by 2^M random mid-point displacement;
      follow the generated path and foreach pixel calculate the
         new chain function value (Knew);
      if Knew>threshold then Kold=Knew;
      if Knew>1 then Kold=1;   (a pixel with an higher lightness value
         has been visited)
   newChromaticPixelValue = average of all N final values Kold;
      (1 foreach random path)

Comparing results with this algorithm to the traditional random path generation approach, we have observed a significant reduction of the number of paths necessary to approximate the lightness value of each pixel. While the random-path algorithm requires about 200 paths, the brownian-path algorithm gives very good results with 20 or sometimes even 4 paths per pixel. So the time complexity spent in generating the brownian path approximation and scan conversion of path edges is improved by the drastic reduction of the total number of paths.

4 Retinex application on color illusions

Besides color constancy, also colour illusions are difficult to explain with classical color models. A well known color illusion derives from what Itten, Albers and others call *simultaneous contrast*, which arises when a small grey square lays over a larger and saturated background color. In this case what the observer perceives is not the grey but a hue which tends to the complemetary color of the background. We have applied the Retinex algorithm to some classical color patches and the results have shown that the Retinex algorithm computes a color triplet which is vey much alike the perceived color in natural conditions.

In Color figure 1 a color illusion by J. Albers is simulated. The configuration shows the effect on the human vision of simultaneous contrast of different color patches. In this example we perceive two different brown patches, the top one with a higher saturation then the bottom one. A similar illusion arises when a gray small square is put on a bigger colored background (see Itten); in this case the perceived color is much more alike the complementary of the background (see Color figure 2). Both examples have been computed using the brownian path algorithm with 20 paths.

In color figure 3 another illusion by Joseph Albers is simulated with Retinex algorithm: the gray background is perceived as non uniform and smoothly changing in an opposite way as the internal squares. The Retinex simulation, using the random path algorithm with 20 paths, shows this effect.

In color figure 4 the Retinex algorithm has been applied to a photograph to equalize colors, using brownian motion approximation on 20 paths.

5   Conclusions and perspectives

We have implemented two algorithms that compute the Retinex model of color vision proposed by Edwin Land in late '60. The algorithms differ in the technique adopted for choosing the paths necessary to explore the image to compute the three Retinex lightness. The first algorithm selects the paths in a pseudo-random way, while the second approach adopts an approximation of a Brownian path by mid-point displacement. The main criticism against Retinex theory concerns its inability to cope with situation with varying backgrounds, therefore reducing predicatability of the results. We consider that this is more a confirmation of the theory rather then a defect: the model emulates human behavior in color perception which is frequently faked by simultaneous contrast effects, so that the perceived color is different form the measured one in a controlled context. This behavior has been simulated in a variety of examples, trypical of color illusions.

Future developments of this research will modify the algorithm to filter digitized images by computing first an approximation of human cones response before the application of the Retinex algorithm. Moreover computational aspects will be explored to improve algorithm efficiency, that in typical images has a computation time ranging from few minutes to half an hour.

6   References

[1] Albers J. *Interaction of Color*, Yale University Press, New Haven (1975)
[2] Brainard D.H. and Wandell B.A., "Analysis of the Retinex theory of color vision", *JOSA-A*, 3(10), 1651-1661 (1986)
[3] D'Zmura M., Lennie P. "Mechanisms of Color Constancy", *Journal of Optical Society of America*, **3**,10, 1662-1672 (1986)
[4] Itten J. *Kunst der Farbe*, Otto Meier Verlag, Ravensburg (1970)

[5] Judd. D.B. "Appraisal of Land's work on two-primary color projections", *Journal of Optical Society of America* **50**, 254-268 (1960)
[6] Land E. and McCann J., "Lightness and Retinex Theory", *Journal of Optical Society of America*, **61**,1, 1-11 (1971)
[7] Land E., "Recent Advances in Retinex Theory and Some Implications for Cortical Computations: Color Vision and the Natural image", *Proc. Natl. Acad. Sci. USA*, Vol. 80, 5163-5169 (1983)
[8] Land E., "The Retinex Theory of Color Vision", *Scientific American*, **237**, 3, 2-17 (1977)
[9] Landy M.S., Movshon J.A. Ed.s *Computational Models of Visual Processing*, The MIT Press, Boston (1991)
[10] Luong Quang-Tuan, "La Couleur en Vision par Ordinateur: 1. une Revue", *Rapports de Recherche INRIA*, n. 1251 (1990)
[11] Maloney L.T., Wandell R.A. "Color Constancy: A Method for Recovering Surface Spectral Reflectance", *Journal of Optical Society of America*, **3**,1, 29-33 (1986)
[12] McCann J.J. & Houston K.L., "Calculating Colour Sensation from Arrays of Physical Stimuli", *IEEE Transaction on SMC*, **SMC-13**, 5, 1000-1007 (1983)
[13] Saupe D. "Algorithms for random fractals" in: Barnsley et alt. *The Science of Fractal Images*, Springer Verlag, New York (1988) IEEE Computer Graphics & Applications, November (1993)
[14] von Helmholtz, H., *Optique physiologique*, Paris, Edition Jacques Gabay, trad. par Javal E. et Klein N.Th. (1989)
[15] Wandell B.A., *Fundations of Vision*, Sinauer Associates Inc. Publishers, Sunderland, Massachusetts (1995)
[16] Wyszecky G., Stiles W.S., *Color Science: Concepts and Methods, Quantitative Data and Formulas*, J. Wiley & Sons, New York (1982) *the II Annual Bionics Symposium*, Vol.1, 126-141 (1961)
[17] Zeki S. *A Vision of the Brain*, Blackwell Scientific Pub., Oxford (1993)

Color Figures

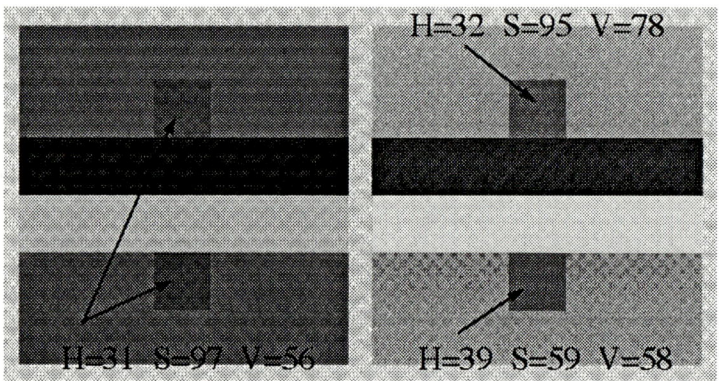

Color Figure 1 - An illusion by Josef Albers

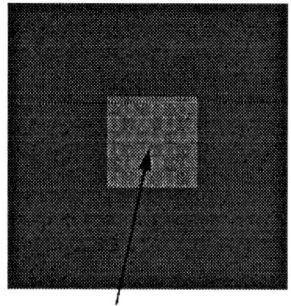 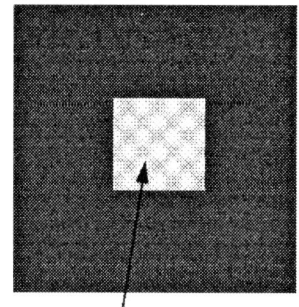

H=0 S=0 Y=50 R=128 G=128 B=128   H=195 S=5 Y=92 R=223 G=232 B=235

Color Figure 2 - The gray foreground is perceived as of the complementary color of the background

 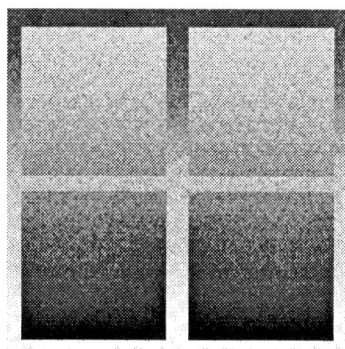

Color Figure 3 - Another illusion by Joseph Albers: left original image, right Retinex filtered image..

 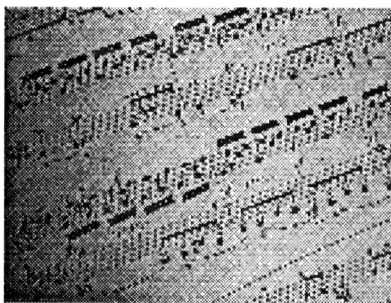

Color Figure 4 - The effect of Retinex algorithm in a non uniform illumination with brown dominant

Improved Textured Images Segmentation Using an Energy Functional

Antoni Grau and Jordi Saludes*

Dept. of Automatic Control and Computer Engineering
* Dept. of Applied Mathematics II
Polytechnic University of Catalonia UPC, Barcelona, Catalonia

Abstract. *In this paper we present a new classification and image segmentation system based on the addition of a variational method to a classic clustering algorithm. This system constitutes an improvement respect traditional segmentation methods. Often due to the nature of the texture features obtained from an image, the segmentation results are not quite precise. If this happens, using the energy functional and its minimization can improve the segmentation. This functional takes into account the information in the feature space and the information in the 2D image domain. The extracted characteristics from the image are texture features that have been defined in order to obtain an admissible trade-off between their discriminant capacity and their effectiveness to be implemented in a vision board to operate at real time. We show some results to appreciate this improvement in the segmentation using the energy functional.*

1 Introduction

Texture perception is an important part of the human vision. The objects can often be discriminated by their texture features despite colors and similar shapes. Despite its ubiquity in scene analysis a precise definition of texture does not exist, but texture is the term used to characterize the surface of a given phenomenon in an image and it is undoubtedly one of the principal features used in image processing and pattern recognition, but in a wide sense, there is no generally accepted definition of texture. In computer vision, all treatments of texture have taken one of the following two approaches [1]. The statistical approach attempts a global characterization of texture. Statistical properties of the spatial distribution of gray levels are used as texture descriptors. The structural approach conceives of texture as an arrangement of a set of spatial subpatterns according to certain placement rules. The first step in objects recognition is at present the segmentation, because normally, each region corresponds to an individual object or each contour corresponds to the border between different objects [2]. For a correct texture discrimination it is necessary to find the differences between them.

In this work we present an image segmentation system which uses texture features and works at high speed, that is, video rate. Due to this reason, the computed texture features are not very sophisticated. We search an admissible trade-off between their effectiveness in terms of quality and their low computational cost. If the features do not have an excessive quality, it will be a hard task for the classifier to segment the parts of the image with certain uniformity: it exists a deficiency in the feature space that the classifier can not solve. It is just in this moment, after classification, when the concept of the energy functional appears: why not take advantage of the information in the feature space together with the information in the 2D image domain?. With this sum of information, the misclassified objects will be reclustered and, at the end of this process, we will find a set of clusters that can be considered correct in terms of visual perception of the image textures. When the final clusters have been obtained from some training images, the system changes of stage and is already prepared to work and classify new objects in an on-line process at high speed. The cluster assignment for every new object entering in the system is achieved in real time and the clusters information can be updated with these news objects. Obviously, a bottle neck in the system is the feature extraction system and we are preparing a specific architecture to implement these texture features. In [3], these features are widely explained and detailed, but we will define them in Section 3. In Section 4 we will see how to classify these texture features in order to create perceptually different texture classes.

2 System overview

For image segmentation using the texture information we propose a system that extracts texture features from the image and groups all the texture elements having something in common. There will be three different and separate steps to achieve the segmentation of the original image. In figure 1 the segmentation system is shown.

Fig. 1. Diagram of the image segmentation system using texture features.

In low-level processing the thinned edges of the image are obtained and the image is prepared for the feature extraction module. In this second module a feature space is generated. Each new point (represented as a vector) in this new space is a texel (texture element) which has texture parameters as components. In the classification step, the texels are grouped in order to obtain texture classes with similar behavior. This classification will be improved with the use of a energy functional. Then, the segmentation output is represented as the borders between texture groups.

3 Feature extraction and feature space

Being perceptible is the principal characteristic of the texture parameters and this perceptible nature of the parameters allow to compute their value with masks created in a perceptible manner too. These masks are local boolean expressions which are applied over each pixel in the image. Due to the masks, it will appear some constraints when the system will try to discriminate the textures, but in the other hand, there is a predisposition for a hardware implementation of the masks to find the texture parameters. The definitions of the parameters are as follows:

1. Straightness. This parameter indicates the straight line density over a region, and it has been derived from the linear regression model.
2. Blurriness. The blurriness is a visual effect where a progressive and slow gray level increasing or decreasing along an image area is noticed. We only consider blur pixels if the change in their intensity is inside a fixed interval, otherwise we consider sudden changes.
3 Abruptness. This parameter indicates sudden changes in the directionality of the texture.
4. Granularity has a high perceptual significance in textures. This value will indicate how many elements in the image are isolated or non-concatenated.
5. Discontinuity. This parameter measures the density of cut edges in the image. When the edges are continuous the value will be high.

Each texel determines a point in the 5-dimension texture space. The images used in this work contain 32-by-32 texels each and this will be the number of points in the feature space. The parameters can be found in a parallel manner because it does not exist any data dependence among them. Then, for an input image, where the minimum element is the pixel, each parameter is calculated in parallel with its masks to produce an output image where the minimum element is the texel (see figure 2). The size of the masks used to calculate the parameters is 4-by-4 elements. A bigger size will generate a large amount of masks for each parameters and the cost in time will be excessive for our intentions. A smaller size of the masks is contradictory with the own definition of the parameters.

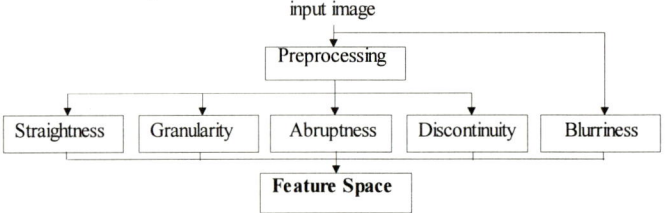

Fig. 2. Feature extraction and generation of the feature space.

The masks are calculated through the image following the raster scan. Therefore, every pixel in the image will be analyzed. These masks are template matching masks.

4 Classification and segmentation

This is the third and last step in the image segmentation system. There are two major techniques in grouping: discriminant analysis and cluster analysis. All varieties of discriminant analysis require prior knowledge of the classes, usually in the form of a sample from each class. In our case, the data do not include information on class membership and moreover we do not know how many classes we are looking for. For this reason, the technique used in this work is cluster analysis and, among the available methods, the algorithm uses a hierarchical technique with an agglomerative method explained below.

A cluster is defined as a collection of points which are close between them [5]. The various clustering methods differ in how the distance between two clusters is computed. There will be two steps in cluster processing: the learning phase and the working phase. In the learning phase, after cluster analysis, the supervisor chooses the number of clusters cutting at specified level of the hierarchical tree. Once the clusters are defined, in the working phase for each new object in the system the centroid distance is computed and the object is assigned to the closest cluster. The learning phase takes a lot of time to compute all the clusters but, luckily, it is an off-line task. The working phase is critical in computing cost, but the system have to compute only several distance measures (as distances as clusters).

In the figure 3 the recognition strategy can be seen. The different phases in recognition are separated. First, from the learning pattern samples the clusters are generated and, in the same phase, by the energy functional a decision rule is derived. The energy functional performs a reclustering of misclassified samples and, in this submodule, the correct clusters are created. In the working phase, the new pattern samples are classified using the clusters generated by the energy functional module. This is the on-line phase.

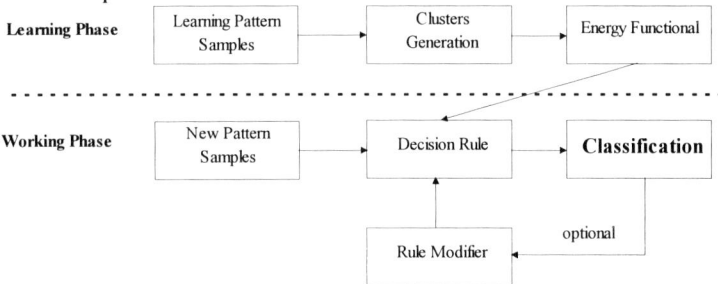

Fig. 3. Diagram of the recognition strategy. Learning phase and working phase.

4.1 Clustering Algorithm

In [6], Ward proposed a method of forming clusters that is based on the loss of information resulting from the grouping of objects into clusters as measured by the total sum of squared deviations of every observation from the mean of the cluster to which it belongs. The assignment rule rests on the increase in the error sum of squares induced from combining every possible pair of clusters. This value, which will be denoted by E.S.S., is used as an objective function. The algorithm also employs a hierarchical grouping procedure. The E.S.S. is computed as

$$E.S.S = \sum_{j=1}^{k} \left(\sum_{i=1}^{C_j} x_{ij}^2 - \frac{1}{C_j} \left(\sum_{i=1}^{C_j} x_{ij} \right)^2 \right) \quad (4.1)$$

where x_{ij} denotes the point value in the feature space (n-dimensional) for the object ith in the jth cluster, k is the total number of clusters at each stage, and C_j is the number of objects in the jth cluster. The E.S.S. is zero at the first stage, since each object constitutes a cluster. For example, we consider the objects shown in figure 4a). After considering all possible clusters of size 2 and combine those objects that yield the smallest E.S.S., we fuse objects E and F. The next stage entails a computation of E.S.S. induced from (1) adding each of the remaining objects to the first cluster and (2) forming all possible pairs of the unclustered objects. Finally, the resulting dendogram is shown in the figure 4b). After the examination of the dendogram, if we choose N3 as the level to cut the tree, the number of total clusters will be 3, as it can be seen in the figure 4c). For simplicity in the figures, we consider a 2-dimensional feature space.

4.2. Energy Functional

The segmentation problem is the problem of subdividing an image into regions so that in each region, the image properties are relatively uniform. When clustering algorithms have been applied to group points in a ^{n} texture feature space, it is not sure that we have arrived at the end of the segmentation process with success and correct results. Therefore, in order to improve the segmentation we will study the problem with a variational approach [4]. This approach is motivated in part by occasional failures of traditional methods, which are based exclusively on clustering objects formed by features extracted from images without taking into account the image domain. The general idea is to define an energy functional over a set of possible segmentations in terms of penalty measures that correspond to various desired properties of a good segmentation. In particular, the functional we have studied is the following. Consider **C** a segmentation, that is, a partition **C** = {C_1, ..., C_M} of the set of texel indexes.

$$J(\mathbf{C}) = \sum_{k=1}^{M} \sum_{j \in C_k} \left\| \vec{x}_j - \overline{x}_k \right\|^2 + \lambda |F| \qquad (4.2)$$

where $\overline{x}_k$ is the center of mass of the texels in C_k, $\vec{x}_j$ is the jth object of the kth cluster, M is the number of clusters, λ is a weight, F is the union of the borders between texture regions in the image and $|F|$ is the length of these borders. The problem is to find a segmentation **C** that minimize the functional $J(\mathbf{C})$. The functional works in two different spaces: R is the image domain and H is the n-dimensional domain of texture features. While the first term of the functional tends to get the maximum number of clusters in the H space, the second term tends to get to minimize the borders between clusters, trying to avoid segmenting the image into too many regions. Thus, the formulation is designed to find a simple segmentation such that in every region the deviation of the objects to the center is minimum. Then, the functional penalties are 1) the minimum number of clusters and 2) the maximum length of their borders. Resuming, this method combines the classification in the H space together with the actual placement of the texels over the image domain R.

The following algorithm finds the pair that minimizes the functional. These are the steps in the algorithm:

1. Choose randomly a texel t_{xy} in the image domain.
2. Seek for any neighbor of texel t_{xy} over the R space that belongs to a different cluster in the H space. If the whole neighbors texels belong to the same clusters of texel t_{xy} go to step 1.
3. Join temporally texel t_{xy} to the clusters of the chosen neighbor.
4. Compute the functional $J(\mathbf{C})$.
5. If the functional value has decreased, then join definitively texel t_{xy} to the cluster of its neighbor. Update $\bar{x}_k$ and go to step 1.
6. If after n iterations the functional value has not decreased we are over a local minima and the algorithm finishes.

Theoretically and practically there are not algorithms that find global minima to this problem due to the nonlinearity and also to the existence of many local minima.

In figures 4c) and 4d) we can see an example of migration of objects to different clusters after minimization of the functional. An object can be moved from a cluster to a different one decreasing the number of borders between regions even if the deviation of this cluster increases.

4.3 Working Phase

In this second phase of grouping, it is required that the clusters be clearly defined. Thus, for each new object entering in the system it will be assigned to a former cluster created in the learning phase with the use of the energy functional. Graphically, this effect is shown in the figure 4e). A new object $\mathbf{X}$ can be placed in any point in the feature space and it is assigned to the closest cluster. The creation of new clusters with new objects is not allowed in this working phase. As assignment rule the minimum distance to the centroids of clusters is chosen.

$$\text{Assignment rule} = \{ \mathbf{X} \in i \mid min\{d(\mathbf{X}, \bar{x}_i)\} \quad (4.3)$$

where i is the cluster to which the object is assigned and $\bar{x}_i$ is the center of mass of the cluster i. For each new object, we compute the distances between the object and the centroids of clusters. The object is assigned to the closest cluster. This process has a low computational cost and it allows its implementability in an on-line system.

In any on-line process the incoming images can vary gradually with time. Therefore, it exists the possibility of updating the centers of mass of the clusters concurrently with the classification of new objects. Each new object in the system is assigned to a cluster by the rule (4.3), and the center of mass of this cluster is computed again. To update the center of mass of a cluster with a new object can be done at real time.

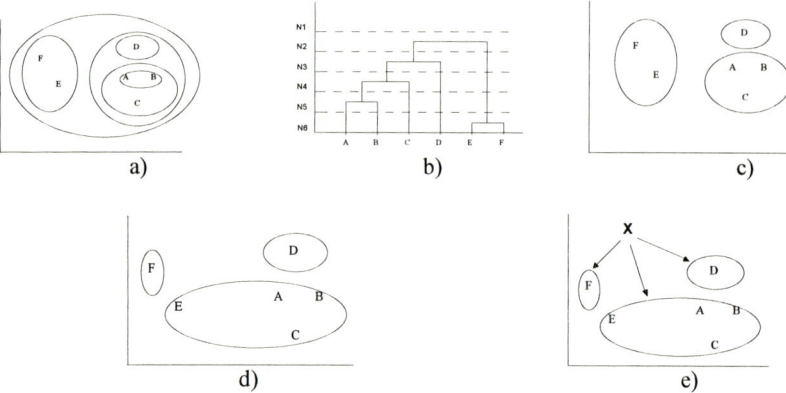

Fig. 4. a) Objects in feature space; b) dendogram; c) clustering process; d) final clustering after energy functional minimization; e) cluster assignment in the working phase.

5 Results and concluding remarks

To show the performance of our system we have chosen as input images some aerial images obtained from a plane.

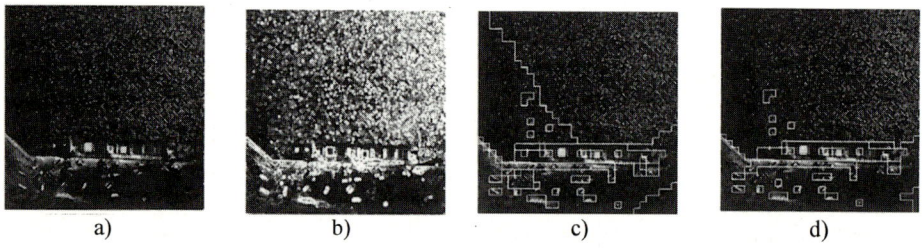

Fig. 5. Learning phase. a) Input; b) Gradient image; c) Clustering; d) Output segmented image.

In the learning and in the working phase, images pass through three different processes: 1) preprocessing: in this stage the gradients of the original image are extracted, then they are binarized and thinned; after this process the image enters in 2) the texture features extractor where five texture parameters are computed generating 32x32 texels per image. The combination of these five parameters forms the feature space with 1024 points with five components each. The next stage is 3) the clustering process: over this 5-dimensional space the points are grouped to form clusters with similar objects. The use of the energy functional will take part only in the learning phase after the classification. All this process can be seen in figure 5 where appears a harbor, the sea, a ship and other objects: from the input image a) the gradients are extracted in b). In order to evaluate the effectiveness of the use of the energy functional versus the traditional clustering methods is interesting to look at the next table (table 1), made up from data in figure 5.c) and 5.d).

		Actual Cluster						Actual Cluster				
		1	2	3	4	Total		1	2	3	4	Total
Assigned Cluster	1	489 71%	10	6	5	510	1	675 98%	0	0	4	679
	2	194	176 76%	4	4	378	2	9	217 94%	7	4	237
	3	6	20	31 72%	0	57	3	5	7	37 86%	0	49
	4	0	24	2	53 85%	79	4	0	6	0	53 85%	59
	Total	689	230	43	62	102	Total	689	230	43	62	1024

Table 1. Clustering results before (left table) and after (right table) energy functional use.

When the texture features have been extracted and hierarchically classified, we cut the tree at a certain level and four clusters have been generated. The cluster #1 corresponds to a image area where it appears the sea. In the cluster #2 are grouped the objects corresponding to the land (in the harbor) and also these elements that will form clusters with a few objects to be considered relevant. In the cluster #3 are grouped those objects with a small area such as trucks, containers in the harbor and also those small components in the ship, while in the cluster #4 are grouped those objects corresponding to the ship.

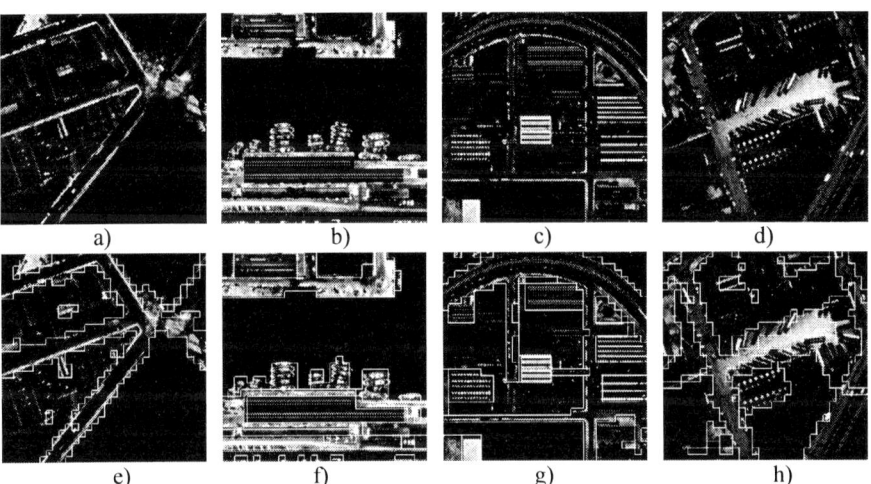

Fig. 6. Results in the working phase. a), b), c) and d) Input images; e), f), g) and h) Output segmented image.

With this information we made two tables: the first table is made after classification but before the use of the energy functional; the second table is made after use it. The table shows the number of objects that appear in the clusters. The columns indicate the objects appearing in the actual clusters while the rows indicate the number of objects appearing in each cluster. The optimal solution is when all the objects appear in the diagonal of the tables, that is, when the assigned objects in the clusters correspond with the actual clusters and the classification is perfect. This is not the case and we indicate the % of correct classification in each cluster. As it can be seen in the tables, the number of objects classified correctly augment in the right table and, specially, in the cluster #1 (corresponding to the sea); practically, the objects that correspond actually to the sea but considered as objects in the land (194 objects in the left table) are moved to their correct clusters after the reclassification with the energy functional. With the cluster #2 there is a similar effect, some objects corresponding actually to land are grouped as ship elements or even as small elements, then many of these elements are moved to its correct cluster. After an evaluation of the system, we pass to the working phase using the data obtained in the learning phase. This can be seen in the figure 6 where some input images are segmented with their texture information.As it is appreciated in the output images, the segmentation is not perfect due to different aspects: first, the input to the image segmentation system is not the gray level image but their edges. That produces an information loss and a degradation in the image. In second place, the parameters computation is not completely accurate because there is a time limitation if we want fast speed, near real time (video rate in this case). Actually, there is a trade-off between the quality of segmentation and the computation time. Our system priorizes the low processing time in front of the high quality in the segmentation not always required in many applications.

6 References

[1] Lipkin, B.C. and Rosenfeld, A., *Picture Processing and Psychopictorics*, Academic Press, New York, 1970.
[2] Tomita, F. and Tsuji, S., *Computer Analysis of Visual Textures*, Kluwer Academic Publishers, 1990.
[3] Casals, A., Amat, J. and Grau, A., "Texture Parametrization Method for Image Segmentation", *Proc. 2nd European Conf. on Computer Vision ECCV'92*, pp. 160-164, 1992.
[4] Mumford, D. and Shah, J., "Optimal Approximations by Piecewise Smooth Functions and Associated Variational Problems", *Comm. Pure Appl. Math.*, Vol. **42**, pp. 577-685, 1989.
[5] Young, T.Y. and Calvert, T.W., *Classification, Estimation and Pattern Recognition*, Elsevier Publishing Company, Inc., 1974.
[6] Ward, J., "Hierarchical Grouping to Optimize an Objective Function", *Journal of the American Statistical Association*, Vol. **58**, pp. 236-244, 1963.

Contribution to the Colour Segmentation by Means of an Algorithm Which Reduces the CCDs Saturation Problems

Jordi Regincós Isern and Joan Batlle Grabulosa

Institut d'Informàtica i Aplicacions, Universitat de Girona,
Av Santaló s/n, E-17003 Girona (Spain)
e-mail:jordir@ima.udg.es jbatlle@eia.udg.es

Abstract. Sometimes, the results provided by colour image processing systems are not accurate enough due to the physical process of image formation. One of that problems is colour clipping, which appear when at least one of the sensor components is saturated. We propose a method to recover the chromatic information of those pixels on which colour has been clipped. The chromatic correction method is based on the fact that some chromatic characteristics are invariant to the uniform scaling of the three RGB components. In this paper we present this method and one study of the chromatic components to which it can be applied.

Key Words : Colour Clipping, Colour recovering, Colour segmentation.

1 Introduction

Colour has been proved to be useful information in computer vision systems, although most of the usual problems in computer vision can be solved on grey-level images. However, in some situations, taking out chromatic information can highly improve the accuracy of the algorithms (see, for instance, Luong [6] and Gershon [3] papers).

There are some steps in the image formation processes which could make changes in the original colour. Actually, if we focus on the first one of this steps, it is obvious that CCD cameras make some important changes in supplied images (see, for instance, Novak [8]).

There are several different ways to model the response of the camera. Shafer notation, shown in [11] is probably one of the most accepted. For a single pixel x, the response of the camera is given by the following equation:

$$C_x = \begin{bmatrix} r_x \\ g_x \\ b_x \end{bmatrix} = \begin{bmatrix} \int X(\lambda)\tau_r(\lambda)s(\lambda)d\lambda \\ \int X(\lambda)\tau_g(\lambda)s(\lambda)d\lambda \\ \int X(\lambda)\tau_b(\lambda)s(\lambda)d\lambda \end{bmatrix} \qquad (1)$$

where $X(\lambda)$ represents the amount of incident light, $s(\lambda)$ corresponds to the camera responsivity and $\tau_r(\lambda)$, $\tau_g(\lambda)$, $\tau_b(\lambda)$ correspond to the filters' transmittance of the three components red, green and blue.

Cameras are finite devices, so there is a maximum value that cameras can provide called the dynamic range. It is usually assumed that camera output ranges from (0,0,0) to (1,1,1). Every value of r_x, g_x or b_x in equation 1 greater than 1 is clipped by the camera to 1 and it is said that this component is saturated.

Some work has been done in order to recover the chromatic information of those pixels of which at least one of RGB components is saturated. This work was based on making assumptions about the actual colour of the pixel from the colour of the neighbouring pixels; see for example the works of Perez and Koch [9] and Klinker et al. [5].

We have proposed a method to recover the chromatic information of a pixel without looking at the neighbouring pixels, and we have studied its applicability to several chromatic characteristics that have been defined on the 0. Section 2 gives an outline of the method, in section 3 we present the chromatic components that have been studied, in section 4 some experimental results are shown and, finally, in section 5 the conclusions to this work are presented.

2 Outline of the proposed method

From equation 1 it can be deduced that for a given colour, the camera response follows a straight line inside the RGB space when light intensity changes. This is illustrated in figure 1. However, due to the limitations of cameras, this response is only linear for one interval of intensities: interval $[a, b)$ in figure 1. For values higher than b, the camera reaches its maximum in at least one of its components. In the interval $[b, c)$ the red component is saturated, and in the interval $[c, d)$ red and green components are saturated. For values higher than d the three components are saturated and white is obtained as the camera response.

This nonlinearity in the camera response produces undesired effects on the algorithms that work with chromatic components such as hue, that are stable under variations in the illuminant intensity. The sketch of the method we propose (which is detailed on our previous work [10]) is:

1. Find the straight line $\rho(t) = (\rho_r(t), \rho_g(t), \rho_b(t))$ that describes the camera response in front of light intensity changes. This line will provide the expected camera response, assuming that the camera has not any physical limitation.
2. From this line $\rho(t)$ it is possible to estimate the actual camera response, which will be the minimum between $\rho(t)$ and the maximum value the camera can return; that is, the actual camera response will be $(\min(\rho_r(t), 1), \min(\rho_g(t), 1), \min(\rho_b(t), 1))$.
3. A mapping is defined from the actual camera response (step 2) to the colour cube in the following way: the expected camera response $(\rho_r(t), \rho_g(t), \rho_b(t))$ is multiplied by a constant k, $0 < k < 1$ and the value obtained is assigned to the actual camera response $(\min(\rho_r(t), 1), \min(\rho_g(t), 1), \min(\rho_b(t), 1))$.
4. The mapping obtained in the previous step is stored in three LUTs, which will be addressed by the camera output. The output of this three LUTs will

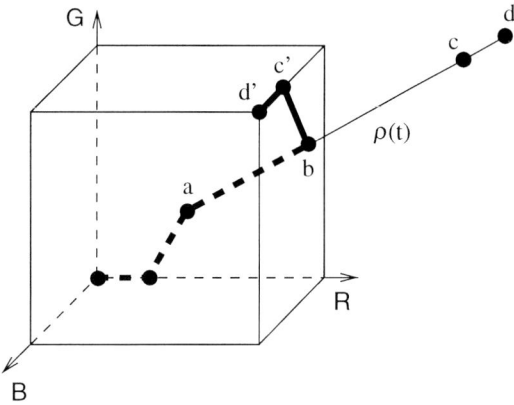

Fig. 1. *Camera response for a given colour when light intensity changes. From a to b the camera response is linear and, from intensity values higher than b, colour is clipped (thick line).*

be the corrected RGB values. Alternatively, the LUTs can be loaded with a transformation to any colour space, such as HSI and, so, it is possible to get the corrected HSI values in real time.

Note that this method works in a two-steps basis: first, colours are characterized offline (the straight lines that describe the colours are estimated) and then, colour correction is performed in real time using LUTs, because the colour correction is defined as a point operation.

3 Chromatic components studied

Our method can be applied to any given chromatic transformation $T(R, G, B)$ which holds the scale invariance property; that is, if $\forall \alpha > 0 \quad T(R, G, B) = T(\alpha R, \alpha G, \alpha B)$. Several chromatic components used on the literature have been studied and the results obtained are summarized in table 1. These transformations from the RGB space to the chromatic components are now presented (for each transformation, a reference to a work in which it has been used is given):

HSI space (Perez and Koch) [9]

$$H(r, g, b) = \arctan \frac{\sqrt{3}(g - b)}{(r - g) + (r - b)} \tag{2}$$

$$S(r, g, b) = 1 - \frac{\min\{r, g, b\}}{(r + g + b)/3} \tag{3}$$

$$I(r, g, b) = \frac{r + g + b}{3} \tag{4}$$

HSI space (Carron and Labert) [2]

$$H(r,g,b) = \begin{cases} \arccos(C_2/S) \text{ if } C_2 \geq 0 \\ 2\pi - \arccos(C_2/S) \text{ if } C_2 < 0 \end{cases} \quad (5)$$

$$S(r,g,b) = \sqrt{C_1^2 + C_2^2} \quad (6)$$

IC1C2 space (Barni et al.) [1]

$$C_1(r,g,b) = \frac{b}{r+g+b} \quad (7)$$

$$C_2(r,g,b) = \frac{2r+b}{r+g+b} \quad (8)$$

HSV space (Smith) [12] Hue is given in the interval $[0, 2\pi]$, although the following hue expression returns a value between $[-pi, pi]$.

$$H(r,g,b) = \begin{cases} \pi/3(g-b)/\Delta(r,g,b) & \text{if } r = \max\{r,g,b\} \\ \pi/3(2+(b-r)/\Delta(r,g,b)) & \text{if } g = \max\{r,g,b\} \\ \pi/3(4+(r-g)/\Delta(r,g,b)) & \text{if } b = \max\{r,g,b\} \end{cases} \quad (9)$$

$$\Delta(r,g,b) = \max\{r,g,b\} - \min\{r,g,b\}$$

$$S(r,g,b) = \frac{\max\{r,g,b\} - \min\{r,g,b\}}{\max\{r,g,b\}} \quad (10)$$

$$V(r,g,b) = \max\{r,g,b\} \quad (11)$$

HLS space (Joblove and Greenberg) [4]

$$L(r,g,b) = (\max\{r,g,b\} + \min\{r,g,b\})/2 \quad (12)$$

$$S(r,g,b) = \begin{cases} \dfrac{\max\{r,g,b\} - \min\{r,g,b\}}{\max\{r,g,b\} + \min\{r,g,b\}} & \text{if } L \leq 0.5 \\ \dfrac{\max\{r,g,b\} - \min\{r,g,b\}}{2 - (\max\{r,g,b\} + \min\{r,g,b\})} & \text{if } L > 0.5 \end{cases} \quad (13)$$

HSV space (Tenembaum) [13]

$$S(r,g,b) = 1 - 3\min\left\{\frac{r}{r+g+b}, \frac{g}{r+g+b}, \frac{b}{r+g+b}\right\} \quad (14)$$

HSV space (Yagi et al.) [15]

$$S(r,g,b) = \max\{r,g,b\} - \min\{r,g,b\} \quad (15)$$

RGB normalized space (Nevatia) [7]

$$Y(r,g,b) = c_1 r + c_2 g + c_3 b \text{ where } c_1 + c_2 + c_3 = 1 \quad (16)$$

$$T_1(r,g,b) = \frac{r}{r+g+b} \quad (17)$$

$$T_2(r,g,b) = \frac{g}{r+g+b}$$

CIEL*u*v* space (Wyscecki and Stiles) [14] CIE components are defined in terms of XYZ components instead of RGB components. There exists a linear transformation from RGB space to XYZ space.

$$h_{uv}(x,y,z) = \arctan \frac{v^*}{u^*} \tag{18}$$

$$L^*(X,Y,Z) = \begin{cases} 116\sqrt[3]{Y/Y_n} - 16 & \text{if } Y/Y_n > 0.008856 \\ 903.3(Y/Y_n) & \text{if } Y/Y_n \leq 0.008856 \end{cases} \tag{19}$$

$$C^*_{uv}(x,y,z) = \sqrt{v^{*2} + u^{*2}} \tag{20}$$

$$s_{uv}(x,y,z) = \frac{C^*_{uv}}{L^*} \tag{21}$$

CIEL*a*b* space (Wyscecki and Stiles) [14]

$$h_{ab}(x,y,z) = \arctan \frac{b^*}{a^*} \tag{22}$$

$$C^*_{ab}(X,Y,Z) = \sqrt{a^{*2} + b^{*2}} \tag{23}$$

4 Results

Figure 2 shows an example of the results obtained with the proposed method applied as a preprocessing step in a hue-based segmentation algorithm. Figure 2.a corresponds to a series of 12 images acquired under illumination changes and figure 2.b shows the same images once they have been corrected. In this case three colours have been corrected: the locomotive colour (orange) and the two wagon colours (green, on the middle, and cyan). Figure 2.c shows the results of the segmentation of the original images and, finally, figure 2.d shows the results of the same segmentation applied to the corrected images. As can be seen in the first eight images of the series, the segmentation has been qualitative improved with the colour correction method.

Quantitative measurements of the segmentation improvement are shown on figure 3. The x-axis corresponds to the percentage of pixels with at least one saturated component and the y-axis shows the percentatge of pixels of the object correctly segmented. Figure 3.a corresponds to the locomotive, figure 3.b corresponds to the green wagon and figure 3.c corresponds to the blue one. As it can be seen, when there are a lot of pixels with some saturated component, the segmentation performs better after colour correction in the locomotive and in the gren wagon. The results obtained with the cyan wagon are similar before and after colour correction: this is due to the fact that cyan composition – in terms of red, green and blue primaries– has similar amounts of green and blue and it nearly has not red.

Component	Chromatic Space	Equation	Invariant
T_1, T_2	normalized RGB	17	Yes
C_1	IC_1C_2	7	Yes
C_2	IC_1C_2	8	Yes
H	HSI	2	Yes
H	HSI	5	Yes
H	HSV	9	Yes
H	$CIEL^*u^*v^*$	18	Yes
H	$CIEL^*a^*b^*$	22	Yes
S	HSI	3	Yes
S	HSI	6	No
S	HSV	10	Yes
S	HSV	14	Yes
S	HSV	15	No
S	HLS	13	Yes if $L \leq 0.5$ No if $L > 0.5$
S	$CIEL^*u^*v^*$	21	Yes
C	$CIEL^*u^*v^*$	20	No
C	$CIEL^*a^*b^*$	23	No
I	HSI	4	No
V	HSV	11	No
L	HLS	12	No
Y	normalized RGB	16	No
L^*	$CIEL^*u^*v^*$	19	No

Table 1. *Chromatic components and their scale invariance property.*

5 Conclusions

We have described a simple preprocessing system, but it is its newness that contributes robustness to segmentation systems based on the hue and saturation components, expanding the range of the virtuality of the camera. Several chromatic components have been analyzed in order to study if the proposed method can be applied to them.

Our system can be applied in areas like the tracking of a known moving object, allowing the effects that changes in the incident light intensity produce on the hue component while the object is moving to be reduced.

We have tested its functioning in different scenes. As was expected, with the method presented the final segmentation is greatly improved when the number of pixels which suffer from colour clipping is high.

References

1. M. Barni, V. Cappellini, and A. Mecocci. A vision system for automatic inspection of meat quality. *8th Int. Conf. on Image Analysis and Processing*, number 974 in Lecture Notes in Computer Science, pages 748–753, 1995.

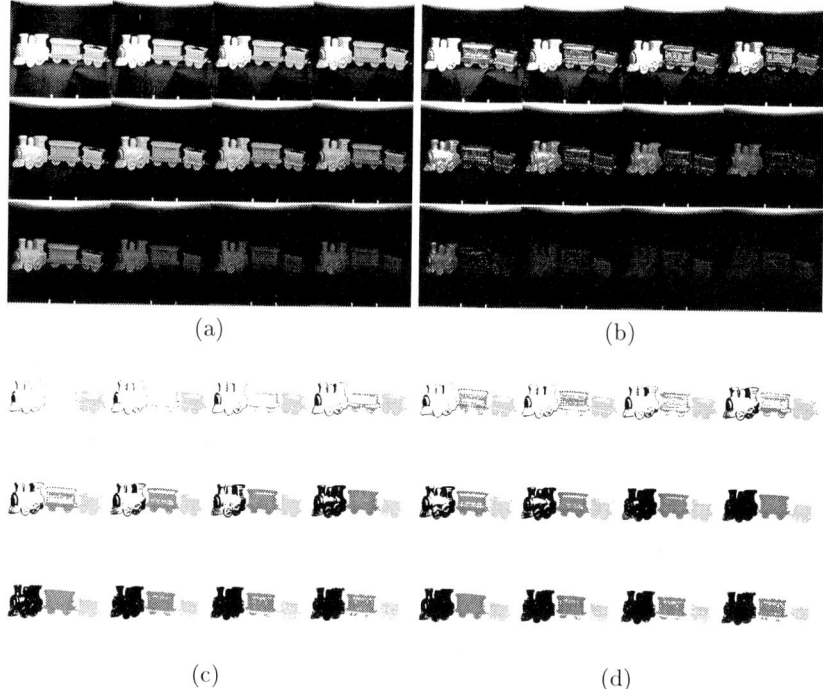

Fig. 2. Some results obtained: Original images (a), corrected images (b), segmentation results in original images (c) and segmentation results in corrected images (d).

2. T. Carron and P. Lambert. Color edge detector using jointly hue, saturation and intensity. *IEEE Int. Conf. on Image Processing*, vol. 3, pages 977–981, 1994.
3. Ron Gershon. Aspects of perception and computation in color vision. *Computer Vision, Graphics and Image Processing*, 32:244–277, 1985.
4. G.H. Joblove and D. Greenberg. Color spaces for computer graphics. *Computer & Graphics*, 12:12–19, 1978.
5. Gudrun J. Klinker, Steven A. Shafer, and Takeo Kanade. The measurement of highlights in color images. *International Journal of Computer Vision*, 2(1):7–32, June 1988.
6. Quang-Tuan Luong. Color in computer vision. In L.F. Pau C.H. Chen and P.S.P. Wang, editors, *Handbook of Pattern Recognition and Computer Vision*, pages 311–368. World Scientific Publishing Company, 1993.
7. R. Nevatia. A color edge detector and its use in scene segmentation. *IEEE Transactions on Systems, Man and Cybernetics*, 7(11):820–826, 1977.
8. Carol L. Novak, Steven A. Shafer, and Reg G. Willson. Obtaining accurate color images for machine vision research. In *Proc. of the Conference on Perceiving, Measuring and Using Color*. SPIE, Volume 1250, February 1990.
9. Frank A. Perez and Christof Koch. Toward color image segmentation in ana-

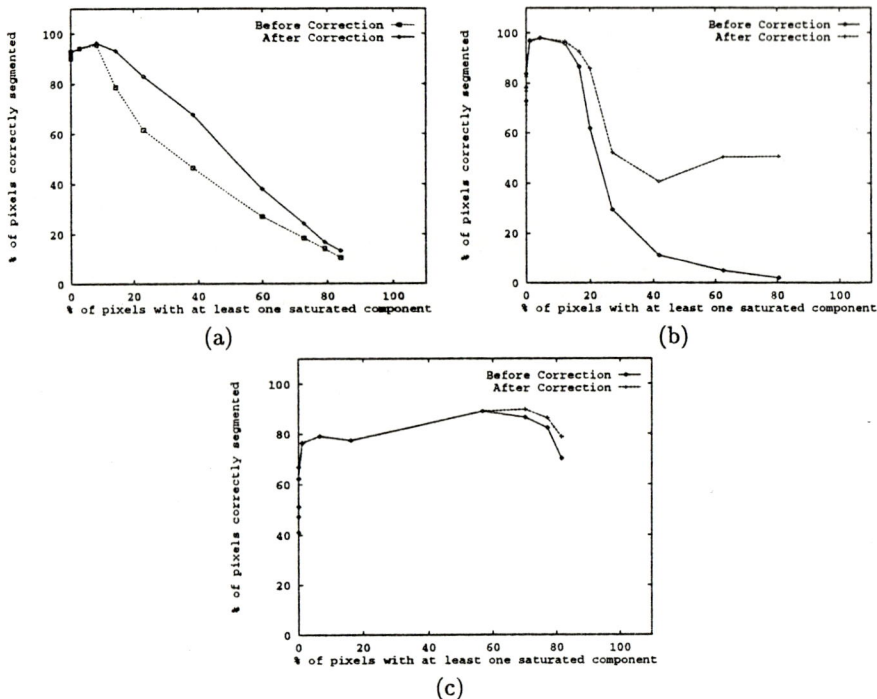

Fig. 3. *Quantitative measurement of the segmentation accuracy improvement after colour correction. Locomotive (a), green wagon (b) and blue wagon (c). Dashed line plots the behaviour of the colour segmentation accuracy before colour correction and the solid line plots the behaviour after colour correction.*

    log VLSI: Algorithm and hardware. *International Journal of Computer Vision*, 12(1):17–42, February 1994.
10. Jordi Regincós and Joan Batlle. A system to reduce the effect of CCD saturation. In *Proc. of the IEEE International Conference on Image Processing*, volume I, pages 1001–1004, 1996.
11. S.A. Shafer. Using color to separate reflection components. *Color Research and Application*, 10(4):210–218, 1985.
12. A.R. Smith. Color gamut transform pairs. *Computer & Graphics*, 12(3):12–19, 1978.
13. J. N. Tenenbaum. An interactive facility for scene analysis research. Technical Note 87, Artificial Intelligence Center, Stanford Research Institute, 1974.
14. Günter Wyszecki and W.S. Stiles. *Color Science (Concepts and Methods, Quantitative Data and Formulae)*. John Wiley & Sons, 1982.
15. Daisuke Yagi, Kejichi Abe, and Hiromasa Nakatami. Segmentation of color aerial photographs using HSV color models. In *MVA '92 Workshop*. IAPR, 1992.

Pyramid-Based Multi-sensor Image Data Fusion with Enhancement of Textural Features

B. Aiazzi°, L. Alparone*, S. Baronti°,
V. Cappellini*, R. Carlà°, L. Mortelli*

°IROE "Nello Carrara" - CNR
Via Panciatichi, 64, 50127 Firenze, Italy
E-Mail: baronti@iroe.fi.cnr.it
*Dip. Ing. Elettronica, University of Florence
Via S. Marta, 3, 50139 Firenze, Italy
E-Mail: alparone@cosimo.die.unifi.it

Abstract. In this work, a multi-resolution procedure based on a generalized *Laplacian pyramid* (GLP) with a *rational* scale factor is proposed to merge image data of any resolution and represent them at any scale. The GLP-based data fusion is shown to be superior to those of a similar scheme based on the discrete *wavelet transform* (WT) according to a set of parameters established in the literature. The pyramid-generating filters can be easily designed for data of any resolutions, differently from the WT, whose filter-bank design is non-trivial when the ratio between the scales of the images to be merged is not a power of two. Remotely sensed images from Landsat TM and from Panchromatic SPOT are fused together. Textured regions are enhanced without losing their spectral signatures, thereby expediting automatic analyses for contextual interpretation of the environment.

1 Multi-sensor Image Data Fusion

The availability of data from many sensors with different characteristics makes data fusion a topic of ever increasing relevance in the field of digital image processing. The main goal of pixel level algorithms [1] is to combine the original images from different sensors in order to synthesize a new set of data whose spatial and/or spectral resolution results to be enhanced, or to concentrate significant features of the various bands in a single image, thus compressing information and enhancing contrast and texture. In some cases processing is made with the main objective of extracting significant features [2] by maximizing the spatial contrast on the basis of the whole data set: distortion measures are not considered. In other applications, such as classification, merging algorithms are requested to maintain the spectral characteristics of the original data as much as possible [3] to avoid misinterpretation and introduction of undesired effects.

Approaches based on *principal component analysis* (PCA), on transformation of the original data in the *hue intensity saturation* (HIS) color space (three bands at a time), and on *high pass filtering* (HPF) [3] have been investigated in the literature to achieve the latter objective: HPF resulted far more efficient than the other algorithms in preserving the spectral features of the enhanced bands. Therefore, such *space-frequency* image representations as *discrete wavelet transform* (WT) and *Laplacian pyramid* (LP) have been recently investigated for image fusion aimed at contrast enhancement [2].

Multi-spectral Earth observations from space exhibit limited spatial resolutions, differently from broad-spectrum imaging sensors, that may be inadequate to specific identification tasks. A typical example of such a situation is represented by Landsat Thematic Mapper (TM) multi-spectral imaging sensor, which has a $30m \times 30m$ ground resolution in seven spectral bands and by the SPOT panchromatic (PAN) sensor, which provides single-band observations on a broad wavelength interval, with a $10m \times 10m$ pixel size. Data fusion of Landsat-TM and SPOT-PAN images have been previously considered in the literature [3] due to their availability and their complementary spatial/spectral features. This paper reports about a *pyramid-based* approach to data fusion of Landsat-TM and SPOT-PAN images, with images previously registered on a common cartographic base, each at its own scale ($30m$ and $10m$, respectively). The proposed algorithm is a variant of the high-pass filter (HPF) method by Chavez *et al.* [3], recognized as one of the most efficient. Its generalization is achieved in a pyramid framework, since a generalize pyramid is an efficient structure by which both the high-pass filtering and the contrast enhancement algorithms can be easily implemented. Images are available at several different spatial scales. The *expansion/reduction* filters can be easily designed to cope with data of any resolutions from different sensors. Once new data from different sensors will be available on the selected test site (e.g., SAR data, digitized aerial photographs, hyperspectral high resolution aircraft data, data from *new-generation* satellites), they will be easily merged with the existing ones in order to assess any advantages occurring from a cooperative analysis based on multiple imaging sources. The algorithm is assessed in terms of both objective scores and visual quality. Spectral feature preservation of Landsat images is evaluated. The performance of the merging procedure is previously discussed and assessed in a comparison with an analogous scheme based on the WT [2], recently established in the literature. The pyramid algorithm is found to be superior on the basis of both subjective and objective criteria.

2 Multi-scale Image Analysis

2.1 Wavelet transform

The wavelet transform provides a multi-resolution representation of continuous and discrete signals [4]. When it is applied to a sequence of discrete data $f(n)$, the original signal can be considered as the coefficients of the projection of a continuous function into the highest resolution subspace: the coefficients relative to the lower resolution subspace and to its orthogonal complement can be obtained through the subsampling of the discrete convolution of $f(n)$ and the coefficients of the impulse response of two digital filters $H(\omega)$ and $G(\omega)$, respectively low-pass and high-pass [4]. The two outcome sequences represent a smoothed version of $f(n)$ and a detail signal, respectively: the latter, being the output of a high-pass filter, highlights the points in which rapid changes of the signal occur. In a similar manner, the higher resolution data can be retrieved from the lower resolution projections by up-sampling and low-pass filtering. Therefore, the wavelet representation is closely related to a two-channel sub-band decomposition scheme.

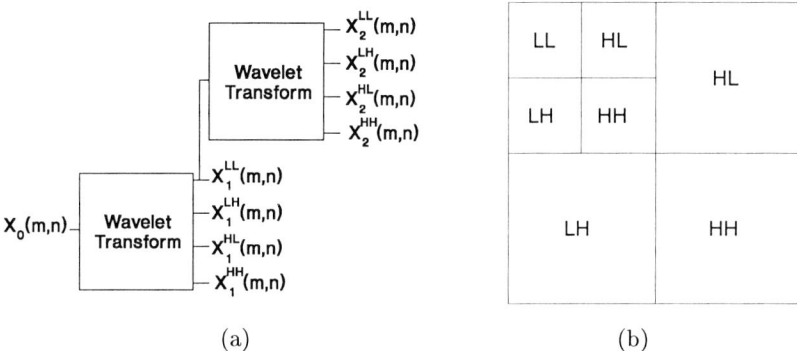

Fig. 1. Two-level 7 sub-bands wavelet transform scheme (a) and associated spatial frequency sub-bands (b).

If two dimension signals are dealt with, a wavelet representation can be obtained by separately processing rows and columns of the array. Let $X_0(m,n)$ be the original image with dimensions $M \times N$, and let $X_1^{LL}(m,n)$, $m = 0,\ldots,M/2-1$, $n = 0,\ldots,N/2-1$ be the lower resolution subsequence obtained by low-pass filtering rows and columns; analogously, let $X_1^{LH}(m,n)$, $X_1^{HL}(m,n)$, and $X_1^{HH}(m,n)$, $m = 0,\ldots,M/2-1$, $n = 0,\ldots,N/2-1$, be the sub-sequences obtained by the combination of low-pass and high-pass filtering along the rows or the columns. Since high-pass filtering highlights edges in an image, $X_1^{LH}(m,n)$ and $X_1^{HL}(m,n)$ will contain information about vertical and horizontal contours, respectively. With analogous considerations $X_1^{HH}(m,n)$ highlights diagonal details. Further splitting of $X_1^{LL}(m,n)$ yields a multi-level decomposition: the signals $X_2^{LL}(m,n)$, $X_2^{LH}(m,n)$, $X_2^{HL}(m,n)$ and $X_2^{HH}(m,n)$, $m = 0,\ldots,M/2^2-1$, $n = 0,\ldots,N/2^2-1$, are produced at the second level of the decomposition and general expressions for the higher levels can easily be derived. Figure 1 shows the scheme for a two-level decomposition yielding a seven sub-bands representation: in the figure the Wavelet Transform block denotes the one-level four sub-bands separable splitting. The low-frequency coefficients $X_i^{LL}(m,n)$, are further decomposed, thus yielding a wavelet space-frequency representation, in which the wavelet coefficients may be accommodated into sub-bands based on their content of spatial frequencies.

2.2 Laplacian pyramid

The Gaussian pyramid (GP) is a multi-resolution image representation obtained through a recursive *reduction*, i.e. low-pass filtering and decimation. Let $G_0(m,n)$, $m = 0,\ldots,M-1$, and $n = 0,\ldots,N-1$, $M = u \times 2^K$, $N = v \times 2^K$, be the input image. The GP [5] is defined with a decimation factor of 2 ($\downarrow 2$) as

$$G_k(m,n) = reduce_2[G_{k-1}](m,n)$$
$$\triangleq \sum_{i=-L_r}^{L_r} \sum_{j=-L_r}^{L_r} r_2(i) \times r_2(j) \, G_{k-1}(2m+i, 2n+j) \quad (1)$$

for $k = 1,\ldots,K$, for $m = 0,\ldots,M/2^k - 1$, $j = 0,\ldots,N/2^k - 1$, in which k identifies the level of the pyramid. The 2D reduction (low-pass) filter is given as the outer product of a linear symmetric odd-sized kernel $\{r_2(i)\}$ which should cut-off at one half of the signal bandwidth, to prevent aliasing.

From the GP, the LP is defined, for $k = 0,\ldots,K-1$, as

$$L_k(m,n) \triangleq G_k(m,n) - expand_2[G_{k+1}](m,n) \quad (2)$$

in which $expand_2[G_{k+1}]$ denotes that the $(k+1)^{st}$ level of the GP is expanded by a factor 2 to match the size of the underlying k^{th} level:

$$expand_2[G_{k+1}](m,n) \triangleq \sum_{\substack{i=-L_e \\ (j+n) \bmod 2=0 \\ (i+m) \bmod 2=0}}^{L_e} \sum_{j=-L_e}^{L_e} e_2(i) \times e_2(j) \, G_{k+1}\left(\frac{i+m}{2}, \frac{j+n}{2}\right) \quad (3)$$

for $m = 0,\ldots,M/2^k - 1$, $n = 0,\ldots,N/2^k - 1$, and $k = 0,\ldots,K-1$. The 2D low-pass filter for *expansion* is given as outer product of a linear symmetric odd-sized kernel $\{e_2(i)\}$, which again should cut-off at one half of the bandwidth. Summation terms are taken to be null for noninteger values of $(i+m)/2$ and $(j+n)/2$, corresponding to interleaving zeroes introduced by up-sampling ($\uparrow 2$).

2.3 Generalized LP with a rational scale factor

The expressions found for (1) and (3) may be generalized to comprise reduction and expansion factors different from 2 [7]. Reduction by q is defined as:

$$reduce_q[G_k](m,n) \triangleq \sum_{i=-L_r}^{L_r} \sum_{j=-L_r}^{L_r} r_q(i) \times r_q(j) \, G_k(qm+i, qn+j) \quad (4)$$

The reduction (low-pass) filter $\{r_q(i)\}$ should be designed to cut-off at one q^{th} of the signal bandwidth. Expansion by a factor p is defined as:

$$expand_p[G_k](m,n) \triangleq \sum_{\substack{i=-L_e \\ (j+n) \bmod p=0 \\ (i+m) \bmod p=0}}^{L_e} \sum_{j=-L_e}^{L_e} e_p(i) \times e_p(j) \, G_k\left(\frac{i+m}{p}, \frac{j+n}{p}\right) \quad (5)$$

The expansion filter $\{e_p(i)\}$ should cut-off at one p^{th} of the signal bandwidth. Summation terms are null for noninteger values of $(i+m)/p$ and $(j+n)/p$.

If $p/q > 1$, p, q integers, is the scale factor between two images to be merged, (1) modifies into the cascade of an expansion by q and a reduction by p

$$G_{k+1} = reduce_{p/q}[G_k] \triangleq reduce_p\{expand_q[G_k]\} \tag{6}$$

while (3) becomes an expansion by p followed by a reduction by q.

$$expand_{p/q}[G_k] \triangleq reduce_q\{expand_p[G_k]\} \tag{7}$$

When (4) is cascaded to (5), convolution can be skipped after up-sampling in (6), as well as before down-sampling in (7).

The Generalized Laplacian Pyramid (GLP) with p/q scale factor between adjacent layers, $\tilde{L}_k$, can thus be defined as:

$$\tilde{L}_k(m,n) \triangleq G_k(m,n) - expand_{p/q}\{reduce_{p/q}[G_k]\}(m,n) \tag{8}$$

The filter design usually is a tradeoff between selectivity (sharp cutoff) and computational cost. Filters with different characteristics have to be designed to cope with bandwidth requirements of data fusion algorithms. In particular for a p/q scale ratio, only low-pass filters with $1/p$ and $1/q$ normalized frequency cut-offs are needed. Instead, the WT requires also a high-pass filter (i.e. a complete filter-bank) which must generally be re-designed for every value of p/q.

3 Multi-resolution Data Fusion Schemes

The idea of the wavelet-based image fusion algorithm developed by Li et al. [2] is to merge couples of sub-bands of corresponding frequency content on the basis of an activity measure locally computed on 2×2 blocks of coefficients. The fused image is produced by taking the inverse transform of the blocks of coefficients chosen as the more *active* between the two images.

The block diagram reported in Figure 2 describes the data fusion algorithm in the general case of two image data sets, preliminarily registered on the same cartographic base, whose scale ratio is p/q. Let S_1 be the data set constituted by a single image having smaller scale and S_2 the data set made up of several multi-spectral observations with larger scale. The goal is to obtain a set of as many multi-spectral images as S_2, each having same spatial resolution as S_1. The upgrade of S_2 to the resolution of S_1 is the zero-mean GLP (8) of S_1, computed for $k = 0$. The high-pass component from S_1 is added to each of the expanded images of S_2 to yield an enhanced set of multi-spectral observations, S_3.

4 Experimental Results

Figures 3(a) and (b) show Band 6 (thermal infrared) and Band 5 (near infrared) of a Landsat TM image portraying a zone of the Elba island, in Tuscany, Italy.

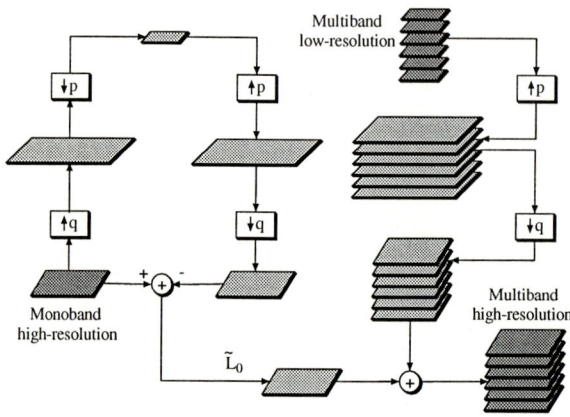

Fig. 2. Outline of data fusion procedure for two images with a p/q scale ratio.

Due to SNR constraints, TM Band 6 is actually sensed with a ground resolution of $120m/pel$ and resampled in order to match the size of the other bands ($30m/pel$). Figures 3(c) and (d) show fusion results of the two algorithms. The results of Fig. 3(c) have been obtained through the FIR implementation of a cubic spline WT [4,6]. Although the wavelet-fused image looks sharper, artifacts are perceivable around edges, due to ringing effects.

Spectral feature preservation is evaluated by taking the pixel differences between any of the merged images and a linearly resampled version of Band 6 (both integer valued). These differences are expected to be either zero or very small on homogeneous areas, and relevant on contours or highly textured areas. The standard deviation of such differences and the number of pixels in which

Table 1. Std. devs. (STD) of the differences obtained by subtracting merged images from expanded TM bands. Percentage of pixels ($P \pm 1$) whose absolute differences are equal to either one or zero. Results reported for wavelet (WT) and pyramid (GLP).

TM Band:	WT: STD	GLP: STD	WT: $P \pm 1$	GLP: $P \pm 1$
1	2.93	2.65	40.91 %	53.63 %
2	2.69	2.32	41.89 %	49.35 %
3	2.62	2.48	41.47 %	53.94 %
4	2.24	2.22	40.96 %	64.09 %
5	2.27	2.33	41.66 %	64.08 %
7	2.39	2.34	41.63 %	65.67 %

they are equal or very close to zero represent two figures of merit for image data fusion [3]. The former should be as close to zero as possible. Due to roundoff to integers, pixel differences are taken to be null if their absolute values do not

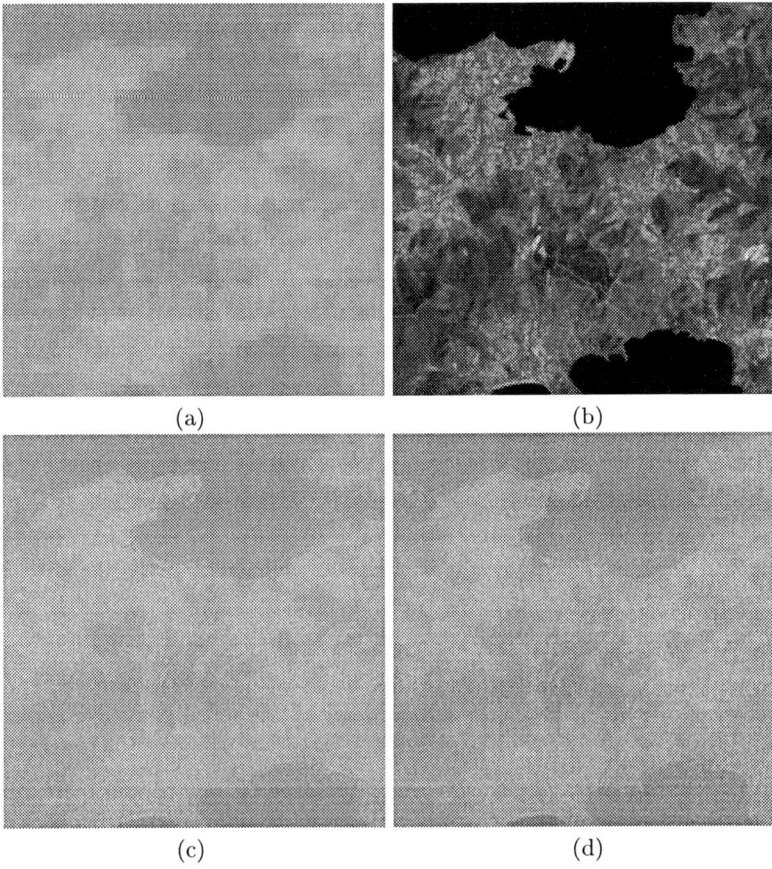

Fig. 3. Fusion of Landsat TM Band 6 (a) with Band 5 (b): (c) wavelet scheme; (d) pyramid scheme. Both images are 256 × 256 details.

exceed unity. Table 1 reports the scores of each TM band. It is apparent that the pixel percentages are far larger for the pyramid scheme. Standard deviations are slightly smaller for the pyramid, with the only exception of Band 5. The values of the parameters reported have been optimized over the pyramid-generating filter (15 taps) and are steady. The results of Table 1 are better than those reported in [3], thanks to the multi-resolution framework which allows a better filter design.

SPOT Panchromatic and Landsat TM data were available for the test area of Metaponto, in Southern Italy. The images were registered on the same cartographic base, each maintaining its own scale. The p/q ratio is 3. Original SPOT-PAN and TM Band-5 images are shown in Figure 4, together with the enhanced TM Band-5 version, in order to visually assess the quality of the results. Contours and textures are highlighted. The local average level is carefully preserved. Such a feature is important in determining spectral signatures, and its alteration may be responsible for misclassification and misinterpretation.

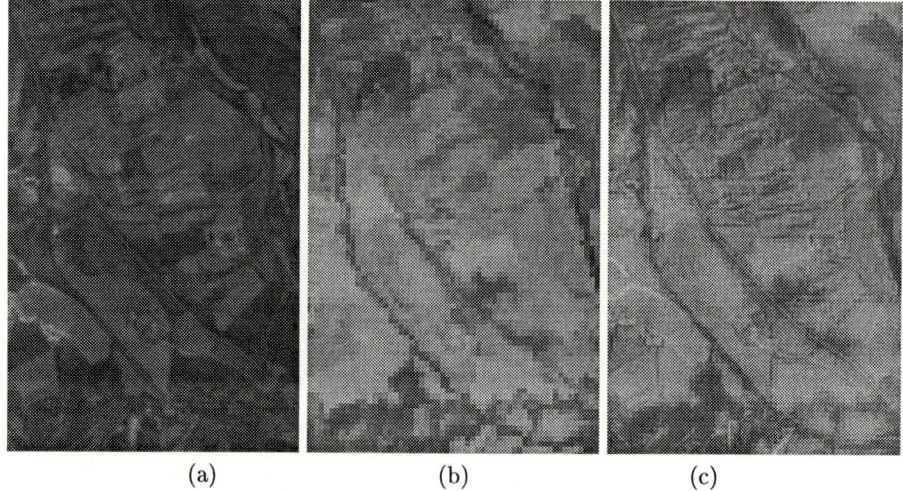

(a) (b) (c)

Fig. 4. 256 × 192 detail of the SPOT-PAN image (a), ground resolution $10m$, and TM-5 image (b) of the test area: resolution is $30m$ and a magnification by 3 has been applied for displaying. TM-5 image pyramid-fused with SPOT-PAN (c). Performance parameters, as defined in Table 1, are $STD = 4.05$ and $P \pm 1 = 36.44\%$.

Acknowledgments

This work was carried out under grants of ASI -Italian Space Agency- within a joint project on multisource classification, and of CNR -National Research Council- in the framework of the nationwide project on Cultural Heritage.

References

1. R. C. Luo and M. G. Kay, "Multisensor integration and fusion in intelligent systems," *IEEE Trans. Systems, Man, and Cybernetics*, **19**(5), 901-931 (1989).
2. H. Li, B. S. Manjunath, and S. K. Mitra, "Multisensor image fusion using the wavelet transform," *CVGIP: Graphical Models and Image Processing*, **57**(3), 235-245 (1995).
3. P. S. Chavez Jr., S. C. Sides, and J. A. Anderson, "Comparison of three different methods to merge multiresolution and multispectral data: Landsat TM and SPOT panchromatic," *Photogram. Engin. Remote Sensing*, **57**(3), 295-303 (1991).
4. S. Mallat, "A Theory for Multiresolution Signal Decomposition: the Wavelet Representation," *IEEE Trans. Pattern Anal. Machine Intell.*, **11**(7), 674-693 (1989).
5. P. J. Burt, "The pyramid as a structure for efficient computation," in *Multiresolution Image Processing and Analysis*, A. Rosenfeld (Ed.), Berlin, Springer-Verlag (1984).
6. M. Unser and A. Aldroubi, "Polynomial Splines and Wavelets- A Signal Processing Perspective", in *Wavelets- A Tutorial in Theory and Applications*, C. K. Chui (Ed.), Academic Press, 91-122 (1992).
7. M. G. Kim, I. Dinstein, and L. Shaw, "A Prototype Filter Design Approach to Pyramid Generation," *IEEE Trans. Pattern Anal. Machine Intell.*, **15**(12), 1233-1240 (1993).

Texture Analysis Using Pairwise Interaction Maps

Dmitry Chetverikov

Computer and Automation Research Institute
1111 Budapest, Kende u.13-17, Hungary
email: mitya@leader.ipan.sztaki.hu

Abstract. Pairwise pixel interactions have proved to be a powerful tool in feature based [3,7] as well model based [9] texture analysis. Different aspects and components of the feature based interaction map (FBIM) approach have already been discussed, but no self-contained description of the FBIM has been published yet. This paper provides a comprehensive up-to-date survey of the approach, including major algorithms and a series of experimental studies that demonstrate the capabilities of the approach.

1 Introduction

Precise directional analysis is essential for the *feature based interaction map* approach proposed recently in [3,7]. Pattern orientation is viewed as direction of maximum statistical symmetry assessed via anisotropy. The term 'interaction map' was originally introduced in the Markov-Gibbs texture model with pairwise pixel interactions [9]. In feature based approach discussed here, this term refers to the structure of statistical pairwise pixel interactions evaluated through the spatial dependence of a texture feature computed from the *extended graylevel difference histogram* (EGLDH) introduced in [2]. The relation between the model based approach [9] and the feature based approach presented in this paper is discussed in [5].

Pairwise pixel interactions described by the EGLDH convey important structural information. Both short- and long-range interactions are relevant. Depending on the scale of the pattern, the range of spacings may include a short-range and a long-range zone. This distinction reflects the difference between the short- and the long-range order which is typical for natural patterns, e.g. crystals. For irregular patterns, the short-range zone is dominant, while structured patterns exhibit their regularity and orientation in the long-range zone.

The power of the FBIM approach is in its selectivity to the fundamental structural properties of textures, such as anisotropy, symmetry and regularity. The FBIM can be implemented as a running *structural filter* [4] that responds to the fundamental structural features of local pattern.

The FBIM approach has been successfully applied to a wide range of texture analysis tasks and applications [3,7,8,4,6]. These studies appeared in the proceedings of diverse conferences, each of them discussing certain aspects of the method. However, no self-contained, comprehensive description of the approach has ever been published. Such presentation is necessary for possible reproduction and use of the algorithms and the results by the members of the computer vision community.

In section 2, we present the structure and the major components of the FBIM approach to texture analysis. Section 3 shows how the fundamental structural features of textures can be recovered using the FBIM approach. Section 4 overviews the current applications of the approach. Finally, future plans and the limits of the approach are discussed.

2 The FBIM approach

The *polar interaction map* $M_{pl}(i,j)$ is the basic entity of the FBIM method. $M_{pl}(i,j)$ is an intensity-coded polar representation of an EGLDH feature, with the rows enumerating the angle α_i, the columns the magnitude d_j of the varying spacing vector. The main steps of the method are as follows.

Step 1 For a discrete set of spacing vectors $\boldsymbol{d}_{ij} = (\alpha_i, d_j)$, $\alpha_i \in [0, 2\pi]$, $i = 0, \ldots, N_a$, $d_j \in [1, d_{max}]$, $j = 0, \ldots, N_d$, compute an EGLDH feature $F(\alpha_i, d_j)$. N_a is the number of angles, N_d the number of displacements.

Step 2 Define polar FBIM as $M_{pl}(i,j) = F(\alpha_i, d_j)$. Each column of $M_{pl}(i,j)$ is treated as a cyclical array and called *anisotropy indicatrix* describing texture anisotropy at the distance d_j: $A^j(i) = M_{pl}(i,j)$.

Step 3 Apply a reflectional symmetry transform to each indicatrix $A^j(i)$, $j = 1, \ldots, N_d$, to obtain the *symmetry indicatrix* $S^j(i)$.

Step 4 Define *polar symmetry map* as $S_{pl}(i,j) = S^j(i)$.

Step 5 Compute the row projections of $S_{pl}(i,j)$ to find the directions of dominant symmetry and evaluate *pattern anisotropy and orientation*.

Step 6 Transform $M_{pl}(i,j)$ to XY coordinates to obtain the *Cartesian (XY) interaction map* $M_{xy}(k,l)$.

Step 7 Use $M_{xy}(k,l)$ to analyze texture structure.

2.1 Computing the interaction maps

The extended graylevel difference histogram shows the frequencies of the absolute graylevel differences between pixels separated by a spacing vector. In the EGLDH, the magnitude and the angle of the spacing vector are independent, arbitrary parameters, while in the conventional graylevel difference histogram (GLDH, e.g., [10]) these parameters are interrelated because of the image raster.

The computation of an EGLDH feature and the layout of $M_{pl}(i,j)$ are illustrated in figure 1. In figure 1a, (m,n) scans the pixels of the image, while (x,y) points at a non-integer location specified by the spacing vector. The intensity $I(x,y)$ is obtained by the linear interpolation of the four neighboring pixels, then truncated to integer. The absolute graylevel difference $|I(m,n) - I(x,y)|$ is used to address and increment a bin of the EGLDH.

The standard GLDH features [10] can be used with the EGLDH as well. In this study, the EGLDH feature F is usually the median of $|I(m,n) - I(x,y)|$. For structural filtering, the mean value of the absolute difference is used as an additive feature suitable for run filtering. Using the more robust median feature results in structurally similar, but somewhat more contrast interaction maps. Examples of median based interaction maps are shown in figure 5.

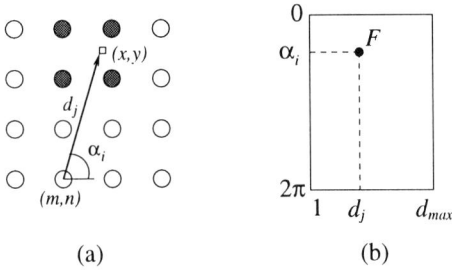

Fig. 1. (a) Computing the EGLDH. (b) The layout of the polar interaction map. F is an intensity-coded EGLDH feature.

2.2 Computing the symmetry map

The anisotropy indicatrix $A^j(i)$ [2] is a column of $M_{pl}(i,j)$ treated as a cyclical array. The indicatrix exhibits texture anisotropy measured at a certain displacement d_j. While the map is intensity-coded, the indicatrix is represented by a polar diagram with the radius proportional to the EGLDH feature value. This diagram is used to compute the symmetry map from $M_{pl}(i,j)$. Figure 2 shows anisotropy and symmetry indicatrices computed for a regular and a linear structure.

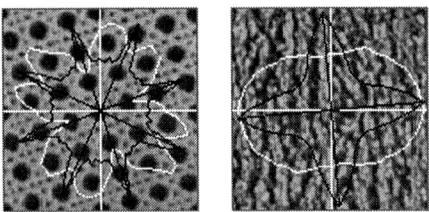

Fig. 2. Anisotropy (light) and symmetry (dark) indicatrices computed for a regular and a linear pattern. For the regular pattern, the displacement $d = 22$ is close to the period. For the linear pattern, a small displacement $d = 2$ indicates the anisotropy of the pattern.

The symmetry map is composed of the columns that are symmetry indicatrices. Each symmetry indicatrix is obtained as a reflectional symmetry transform of the corresponding anisotropy indicatrix. For these purpose, a symmetry value is assigned to each direction α_i, as illustrated in figure 3. The two halves of the anisotropy indicatrix $A(i)$ are matched to yield the symmetry indicatrix $S(i)$.

The symmetry indicatrix $S(i)$ is defined as follows:

$$S(i) = \left(\frac{2}{i_{max}} \sum_{k=1}^{i_{max}/2} \left[1 - \left| \frac{A(i+k) - A(i-k)}{A(i+k) + A(i-k)} \right| \right] \right)^{\gamma}$$

Here $i = 1, \ldots, i_{max}$, where $i_{max} = N_a - 1$; k is defined (mod i_{max}). The power $\gamma = 5$ is used to make the symmetry measure more sensitive.

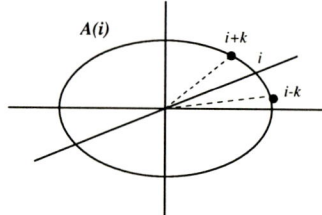

Fig. 3. Computing reflectional symmetry of the anisotropy indicatrix $A(i)$ for the angle α_i.

2.3 Analyzing texture symmetry and orientation

An algorithm for analysis of the symmetry map was proposed in [3]. The analysis is aimed at the evaluation of pattern anisotropy and orientation defined by the axes of maximum symmetry. This is done for the short- and the long-range zones separately. For regular patterns, the short-range zone is not specific. For irregular pattern, this zone is relevant, while the decorrelated long-range zone only distorts the results by adding high but non-informative symmetry values. The procedure [3] tries to separate the two zones in the symmetry map. The border between the zones is detected as the minimum correlation between pairs of columns separated by a small displacement. The operation of this algorithm is illustrated in figure 4.

Once the zones have been identified, the orientation analysis in each zone involves computing the row projections for each zone separately. The directions of the maximum symmetry are the peaks of the projections. These directions indicate the orientation of texture. In figure 4, the row projections are shown as polar diagrams overlaid on the pattern.

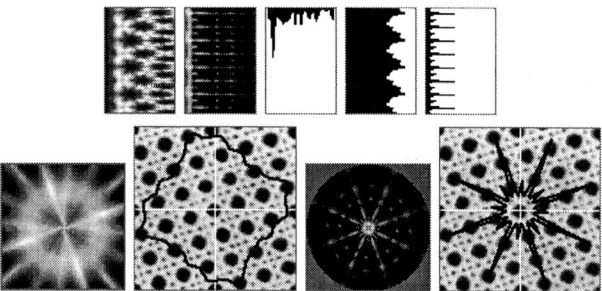

Fig. 4. Computing anisotropy and orientation of a texture pattern. Upper row: polar interaction map; polar symmetry map; correlations between columns of symmetry map; row projections of short-range zone; row projections of long-range zone. Lower row: short-range symmetry map; short-range anisotropy; long-range symmetry map; long-range anisotropy.

3 Recovering the basic structural features of textures

Figures 5 exemplifies the FBIM analysis of different natural textures from the album [1]. The first pattern is an irregular linear structure. The rest of the textures are structured patterns with various degrees of regularity. For the first pattern the short-range zones of $M_{xy}(k,l)$ is displayed enlarged. The structures of the other patterns are represented by the dark blobs of the interaction maps.

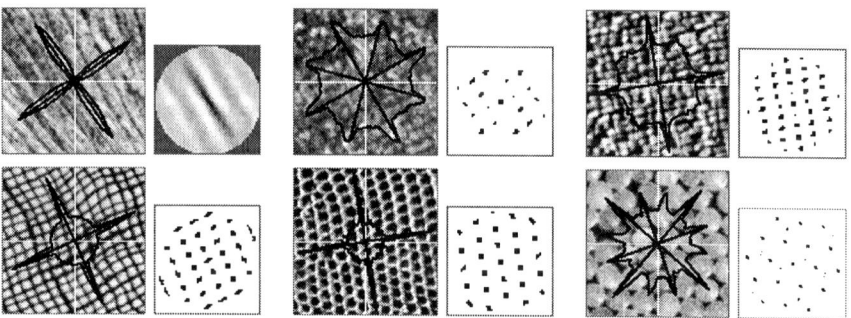

Fig. 5. Texture patterns with anisotropies overlaid and structures shown separately.

The anisotropies, orientations and structures of the patterns are correctly indicated even for weak structures. The texture orientation results by the FBIM approach were compared in [3] to the results obtained by an alternative, Fourier based method. The superiority of the FBIM over the spectral domain was further discussed in [4].

4 Applications of the FBIM approach

4.1 Rotation-invariant texture classification

To demonstrate the texture discriminating power of the FBIM approach, rotation-invariant texture classification experiments were carried out in [7]. Thirty textures were selected and digitized to $512 \times 512 \times 8$ resolution test images.

The test set contained patterns with different degree of regularity as well as non-regular directional patterns. No isotropic irregular textures were included since they do not have characteristic interaction maps suitable for discrimination. This is a limitation of the proposed approach as far as classification is concerned. Another condition is a relatively large size of the texture patches to be classified: the patch size should be large enough to contain two or more periods of any structured texture included in the test set. Otherwise, the interaction map does not represent the structure of the pattern.

Under these conditions, an accuracy of 96% was achieved using a simple but strict classification procedure that assigns to the test patch the class of the learning sample whose map is most similar to the map of the test patch.

4.2 Using the FBIM filter to detect texture defects

Detection of structural imperfections in textures is a specific task of texture analysis related to automated inspection of materials and surfaces. Recently, we have proposed a uniform approach to texture defect detection based on the concept of structural filtering introduced in [4]. When the polar maps of a reference sample and a sliding window are matched, $M_{pl}(i,j)$ becomes a powerful structural filter selective to anisotropy and regularity. Both rotation and scaling are easily incorporated into $M_{pl}(i,j)$ making the polar map suitable for orientation- and size-adaptive filtering. Potential applications and limits of the FBIM filter were discussed in [4] where a fast running implementation of the filter was proposed.

Many natural textures contain gradual variations of orientation and size that may be treated as tolerable or intolerable depending on the application. The FBIM filter can cope with both types of variations by specifying the degrees of variations that should be tolerated.

Figure 6 shows the four 256×256 pixel size grayscale test images used in the defect detection experiments, with sample detection results overlaid. Each of the images contains a single structural imperfection. The defect detection tests were run for different ranges of the algorithm parameters. More details of this experimental study are given in a forthcoming paper [6].

4.3 Applications to document analysis

A typical problem of document image analysis is the text/non-text separation. In a recent study [8], we applied the FBIM approach to the problem of *zone classification* in document image processing. Document blocks were labeled as

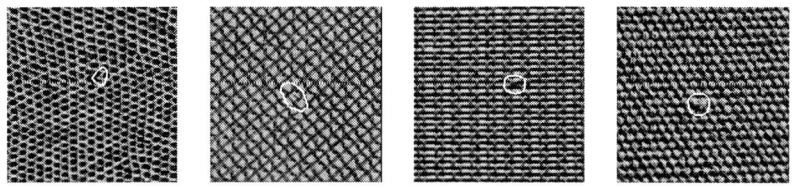

Fig. 6. The test textures with sample detection results overlaid.

text or non-text using texture features derived from a feature based interaction map.

The zone classification method proposed was tested on the comprehensive document image database UW-I created at the University of Washington in Seattle. This database comprises about 1000 digitized pages with approximately 15000 zones. Different classification procedures were considered. The performance ranged from 96% to 98% using 6 FBIM texture features only.

Another application is the determination of the *document skew* for subsequent skew normalization. The skew angle was computed as the orientation of the texture composed of the text lines. The original resolution of the UW-I document images was reduced by a factor of 8. The angular resolution of the FBIM was 0.5°. The row projection array of the symmetry map was interpolated to calculate more accurate position of the maximum. The probability that the estimated skew angle lies within 0.5° from the ground truth was 96%.

5 Conclusion

FBIM is an efficient tool for texture analysis whose capability to solve different tasks has been demonstrated by numerous experiments in various applications. Currently, a basic limitation of the approach seems to be its unsatisfactory performance in the case of irregular patterns whose interaction maps do not show specific structure. We are now investigating this problem in order to find a solution based on a more detailed analysis of the short-range zone of the maps.

A related open problem is that of finding textured regions in images and roughly classifying the regions as irregular, linear, or structured. Interaction maps of each category have specific features that should be detected and analyzed. When the rough classification is done, the structural analysis should proceed depending on the category.

Other plans for future research and application of the FBIM approach include creation of a local symmetry filter, invariant detection of textured objects, shape from texture and query-by-texture in image databases.

6 Acknowledgment

This work was supported in part by the grants OTKA T14520 and EU INFO COPERNICUS IC15 CT94 0742.

References

1. P. Brodatz. *Textures: a photographic album for artists and designers*. Dover, New York, 1966.
2. D. Chetverikov. GLDH based analysis of texture anisotropy and symmetry: an experimental study. In *Proc. International Conf. on Pattern Recognition*, pages 444–448. Vol.I, 1994.
3. D. Chetverikov. Pattern orientation and texture symmetry. In *Computer Analysis of Images and Patterns*, pages 222–229. Springer Lecture Notes in Computer Science vol.970, 1995.
4. D. Chetverikov. Structural filtering with texture feature based interaction maps: Fast algorithm and applications. In *Proc. International Conf. on Pattern Recognition*, pages 795–799. Vol.II, 1996.
5. D. Chetverikov. Texture feature based interaction maps: Potential and limits. In *Proc. of Seminar on Theoretical Foundations of Computer Vision*, Dagstuhl, Germany, 1997 (to appear). Springer Verlag.
6. D. Chetverikov and K. Gede. Textures and structural defects. In *International Conference on Computer Analysis of Images and Patterns*, Kiel, Germany, 1997 (to appear).
7. D. Chetverikov and R.M. Haralick. Texture anisotropy, symmetry, regularity: Recovering structure from interaction maps. In *Proc. British Machine Vision Conference*, pages 57–66, 1995.
8. D. Chetverikov, J. Liang, J. Kőműves, and R.M. Haralick. Zone classification using texture features. In *Proc. International Conf. on Pattern Recognition*, pages 676–680. Vol.III, 1996.
9. G. Gimel'farb. Non-Markov Gibbs texture models with multiple pairwise pixel interactions. In *Proc. International Conf. on Pattern Recognition*, pages 591–595. Vol.II, 1996.
10. R. M. Haralick and L. G. Shapiro. *Computer and Robot Vision*, volume I. Addison-Wesley, 1992.

Estimation of the Color Image Gradient with Perceptual Attributes

[1]Philippe Pujas , [2]Marie-José Aldon

[1]Institut Universitaire de Technologie de Montpellier, Université Montpellier II,
17 quai Port Neuf, 34500 Béziers, FRANCE
[2]LIRMM - UMR C55060 - CNRS / Université Montpellier II
161 rue ADA, 34392 Montpellier Cedex 05, FRANCE

Abstract
Classical gradient operators are generally defined for grey level images and are very useful for image processing such as edge detection, image segmentation, data compression and object extraction. Some attempts have been made to extend these techniques to multi-component images. However, most of these solutions do not provide an optimal edge enhancement.
In this paper we propose a general formulation of the gradient of a multi-image. We first give the definition of the gradient operator, and then we extend it to multi-spectral images by using a metric and a tensorial formula. This definition is applied to the case of RGB images. Then we propose a perceptual color representation and we show that the gradient estimation may be improved by using this color representation space. Different examples are provided to illustrate the efficiency of the method and its robustness for color image analysis.

1 Introduction

This paper addresses the problem of detecting significant edges in color images. More specifically, given a scene including objects which are characterized by homogeneous colors, we want to detect and to extract their contours in the image. Our objective is to propose a solution which overcomes the problems of shades and reflections due to lighting conditions and objects surface state, in order to achieve an adequate image segmentation.

The paper is divided into four parts. In the first part, we propose a general formula to define the gradient of a multi-spectral image. In the second part, we apply this definition to color images whose components are described in the RGB space. We present a brief overview of the classical approaches consisting of an evaluation of the gradient as a simple function of its three components. Thirdly, we show that it is possible to improve this gradient estimation by using an Euclidean metric in the RGB space.

We then introduce the perceptual color space representation which separates the intensity and chromatic components. We improve the classical HSV model by introducing a new parameter, the "chromaticity degree" γ, which allows the separation of chromatic and achromatic areas within the image. In the fourth part of the paper, we use the HVγ representation to propose a gradient estimator which avoids enhancement of non significant edges created by object shades or reflections. Results obtained with true images illustrate the advantages of this approach.

2 Multi-spectral image gradient

2.1 Gradient definition

Let f(M) be a scalar potential field in R^2 (for instance a monochrome image), with $M = M(x,y)$. The gradient of f is referred to as ∇f:

$$\nabla f = \begin{pmatrix} \dfrac{\partial f}{\partial x} \\ \dfrac{\partial f}{\partial y} \end{pmatrix} \qquad (1)$$

In the image plane the gradient ∇f may be represented by a 2D vector; its orientation corresponds to the direction along which f has the maximum rate of change and its magnitude is the absolute value of this maximum rate of change. Let . denote the dot product in R^2 and : $dM = (dx, dy)^T$. Then we can write:

$$\nabla f . dM = df \qquad (2)$$

Currently, different kinds of methods are available to compute the gradients' components in a monochrome image. The earliest of them are based on simple discrete approximations of continuous derivatives [1,2]. More recent operators have been designed by taking into account a model of the edges to be detected and a quantitative definition of the performance of the edge detector [3,4,5].

2.2 Application to multi-images

The previous gradient definition (1) cannot be applied to a multi-image such as a color image. Such an image being described by a set of components cannot be modelled by a scalar potential field but by a 2D vector field [6]. Let C be the multi-spectral image and ∇C its gradient. Let $K=(k_1,k_2,...k_n)^T$ be the image representation in a given space. The problem discussed in this section is the way to combine the elementary gradient components ∇k_i $i \in [1,n]$ in order to obtain the best gradient estimation ∇C :

$$\nabla C = F(\nabla k_1, \nabla k_2, ... \nabla k_n) \qquad (3)$$

Several solutions have been proposed to deal with color images (n=3). The straightforward approaches make use of linear combinations of the ∇k_i, the simplest method consisting in a vectorial sum of the ∇k_i. Another approach is to estimate the resultant gradient magnitude at point M(x,y) as an Euclidean distance between the color vectors K(M) and K(M'). More sophisticated methods make use of distance between averaged vector values. As explained in [7], in all of these approaches the image components k_i do not cooperate with one another. To avoid this drawback, Di Zenzo proposes a solution based on the use of a tensor gradient of the multi-image which is defined as a vector field. This formula has been adopted by Chapron [8].

We propose here a general gradient definition which can be applied to a color image, or to any kind of multi-image. Given a point $K=(k_1,k_2,...k_n)$ in the image space, we define a metric dC^2 in this space such as:

$$dC^2 = dK^T g \, dK \qquad (4)$$

where g represents the metric tensor and dK an elementary displacement in the considered space. Notice that g is a non negative defined tensor, in order to define a positive metric. K being a function of the pixel position M, it may be considered as a vector field. Then:

$$dK = \frac{\partial K}{\partial M} dM \qquad (5)$$

$$dC^2 = dM^T \left(\frac{\partial K}{\partial M}\right)^T g \left(\frac{\partial K}{\partial M}\right) dM \qquad (6)$$

We note $\tilde{g}$ the non negative defined tensor associated to this metric:

$$\tilde{g} = \left(\frac{\partial K}{\partial M}\right)^T g \left(\frac{\partial K}{\partial M}\right) \qquad (7)$$

Let θ be the direction of the maximum image change, i.e. the gradient direction. An elementary displacement along this direction can be noted:

$$dM = dl \begin{pmatrix} \cos(\theta) \\ \sin(\theta) \end{pmatrix} \qquad (8)$$

Equation (6) can be reformulated as:

$$dC^2 = dl^2 (\tilde{g}_{11} \cos^2(\theta) + (\tilde{g}_{12} + \tilde{g}_{21}) \cos(\theta) \sin(\theta) + \tilde{g}_{22} \sin^2(\theta)) \qquad (9)$$

The gradient direction corresponds to the value of θ which maximizes the function $F(\theta)$, such as:

$$F(\theta) = \tilde{g}_{11} \cos^2(\theta) + (\tilde{g}_{12} + \tilde{g}_{21}) \cos(\theta) \sin(\theta) + \tilde{g}_{22} \sin^2(\theta) \qquad (10)$$

The gradient norm is obtained from equation (2) and can be expressed as a function of $F(\theta)$ [9].

3 Application to RGB Images

3.1 Gradients combination

In the (R,G,B) color space, the color image is noted:

$$f(x,y) = (R(x,y), G(x,y), B(x,y)) \qquad (11)$$

As seen in section 2.2, the simplest method to obtain the gradient of a color image consists in adding the elementary gradient components:

$$\nabla C = \nabla R + \nabla G + \nabla B \qquad (12)$$

Using equation (2), we obtain the color differential :

$$dC = dR + dG + dB \qquad (13)$$

This equation expresses the fact that points of the (R,G,B) space which belong to a plane parallel to the R+G+B=0 plane have the same color. In other words, the iso-chroma (r,g,b) is a plane whose equation is R+G+B=r+g+b. For instance, grey (1,1,1) and red (3,0,0) are identical colors! This is always true for all linear transforms.

We have applied this gradient definition to the image of figure 1a. Figure 1b presents the resultant gradient magnitude whose components have been obtained with a Sobel operator [1]. This result show that (18) can provide an over-segmentation. For instance internal edges appear in the front face of the upper orange cube. Due to shadows, the lower left purple cube has been separated in two regions. One of these regions has been merged with another one belonging to the green neighbor cube.

3.2 Euclidean metric

In [7] and [8], the gradient estimator is implicitly based on the use of an Euclidean metric in the RGB space. Here, the metric tensor is a unit tensor : $g = I_d$. Then, $\tilde{g}$ is a symmetric tensor such as:

$$\tilde{g} = \begin{pmatrix} (\frac{\partial R}{\partial x})^2 + (\frac{\partial G}{\partial x})^2 + (\frac{\partial B}{\partial x})^2 & \frac{\partial R}{\partial x}\frac{\partial R}{\partial y} + \frac{\partial G}{\partial x}\frac{\partial G}{\partial y} + \frac{\partial B}{\partial x}\frac{\partial B}{\partial y} \\ \frac{\partial R}{\partial x}\frac{\partial R}{\partial y} + \frac{\partial G}{\partial x}\frac{\partial G}{\partial y} + \frac{\partial B}{\partial x}\frac{\partial B}{\partial y} & (\frac{\partial R}{\partial y})^2 + (\frac{\partial G}{\partial y})^2 + (\frac{\partial B}{\partial y})^2 \end{pmatrix} \quad (14)$$

Computing the maxima of the function $F(\theta)$ requires the estimation of the partial derivatives of each component of $f(x,y)$. These derivatives are obtained using a classical operator [1,2,3,4,5].

Results of the algorithm using this Euclidean metric are illustrated in [9]. Compared with the previous results, it provides a best edge enhancement; however, there are always over-segmentations. The drawbacks of this solution are partly due to the used color space representation. It cannot represent color information like it is perceived by a human. Moreover, this space is not adapted to direct color comparison, because equal geometric distances in the RGB frame does not correspond to equal perceptual changes in color [10].

4 Perceptual Color Space Representation

In [11], Nevatia noticed that most of the edge information was in the intensity component of the image. The RGB model being enable to separate the chromatic and luminance information in a color image, Nevatia proposed an edge detector in a new space defined by the intensity. Later, other perceptual spaces have been defined [10,12]. The classical HSV representation [12] makes use of three attributes to describe a color: the hue H, the saturation S and the value (or intensity) V.

4.1 The HSV frame

All the perceptual frames are defined by a non-linear diffeomorphism. Our HSV frame is obtained by the following equations:

$$V = \sup(R,G,B)$$

$$S = 256 \frac{V - \inf(R,G,B)}{V}$$

$$H = \begin{bmatrix} \frac{G-B}{V-\inf(R,G,B)} & \text{if } V = R \\ \left(2 + \frac{B-R}{V-\inf(R,G,B)}\right) & \text{if } V = G \\ \left(4 + \frac{R-G}{V-\inf(R,G,B)}\right) & \text{if } V = B \end{bmatrix} \quad (15)$$

In these equations we suppose the dynamic range of the image signal to be 256. This coordinate transform requires few operations and is simpler than most of the other perceptual representations. For instance, the Luv transform [13] is five time more consuming.

4.2 Some difficulties with the perceptual frames

In a previous paper [15], we specified two kinds of problems connected with the definition of the hue:
- H is 6-periodic
- H is undefined if S=0 (15).

The first problem leads to computational difficulties for estimating hue averages or differences. It is easily solved using a suitable algebra [15]. The second problem makes the interpretation more difficult when using the hue to describe the image. We call this case the "low saturation effect" because in experiments, it begins to appear when the value of S is low. In order to solve this problem, we consider the existence of achromatic and chromatic areas in the perceptual color space [15]:
- In the achromatic area, H is undefined (or badly defined), so that it should not be used; the color is only described by S and V.
- In the chromatic area, previous research works [14] confirm that hue is generally the most discriminant attribute. It can be used for an efficient image analysis.

In order to optimize the analysis algorithms, it is necessary to define a way for classifying pixels into chromatic and achromatic zones. We have proposed the concept of "chromaticity degree" [15]. The chromaticity degree γ of a pixel is a scalar function depending on S and V in the range [0,1]. The closer γ is to one, the more chromatic the pixel is; the closer γ is to zero, the more achromatic the pixel is. The identification of γ is done by manual classification of a representative set of pixels. We have proposed in [15] different identification methods.

4.3 The HVγ representation

γ being a function of S and V and $\frac{\partial \gamma}{\partial S}$ being different of zero, the HVγ transform is a diffeomorphism. Consequently, it defines a new perceptual color space. In this frame, H is still periodic and undefined if $\gamma = 0$. The interest of the HVγ representation is that it does not require any external variable to perform the analysis of the color image, unlike HSV which also needs the chromaticity degree.

5 Image Gradient in a Perceptual Color Frame

One of the most important advantages of a perceptual space is that it allows direct color comparison based on geometric distance estimation.

5.1 Linear gradients combination

Using a linear combination of the three gradient components in the HSV or in the HVγ space cannot provide a correct gradient estimation. In fact, hue being not defined inside achromatic areas, the component ∇H has no significance here. Moreover, across transitions between chromatic and achromatic zones, a large hue variation induces a wrong edge.

Figure 1c illustrates this problem. We present the magnitude of the ∇H gradient component which has been obtained with the Sobel operator. Erroneous edges appear in the achromatic areas and along the reflections boundaries. In the next section we propose a solution to this problem by taking into account the chromaticity degree.

5.2 Non Linear gradients combination

In the HSV space, we propose the following definition of the resultant gradient:
$$\nabla C = \gamma(\nabla H + \nabla S) + (1-\gamma)\nabla V \qquad (16)$$

This definition may be improved to solve the reflections problem using:
$$\nabla C = \gamma \nabla H + (1-\gamma)\nabla V \qquad (17)$$

In the HVγ space the non linear gradient combination takes the form:
$$\nabla C = \gamma \nabla H + (1-\gamma)\nabla V + \nabla \gamma \qquad (18)$$

We have shown, in [9], that this solution improves the previous results. However some edges are not correctly enhanced.

5.3 A metric for the perceptual space

We have defined a new metric in the perceptual space which is based on a tensorial formula:
$$dC^2 = \gamma\, dH^2 + (1-\gamma)\, dV^2 + d\gamma^2 \qquad (19)$$

This definition means that in chromatic zones the distance between two pixels is a function of the hue H, while it is a function of the intensity V in achromatic zones. The last term $d\gamma^2$ allows the separation of chromatic and achromatic zones. This yields the metric tensor:

$$\tilde{g} = \begin{pmatrix} \gamma(\frac{\partial H}{\partial x})^2 + (1-\gamma)(\frac{\partial V}{\partial x})^2 + (\frac{\partial \gamma}{\partial x})^2 & \gamma\frac{\partial H}{\partial x}\frac{\partial H}{\partial y} + (1-\gamma)\frac{\partial V}{\partial x}\frac{\partial V}{\partial y} + \frac{\partial \gamma}{\delta x}\frac{\partial \gamma}{\delta y} \\ \gamma\frac{\partial H}{\partial x}\frac{\partial H}{\partial y} + (1-\gamma)\frac{\partial V}{\partial x}\frac{\partial V}{\partial y} + \frac{\partial \gamma}{\partial x}\frac{\partial \gamma}{\partial y} & \gamma(\frac{\partial H}{\partial y})^2 + (1-\gamma)(\frac{\partial V}{\partial y})^2 + (\frac{\partial \gamma}{\partial y})^2 \end{pmatrix} \qquad (20)$$

In figure 1d we give the multilevel gradient image obtained with this new metric. Now, homogeneous color regions are well separated, with few over-segmentations. Their edges are correctly enhanced. This gradient estimator is less sensitive to shadow and reflection effects.

In figure 2a we present the original color image of a natural landscape. The following normalized gradient images have been obtained respectively with a sum of the elementary gradient components in the RGB space (18) (Fig. 2b), and by using the perceptual metric in the HVγ space (20) (Fig. 2c). We can see that with this second metric, the road and mountains edges are more correctly enhanced. We have also compared these gradient estimators with the color image of two birds (fig. 3a). The same performance can be observed for edge detection (fig. 3b, fig 3c).

6 Conclusion

In this paper we have proposed a new definition based on a spatial metric, for the gradient of a multi-spectral image. Our main goal being to obtain a perceptual gradient, we have defined a new color representation based on the concept of chromaticity degree. We have proposed a metric adapted to this color space and deduced a perceptual gradient formulation. Experimental results show the performance of this gradient estimation. They illustrate the interest of the chromaticity degree concept and its robustness. Moreover, this concept can be extended with other perceptual color representations in order to define a perceptual gradient. It is obvious that such a color gradient definition may be applied in classical gradient-based edges detectors or in segmentation methods like those using region growing [15].

References

1. I. Sobel, "Neighbourhood coding of binary images for fast contour following and general array binary processing", Computer Graphics and Image Processing, vol. 8, 1978, pp. 127-135.
2. J.M.S. Prewitt, "Object enhancement and extraction", Picture Processing and Psychopictorics, B.S Lipking and A. Rosenfeld, editors", Academic Press, New York, 1970, pp. 75-149.
3. J.F. Canny, "A Computational approach to edge detection", IEEE Transactions on Pattern Analysis and Machine Intelligence, vol. 8, November 1986, pp. 769-798.
4. R. Deriche, "Using Canny criteria to derive an optimal edge detector recursively implemented, The International Journal of Computer Vision, vol. 2, April 1987, pp. 167-187.
5. J. Shen and S. Castan, "An Optimal linear operator for edge detection", Conference on Computer Vision and Pattern Recognition, Miami Beach, Florida, USA, 1986, pp. 109-114.
6. R. Machuca and K. Philips, "Application of Vector Fields to Image Processing", IEEE Transactions on Pattern Analysis and Machine Intelligence, vol. 5, n°3, may 1983, pp. 316-329.
7. S. Di Zenzo, "Note on the gradient of a multi-image", Computer Vision, Graphics, And Image Processing, vol. 33, 1986, pp. 116-125.
8. M. Chapron, "A New Chromatic Edge Detector Used for Color Image Segmentation", 11th IAPR International Conference on Pattern Recognition, The Hague, vol. 3, 1992, pp. 311-314.
9. P. Pujas: Analyse de scènes exploitant des images couleur et 3D, PhD Thesis, University Montpellier II, France, February 1996.
10. R. Taylor and P. Lewis, "Color Image Segmentation Using Boundary Relaxation", 11th IAPR International Conference on Pattern Recognition, The Hague, vol. 3, 1992, pp. 721-724.
11. R. Nevatia, "A Color Edge Detector and Its Use in Scene Segmentation", IEEE Trans. on Systems, Man, and Cybernetics, vol. 7, 1977, pp. 820-826.
12. A.R. Smith", "Color gamut transform pairs", SIGGRAPH'78, Atlanta, USA, august 1992, pp. 721-724.
13. Y. Ohta and T. Kanade and T. Sakai, "Color Information for Region Segmentation", Computer Graphics And Image Processing", vol. 13, 1980, pp. 222-241.
14. D. Tseng and C. Chang, "Color Segmentation Using Perceptual Attributes", 11th IAPR International Conference on Pattern Recognition, The Hague, vol. 3, August 1992, pp. 228-231.
15. P. Pujas and M.J. Aldon, "Robust Color Image Segmentation", 7th ICAR, Sant Feliu de Guixols, Catalonia, Spain, September 1995, pp. 145-155.
16. S. Tominaga, "Color Image Segmentation Using Three Perceptual Attributes", CVPR, 1996, pp. 628-630.

Fig. 1. (a) Manufactured objects. (b) Gradient magnitude computed with (12) in the RGB space. (c) Hue gradient computed with a Sobel operator. (d) Gradient magnitude computed with (19) in the HVγ space.

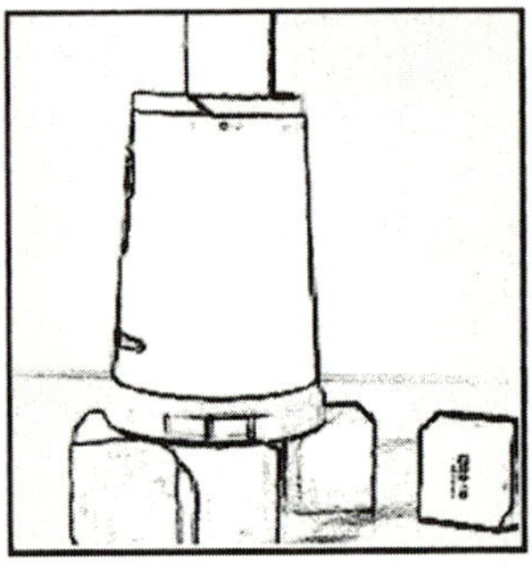

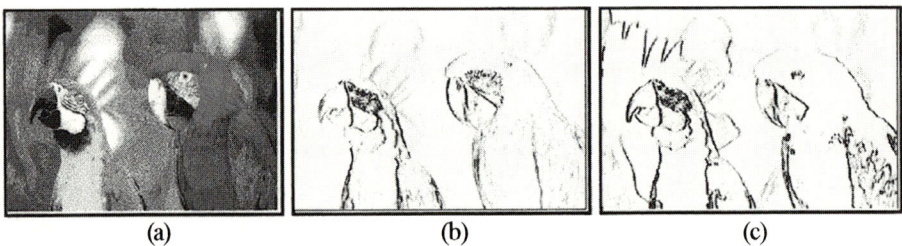

Fig. 2. (a) Landscape. (b) Gradient magnitude computed with (12) in the RGB space. (c) Gradient magnitude computed with (19) in the HVγ space.

Fig. 3. (a) Birds. (b) Gradient magnitude computed with (12) in the RGB space. (c) Gradient magnitude computed with (19) in the HVγ space.

Contour Line Extraction from Color Images of Scanned Maps

Marc Lalonde and Ying Li

Knowledge-Based Systems Group, Centre de recherche informatique de Montréal,
1801 McGill College Avenue, Suite 800, Montréal, Québec, Canada H3A 2N4

Abstract. This paper reports our work on contour line extraction from color images of scanned maps. Color processing extracts the basic colors of an image by switching from RGB to L*a*b color space, projecting the image on its principal axes and using modified histogram splitting. After the user has selected one of the extracted colors, pixel color classification yields a binary image that is thinned for connected-component analysis. Line extraction and merging follows, along with contour line formation based on connectability between the line segments.

1 Introduction

Contour lines in topographic maps provide terrain information for a geographical information system (GIS) application such as constructing 3D models of strategic areas for flight or navigation simulation. Current map digitization process requires hours of an operator's manual work on a single map, and consequently, there is a need for an automatic or semi-automatic data collection tool as a preprocessing module in a generic GIS application design environment.

Automating contour line extraction from color images is not a trivial task. Color analysis is complicated by the poor quality of the images produced by the scanner and the use of dithering in the map printing process. The contour lines are often broken due to the overlaying of other map objects. Our contributions are twofold. 1. Colors in a color image are extracted by switching from RGB to L*a*b color space, projecting the image on its principal axes, and performing modified histogram splitting; 2. Line segments are extracted and merged to form contour lines based on connectability between them.

In the following, Section 2 reviews some related work on contour line extraction. Section 3 describes the color processing algorithms applied to color map images. Section 4 describes some line manipulation algorithms working on the binary images produced by the color module. In Section 5, contour line formation is explained, followed by preliminary results presented in Section 6. Finally, Section 7 outlines the future work required to complete the project.

2 Related Work on Contour Line Extraction

Work on automatic contour line extraction is relatively limited, and it is usually presented as an extra feature of a basic map processing package. For example, the recognition system of Suzuki et al. [1] can perform extraction of buildings and various other constructs made of lines such as railways, roads, and contour lines. After some preprocessing, thinning of the original binary image is performed and lines of all widths are extracted. Of the extracted lines, contour lines are isolated using the knowledge about their widths and layout. The extraction rate (the ratio of the total length of correctly extracted lines to the total length of the lines to be found) was 82%. This work led to the development of a complete map recognition system, MARIS [2], which has a recognition rate of about 90% for contour lines.

Ebi et al. [3] present their system designed for extracting GIS information from color maps. A map image is sliced into layers, each corresponding to one color, and map objects (including contour lines) are extracted from each layer. To separate color layers, color pixels are transformed from RGB to the L*u*v color space, and the color clusters are determined through the detection of peaks in the histogram of the $u'v'$ coordinates.

Shimada et al. [4] adopt an agent-based method, where each agent is responsible for detecting a contour line, and a supervisor agent makes sure that all agents complete their tasks and talk to each other in case of ambiguous situations (e.g. touching lines). The system is semi-automatic since an operator is asked to draw cutting lines between ridges and valleys so that starting points could be assigned to the agents. No recognition results are given.

In a closely related domain, Taniguchi et al. [5] present a system for the automatic interpretation of weather reports, in which a sub-module is in charge of extracting lines of constant pressure from weather maps.

3 Color Processing

Our color processing stage for dividing a map image into color layers has two modules: 1. color extraction that analyses an image and extracts its colors; 2. color segmentation that classifies image pixels according to a selected color. The interface to the system is such that a color palette is displayed to the user who selects a color found in the map image. This allows the user to easily input the color of the contour lines to be extracted by clicking on the relevant color patch.

3.1 Extraction of the Colors

The low quality of the scanner combined to the use of dithering in the map printing process complicate the design of a reliable color extraction module. The very high spatial frequencies on the map cause aliasing during the pixel sampling, which results in the production of new colors around the dithering dots and along edges. Smoothing the image with a three-dimensional median filter [6] would remove some noise but also destroy the thin contour lines.

Our approach is based on the histogram splitting algorithm of Tominaga [7] with the following measure to reduce the dithering effect.

- The image is first thresholded with respect to its saturation component. The saturation of a pixel with RGB color values (r,g,b) is defined as

$$\text{Saturation} = 1 - 3 \cdot \frac{\min(r,g,b)}{r+g+b} \quad . \tag{1}$$

The resulting image is interpreted as a mask that is used for processing the chromatic and achromatic layers of the map separately.
 - A connected-component analysis is performed over the color mask image. When small blobs are considered to be part of dithering regions, their pixels are also masked.
 - The remaining pixels are remapped from the RGB to the $L^*a^*b^*$ color space and passed to the histogram splitting process:
   - the L^*a^*b image is decomposed into its principal components. Each color vector $\mathbf{c}$ is recomputed as $\mathbf{c}' = \mathbf{U}^t(\mathbf{c} - \mathbf{m})$, where $\mathbf{m}$ is the mean color vector computed over all non-masked pixels, and $\mathbf{R}$ is the covariance matrix of $\mathbf{c}$, while $\mathbf{U}^t$ is the result of the decomposition of $\mathbf{R}$ into its eigenvectors.
   - The histogram of the first principal component is constructed. The contribution to the histogram is weighted according to the neighborhood of the pixel and the histogram is smoothed.
   - Upon determination of the dominant peak, the pixels falling into the corresponding region are given a unique label and are also masked to withdraw them from the process.
   - If the histogram of the first principal component is noisy or is unimodal, the second, and possibly third, principal component is analyzed. Histogram splitting ends when the histograms are either unimodal or noisy.

Extraction of the dominant grey levels (using the achromatic mask image) follows the same idea. Only a section of the image with a rich color contents is sufficient to capture the dominant colors.

3.2 Pixel Classification

A classification rule evaluates if a pixel belongs to the class of interest. We use a nearest-neighbor rule in the RGB color space with a rejection threshold set empirically to 2σ, where σ is the standard deviation of the color cluster center in the RGB space as measured after color extraction. Region growing refines the classification by merging unlabeled pixels whose hue difference with respect to the mean hue of the neighboring blob is lower than a threshold. The hue value of a pixel with RGB values (r,g,b) is an angle between $-\pi$ and π and is given by:

$$\text{hue} = \operatorname{atan}(\frac{\sqrt{3} \cdot (g-b)}{2r - g - b}) \quad . \tag{2}$$

Although sharing elements with Ebi et al. [3], our algorithm has advantages: 1. Separation of the image into chromatic and achromatic planes increases resolution; 2. The use of principal component analysis yields more discriminant images, and thus richer histograms; 3. The introduction of weights on pixels according to their neighborhood is new.

4 Line Extraction and Merging

Line extraction is performed on the binary image yielded by the color segmentation module. The image is thinned using Guo and Hall's algorithm [8] and the pixels are labeled as LINELINK (pixels in the middle of a line segment), FREEOL (pixels at endpoints), or LINEJUNC (pixels at junctions). To find the connected pixels that form lines, the image is scanned until a pixel marked as LINELINK is found. The neighboring LINELINK pixels are recursively extracted from the image until FREEEOL or LINEJUNC pixels are reached.

Line merging is needed to amend either imprecise segmentation or line breaks produced by other map items overlaying the contour lines. Two cases for merging are depicted in Fig. 1. The first rule (Fig. 1a) merges L_1 with L_2 instead of L_3 if the ratio of line_dist(L_1, L_2) to line_dist(L_1, L_3) is within a proximity threshold, where the line distance line_dist(L_i, L_j) is the minimum of the distances between the endpoints of L_i and L_j. This rule would not merge A to B in Fig. 1b because too many possible connections are allowed in the restricted radius around A. Let point_dist(p, q) be the Euclidean distance between points p and q. We add a second rule to merge between A and B if

$$\text{point\_dist}(A, B) < \text{point\_dist}(A, C) \text{ , and} \tag{3}$$
$$\text{point\_dist}(A, D) > \text{point\_dist}(A, B) + \text{point\_dist}(B, C) \text{ .} \tag{4}$$

These merging rules usually reduce the number of lines found in the image by one half.

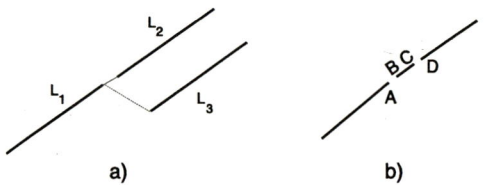

Fig. 1. Merge cases

5 Contour Line Formation

To form contour lines, the line segments obtained from line extraction and merging are chained into closed curves, for which confidence values are computed for determining whether they form valid contour lines.

To chain segments into closed curves, candidate connections for each line end are selected based on a smoothness index. A recursive algorithm selects a line with length above a threshold as the top node and builds a connection tree by connecting more line segments until a closed curve is obtained. When a closed contour is found, i.e. when the free endpoint of the last concatenated segment is close to the free endpoint of the starting line, the child node in the tree is labeled accordingly. Retrieving all combinations of closed contours in the tree is thus easy (tests are made to ensure that all contours are unique).

The smoothness index for each possible connection is computed based on the lines' lengths, the tangent vectors at the endpoints to be connected, and the gap between their endpoints. In cases where a line end is too noisy, it is marked as unreliable and a crude approximation is used for the tangent vector. When the gap is small (a few pixels, as in Fig. 2a), the smoothness is measured as

$$\text{smoothness} = \mathbf{tg}_{L_i} \cdot \mathbf{tg}_{L_j} , \qquad (5)$$

if the two line ends are reliable. Otherwise, the smoothness is set to 1 (the maximum value). When the gap is large, then an additional vector $\mathbf{v}_{ij}$ joining the two endpoints (Fig. 2b) is used:

$$\text{smoothness} = \begin{cases} \min(\mathbf{tg}_{L_i} \cdot \mathbf{v}_{ij}, \mathbf{v}_{ij} \cdot \mathbf{tg}_{L_j}) & L_1 \text{ and } L_2 \text{ reliable} \\ \mathbf{tg}_{L_i} \cdot \mathbf{v}_{ij} & L_1 \text{ reliable} \\ \mathbf{v}_{ij} \cdot \mathbf{tg}_{L_j} & L_2 \text{ reliable} \\ \min(\mathbf{tg}_{L_i} \cdot \mathbf{v}_{ij}, \mathbf{v}_{ij} \cdot \mathbf{tg}_{L_j}) & L_1 \text{ and } L_2 \text{ unreliable} \end{cases} \qquad (6)$$

The last rule states that it is preferable to resort to bad estimates of the tangent vectors than to reject the possibility of a connection. The general idea is similar to that of Taniguchi et al. [5] with the addition of the 'reliability' qualifier.

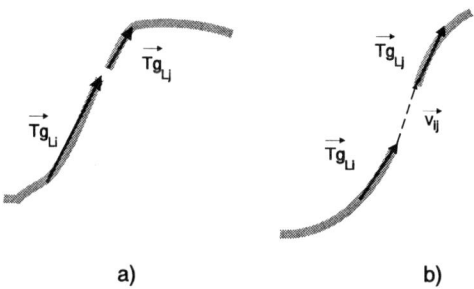

Fig. 2. Definition of the vectors

The confidence that a closed line might be a valid contour line is a function of the quality of connection along the line: for each connection, the ratio (smoothness of connection)/(best smoothness from the starting point) is computed, and the confidence of the closed contour is the minimum ratio found. Connection trees can be reused for interactively extracting polylines from the image. When the user clicks on a line segment, it triggers a search in the corresponding connection tree and all segments logically attached to the selected line segment (in a 'smoothness' sense) are selected as well. This allows the user to add pieces of lines to the current contour model built with the closed lines already isolated.

6 Results

Experiments have been conducted on maps of varying quality. We show our results on two pieces of maps: one from the Salzburg area and one from the Montreal area. The maps are digitized at 600 dpi using an off-the-shelf true color (24-bit RGB) scanner. Figures 3 and 4 show the result of extracting the colors for the contour lines. The Montreal map (Fig. 4) has shown to be more difficult to handle because of the greater use of dithering in the map. The processing time for a 1000x900 image is 55 seconds on an SGI Indigo2 workstation.

Figure 5 shows the best contour lines (of confidence value over 0.90) found by the system. The other line segments are not displayed since the system does not have enough knowledge about them being parts of contour lines. Either more research is needed to improve the construction of the model by automatically merging these line segments into the solution, or some interface functionality is to be provided to the user for selecting polylines and merging them manually. Despite some verifications that no two closed contours are perfectly equal, it sometime happens that many instances of the same contour line are found, with each being slightly different from its 'siblings' (e.g. closed contour A has a one more small segment than closed contour B). Future work will provide a function that picks the best closed line among those that share the same basic segments. The line processing operation takes about 30 seconds where more than half of the time is spent on thinning.

7 Conclusions

This paper described our method for extracting contour lines from color images of scanned maps. The quality of line extraction depend mostly on the result of the color processing, which has been giving acceptable results in spite of the color noise generated by the scanner in the dithered images. We hope more tests will reveal the sensitivity of the quality of the segmentation to the saturation threshold (see Section 3.1). We expect low sensitivity when dealing with bright colors, whereas problems might arise when extracting low saturated colors (such as light yellow or dark brown).

Future work will involve integration of additional algorithms such as functions for line thickness measurement and verification of line inclusion. We plan to build a user interface that lets the user browse the list of found closed contours, edit them or aggregate line segments or polylines to the current set.

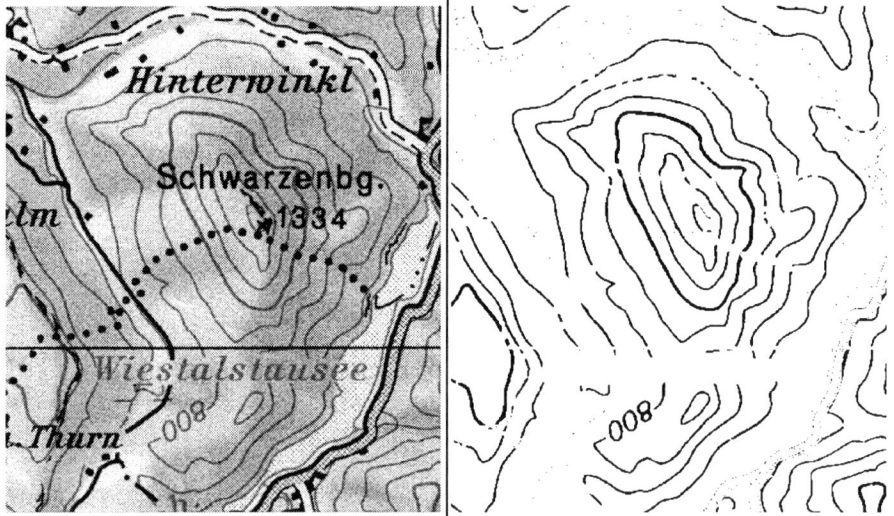

Fig. 3. The Salzburg map. Left: Original color image (printed in grey scale). Right: Result of the color segmentation; the user has selected the color *orange* for pixel classification.

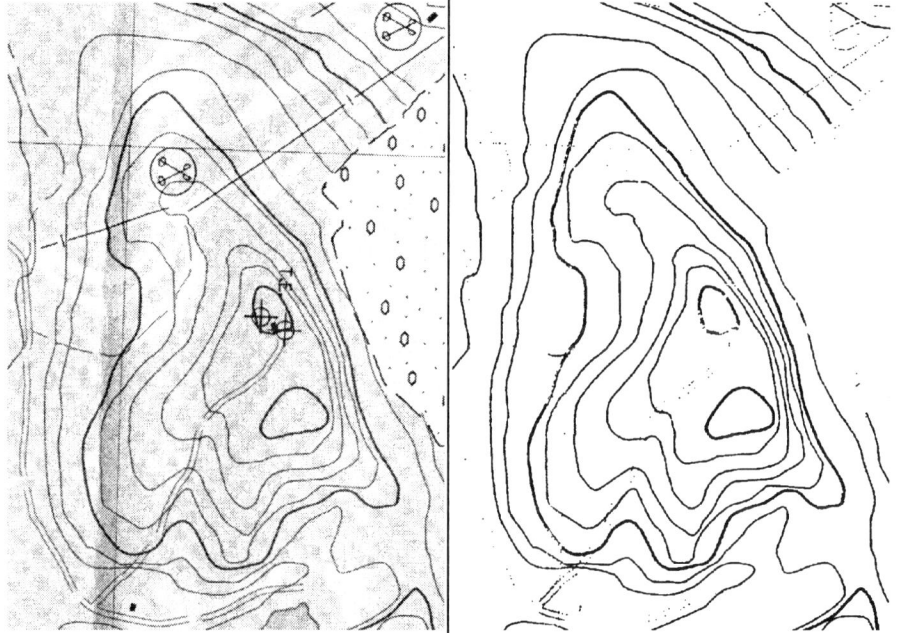

Fig. 4. The Montreal Map. Left: Original color image (printed in grey scale). Right: Result of the color segmentation; the user has selected the color *brown* for pixel classification.

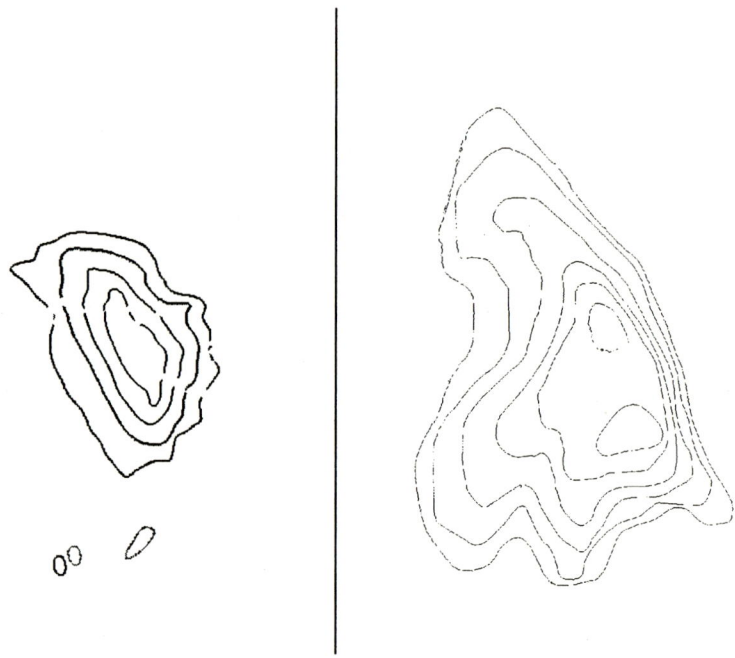

Fig. 5. Extracted closed contours for the Salzburg map (left) and the Montreal map (right).

References

1. Suzuki, S., Kosugi, M., Hoshino, T.: Automatic line drawing recognition of large-scale maps. Optical Engineering **26** (1987) 642–649
2. Suzuki, S., Yamada, T.: MARIS: map recognition input system. Pattern Recognition **23** (1990) 919-933
3. Ebi, N., Lauterbach, B., Anheier, W.: An image analysis system for automatic data acquisition from colored scanned maps. Machine Vision and Applications **7** (1994) 148–164
4. Shimada, S., Maruyama, K., Matsumoto, A.: Agent-based parallel recognition method of contour lines. Proc. of the 3rd Int. Conf. on Document Analysis and Recognition **1** (1995) 154–157
5. Taniguchi, R., Yokota, M., Kawaguchi, E., Tamati, T.: Knowledge-based picture understanding of weather charts. Pattern Recognition **17** (1984) 109–123
6. Pitas, I., Venetsanopoulos, A.N.: Nonlinear Digital Filters: Principles and Applications. Kluwer Academic Publishers, 1990
7. Tominaga, S.: A Color Classification Method for Color Images Using a Uniform Color Space. Proc. 10th Int. Conf. Patt. Recog. 1990 803–807
8. Guo, Z., Hall, R.W.: Parallel thinning with two-subiteration algorithms. Comm. of the ACM **32** (1989) 359–373

Subjective Analysis of Edge Detectors in Color Image Processing

P. Androutsos*, D.Androutsos, K.N. Plataniotis and A.N. Venetsanopoulos

Department of Electrical and Computer Engineering
University of Toronto
Toronto, Ontario, M5S 3G4, CANADA
http://www.comm.toronto.edu/~dsp/dsp.html

Abstract. There exist many experimental situations in which a subjective rather than an objective test of a specific variable proves to be a much more relevant method of investigation. Examples of such cases abound in experiments involving human perception or human interaction. When performing tests of the human visual process, one particular subject may view something differently than another. In such situations, objective tests are very difficult to generate and often completely unfeasable due to the fact that they do not accurately model human perception. Because of the intimate relationship between image processing and the human eye, subjective tests are extremely important when the final judgement if an image is passed by the human eye. In this paper insight into what method of colour edge detection results in edgemaps which are in best accordance with what the human eye sees. In particular, this paper presents a comparison of the relative subjectively based performances of a group of basic order statistic and difference vector operator detectors.

1 Introduction

Order Statistics have recently gained much attention in the field of image processing, and have proved to be extremely powerful when applied to the area of colour edge detection [1]. Order statistics provide elegant methods of analyzing the vast amounts of data within digital images, followed by straightforward means by which various features (e.g. edges) can be extracted. Unfortunately, numerical data useful in providing an effective means to compare edge detectors is very hard to come by. As a result, it is often more convenient to rate various detectors using subjective tests. Subjective testing provides insight into how specific detectors fare when compared to that being modeled, namely the human eye. This paper attempts to address the relative performance of various colour edge detectors such as the Vector Range (VR) and the Difference Vector Median (DVMedian), based on a subjective analysis of visual inspection by several test subjects.

---

* The authors can be reached at the University of Toronto via electronic mail at the following addresses: androup@dsp.toronto.edu, zeus@dsp.toronto.edu, kostas@dsp.toronto.edu and anv@dsp.toronto.edu respectively.

2   Colour Edge Detection via Oder Statistics

The detectors considered here can be separated into two groups. Group 1 is comprised of order statistic based edge detectors [1], [2]. Namely, they are the Vector Range detector (VR), the Minimum Vector Dispersion detector (MVD), and the Nearest Neighbor detector (NN) which has two different forms. First, the definition for the Vector Range detector can be expressed as

$$VR = |v_N - v_1|, \qquad (1)$$

where the distance metric used is the Euclidean distance or $\ell 2$ norm. The vectors v_1 and v_N represent the first and last members in the set of R-Ordered vectors [3] derived from the 5x5 pixel window. The MVD detector has a more complex form than that of the simple VR detector, and is defined as

$$MVD = min_j(|v_{(N+1)-j} - \sum_{i=1}^{l} \frac{v_i}{l}|), \qquad (2)$$

with delimiters on the parameters $i = 1, 2, ...N, j = 1, 2, ...k, k < N$, and $l < N$, where N is the number of pixels in the window. After much experimentation, it was found that using $l = 10$, and $k = 6$ provided satisfactory results. Lastly, the Nearest Neighbor method expands on the ideas of the VR and MVD edge detectors. In the first variant, the VR detector is altered to determine the range between the maximum outlier and a vector obtained by use of the adaptive Nearest-Neighbour filter [2]. This edge detector is defined in the following equation:

$$NNVR = |v_N - \sum_{i=1}^{N-1} w_i v_i|. \qquad (3)$$

The term w_i is a weighting constant which is inversely proportional to the divergence of the vector v_i from the group. In other words, outliers which differ very much from the reference vector are apportioned a smaller percentage of the total weight available than those vectors which are close to the reference. This weighting constant is characterized by the following:

$$w_i = \frac{d_N - d_i}{Nd_N - \sum_{i=1}^{N} d_i}. \qquad (4)$$

The denominator term is used to normalize all the weights such that their sum is equal to unity. The distances, d_i, and d_N, are determined from the method of reduced ordering [3], where the $\ell 2$ norm, and a median vector are used as the distance metric and reference vector respectively. A special case exists for this weighting function. Specifically, in a highly uniform pixel window where there are no edges, all the Neighbouring pixels have the same value as the centre pixel. This results in the denominator of the weighting function equating to zero, thereby causing the weights to approach infinity. In such a scenario, amends can be made by replacing the summation in (4) with the vector v_0 (the centre pixel). Since all pixels are identical, the distance between the vector v_0 and the outlier v_N will be zero, and no edgemap will be created.

In the second variant, a method similar to the MVD detection technique is used. The NNMVD detector can be defined as:

$$NNMVD = min_j(|v_{(N+1)-j)}| - \sum_{i=1}^{l} w_i d_i|). \quad (5)$$

The four edge detectors that constitute Group 2 are based on the concept of Difference Vector Operations [4]. The difference vector operation uses opposite vector pairs within the pixel window along with the Euclidean distance to determine the gradient in each of four possible directions $(0°, 45°, 90°, 135°)$.

v_4	v_3	v_2
v_5	v_0	v_1
v_6	v_7	v_8

Fig. 1. 3x3 pixel window.

An example of a gradient resulting from the opposite pair in the direction $0°$ is:

$$d_0 = |v_1 - v_5|. \quad (6)$$

The vector gradient can then be determined as the maximum resulting from the set of all such possible gradients.

$$VG = max(d_0, d_{45}, d_{90}, d_{135}). \quad (7)$$

Changes must be made to this edge detector to enable us to work with a 5x5 pixel Window. First, the window must be partitioned into a set of four subwindows each corresponding to the same four directions as the in case of the 3x3 window. This is shown in Figure 2.

Each subwindow contains eleven vectors [4]. The vector for the centre pixel is common to each subwindow. By applying each detector's characteristic filtering algorithms to each of the eight subwindows (e.g. Median or Mean filter), a total of eight distinct vectors $(s_{0+}, s_{0-}, s_{45+}, ..., s_{135-}, s_{135+})$ can be obtained to reduce the 5x5 window to a 3x3 window. Four filtering algorithms were used in the reduction of the 5x5 window, resulting in four different edge detectors. Specifically, detectors were realized with the Mean filter, the Median filter, the α-trimmed Mean filter (α=0.3), and the Nearest Neighbor filter [5].

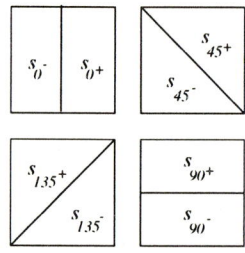

Fig. 2. Subwindow configurations.

3 Experimental Procedure and Subjective Testing

In this experiment, the aforementioned edge detectors were compared in four different scenarios. All eight algorithms were applied to the original image, and corrupted images using Gaussian, Impulsive, and mixed noise composed of a combination of impulsive and Gaussian noise. After detection, the outputs of all the algorithms were thresholded using a histogram. Through experimentation, a ten percent threshold value was found to provide adequate results.

Fig. 3. Gaussian corrupt *lenna*.

The criteria used for subjective testing were as follows:

Detection - This criterion measured how well a particular detector performed considering the noise environment within which it operated. It was explained to the subjects that good detection meant that the algorithm had to create an edgemap which effectively delineated actual edges in comparison to the original, while suppressing noise and not generating false edges.

Continuity - Subjects rated each detector on the consistency of the lines outlining the edges. Edge lines that were not broken were considered very good, and lines with absolutely no breaks were considered ideal.

Thinness - This criterion provided a measure of how much uncertainty was created by a particular detector when creating an output. Thin edgemaps (one pixel in width deemed ideal) meant that the edge detector could precisely find a particular edge with very little ambiguity.

Visual Appeal - This category was used to take into account any human factors involved with the rating of the edgemaps. Persons participating were told that scores in this category could be based under whatever criteria they deemed necessary or important for an overall good edge detector output.

Each output was subjectively rated by 19 persons on a scale of 0(poor), 1(below average), 2(average), 3(above average) and 4(good). The rationale for using this particular scoring system was that it permitted the rating of outputs that were extremely good or bad, and outputs that ranked somewhere in the middle. In addition, it provided a means for the subjects to assign scores in cases where they were undecided on how to rate the output. Each subject was presented with the entire set of eight edge detector outputs, as well as the original colour image for comparison in each case (noiseless, Gaussian, impulsive, and mixed cases). Each subject was permitted to examine all edgemaps at their own leisure. No time limit was set for output examination, and the changing of edgemap scores as desired by the subject was permitted All subjects used were volunteers, and non-experts in the fields of edge detection and image processing. Data values were tallied in tabular form and the averages were calculated for each detector for all criteria in each scenario.

4 Experimental Results

Table 1 gives an overview of the various cases of noise environments, whereas Table 2 shows the average scores of each detector for all four noise scenarios. There is a marked division between the overall performances of the difference vector detectors and the basic order statistic detectors. All of the difference order detectors perform very similarly, with slight variations in score and some poor performances in specific noise scenarios. The basic order statistic detectors behave analogously, but with clearly inferior scores.

NOISE TYPE	AMOUNT	VARIANCE
Gaussian	30%	5%
Impulsive	4%	5%
Mixed	2% Imp, 15% Gau	5%

Table 1. Noise Distributions.

Throughout all the tests, two particular edge detectors consistently attained high performance indexes. Namely, the Difference Vector Mean, and the Difference Vector Nearest Neighbour detectors produced outputs that agreed with the subjects in all cases. While only attaining only moderate and low scores,

Fig. 4. Output from Difference Vector Nearest Neighbor Detector.

Fig. 5. Output from Minimum Vector Dispersion Detector.

the Nearest Neighbour Minimum Vector Dispersion detector achieved consistent scores in all scenarios with only small variations, and justifiably provided the most reliable output among the basic order statistic edge detectors.

The minimum vector dispersion detector provided edgemaps that delineated edge locations, but did not suppress noise very well. The detector which received the highest performance index scores was the Difference Vector Nearest Neighbour detector. For all input images, this algorithm effectively detected the true image edges, and suppressed all three different types of noise. In addition, the edgemap lines it created were thin, and continuous. The Difference Vector Mean provided comparable edgemaps. The only differences between it and the DVNN detector were in the noisy cases, where it did not suppress noise as effectively.

NOISELESS	VR	MVD	NNVR	NNMVD	DVMean	DVMed	DVαMean	DVNN
detection	1.21	2.11	0.89	0.89	3.00	2.68	2.47	3.21
continuity	1.00	1.79	1.00	1.26	3.21	2.63	2.68	2.84
thinness	2.31	2.21	2.21	1.89	2.68	2.52	2.58	2.11
visual appeal	0.79	1.79	1.32	1.21	3.21	2.42	2.79	2.89
OVERALL	1.32	1.97	1.36	1.31	3.03	2.56	2.63	2.76
GAUSSIAN	VR	MVD	NNVR	NNMVD	DVMean	DVMed	DVαMean	DVNN
detection	0.32	0.21	0.47	1.79	2.58	0.10	0.89	3.00
continuity	0.21	0.42	0.68	1.42	3.00	0.21	0.79	3.00
thinness	0.89	1.11	1.21	1.00	2.21	0.32	1.32	2.68
visual appeal	0.21	0.21	0.47	1.11	2.68	0.00	0.58	3.32
OVERALL	0.41	0.49	0.71	1.33	2.62	0.14	0.90	3.00
IMPULSIVE	VR	MVD	NNVR	NNMVD	DVMean	DVMed	DVαMean	DVNN
detection	0.00	0.79	0.00	1.11	2.32	2.32	0.00	3.68
continuity	0.00	1.32	0.21	1.42	2.00	1.42	0.32	3.42
thinness	0.05	1.58	0.00	1.11	2.47	2.11	0.32	2.79
visual appeal	0.00	0.68	0.00	0.89	2.00	1.79	0.32	3.58
OVERALL	0.01	1.09	0.05	1.08	2.20	1.91	0.24	3.37
MIXED	VR	MVD	NNVR	NNMVD	DVMean	DVMed	DVαMean	DVNN
hline detection	0.00	1.21	0.00	1.74	2.53	1.21	0.37	3.52
continuity	0.00	1.21	0.00	1.58	2.16	0.94	0.16	3.21
thinness	0.00	1.16	0.00	1.58	1.16	1.53	0.53	2.74
visual appeal	0.00	0.74	0.16	1.21	2.68	1.32	0.21	3.47
OVERALL	0.00	1.08	0.04	1.54	2.13	1.25	0.32	3.24

Table 2. Edge detector scores for various noise.

5 Conclusions

Edgemaps can be considered to be one of the simplest representations of an image. Since they are monochromatic, and because like hand drawn sketches, they contain the most crucial information of an image, they can be used for object recognition, and robot vision. Also, since edgemaps are small, and easily compressed they can be efficiently transmitted with very little bandwidth. This is almost an ideal scenario, but due to the fact that edge detectors behave differently in various noise situations (and sometimes are specifically suited to perform well in particular noise distributions), discretion must be used when choosing a particular edge detector. The data presented suggests that from a subjective standpoint, the most effective edge detectors for the cases studied were clearly the Difference Vector Mean, and the Difference Vector Nearest Neighbour detectors. This leads to the conclusion that the Difference Vector Nearest Neighbour detector provides results which are perceived as favorable by the human eye. A similar statement can be made for the Difference Vector Mean detector but to a lesser degree as indicated by its performance indeces. The clear dominance of the difference vector operators over the basic order statistic detectors, ushers in the conclusion that the use of the directional gradient along with application of the nearest neighbour filter provides edgemaps that are in accordance with what the human eye sees. In addition, the nearest neighbour filter has a limited ability to compensate for the type of environment in which it operates, meaning that it can effectively suppress noise, and may be the reason why so many of the subjects gave it high scores. By subjectively testing edge detectors, insight into the way in which the eye perceives edges can be obtained. Development of an edge detector which directly addresses the issue of human perception and has the ability to adapt to the noise environments which it is placed is a direct consequence of this experiment. Future research is being driven by fuzzy and multispectral techniques that show great promise in such a pursuit. It is notable to mention that a far greater number of persons for subjective testing would be extremely useful in the sense that experimental data would become much more statistically significant, and would cover a more diverse set of people.

References

1. P.E. Trahanias and A.N. Venetsanopoulos "Colour edge detection using vector order statistics," *IEEE Transaction on Image Processing, pp1-18*, Sept. 1995.
2. I. Pitas and A.N. Venetsanopoulos, "Nonlinear Digital Filters: Principles and Applications" *Kluwer Academic Publishers*, 1990.
3. S. Sanwalka "Vector order statistic filters for colour image processing," *University of Toronto M.A.Sc. Thesis*, Sept. 1992.
4. Y. Yang "Colour edge detection and segmentation using vecotr analysis," *University of Toronto M.A.Sc. Thesis*, Sept. 1995.
5. K.N. Plataniotis, D. Androutsos, A.N. Venetsanopoulos "Nearest Neighbour multichannel filters for image processing," *Signal Processing VIII, Theories and Applications, vol 3*, Sept. 1996.

Similarity Measures for Binary and Gray Level Markov Random Field Textures

Abdurrahman Çarkacıoğlu, and Fatoş T. Yarman-Vural
Department of Computer Engineering
Middle East Technical University
E-mail: vural@ceng.metu.edu.tr

ABSTRACT

In this study a new set of texture measures, namely, *Clique Length* and its *moments are* introduced. These measures are defined employing new concepts which agrees with the human visual system. The simulation experiments are performed on binary and gray level MRF texture alphabet to quantify the data by the kth moment of Clique Length. Experimental results indicate that the introduced measures identify the visually similar textures much better than the mathematical distance measures.

1. INTRODUCTION

Loosely speaking, texture can be defined as a stochastic, possibly periodic, two dimensional image field. In recent computer vision literature there has been an increasing interest in the use of statistical techniques for modeling and processing the textured image. Some of this work has been directed towards the application of Markov Random Fields (MRF) in texture modeling, classification and restoration of noisy and textured images [1],[2].

Texture generation using the MRF model is a classical problem. Using different set of parameters, it is possible to generate extremely wide class of textures. However, it is not easy to generate a desired form of texture, since the relationship between the model parameters and certain features of texture is not a linear one.

In this study, a set of measures is introduced to quantify the similarity of textures which can be modeled by Markov Random Fields (MRF). First, a brief explanation of MRF model is given in Section 2. Texture realization problem is addressed in Section 3. Then, based on the theoretical studies and various observations, a set of texture measures are defined in Section 4. In Section 5, simulation experiments are performed to determine the relation between the data type, model parameters and the texture measures defined in Section 4. Section 6, concludes the paper by discussing the proposed texture measures and commenting on the directions of the feature research.

2. MARKOV RANDOM FIELDS (MRF) FOR TEXTURE MODELING

Consider the random field defined over a finite lattice of points (i, j), $L = \{(i,j): 1 \leq i \leq N_1, 1 \leq j \leq N_2\}$. A collection of subsets of L described as, $\eta = \{\eta_{ij} : (i,j) \in L, L \supset \eta_{ij}\}$, is a neighborhood system on L iff η_{ij}, the neighborhood of pixel (i,j), is such that (a) $(i,j) \notin \eta_{ij}$, (b) $(k,l) \in \eta_{ij}$, for any $(i,j) \in \eta_{kl}$.

A random field $X = \{X_{ij}\}$, defined over the lattice L, is a Markov Random Field (MRF) with respect to the neighborhood system η, iff the distribution of X, $P(X=x) \geq 0$, $\forall x$ and $P(X_{ij} = x_{ij} | \{X_{kl} = x_{kl}, (k,l) \in L, (k,l) \neq (i,j)\}) =$
$P(X_{ij} = x_{ij} | \{X_{kl} = x_{kl}, (k,l) \in \eta_{ij}\})$, for all $(i,j) \in L$.

The above theoretical background is used to generate textures and estimate the model parameters. The reader is referred to [1], [2], [3] for a more detailed description of Markov Random Field texture realization problem.

3. TEXTURE SIMILARITY MEASURES

Because of its complexity and tremendous amount of variations, there is no clear-cut mathematical definition of texture. Therefore, a measure of texture which gives an idea about the data type is highly desirable. This measure would be very useful to give a quantitative idea about the texture similarity among the textures, in texture classification problem. In the following, behind a series of definitions, a new set of texture measures is introduced. First, a simple tool for measuring various properties of texture, called base clique is defined. For this purpose, the general concept of clique [4], [5] and neighborhood is utilized.

Definition 1: Base Clique: Given a seed pixel (i,j) in a neighborhood system η, the *base clique* of (L, η) denoted by a pair of pixels $B_p(ij,kl)$ where $p = 1..P$, is a subset of the Lattice L such that (a) $(i,j) \neq (k,l)$, ij $\in \eta_{kl}$ and (b) Pixels (i,j) and (k,l) satisfies $|x_{ij} - x_{kl}| < \varepsilon$, for a given $\varepsilon > 0$, where P indicates the number of distinct base cliques (see: Fig. 1)

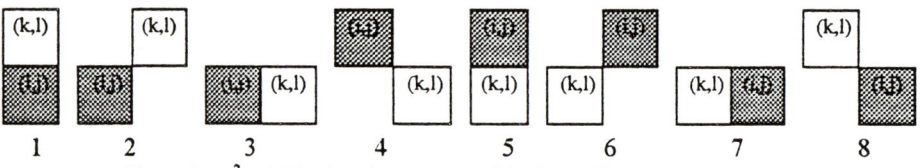

Figure 1. η^2 neighborhood systems and its base cliques B_P, $P = 1,2..8$.

Definition 2: Base Clique Chain: Given pixel (i,j) as a seed, the *base clique chain* $C_P(ij)$ is the connected chain of pixels with the same base clique,

$$\forall p, \quad C_p(ij) = \{B_p(ij,kl) \cup B_p(kl,mn) \cup B_p(mn,..) \cup .. B_p(qr,st)\}$$

where (qr,st) is the last connected pair of pixels, with the base clique B_p. Briefly base clique chains are the line likes starting from the seed pixel and ending at the last connected pair of pixel.

Definition 3: Given (i,j) as a seed, n^{th} *Order Clique Chain* is

$$O^n_z(ij) = \bigcup_{k \in z} C_k(ij)$$

where z is the n-combination of the integers 1,...,P. Note that the number of nth order clique chain is $P!/[(P-n)! \cdot n!]$ where P indicates the number of distinct base cliques.

n^{th} *Order Clique Chain measures* give us cornering information in a given texture.

For example, a *second order clique chain* is a combination of two base clique chains and defined as $O^2_{pq}(ij) = C_p(ij) \cup C_q(ij)$.

Note that there are 28 distinct second order clique chains for a seed pixel (i,j). Figure 2 illustrates some of the second and third order clique chains.

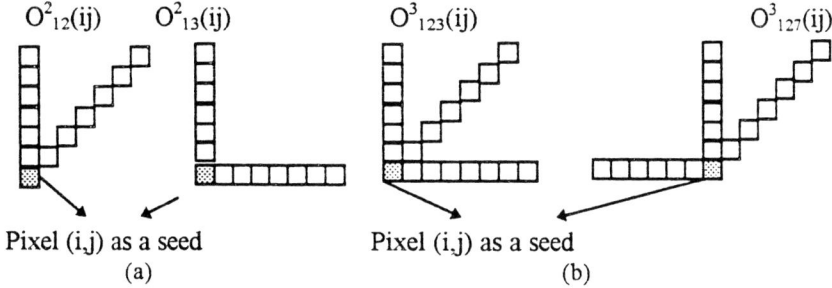

Figure 2: Some of the a) second and b) third order clique chains, respectively.

Definition 4: Given pixel (i,j) as a seed, and its base clique chain $C_p(ij)$, the **base clique length $L_p(ij)$** is the number of the pixels in $C_p(ij)$.

In other words, the base clique length, $L_p(ij)$, is the number of connected chain of pixels formed by the elements of the base clique chain, $C_p(ij)$.

In a textured image, the base clique length, $L_p(ij)$ of each base clique chain is computed by counting the pair of pixels as long as they belong to the same **base clique chain** set, $C_p(ij)$.

Definition 5: Given a seed pixel (i,j) and its nth order clique chain $O^n_z(ij)$, the nth **order clique length $L^n_z(ij)$**, for n>1, is defined as $L^n_z(ij) = \min\{L_p(ij)\}$ where $p \in Z$.

The effect of nth order clique on the human visual system can be measured by considering the minimum length of the nth order clique chain components, $L_p(ij)$. Suppose that, we have $O^2{}_{12}(ij)$ with base clique lengths $L_1(ij)=1, L_2(ij)=8$ and $O'^2{}_{12}(ij)$ with base clique lengths $L'_1(ij)=4$ and $L'_2(ij)=4$. For our visual system, $O'^2{}_{12}(ij)$ caries more second order information than $O^2{}_{12}(ij)$.

Due to the stochastic nature of the texture, $L_z{}^n(ij)$ can be considered as a random variable. Therefore, moments, especially, second order statistics of $L_z{}^n(ij)$ can give us an idea about the appearance of the texture.

Definition 6: $\mu^n{}_z$ is the second order moment of $L^n{}_z(ij)$ (Clique Length)

$$\mu^n{}_z = E\,[[L^n{}_z(ij)]^2] \approx (1/N) \sum [L^n{}_z(ij)]^2$$

where E indicates the expected value operator and N indicates the total number of clique chain sets in Clique type $O^n{}_z(ij)$. Note that the number of the $\mu^n{}_z$ is the same as the number of the nth order clique chains in a given texture.

Notice that each clique type is represented by a single second order moment. $\mu^{k,n}{}_z$ is the kth order moment of $L^n{}_z(ij)$ (Clique Length)

$$\mu^{k,n}{}_z = E\,[[L^n{}_z(ij)]^k] \approx (1/N) \sum [L^n{}_z(ij)]^k$$

Definition 7: For a given texture T, the Nx1 vector $\underline{M}^n(T)$, where the entries are the $\mu^n{}_z$ (second order moment of $L^n{}_z(ij)$), for all the nth order clique chains, are called the **second order moment vector**.

$\underline{M}^n(T)$ is used to represent texture T. At this point, any distance (Euclidean, Mahalanobis, Yule, Jakard etc.) can be used to measure the similarity between the textures by using the $\underline{M}^n(T)$ vectors.

For examples, the second order moment vector of the second order clique chains for texture T is $\underline{M}^2(T) = \{\mu^2{}_{12}, \mu^2{}_{13}, \mu^2{}_{14}, \ldots, \mu^2{}_{78}\}$. $\underline{M}^2(T)$ gives only a rough idea about the texture appearance. However, as the order n gets larger, the second order moment vector provides a more detailed information, with the price of increasing the dimension of $\underline{M}^n(T)$. In the following, an algorithm is proposed to compute $\mu^n{}_z$ or each nth order clique chain. It is clear that $O^2{}_{15}, O^2{}_{26}, O^2{}_{37}, O^2{}_{48}$ clique chains are line like. For these types of clique chains, we replace the length of clique chain formulas to $L_{PQ}{}^2(ij) = L_P(ij) + L_Q(ij)$.

4. EXPERIMENTAL RESULTS

In our simulation experiments, first, second, third and fourth order MRF models are used for modeling binary and gray level images. Textures of size 64X64 are generated according to various settings of MRF parameters. The algorithm is implemented under C programming language.

The second order moments for a given texture are scaled to 1 by dividing all the second order moments to the value of the maximum second order moments of the texture. This scaling provides invariancy in size and dimension of texels. As a next step, the distance of second order moments are used to calculate the distance between the textures T1 and T2 is, then, given as d $(x, y) = |\underline{M}^n(T1) - \underline{M}^n(T2)|$

The above distance gives us a measure for the similarity of texture T1 and T2. Using this distance, it is possible to classify the textures. Experimental results show that μ^n_z very successful for identifying visually similar textures. Table 2 shows the most similar five textures according to the Euclidean distance of second order moments. Visual inspection of the textures 1-30 (Figure 3) and the quantitative analysis of table 2 indicate that the proposed distance is highly consistent with our human visual system in measuring the texture similarity.

5. CONCLUSIONS

In this paper, second order *moments* of $L^n_z(ij)$, are used to measure the similarity of binary and gray level Markov Random Fields textures. It is intuitively clear that the distribution of different clique types in the image defines texture. Therefore, by using the clique type statistics on a given texture, we can get an idea about the appearance of the texture.

Based on the above argument, we introduce the definitions for base clique, n^{th} order clique, second order *moments with respect to* origin of $L^n_z(ij)$. μ^n_z is used to obtain quantitative values about the distribution of the different base cliques. Simulation examples demonstrate that μ^n_z provide most of the information about the appearance of the texture of an image. The results are very consistent with the human visual system. In conclusion, using the second order statistics of the length of nth order clique chains, visually similar textures can be easily identified.

REFERENCES
1. Derin and H. Elliott, "Modelling and segmentation of noisy and textured images using Gibbs random fields", *IEEE Trans. Pat. Anal. Mach. Intel.*, vol. PAMI-9, pp. 39=55, 1987.
2. S.Lakshamanan and H. Derin, "Simultaneous parameter estimation and segmentation of Gibbs random fields simulated annealing", *IEEE Trans. Pat. Anal. Mach. Intel.*, vol. 11., pp. 799 - 813, 1989.
3. S. Cohen and D. B. Cooper, "Simple parallel hierarchical and relaxation algorithms for segmenting noncausal Markovian random fields", *IEEE Trans. Pat. Anal. Mach. Intel.*, vol. PAMI-9, pp. 195 - 219, 1987.

4. Derin, "The use of Gibbs distributions in image processing", in *Communications and Networks*: A Survey of Recent Advances, I. Blake and V. Poor, Eds. New York: Springer - Verlag, 1985.
5. A. Carkacioglu and F.T. Yarman-Vural, "Set of Texture Similarity Measures", Electronic Imaging 97, Machine Vision Applications in Industrial Inspection, SPIE Proceedings Vol 3029, pp76-84 San Jose, 1997.

Table 1: The most similar three textures shown in Figure 3

Texture No	Texture No	Distance	Texture No	Distance	Texture No	Distance
1	2	0.13	3	0.28	12	1.97
2	1	0.13	3	0.20	5	1.95
3	2	0.20	1	0.28	5	1.94
4	6	0.28	5	0.29	11	1.96
5	6	0.19	4	0.29	3	1.94
6	5	0.19	4	0.28	12	1.95
7	20	0.77	19	0.81	21	0.81
8	19	0.30	20	0.35	21	0.35
9	8	0.58	19	0.76	21	0.80
10	12	0.77	11	0.80	8	1.94
11	12	0.09	10	0.80	6	1.95
12	11	0.09	10	0.77	6	1.95
13	14	0.69	15	0.73	28	1.11
14	13	0.69	15	1.00	28	1.31
15	13	0.73	28	0.76	14	1.00
16	17	0.87	18	0.90	25	2.46
17	16	0.87	18	1.04	3	2.41
18	16	0.90	17	1.04	5	2.42
19	21	0.23	20	0.23	8	0.30
20	21	0.18	19	0.23	8	0.35
21	20	0.18	19	0.23	8	0.35
22	24	2.21	23	2.61	26	3.97
23	24	2.06	22	2.61	26	2.85
24	26	1.96	23	2.06	22	2.21
25	27	0.82	26	1.34	16	2.46
26	25	1.34	27	1.51	24	1.96
27	25	0.82	26	1.51	16	2.48
28	15	0.76	30	0.91	13	1.11
29	30	0.84	28	1.25	15	1.71
30	29	0.84	28	0.91	15	1.28

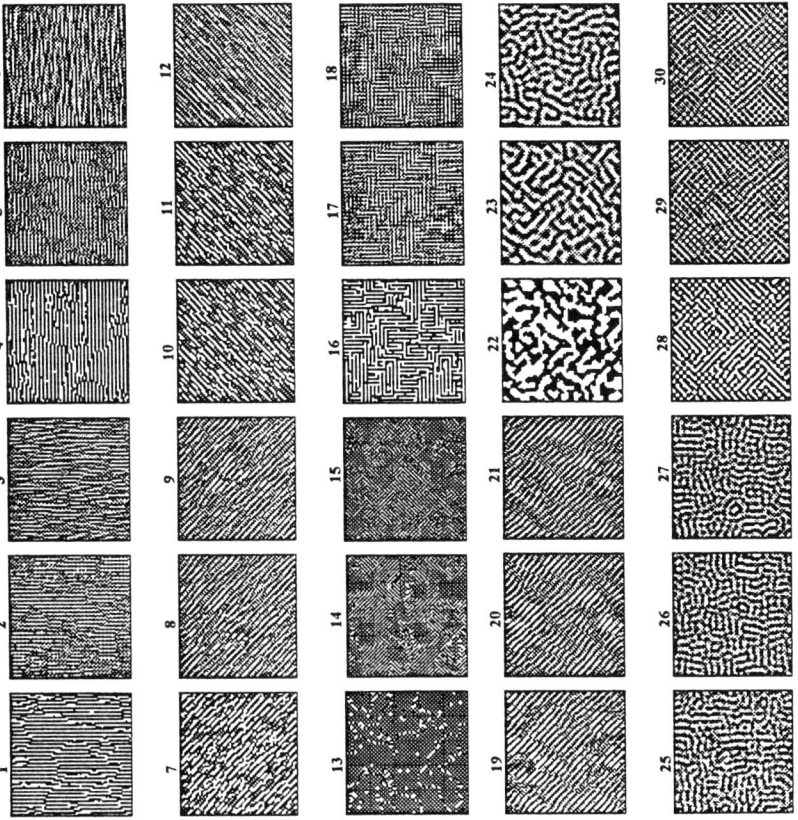

Figure 3. TEXTURES (1-30)

A Simple and Effective Edge Detector

C. Cafforio, E. Di Sciascio, C. Guaragnella and G. Piscitelli

Dipartimento di Elettrotecnica ed Elettronica, Politecnico di Bari
Via E.Orabona, 4 I-70125 Bari, Italy

Abstract. Nonlinear filtering based on a two concentric circular windows operator is introduced as a simple and effective way to find edges in image processing. The dual windows edge detector can operate with no fixed threshold, is isotropic, i.e. its response does not depend on edge orientation and has good noise immunity in the vicinity of edges. Computational requirements are limited and experimental results are good enough.

1 Introduction

Many image processing applications require edge detection. Edges are of extreme importance, as they convey essential information in a picture. Accurate edge detection is necessary for a number of image analysis and recognition techniques. A lot of research has been done on this topic in the past, yet a completely satisfactory algorithm is still to be devised.
An efficient edge detection procedure should comply with some basic requirements: edges should be found with a low probability of false detection due to noise; edges should not be misplaced; algorithms should perform similarly with any image; procedures should lead to an efficient implementation.
Ideally edges are steep transitions between smoothly varying luminance areas. In real scenes transitions are not so steep, thus their extraction requires careful processing.
Several edge detection techniques exist and a number of modifications to basic algorithms have been reported in the literature. Usually these algorithms use first or second derivatives of image luminance and are forced to adopt some regularization techniques to control the effects of noise. Marr and Hildreth [1] proposed an algorithm that finds edges at zero-crossing of the image Laplacian, and several modification to their basic approach have been reported. This algorithm guarantees that continuos contours are extracted. The Canny algorithm [2] finds edges at peaks of the first difference in the local gradient direction. A variation on the basic algorithm, due to Boie-Cox [3], uses the maximum first difference and localizes edges with the aid of a directional second difference. A number of other modifications to the approach of Canny also exist [4-8]. Edge detection algorithms based on non linear filters have also been presented in the literature [9]. More recently, techniques have been proposed that characterize edge detection as a fuzzy reasoning problem [10]. There is much literature on the subject and the above references are only a few, not exhaustive, examples.
Most techniques heavily rely on low-pass filtering and the use of thresholds.

Work supported in part by CNR grant n. 95.00439.CT12.

Although this is a sure way to drastically reduce noise effects, it leads, most likely, to discontinuities in the detected edges and to contour misplacement. To partially avoid the negative effects of filtering, adaptive directional filters that smooth the image only in the direction orthogonal to local image gradient have been proposed. These filtering schemes are effective, but lead to an increase in the algorithmic complexity without the possibility to assure precise edges localization.

In this paper we present a simple operator that uses two concentric windows. The operator does not require fixed thresholds, is isotropic, i.e. its response is almost independent on contours orientation, and guarantees a good edge localization. Computational requirements are limited.

The proposed algorithm originates from a previous work where no known edge extraction algorithm was accurate enough to allow an efficient extrapolation of high frequency terms from a low passed image [12].

Next section presents the proposed edge detection approach. In section III the obtained results are shown and discussed. The arguments that call for further research are outlined in the final section.

2 Nonlinear Edge Detection

Let X be a gray level image and $x(i,j)$, $i=1,..,M$, $j=1,..,N$ its samples. It is reasonable to consider contours placed exactly half way along the slopes of transition zones between almost uniform luminance areas. This is a simple task in one dimension; in two dimensions the need for dealing with the orientation of the transition zone makes things more difficult. It is much more so because of the rectangular sampling grid. To correctly locate a contour, it is necessary to estimate the width of the transition zone and to track the position within this zone. This is not achievable with just one local measure.

The proposed algorithm uses two concentric circular windows W_1 and W_2 (as much circular as the rectangular sampling grid allows). The window shape is not an essential feature for our approach: a square window gives similar results, but the response of the operator is no longer isotropic.

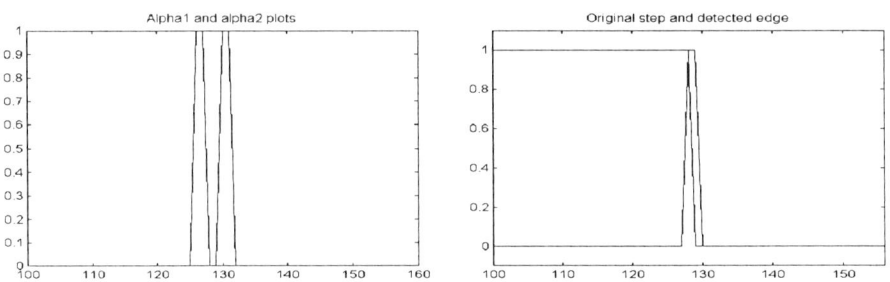

Fig. 1. a) A sharp profile and the extracted contour; b) corresponding α_1 and α_2 plots. $L_{W_2} = 7$, $L_{W_1} = 3$ pixels.

Fig. 1 shows the edge detection operator applied on sharp profile. It can be noticed that whatever the external window size L_{W_2} the detected edge width is determined by the inner window size L_{W_1}, in fact it is equal to (L_{W_1} - 1).

The maximum and minimum pixel values are measured within both considered windows:

$$min = \min_{i,j \in W_1} x(i,j) \qquad MIN = \min_{i,j \in W_2} x(i,j)$$
$$max = \max_{i,j \in W_1} x(i,j) \qquad MAX = \max_{i,j \in W_2} x(i,j) \qquad (1)$$

and the differences between maxima and minima in respective windows are then computed:

$$\alpha_1 = MAX - max \qquad \alpha_2 = min - MIN \qquad (2)$$

A contour is considered present wherever $\alpha_1 = \alpha_2 \neq 0$, i.e. when the luminance ranges in W_1 and W_2 coincide.

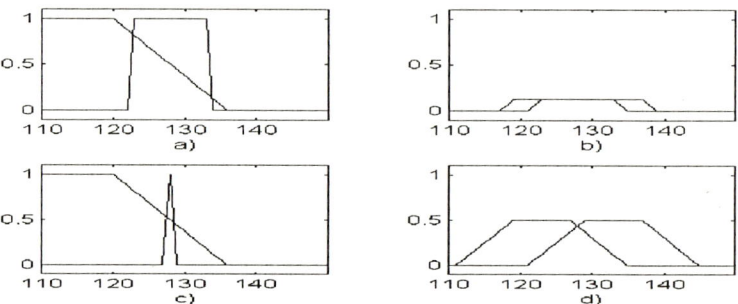

Fig. 2. A linear profile and the operator response for different outer window sizes: a) $L_{W_2} = 7$, $L_{W_1} = 3$ pixels; b) corresponding α_1 and α_2 plots; c) $L_{W_2} = 19$, $L_{W_1} = 3$ pixels; d) corresponding α_1 and α_2 plots.

The rationale of this procedure is that the larger window identifies the transition zone, while the smaller window, placed exactly at the center of the larger one, tracks the half height point. If the external window is larger than the transition zone, the contour is exactly located. If the transition zone is larger, the condition $\alpha_1 = \alpha_2$ is met, assuming the slope is constant, until one of the extrema of the outer window leaves the discontinuity area.

To clarify how the operator works, let us consider its application in the one dimensional case for two further ideal signals: a linear and a cosine like degrading profiles.

For a linear degrading profile the detected edge width may result thick (see figs. 2.a and 2.b) when a small outer window is used.

A single pixel edge can be obtained if the whole constant slope luminance transition is included within the outer window. This behavior clearly appears in figs. 2.c and 2.d.

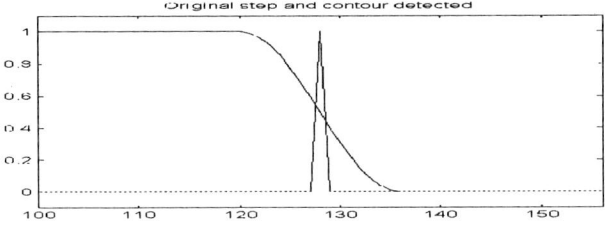

Fig. 3. A cosine profile and the detected edge $L_{W_2} = 11, L_{W_1} = 3$ pixels

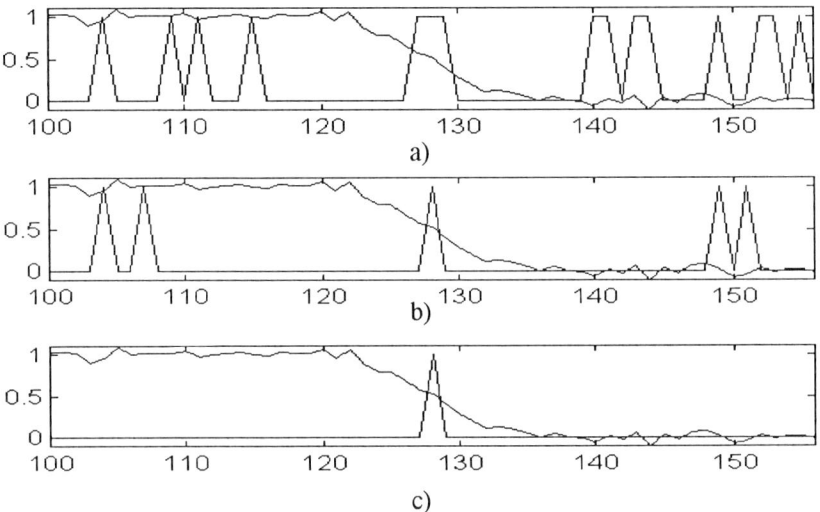

Fig. 4. A cosine profile with added Gaussian noise ($\sigma = 0.05$, PSNR = 26 dB): a) $L_{W_2} = 11, L_{W_1} = 3$ pixels; b) $L_{W_2} = 31, L_{W_1} = 3$ pixels; c) $L_{W_2} = 41, L_{W_1} = 3$ pixels.

In fig. 3 a cosine degrading 15-pixels wide profile is considered. With this type of signal a single pixel edge is obtainable even with small outer window size. This is because the luminance profile has not an exactly constant slope. In fact the difference $\alpha_1 - \alpha_2$ zeroes only exactly in the middle of the luminance transition, where the contour should be placed.

Condition $\alpha_1 = \alpha_2$ allows edge detection only conceptually. As a matter of fact, in a real environment, a thresholding strategy has to be adopted.

The difficulty lies in the image sampling: it is extremely improbable that the difference $\alpha_1 - \alpha_2$ exactly zeroes on a sampling point. The change in the sign of $\alpha_1 - \alpha_2$ is a much more effective indicator of the presence of a contour. If a two dimensional search of sign changes in $\alpha_1 - \alpha_2$, which can allow a fractional pixel positioning of contours, is considered a complication, a simplified procedure can be used instead.

a) b)

Fig. 5. a) Original test image "building". b) Transition areas in "building". $\alpha_1 > \alpha_2$ black; $\alpha_1 < \alpha_2$ gray, $\alpha_1 = \alpha_2$ white.

The equality of the two values α_1 and α_2 can be considered verified when their absolute difference is smaller than the estimated luminance change between contiguous pixels or a fraction thereof:

$$\varepsilon = \frac{(max - min)}{W_1} \qquad (3)$$

This is a thresholding operation and, though adaptive, it has drawbacks. In fact, it is no longer possible to assure that the extracted contours are closed. In some application this is unacceptable, but discontinuities in detected contours are always an unpleasant effect to consider. On the other hand, continuous contours do not necessarily mean physically meaningful contours.

In fig. 4 the previous cosine degrading profile is considered when an added Gaussian noise is introduced and the adaptive threshold is implemented. By increasing the outer window size it can be noticed that, while the edge position remains unchanged, false edges due to noise are drastically reduced in the area around the true edge position.

The proposed operator can be considered an application to contour detection of the principles of morphological filtering. However, it is possible to give a simpler interpretation of its operation.

The external window is used to circumscribe the area in which the contour is to be looked for, and it determines the local luminance change. A luminance threshold put exactly half-way on the measured luminance change would reproduce ancient techniques used in image binarization. Here a slightly more sophisticate approach is proposed looking for the position where the luminance change within the inner window is centered with respect to that in the outer one. This provides a little (the only one) rejection of the noise inevitably superimposed on the image.

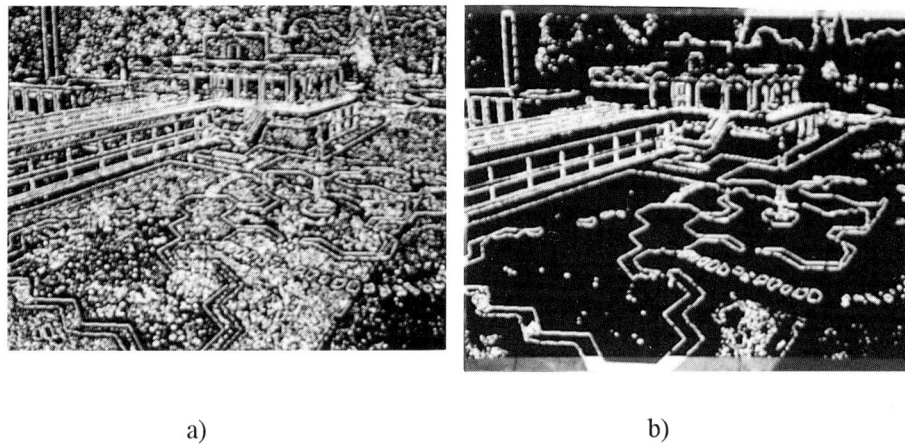

Fig. 6. a) Edge detector output with $L_{W_1}=3$ and $L_{W_2}=11$. b) Edge detector output with $L_{W_1}=3$ and $L_{W_2}=61$.

The noise rejection is more effective with a larger inner window (but this has other drawbacks, as already pointed out) and with a higher local gradient. The noise rejection is maximum when a contour is present: this results in better contour detection and localization. However, the algorithm is completely ineffective in flat luminance areas, where noise can detect a lot of false contours. A threshold on *MAX-MIN* or *max-min* is in some way unavoidable.

The size of the external window, i.e. W_2, affects the sensitivity of the operator: small external windows will result in extremely accurate contours, but also in a high sensitivity to noise. Larger windows will extend the effect of high contrast contours over a larger area, preventing smaller contrast contour from being detected near higher contrast ones (see fig. 4). As a matter of fact, a larger window provides the large support that is needed to suppress noise, while the inner window has the small support required for edge localization [11].

An effective way to combine the two windowing strategies is possible adopting a simple tracking algorithm. Let us suppose to apply the edge detection algorithm to the same image, with different window sizes. Let image Y_1 contain the contours obtained with the larger external window and image Y_2 contain those obtained with the smaller external window. The algorithm tracks edge samples in Y_1 and, when a break is observed, it checks image Y_2 for contour points that allow to proceed in edge following. This is done with a window 5 pixels wide having at least 3 pixels in Y_1 already marked as edge samples. When such a condition is met other edge samples in Y_2 are marked as edge pixels also in the new edge image. The algorithm obviously avoids to consider isolated edge pixels, i.e. pixels having no edge neighbors within the window. It is clear that the current implementation of the tracking algorithm is extremely simple, although effective in reducing contours discontinuities.

Similar or better results could be obtained with more refined techniques that could operate without the two passes with different window sizes. A dynamically varying external window size could be a better choice for a single pass algorithm.

Fig. 7. a) Edge image obtained with the application of the tracking algorithm on the previous two edge images. b) Edges detected with $L_{W_1}=3$, $L_{W_2}=5$ and $\varepsilon > 5$ for test image building.

3 Experimental Results

All experiments were based on 720×560 pixels gray level images quantized with 8 bits. No filtering whatsoever was used to reduce noise effects, in order to check the noise sensitivity of the algorithm.

Fig. 5.a shows the original test image named "building". Figure 5.b shows the image obtained considering the difference ($\alpha_1 - \alpha_2$) in the test image. The image is obtained using $L_{W_1}=3$ and $L_{W_2}=11$, respectively. It is evident that where transitions take place, i.e. where the difference ($\alpha_1 - \alpha_2$) changes sign, contours should be placed.

The application of the nonlinear operator on the test image produces edge images shown in figs. 6.a and 6.b. The first one is obtained with window sizes $W_1=3$ and $W_2=10$, the second one is obtained with $W_1=3$ and $W_2=61$. The difference is noticeable; figure 6.b has all higher energy contours detected, but some of them are incomplete, while figure 6.a, though having all edges detected, has a lot of noise. The application of the tracking algorithm on images in figures 6.a and 6.b produces the image in figure 7.a, where various contours become closed. Edges are correctly positioned even after the modification of the window size and this is what allows a correct edge tracking. The proposed algorithm is able to detect multiple converging edges, as no assumption on edge direction is made.

Other experiments have been performed introducing a further fixed threshold on ε to reduce the noise effect. The introduction of this threshold allows to reduce used window sizes also producing accurate but discontinuous contours. Fig. 7.b has been obtained using $L_{W_1}=3$, $L_{W_2}=5$ and a fixed threshold $\varepsilon > 5$.

4 Conclusions

Edge detection is a crucial stage in almost any image processing application. This work offers a contribution to the design of simple and reliable edge detectors. The basic nonlinear operator we propose uses two concentric windows; the outer one identifies a transition zone, while the inner one tracks the half point that can be assumed to be an edge sample. A noteworthy advantage of this method is the edge detector requires no fixed threshold; an adaptive threshold, proportional to the luminance change in the inner window has been introduced.

Experimental results obtained with various test images seem to support the theoretical background. Further work is needed to make the algorithm more robust in low contrast areas. Adaptive window size and a fuzzy logic approach to thresholding will be considered.

References

1. D. Marr, E. C. Hildreth, "Theory of edge detection", Proc. Royal Soc. London, B, vol. 207, pp. 187-217, 1980.
2. Canny, "A computational approach to edge detection", IEEE Trans. on PAMI, vol. 8, pp. 679-698, Nov. 1986.
3. A. Boie, I. J. Cox, "Two dimensional optimum edge recognition using matched and Wiener filters for machine vision", Proc. of Int. Conf. On Computer Vision, London, pp. 450-456, 1987.
4. Deriche, "Using Canny's criteria to derive a recursively implemented optimal edge detection, Int. Journal of Computer Vision, pp. 167-187, 1987.
5. Petrou, J. Kittler, "Optimal edge detectors for ramp edges", IEEE Trans. on PAMI, vol. 13, pp. 1154-1171, Nov. 1991.
6. Sarkar, K. L. Boyer, "On optimal infinite impulse response edge detection filters", IEEE Trans. on PAMI, vol. 13, pp. 483-491, May, 1991.
7. A. Spacek, "Edge detection and motion detection", Image and vision computing, vol. 4, pp. 43-56, Feb. 1986.
8. Ziou, "Line detection using an optimal IIR filter", Pattern Recognition, vol 24, pp. 465-478, 1991.
9. I. Pitas, A. N. Venetzanopulos, "Edge detectors based on nonlinear filters", IEEE Trans. on PAMI, vol. 8, N0. 4, pp. 538-550, 1986.
10. Law, H. Itoh, H. Seki, "Image filtering, edge detection, and edge tracing using fuzzy reasoning, IEEE Trans. on PAMI, vol. 18, No. 5, pp. 481-491, 1996.
11. Gokmen, C. Li, "Edge detection and surface reconstruction using refined regularization", IEEE Trans. on PAMI, vol. 15, No. 5, pp.492-499, 1993.
12. C. Cafforio, E. Di Sciascio, C. Guaragnella, "Spectral extrapolation in sub-band coding", Proceedings of IEEE Workshop on Digital Signal Processing, pp. 13-16, Loen (Norway), 1996.

Improvements to Image Magnification

A. Biancardi, L. Lombardi and V. Pacaccio

Università di Pavia, DIS, Via Ferrata 1, I-27100 Pavia, Italy

Abstract. The main limitation of current magnifying techniques is that they do not introduce any new information into the original image. This lack of information is responsible for the perceived degradation of the enlarged image. The idea underlying this work is to estimate missing frequencies from the original low resolution image and to synthesize them. Sub-pixel edge estimation and a polynomial interpolation step are the key techniques of the proposed method. Furthermore, a new extension to color images is presented. Results are encouraging even if they suggest that further effort should be spent in improving edge localization accuracy.

1 Introduction

Image definition is the amount of performable detail with a given resolution and it is a key factor in perceived-quality assessment. Preserving definition in image transforms is a demanding task because it sits on the overlapping between two domains: the physical and the perceptual ones.

Interpolation methods, usually employed in image magnification, cause a degradation in the enlarged image due to definition loss. In this paper we developed a new magnifying detail-preserving technique. The input of the problem is a low resolution (LR) image, with its finite information content, and the output is a high resolution (HR) version of the same image. The research goal is to produce a perceived high quality in the high resolution image, i.e. an image that should be perceived at least as nicely as the low resolution one by a human observer.

As edges are high spatial frequency features, they strongly affect the perceived image sharpness and quality. If we assume that the low resolution image is as a subsampled version of the high resolution one, we can estimate where the edges "were" before subsampling and then use this spatial information to reconstruct as exactly as possible the missing information. Therefore, edge estimation must be sub-pixel and the reconstruction of the HR image must keep into account the estimation results. The idea consists in modifying a usual intepolation scheme, e.g. bilinear or bicubic interpolation, in order to prevent interpolation across edges.

2 Sub-Pixel Edge Estimation

Our sub-pixel edge estimation is based on the Allelbach and Wong's work [1]. Problem analysis revealed that estimation accuracy is the most important problem in image magnification because edge localization strongly affects the enlarged image quality. The authors proposed a linear approach which we have found to be error prone. Starting from a center-on-surround-off filtered LR image they linearly interpolate the zero-crossing (ZC) points by a heuristic procedure which depends on the sign geometry of the low resolution pixels; then they quantize the analytical estimated edge to the high resolution grid.

Contrary to Allelbach and Wong, we found that both rotational invariance and localization of the filter strongly affect the edge-map accuracy, and hence we looked for a trade-off between these two properties, developing a 5×5 filter which we called ROT. This filter, built by discretization of a positive constant disk-shaped region with a constant negative surround ring, mimics the well known LOG, which is rotationally invariant, but with more localized response due to its smaller dimension. Of course the non-directional property of the Laplacian is lost in discretization; thus also the ROT filter is not non-directional, but we found that it has an increased sensitivity along diagonals and other angles if compared to the simple rectangular filter proposed in [1].

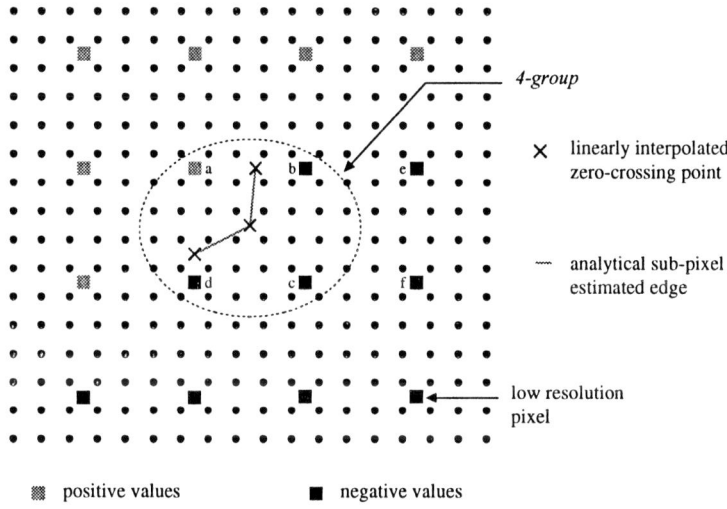

Fig. 1. Linear sub-pixel edge estimation.

Once the low resolution image has been ROT-filtered the ZCs are estimated according to the Allelbach and Wong's idea but using bicubic interpolation instead of the linear one. Referring to figure 1, the linear sub-pixel estimation procedure may be summarized as follows: for each group of 4 LR pixels (*the 4-group*) of the filtered image, according to their sign geometry, sub-pixel ZCs are estimated by linear interpolation between LR pixel pairs a-b, a-c and a-d.

Unfortunately the simple linear approach is not accurate, thus we modified the estimation scheme considering the bicubically interpolated filtered image.

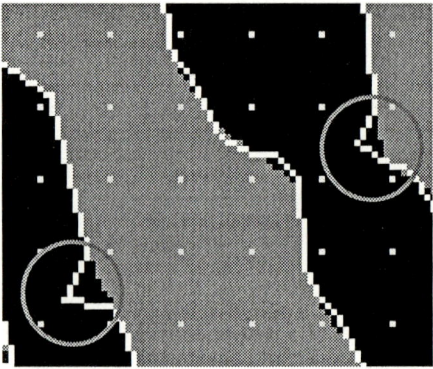

Fig. 2. Comparison between linear estimation approach and bicubically interpolated filtered image.

Edges are still represented as joint segments as well as in Allelbach and Wong's approach, but ZCs are computed by mean of bicubic interpolation which is much more accurate mainly along diagonals. Figure 2 shows the comparison between the linear estimated edge-map (white lines), and the bicubically interpolated ROT-filtered image (regions); the isolated points are the LR pixels. Circles point out that significant differences are prominent in diagonal directions, while horizontal and vertical estimates are substantially good. The figure also shows how, where linear boundary estimations conincide to the bicubic ones, the piecewise linear boundary representation appears somehow smooth, while preserving all the advatages of a representation by 1st order curves.

A great advantage of such a piecewise linear boundary description is its flexibility, which may be exploited in order to solve another important problem in edge-map estimation, i.e. the disagreement between the estimated edge positions and real region boundaries (true edges) in the LR image. Unmatching information leads to undesired resuts in interpolation. Allelbach and Wong's preprocessing step tries to solve this problem modifying the original image according to edge-map estimations. We found that this method creates artifacts in many real images. Furthermore, modifying the original image according to edge estimation results introduces the estimation error inside the original data, so that no correction will be possible. Unfortunately an edge correction scheme is not yet available due to the non-local nature it should have, so, while investigating on such a global corrector, we improved the preprocessing step: we exploited edge-strength information from first derivative edge detection in order to replace only pixels on steep edges.

3 Rendering the Surface

Rendering is the final step in which we reconstruct the high resolution image filling the absent information inside each 4-group by an EDI scheme. The essential feature of this strategy is that the interpolation process is modified to handle interpolation across edges, while, when no edges are found, a simple bilinear interpolation is performed.[1] When there is an edge inside the current 4-group, the rendering procedure attempts to reconstruct the surface according to a slope-controllable edge model.

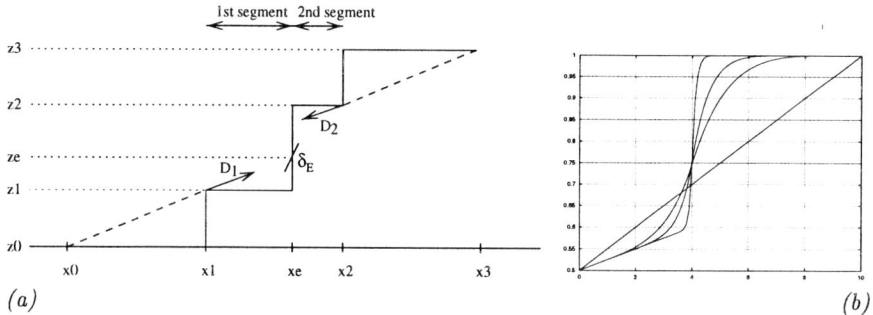

Fig. 3. (a) General 1-D situation of polynomial interpolation; (b) Example plot of the polynomial curve at different values of n.

The figure 3.a describes the typical 1-D situation. x_0, x_1, x_2, x_3 are LR pixel positions and z_0, z_1, z_2, z_3 are their respective gray levels. x_e is the estimated ZC point. Our idea is to fit a polynomial curve that has a controllable slope δ_E across the edge transition, and that must interpolate the points (x_1, z_1) and (x_2, z_2). The point z_e is the HR image value, i.e. the interpolating polynomial value, across the edge, and it represents a free parameter of the algorithm. In order to have some correlation between the reconstructed image in the $[x_1, x_2]$ range and neighbor LR pixels, we choose to drive the interpolating polynomial shape by its first derivatives D_1 and D_2 in the range extrema x_1 and x_2 respectively. These derivatives are computed as $D_1 = z_1 - z_0$ and $D_2 = z_2 - z_3$.

3.1 Polynomial Interpolation

The essential feature that the interpolating polynomial should have is the absence of oscillations whatever values the edge slope may have. Observing that the maximum slope depends on polynomial degree and oscillations depend on intermetiate degree terms, we use the following polynomial curves

$$Z_1(t) = \Delta_1 \left[\hat{D}_1 t + (1 - \hat{D}_1) t^n \right] + z_1, \text{ for } x \in [x_1, x_e] \qquad (1)$$

$$Z_2(s) = \Delta_2 \left[\hat{D}_2 s + (1 - \hat{D}_2) s^n \right] + z_2, \text{ for } x \in [x_e, x_2] \qquad (2)$$

---

[1] Higher degree interpolation is not needed because no high frequency components are present (edges) so that the bilinear interpolation, which is low-pass in its nature, does not cause blurring.

which represent the two joint curve segments in $[x_1, x_2]$, and where

$$x \in [x_1, x_e] \Leftrightarrow x = x1 + t(x_e - x_1) \text{ and } t \in [0, 1]$$
$$x \in [x_e, x_2] \Leftrightarrow x = x2 + s(x_2 - x_e) \text{ and } s \in [0, 1]$$

Let $\Delta_E = z_2 - z_1, \hat{D}_1 = D_1/\Delta_1$, and $\hat{D}_2 = D_2/\Delta_2$; it may be easily proved that interpolation conditions are satisfied. Then we assign C^0 and C^1 continuity on the common point (x_e, z_e) by fixing $\Delta_1 = \Delta_2 + \Delta_E$ for C^0 and $Z_1'(1) = -kZ_2'(1)$ for C^1, where $k = (x_e - x_1)/(x_2 - x_e)$ is a normalization factor. Finally, computing first derivatives and doing substitutions, we get

$$\Delta_2 = \frac{(n-1)(D_1 + kD_2)}{(1+k)n} - \frac{\Delta_E}{1+k} \quad (3)$$

$$\Delta_1 = \Delta_2 + \Delta_E. \quad (4)$$

It should be noted that the continuity implies that the value z_e depends on the other parameters; this is not a problem because z_e is a dummy point and it does not affect the perceived final image quality.

The polynomial degree n must somehow be bounded to the LR edge slope in order to exactly reproduce the HR transition. Roughly speaking the higher is the edge step Δ_E the higher is the probability the edge to be sharp, then the higher should be n. Anyway, we must consider that the effect of varying the exponent depends also on the magnification factor, because the polynomial curve with a low value of n affects a larger number of HR pixels, due to its flatter shape. By now we employ an heuristic relationship to bound n to the edge step height Δ_E and to the magnification factor $x_2 - x_1$, but we plan to gather and exploit some other information about the LR edge slope to improve the reproduction fidelity.

3.2 2-D Extension

Referring to figure 4.a and 4.b we compute the interpolating surface according to an heuristic procedure which ensures the desired slope across the edge and C^0 continuity among patches. Given the the HR pixel position, the interpolating segment through it, which is parallel to the edge, intercepts the border and the diagonal in points p and q respectively, hence the 1-D polynomial curve is evaluated in those points and computed values are used to linearly interpolate the HR pixel value.

4 Mixing Edges and Smooth Areas

A great disadvantage of ZC edge maps is that they do not carry any information about the edge strength, i.e. its slope. Roughly speaking the underlying assumption about the step-like shape of edges we attempt to reconstruct might be not verified where the real edge is more than one pixel wide. On the other hand we found that bicubic interpolation is suitable in such a case. Hence we compute

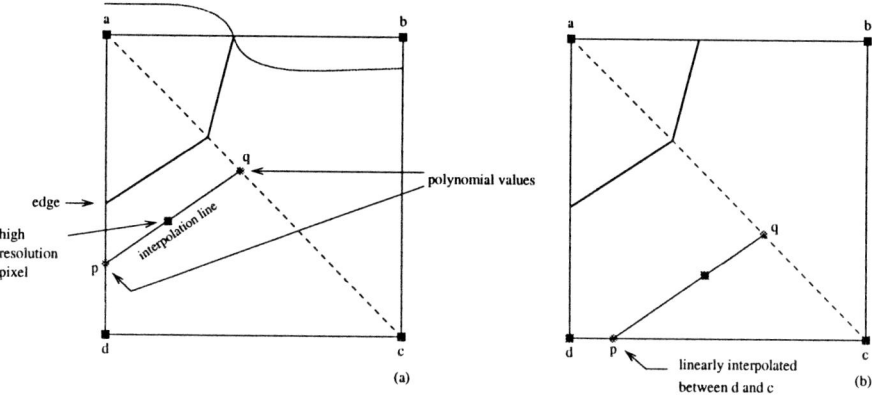

Fig. 4. 2-D extension of polynomial interpolation, example of geometry. A 4-group with its piecewise linear edge inside is shown. (a) Intercepted points (e.g. p and q in figure) are both on a border where a polynomial curve lies. (b) When an intercepted point is on a border where the polynomial curve does not lie (e.g. p in figure), its value is linearly interpolated between LR pixel values (e.g. c and d in figure).

the final result, say $R(i,j)$, by linearly mixing the polynomial synthesized image, say $I(i,j)$, and the bicubically interpolated one, say $B(i,j)$. Weights, between 0 and 1, are choosen to be a HR version of the first derivative $FD(i,j)$ of the LR image:

$$R(i,j) = FD(i,j) \cdot I(i,j) + [1 - FD(i,j)] \cdot B(i,j)$$

foreach i,j in the HR space.

5 Extension to Color Images

There is evidence that the human visual system's (HVS) color coding process is based on three visual channels, one type is independent on the wavelenght, yet the other two have a chromatic sensitivity. Moreover, there seems to be general agreement that spatial resolution is markedly lower in chromatic channels than in the achromatic one, hence high frequency informations, i.e. edges, come mainly from this channel [5-7].

Another important consideration is that, in order to avoid chromatic artifacts in the enlarged image, a non-linear operator cannot be applied to each RGB component separately. The proposed technique is strongly non-linear due to the edge detection procedure.

Hence the edge-map is built from the luminance image according to the HVS's spatial sensitivity, and directs the subsequent interpolation step on each RGB component separately. Artifacts due to color aberrations are negligible. The method can be improved by exploiting color information for a more accurate segmentation process and using perceptually uniform color spaces but which separate chrominance from luminance [6,4].

6 Results

We tried our algorihm with many real images and results show that image definition is well preserved, even if magnification factors are quite large, i.e. 4-8 along each dimension. Figure 5 shows a particular of a gray scale real image: the original image was magnified 8 times along each dimension, 64 times the surface, by bicubic interpolation and by our proposed method. The comparison between the two enlarged images exhibits a noticeable difference: boundaries in our image are quite sharp while a strong blurring effect is present in the bicubically interpolated one.

Fig. 5. (a) Proposed magnifying method; (b) Bicubic interpolation.

7 Conclusions

The problem of recovering the high frequency (HF) contents of a magnified image was tackled at first by reviewing existing proposals and learning about the models of human visual perception. The procedural starting point was chosen to be the work by J. Allelbach and P. W. Wong. Their methodology was studied and verified on some non-synthetic images. Main point of this methodology is the idea of using sub-pixel estimations of edges as the most probable locations with high HF contents. This estimation is carried out by interpolating the result of a center-on-sorround-off filter and then looking for ZCs in the filtered image.

Unfortunately the filter design is very critical because it must be as irrotational as possible and as insensitive to noise as possible. A new version of such a filter is proposed in order to minimize edge displacements caused by the space discretization of digital images. Additionally some pre-processing filters were tried with some improvements on most images.

Secondly, the comparison between a bicubically interpolated filtered image and the linear sub-pixel extimation showed that the linear approach is error prone, leading to artifacts in the magnified image.

The methodology has been extended to color images thanks to a unified (monochrome) edge map; artifacts due to color aberrations are negligible.

A great advantage of our approach is the possibility to further sharpen the magnified image with an edge-enhancing filter [8] processing that is not possible with standard (e.g. bilinear or bicubic) interpolations.

Current results are encouraging (when magnifying each image dimension by 4-8 times, i.e. 16-64 times the surface). Once the remaining sources of artifacts will be removed, this scheme will lead to a crisp and pleasing final result.

Acknowledgments

We wish to thank Dr. Lorenzo Coslovi (Hewlett-Packard Italia) for his kind support.

References

1. J. Allelbach, P.W. Wong, *Edge-Directed Interpolation*, Proc. ICIP-96, IEEE Press, Lausanne CH, 1996, vol. III, pp. 707-710.
2. J.D. Fahnestok, B.R. Hunt, *The Maintenance of Sharpness In Magnified Digital Images*, CVGIP 27, 1984, pp. 32-45.
3. R.G. Keys, *Cubic Convolution Interpolation for Digital Image Processing*, IEEE Trans. ASSP, vol. 29, no. 6, December 1981, pp. 1153-1160.
4. J.D. Foley, A. van Dam, S.K. Feiner, J.F. Hughes, *Computer Graphics, principles and practice*, second edition, Addison-Wesley, 1992.
5. A.B. Watson, *Perceptual-components architecture for digital video*, J. Opt. Soc. Am. A, vol. 7, no. 10, October 1990, pp. 1943-1954.
6. M. Gross, *Visual Computing,* Springer-Verlag, Berlin, 1994.
7. D.J. Granrath, *The role of human visual models in image processing,* Proc. of the IEEE, vol 69, no. 5, May 1981, pp. 552-561.
8. D.E. Knuth *Digital Halftones by Dot Diffusion*, ACM Transactions on Graphics, vol. 6, no. 4, October 1987, pp. 245-273.

Refining Surface Curvature with Relaxation Labeling

Richard C. Wilson and Edwin R. Hancock

Department of Computer Science
University of York, York, Y01 5DD, UK.

Abstract. Our main contributions in this paper are twofold. In the first instance, we demonstrate how $H - K$ surface labelling can be realised using dictionary-based probabilistic relaxation. To facilitate this implementation we have developed a dictionary of feasible surface-label configurations. These configurations observe certain constraints on the contiguity of elliptic and hyperbolic regions, and, on the continuity and thinness of parabolic lines. The second contribution is to develop a statistical model which allows scheme to be initialised using the probabilities of the different $H - K$ labels to be estimated from surface normal information.

1 Introduction

Curvature labels provide a natural way of describing the intrinsic differential structure of surfaces. Mean and Gaussian curvature labels derived from the eigen-structure of the Hessian matrix allow surfaces to be segmented into meaningful structures such as ridges or valleys, saddle points or lines, and, domes or cups. These structures can be further organised into simply connected elliptical or hyperbolic regions which are separated from one-another by parabolic lines. Unfortunately, because the Hessian matrix is based on second-derivatives the reliable estimation of surface curvature has proved to be a task of notorious difficulty in the analysis and range or volumetric imagery [8,9]. Some of the limitations of the alternative strategies for curvature estimation were unearthed in the comparative study of Flynn and Jain [3].

It is for these reasons that strategies aimed at circumventing the direct estimation of second-derivatives have been developed. One of the most popular approaches is to approximate the surface by a low-order piecewise continuous surface [5,2]. For instance, Besl and Jain adopt a hierarchical fitting technique [1]. Firstly, a local tangent plane is extracted by identifying the principal component axes for the surface point distribution over a support neighbourhood, Next, the plane-fit is refined using a cubic patch. Hilton, Illingworth and Windeat [6] have addressed the issue of analysis of variance to improve the statistical fidelity of the fitting process.

Despite these efforts at improving the reliability of curvature estimation the problem of how to refine inconsistent curvature estimates has received less attention. In essence, surface-fitting does not guarantee that the extracted estimates of the Hessian are consistent when viewed from the requirements that elliptic and hyperbolic regions should be simply connected, or, that parabolic lines should be thin and continuous. Interrogation of the literature reveals that it is only Sander and Zucker [10] who have made any serious attempt at exploiting the idea of curvature consistency to improve the recovery of a consistent surface description. Their idea has been to iteratively update Darboux frames by imposing the constraint that the principal curvature directions should vary smoothly across the surface. The initial estimates of the Hessian required in this analysis are derived from the least squares fitting of bi-quadric patches. However, there is no attempt to reconcile the quality of the recovered surface description with the underlying statistical uncertainties in the raw surface data. Neither is there any attempt to exploit the structured nature of the $H - K$ surface-labels in improving curvature consistency.

Our overall aim in this paper is to present a statistical framework for surface curvatur labelling. We commence by showing how the Hessian matrix can be directly estimated using statistics derived from surface normals. Our motivation in embarking on this statistical analysis is to realise the process of consistent curvature-label refinement using probabilistic relaxation labelling. The framework adopted in this study is the dictionary-based relaxation scheme of Hancock and Kittler [4]. The critical ingredient is a dictionary which represents the valid configurations of HK curvature labels that can be consistently assigned to neighbouring sites on the surface. In this way we tap the rich source of constraints provided by the highly structured nature of surface curvature labels.

2 Representing Differential Surface Structure

In this paper we are interested in estimating the local differential structure of surfaces using computed estimates of the surface normal directions. This is to be contrasted with the fitting of a local surface patch and estimating curvature from the computed parameters of the patch We commence by providing some of the formal ingredients of our surface representation. The local surface orientation is determined by the direction of the surface normal $\mathbf{n} = (n_x, n_y, 1)^T$. When the surface is represented by a twice differentiable function $z = f(x, y)$, then the components of the normal are related to the surface gradient, i.e. $\mathbf{n} = (\frac{\partial f}{\partial x}, \frac{\partial f}{\partial y}, 1)^T$. In this continuous case, the differential structure of the surface is captured by the Hessian matrix

$$\mathcal{H} = \begin{pmatrix} \frac{\partial^2 f}{\partial x^2} & \frac{\partial^2 f}{\partial x \partial y} \\ \frac{\partial^2 f}{\partial x \partial y} & \frac{\partial^2 f}{\partial y^2} \end{pmatrix} \tag{1}$$

The eigen-structure of the Hessian matrix can be used to gauge the curvature of the surface. The two eigen-values of $\mathcal{H}$ are the maximum and minimum curvatures. The orthogonal eigen-vectors of $\mathcal{H}$ are known as the principal curvature directions. The mean-curvature of the surface is found by averaging the maximum and minimum curvatures. The Gaussian curvature is equal to the product of the two eigenvalues.

In the case when surface normal information is being used to characterise the surface, then the Hessian matrix takes on the following form

$$\mathcal{H} = \begin{pmatrix} \alpha & \beta \\ \beta & \gamma \end{pmatrix} \qquad (2)$$

The diagonal elements of the Hessian are related to the rate-of change of the surface normal components, i.e., $\alpha = \frac{\partial n_x}{\partial x}$ and $\gamma = \frac{\partial n_x}{\partial x}$. Treatment of the off-diagonal elements is more subtle. However, under the assumption that the surface is locally developable then we can write $\beta = \frac{\partial n_x}{\partial y} = \frac{\partial n_y}{\partial x}$. In the next Section we will describe how the elements of the Hessian, i.e. α, β and γ, can be estimated from raw surface normal data using the method of least-squares.

With estimates of the elements of the Hessian to-hand, we can compute the mean(K) and Gaussian(H) curvatures of the surface. According to the definitions given above $K = \frac{1}{2}(\alpha + \gamma)$ and $H = \alpha\gamma - \beta^2$. The signs and zeros of these two quantities can be used to label the surface according to curvature class. The different classes are defined in Table 1. It is important to stress that there are adjacency constraints applying to the curvature labels. In particular, the the cup (C) and dome (D) surface types may not appear adjacent to each other on a surface. Moreover, elliptic regions on the surface (those for which H is positive) must be separated from hyperbolic regions (those for which H is negative) by a parabolic line (where $H=0$). In other words, domes and cups are enclosed by ridge or valley-lines. Moreover, domes or cups can not be adjacent to saddle-structures. In Section 5 we will exploit these constraints to construct a dictionary for the $H - K$ curvature labels.

Class	Symbol	K	H	Region-type
Dome	D	-	+	Elliptic
Ridge	R	-	0	Parabolic
Saddle ridge	SR	-	-	Hyperbolic
Plane	P	0	0	Hyperbolic
Saddle-point	S	0	-	Hyperbolic
Cup	C	+	+	Elliptic
Valley	V	+	0	Parabolic
Saddle-valley	SV	+	-	Hyperbolic

Table 1. Curvature classes

3 Computing the Hessian using Sampled Normals

In this section we describe how to make a statistical estimate of the Hessian matrix from a sample of surface normals. Specifically, we use the method of least squares to estimate the elements of $\mathcal{H}$ and to compute the errors associated with these estimates.

We commence by assuming that we have a set of surface normal measurements associated with a tentative surface. Moreover, we assume that the variance in the surface normal components is known. For instance, in the case of volumetric intensity images with the surface normals estimated using directional edge-detection operators, then Sharp and Hancock [11] have shown how the variance-covariance matrix for the surface normals is determined by the intensity noise-variance together with the autocorrelations of the filter kernels. Let $\mathbf{n}_0$ represent the surface normal at the position (x_0, y_0, z_0). and let $\mathbf{n}_m$ be a neighbouring surface normal with position (x_m, y_m, z_m). If the normals are close to each other, then we can approximate the change in the components of the surface normal using a first-order Taylor expansion. Accordingly, $\Delta n_x = \frac{\partial n_x}{\partial x}\Delta x + \frac{\partial n_x}{\partial y}\Delta y$ and $\Delta n_y = \frac{\partial n_y}{\partial x}\Delta x + \frac{\partial n_y}{\partial y}\Delta y$, where the measured change in the components of the surface normal is given by $\mathbf{n}_m - \mathbf{n}_0 = (\Delta^m n_x, \Delta^m n_y, 0)^T$. The displacements in point co-ordinates are $\Delta^m x = x_m - x_0$ and $\Delta^m y = y_m - y_0$. With these relationships to hand we can rewrite the Taylor expansion in terms of elements of the Hessian matrix, i.e. $\Delta^m n_x = \alpha \Delta^m x + \beta \Delta^m y$ and $\Delta^m n_y = \beta \Delta^m x + \gamma \Delta^m y$. These equations govern the parallel transport of the vector across the curved geometry of the surface. So, to first-order, the change in the normal is linear in the elements of the Hessian matrix. Unfortunately, for the single neighbouring normal these equations are under-constrained and we can not recover the Hessian. However, if we have a sample of N neighboring surface normals, then there are $2N$ linear homogenous equations in the elements of $\mathcal{H}$ and the problem of recovering differential structure is no-longer under-constrained. We make the homogeneous nature of the equations more explicit by writing

$$\begin{aligned} \Delta n_x^{(m)} &= \Delta x^{(m)} \cdot \alpha + \Delta y^{(m)} \cdot \beta + \quad 0 \cdot \gamma \\ \Delta n_y^{(m)} &= \quad 0 \cdot \alpha + \Delta x^{(m)} \cdot \beta + \Delta y^{(m)} \cdot \gamma \end{aligned} \quad (3)$$

In order to simplify notation, we can write the full system of $2N$ equations in matrix form as $\mathbf{N} = \mathbf{XP}$, where $\mathbf{N}$ is an aggregated column-vector of normal components, i.e. $\mathbf{N} = (\Delta n_x^{(1)}, \Delta n_y^{(1)}, \Delta n_x^{(2)})^T$. The design matrix $\mathbf{X}$ represents the co-ordinate displacements

$$\mathbf{X} = \begin{pmatrix} \Delta x^{(1)} & \Delta y^{(1)} & 0 \\ 0 & \Delta x^{(1)} & \Delta y^{(1)} \\ \Delta x^{(2)} & \Delta y^{(2)} & 0 \\ & \vdots & \end{pmatrix}$$

and the parameter vector $\mathbf{P} = (\alpha, \beta, \gamma)^T$. When the system of equations is overspecified in this way, then we can extract the set of parameters that minimises the vector of error-residuals $\mathbf{N} - \mathbf{XP}$. We pose this parameter recovery process as a least-squares estimation problem. In other words we seek the set of estimated parameters $\hat{\mathbf{P}} = (\hat{\alpha}, \hat{\beta}, \hat{\gamma})^T$ which satisfy the condition $\hat{\mathbf{P}} = \arg\min_\mathbf{P} (\mathbf{N} - \mathbf{XP})^T (\mathbf{N} - \mathbf{XP})$. The solution-vector is given by $\hat{\mathbf{P}} = (\mathbf{X}^T\mathbf{X})^{-1}\mathbf{X}^T\mathbf{N}$. The least-squares estimates of the parameters can then be used to compute the set of surface labels.

4 Labelling the surface

In order exploit the highly structured nature of the surface labelling constraints, we have chosen to employ the technique of dictionary-based probabilistic relaxation. Our reasons for this are twofold. In the first instance, the method draws on labelling constraints using an exhaustive list or compilation of valid neighbourhood label configurations. This list is referred to as a dictionary. The second reason is that the framework is Bayesian and combines evidence for label assignments. Rather than commencing from hard and potentially erroneous label assignments, the initial characterisation is in terms of *a posteriori* label probabilities. These initial probabilities are computed from distribution functions which characterise uncertainties in the unary attributes from which label decisions are to be derived. In other words, dictionary-based relaxation allows us to exploit both consistent surface label structure and the covariance structure of the H-K curvatures.

According to the original formulation of Hancock and Kittler [4], the local labelling is described by a set of probabilities. Specifically, $P^{(n)}(S_i = \omega)$ is the weight of evidence assigned to the label assignment ω at site S_i for the iterative epoch n of the algorithm. These weights are initialised with the probability that label ω takes on one of the eight possibilities from Table 1. Initially, these probabilities are calculated using the computed values of H and K, together with their known covariance structure.

These label-probabilities are iteratively updated using the non-linear relaxation rule

$$P^{(n+1)}(S_i = \omega) = \frac{P^{(n)}(S_i = \omega)Q^{(n)}(S_i = \omega)}{\sum_{\omega' \in \Omega} P^{(n)}(S_i = \omega')Q^{(n)}(S_i = \omega')} \quad (4)$$

The critical ingredient in the update formula is the support function $Q^{(n)}(S_i = \omega)$ which combines evidence from the context-conveying neighbourhood K_i of the surface-site S_i for the label assignment $\omega \in \Omega$. Here $\Omega = \{D, R, SR, S, SV, V, C\}$ denotes the complete set of curvature labels. According to Hancock and Kittler [4], the support function takes on the following product-form

$$Q^{(n)}(S_i = \omega) = \sum_{\Lambda \in \Theta(\omega)} \left\{ \prod_{k \in K_i} \frac{P^{(n)}(S_k = \lambda_k)}{P(S_k = \lambda_k)} \right\} P(S_l = \lambda_l \; \forall l \in K_i) \quad (5)$$

In the above formula the dictionary $\Theta(\omega)$ contains a set of legitimate labellings over this neighbourhood K_i. The dictionary item $\Lambda = \{\lambda_k; k \in K_i\}$ is a configuration of valid curvature labels on the neighbourhood K_i.

5 Dictionary

As mentioned above, there are strong adjacency constraints on the valid label configurations appearing on the neighbourhoods of some classes of surface. One of the goals of the work reported in this paper is to describe a methodology for enumerating and encoding these constraints in a dictionary for $H-K$ labels. Of course, the construction of a dictionary depends critically on the choice of neighbourhood topology. In the experimental section later we will be demonstrating the utility of the method on triangular surface meshes. In this case the natural neighbourhood consists of a triangle and it's three directly adjoining elements. We will therefore confine our attention exclusively to this arrangement of objects in constructing the dictionary.

We commence by considering the dome-class (D) for which $H>0$ and $K>0$. This is an example of an elliptic region. It can therefore be connected to other dome labels. It can not co-occur with any class other than the ridge for which $H=0$ and $K>0$. The configurations satisfying these two constraints are shown in Figure 1. It should be noted that all sections of the dictionary are rotation invariant. The cup-class (C) is symmetric with the dome under reversal of the sign of the mean curvature (K). Under this transformation, the ridge-label is replaced with the valley label. The dictionary for the cup class is therefore constructed by performing the mappings $D \to C$ and $R \to V$ in Figure 1.

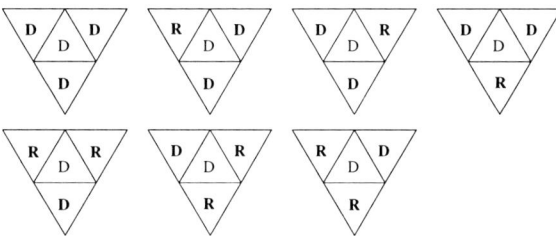

Fig. 1. Elements of the dictionary for the D-label

The two hyperbolic region labels have a more complicated neighbourhood structure. The saddle-valley and the saddle-ridge labels can again form contiguous patches. However, they can be bounded by both the saddle-point and the parabolic line of appropriate mean-curvature. For instance the saddle-ridge can be adjacent to the ridge and the saddle-point. The ridge and valley labels fall into the category of parabolic lines. In other words, they must form the boundaries between hyperbolic regions and elliptical regions. Specifically, they are effectively zero crossings of Gaussian curvature. In consequence the ridge label intercedes between elliptical domes and hyperbolic saddle-ridges. The dictionaries for these hyperbolic and parabolic label classes can be compiled in an analogous manner to Figure 1, but are omitted here due to space limitations.

6 Experimental evaluation

In this section we offer some experimental validation of our surface-labelling algorithm. In order to evaluate the method under controlled conditions, the experiments are conducted using synthetic surfaces. There surfaces are subjected to controlled levels additive Gaussian noise. The surfaces simulate range images. The additive Gaussian noise models sensing errors in the surface height distribution. The raw height data is triangulated using sample points from the surface. These triangles then provide surface normal information.

Figure 2a shows the initial labelling of a damped-cosine surface. The surface labels are shaded according to a convention in which cups, domes, saddle valleys and saddle ridges apear as progressively light regions. Notice that the initial labelling contains no parabolic line-structure. In other words, the abutting boundaries of the different regions are intrinsically inconsistent when viewed from the perspective of the H-K label-set. Figure 2b shows the labelling of the surface after 6 iterations of the dictionary-based relaxation scheme. The green triangles appearing in the updated labelling are parabolic lines (i.e. either ridges or valleys). Although there are sampling artifacts due to the triangular elements used in our labelling scheme, the parabolic lines are thin and continuous. More significantly, they delineate elliptic and hyperbolic regions. Finally, Figure 3 shows the labelling of two noisy damped-cosine surfaces ($\sigma = 0.5$ and $\sigma = 1.0$ respectively).

Fig. 2. Initial (left) and final (right) labellings of damped-cosine surface

7 Conclusions

Our main contributions in this paper are twofold. In the first instance, we have demonstrated how $H - K$ surface labelling can be realised using dictionary-based probabilistic relaxation. To facilitate this implementation we have developed a dictionary of feasible surface-label configurations. These configurations observe certain constraints on the contiguity of elliptic and hyperbolic regions, and, on the continuity and thinness of parabolic lines. The second contribution has been to develop a statistical model which allows curvature lables to be estimated using surface normal statistics.

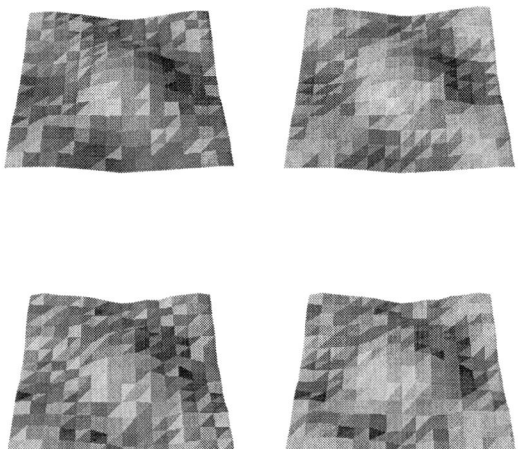

Fig. 3. Initial and final labelling two noisy damped-cosine surfaces

References

1. Besl P.J. and Jain R.C., "Segmentation through variable order surface fitting", IEEE PAMI, PAMI 10, pp 167-192, 1988.
2. Bolle R. M. and Cooper D.B, "Bayesian recognition of of local 3D shape by approximating image intensity functions with quadric polynomials", *IEEE PAMI*, **6**, pp. 418–429, 1984.
3. Flynn P.J., and Jain A.K., "On reliable curvature estimation", IEEE Computer Vision and Pattern Recognition Conference, pp. 110-116, 1989.
4. Hancock E.R. and Kittler J., "Edge labelling using dictionary based relaxation", *IEEE PAMI*, **PAMI 12**, pp.165–181, 1990.
5. Haralick R.M. and Watson L., "A facet model for image data", *Computer Graphics and Image Processing*, **15**, pp. 115–129, 1981.
6. Hilton A., Illingworth J. and Windeatt T., "Statistics of surface curvature estimates", *Pattern Recognition*, **28**, pp. 1201–1221, 1995.
7. Koenderinck J.J. and Van Doorn A., "Surface shape and curvature scales", *Image and Vision Computing*, **10**, pp. 557–564, 1992
8. Monga O., Deriche R., and Malandain G., "Recursive Filtering and edge closing; two primary tools for 3D edge detection", Image and Vision Computing, 9, pp. , 1991.
9. Monga O. and Benayoun S., "Using differential geometry in R4 to extract typical features in 3D density images", 11th International Conference on Pattern Recognition, Volume 1, pp 379-382, 1992.
10. Sander P.T. and Zucker S.W., "Inferring surface structure and differential structure from 3D images", IEEE PAMI, PAMI 12, pp 833-854, 1990.
11. Sharp N.G. and E. R. Hancock, " Feature Tracking by Multi-frame Relaxation", *Image and Vision Computing*, **13**, pp. 637–644, 1995.
12. Wilson R.C. and Hancock E.R., "A minimum variance surface mesh", *submitted*, 1997.

Dynamic Scale-Space Theories

Alfons H. Salden*

INRIA Sophia Antipolis, 2004 route des Lucioles, BP 93,
F-06902 Sophia Antipolis Cedex, France

Abstract. In this paper image formation is quantised by imposing an image induced connection and computing the associated torsion and curvature. Next dynamic scale-space theories are proposed that couple to and smooth the image formation itself.

1 Introduction

A major reason for developing scale-space theories [1] is to find a stable and reproducible image by sieving noise, by detection, conservation, restoration and/or enhancement of topologically interesting objects like edges, corners, and more global entities like volumes, and by data compression and reduction. A requirement imposed on all these theories is that the so-called scale-space operator causes a smoothing of the image that is invariant or slightly affected by certain sets of transformations such as the group of anamorphoses and the group of diffeomorphisms of the image, and transformations caused by noise. Subsequently the image can initiate and control the dynamics and architecture of autonomous artificial systems like robots that, of course, influences the image formation itself. In these contexts we'll restrict ourselves to presenting so-called dynamic scale-space theories [1], in which dynamic refers to the coupling of the paradigm to the image formation process.

Normally one assumes in the scale-space theories that the connection on the image domain is flat. This implies that the image domain or the objects defined on them are neither curved nor twisted in a modern geometric sense. For example, in linear scale-space theory one takes as the base-manifold just Euclidean space. Another example is shortening flow theory for curves and surfaces solved by means of the level set method. Here one encounters notions like curvature and torsion coinciding with a classical geometry living on e.g. the isophotes and flowlines, but only in the context of zero-forms being invariant under classical transformation groups and being factors in the connection coefficients of a flat connection on these objects. The resulting smoothing schemes lead to a family of images in which non-isolated singularity and discontinuity sets in the grey-valued image are destroyed instantaneously.

---

* This work was supported by the Netherlands Organisation of Scientific Research, grant nr. 910-408-09-1, and by the European Communities, H.C.M. grant nr. ERBCHBGCT940511.

Simultaneously anisotropic scale-space theories were developed to retain discontinuity sets of the grey-valued image as much as possible [2]. The latter theories found a solid foundation in [1] where it was pointed out for the first time that Alvarez and Weickert used without realising it modern geometry. Measuring the homogeneity of the image gradient field to steer the diffusion process on the grey-valued image is equally well expressed as measuring the inhomogeneity of the image gradient field to control it.

Introducing an image induced connection [1] characterised by a non-zero torsion and curvature the image formation can be captured in terms of so-called translation and/or rotation vector density fields. Such vector fields one identifies in defect theory and gauge field theories with a so-called Burgers and Frank vector fields, respectively. These vector fields are in particular non-zero at so-called defect or cut lines where the physical properties of media are multi-valued. The reason for these properties being multi-valued lie in the fact that actually different media touch each other or that material has been inserted or removed causing disclinations and dislocations of e.g. the crystal lattice. Such lines happen to be also retrievable in images by the same geometric expertise. They are normally called ridges and ruts in cartography. Having quantised the inhomogeneity in image formation in terms of vector density fields the relaxation of the image or its formation coupled to those fields comes into sight.

The organisation of this paper is as follows. In section 2 we will very briefly treat modern geometry and illustrate its use in quantising the image formation at ridges, ruts and other type of topological objects. Finally, in section 3 the image induced connections are used to formulate dynamic scale-space theories.

2 Modern Geometry of Image Formation

As in the sequel the presentation of the dynamic scale-space theories heavily relies on modern geometry its most important ingredients are summarised. In subsection 2.1 differential geometry and in subsection 2.2 integral geometry are briefly treated. For a more thorough treatment of modern geometry the reader is referred to [1] and the references therein. In subsection 2.3 modern geometry is applied to the problem of finding the essential physical objects in images.

2.1 Differential Geometry

Let M be a D-dimensional image domain parametrised by canonical coordinates $p = (p^1, \ldots p^D)$. Now consider the frame bundle $F \equiv P(M, \pi, A(D, \mathbb{R}))$ where P is the total space consisting of all frames Φ_p at each point $p \in M$, $\pi : P \to M$ is the projection and $A(D, \mathbb{R}) = GL(D, \mathbb{R}) \rhd T(D, \mathbb{R})$ the full affine group, where $Gl(D, \mathbb{R})$ is the general linear group and $T(D, \mathbb{R})$ the translational group. In this context let's define a local frame as follows.

Definition 1. A local frame Φ_p is defined by:

$$\Phi_p = (x; e_1, \ldots, e_D)(p),$$

where the vectors $(x, e_1, \ldots e_D)(p)$ span the local tangent space $T_p A(D, \mathbb{R})$.

Now an affine connection Γ in the frame bundle F is defined as follows.

Definition 2. An affine connection Γ in the frame bundle F is defined in terms of the Lie algebra $\mathcal{G}(D, \mathbb{R})$-valued connection one-forms (ω^i, ω_i^j) and the frame vectors $(x, e_1, \ldots e_D)$ through the following equality:

$$\nabla x = \omega^i e_i, \quad \nabla e_i = \omega_i^j e_j,$$

where ∇ is the covariant differential operator.

The affine connection Γ satisfies so-called structure equations:

Theorem 3. *Given an affine connection Γ in the frame bundle F, defined in (2), then the connection one-forms satisfy the following structure equations:*

$$D\omega^i = d\omega^i + \omega_k^i \wedge \omega^k = \Omega^i, \quad D\omega_j^i = d\omega_j^i + \omega_k^i \wedge \omega_j^k = \Omega_j^i,$$

where d the ordinary exterior derivative, $\wedge$ is the wedge product, D the covariant derivative, Ω^i is the torsion 2-form and Ω_j^i is the curvature 2-form.

In turn the torsion and the curvature 2-form satisfy so-called Bianchi identities:

Theorem 4. *Let Γ be an affine connection in the frame bundle F with torsion 2-form Ω_0^i and curvature 2-form Ω_j^i. The integrability conditions for the structure equations, that are the Bianchi identities, are given by:*

$$D\Omega^i = \Omega_j^i \wedge \omega^j, \quad D\Omega_j^i = 0,$$

2.2 Integral Geometry

Following Cartan [1] one can apply a displacement to determine the translation vector field and the rotation vector fields to operationalise the torsion and the curvature of the frame bundle F with connection Γ.

Definition 5. Let Γ be a connection in the frame bundle F. The translation vector field b and the rotation vector fields f_i determined by the connection are defined by:

$$b = \oint_C \nabla x, \quad f_i = \oint_C \nabla e_i,$$

where C is an infinitesimally small closed loop and boundary of a 2-dimensional submanifold S of M with the same induced connection Γ. The sense of traversing the loop is chosen such that the enclosed submanifold is to the left.

On the basis of the connection one forms ω_0^i a foliation of the manifold (M, Γ) can be realised and choosing $\frac{1}{2}D(D-1)$ pairs of them will yield submanifolds containing the desired submanifold S. These integral invariants are intrinsic vectors of the submanifold (S, Γ) and also of the manifold (M, Γ). Using Stokes' theorem the translation and rotation vector fields can be expressed as [1]:

$$b = \int_S \Omega^i e_i, \quad f_i = \int_S \Omega^j_i e_j.$$

At branching points the translation and rotation vector fields satisfy the following superposition principles (conservation laws for "topological" currents):

$$B = \sum b, \quad F_i = \sum f_i.$$

The latter principles can be conceived as the integral geometric formulations of the Bianchi identities in the previous subsection.

2.3 Application

Let's find the essential physical objects of an image that are invariant under the group of (not necessarily total grey-value preserving) diffeomorphisms of the image caused by active transformations of the scene. The latter active transformations may lead to e.g. anamorphoses of the image or integrable deformations of the net of flowlines and isophotes. It's clear that the set of (non)-isolated singularities and the set of discontinuities of the image remain the same topologically equivalent sets under these transformations. The vanishing of the image gradient is not affected, neither are discontinuities. A set of nonisolated singularities occurs, for example, for images like $L_0^n(x,y) = -\Re\left((x+iy)^{\frac{2n+1}{2}}\right)$, $n \in \mathbb{N}$. It is not so obvious that this invariance also holds for the landscape of ridges and ruts of a smooth image [1]. The latter topological equivalence of non-isolated singularity sets, ridges and ruts can be explained by the fact that across them either the flowlines or the isophotes have opposite convexity. In order to detect subsesequently e.g. ridges and ruts in a two-dimensional grey-valued image a non-local integral geometric difference operation is needed along isohotes with respect to the normalised curvature vector field. At ridges and ruts application of this curvature operator then yields a vectorial Dirac pulse.

Now let us demonstrate that an integral geometric operation suffices to detect certain types of singularity sets. In figure 1 the length of the translation vector field b for a discretised input image L_0 on a two-dimensional Euclidean space E^2 is computed by means of linear scale-space theory [1]. The set of non-isolated singularities will instantaneously disappear upon linear scaling, but the "apparent ridge" and endpoint occurring normally in e.g. fingerprint images can be nicely detected. Note that here we only imposed invariance under the group of Euclidean movements allowing us to consider e.g. the Euclidean differential geometry of the net of isophotes and flowlines of the smoothed images. This geometric and topological expertise enables us to come up with the necessary instructions for the construction of an image.

The proposed non-local topological and integral geometric analysis can be extended to grey-valued images defined on higher dimensional image domains.

For example in the case of three-dimensional grey-valued images normally the unit normal frame field to an isophote will be unique but for some singular ones they will be multi-valued. Again topological operations and slot-machines for reading out torsion or curvature can be used to locate singularity or discontinuity sets of non-constant co-dimension.

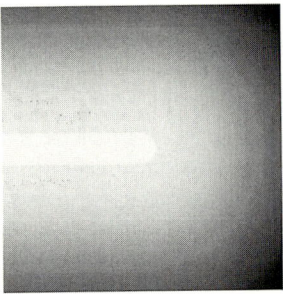 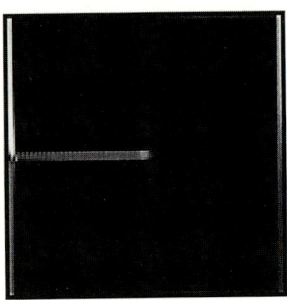

Fig. 1. Left frame: a 256×256 pixel-resolution discrete input image $L_0(x,y) = L_0^0(x,y)$. Right frame: the Euclidean length of the translation vector $|b|$ for a linearly scaled version of that image.

3 Dynamic Scale-Space Theories

In subsection 3.1 dynamic scale-space theories for smoothing the input image are derived by determining the Beltrami-Laplace operator consistent with a particular choice of an image induced connection and metric. In subsection 3.2 dynamic scale-space theories for smoothing the image formation are derived by coupling the smoothing of the image formation to itself.

3.1 Input Images

First let us denote with label ref and ind aspects related to the geometry of the image domain (the camera system) and the induced geometry by the input image, respectively. Assume that dynamic scale-space theories for the input image can be based on the following conservation law on a region μ_{ind} of the image domain M_{ind} with boundary $\partial \mu_{ind}$:

$$\int_{\mu_{ind}} \frac{\partial L}{\partial s} \eta(\mu_{ind}) = \int_{\partial \mu_{ind}} \gamma_{ind}(\nabla_{ind} L, \partial^n \mu_{ind}) \eta(\partial \mu_{ind}),$$

where $\partial^n \mu_{ind}$ is a normal field, ∇_{ind} is the induced covariant derivative constructed on the basis of the induced connection ω_{ind}, γ_{ind} is an induced metric tensor, $\eta(\mu_{ind})$ and $\eta(\partial \mu_{ind})$ are volume measures on the interior μ_{ind}^0 and the boundary $\partial \mu_{ind}$, respectively. Nothe that the metrics need not to be compatible with the connections [1]. Using the divergence theorem and reflective boundary conditions these dynamic scale-spaces are governed by the following Cauchy problem:

$$\frac{\partial L}{\partial s} = \gamma_{ind}^{\alpha\beta} \nabla_{ind,\alpha} \nabla_{ind,\beta} L, \qquad (1)$$

$$\frac{\partial L}{\partial n_{ind}} = 0 \text{ on } \partial \mu_{ind} \times \mathcal{R}_s, \ n_{ind} \in \partial^n \mu_{ind}, \qquad (2)$$

$$L(x,0) = L_0(x), \qquad (3)$$

in which the first equation represents the scaling operation, the second equation a reflective boundary condition ensuring the conservation of total flux and the third equation states the initial condition.

Requiring the dynamic scale-spaces for the input image to be invariant under spatially homogeneous grey-value transformations and applying a similar conservation principle as above for geometries induced by the input image, the divergence theorem and a variational principle the dynamic scale-spaces can be shown to be governed by the following Cauchy problem [1]:

$$\frac{\partial L}{\partial s} = -\gamma_{ref}^{\alpha\beta} \nabla_{ref,\alpha} L \triangle_{ind} x_\beta, \qquad (4)$$

$$\frac{\partial L}{\partial n_{ind}} = 0 \text{ on } \partial \mu_{ind} \times \mathcal{R}_s, \ n_{ind} \in \partial^n \mu_{ind}, \qquad (5)$$

$$L(x,0) = L_0(x), \qquad (6)$$

with $\triangle_{ind}$ the induced Beltrami-Laplace operator given by:

$$\triangle_{ind} = -\gamma_{ind}^{\alpha\beta} \nabla_{ind,\alpha} \nabla_{ind,\beta}. \qquad (7)$$

Example 6. The smoothing of a grey-valued input image on two-dimensional Euclidean space E^2 is normally steered by the image gradient through the conductivity tensor [2]. Alternatively, one could steer the heat capacity on the basis of the image gradient. Choosing a flat connection and a metric equal to

$$\gamma = \exp\left(-\frac{L_i L_i}{\lambda^2}\right) dx^p \otimes dx^p,$$

one obtains, upon substitution into equation (1) and (3), and assuming the image domain not to be bounded, as Cauchy problem for the input image:

$$\frac{\partial L}{\partial s} = \exp\left(-\frac{L_i L_i}{\lambda^2}\right) \triangle_{ref} L, \ L(\cdot,0) = L_0(\cdot),$$

where λ a contrast parameter.

Example 7. Instead of steering the smoothing of a grey-valued input image on two-dimensional Euclidean space directly as in the above example one can also prefer to control the smoothing indirectly choosing the connection and the metric as follows:

$$\omega^1 = \exp\left(\frac{L_w^2}{2\lambda^2}\right) dv, \quad \omega^2 = 0, \quad \omega_j^i = 0, \quad \gamma = \eta_{ref},$$

where v and w are the coordinates with respect to the orthonormal frame field to the isophotes and substituting these choices into equations (4), (6) and (7). The dynamic scale-spaces of the input image are then governed by the following Cauchy problem:

$$\frac{\partial L}{\partial s} = \exp\left(-\frac{L_w^2}{\lambda^2}\right)\frac{\partial^2 L}{\partial v^2}, \quad L(\cdot, 0) = L_0(\cdot),$$

which is just an alternative to the controlled Euclidean shortening flow considered in e.g. [2].

3.2 Image Formation

In anisotropic scale-space theories [2] one is concerned in retaining the discontinuity sets as much as possible under the smoothing. It is demonstrated in [1] that these theories have a nice geometric foundation. Instead of smoothing the *input image* by means of the inhomogeneous group actions as proposed in the first reference in the sequel the smoothing of the *image formation* is presented as proposed in [1]. Because the image formation of a grey-valued input image on a two-dimensional Euclidean space E^2 can be described as an inhomogeneous Euclidean group action induced by the input image it is natural to diffuse the translation vector field and the rotation vector fields (see section 2). As noted in geometric formulations of the anisotropic scale-space theories [1] the smoothing of the input image can be suppressed at discontinuity sets, such as ridges and ruts. Analogously, the smoothing of the image formation can be suppressed at such objects. Leaving them as much as possible untouched the relaxation of the translation vector field b and the rotation vector fields f_i can be defined as follows.

Definition 8. A dynamic scale-space theory for the translation vector field b and the Frank vector fields f_i is governed by the following Cauchy problem:

$$\frac{\partial \psi}{\partial \tau} = \gamma^{ij} \nabla_i \left(O_\psi \psi\right)_j, \quad \text{on } \Omega \times \mathbb{R}^+$$
$$\psi(x, 0) = \psi_0(x), \quad \text{on } \Omega$$
$$\gamma^{ij} \left(O_\psi \psi\right)_j n_j = 0, \quad \text{on } \partial\Omega \times \mathbb{R}^+$$

with ψ the translation vector field or the matrix of rotation vector fields, O_ψ a diffusion operator consistent with ψ, and the induced (not necessarily) metric connection (Γ, γ) such that the components of the metric tensor are given by $\gamma_{ij} = \gamma_{ref}(\hat{\psi}_i, \hat{\psi}_j)$ in which $\hat{\psi}_i$ are so-called canonical fields.

Note that in the case of smoothing of the translation vector field b the metric becomes degenerate. Furthermore, in analogy with the anisotropic scale-space theories [2] the flow $j = -O_\psi \psi$ can be set equal to:

$$j = -d\nabla\psi, \ d = \exp\left(-\alpha_\phi \psi_i \psi_j\right),$$

with α_ϕ a contrast parameter. On the basis of the divergence formulation of these dynamic scale-space theories the evolution of the image formation in terms of the translation vector field and the rotation vector field can readily be computed [2]. Concluding these dynamic scale-space theories enable a multi-scale description of the image formation in terms of scaled measures of torsion and curvature. Note that similar dynamic scale-space theories can be formulated, if the image formation should be invariant under the group of monotonic grey-value transformations.

4 Discussion

Modern geometry is demonstrated to be useful in describing the image formation of a grey-valued input image. Furthermore, it enables us to come up with dynamic scale-space theories in particular for the image formation.

The acquired geometric expertise allows us to conceive an image as a finite CW-complex in which the ridges, ruts and other type of singularity sets are the essential physical objects bordering different image formation processes. Subsequently, the topological aspects of image induced paths on such a CW-complex can be quantified in terms of so-called (generalised) Vassiliev invariants characterising dynamical processes involved in the image formation.

In the context of dynamic scale-space theories it might be ultimately interesting to formulate also theories that are topologically equivalent or, as physicists say, that are covariant. According to this author one then has to turn to theories taking the landscape of ridges and ruts, and the total grey-values in between as a finite CW-complex. The total grey-values attained at the singularities can savely be redistributed according to the valencies and the couplings on the CW-complex. One might conjecture that, depending on the particular paradigm chosen to achieve a task certain types of dynamical processes will survive, whereas others definitively will fade out. The outcomes of further research in this area might have some considerable impact in autonomous system research, cognitive sciences and the field of artificial intelligence.

References

1. Salden, A.H.: *Dynamic Scale-Space Paradigms*. PhD thesis, Utrecht University, The Netherlands, November, 1996, http://www.ceremade.dauphine.fr/ cohen/MSPCV/pgm961128.html.
2. Weickert, J.: *Anisotropic Diffusion in Image Processing*. PhD thesis, Dept. of Mathematics, University of Kaiserslautern, Germany, January, 1996.

Reconstructing Digital Sets from X-rays

E. Barcucci[1], A. Del Lungo[1], M. Nivat[2], R. Pinzani[1], A. Zurli[1]

[1] Dipartimento di Sistemi e Informatica, Via Lombroso 6/17, Firenze, Italy,
pire@ingfi1.ing.unifi.it
[2] LITP Institut Blaise Pascal, Université Paris 7 "Denis Diderot", 2 Place Jussieu, Paris Cedex 05, France, Nivat@litp.ibp.fr

Abstract. In this paper, we study the problem of determining digital sets by means of their X-rays. An X-ray of a digital set F in a direction u counts the number of points in F on each line parallel to u. A class Φ of digital sets is characterized by the set U of directions if among all Φ's elements, each element in Φ is determined by its X-rays in U's directions. This discrete tomography's problem is of primary importance in reconstructing three-dimensional crystals from two-dimensional images taken by an electron microscope by measuring the number of atoms lying on each line in some directions (see [16]). There are some classes of digital sets that satisfy some connection and convexity conditions and that cannot be characterized by any set of directions [2]. Gardner and Gritzmann [10] show that there are some sets of four prescribed directions that characterize the class of totally convex sets (sets which are convex with respect to all the directions). We make the conjecture that there is a certain set U of four directions that characterizes the class of convex polyominoes (sets which are convex with respect to only two directions: horizontal and vertical). In order to give experimental evidence of this conjecture, we present a polynomial algorithm that reconstructs convex polyominoes from their X-rays.

1 Introduction

A *digital set* is a finite subset of the integer lattice $\mathbb{Z}^2$ defined up to a translation; it can be represented by a binary matrix or a set of *cells* (unitary squares), as shown in figure 1. An *X-ray* of a digital set in a direction u is a function giving the number of its points on each line parallel to u.

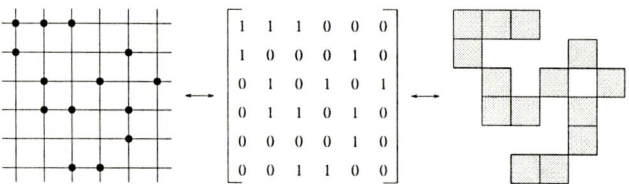

Fig. 1. Correspondence between a digital set, a binary matrix and a set of cells.

The reconstruction of a digital set from X-rays is of primary importance in computer-aided tomography, pattern recognition, image processing and data compression [5, 6, 7, 11, 12, 13, 17]. In detail, it can also be of help in reconstructing images taken by an electron microscope. During the symposium on *Discrete Tomography: Algorithm and Complexity* organized by G. Gritzmann and M. Nivat in Dagstuhl, Peter Schwander (a physicist at the Institute for semiconductor Physics in Frankfurt), had asked discrete tomography's experts for help in reconstructing three-dimensional crystals from two-dimensional projections taken by an electron microscope. He uses a new technique based on high resolution transmission electron microscopy (HRTEM) that can effectively measure the number of atoms lying on each line in some directions (see [16]).

In studying the problem of determining digital sets by X-rays, many authors only use the horizontal and vertical ones [1, 4, 5, 7, 8, 12, 14]. Given a digital set F and its matrix representation $A = (a_{ij})$, the X-rays in the horizontal and vertical directions are defined as $H = (h_1, h_2, ..., h_m)$ and $V = (v_1, v_2, ..., v_n)$ respectively, where:

$$h_i = \sum_{j=1}^{n} a_{ij} \quad 1 \leq i \leq m, \qquad v_j = \sum_{i=1}^{m} a_{ij} \quad 1 \leq j \leq n.$$

In this case, the number h_i is the number of 1 in the ith row, and v_j is the number of 1 in the jth column of A. The main problem met with in the reconstruction from horizontal and vertical X-rays is the "ambiguity" involved because, in some cases, a great many sets have the same pair (H, V). The following two strategies have been adopted to reduce this ambiguity:

- more than two X-rays are assigned (see [6, 15]);
- some properties of the set to be reconstructed are given "a priori" (for example: convexity, connection and symmetry) and the algorithms take advantage of this further information to reconstruct the set (see [1, 5, 8, 12]).

As far as the latter method is concerned, some properties imposed on the set completely eliminate the ambiguity, while some others only partially reduce it. It is shown in [9] that there is an exponential number of convex polyominoes having the same horizontal and vertical X-rays. Therefore, when convexity conditions are imposed on the sets, the ambiguity is usually reduced but not eliminated. We also wish to point out that some convexity conditions do not help us to reconstruct the digital set. For example, the problem of reconstructing a column-convex polyomino (i.e. a digital set which is convex with respect to only vertical direction) from horizontal and vertical X-rays (H, V) is NP-complete [1]. In this paper, we begin by examining the problem of reconstructing a digital set from horizontal and vertical X-rays (H, V) and its extension to the problem of reconstructing a pattern from more than two X-rays' sets. We study the ambiguity problem with respect to some classes of digital sets on which some connection constraints are imposed. In particular, given a class Φ of digital sets, we want to know if a set U of directions exists such that among all Φ's elements, each element in Φ is determined by its X-rays in U's directions. If the set U exists, we say that the class Φ is characterized by U. We then go on to establish the existence of a set of directions that completely eliminates ambiguity on Φ.

By extending the concept of switching component introduced by Chang and Ryser [4, 14], we can prove that there are some classes of digital sets that satisfy some connection and convexity conditions and that cannot be characterized by any set of directions. One of these classes is the set of column-convex polyominoes. We go on to study the ambiguity problem with respect to the class of convex polyominoes (i.e. digital sets which are convex with respect to only horizontal and vertical directions). We theorize that a certain set U of four directions exists which allows us to eliminate ambiguity from convex polyominoes. We made an experimental verification of this conjecture that consists of two steps:

1. the uniform random generation of a convex polyomino F having a specified number of rows and columns and the determination of F's X-rays in the four directions U;
2. the reconstruction of F by means of the four discrete X-rays determined in the previous step.

The second step is performed by means of an algorithm that starts out from the four X-rays obtained in the second step and reconstructs the convex polyominoes having these X-rays in U's directions. This algorithm takes advantage of the convexity properties and is very efficient. It is described in section 3.

2 Preliminaries

A *direction* is an unit vector of the Euclidean bidimensional space $\mathbb{E}^2$. If u is a direction, we denote the line through the origin parallel to u by l_u. A *discrete direction* is a direction $u = (u_x, u_y)$ such that $\frac{u_y}{u_x}$ is a rational number. If F is a digital set and u is a discrete direction, then the *discrete X-ray of F in the direction u* is the function $X_u F$, defined as:

$$X_u F(x) = |F \cap (x + l_u)|$$

for $x = (n, m)$ with $n, m \in \mathbb{Z}$ (see figure 2).

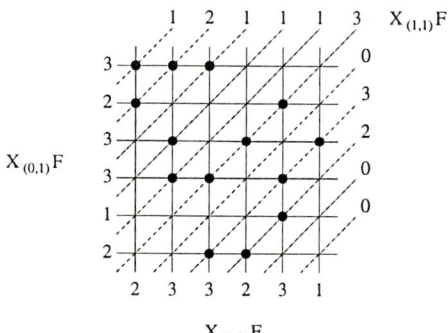

Fig. 2. Example of X-rays in the directions $u_h = (1, 0)$, $u_v = (0, 1)$, $u_1 = (1, 1)$.

Let U be a finite set of discrete directions and Φ be a class of digital sets. We say that $F \in \Phi$ is *determined* by the discrete X-rays in the directions of U if whenever $F' \in \Phi$ and $X_u F = X_u F'$ for all $u \in U$, we have $F = F'$. Otherwise, F is *ambiguous* with respect to U in Φ. In both cases, we say that F satisfies the set of X-rays $X_u F$. We say that the class Φ is *characterized* by the set U of directions if each set in Φ is determined by the discrete X-rays in the directions of U.

We can now define the following problem: " Given a class Φ of digital sets, is there a finite set U of discrete directions characterizing Φ ?"

We approach this problem by defining some classes by means of the correspondence among the digital sets, the binary matrices and the sets of cells whose cardinality is unrestricted. A *polyomino* is a connected finite set of adjacent cells lying two by two along a side, and defined up to a translation (see figure 3).

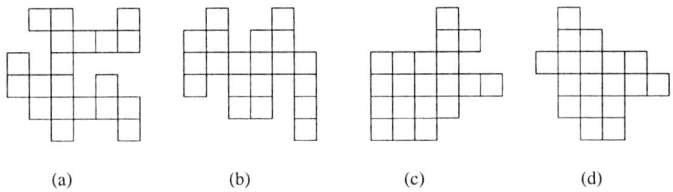

Fig. 3. A polyomino (a) and column-convex (b), row-convex (c), convex polyomino (d).

Let F be a digital set. A *column (row)* of F is the intersection of F with a line $x = n$ ($y = n$), $n \in \mathbb{Z}$. We say that a digital set F verifies the properties **p**, **v** and **h** if:

- **p**: F is a polyomino;
- **h**: every row of F is a connected set;
- **v**: every column of F is a connected set.

A set F belongs to the class (**x**) ($F \in $ (**x**)) if and only if it verifies the property **x**. We denote the set of all the digital sets by ($\emptyset$). We now introduce the following definitions:

- F is a column-convex (row-convex) polyomino if $F \in$ (**p,v**), ($F \in $ (**p,h**)),
- F is a convex polyomino if $F \in$ (**p,h,v**).

We wish to point out that a convex polyomino can be considered to be the intersection between a set $P \in \mathbb{E}^2$, which is convex with respect to vertical and horizontal directions, and the lattice $\mathbb{Z}^2$.

In [2] we showed the following properties:

Theorem 1. *There is not a finite set of discrete directions U that characterizes the following classes: ($\emptyset$), (**p**), (**v**), (**h**), (**p,v**), (**p,h**), (**h,v**).*

Theorem 2. *The class (**p,h,v**) cannot be characterized by any set of three discrete directions.*

Gardner and Gritzmann [10] proved that totally convex sets (i.e., which are convex with respect to all the directions) are characterized by any set U of four discrete directions having *cross ratio* $\rho(U) \notin Q = \{\frac{4}{3}, \frac{3}{2}, 2, 3, 4\}$. Given a set $U = \{u_1, u_2, u_3, u_4\}$ of discrete directions, the cross ratio $\rho(U)$ is defined as $\rho(U) = \frac{(h_3-h_1)(h_4-h_2)}{(h_3-h_2)(h_4-h_1)}$, where $u_i = (u_{ix}, u_{iy})$, $h_i = \frac{u_{iy}}{u_{ix}}$, for $i = 1, 2, 3, 4$ and $h_1 < h_2 < h_3 < h_4$. If $u_i = (0,1)$, then $\rho(U) = \frac{(h_j-h_k)}{(h_j-h_l)}$, where $j, k, l \neq i$. In [2] we showed that convex polyominoes are not characterized by $U = \{u_h, u_v, u_3, u_4\}$ when $\rho(U) \in Q$. Furthermore, for any set of three discrete directions $\{u_h, u_v, u_3\}$ there are 15 directions u_4 such that $\rho(\{u_h, u_v, u_3, u_4\}) \in Q$. These results can be easily extended to any set of directions $\{u_1, u_2, u_3, u_4\}$ and to the class of totally convex sets. On the basis of the previous results we make the following conjecture:

Conjecture 1 *The class* (**p,h,v**) *is characterized by* $U = \{u_h, u_v, u_3, u_4\}$, *where* $\rho(U) \notin Q = \{\frac{4}{3}, \frac{3}{2}, 2, 3, 4\}$.

In order to obtain an experimental verification of our conjecture, we used an algorithm that reconstructs convex polyominoes from their discrete X-rays and considered the directions $U = \{u_h, u_v, (2,1), (-1,2)\}$, which give the minimun size X-rays such that $\rho(U) \notin Q$. Our experiments consist of two steps:

- the uniform random generation of a convex polyomino F having a specified number of rows and columns and the determination of F's X-rays in the four directions $U = \{u_h, u_v, (2,1), (-1,2)\}$;
- the reconstruction of F by means of the four discrete X-rays determined in the previous step.

The first step is carried out by using *Polylab* [3]. We obtain a random generation of four vectors that are consistent on the class (**p,h,v**). The second step is performed by means of an algorithm that starts out from the four X-rays obtained in the first step and reconstructs the convex polyominoes having these X-rays in U's directions. If $U = \{(1,0), (0,1), (2,1), (-1,2)\}$ characterizes (**p,h,v**), then, for each randomly generated convex polyomino F, the algorithm reconstructs F from its X-rays in U's directions. In order to verify our conjecture experimentally, we verify that the algorithm only reconstructs the randomly generated convex polyomino F.

3 Reconstruction algorithm

Here we present a reconstruction algorithm derived from the one in [1]. Let us consider the X-rays along the set of directions $U = \{u_h, u_v, u_3, u_4\}$ with $u_3 = (2,1)$ and $u_4 = (-1,2)$. A convex polyomino Π that satisfies them is contained in a rectangle R of size $n \times m$, where $m = |H|$ and $n = |V|$. We call *kernel* any set α of cells such that $\alpha \subseteq \Pi \subseteq R$, and we call *shell* any set β of cells such that $\Pi \subseteq \beta \subseteq R$. Consequently, the shell contains Π, while the kernel is contained in Π. Assuming that $\alpha = \emptyset$ and $\beta = R$, the basic idea of the algorithm that reconstructs Π by starting out from the empty set $\emptyset$ is to reduce the shell and expand the kernel by means of some filling operations that take advantage of both the convexity constraint and the X-rays along u_h, u_v, u_3, u_4. The shell is reduced by eliminating the cells not belonging to Π from β. Vice versa, the kernel is expanded by putting the cells belonging to Π into α. If Π does not exist, the reconstruction fails, that is, the filling operations produce a kernel α and a shell β such that $\alpha \not\subset \beta$. On the other hand, if only one polyomino Π exists, that satisfies the given X-rays, the reconstruction algorithm stops when $\alpha = \beta = \Pi$. However, if the quadruple $U = \{u_h, u_v, u_3, u_4\}$ does not characterize the class of convex polyominoes, it is likely that the reconstruction algorithm stops when $\alpha \subset \beta$ and $\alpha \neq \beta$, for a certain set of X-rays along U. Consequently, Π is indeterminate. In our tests, this never occurred. We use the following notations to describe the filling operations:

- $c(i,j)$ is the cell of R's in the ith row and jth column;
- R^i is R's ith row (the intersection of R with the strip R $\times [i-1, i]$);
- R_j is R's jth column (the intersection of R with the strip $[j-1, j] \times$ R);
- α^i, β^i are α's ith row ($\alpha^i = \alpha \cap R^i$) and β's ith row ($\beta^i = \beta \cap R^i$) respectively;
- α_j, β_j are kernel α's jth column ($\alpha_j = \alpha \cap R_j$) and shell β's jth column ($\beta_j = \beta \cap R_j$) respectively;
- $R_i^{(2,1)}$, $\alpha_i^{(2,1)}$, $\beta_i^{(2,1)}$ are R's, α's and β's cells, respectively, that lie on the ith parallel to the $(2,1)$ direction;
- $R_i^{(-1,2)}$, $\alpha_i^{(-1,2)}$, $\beta_i^{(-1,2)}$ are R's, α's and β's cells, respectively, that lie on the ith parallel to the $(-1, 2)$ direction;
- h_i and v_j are the ith and the jth components of the X-rays $X_{u_h} F(x)$ and $X_{u_v} F(x)$;
- $p_i^{(2,1)}$ e $p_i^{(-1,2)}$ are the ith component of the X-rays $X_{u_1} F(x)$ and $X_{u_2} F(x)$;
- $l(\alpha^i) = min \{j \in [1..n] : c(i,j) \in \alpha^i\}$, $r(\alpha^i) = max \{j \in [1..n] : c(i,j) \in \alpha^i\}$;
- $d(\alpha_j) = min \{i \in [1..m] : c(i,j) \in \alpha_j\}$, $u(\alpha_j) = max \{i \in [1..m] : c(i,j) \in \alpha_j\}$.

Kernel expansion

- *Connecting operation Γ on row R^i:*
  1. if $\alpha^i = \emptyset$, then $\Gamma(\alpha^i) = \alpha^i$;
  2. if $\alpha^i \neq \emptyset$, then $\Gamma(\alpha^i) = \{c(i,j) : l(\alpha^i) \leq j \leq r(\alpha^i)\}$.
- *Coherence operation Λ on row R^i:*
  1. if β^i is disconnected, then $\Lambda(\alpha^i) = \alpha^i$;
  2. if $\beta^i = \{c(i,j) : j_1 \leq j \leq j_2\}$ (that is, connected) and $\alpha^i = \emptyset$, then $\Lambda(\alpha^i) = \{c(i,j) : j_2 - h_i + 1 \leq j \leq j_1 + h_i - 1\}$;
  3. if $\beta^i = \{c(i,j) : j_1 \leq j \leq j_2\}$ and $\alpha^i \neq \emptyset$, then $\Lambda(\alpha^i) = \alpha^i \cup \{c(i,j) : l(\alpha^i) \leq j \leq j_1 + h_i - 1$ or $j_2 - h_i + 1 \leq j \leq r(\alpha^i)\}$.
- *Coherence operation Λ on cells $R_i^{(p,q)}$ for $(p,q) = (2,1)$, $(p,q) = (-1,2)$:*

  if the number of cells belonging to $\beta_i^{(p,q)}$ is equal to $p_i^{(p,q)}$, then $\Lambda\left(\alpha_i^{(p,q)}\right) = \beta_i^{(p,q)}$.

Shell reduction

- *Connecting operation Φ on row R^i:*
  1. if $\alpha^i = \emptyset$, then $\Phi(\beta^i) = \beta^i$;
  2. if $\alpha^i \neq \emptyset$ and $c(i,k) \notin \beta^i$ and $k < l(\alpha^i)$, then $\Phi(\beta^i) = \beta^i - \{c(i,j) : j \leq k\}$;
  3. if $\alpha^i \neq \emptyset$ and $c(i,k) \notin \beta^i$ and $k > r(\alpha^i)$, then $\Phi(\beta^i) = \beta^i - \{c(i,j) : j \geq k\}$.

- *Coherence operation Θ on row R^i:*
  1. if $\alpha^i = \emptyset$, then $\Theta(\beta^i) = \beta^i$;
  2. if $\alpha^i \neq \emptyset$ then $\Theta(\beta^i) = \beta^i - \{c(i,j) : j \leq r(\alpha^i) - h_i \text{ or } j \geq l(\alpha^i) + h_i\}$.

- *Coherence operation Θ on cells $R_i^{(p,q)}$ for $(p,q) = (2,1)$, $(p,q) = (-1,2)$:*

  if the number of cells belonging to $\alpha_i^{(p,q)}$ is equal to $p_i^{(p,q)}$, then $\Theta\left(\beta_i^{(p,q)}\right) = \alpha_i^{(p,q)}$.

As far as columns are concerned, connecting and coherence operations $\Gamma, \Lambda, \Phi, \Theta$ are defined in the same way as for the rows, by considering $\alpha_j, \beta_j, d(\alpha_j), u(\alpha_j), v_j$ instead of $\alpha^i, \beta^i, l(\alpha^i), r(\alpha^i), h_i$.

Assuming that a convex polyomino Π exists that satisfies the quadruple of X-rays along U, kernel α and shell β are two sets such that: $\alpha \subseteq \Pi \subseteq \beta$. In this situation, if we perform operations Γ, Λ, Φ and Θ on R^i and R_j, we can easily prove that:

$$\alpha^i \subseteq \Gamma(\alpha^i) \subseteq \Pi^i \quad \alpha^i \subseteq \Lambda(\alpha^i) \subseteq \Pi^i \quad \Pi^i \subseteq \Phi(\beta^i) \subseteq \beta^i \quad \Pi^i \subseteq \Theta(\beta^i) \subseteq \beta^i$$
$$\alpha_j \subseteq \Gamma(\alpha_j) \subseteq \Pi_j \quad \alpha_j \subseteq \Lambda(\alpha_j) \subseteq \Pi_j \quad \Pi_j \subseteq \Phi(\beta_j) \subseteq \beta_j \quad \Pi_j \subseteq \Theta(\beta_j) \subseteq \beta_j$$

Consequently, operations Γ and Λ expand α, while Φ and Θ reduce β. It follows that the algorithm that reconstructs Π initially sets $\alpha = \emptyset$, $\beta = R$ and performs the filling operations on R's rows and columns in Γ, Λ, Φ and Θ order and sets $\alpha = \Gamma(\alpha)$, $\beta = \Phi(\beta)$, $\alpha = \Lambda(\alpha)$ and $\beta = \Theta(\beta)$ at each row or column.

4 Conclusions

As stated in section 2, we verify our conjecture experimentally in two steps:

1. the random generation of four vectors that are consistent on (**p,h,v**);
2. the execution of the algorithm performing on the vectors of step (1).

We ran the algorithm up to thousands of cases and the algorithm always reconstructed one convex polyomino (the one randomly generated at step (1)) without any ambiguity. The results of these tests gave us an experimental evidence of both our conjecture and the good performances of the reconstruction algorithm.

References

1. E. Barcucci, A. Del Lungo, M. Nivat and R. Pinzani, Reconstructing convex polyominoes from horizontal and vertical projections, *Theor.Comp. Sci.*155 (1996) 321-347.
2. E. Barcucci, A. Del Lungo, M. Nivat, R. Pinzani, X-rays characterizing some classes of digital pictures, Research report RT 4/96, Dipartimento di Sistemi e Informatica, Università di Firenze (1996).
3. E. Bertoli, A. Del Lungo and F. Ulivi, POLYLAB, A package for the study of polyominoes, *Proceedings of the 7th FPSAC*, eds. B. Leclerc and J. Y. Thibon, Marne-la-Vallé, (1995) 65–74.
4. S. K. Chang, The reconstruction of binary patterns from their projections, *Comm. ACM* 14 (1971) 21-25.
5. S. K. Chang and C. K. Chow, The reconstruction of three-dimensional objects from two orthogonal projections and its application to cardiac cineangiography, *IEEE Trans. Comput.* C-22 (1973) 661-670.
6. S. K. Chang and G. L. Shelthon, Two algorithms for multiple-view binary pattern reconstruction, *IEEE Trans. Syst., Man., Cybern.* SMC-1 (1971) 90-94.
7. S. K. Chang and Y. R. Wang, Three-dimensional objects reconstruction from two orthogonal projections, *Pattern Recognition* 7 (1975) 167-176.
8. A. Del Lungo, Polyominoes defined by two vectors, *Theor. Comp. Sci.* 127 (1994) 187-198.
9. A. Del Lungo, M. Nivat and R. Pinzani, The number of convex polyominoes reconstructible from their orthogonal projections, *Disc. Math* 157 (1996) 65-78.
10. R. J. Gardner and P. Gritzmann, Discrete tomography: determination of finite sets by X-rays, to appear in *Transactions of the AMS*.
11. R. Gordon and G. T. Herman, Reconstruction of pictures from their projections, *Graphics and Image Processing* 14 (1971) 759-768.
12. A. Kuba, The reconstruction of two-directionally connected binary patterns from their two projections, *Comp. Vision, Graphics, and Image Proc.* 27 (1984) 249-265.
13. D. G. W. Omnasch and P. H. Heintzen, A new approch for the reconstruction of the right or left ventricular form from biplane angiocardiographic recordings, *Digital Imaging in Cardiovascular Radiology*, Georg Thieme Verlag, Stuttgart-New York, 1983, 151-163.
14. H. Ryser, *Combinatorial Mathematics*, The Carus Mathematical Monographs No. 14, The Mathematical Association of America, Rahway 1963.
15. A. R. Shliferstein and Y. T. Chien, Switching components and the ambiguity problem in the reconstruction of pictures from their projections, *Pattern Recognition* 10 (1978) 327-340.
16. P. Schwander, C. Kisielowski, M. Seibt, F. H. Baumann, Y. Kim and A. Ourmazd, Mapping projected potential, interfacial roughness, and composition in general crystalline solids by quantitative transmission electron microscopy, *Physical Review Letters* 71 (1993) 4150-4153.
17. C. H. Slump and J. J. Gerbrands, A network flow approach to reconstruction of the left ventricle from two projections, *Comp. Graphics Image Proc.*18 (1982) 18-36.

Pattern Recognition from Compressed Labelled Trees of Fuzzy Regions

Laurent WENDLING[1], Jacky DESACHY[1], and Alain PARIES[2]

[1] Université Paul Sabatier (Toulouse III) IRIT 118, route de Narbonne 31062 Toulouse Cedex, France tel: (+33)561.556.599; fax: (+33)561.556.258
E-mail: {wendling,desachy}@irit.fr
[2] Université Bordeaux I, LaBRI 351, avenue de la Libération 33405 TALENCE CEDEX tel: (+33)556.846.916; fax: (+33)556.846.669
E-mail: Alain.Paries@INSAT.fr

Abstract. In this paper, a method of pattern recognition based on images split into a set of trees composed of fuzzy regions is presented. First, a fuzzy segmentation based on possibilistic c-means is carried out in the raster image. Fuzzy support have been defined from a first level cut. On each cluster, a fuzzy region is assumed to be a convex combination of sets with associated features. A set of sample trees is achieved from the application of the segmentation algorithm on characteristic objects. Then, a tree isomorphism to recognize is defined to recognize an object. At last, a new tree compression method is introduced to decrease the complexity when we have to manage with a large set of trees.

Keywords: Fuzzy Regions, Tree Isomorphism, Directed Acyclic Graph Compression, Pattern Recognition.

1 Introduction

This paper deals with both a new method of pattern recognition based on a tree description of sample images and an optimization approach managing with labelled trees isomorphisms. Currently pattern recognition methods are based on Segmentation – Interpretation cycles which have high processing time costs. The presented approach allows to convert these cycles into a linear application: only one segmentation and one interpretation. First, a fuzzy segmentation based on possibilist c-means is applied on the raster image to obtain fuzzy sets supports. Using fuzzy sets theory, a "convex combination of sets" (also called random sets) is defined, which is composed of a set of crisp regions (defined by successive level-cuts) and a positive weight linked to it. An object is defined by a set of samples (characteristic images) given by an expert. These samples are split with the fuzzy segmentation algorithm into a set of fuzzy regions hierarchical trees. Then, we define the tree isomorphism which links the tree representation of the global image to these sample trees descriptions in order to recognize a particular object. The application complexity increases with the samples set size. So, a new tree compression method is introduced to reduce the processing complexity.

2 Pattern Recognition System

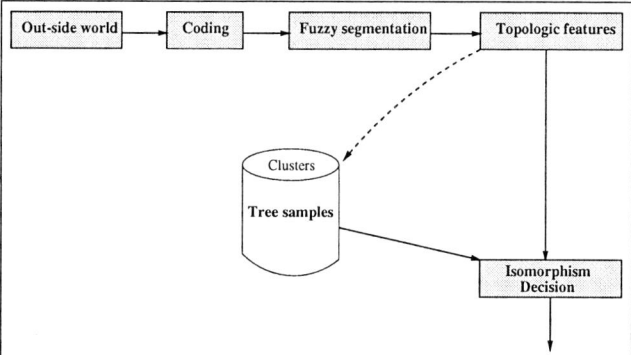

Fig. 1. System Description

The pattern recognition system consists in five parts:
- **Fuzzy Segmentation:** To split the raster image into a set of fuzzy regions.
- **Topological Features:** To Compute features associated to each region.
- **Tree samples:** Each cluster is composed of a set of tree samples.
- **Decision part:** To Perform a tree isomorphism combined with a similarity ratio in order to find an object with a recognition rate.

3 Fuzzy Segmentation

3.1 Fuzzy Partition Definition

The goal of fuzzy segmentation method is to manage with vagueness boundaries [14] [8] and to allow a pixel to belong more ore less to a given region. Most of fuzzy segmentation methods are based on defining fuzzy partitions by using fuzzy c-means algorithms [3]. Nevertheless these approaches give noisy and non totally coherent results [10] [5]. A recent approach proposed by Krishnapuram [9] seems to overcome these problems. His method is independent of the interclass distance and is based on a "good" membership profile [18]. The initialisation of the defined algorithm is fundamental. Barny & al. [2] have shown that the use of c-means algorithm to define the input partition can fail by defining coincident clusters. This problem can be overcome if possibilistic c-means algorithm [12] is used with a number of classes equal to 1. First any partition point is set at 1. Then the most favourable cluster (validity criteria and partitions variations) is carried out. Then points of cluster data which most verify this new achieved cluster are removed from the image. Processing is runned again until the achievement of unconsistent partitions (too small for example) is performed from the image. Currently clusters validity algorithms have generally high processing times with sometimes unconsistent results [10]. In the present system this algorithm has been applied with a level cut criteria to decrease processing time (0.9 s. on a 100 MHz SUN SPARC 4). Then a set of partitions composed of fuzzy clusters is achieved. On each cluster fuzzy regions are defined from their support (first level-cut). The hierarchical structure is defined from support inclusions.

3.2 Convex Combination of Sets

Using fuzzy sets theory [17], a fuzzy set can be defined by a convex combination of sets representation [6]. This combination is composed of n included crisp sets $A_i(A_1 \supset A_2 \supset \ldots \supset A_n)$ with $m(A_i)$ the associated positive weight [15] such as: $\sum_{i=1}^{n} m(A_i) = 1$. Assume n values $\alpha_i \in\ ]0,1]$ and $\alpha_1 < \alpha_2 < \cdots < \alpha_n$. Let $A = \{A_1, A_2, \cdots, A_n\}$ be the fuzzy region where A_i is the crisp set obtained from the level-cut α_i. A level-cut, for the value α_i, is defined as the set of pixels such as membership degree is greater than α_i. For all crisp set A_i, the associated weight is (assuming that $\alpha_0 = 0$): $\forall A_i \subset A, m(A_i) = \alpha_i - \alpha_{i-1}$. The membership function of a fuzzy set is obtained from its convex combination of sets representation. This property allows to manage with fuzzy regions by using their level-cuts.

Fuzzy Parameters. A crisp region A_i is a set of connected pixels with a non-zero membership degrees value. Features of a region such as perimeter P, surface S, compactness degree $\frac{S}{P^2}$, moments of order one and two... can be computed. The valuation of one particular topologic feature is performed separately on each crisp region by using a measurable function F. The final measurable function $\tilde{F}$ of the fuzzy region $A = \{A_1, A_2, \cdots, A_n\}$, characterising the topologic features, is achieved by the following formula [7]:

$$\tilde{F}(A) = \sum_{i=1}^{n} m(A_i) \cdot F(A_i)$$

4 Tree Isomorphism

The Euclidian plane of an image can be assumed to the fuzzy sets referential. The first level cut give the support of each fuzzy region. A tree based on the included notion is defined: i.e. let A and B be two fuzzy regions, s and t their corresponding nodes in the tree. t is a successor of s (denoted by $s \rightarrow t$) if $B \subset A$ and there is no fuzzy region C such as $B \subset C \subset A$. In fact, fuzzy segmentation splits the image into a hierachical tree.

4.1 Definitions

If there exists an arc, in a graph G, between s and t, then t is a **successor** of s in G. We denote $Succ_G(s)$ the set of all successors of s in G. If $t \in Succ_G(s)$, s is the **father** of t, $s \rightarrow_G t$. A graph $G = <V_G, M_G>$ is defined by the set of its vertice V_G and the function M_G which, for all $s \in V_G$, associates $Succ_G(t)$. $deg_G(s)$ is the degree of s, i.e. $|Succ_G(s)|$. A graph T is a tree iff.:

- There exists a unique vertex, called the root, which is never the ending of an arc. The root of T is denoted by $root_T$.
- Each vertex is accessible from the root through a unique path.
- There is no cycle in the underlying undirected graph.

T/s is the sub-tree of T rooted in s and $height_T(s)$ is the length of the longest finite path starting in s.

4.2 Isomorphism Search

Algorithm Description. A depth first search algorithm is applied on the tree image I. Let s be the current node with $deg_I(s) = k$, $Succ_I(s) = \{s_1, s_2, \ldots, s_k\}$. For each node t, let $\{t_1, t_2, \ldots, t_k\}$ be its set of successors. Then, a classical isomorphism is used which stores each couple $\{s, t\}$ performed from a tree searched in a list V. If a tree of O_i is found in the image I tree, a recognition rate is computed. As the number of parameters can be large and as the tree isomorphism has a linear complexity, these two operations can be separated. This method takes into account images composed of objects set at top level. A basic algorithm [1] which allows to define a tree isomorphism in $\mathcal{O}(n)$ time is used and extended to the present application. Two trees T_1 and T_2 are said isomorphic if a tree can be mapped into the other one by permuting the order of the sons of the vertice. The aim of this algorithm is to assign integers to the vertice of the two trees. The processing starts from the leaves and works up towards the roots. Two trees are isomorphic iff. the same integer is assigned to their roots.

Similarity Ratio. A simple mean scheme has been used to carry out the nodes comparison. Other distance measure can obviously be used... Let s be a node of the image tree I and t be a node of the object tree O_i, with p features:

$$C(s,t) = \frac{1}{p} \cdot \sum_{i=1}^{p} S(P_i, P_i') \text{ with: } \begin{cases} S(0,0) = 1 \\ S(P_i, P_i') = \frac{min(P_i, P_i')}{max(P_i, P_i')} \end{cases}$$

A feature can have a greater effect than the others. In this case, a constant λ is introduced. Then, the similarity ratio between nodes becomes:

$$C(s,t) = \frac{1}{\sum_{i=1}^{p} \lambda_i} \cdot \sum_{j=1}^{p} S(P_j, P_j) \cdot \lambda_j$$

And, for all couples of nodes in $V = \{\{s^{N1}, t^{N1}\}, \{s^{N2}, t^{N2}\}, \ldots, \{s^{Nm}, t^{Nm}\}\}$ with $|I/s| = |O_i| = m$ nodes. Assume $\{\mu_1, \ldots, \mu_m\}$ is a set of importance degrees with the condition: $\sum_{i=1}^{m} \mu_i = 1$. Then, an importance degree is assigned to each node as follows:

$$C\_Tree(I/s, t = O_i) = \sum_{k=1}^{m} C(s^{Nk}, t^{Nk}) \cdot \mu_k$$

The final recognizing rate is: $Reco(I/s, t = O_i) = 100 \cdot C\_Tree(I/s, t = O_i)$

If an expert gives a large number of characteristic sample images for a particular cluster, this application may have a high processing time. So a new labelled tree compression method is introduced to decrease the time complexity. Moreover, this compression allows to perform typical objects (most frequent objects or object parts) and a frequency degree can also be given to each edge in order to apply A^* algorithm.

5 Tree Compression

5.1 Labelling

Each node of the tree samples can be assumed to be a $\Re^p$ value with p the number of topologic parameters associated to the node. The set of all the nodes defines a set of points in a $\Re^p$ space. We can suppose that the probability to find two identical $\Re^p$ points is very low, but as the expert gives representation of a typical object, the probability to find two nearest $\Re^p$ points is high. The idea is to compute the clusters defined by these sets of points. An unsupervised method must be used because, at the begining, the number of clusters is unknown. We have also used the algorithm presented subsection 3.1. A label corresponding to a cluster center is assigned to each cluster. A set of labelled trees $\{T_1, T_2, \ldots, T_k\}$ is achieved. At last, this set of trees is grouped into a single tree T such as: $Succ_T(root_T) = \{root_{T_1}, root_{T_2}, \ldots, root_{T_k}\}$ and $Label(root_T) = 0$ [16].

5.2 Directed Acyclic Graph Compression

Let $G = <V_G, M_G>$ be a DAG (i.e. a Directed Acyclic Graph) and $s \in V_G$. $Tree(G, s)$ is the tree of all finite paths starting from s. If $T = <V_T, M_T>$ is a labelled tree, $Label(t)$ is noted as the label of the node $t \in V_T$. As described in [4], a unique DAG can be built: $G = Dag(T)$ such as T and $Tree(G, root_G)$ are isomorphic. In this paper, this approach is extended to labelled tree. An equivalence relation $\approx_L$ is defined such as: $\forall t, s \in V_T$, $t \approx_L s$, iff. T/t and T/s are isomorphic and $Label(t) = Label(s)$. So, for all tree $T = <V_T, M_T>$, $Dag(T)$ is the graph $<V', M'>$ such as:

- $V' = V/\approx_L$
- $M'([s]) = \{[s_1], \ldots, [s_k]\}$ with $[s_i]$ the cluster number of s_i, $Succ_T(s) = \{s_1, \ldots, s_k\}$ and $Label([s_i]) = Label(s_i)$.

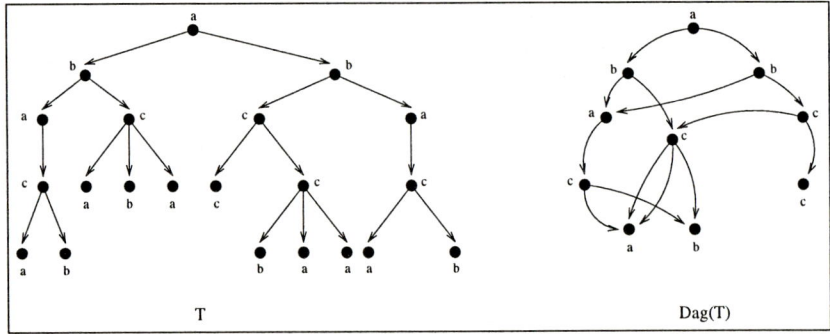

Fig. 2. DAG compression (example)

For all tree $T = <V_T, M_T>$, $Dag(T)$ can be built in $\mathcal{O}(n)$ with $|V_T| = n$. Let $T = <V_T, S_T>$ be a tree. $D = <V_D, S_D>$ is going to be built with $V_D = \{1, 2, \ldots, n\}$, and $p : V_T \to V_D$ such as $p(v)$ is the element of V_D which represents the equivalence cluster of v for $\approx_L$. S_D is such that $S_D(p(v)) = \{p(v_1), p(v_2), \ldots, p(v_k)\}$ with $S_T(v) = \{v_1, v_2, \ldots, v_k\}$. For all $v, v' \in V_T$, $v \approx_L v' \Rightarrow height_T(v) = height_T(v')$. Let $h \in N$, $B_h = \{v/v \in V_T, height_T(v) = h\}$ is defined. For all $v, v' \in B_h$: $v \approx_L v' \leftrightarrow S_D(v) = S_D(v')$ and $Label(v) = Label(v')$. For all $w \in V_T$, the serie $s(w)$ is associated to the integer representing $S_D(p(w))$, in increasing order. On every set B_h, a global order $v \prec v' \Leftrightarrow p(S_T(v)) \leq_{lex} p(S_T(v'))$ is computed. $\leq_{lex}$ is the lexicographical order on the vertex series of T associated wih $\leq$. $v \approx_L v' \Leftrightarrow v \prec v'$, $v' \prec v$ and $Label(v) = Label(v')$. On the set of labels, the global order $<_L$ is defined.

6 Achieved Results and Discussion

6.1 Elliptical Biologic Cells Recognition

• **Elliptical Cells Cluster.** These little images (30 in the present application) are selected on cells images and represent prototypes to find.

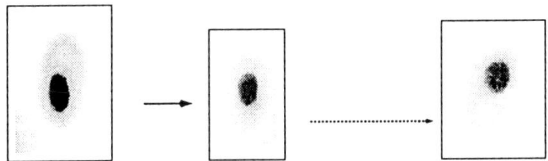

• **Fuzzy Segmentation.** A fuzzy segmentation is carried out from possibilistic c-means algorithm (subsection 3.1). Compactness and ellipticity degree (based on moments form 0 to 2 [13]), features are computed for each fuzzy region. A 0.75 weight is, a priori, attributed to the cytoplasm node, and a 0.25 weight is attributed to the kernel node because the main information (elliptical cell) is given by the cytoplasm.

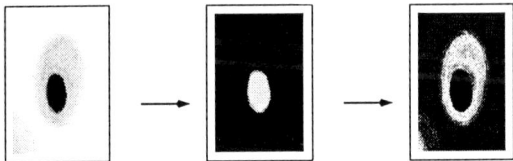

• **Application.** A test image composed of typical cells which are usally found in cells images is defined. Cells which reach at least a 90% recognition rate are recognized (the following images represent the basic image and its fuzzy partition). After applying the isomorphism algorithm, we achieve there is two elliptical cells with a 94.37%, for the best one, recognition rate in the image.

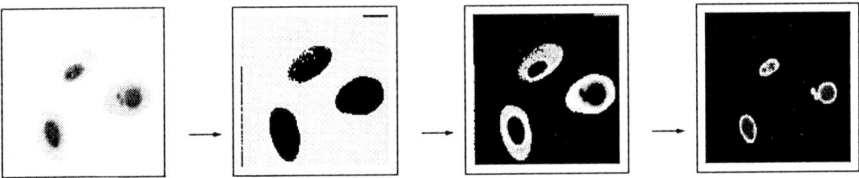

6.2 Second Application of the Method: Mice Detection

- **Mice Cluster.** A set of characteristic mice (catched in the lab) is defined.

- **Fuzzy Segmentation.** A fuzzy segmentation is achieved. For each fuzzy region compactness feature is computed (image and its fuzzy partition).

- **Test Image.**

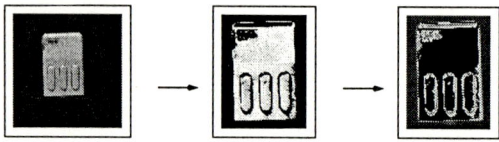

After searching the tree isomorphism, the associated similarity degree is computed. A little default (upper left corner of the test mouse) is found in the test image (maximum recognition rate is 94% for the sample type number one).

6.3 Remarks

A compression rate around 75% is reached for the previous applications. Results have been achieved 3.5 times faster due to the compression method. These examples are too simple to show the power of the proposed compression approach. Currently, studies are in progress with more complex scenes. Nevertheless, 4 sets A_1, A_2, A_3 and A_4 of random trees have been generated using a well known one-to-one correspondance between Dyck language and trees. Each set contains 100 trees of size 50 for A_1, 60 for A_2, 100 for A_3, and 200 for A_4. A comparison of compression rates can be found on the table bellow:

Nb. of labels	1	2	3	...	10
A_1	62%	53%	49%	...	34%
A_2	64%	54%	49%	...	35%
A_3	67%	59%	54%	...	43%
A_4	71%	63%	59%	...	47%

This table does not take into account objects likeness. If an expert gives several objects belonging to the same cluster, the compression rate would be higher. A frequency degree can be defined as follows: $\frac{\text{duplicate edges}}{\text{total number of edges}}$.

7 Conclusion

A new method of pattern recognition based on a tree description of sample images from a fuzzy segmentation method and an optimization method for managing labelled trees isomorphisms (isomorphism and labelled tree compression) has been introduced. This method has been successfully applied on different kinds of application. Currently, we try to generalize this method to more complex scenes with graphs ismorphisms involving relations between nodes like distance between two fuzzy regions, orientation degree between two fuzzy regions...

References

1. A.V. Aho, J. E. Hopcroft and J. D. Ullmann, "The Design and Analysis of Computer Algorithms", book, ed. Addison Wesley, 1974.
2. M. Barni, V. Cappellini and A. Mecocci, *Comments on "A Possibilistic Approach to Clustering"*, IEEE Transactions on Fuzzy Systems, vol 4:3, 1996, pp 393-396.
3. J.C. Bezdek, *Pattern Recognition with Fuzzy Objective Function Algorithms*, Plenum Press, New York, 1981.
4. B. Courcelle and A. Pariès, *"Mineur d'arbres avec racines"*, Informatique Théorique et Applications, 1995.
5. R.N. Dave, *Characterization and Detection of Noise in Clustering*, Pattern Recognition Letters, vol 12, 1991, pp 657-664.
6. D. Dubois and M.C. Jaulent, *"A General Approach to Parameter Evaluation in Fuzzy Digital Pictures"*, Pattern Recognition Letters 4, 1987, pp 251-261.
7. D. Dubois and H. Prade, *"Possibility Theory, an Approach to the Computerized Processing of Uncertainty"*, Plenum Press, New-York, 1988.
8. J.M. Keller and C. L. Carpenter, *"Image Segmentation in the Presence of Uncertainty"*, International Journal of Intelligent Systems, Vol. 5, Iss 2, 1990, pp. 193-208.
9. Raghu Krishnapuram, *"Fuzzy Clustering Methods in Computer Vision"*, EUFIT'93, Aachen, Sep. 7, 1993, pp 720-730.
10. R. Krishnapuram, *"Generation of Membership Functions via Possibilistic Clustering"*, 3rd IEEE Conference on Fuzzy System, vol 3, 1994, pp 902-908.
11. R. Krishnapuram and J.M. Keller, *"A Possibilistic Approach to Clustering"*, IEEE Transactions on Fuzzy Systems, vol 1:2, 1993, pp 98-110.
12. R. Krishnapuram and J.M. Keller, *"The Possibilistic C-Means Algorithm: Insights and Recommendations"*, IEEE Transactions on Fuzzy Systems, vol 4:3, 1996, pp 385-393.
13. Reed Teague, *"Image Analysis via the General Theory of Moments"*, J. Optical Society of America, Vol. 70, N. 8, 1980, pp 920-921.
14. E.H. Ruspini, *"A New Approach to Clustering"*, Inform. and Control, vol 15, 1969, pp 22-32.
15. G. Shafer, *"A Mathematical Theory of Evidence"*, Princetown University Press, 1976.
16. L. Wendling, J. Desachy and A. Paries *"Pattern Recognition by Splitting Images Into Trees of Fuzzy Regions"*, International Journal of Intelligent Data Analysis, Elsevier, 1997.
17. L. A. Zadeh, *"Fuzzy Sets"*, Information and Control 8, pp 338-353, 1965.
18. H. J. Zimmerman and P. Zysno, *"Quantifying Vagueness in Decision Models"*, European J. Operational Res., vol 22, 1985, pp 148-158.

Optimality Analysis of Edge Detection Algorithms for Range Images

Xiaoyi Jiang

Department of Computer Science, University of Bern
Neubrückstrasse 10, CH-3012 Bern, Switzerland
jiang@iam.unibe.ch

Abstract. In this paper we propose a methodology to measure the optimality of edge detection algorithms for range images. Our optimality analysis is based on the potential of an edge detection algorithm to recover intrinsic surface information. We also propose an algorithm for actually performing the recovery. An optimality analysis on four selected edge detection algorithms demonstrate the usefulness of our approach.

1 Introduction

In the past years there has been an increasing interest in both theoretical and experimental evaluation of vision algorithms. In the field of range image analysis, the accuracy of curvature estimates has been studied by some researchers [1, 4, 6]. In [7, 9] techniques for experimental comparison of edge- and region-based range image segmentation algorithms have been proposed.

In the present paper we consider the optimality of edge detection algorithms for range images. An edge detector usually provides edge strengths that are strongly related to the local surface configuration of edge points. Thus, an edge detector generally possesses the ability of recovering intrinsic surface information (to be defined in Section 2) of edge points besides their detection. So far this surface information recovery aspect of edge detection algorithms has been entirely ignored in the literature. We define an optimal edge detector as one that allows a perfect recovery of intrinsic surface information. Measures are suggested to characterize the potential of surface information recovery of edge detection algorithms. Moreover, we propose an algorithm for actually performing this recovery.

An optimality analysis is carried out on four selected edge detection algorithms to demonstrate the usefulness of our approach by analyzing their relative performance. Such an optimality analysis can help us deepen our understanding of known algorithms. In addition, it can potentially be used to guide the design of improved edge detection algorithms. We believe that both aspects are useful to solving the important edge-based range image segmentation problem.

2 Optimal edge detection in range images

In range images we can distinguish between *jump*, *crease*, and *smooth* edges. Jump edges are relatively easy to detect. The detection of smooth edges is still an unsolved problem. In this paper we only consider crease edges.

The general approach to the detection of crease edges can be stated as follows. An edge detector assigns an edge strength to each pixel. Then, a thresholding operation is performed such that pixels with an edge strength higher than a specified threshold T are considered as edge points.

We model the local environment of a crease edge point by two planar surfaces $z = a_1 x + b_1 y + c_1$ and $z = a_2 x + b_2 y + c_2$. Notice that since a small local environment can always be reasonably well approximated by a planar surface patch, this model is useful for curved surfaces as well. Then, the angle between the normals of the two surfaces (shortly angle of normals or AON):

$$\cos^{-1} \frac{(-a_1, -b_1, 1) \cdot (-a_2, -b_2, 1)}{||(-a_1, -b_1, 1)|| \cdot ||(-a_2, -b_2, 1)||} \qquad (1)$$

can be regarded as an ideal edge strength. It is independent of the position and orientation of the scene relative to the range scanner. In addition, it is invariant to changes of the coordinate system. Therefore, the AON represents an intrinsic property of edges. If it is provided by some edge detector as edge strength, the threshold T needed by the thresholding operation is theoretically simply zero. In practice, however, we have to select a small angle value as threshold in order to tolerate low responses caused by noise. In [10] this discussion has motivated the definition of an *optimal* edge detector for range images as one that supplies the AON or a monotonic function of the AON as edge strength.

The usefulness of this definition, however, is very limited. The reason is that the majority of edge detection algorithms provides edge strengths that are not directly comparable to the AON. Moreover, the same AON can be generated by many possibilities of two intersecting planes, resulting usually in different edge strengths. In the general case, thus, an edge detector can actually be considered as a function $f(\alpha)$ that maps a value α of AON as defined in (1) to an interval $[\min_\alpha, \max_\alpha]$. The definition of optimal edge detectors given in [10] must be extended to treat this general case. For this purpose we require that an optimal edge detector fulfill the following two conditions:

- $\max_\alpha < \min_\beta$ for $\alpha < \beta$, i.e., the edge strength response increases with the angle of normals;
- $O_{\alpha,\beta} \equiv [\min_\alpha, \max_\alpha] \cap [\min_\beta, \max_\beta] = \emptyset$ for $\alpha \neq \beta$, i.e., the intervals corresponding to two different values of AON don't overlap.

The earlier definition of optimal edge detectors given in [10] is a special case of this general definition. The edge strength function $f(\alpha) = \alpha \equiv [\alpha, \alpha]$ obviously fulfills the two conditions above.

The following rationale lies behind the general definition of optimal edge detectors above. Although an optimal edge detector maps a value α of AON not to a single edge strength but to an interval, we can still distinguish edge points originating from different angles of normals. As a direct application of this fact, we can simply select $\min_\alpha$ as the threshold T for generating a binary edge map if we want to tolerate responses caused by noise that is below the response of some small angle of normals α. Moreover, we can uniquely map the edge strength of edge points back to AON and therefore recover the intrinsic surface information.

3 Optimality measures

Given an edge detection algorithm that maps a value α of AON to the interval $[\min_\alpha, \max_\alpha]$, we define now measures to express its optimality according to the definition of optimal edge detectors given in the last section. Without loss of generality we consider only a discrete set of AON values, say $G = \{0^\circ, 1^\circ, 2^\circ, \cdots\}$. For an optimal edge detector, we have $O_{\alpha,\beta} = \emptyset$ for $\alpha \neq \beta$. To which extent this condition is fulfilled can be measured by the quantity $|[\min_\alpha, \max_\alpha] \cap [\min_\beta, \max_\beta]|$ where $|[a,b]| = b - a$ means the length of the interval $[a, b]$. Generally, edge detection algorithms provide different value domains of edge strengths. In order to make them directly comparable we actually use the relative measure $|[\min_\alpha, \max_\alpha] \cap [\min_\beta, \max_\beta]|/|[\min_\alpha, \max_\alpha]|$. Considering the combination of all possible values of α and β, we receive the optimality measure

$$M_1 = \sum_{\alpha \in G, \beta \in G, \alpha \neq \beta} \frac{|[\min_\alpha, \max_\alpha] \cap [\min_\beta, \max_\beta]|}{|[\min_\alpha, \max_\alpha]|}.$$

In case of optimal edge detectors this measure takes the value zero.

The optimality measure M_1 has the unwanted effect that the same amount of relative overlap contributes equally to the measure, independent of the difference between α and β that produce the overlap. In general, however, we prefer small differences since otherwise it becomes difficult to distinguish between edge points caused by very different angles of normals. To take this aspect into account we add a penalty term $|\alpha - \beta|$ and receive the second optimality measure

$$M_2 = \sum_{\alpha \in G, \beta \in G, \alpha \neq \beta} \frac{|[\min_\alpha, \max_\alpha] \cap [\min_\beta, \max_\beta]|}{|[\min_\alpha, \max_\alpha]|} \cdot |\alpha - \beta|.$$

The optimality measures M_1 and M_2 are related to the recovery of intrinsic surface information, i.e., the angle of normals of edge points. In case of optimal edge detectors with both M_1 and M_2 being zero, the function $f(\alpha)$ is invertible and thus a perfect recovery is possible. The recovery accuracy decreases with increasing values of M_1 and M_2.

Besides surface information recovery there is another aspect that is relevant to the characterization of edge detection algorithms. Let's consider a non-optimal edge detector, and a small α and large β with $O_{\alpha\beta} \neq \emptyset$. In this case we should choose the threshold T smaller than the lower bound of $O_{\alpha\beta}$ in order to detect all edge points caused by the significant angle of normals β. This way we regard all edge candidates with edge strength in the range $O_{\alpha\beta}$ as true edge points. These points, however, partly correspond to the weak angle of normals α likely produced by noise, thus generating spurious edge points. If we set T to be larger than the upper bound of $O_{\alpha\beta}$, on the other hand, we reduce the detection of spurious edge points at the risk of missing edge points produced by the significant angle of normals β. Generally, an overlap $O_{\alpha\beta} \neq \emptyset$ between a small α and a large β indicates difficulties in the balance of reducing spurious and missing edge points. In some sense the ease of this balance is expressed in the optimality measure M_1. In order to emphasize that only situations are relevant where small values of AON are involved, we add a penalty term $(180 - \min(\alpha, \beta))$ to M_1 and receive the third optimality measure

$$M_3 = \sum_{\alpha \in G, \beta \in G, \alpha \neq \beta} \frac{|[\min_\alpha, \max_\alpha] \cap [\min_\beta, \max_\beta]|}{|[\min_\alpha, \max_\alpha]|} \cdot (180 - \min(\alpha, \beta)).$$

Here 180 corresponds to the maximal possible value (in degrees) of α and β.

The monotony requirement of optimal edge detectors can be roughly measured as follows. For each $\alpha \in G$ we consider the center c_α of the interval $[\min_\alpha, \max_\alpha]$, i.e., $c_\alpha = (\max_\alpha + \min_\alpha)/2$. We count for pairs $(\alpha, \beta), \alpha < \beta$, with $c_\alpha \geq c_\beta$. This way a fourth optimality measure

$$L = \sum_{\alpha \in G, \beta \in G, \alpha < \beta} l_{\alpha\beta}, \quad l_{\alpha\beta} = \begin{cases} 0, c_\alpha < c_\beta \\ 1, c_\alpha \geq c_\beta \end{cases}$$

is defined. A high value of L indicates difficulties in distinguishing between edge points caused by different angles of normals.

In summary we have defined four optimality measures for edge detection algorithms. While M_1 and M_2 are related to the recovery of intrinsic surface information, M_3 is concerned with the balance of spurious and missing edge points. The monotony requirement of optimal edge detectors is measured by L. All the optimality measures have the property that an edge detector with the mapping function $f'(\alpha) = af(\alpha) + b$ is characterized by the same measures as one with the mapping function $f(\alpha)$. Our optimality measures are therefore invariant to linear transformations of the mapping function.

4 Computation of optimality measures

For the computation of the optimality measures of an edge detection algorithm the mapping function $f(\alpha)$ is assumed to be known. This function can be determined by simulation as follows. All possibilities of two intersecting planar surfaces at an edge point are considered. Theoretically, each of the intersecting planes can take an arbitrary orientation as long as both are visible, i.e., their slant angle (with the z-axis) is in the range $[0^o, 90^o)$. For practical reasons the plane orientation is limited such that its slant angle is smaller than 70°, since other orientations are rarely observed in range images.

A plane orientation corresponds to a point on the unit Gaussian sphere. In order to discretely enumerate all possible plane orientations we need a uniform tessellation of the unit sphere. In our simulation we apply a tessellation method based on the well-known geodesic dome constructions [8]. Starting with a regular icosahedron, each of its edges is divided into f equal sections, where f is called the frequency of the geodesic division. This results in f^2 triangles for each face and totally $20f^2$ triangles. Then, this divided icosahedron is projected onto the unit sphere and the centers of all cells define an approximate uniform tessellation of the unit sphere. We have used a division of frequency 8. Those orientations with a slant angle larger than $70°$ are excluded from further consideration.

We consider all combinations of two plane orientations and compute for each such combination the (discrete) angle of normals $\alpha \in G$ and the edge strength supplied by the edge detection algorithm under study. This way we obtain for each value $\alpha \in G$ a set SL_α of possible edge strengths.

Let A_α and S_α be the average and standard deviation of the set SL_α. We model the edge strengths resulting from edges of AON α by a normal distribution. Then, more than 95% of the responses caused by these edges lies in the range $A_\alpha \pm 2S_\alpha$. Therefore, we define $f(\alpha) = [A_\alpha - 2S_\alpha, A_\alpha + 2S_\alpha]$.

5 Recovery of intrinsic surface information

As discussed in Section 2, the angle of normals is an intrinsic property of the surfaces around an edge point. Our optimality measures are mainly concerned with the potential of an edge detection algorithm to recover this intrinsic surface information. In this section we give an algorithm for actually doing the recovery.

If an edge detector is not optimal, there is no unique answer to the question which angle of normals produces a particular edge strength value. In the following we try to find out the most likely answer in a probabilistic sense.

Each AON value $\alpha \in G$ is associated with a normal distribution that is determined by simulation described in the last section and represents the probability of edge strength values being caused by edges of AON α. For a particular edge strength value we compute its probabilities with regard to the normal distributions of all AON values out of G. The AON value that produces the maximal probability is considered as the most likely recovery solution.

Given the recovery algorithm above we are naturally interested in the recovery accuracy of an edge detection algorithm. For this purpose we perform a simulation, considering again all possibilities of two intersecting planar surfaces at an edge point. For each the edge strength provided by the edge detection algorithm is computed. Then, the recovered AON value is determined and the absolute difference to the ground truth is recorded. Assuming that the absolute differences obey a positive normal distribution (the negative side is meaningless in this context), the standard deviation of this distribution gives an overall characterization of the recovery accuracy.

6 Edge detection algorithms under study

The optimality measures and the recovery algorithm described in the previous sections are useful for all edge detection algorithms that provide quantitative edge strengths. In this work we have performed a concrete optimality analysis on four selected edge detection algorithms. A popular class of edge detection algorithms applies step edge detection operators developed in the intensity image domain independently to the three components of the normals of the imaged surfaces and then combine the three results. Typical examples of this class are described, for instance, in [3, 9] based on a morphological edge detection method and the Canny operator. Let (a_1^*, b_1^*, c_1^*) and (a_2^*, b_2^*, c_2^*) be the unit surface normals of the two intersecting planes of an edge point. The edge strength provided by this class of edge detection algorithms is proportional to $\max(a_1^* - a_2^*, b_1^* - b_2^*, c_1^* - c_2^*)$ that is used in our evaluation.

In [10] an edge detection method based on scan line approximation is proposed. Four directional curves (horizontal, vertical, and two diagonals) centered at a pixel on the imaged surfaces are considered. For each direction the angle between the normals of the two sides on the curve is computed. The maximal angle of the four directions gives the final edge strength.

The approach of Al-Hujazi and Sood [2] is based on residual analysis. Again, four directional curves are investigated. Both sides on a directional curve centered at a pixel are represented by a straight line and the difference between their slopes indicates the edgeness for that direction. The overall edge strength is given by the maximal slope difference of the four directions.

In [5] Zernike-moments are used to compute an edge strength that is roughly the difference between the slopes of the two intersecting planes around an edge point.

7 Simulation results

In the following the four algorithms used in our study will be called algorithm A-D in the order as described in the last section. For these algorithms we have computed the optimality measures and the recovery accuracy, see Table 1. In addition, Figure 1 shows the mapping functions $f(\alpha)$. The x-axis represents the angle of normals α. For each α, the interval $[\min_\alpha, \max_\alpha]$ is drawn as a vertical line with the center (average) point marked as well. In order to make a direct comparison possible, the mapping functions have been normalized to an overall edge strength range $[0, 1]$. In all algorithms the measure L is very small, indicating the general monotonic trend of the edge strength. This can be easily verified by the average curves drawn in Figure 1. The measures M_1, M_2 and M_3 suggest an optimality ranking Alg. B > Alg. A > Alg. C > Alg. D. But the recovery accuracy reveals that algorithm B is slightly worse than algorithm A, while the other two algorithms demonstrated a high inaccuracy. It seems that the recovery accuracy is dependent on the shape of the average curve. In general, a linear average curve indicates the potential of an accurate recovery, as exemplified by the first two algorithms in Figure 1. The mapping function of algorithm D is particularly interesting. The average curve is very flat in the first half of AON values. The response intervals overlap even for significantly different AON values, say $10°$ and $70°$. This reveals the general difficulty of this edge detection approach in balancing spurious and missing edges. Overall, our measures are able to characterize the optimality of edge detection algorithms.

	M_1	M_2	M_3	L	Recovery accuracy
Alg. A	0.22	2.92	21.73	91	$8.3°$
Alg. B	0.18	1.82	20.79	5	$8.9°$
Alg. C	0.28	4.30	34.41	19	$14.6°$
Alg. D	0.33	6.04	41.84	26	$17.4°$

Table 1. Optimality measures and recovery accuracy.

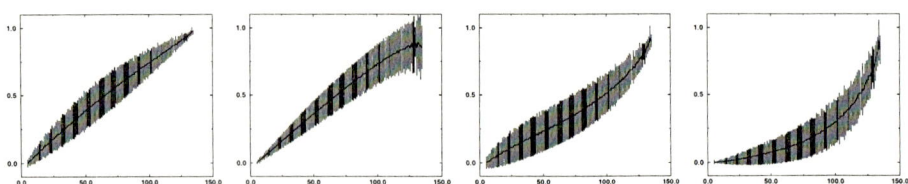

Fig. 1. From left to right the mapping functions of algorithms A-D are drawn.

Now we look at the recovery accuracy in more detail. Figure 2(a) shows the recovery accuracy as a function of AON. The behavior of algorithms C and D are quite similar. For algorithm A the recovery accuracy decreases with AON so that in the area of large AON values (about $\geq 80°$), algorithm B becomes better. The overall recovery accuracy is certainly affected by the occurrence frequency of AON that is drawn in Figure 2(b).

There is another way to look at the recovery accuracy. The plane containing the surface normals of the two intersecting planes of an edge point makes an angle θ with the z-axis. Even for the same angle of normals many edge detectors produce quite different edge strengths dependent on the angle θ. It is thus interesting to see the recovery accuracy as a function of θ. This relationship is shown in Figure 2, together with the occurrence frequency of θ. Again, algorithms C and D demonstrated a similar behavior. Interestingly, algorithm B is superior to algorithm A for θ smaller than about $35°$. But afterwards the accuracy decreases steeply while the recovery accuracy of algorithm A increases slightly.

8 Discussions and conclusion

In the present paper we have proposed a methodology to measure the optimality of edge detection algorithms for range images. Our optimality analysis is based on the potential of an edge detection algorithm to recover intrinsic surface information. We have also proposed an algorithm for actually performing the recovery.

The proposed optimality analysis is useful in two ways. It can help us deepen our understanding of known algorithms. The optimality measures and the detailed analysis of the simulation results enable us to study the performance of recovering intrinsic surface information of edge detection algorithms from different views, as exemplified in this work for four selected edge detection algorithms.

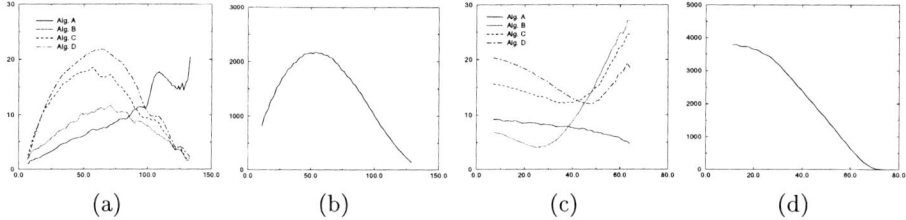

Fig. 2. Recovery accuracy as a function of AON (a) and the occurrence frequency of AON (b). Recovery accuracy as a function of θ (c) and the occurrence frequency of θ (d).

Our optimality analysis can be potentially used to design new algorithms with better performance. As an example, let's consider the algorithm B. In the original work [10] the maximal angle resulting from the four directions has been suggested. However, it turned out that we get an improved recovery accuracy of 6.3^o by using the second largest angle. We believe that both aspects are useful to solving the important edge-based range image segmentation problem, making the proposed approach in the present paper a valuable contribution to the increasing efforts in the characterization of vision algorithms.

References

1. N.N. Abdelmalek, Algebraic error analysis for surface curvature segmentation of 3-D range images, Pattern Recognition, 23(8): 807–817, 1990.
2. E. Al-Hujazi and A. Sood, Range image segmentation with applications to robot bin-picking using vacuum gripper, IEEE Trans. on SMC, 20(6):1313–1325, 1990.
3. P. Boulanger et al., Detection of depth and orientation discontinuities in range images using mathematical morphology, Proc. of 10th Int. Conf. on Pattern Recognition, Atlantic City, 729–732, 1990.
4. P.J. Flynn and A.K. Jain, On reliable curvature estimation, Proc. of Computer Vision and Pattern Recognition, 110–116, 1989.
5. S. Ghosal and R. Mehrotra, Detection of composite edges, IEEE Trans. on Image Processing, 3(1): 14–25, 1994.
6. A. Hilton et al., Statistics of surface curvature estimates, Pattern Recognition, 28(8): 1202–1221, 1995.
7. A. Hoover et al., An experimental comparison of range image segmentation algorithms, IEEE Transactions on PAMI, 18(7): 673–689, 1996.
8. B.K.P. Horn, *Robot Vision*, The MIT Press, 1987.
9. X.Y. Jiang et al., A methodology for evaluating edge detection techniques for range images, Proc. of 2nd Asian Conf. on Computer Vision, Singapore, Vol.II, 415–419, 1995.
10. X.Y. Jiang and H. Bunke, Robust and fast edge detection and description in range images, Proc. of IAPR Workshop on Machine Vision Applications, Tokyo, 538–541, 1996.

Analysis Situs and Image Processing

Fridrich Sloboda and Bedrich Zaťko

Institute of Control Theory and Robotics, Slovak Academy of Sciences, Dúbravska cesta 9, 842 37 Bratislava, Slovakia

Abstract. In the paper a topological approach to approximation of planar Jordan curves and arcs is described. The approximation is based on the basic notions of intrinsic geometry of metric spaces: on the notion of a shortest path in a polygonaly bounded compact set and on the notion of a geodesic diameter of a polygon. Furthermore, a new linear time algorithm for the shortest path problem solution is described, and the approximation of the most important characteristic set in image processing is shown.

1 Introduction

Analysis situs, i.e. topology, is a part of geometry related to investigation of invariants of connected compact sets. In two–dimensional case the boundaries of connected compact sets represent planar curves. The notion of a curve belongs to the hardest problems in the history of mathematics. The modern history is related to C. Jordan. A planar Jordan curve is a simple closed curve, that is a curve which belongs to a parametrized path $\phi : [a, b] \to R^2$ with $a \neq b$, $\phi(a) = \phi(b)$, $\phi(s) \neq \phi(t)$ for all $a \leq s < t < b$. Though the class of planar curves which possess parametric forms is large, not all planar curves can be expressed in this form. Therefore there was an interest to find a more general definition of a curve. The first topological definition of a planar curve was given by G. Cantor : a planar curve is a connected compact set of points, which does not possess internal points [11]. The most general topological definition was introduced by P. Urysohn and K. Menger : a curve is an one-dimensional connected compact set, whereby a connected compact set $S \subset R^n$ is one-dimensional, if $\forall s \in S \; \exists \bar{\delta} = \bar{\delta}(s) > 0 : \forall \delta \leq \bar{\delta}$:

$$M(s, \delta) = \{x \in R^n | \text{dist}(x, s) = \delta\} \cap S \tag{1}$$

does not possess a connected compact component, which consists of more than one point [9, 11, 17]. It has been shown that in the case of planar curves both topological definitions are equivalent [11]. A connected compact set $S \subset R^2$ is a simple closed Urysohn curve if $\forall s \in S \; \exists \bar{\delta} = \bar{\delta}(s) > 0 : \forall \delta \leq \bar{\delta} : M(s, \delta)$ defined by (1) possesses exactly two connected compact components, which consist of one point. A connected compact set $S \subset R^2$ is a simple Urysohn arc, if $\forall s \in S \setminus \{s_1, s_2\} \; \exists \bar{\delta} = \bar{\delta}(s) > 0 : \forall \delta \leq \bar{\delta} : M(s, \delta)$ defined by (1) possesses exactly two connected compact components, which consist of one point, and $\forall s \in \{s_1, s_2\} \in S$

$\exists \bar{\delta} = \bar{\delta}(s) > 0 : \forall \delta \leq \bar{\delta} : M(s,\delta)$ defined by (1) possesses exactly one connected compact component, which consists of one point. Not all planar Cantor (Urysohn) curves possess a parametric form. A planar Cantor (Urysohn) curve is a continuous image of a linear segment iff it is locally connected [9, 11]. The notion of local connectedness was introduced by S. Mazurkiewicz [8] and H. Hahn [5]. A set is locally connected if each point of the set possesses an arbitrary small connected neighborhood. A simple planar Urysohn curve (a simple planar Urysohn arc) is locally connected, so that it is parametrizable [9, 11]. But if a curve is parametrizable it does not imply that the parametric form itself will be ever found. W. Sierpinski has shown an example of a planar curve, so called Sierpinski carpet, which has a unique property: it contains all planar curves. To each planar Cantor (Urysohn) curve C there exists a subset C' of the Sierpinski carpet which is homeomorph with C [11].

The trace of one–dimensional continua which do not possess a parametric form or if this form is not available is "untouchable". Untouchable in the sense, that it is not possible to generate points which lie on the trace of such continua. How to measure the length of untouchable one–dimensional continua? How to represent such continua? These questions are related to the basic problems of set theoretical topology, but they are related also to the basic problems of image processing.

2 Inner and Outer Jordan Content

In order to answer above mentioned questions let us consider a regular grid in R^2. The grid itself represents a theoretical tool of set–theoretical topology and was introduced by C. Jordan and G. Peano by the end of last century in order to define measurable sets. More formally, for $p = 0, 1, 2, ...$ and for each couple (w_1, w_2) of integer numbers let

$$N^p_{(w_1,w_2)} = \{x \in R^2 \mid w_i \, 2^{-p} \leq x_i \leq (w_i + 1) \, 2^{-p}, \ i = 1, 2\}. \tag{2}$$

$N^p_{(w_1,w_2)}$ represents the topological unit of an orthogonal grid.

Let $M \subset R^2$ be a simply connected compact set, and let us define

$$I^+_p(M) = \sum I(N^p_{(w_1,w_2)}), \quad I^-_p(M) = \sum I(N^p_{(w_1,w_2)}), \tag{3}$$

where I denotes the content of (.), and the sum defining I^+_p is taken over all squares for which $N^p_{(w_1,w_2)} \cap M \neq \emptyset$ and I^-_p corresponds to squares for which $N^p_{(w_1,w_2)} \subset M^\circ$. M is measurable if

$$I(M) = \inf_p I^+_p(M) = \sup_p I^-_p(M) = \lim_{p \to \infty} I^\pm_p(M), \tag{4}$$

where $I(M)$ is the Jordan content of M. M is measurable iff the boundary of M, ∂M, has measure zero. A set which has measure zero does not possess internal points, i.e., it does not possess a square $N^p_{(w_1,w_2)}$ internal to this set.

Squares with $N^p_{(w_1,w_2)} \subset M^o$ will be called inner elements of M and squares with $N^p_{(w_1,w_2)} \cap M \neq \emptyset$ boundary elements of M. Let

$$^+M_p = \bigcup_{I^+_p} N^p_{(w_1,w_2)}, \quad ^-M_p = \bigcup_{I^+_p} N^p_{(w_1,w_2)},$$

where $^+M_p, ^-M_p$, corresponds to those $N^p_{(w_1,w_2)}$ which belong to I^+_p, I^-_p, respectively. Note that there exists p such that ^+M_p and ^-M_p are simply connected. Further, $^+M_p(^-M_p)$ will be called *edge connected* if each element of $^+M_p(^-M_p)$ shares a common edge with some other element of $^+M_p(^-M_p)$.

The approximation of a smooth planar Jordan curves and arcs is given by the following [12, 13]

Theorem 1: Let $\gamma : [0, d(\gamma)] \to R^2$ be a smooth Jordan curve with bounded length $d(\gamma)$. Let $G_p = {^+M_p} \setminus {^-M^o_p}$ be edge connected, $p = 0, 1, \ldots$. Let $p_p : [0, d(p_p)] \to R^2$ denote the shortest Jordan curve in G_p containing ^-M_p, $p = 0, 1, \ldots$. Then

$$\lim_k d(p_k) = d(\gamma).$$

Theorem 2: Let $\gamma : [0, d(\gamma)] \to R^2$ be a smooth Jordan arc with bounded length $d(\gamma)$. Let

$$G_p = \cup N^p_{(w_1,w_2)},$$

where G_p corresponds to squares for which $N^p_{(w_1,w_2)} \cap M \neq \emptyset$. Let G_p be edge connected, $p = 0, 1, \ldots$, and let $g_p : [0, d(g_p)] \to R^2$ be a geodesic diameter corresponding to G_p, $p = 0, 1, \ldots$. Then

$$\lim_p d(g_p) = d(\gamma).$$

According to these theorems the length of a planar Jordan curve and the length of a planar Jordan arc is defined on basis of the basic notions of intrinsic geometry of metric spaces [1]: on the notion of the shortest Jordan curve in a polygonally bounded compact set and on the notion of a geodesic diameter in a polygon. The approximating curves are represented by grid points, i.e., by points whose coordinates are integer numbers. The achievements in high technology have allowed to produce high resolution monitors, plotters, scanners and CCD cameras. All of them are built on the basis of an orthogonal grid. The resolution has been increased up to 2 microns at the present linear scanners. The achievements in high technology have set up new demands on algorithms and their complexity. They have created new scientific directions, such as computer aided geometric design, computer graphics and image processing.

3 Implicit Forms and Characteristic Sets

In the previous section theoretical gridding technique was considered. The practical gridding is related to the approximation of planar Jordan curves and arcs in implicit forms [15]:

Theorem 3: Let $D \subset R^2$ be connected and compact, and let $f : D \subset R^2 \to R^1$ be a continuously differentiable function: $\exists x^* \in D^0 : \forall x \neq x^* \in D : f(x) < f(x^*)$, where $f(x^*)$ is the only local maximum of f in D. Let

$$L(z) = \{x \in D \mid f(x) - z = 0\}$$

be a simple planar Urysohn curve. Let $p_p : [0, d(p_p)] \to R^2$ be the noncontractible shortest Jordan curve in

$$G'_p = \cup N^p_{(w_1, w_2)},$$

where $N^p_{(w_1, w_2)}$ are squares, which have the property that not all of their vertices have the same value of sign $[f(x) - z = 0]$, and $\partial G'_p = L_1^{(p)} \cup L_2^{(p)}$, $L_1^{(p)} \subset I(L_2^{(p)})$. Then

$$\lim d(p_p) = d(L),$$

where $d(L)$ is the length of $L(z)$.

Proof. According to the assumption

$$L(z) = \{x \in D \mid f(x) - z = 0\} \tag{5}$$

is a simple planar Urysohn curve. Any simple planar Urysohn curve is locally connected, so that it is parametrizable, and it is homeomorph with the unit circle. Because f is continuously differentiable $L(z)$ is rectifiable.

Let us consider the orthogonal grid defined by (2). Because $f : R^2 \to R^1$ is continuously differentiable, there exists p such that G'_p consists of squares, which have the property that $L(z)$ enters and leaves a square exactly once. According to the definition G'_p consists of squares, which have the property that not all of their vertices have the same value of sign $[f(x) - z = 0]$. But for any grid points resolution there might exist squares, which were entered and left by the trace of $L(z)$ through the same edge. These squares do not belong to G'_p and can not be identified. Let $p_p : [0, d_p] \to R^2$ denote the shortest Jordan curve in G'_p, encircling $L_1^{(p)}$. Because $f : R^2 \to R^1$ is continuously differentiable, there exists p such that the squares which were entered and left by the trace of $L(z)$ through the same edge, share exactly one edge with a square belonging to G'_p. In this situation the shortest polygonal path in G'_p can not pass through such squares, otherwise it can be shortened. According to this the shortest polygonal path in G'_p is identical to the shortest polygonal path in G_p, so that Theorem 1 applies.

A similar theorem holds for length approximation of smooth planar Jordan arcs in implicit forms [15].

Comment: The consequence of Theorem 3 is, that if the inner content elements are given, for the approximation of the boundary it is sufficient to border the inner content elements by elements which share one edge or one vertex with an inner content element, and in this set to find the shortest Jordan curve.

Let $f : D \subset R^2 \to R^1$ be twice continuously differentiable. Surface analysis of f is related to approximation and representation of the following planar curves and arcs in implicit forms

$$L(z) = \{x \in D \mid f(x) - z = 0\}, \tag{6}$$

$$det\, H = 0, \text{ for } (g,g) \neq 0, \tag{7}$$

where H is the Hessian matrix of f and g is the gradient vector of f.

(6) represents equiconstant level sets of f, and enables to visualize f. (7) represents the boundary of convexity (pseudoconvexity), or concavity (pseudoconcavity). Let $y \in R^2$ be a vector such that $\|y\| = 1$. Then $\partial f/\partial y = (g, y)$, where g is the gradient vector of f and $\partial^2 f/\partial y^2 = (Hy, y)$, where H is the Hessian matrix of f [10]. If $y = g^N$, where g^N is the normalized gradient vector, then $(g, y) = \|g\|\|y\| \cos(g,y) = \|g\|$. Suppose that for a given $f : D \subset R^2 \to R^1$ there is $\epsilon > 0$, such that the set

$$L^* = \{x \in D \mid \max_{-\epsilon < \alpha < \epsilon} (g(x + \alpha y), y) = (g(x), y)\}$$

is nonempty. The first directional derivative possesses its maximum where the second directional derivative vanishes, which corresponds to the set

$$L^* = \{x \in D \subset R^2 \mid (Hy, y) = 0\}.$$

Because $\|y\| = 1$, $(Hy, y) = 0$ if and only if $det\, H = 0$. $f : R^2 \to R^1$ is convex on a convex set $D \subset R^2$ if and only if the Hessian matrix is positive semidefinite. Moreover, f is strictly convex on D if H is positive definite [10]. A 2x2 Hessian matrix is positive definite if $\partial^2 f/\partial x_1^2 > 0$ and $det\, H > 0$ and is negative definite if $\partial^2 f/\partial x_1^2 < 0$ and $det\, H > 0$. It means that the boundary of a characteristic set on which f is convex or concave, respectively, is related to the set $L^* = \{x \in D \subset R^2 \mid det\, H = 0\}$. In the case, the boundary of a compact set, where $det\, H = 0$ is not convex, we speak about pseudoconvexity, pseudoconcavity, respectively.

Comment: In image processing, the set where the norm of the gradient of $f : R^2 \to R^1$ is locally maximal, is related to the set where the Laplace operator vanishes [7]. The Laplace operator is the trace of the Hessian matrix of f, and is by no way related to the set where the norm of the gradient vector is locally maximal. The set where the norm of the gradient vector is locally maximal is related to the set where $det\, H = 0$, which is well known in the theory of convexity, and this set is the most important set according to which a segmentation of $f : R^2 \to R^1$ can be performed [4, 10, 18, 19]. In some cases this set represents an equiconstant level set of f.

4 Algorithm

Let $G = P_{L_2} \setminus P_{L_1}^o$, where P_{L_1}, P_{L_2} are polygons, $P_{L_1} \subset P_{L_2}^o$. The vertices of the shortest polygonal path in G have the following property [14]:

Theorem 4: Let P_{L_1}, P_{L_2}, be polygons: $P_{L_1} \subset P_{L_2}^o$, $\partial P_{L_1} = L_1$, $\partial P_{L_2} = L_2$. A vertex of L_1, L_2, is a vertex of the shortest polygonal curve in $G = P_{L_2} \setminus P_{L_1}^o$ encircling L_1 iff there exists a linear segment S passing through this vertex, endpoints of which belong to L_2, L_1, respectively.

The linear segments related to Theorem 4 are called relative support lines, and represent a generalization of the notion of a support line introduced by G. Minkowski. The shortest polygonal path itself has the property [14]:

Theorem 5: Let $p_0, p_1, \ldots p_{n-1}$, $p_0 \equiv p_{n-1}$ be a polygonal curve in $G = P_{L_2} \setminus P_{L_1}^o$, $P_{L_1} \subset P_{L_2}^o$, $\partial P_{L_1} = L_1, \partial P_{L_2} = L_2$, encircling L_1 in the positive sense. Let p_0 be a vertex of the shortest polygonal curve in G. Then $p_0, p_1, \ldots, p_{n-1}$ is a shortest polygonal curve in G encircling L_1 iff $\overrightarrow{p_i p_{i+1 (\text{mod } n)}}$ points on L_2, L_1 if $p_{i+1(\text{mod } n)} \in L_1, p_{i+1(\text{mod } n)} \in L_2$, respectively.

Almost all algorithms for the shortest path problem solution are based on partition of G [3, 6]. The most popular partition of G is the triangulation of G [2]. According to Theorem 4 the pseudomonotone polygon partition of G is related only to extremal vertices of L_1, L_2, respectively, which is the smallest subset of all vertices of G according to which a partition of G can be performed.

The algorithm for the shortest path problem solution in G, related to image processing problems, is shown on Fig. 1 and described as follows [14]:

Algorithm

Step (1): Find all extremal convex vertices of P_{L_1} and all extremal concave vertices of P_{L_2}.

Step (2): Find all x-axis parallel pseudodiagonals of G related to extremal convex vertices of P_{L_1} and extremal concave vertices of P_{L_2}.

Step (3): Eliminate all monotone polygons whose boundaries contain only vertices of L_1, L_2, respectively; and denote the resulting polygonally bounded set by G'.

Step (4): Find a shortest path in each pseudomonotone polygon between two extremal vertices of G'.

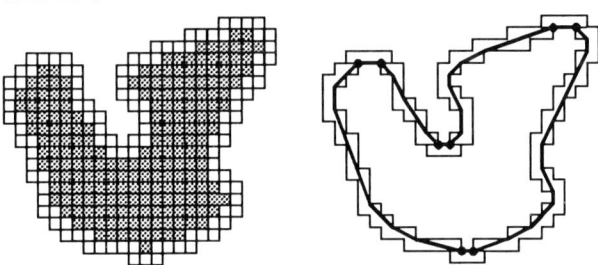

Fig. 1.

Triangulation and pseudomonotone polygon partition of G are based on trapezoidation, which can be performed in linear time [2]. In view of Theorem 5, Step (4) has linear time complexity which implies that Algorithm has also linear time complexity.

Comment: In image processing trapezoidation does not require a special procedure, because it is a part of the connectivity analysis. In this case implicit forms are related to $f : R^2 \to R^1$ which is given by discrete function values.

Geodesic diameter calculation has $O(n \ log \ n)$ time complexity, where n is the number of the polygon [16].

5 Examples

Let us consider the following (see Fig. 2)

Example: $f(x,y) = \exp\left(-\frac{1}{30}(x-2)^2 - \frac{1}{20}(x-2)(y-2) - \frac{1}{20}(y-2)^2\right) -$
$- \exp\left(-\frac{1}{40}x^2 - \frac{1}{20}(y+2)^2\right) + \exp\left(-\frac{1}{20}(x+4)^2 - \frac{1}{20}(y-2)^2\right).$

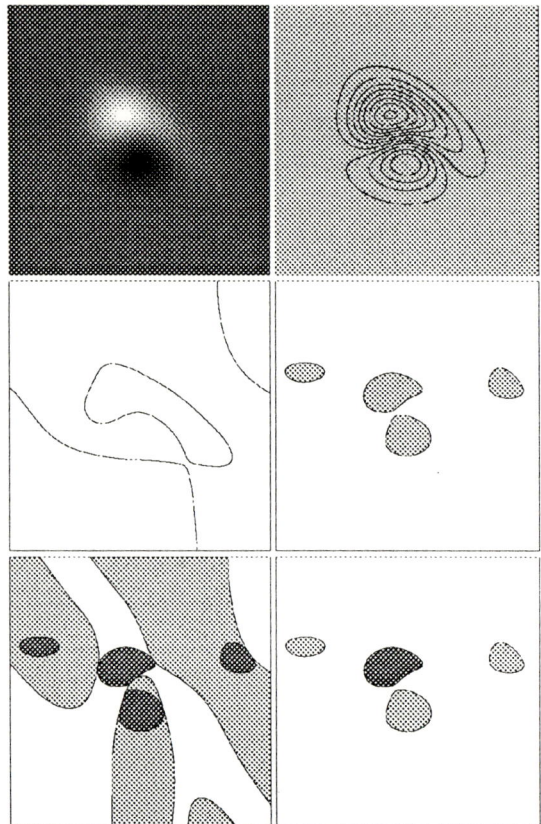

Fig. 2.

Fig. 2 shows in the first rows the function values of f and the corresponding equiconstant level sets of f. In the second rows are shown sets where Laplacian operator vanishes and $\det H \geq 0$, and in the third rows are shown sets where the second partial derivative $\partial f/\partial x^2$ is positive or negative, and sets where f is concave (pseudoconcave) or convex (pseudoconvex).

References

1. A.D. Alexandrov, Intrinsic Geometry of Convex Surfaces, (in Russian) OGIZ, Moscow (1948).
2. B. Chazelle, Triangulating a Simple Polygon in Linear Time, Discrete Comput. Geom. 6 (1991) 485–524.
3. B. Chazelle, A theorem on polygon cutting with applications, Proc. 23-rd Annual Symp. on Found. Comput. Sci. (1982) 339–349.
4. H. Enomoto, T. Katayama, Structure Lines of Images, Proc. of 3rd Int. Joint Conf. On Pattern Recognition, (1976) 811–815.
5. H. Hahn, Über die Komponenten offener Mengen, Fundamenta Math. (1921) 189–192.
6. D. T. Lee, F. P. Preparata, Euclidean Shortest Paths in the Presence of Rectilinear Barriers, Networks 14 (1984) 393–410.
7. D. Marr, E.C. Hildreth, Theory of Edge Detection, Proc. Rog. Soc. London (1980) 187–217.
8. S. Mazurkiewicz, Sur les lignes de Jordan, Fundamenta Math. (1920) 166–209.
9. K. Menger, Kurventheorie, B.G. Teubner Press, Leipzig, (1932).
10. J.M. Ortega, W.C. Rheinboldt, Iterative Solution of Nonlinear Equation in Several Variables, Academic Press, New York, London (1970).
11. A.S. Parchomenko, Was ist eine Kurve, VEB Deutscher Verlag der Wissenschaften, Berlin (1957) (original Moscow (1954)).
12. F. Sloboda, J. Stoer, On Piecewiese Linear Approximation of Planar Jordan Curves, J. Comp. Applied Mathematics 55 (1994) 369–383.
13. F. Sloboda, J. Stoer, On Piecewise Linear Approximation of Planar Jordan Arcs., Tech. Report of the Inst. of Control Theory and Robotics, Bratislava (1996).
14. F. Sloboda, B. Zaťko, On Linear Time Algorithm for the Shortest Path Problem in a Polygonally Bounded Compact Set, Techn. Report of the Inst. of Control Theory and Robotics, Bratislava (1996).
15. F. Sloboda, B. Zaťko, On Approximation of Planar Jordan Curves and Arcs in Implicit Forms, Techn. Report of the Inst. of Control Theory and Robotics, Bratislava (1997).
16. S. Suri, The All Geodesic–Further Neighbours Problem for Simple Polygons, Proc. Third Ann. Symp. on Comp. Geom. (1987) 64–75.
17. P. Urysohn, Mémoire sur les multiplicités Cantoriennes, Fundamenta Math. (1925), 30–130.
18. A. H. Wallace, Differential Topology, N. Y., W. A. Benjamin, (1968).
19. Y. Watanabe, Structural Features of Three–Dimensional Images, Proc. 1st Int. Symp. for Science on Form, KTK Scientific Publ. Tokyo, (1986) 247–254.

Defining Cost Functions and Profitability Measures for Digraphs Associated with Raster DEMs

Pascal Matsakis[*†], Julien Gadiou[*], Jacky Desachy[*]
matsakis@irit.fr

[*]Université Paul Sabatier - IRIT
118, route de Narbonne
31062 Toulouse Cedex, France

[†]CRIL Ingenierie - Groupe Coritec
5, avenue Marcel Dassault
31500 Toulouse, France

Abstract. With a raster Digital Elevation Model, it is usual to associate a directed graph. Firstly, the problem of defining cost functions for such digraphs is discussed in a general and formal framework, and a particularly simple and natural way to tackle this problem is proposed. Secondly, the notion of profitability, which is commonly linked with the notion of cost, is put forward. Thus, profitability measures are introduced. In particular, the profitability of a point according to a region is defined. Finally, it is shown that profitability measures and cost functions provide complementary information.

Keywords. *DEMs, directed graphs, cost functions, problems of optimal paths.*

1 Introduction

With a raster DEM, it is usual to associate a digraph. To each path of this digraph a cost may be attached. Typically, a DEM represents a part of the surface of the earth. A path of the associated digraph then corresponds to a path on the surface. And the cost of the digraph path may correspond to the (euclidean) length of the surface path. It may also correspond to a time or a gasoline consumption. In this paper, only digraphs associated with raster DEMs are considered. Firstly, the problem of defining cost functions for these graphs is discussed in a general and formal framework. A particularly simple and natural way to tackle this problem is proposed in section §2. The presented approach develops from euclidean to discrete geometry. Thus, cost functions defined on paths of the affine euclidean space are introduced. As shown in §4, cost functions of graphs are to space cost functions, as a raster DEM is to the represented surface. Secondly, the notion of profitability is put forward. It is in the habit of saying that such a place is near and of easy access, or near but inaccessible, distant but quite accessible, etc. Distance and accessibility can thereby be complementary criteria for the research of particular spots, for instance within the framework of regional development. Profitability is to cost as accessibility is to distance. Profitability measures on paths of the affine euclidean space are introduced in §3. As shown in §4, they enable to define on a DEM the profitability of a point according to a region. By way of conclusion, §5, experimental results illustrate how profitability measures and cost functions provide complementary information. Note that **N** is the set of natural numbers, **Z** is the one of relative numbers and **R** the real numbers one. a..b, where a and b are relative numbers, is $\{n \in \mathbf{Z} \ / \ a \leq n \leq b\}$ and $\overline{\mathbf{R}}$ is $\mathbf{R} \cup \{-\infty, +\infty\}$.

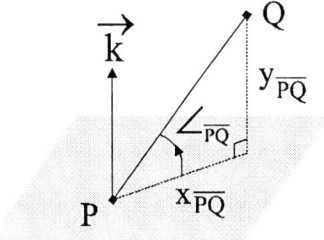

Fig. 1. Linear paths of C.

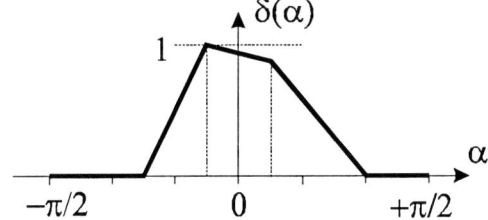

Fig. 2. Generating a hill-climbing energetic function. The hill-climber feels mostly at ease on a 15° slope. He refuses slopes up over 60° and slopes down over 45°.

2 Space Cost Functions

2.1 Terminology and Notations

The *space* is a directed affine euclidean space referred to the direct orthonormal frame $(O, \vec{i}, \vec{j}, \vec{k})$. The *plane* is the affine subspace referred to the direct frame $(O, \vec{i}, \vec{j})$. In this section §2, a *path* is an oriented geometrical arc of space. The *length* of a path is its euclidean length. A *linear path* is a path of space without double point and whose support is a segment. $\overline{P_1P_2}$, where P_1 and P_2 are two points of space, denotes the linear path joining P_1 to P_2. Let n be an integer such that n≥3, $(P_i)_{i \in 1..n}$ a sequence of n points of space. $\overline{P_i}^{\,i \in 1..n}$ denotes the path obtained by juxtaposition of the arcs $\overline{P_i}^{\,i \in 1..n-1}$ and $\overline{P_{n-1}P_n}$. Its length is $\Sigma_{i \in 1..n-1}\ P_iP_{i+1}$. Now consider the set of the paths which benefit by the recurrent notation that has been introduced. C will represent the part of this set including following paths only: $\overline{P_i}^{\,i \in 1..n}$ such that for any element i of 1..n−1 the orthogonal projections of P_i and P_{i+1} onto the plane are distinct. From now on, the term *path* will be applied to the elements of C only. Let $\overline{PQ}$ be a path of C (more precisely, "let P and Q be two points of space such that: $\overline{PQ} \in C$"). Let $\vec{u}$ be the orthogonal projection of $\overrightarrow{PQ}$ onto the plane $(\vec{i}, \vec{j})$ and let $\vec{v}$ be the unit vector $\vec{u}/|\vec{u}|$. Defined relatively to the direct vector plane $(\vec{v}, \vec{k})$: $\angle\overline{PQ}$ is the measure of the oriented angle between $\vec{v}$ and $\overrightarrow{PQ}$, $x\overline{PQ}$ and $y\overline{PQ}$ are the coordinates of $\overrightarrow{PQ}$ (Fig.1). $\angle\overline{PQ} \in\ ]-\pi/2, \pi/2[$ and $x\overline{PQ} \in \mathbf{R}^*_+$ and $y\overline{PQ}/x\overline{PQ} = \tan(\angle\overline{PQ})$. $\angle\overline{PQ}$ measures the angle between the $\overline{PQ}$ and the plane. It is the *slope* of the path.

2.2 Slope-Dependent Cost Functions

Definition 1. A *space cost function* is a map C from C into $\bar{\mathbf{R}}^*_+$ satisfying [A1]:
[A1] Let $\overline{P_i}^{\,i \in 1..n}$ be a path of C: $C(\overline{P_i}^{\,i \in 1..n}) = \Sigma_{i \in 1..n-1}\ C(\overline{P_iP_{i+1}})$.
Given μ a path of C. $C(\mu)$ is the *cost* of μ. If $C(\mu)$ is finite, μ is *potentially profitable*.

Any path of C hence admits a cost (C is a map) and may absolutely not be profitable (a cost may be infinite). Moreover, travelling is always costly (a cost is a strictly positive value). Note that certain problems of optimal paths call for replacing the sum calculation in [A1] by the calculation of the minimum [Gon84] or of the average [Ahu93], or even of the product [Pri94]. And the matter may be then to research max-cost paths and no more min-cost paths. The previous definition is however adapted to the majority of the practically encountered problems.

Definition 2. A space cost function C is said to be *slope-dependent* iff it satisfies [A2]:
[A2] Let $\overline{PQ}$ and $\overline{RS}$ be two paths of C: [PQ=RS and $\angle\overline{PQ}=\angle\overline{RS}$] $\Rightarrow$ C($\overline{PQ}$)=C($\overline{RS}$).

The slope and length of a linear path then determine its cost. In other words, the exact localization of the path in space is not important: space is considered *homogeneous*. In practice, it is far from being always the case. For instance, a 4×4 vehicle can travel on road or on uneven ground, on dry or sodden soil, through a thick or scattered vegetation, along or against the wind direction, etc. ([Mit91], [Zha93], [Kre94], [Dub95]...). Slope-dependent cost functions are then not adapted to cost modelization (supposing that the available data do not only consist in topographical ones!). However that may be, these are fundamental functions because the cost function associated with a non-homogeneous space can be defined from a parametrized family of slope-dependent functions (one corresponding for instance to the travel on tarred road, the other on stony path, etc.). Moreover, the profitability measures (see §3) need to be based on such cost functions, representing ideal spaces. Proposition 1 expresses that the cost of a linear path of given slope is proportional to the length of this path.

Proposition 1. Let C be a slope-dependent cost function, $\overline{PQ}$ and $\overline{RS}$ two paths of C and k a strictly positive real number: [PQ=k.RS and $\angle\overline{PQ}=\angle\overline{RS}$] $\Rightarrow$ C($\overline{PQ}$)=k.C($\overline{RS}$)

2.3 Generator of a Space Cost Function

Definition 3. Let θ be an element of $]-\pi/2,\pi/2[$ and let C be a slope-dependent space cost function. C is said to be *minimal at* θ iff it satisfies the following properties:
[A3] Let $\overline{PQ}$ and $\overline{RS}$ be two paths of C: (PQ = RS and $\angle\overline{RS} = \theta$) $\Rightarrow$ C($\overline{PQ}$) $\geq$ C($\overline{RS}$)
[A4] Let $\overline{RS}$ be a path of C: [RS = 1 and $\angle\overline{RS} = \theta$] $\Rightarrow$ C($\overline{RS}$) = 1

From [A3], it derives that among all linear paths with length 1, those with slope θ have the lowest cost. [A4] sets this minimal cost to 1. The only real contribution of [A4] is to guarantee the existence of potentially profitable paths. Remark that the map from C into $\vec{R}_+^*$ which associates each path with its length is minimal at any element of $]-\pi/2,\pi/2[$.

Proposition 2. $\Rightarrow$ Let θ be an element of $]-\pi/2,\pi/2[$ and let C be a slope-dependent space cost function. If C is minimal at θ, there exists a map δ from $]-\pi/2,\pi/2[$ into [0,1] such that for any $\overline{PQ}$ of C: C($\overline{PQ}$) = x$_{\overline{PQ}}$ / [cos($\angle\overline{PQ}$).δ($\angle\overline{PQ}$)]. This map δ is unique and takes the value 1 at θ. $\Leftarrow$ Let θ be an element of $]-\pi/2,\pi/2[$ and let δ be a map from $]-\pi/2,\pi/2[$ into [0,1] taking the value 1 at θ. There exists a space cost function C such that for any path $\overline{PQ}$ of C: C($\overline{PQ}$) = x$_{\overline{PQ}}$ / [cos($\angle\overline{PQ}$).δ($\angle\overline{PQ}$)]. This cost function is unique, slope-dependent and minimal at θ.

Definition 4. Let θ be an element of $]-\pi/2,\pi/2[$. According to proposition 2, the datum of a space cost function, slope-dependent and minimal at θ, is equivalent to the one of a map δ from $]-\pi/2,\pi/2[$ into [0,1] taking the value 1 at θ: δ is the cost function *generator*.

$\delta(\alpha)$, for any element α of $]-\pi/2,\pi/2[$, is the distance a cost unit enables to cover on a linear path of slope α. The interest of expressing C in terms of δ lies in this simple interpretation. From a practical point of view, defining a cost function by means of its generator is particularly convenient and natural (Fig.2).

2.4 A Typical Family of Space Cost Functions

Definition 5. Let θ be an element of $]-\pi/2,\pi/2[$ and C a space cost function, slope-dependent and minimal at θ. C is called a *hill-climbing energetic function* iff it satisfies [A5]:
[A5] Let $\overline{PQ}$ and $\overline{RS}$ be two paths of C:
$[PQ = RS$ and $(\angle \overline{PQ} \leq \angle \overline{RS} \leq \theta$ or $\theta \leq \angle \overline{RS} \leq \angle \overline{PQ})] \Rightarrow C(\overline{PQ}) \geq C(\overline{RS})$

A hill-climber gets less tired on a linear path of slope θ. The further from θ the slope is (i.e. the more abrupt the slope up or the steeper the slope down), the more energy is consumed. The idea is to penalize abrupt ways (Fig.2). The map from C into $\overline{\mathbb{R}}_+^*$ which associates each path with its length is a hill-climbing energetic function.

3 Space Profitability Measures

In this section, profitability measures on paths of the space are introduced. The calculation of the profitability of a path joining P to Q is based on the estimate, drawn from a priori knowledge, of the travel cost from P to Q. The knowledge at stake are voluntarily limited: for instance to the position of the orthogonal projections of P and Q onto the plane, or to the position of P and Q in space, the length of the min-cost path from P to Q, etc. Each case leads to a particular profitability measure. As illustrations, two measures are briefly described here. C denotes a space cost function, D_P and D_Q the lines directed by $\vec{k}$ and running through P and Q, P@Q the set of paths belonging to C and joining P to Q, $D_P @ D_Q$ the set of paths belonging to C and joining one point of D_P to one point of D_Q: $D_P @ D_Q = \bigcup_{(P',Q') \in D_P \times D_Q} P'@Q'$

3.1 The 2D-Profitability Measure

Definition 6. The *2D-profitability measure* is a function A^{2D} from C into $[0,1]$. Let μ be a path of C joining a point P to a point Q. A^{2D} is defined at μ iff $\inf_{v \in D_P @ D_Q} C(v)$ is finite (it is especially the case when μ is potentially profitable). It is then set that: $A^{2D}(\mu) = [\inf_{v \in D_P @ D_Q} C(v)] / C(\mu)$. $A^{2D}(\mu)$ is the *2D-profitability* of μ (relatively to C).

Forecasting to spend $\inf_{v \in D_P @ D_Q} C(v)$ to join Q from P, means at the same time to be economical, pragmatic, very optimistic and (really) misinformed. As if a hill-climber, in order to assess the distance he still has to cover, would draw a segment on a rudimentary touristic map and consider the relief to be certainly as he hopes to be. So is A^{2D}. And its judgement gets more severe: if somebody is advised to follow μ to get from P to Q, the measure will probably assess that, comparing $C(\mu)$ with the cost initially forecast, the suggestion was not the best (and even dishonest). In the case where C is the map from C into $\overline{\mathbb{R}}_+^*$ which associates each path with its length, the forecast cost is the distance between lines D_P and D_Q, i.e. $x_{\overline{PQ}}$. Consequently, measure A^{2D} is a map and value $A^{2D}(\mu)$ is simply $x_{\overline{PQ}} / \lambda$ — where λ denotes the length of μ. The following proposition gives a practical means to characterize A^{2D} in a more general case.

Proposition 3. If the cost function C is generated by a continuous map δ, the profitability measure A^{2D} is an everywhere defined measure. Moreover, for any couple (P,Q) of points:
$[\overline{PQ} \in C \Rightarrow \inf_{v \in D_P @ D_Q} C(v) = x_{\overline{PQ}} / \max_{\alpha \in]-\pi/2,\pi/2[} (\delta(\alpha).\cos(\alpha))]$ and $[\overline{PQ} \notin C \Rightarrow \inf_{v \in D_P @ D_Q} C(v) = 0]$

3.2 The 3D-Profitability Measure

Definition 7. The *3D-profitability measure* is a function A^{3D} from C into $[0,1]$. Let μ be a path of C joining a point P to a point Q. A^{3D} is defined at μ iff $\inf_{v \in P@Q} C(v)$ is finite (it is especially the case when μ is potentially profitable). It is then set that: $A^{3D}(\mu) = [\inf_{v \in P@Q} C(v)] / C(\mu)$. $A^{3D}(\mu)$ is the *3D-profitability* of μ (relatively to C).

Forecasting to spend $\inf_{v \in P@Q} C(v)$ to join Q from P, means at the same time to be economical, pragmatic, very optimistic and (rather) misinformed. As if a motorist, in order to assess the distance he still has to cover, would scan an ordnance map and consider that tunnels have certainly been excavated and bridges erected. A^{3D} sticks more to realities than A^{2D}. Its judgement gets less severe. If C associates each path with its length, then the forecast cost is PQ. In the general case, $\inf_{v \in P@Q} C(v)$ may be difficult to calculate. This point will not be tackled here.

4 Back to the Discrete Space

After a quick reminder about cost functions of graphs and optimal paths, in §4.1, it is shown in §4.2 how to associate a weighted digraph with a raster DEM and a space cost function. Moreover, in §4.3, profitability measures are defined on the graph vertices.

4.1 Cost Functions of Graphs and Optimal Paths

Let (X,U,V) be any weighted and directed graph. Assume that V is a map from U into $\overline{R}_+^*$. The function from the set of graph paths into $\overline{R}_+^*$, which associates each path $(u_i)_{i \in 1..n}$ with the value $\Sigma_{i \in 1..n} V(u_i)$, is the *cost function* of (X,U,V). $\Sigma_{i \in 1..n} V(u_i)$ is the *cost* of $(u_i)_{i \in 1..n}$. Now, let p and q be two vertices and let μ be a path from p to q. If the cost of any path from p to q is greater than or equal to the cost of μ then μ is an *optimal path* — or a *min-cost path* — from p to q. The cost of μ is the *min-cost* to reach q from p. Finally, let Δ be the function from X^2 into $\overline{R}_+^*$ which associates each couple (p,q) with the min-cost to reach q from p, let Y be a non-empty subset of X and let Δ_Y be the function from X into $\overline{R}_+^*$ defined by: $\forall p \in X$, $\Delta_Y(p) = \min_{q \in Y} \Delta(q,p)$. $\Delta(q,p)$ is the min-cost to reach p from q. A path from q to p is optimal iff its cost is $\Delta(q,p)$. $\Delta_Y(p)$ is the *min-cost* to reach p from Y. Δ_Y is the *min-cost function* according to Y.

4.2 Weighted Digraphs, Raster DEMs and Space Cost Functions

With a numerical image, it is usual to associate the directed graph whose vertices are the image pixels and the arcs are, generally, the couples of 8-adjacent vertices. Let hence I be a raster DEM and let (X,U) be the digraph associated with I. The pixels of I are assimilated to points of the discrete space Z^2. This space itself is embedded into the affine euclidean xy-plane. The choice of U incites to provide Z^2 with a discrete distance, and more precisely with a ponderate distance d defined by a 3×3 mask. In practice, in order to approach the euclidean distance, the chamfer distance 3-4 is generally recommended [Bor86]. Now, it is considered that I is a raster model of a surface of the affine euclidean space. To any pixel of I consequently corresponds a point of this surface. Pixel and associated point will be named by the same letter:

small for the first and capital for the second. To pixel p of I thus corresponds a point P of the affine space: p is the orthogonal projection of P onto the xy-plane and the z-coordinate of P is the gray-level of p in I (up to a scale factor). Finally, let C be a space cost function, as defined in §2. C enables to weight the digraph (X,U) by means of the map V from U into $\overline{\mathbb{R}}_+^*$ which associates each arc (p,q) with $C(\overline{PQ})$ (we will hark back to this point in §4.3). Consider a path μ of the weighted digraph (X,U,V). It can be represented by a sequence $(p_i)_{i \in 1..n}$ of vertices such that: $\forall i \in 1..n-1, (p_i,p_{i+1}) \in U$. The cost of μ is the value $\Sigma_{i \in 1..n-1} V(p_i,p_{i+1})$, or also $\Sigma_{i \in 1..n-1} C(\overline{P_i P_{i+1}})$, i.e. $C(\overline{P_i}^{i \in 1..n})$. It then appears natural that the cost function of the graph, as well as the space cost function, should be denoted C.

4.3 Profitability of a Point According to a Region

Let I be a raster DEM, C a space cost function and (X,U,V) the graph associated with (I,C). Consider a non-empty subset Y of X. Here will be defined a function from X into [0,1] called profitability measure according to Y. The notion of space profitability measure, developed in §3, will of course contribute to this end. Within the framework of this paper, it will exclusively be referred to measure A^{2D}. Moreover, it will from now on be supposed that C is generated by a continuous map δ. Consider, for a given vertex p of X–Y, the expression: $\min_{q \in Y} [\inf_{v \in D_Q @ D_P} C(v)]$. As a direct extension of §3.1, it appears natural to interpret this value as the cost forecasted by A_Y^{2D} to reach p from Y — where A_Y^{2D} denotes the profitability measure we want to define. It also appears natural to welcome p in the definition domain of A_Y^{2D} iff this cost is finite. In this case: $A_Y^{2D}(p) = (\min_{q \in Y} [\inf_{v \in D_Q @ D_P} C(v)]) / \Delta_Y(p)$. The path that will be "really taken" is indeed the optimal path, whose cost is $\Delta_Y(p)$. Now, according to proposition 3: $\inf_{v \in D_Q @ D_P} C(v) = qp / \max_{\alpha \in]-\pi/2,\pi/2[} (\delta(\alpha).\cos(\alpha))$. Where qp is obviously the euclidean distance between q and p. Consequently:

$$\min_{q \in Y} [\inf_{v \in D_Q @ D_P} C(v)] = (\min_{q \in Y} qp) / \max_{\alpha \in]-\pi/2,\pi/2[} (\delta(\alpha).\cos(\alpha))$$

The discrete transcription of the numerator is $\min_{q \in Y} d(q,p)$ or also $d_Y(p)$ — by denoting d_Y the distance image according to Y (remember that d is a chamfer distance, see §4.2). For obvious practical reasons, it is tempting to adopt it. But, to this end, the discrete transcription of the cost calculation must be operated. In §4.2, we had weighted each arc (p,q) by $C(\overline{PQ})$, i.e., according to proposition 2, by: $C(\overline{PQ}) = x\overline{PQ} / [\cos(\angle \overline{PQ}).\delta(\angle \overline{PQ})]$. Coming back to the definition of V, we set:

$$\forall (p,q) \in U, \ V(p,q) = d(p,q) / [\cos(\angle \overline{PQ}).\delta(\angle \overline{PQ})]$$

Remark that C(p,q) — where C denotes the cost function of the digraph — is not exactly equal to $C(\overline{PQ})$ any more — where C now denotes the space cost function.

Definition 8. The *2D-profitability measure* A_Y^{2D} *according to* Y — or *2D-profitability image according to* Y — is the map from X into [0,1] which takes the value $(d_Y(p) / \Delta_Y(p)) / \max_{\alpha \in]-\pi/2,\pi/2[} (\delta(\alpha).\cos(\alpha))$ at each vertex p of X–Y and takes the value 1 at each vertex of Y. $A_Y^{2D}(p)$ is the *2D-profitability of p according to* Y.

If C is the hill-climbing energetic function which associates each path of $\mathcal{C}$ with its euclidean length, then A_Y^{2D} is defined on X by: $\forall p \in X, A_Y^{2D}(p) = d_Y(p) / \Delta_Y(p)$.

Gaussian Hill. Costs and profitabilities are according to the upper-left corner.

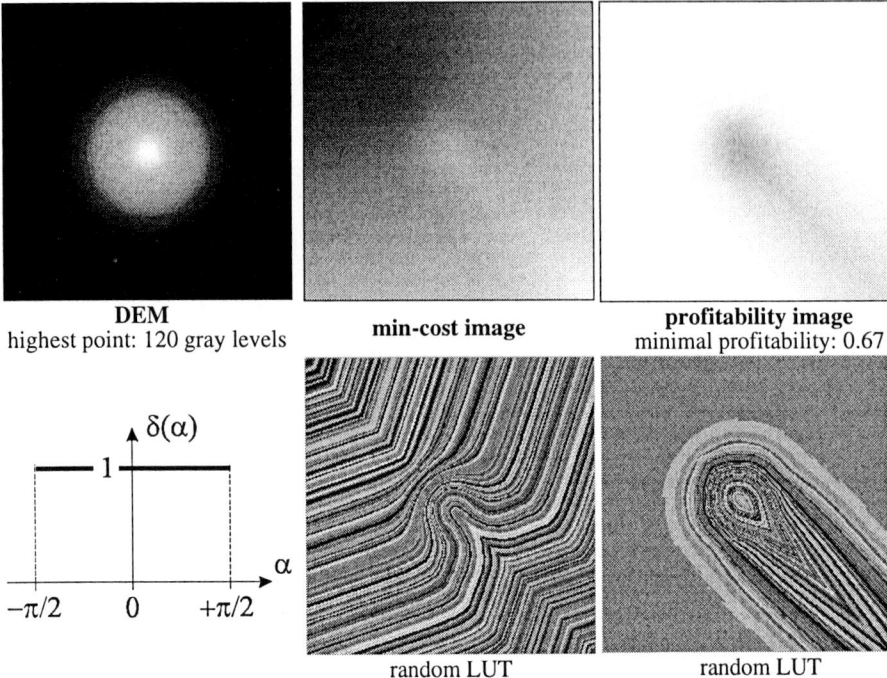

Mont Ventoux (France). Covered surface: 100 km². Maximal difference in level: 600 m. Costs and profitabilities are according to a point situated amid the upper part of the image.

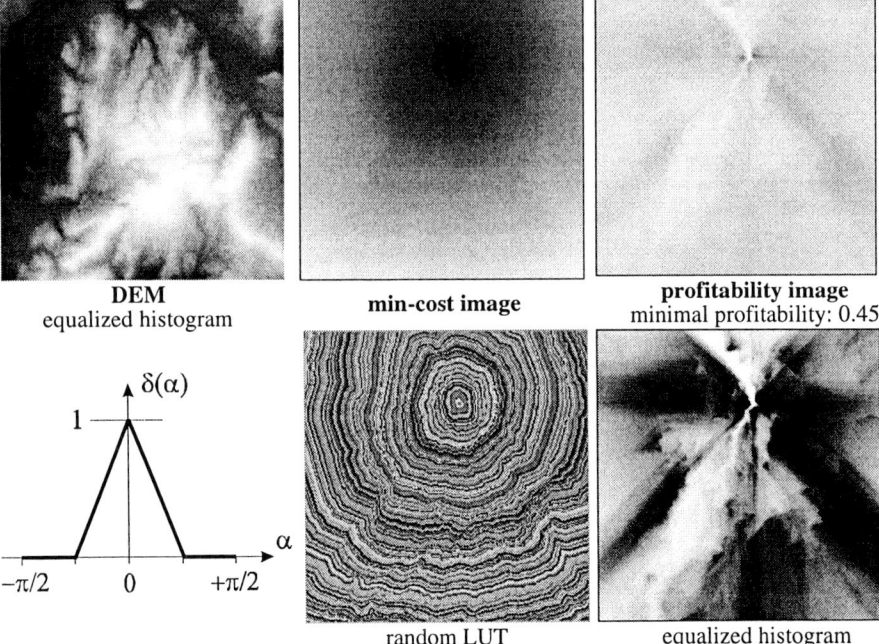

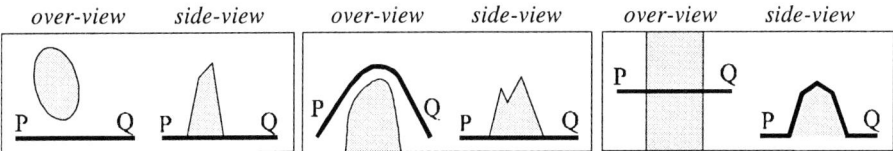

Fig. 3. 2D-profitability measures. Some characteristics.
The represented paths are the optimal paths from P to Q. In the first case (on the left), point Q may be assessed totally profitable according to {P}. Depending on δ, Q may be assessed more profitable in the second case than in the third one, even if the PQ distances are identical and also the lengths of the optimal paths.

5 Experimental Results and Conclusion

In this paper, the problem of defining cost functions for digraphs associated with raster DEMs has been discussed in a general and formal framework. A particularly simple and natural way to tackle this problem has been proposed. Moreover, the notion of profitability has been put forward and profitability measures have been defined. The calculation of profitabilities is based on the datum of a space cost function — representing an ideal homogeneous space — and consists in drawing estimates from a priori knowledge of travel costs. Profitability measures and cost functions provide useful and complementary information. The results of two experiments are given here in order to illustrate this point. The min-cost images Δ_Y have been computed by means of the well-known Bellman's algorithm [Bel58] and the distance images d_Y by means of a very efficient algorithm [Ros66] [Bor84] which needs exactly two passes over the data set. High elevations, profitabilities and costs are represented in light gray. All images are 256×256.

References

[Ahu93] Ahuja, R.K., T.L. Magnanti, and J.B. Orlin, *Network flows*, Prentice Hall, 1993.
[Bel58]   Bellman, R., "On a toutin problem", *Quaterly of Applied Math.*, 16(1), pp.87-90, 1958.
[Bor84]   Borgefors, G., "Distance transformations in arbitrary dimensions", *CVGIP* 27, pp.321-345, 1984.
[Bor86]   Borgefors, G., "Distance transformations in digital images", *CVGIP* 34, pp.344-371, 1986.
[Dub95] Dubois, N., F. Semet, "Estimation and determination of shortest path length in a road network with obstacles", *European Journal of Operational Research* 83, pp.105-116, 1995.
[Gon84] Gondran, M., and M. Minoux, *Graphs and Algorithms*, Wiley, Chichester, 1984.
[Kre94]   van Kreveld, M., "On Quality Paths on Polyhedral Terrains", *IGIS'94*, LNCS 884, Springer-Verlag, pp.113-122, 1994.
[Mit91]   Mitchell, J.S.B., C.H. Papadimitriou, "The weighted region problem : finding shortest paths through a weighted planar subdivision", *Journal of ACM* 38(11), pp.18-73, 1991.
[Pri94]   Prins, C., *Algorithmes de graphes*, Eyrolles, 1994.
[Ros66]   Rosenfeld, A., and J.L. Pfaltz, "Sequential operations in digital picture processing", *Journal of ACM* 13(4), pp.471-494, 1966.
[Zha93] Zhan, C., S. Menon, and P. Gao, "A Directional Path Distance Model for Raster Distance Mapping", *COSIT'93*, LNCS 716, Springer-Verlag, pp.434-443, 1993.

Using Proximity and Spatial Homogeneity in Neighbourhood-Based Classifiers *

J.S. Sánchez[1], F. Pla[1] and F.J. Ferri[2]

[1]Departament d'Informàtica. Universitat Jaume I
Campus Penyeta Roja, E-12071 Castelló. SPAIN.
{sanchez,pla}@inf.uji.es, +34 64 345676
[2]Departament d'Informàtica i Electrònica. Universitat de València
Dr. Moliner 50, E-46100 Burjassot (València). SPAIN.
ferri@uv.es, +34 6 3864768

Abstract

In this paper, a set of neighbourhood-based classifiers are jointly used in order to select a more reliable neighbourhood of a given sample and take an appropriate decision about its class membership. The approaches introduced here make use of two concepts: proximity and symmetric placement of the samples.

1 Introduction

The ultimate aim of any pattern recognition system is to achieve the best possible classification accuracy for the problem to be solved. In practical applications, due to the availability of a wide variety of classifiers, it is logical to test only a number of them and choose the algorithm with highest reliability as a final solution to the problem. Nevertheless, although one of the classifiers would yield the best performance, the other tested schemes can contribute complementary information about the samples to be classified and therefore, it seems obvious that those could be efficiently used to improve the performance of the selected classifier.

A traditional strategy to obtain an increase in performance consists of using some form of voting with the results of several single classifiers in order to arrive at a combined decision [4] [5]. In addition to simple majority vote (in which all votes have equal weight), the votes can be weighted for each class or even, according to the performance of each classifier on each class [8]. A common theoretical framework for combining classifiers which use distinct pattern representations is given in [7].

In a similar way, it is also possible to improve the classification rate by means of some kind of cooperation between different classifiers. In this case, there is not a combination of the decisions of various classifiers but, on the contrary, all the information given by a set of classifiers is used to arrive at a unique decision. In this paper, several "cooperating" neighbourhood-based classifiers are introduced and implemented in order to apply them to some real problems. The single classifiers used to obtain the proposed schemes are the well-known *k-Nearest Neighbours* decision

---

* This work was partially supported by grants P1A94-23, P1B96-13 (Fundació Caixa-Castelló-Bancaixa), GV2110/94 (Conselleria d'Educació i Ciència), AGF95-0712-C03-01 and TIC95-676-C02-01 (Spanish CICYT).

rule [3] (or *k*-NN, in which the *k* closest neighbours vote for the label of the sample to classify) as well as two novel non-parametric approaches [10], namely the *k-Nearest Centroid Neighbours* (*k*-NCN) and the *Graph Neighbours* (GN) rules, whose general idea consists of classifying a sample by the prototypes placed around it (instead of close to it). The reason to use these neighbourhood-based classifiers is that, while the *k*-NN provides a suitable notion of the statistical distribution in the proximity of a sample, the *k*-NCN and GN rules mainly take into account geometric information, such as the spatially homogeneous placement of prototypes around a sample.

2 Neighbourhood-based classifiers

Classification in pattern recognition has traditionally been tackled through two alternative approaches; namely, *parametric* and *non-parametric* methods [3]. The parametric classifiers assume a functional distribution of given samples. On the other hand, the non-parametric do not assume any functional distribution of the set of prototypes. While the parametric approach has theoretically been shown to be potentially capable of yielding optimal results, in practice, it often tends to actually fail because of inappropriate assumptions of a priori distributions.

Among non-parametric methods, those which are based on sample-to-sample distances are particularly remarkable; namely, *k*-NN techniques. When applied to classification, these schemes require the classes to be represented by appropriate sets of prototypes and the decision rule is generally reduced to label each given sample with the class that contains most of its *k* nearest neighbours. It is the conceptual simplicity of such a rule, along with its asymptotical tendency towards the *Bayes* rule in terms of minimum classification error, what makes the *k*-NN approach particularly appealing in many practical situations. Nevertheless, when the number of prototypes in the training set is not large enough, the *k*-NN rule is no longer optimal. This problem becomes more relevant when having few prototypes compared to intrinsic dimensionality of the feature space, which is a very common practical situation.

Some alternative definitions of neighbourhood have been used to obtain other non-parametric classifiers, trying to partially overcome the practical drawback pointed out above for the *k*-NN rule. In particular, the recently introduced concept of *Nearest Centroid Neighbourhood* [2] along with the neighbourhood relation derived from the *Gabriel Graph* (GG) and the *Relative Neighbourhood Graph* (RNG) [6], have been used to obtain the so-called *k*-NCN and GN rules [10], respectively.

As mentioned before, the *k*-NN rule consists of estimating the class of a given sample through its *k* closest prototypes. In other words, this classifier considers that all the information required to classify a new sample can be obtained from a small subset of prototypes close to the sample. However, it does not take into account the geometrical distribution of those *k* prototypes with respect to the given sample (that is, in general the nearest prototypes do not completely surround the sample since the *k*-NN rule considers the neighbourhood only in terms of Euclidean distance).

The non-parametric *k*-NCN and GN approaches are also based on the general idea of estimating the class of a sample from its neighbours, but considering a different kind of neighbourhood which allows to inspect a sufficiently small area *around* the sample, in such a way that all prototypes surrounding that sample take part in the classification. As already pointed out, this is accomplished by using two different concepts about surrounding a sample with nearby prototypes: firstly, the Nearest Centroid Neighbourhood [2], which tries to surround a sample by taking prototypes in such a way that *(a)* they are as near as possible to the sample, and *(b)* their centroid is also as close as possible to the sample. Secondly, the Graph Neighbourhood (i.e., GG and RNG-based neighbourhoods) of a sample, defined as the union of all its graph neighbours. Taking into account that two points are graph neighbours if no other point lies inside a certain region of influence between them [6], it is possible to surround completely a sample by means of its graph neighbours.

Bearing this in mind, the Nearest Centroid Neighbourhood and Graph Neighbourhood concepts can be used to obtain two alternative non-parametric classifiers; namely, the k-NCN and GN rules, respectively. Both of them have in common the fact of considering a number of prototypes *around* (instead of *close to*) a sample to estimate its class. Given a set of labelled prototypes $X = \{x_1, ..., x_n\}$ and a new unknown sample q, the k-NCN and GN classification rules assign to q the class with majority of votes among its k nearest centroid neighbours or its graph neighbours, respectively. In the case of the GN rule, considering the GG and the RNG, two different approaches are defined: the *Gabriel Graph Neighbours* (GGN) and the *Relative Neighbourhood Graph Neighbours* (RNGN) rules, respectively.

Note that the surrounding decision rules (i.e., the k-NCN and the GN rules) can take prototypes which are not sufficiently close to the given sample. Obviously, this constitutes a relative drawback for those rules with respect to the k-NN classifier in some practical problems. Therefore, it could be interesting to look for a certain balance between both categories of non-parametric classifiers, trying to overcome as far as possible some of the mentioned disadvantages of each one.

3 Proposed classification schemes

From the discussion made in Section 2, the neighbourhood-based classifiers considered here can be divided into two groups according to the information used by each one: the *proximity-based* classifiers (the k-NN rule) and the *surrounding* ones (the k-NCN and GN rules). Since our general purpose is to obtain alternative decision rules with the advantages of both groups, it is possible to relate their information in some effective way to derive a single decision, which should presumably improve the corresponding performance. Here, we propose a way to obtain a classification scheme from the cooperating work between the proximity-based classifiers and the surrounding ones. In a few words, it consists of using all the information given by prototypes which belong to a relatively small neighbouring region around a sample.

Let $X = \{x_1, ..., x_n\}$ be a set of n labelled prototypes and let q be a new sample to classify. The set of surrounding neighbours of q (that is, the nearest centroid neighbours or the graph neighbours) can be represented as $SN_q = \{s_1, ..., s_m\}$, $m < n$. Now, let $k = |SN_q|$ in such a way that the set $k\text{-}NN_q = \{p_1, ..., p_k\}$ constitutes the set of its k-proximity-based neighbours (or simply, k-nearest neighbours). Thus, we can define the set of the *nearest surrounding neighbours* of q (NSN_q) as the subset of surrounding neighbours which are also nearest neighbours. Consequently, this subset can be represented as the intersection of both sets of surrounding and nearest neighbours of q:

$$NSN_q = SN_q \cap k\text{-}NN_q$$

Obviously, the set of the nearest surrounding neighbours, NSN_q, contains the prototypes of X which satisfies both conditions pointed out before: they must be around the sample q (*surrounding neighbourhood*) and, they must be sufficiently close to it (*nearest neighbourhood*). In a first step, the surrounding classifier (i.e., the k-NCN or the GN) takes prototypes around q and in a further step, the k-NN rule constrains that initial set of neighbours (the set of surrounding neighbours SN_q) to only those prototypes which are really close to q. In such a way, we are trying to correct the fact that some surrounding neighbours can be too far from the sample q. There clearly exists a relationship among the size of those sets of neighbours of q:

$$|NSN_q| \leq |k\text{-}NN_q| = |SN_q| < |X|$$

The geometrical meaning of the set NSN_q is as follows: the prototypes of NSN_q constitute a small region (namely, the *nearest surrounding region*) close to the sample q in such a way that all prototypes are distributed around q. It is worth noting that this region does not contain the surrounding neighbours which are too far from q since they are eliminated during the *nearest neighbourhood* test. Formally, the general decision rule can be formulated in the following way:

Step 1: Find the surrounding neighbours of q: $SN_q = \{s_1, ..., s_m\}$, $m < n$.
Step 2: Let $k = SN_q$.
Step 3: Find the k nearest neighbours of q: $k\text{-}NN_q = \{p_1, ..., p_k\}$.
Step 4: Compute the nearest surrounding neighbours of q:
$$NSN_q = SN_q \cap k\text{-}NN_q, \quad j = |NSN_q| \leq k$$
Step 5: Assign to q the class with majority of votes among its j nearest surrounding neighbours in the set NSN_q (resolve ties randomly).

Note that this scheme can lead to three alternative but analogous classification rules, simply varying the surrounding approach used in Step 1 of the algorithm: we can apply either the k-NCN rule or the GN rule and, in the later case, we can choose between the GG or the RNG. A priori, almost nothing can be said about the relative merits of each one of these approaches in terms of classification accuracy. However, it has a practical interest in evaluating other issues related to the surrounding rules: first, according to [11], the averaged computational cost to search for the graph neighbours (i.e., Gabriel neighbours or relative neighbours) of a sample in a d-dimensional feature space is close to $O(dn)$, while k nearest centroid neighbours can be found in $O(kn)$ time [2]. Second, while the GN rules do not require any tuning or external parameter, the k-NCN rule needs the number of neighbours k.

4 Description of the databases

Three different real databases are used to investigate the performance of those classification schemes. For the first and third databases, the *Holdout* method averaged over five different random partitions (half for training and half for testing purposes) of each original set, has been used to obtain the results reported. For the second database [9], only one unbalanced partition of the initial set has been considered to compare the performance achieved here to that obtained in previous related works.

The first database

The aim of the experiments carried out over this database is to distinguish among 11 different textures, each pattern (pixel) being characterised by 40 attributes built by the estimation of fourth order modified moments in four different orientations: 0, 45, 90 and 135 degrees. There are a total number of 5,500 patterns.

The second database

This database was already considered in [9] in order to study colour segmentation to locate oranges in outdoor scenes under daylight conditions. It consists of values from RGB colour images with a resolution of 256 x 256 pixels. There are 19,164 patterns (6,386 samples for training and 12,778 for testing) with two attributes (coordinates φ and θ of the colour vectors in the RGB space) and three classes.

The third database

This database was used in [1] about the development of a real time analytical system for French and Spanish speech recognition. There are two different classes: the nasal vowels and the oral vowels. It contains samples coming from 1,809 different isolated syllables. Five different attributes were chosen to characterise each vowel: the amplitudes of the five first harmonics, normalised by the total energy. There are 5,404 patterns: 3,818 for nasal vowels and 1,586 for oral ones.

5 Experimental results

Some experiments by using four single neighbourhood-based classifiers (k-NN, k-NCN, GGN, and RNGN) and the three schemes proposed here (k-NN with k-NCN, k-NN with GGN, and k-NN with RNGN) have been carried out. For the k-NN and k-NCN rules, typical values for the parameter k (ranging from 1 to 11) have been tried in each experiment.

Table 1 shows the accuracy levels obtained by the four single classifiers. Note that the results for the first and third databases correspond to the averaged classification rates and the standard deviations over the five random partitions of the original data sets. These results should be taken as illustrative in the sense that they will be used to compare the single classifiers to the schemes proposed here.

	Database 1	Database 2	Database 3
1-NN (1-NCN)	98.38 (± 0.04)	97.56	88.41 (± 0.21)
3-NN	98.03 (± 0.18)	97.61	87.41 (± 0.10)
5-NN	97.75 (± 0.14)	97.61	86.33 (± 0.12)
7-NN	97.65 (± 0.18)	97.72	86.12 (± 0.12)
9-NN	97.44 (± 0.16)	97.60	85.87 (± 0.06)
11-NN	97.33 (± 0.17)	97.46	85.13 (± 0.19)
3-NCN	98.14 (± 0.05)	97.64	88.11 (± 0.09)
5-NCN	98.20 (± 0.11)	97.75	86.97 (± 0.13)
7-NCN	98.48 (± 0.08)	97.71	87.02 (± 0.09)
9-NCN	98.58 (± 0.04)	97.60	87.00 (± 0.18)
11-NCN	98.73 (± 0.07)	97.68	86.34 (± 0.15)
GGN	80.01 (± 0.08)	97.68	84.85 (± 0.20)
RNGN	97.90 (± 0.14)	98.08	88.03 (± 0.12)

Table 1. Accuracy levels by using single neighbourhood-based classifiers

It is also important to point out the large difference in performance between the GGN rule versus the other ones for the first and third databases. This is in contrast to the RNGN (the other GN classifier), which obtains rates comparable to the k-NN and k-NCN rules, specially in the case of the last two databases. The "surprising" differences between the GGN and the RNGN rules can be interpreted as follows: the number of Gabriel neighbours for a given sample is much bigger than the number of its relative neighbours (i.e., the neighbours in the corresponding RNG), so the GGN is inspecting into a too large region while in practice, the neighbourhood of a sample should be relatively small. This can be assessed in Table 2 where the averaged number of graph neighbours as well as the percentage of prototypes in the test set with respect to the neighbourhood size for each database are reported. Note that the cases for which the GGN obtains a sufficiently high performance correspond to the second and third databases, where the percentage is smaller than 1.

	Database 1		Database 2		Database 3	
	Size	%	Size	%	Size	%
GGN	33.30	1.21	77.72	0.60	10.21	0.37
RNGN	3.35	0.12	50.23	0.39	2.87	0.11

Table 2. Neighbourhood size for the GGN and RNGN rules

The classification accuracy and the standard deviations achieved with the schemes proposed are summarised in Table 3. As expected, the approaches introduced here generally present a certain improvement with respect to the single classifiers. This confirms that the use of the information given by both groups of neighbourhood-based classifiers really manages to bound better the more reliable neighbourhood of a sample (i.e., the decision boundaries among classes).

	Database 1	Database 2	Database 3
1-NCN/1-NN	98.38 (± 0.04)	97.56	88.41 (± 0.21)
3-NCN/3-NN	98.34 (± 0.03)	97.54	88.15 (± 0.27)
5-NCN/5-NN	98.25 (± 0.08)	97.44	88.29 (± 0.30)
7-NCN/7-NN	98.02 (± 0.10)	97.39	87.82 (± 0.29)
9-NCN/9-NN	98.00 (± 0.16)	97.58	87.65 (± 0.13)
11-NCN/11-NN	98.03 (± 0.23)	97.57	87.44 (± 0.21)
GGN/k-NN	92.36 (± 0.06)	97.80	87.21 (± 0.28)
RNGN/k-NN	98.57 (± 0.06)	98.09	88.70 (± 0.21)

Table 3. Percentage of test cases correctly classified by the proposed schemes

Note that the GGN/k-NN rule achieves the highest improvement rates on the first database, in fact for the one in which the single application of the GGN classifier obtains the worst results. In this case, it seems that the k-NN rule adjusts as far as possible the disproportionate neighbourhood given by the GGN classifier. On the other hand, it is clear that the cooperation between the GGN and the k-NN rules achieves the highest increase in performance in all experiments.

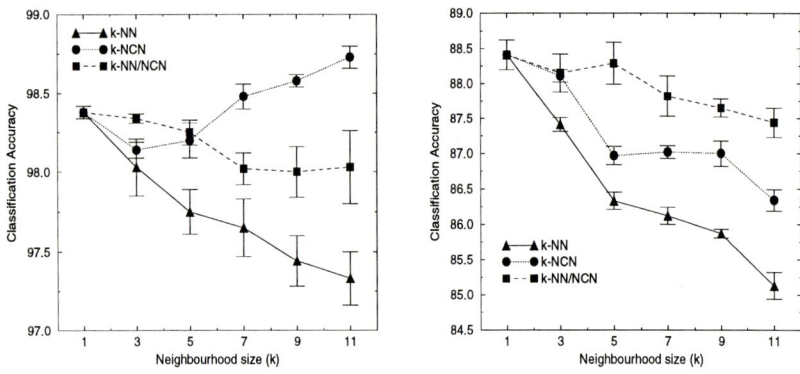

Fig. 1. Classification accuracy for the first and third databases

Fig. 1 shows the performance of the k-NN and k-NCN classifiers, as well as the one achieved by the cooperation between them on the first and third databases. As can be seen, the combined application of those schemes achieves more significant differences when the single k-NCN rule performs worse (that is, from $k = 5$). In fact, these results are also consistent with the discussion made for the GG-based classifier.

6 Concluding remarks

A way of cooperation between different neighbourhood-based classification schemes has been introduced in order to increase the performance achieved by using single classifiers. The general idea of this approach consists of using two complementary neighbourhood definitions: the *nearest* neighbourhood and the *surrounding* neighbourhood. The former is represented by the well-known k-NN decision rule [3] and, the latter by means of the recently introduced k-NCN [10] and graph-based classifiers [10]. The aim of using jointly the information obtained from different neighbourhood-based classifiers is to benefit from the advantages of both groups of neighbourhood defined in this paper. On the one hand, the k-NN rule gives a suitable measure of distance-based proximity with respect to a sample. On the other hand, the surrounding approaches conveniently define the geometrical distribution of prototypes around the given sample.

From the experimental results, it is possible to draw some conclusions. First, the schemes proposed really achieve a relative improvement in terms of classification accuracy with respect to the single neighbourhood-based rules. Second, it seems that the combination of the k-NN and the GGN classifiers is not appropriate since it systematically achieves the lowest performance in all experiments. Finally, with regard to the jointly use of the information obtained from the k-NN and the k-NCN and RNG-based rules, differences between them are not significant enough and, it is not possible to make a final statement about the suitability of either of them. In this case, one should weigh other related issues such as the computing time.

References

1. Alinat, P. (1993). Periodic Progress Report 4: ROARS Project ESPRIT II 5516.
2. Chaudhuri, B.B. (1996). A new definition of neighbourhood of a point in multi-dimensional space, *Pattern Recognition Letters* **17,** 11-17.
3. Duda, R. and P.E. Hart (1973). *Pattern classification and scene analysis.* John Wiley & Sons, New York.
4. Franke, J. and E. Mandler (1992). A comparison of two approaches for combining the votes of cooperating classifiers, *Proc. of 11th. International Conference on Pattern Recognition*, 611-614.
5. Ho, T.K., J.J. Hull and S.N. Srihari (1994). Decision combination in multiple classifier systems, *IEEE Trans. on Pattern Analysis and Machine Intelligence* **16**, 66-75.
6. Jaromczyk, J.W. and G.T. Toussaint (1992). Relative neighbourhood graphs and their relatives, *Proc. of IEEE* **80**, 1502-1517.
7. Kittler, J., M. Hatef and R.P.W. Duin (1996). Combining classifiers, *Proc. of 13th. International Conference on Pattern Recognition*, 897-901.
8. Lam, L. and C.Y. Suen (1995). Optimal combinations of pattern classifiers, *Pattern Recognition Letters* **16**, 945-954.
9. Pla, F., F. Juste, F.J. Ferri and M. Vicens (1993). Colour segmentation based on a light reflection model to locate citrus fruits for robotics harvesting, *Computers and Electronics in Agriculture* **9**, 53-70.
10. Sánchez, J.S., F. Pla and F.J. Ferri (1997). On the use of neighbourhood-based non-parametric classifiers, *Proc. of Pattern Recognition in Practice V* (In press).
11. Toussaint, G.T., B.K. Bhattacharya and R.S. Poulsen (1985). The application of Voronoi diagrams to nonparametric decision rules, *Computer Science and Statistics*, 97-108.

Image Segmentation by Means of Fuzzy Entropy Measure

C. Di Ruberto, M. Nappi, S. Vitulano

Istituto di Medicina Interna - Policlinico Universitario
Via S. Giorgio, 12 - 09124 Cagliari - ITALY
tel. (39) (70) 6028209 - fax (39) (70) 663651
E-Mail vitulano@vaxca1.unica.it

Abstract
The paper describes an algorithm for image segmentation using fuzzy entropy measure. The relation between the fuzzy entropy of an image domain and the fuzzy entropy of its subdomains is explored as a uniformity predicate. With the aim of implementing the model, we have introduced a well known technique of Problem Solving. The most important roles of our model are played by the Evaluation Function (*EF*) and the Control Strategy. So the *EF* is related to the ratio between the fuzzy entropy of one region or zone of the picture and the fuzzy entropy of the entire picture. The Control Strategy determines the optimal path in the search tree (quadtree) so that the nodes of the optimal path have minimal fuzzy entropy. The paper shows some comparisons between the proposed algorithm and classical edge detection techniques.

1. INTRODUCTION

Segmentation is a significant issue in the field of image processing and image understanding. The segmentation is the process, both human and automatic, that isolates in a pictorial scene zones or regions, edges or contours and angles with respect to a certain uniformity predicate.
In the last years several approaches have been proposed in the literature and may be classified as follows:
- Local (or atomistic) approach:

Such a method takes advantages of grey level discontinuities that are considered relevant features of image. In order to extract these features several local operators have been introduced and the most important are Sobel [1, 2], Marr-Hildreth [3]. In this case the discontinuities assume the shape of monodimensional step function. In [1] Gonzales and Wintz propose an effective algorithm that looks at regions in image with non-unimodal histogram, using local thresholds.
To the aim of resolving the strong simplification that associates edges with a monodimensional step function, a lot of corrections have been introduced. In this manner local operators can process image with smoothing and shading effects [2, 4, 5].
In [6, 7] any angle of a region in a digital image is approximated by means of a series of segments or half edges, where a half edge can be characterised by its position and orientation angle. So the angle is a point at which two (or more) half edges cross [7-9].

Recently some techniques based on image iterated smoothing have been proposed. Such a method eliminates higher frequencies of image by iterated sampling, preserving -as much as possible- the shape and the edge positions [10, 11]. This approach presents some obvious limits such as an high computing time, and the strictly depending on threshold and on smoothing parameters. For this reason usually local operators are not utilised to process medical images.

In [12] Higgins presents and compares three different methods to detect grey level discontinuities by utilising structural local informations.

- Global (or structural) approach:

For its simplicity the global threshold is the oldest technique for image segmentation. In this approach it is customary to utilise one threshold for the whole image (global information) or one threshold for each image region (contextual method) [13-16]. The threshold is based on the following rules: each region in image may be associated with a peak histogram [17-20]. Of course such a rule is very restrictive.

In order to extract the objects that make up an image, entropy -high order entropy or conditional entropy- is often adopted as uniformity predicate [21,22].

The Gestalt theory and Neurophysiologic theory affirm that in the human perceptive process the eye aims to minimise objects and background variations -homogenise- enhancing transitions regions, i.e. edges.

The goal of this paper is to propose a new algorithm for image segmentation using the fuzzy entropy, namely FISE, exploiting the ratio between each image region and the whole image [23, 24, 32]. In other words such a ratio is minimum on the regions and maximum on the edges.

2. FUZZY ENTROPY MEASURE FOR EDGE DETECTION

The task of the segmentation is to enhance the different regions of the image. In order to segment the image into regions each of them has to satisfy the uniformity predicate. Now it is necessary to analyse some aspects of the uniformity predicate before we introduce it.

In our opinion the uniformity predicate couldn't be determined without considering the features of the whole image. In other words the measures calculated to determine the uniformity predicate have to be related to the same measures calculated on the whole image. In this paper the fuzzy entropy has been chosen as the uniformity predicate. In particular the fuzzy entropy of a region (or a background) -that is a region too- is always lower than the entropy of whole image or, in other words, the fuzzy entropy of a region is always greater or equal than the entropy of its subdomains.

The FISE adopts a strategy to segment the image similar to that introduced in [25]: the merge and the split are applied in cascade on the image using a region growing algorithm.

A search tree of any hierarchical structure, as for example t-ari trees, can be utilised. In our work we have chosen the quadtree.

The strategy we'll propose segments into regions whose fuzzy entropy is nearly zero. More precisely, the zones extracted as regions are those whose evaluation function is gradually goes to zero.

We have defined the evaluation function as a measure of the fuzzy entropy of a region according to the theorem of fuzzy entropy [33].

The theorem of fuzzy entropy defines the entropy of a fuzzy set A as a measure of the subset A and $notA$ (A and A^c) in relation to the subset A or $notA$ (A or A^c) as showed in Figure 1.

Let a_i be an internal node of quadtree whose grey tones are $g_1, g_2, ..., g_n$ ($g_i < g_j$, $i < j$, $i,j = 1,...,n$) and frequencies $f_1, f_2, ..., f_n$, respectively.

Let a_{4i+t}, $t \in [1,4]$, a child node of a_i, whose greytones are $g'_1, g'_2, ..., g'_m$ ($g'_i < g'_j$, $i < j$, $i,j = 1,...,m$) and frequencies $f'_1, f'_2, ..., f'_m$, respectively.

For each grey tone g_i, $i = 1,...,n$, of the node a_i we define

$$\Delta f_i = |f_i - f'_j|$$

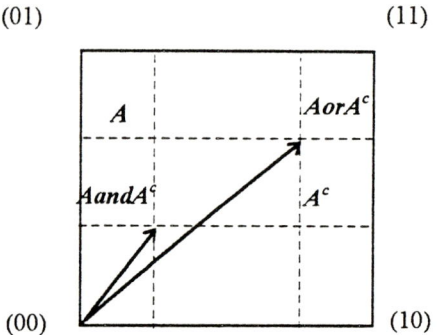

Figure 1 - The entropy fuzzy theorem.

where f_i is the frequency of the grey tone g_i while f'_j is the frequency of the grey tone of the node a_{4i+t}, with $g_i = g'_j$.

Then we evaluate

$$\Delta^{100} f_i = \sqrt{(100 - \Delta f_i)^2 + 100^2}$$

and finally we define the fuzzy entropy of the greytone g_i as:

$$e_i = \frac{\Delta f_i}{\Delta^{100} f_i}.$$

So, the fuzzy similarity of the node a_i respect to its child node a_{4i+t} is:

$$E(a_i, a_{4i+t}) = \frac{1}{K} \sum_{i=1}^{K} e_i \quad (1)$$

where $K = \max(n,m)$.

Now we can define the fuzzy homogeneity of the node a_{4i+t}, $t \in [1,4]$ as the average fuzzy similarity of a_{4i+t} and its four children nodes:

$$EM(a_{4i+t}) = \frac{1}{4}\left(E(a_{4i+t}, a_{(4i+t)+1}) + E(a_{4i+t}, a_{(4i+t)+2}) + E(a_{4i+t}, a_{(4i+t)+3}) + E(a_{4i+t}, a_{(4i+t)+4})\right)$$

We can observe that if $EM(a_{4i+t})$ is nearly zero then the node a_{4i+t} is "totally homogeneous".

So, the evaluation function of the node a_{4i+t} is given by:

$$F(a_{4i+t}) = \frac{EM(a_{4i+t})}{E(a_i, a_{4i+t})}.$$

$F(a_{4i+t})$ evaluates the fuzzy homogeneity of the node a_{4i+t} respect to the fuzzy similarity to its father node a_i, assuming minimal value inside a region and maximal value along the contour of a region.

So, if a_i is an internal node of a quadtree and the evaluation functions of its four children nodes are $F(a_{4i+1})$, $F(a_{4i+2})$, $F(a_{4i+3})$, $F(a_{4i+4})$, the control strategy expands the child node a_s, with $s = 4i+1, ..., 4i+4$, whose $F(a_s)$ is minimal:

$$F(a_s) = \min\{F(a_{4i+1}), F(a_{4i+2}), F(a_{4i+3}), F(a_{4i+4})\}.$$

According this rule, along the search path of the quadtree, $F(a_s) = \dfrac{EM(a_s)}{E(a_i, a_s)}$ is different from zero, in particular $EM(a_s) \neq 0$ and $E(a_i, a_s) \neq 0$.

Applying this rule at each level of the search tree, we'll individuate a node a_j whose $EM(a_j) \cong 0$ and so $F(a_j) \cong 0$. Then we can state that the subdomain relative to the node a_j belongs to a region (or an object) R.

If, inside the region R the evaluation function $F(a_{i'})$ of an expanded node $a_{i'}$ is nearly zero and the evaluation function $F(a_{j'})$ of its child node $a_{j'}$ to expand is different from zero then we can state that the subdomain relative to the node $a_{i'}$ is the smallest subdomain including the partition element of the region. In other words, if the dimension of the subdomain relative to the node $a_{i'}$ is $k_{i'} \times k_{i'}$ and its initial coordinates in the entire image domain are $(x_{i'}, y_{i'})$, then the dimension of the partition element P of the region R will be $dx \times dy$, with $k_{i'/2} \leq dx \leq k_{i'}$, $k_{i'/2} \leq dy \leq k_{i'}$ and $dx = \min\{x\}$, $k_{i'/2} \leq x \leq k_{i'}$ and $dy = \min\{y\}$, $k_{i'/2} \leq y \leq k_{i'}$ such that the evaluation function computed on the subdomain with dimension $x \times y$, included in the domain of the node $a_{i'}$, and initial coordinates $(x_{i'}, y_{i'})$ is nearly zero. So, the region R of the image will be the union of the elements (or subdomains) $P_1, P_2, ..., P_N$ fuzzy similar to the partition element P:

$$R = \bigcup_{i=1}^{N} P_i$$

where P_i has coordinates (px_i, py_i) and dimension $dx_i \times dy_i$, with $dx_i = dx$ and $dy_i = dy$.

3. EXPERIMENTAL RESULTS AND DISCUSSIONS

In order to evaluate the performance of the proposed edge detection algorithm we have tested it on several theoretical and real 512x512x8 bits images. For the sake of brevity in this Section objective results obtained on theoretical images are presented. Moreover, FISE has been widely compared with the segmentation algorithm based on entropy (namely ISE) proposed by Vitulano et alt. in [32] and some of the most useful local operators, i.e. Sobel, DOG zero-crossing, Haralick [1, 2], Anisotropic diffusion and three methods proposed by Higgins [12].

We describe briefly ISE segmentation method.

Let a_s be a node at level z of quadtree and a_{4s+t}, $t \in [1,4]$ a child node of a_s. Let $f_k(a_s)$ with $k \in [1,n]$ be the maxima frequencies of the histogram of the image area relative to the node a_s and $f_l(a_{4s+t})$ with $l \in [1,m]$ be the maxima frequencies of the histogram of the image area relative to the node a_{4s+t}.

For simplicity we denote the set of these frequencies with $\{f_k\}$ and $\{f_l\}$. With each maximum frequency f_i in $\{f_k\}$ (or in $\{f_l\}$) we associate a gaussian function whose standard deviation is defined as:

$$\sigma_i = \frac{1}{\sqrt{2\pi f_i}}. \qquad (2)$$

So, with each frequency f_i in $\{f_k\}$ (or in $\{f_l\}$) we can associate an interval of graytones of the histogram of a_s (or of a_{4s+t}):

$$[g_i - \sigma_i, g_i + \sigma_i]$$

where g_i is the greytone with frequency f_i and standard deviation σ_i defined as in (2).

If g_i is the graytone with frequency $f_i \in \{f_k\}$ and g_j is the graytone with frequency $f_j \in \{f_l\}$ then the following relations could be satisfied:

$$|g_i - g_j| \leq |a - b| \leq 2\sigma * \qquad (3)$$

where a and b are the extrema of the intersection interval $[g_i - \sigma_i, g_i + \sigma_i] \cap [g_j - \sigma_j, g_j + \sigma_j]$ and $\sigma^* = \min(\sigma_i, \sigma_j)$,

otherwise

$$[g_i - \sigma_i, g_i + \sigma_i] \cap [g_j - \sigma_j, g_j + \sigma_j] = \emptyset \qquad (3')$$

We consider now the sets $\{f_k\}$ and $\{f_l\}$ sorted according increasing graytones and we state that if $n = m$ and $|g_i - g_j| \leq 2\sigma *$, where g_i is the greytone with frequency $f_i \in \{f_k\}$ and standard deviation σ_i and g_j is the greytone with frequency $f_j \in \{f_l\}$ and standard deviation σ_j, and $\sigma^* = \min(\sigma_i, \sigma_j)$, then the histograms of a_s and a_{4s+t}, $t \in [1,4]$, have the maxima frequencies in correspondence to the same graytones. In terms of multivariate analysis the histograms have the same variables. Moreover, if

$$\forall i = j \in [1,n], \quad f_i = f_j \text{ and } |g_i - g_j| = 0 \qquad (4)$$

then the histograms of a_s and a_{4s+t} have the same graytones distribution.

If (3) and (4) are satisfied we define the domain relative to a_s "totally homogeneous" to the domain relative to a_{4s+t}.

Obviously, the relations (3) and (4) are valid for theoric images and for a limited number of nodes of the quadtree. Instead for real images it's necessary the introduction of an evaluation function to determine the homogeneity degree of two different domains.

The control strategy prefers the nodes with minimum E.F. along the search tree as in FISE method.

So, for each $f_i \in \{f_k\}$ and $f_j \in \{f_l\}$ satisfying the relation (3), we define the entropy e_i relative to the frequencies f_i and f_j as:

$$e_i = \frac{|a-b|}{|a'-b'|}$$

where a and b are the extrema of the intersection interval $[g_i - \sigma_i, g_i + \sigma_i] \cap [g_j - \sigma_j, g_j + \sigma_j]$, a' and b' are the extrema of the union interval $[g_i - \sigma_i, g_i + \sigma_i] \cup [g_j - \sigma_j, g_j + \sigma_j]$ and g_i, σ_i, g_j and σ_j are defined as in (3). So, we define the E.F. of a_s in relation to a_{4s+t} as:

$$E(a_s, a_{4s+t}) = \frac{1}{p}\left(\sum_{i=1}^{p}(1-e_i) + \alpha\right) \quad (5)$$

where a is the number of frequencies of $\{f_k\}$ and $\{f_l\}$ not satisfying the relation (3) and $p = \frac{n+m-\alpha}{2}$.

For the test cases we assume the image f is made up of disjoint constant intensity region corrupted by additive Gaussian noise (AWGN); i.e. if point $(x,y) \in$ region R_k, than $f(x,y) = \mu_k + \eta_k(x,y)$, where μ_k is the constant intensity value for points in R_k and $\eta_k(x,y)$ is a sample of AWGN having statistics $N(0, \sigma_k^2)$.

In order to determine an objective parameter to test the quality of processed images, as proposed by Haralick [31], we use two performance metrics to compare the various methods: P(AETE) and P(TEAE). The first metric is the conditionally probability of a point being assigned as an edge point, given that the point is a true edge point, while the second one the conditional probability of a point being a true edge point, given that the point is assigned as an edge point. Assigned edge points are those points that a particular edge detection method assigns. True edge points are defined to be the points within the two-point wide region in which each point is adjacent to some point having a value different from it on the uncorrupted checkerboard. For each method, except for our algorithm, a threshold is adjusted until $P(AETE) \approx P(TEAE)$. This equalisation in

practice represents an even better trade-off between detecting true edge points and rejecting non-edge points.

The numerical comparison results proposed by Higgins [12], ISE and FISE are illustrated in Table 1.

Table 2 presents the numerical results obtained by applying ISE and FISE on a checkerboard image corrupted by AWGN, mean=0 and standard deviation=60.

	P(TEAE)	P(AETE)	Average
Sobel gradient	0.660	0.656	0.658
DOG	0.865	0.833	0.849
Haralick operator	0.760	0.759	0.759
Anisotropic Diffusion	0.889	0.898	0.894
Higgins-Method 1	0.948	0.920	0.934
Higgins-Method 2	0.828	0.823	0.825
Higgins-Method 3	0.866	0.847	0.857
FISE Method	0.457	0.598	0.528
ISE Method	0.974	1.000	0.987

Table 1 - Results for noisy checkerboard using local structure operators, ISE and FISE.

	P(TEAE)	P(AETE)	Average
FISE Method	0.579	0.518	0.549
ISE Method	0.669	0.956	0.812

Table 2 - Results for noisy checkerboard using ISE and FISE methods.

4. CONCLUDING REMARKS

The FISE method seems that it should be helpful both for edge detection and segmentation of regions.
The most important features of such method could be summarised as follows:
- high noise tolerance both for real and theoretic image;
- threshold independence;
- low computing time ($n \log n$) where n is the size of the image;
- the edges detected don't present steps usually introduced by local operators.

We underline that the method doesn't need the choice of thresholds or operators dimensions and requires short consuming time. The control strategy makes totally automatic the entire process.

The checkerboard image represents the most significant test image. Infact, since it is a theoric image we know all its edge points and the regions contained in it. The theoretic image has been corrupted by additive Gaussian noise (AWGN) in order to test the goodness of different algorithms. Local operators require high computing time ([12]) and the choice both of thresholds and of operators dimensions (i.e. DOG 21x21, DOG 69x69) doesn't correspond to deterministic criteria. Through a visual analysis we have observed that the contours are irregular and present many breakings. Many points that are not true edge points are assigned as edge points. In general the edge image is smaller than the input image: this decreasing depends on the dimensions of the local operator. The numerical results confirm the limits of the local operators and the goodness of the global FISE and ISE methods. We have observed that FISE method works better as the complexity of the image increases. So, in the next future we are going to improve the experimentation of the method on complex real test images.

5. REFERENCES

1. R.C. Gonzales and P. Wintz, *Digital Image Processing*, 2nd Edn. Addison-Wesley, Reading Mass. (1987).
2. R. Rosenfeld and A.C. Kak, *Digital Picture Processing*, 2nd Edn. v.2, Academic Press, N.Y. (1982).
3. D.C. Marr and E. Hildreth, "Theory of edge detection", *Proc. R. Soc. London B 207*, 187-217 (1980).
4. B. Born, *Robot Vision*, MIT Press, Cambridge, Mass. (1986).
5. J. Canny, "A computation approach to edge detection", *IEEE Trans. on Pattern Analysis Mach. Intell.*, PAMI-8, 679-698 (1986).
6. M.A. Gennert, "Detecting half-edges and vertices in images", *Proc. IEEE Int. Conf. Comput. Vision Pattern Recogn.*, 552-557 (1986).
7. V.S. Nalwa and T.O. Binford, "On detecting edges", *IEEE Trans. on Pattern Analysis Mach. Intell.*, PAMI-6, 699-714 (1986).
8. M.M. Fleck, "Some defects in finite-difference edge finders", *IEEE Trans. on Pattern Analysis Mach. Intell.*, PAMI-14, 337-345 (1992).
9. Y.G. Leclerc, "Capturing the local structure in image discontinuities in two dimensions", *Proc. IEEE Int. Conf. Comput. Vision Pattern Recogn.*, 34-38 (1985).
10. P. Perona and J. Malik, "Scale-space and edge detection using anisotropic diffusion", *IEEE Trans. on Pattern Analysis Mach. Intell.*, PAMI-12, 629-639 (1990)
11. P.Saint Marc, J.Chen and G.Medioni, "Adaptive smoothing: a general tool for early vision", *IEEE Trans. on Pattern Analysis Mach. Intell.*, PAMI-13, 514-530 (1991).
12. W.E. Higgins and C. Hsu, "Edge detection using two dimensional local structure information", *Pattern Recognition*, 27, 277-294 (1994).
13. S.D. Yanowitz and A.M. Brukstein, "A new method for image segmentation", *Comput. Vision Graphics Image Processing*, 46, 82-95 (1989).
14. T. Taxt, P.J. Flynn and A.K. Jain, "Segmentation of document images", *IEEE Trans. on Pattern Analysis Mach. Intell.*, PAMI-11, 1322-1329 (1989).
15. T.W. Ridler and S. Calvard, "Picture thresholding using an iterative selection method", *IEEE Trans. on System Man and Cybern.*, SMC-8, 630-632 (1978).
16. P.K. Sahoo, S. Soltani, A.K.C. Wong and Y.C. Chen, "A survey of thresholding techniques", *Comput. Vision Graphics Image Processing*, v.41, 233-260 (1988).
17. N.R. Pal and S.K. Pal, "Object-background segmentation using a new definition of entropy", *IEE Proc.*, Pt, 136, 284-295 (1989).
18. N.R. Pal and D. Bhandari, "On object-background classification", *Int. J. Syst. Sci.*, 23, 1903-1920 (1992).
19. A.S. Abutaleb, "Automatic thresholding of grey level pictures using two-dimensional entropy", *Comput. Vision Graphics Image Processing*, 47, 22-32 (1989).
20. P.J. Burt, T.Hong and A.Rosenfeld, "Segmentation and estimation on image properties through cooperative hierarchical computation", *IEEE Trans. on System Man and Cybern.*, v.11, n.12 (1981).
21. Y. Nakagawa and A. Rosenfeld, "Some experiments on variable thresholding", *Pattern Recognition*, 11, 191-204 (1979).
22. J. Kittler and J. Illingworth, "Minimum error thresholding", *Pattern Recognition*, 19, 41-47 (1986).

23. S.Vitulano, C.Di Ruberto and M.Nappi, "A.I. based image segmentation", *Lectures Notes in Computer Science*, Proc. of Int. Conf. of Image Analysis and Processing, C. Braccini, L. De Floriani, G. Vernazza (Eds.), Springer Verlag, 974, 429-434 (1995).
24. S. Tanimoto, M. Savini and S. Vitulano, "Allocation in attention vision", *Human and Machine Vision*, Proc. of the Third Int. Workshop on Perception, V. Cantoni (Ed.), Plenum Press, 171-180 (1994).
25. S.L. Horowitz and T. Pavlidis, "Picture segmentation by a directed split and merge procedure", *Proc of the Second Intern. Joint Conf on Pattern Recognition*, pp.424-433, Copenhagen, Aug. (1974).
26. N.J. Nilsson, *Principles of Artificial Intelligence*, Springer Verlag (1982).
27. S.Z. Selim and M.A. Ismail, "Soft clustering of multidimensional data: a semi-fuzzy clustering approach", *Pattern Recognition*, 15, 559-568 (1984).
28. M.S. Kamel and S.Z. Selim, "A thresholded fuzzy k-means algorithm for semi-fuzzy clustering", *Pattern Recognition*, 24, 825-833 (1991).
29. Q. Zhang and R.D. Boyle, "A new clustering algorithm with multiple runs of iterative procedures", *Pattern Recognition*, 24, 835-848 (1991).
30. M. Nappi and S. Vitulano, "A new more efficient k-means algorithm for clustering", *Proc. of the Thirteenth IASTED Int. Conf., Applied Informatics*, ed. M. H. Hamza, pp. 324-327 (1995).
31. R.M. Haralick, "Digital step edges from zero crossing of second directional derivatives", *IEEE Trans. on Pattern Analysis Mach. Intell.*, PAMI-6, 58-68 (1984).
32. S. Vitulano, C. Di Ruberto, M. Nappi, "Biomedical Image Processing", *Proc. III IEEE ICECS '96*, Oct. 13-16, Rodos, v.2, pp.1116-1119 (1996).
33. B.Kosko, "Fuzzy Thinking: the New Science of Fuzzy Logic", Hyperion Ed. (1993).

Efficient Region Segmentation through 'Creep-and-Merge'

Antranig Basman, Joan Lasenby and Roberto Cipolla

University of Cambridge, Department of Engineering, Cambridge CB2 1PZ, England

Abstract. We present a novel architecture for region-based segmentation of stationary and quasi-stationary statistics, which is designed to function correctly under the widest range of conditions. It is robust to the extremes of region topology and connectivity, and automatically maintains region boundaries sampled to the minimum scale at which the region configuration can be determined with statistical confidence. The algorithm is deterministic, and when operating on images from within its domain of validity, contains no adjustable parameters. In contrast to most other techniques directed at the same problem, the progress of the algorithm cannot be described by the optimisation of a global energy criterion.
We describe a specific implementation using Gaussian stationary statistics, and present test results which demonstrate superior performance to a collection of other systems.

1 Introduction

Many areas of computer vision could benefit from the replacement of its long–standing preference for local edge and corner detectors, based on linear correlation and convolution operations by more global methods, making use of region–based information. However, current region–based schemes suffer from drawbacks that make them unattractive for general use. These include:

Inflexibility : these schemes are often tied to a particular region modelling framework (MRF, piecewise polynomial, wavelet, etc.), which makes it difficult to adapt them for general use, or as new models arise. Schemes are also frequently [7, 3, 6] restricted to grayscale images.

Inefficiency : either through the performance of unnecessary extensive search through small scales of the image, or due to non-deterministic elements of the optimisation [1], region–based schemes traditionally take many orders of magnitude longer to run than localised feature detectors.

Non-regularisation : arguably the greatest challenge facing region optimisation is the regularisation of parameters, e.g. the boundaries, size and number of regions. Current schemes have difficulty accommodating regions with greatly varying size [12, 15], and often include explicit penalties for increasing the length of boundaries (a 'smoothing' regularisation) or increasing the number of regions. These introduce arbitrary parameters without supplying satisfactory algorithms to compute them for different situations.

The scheme we present here is flexible, efficient, and intrinsically regularises the necessary parameters in a dynamic manner during the course of optimisation; it thus requires no manual adjustments.

1.1 Motivation

A popular region–based segmentation paradigm (the "snakes/balloons" framework put forward in [9, 5], and generalised in [15]) idealises the image as a continuous field, and region boundaries as differentiable contours; optimisation then proceeds by steepest descent deformation of the contours with respect to some global energy criterion.

Another paradigm (split–and–merge/link, presented in [11, 2]) subdivides the image into discrete nested (square) cells, which are then recursively connected/split apart according to some homogeneity predicate [14].

The current scheme avoids the problems associated with these systems. Contour frameworks suffer from excessive locality (the image is only examined in a curve neighbourhood), troublesome discretisation – image 'forces' often involve curvature and gradient terms that are sensitive to quantisation artifacts – and topological difficulties which all restrict the scope of deformations.

On the other hand, splitting frameworks can overlook important features, since they proceed through a fixed number of recursive passes over the image.

Our framework may either be viewed as an explicit discretisation of a contour–based system, or as an iterative enhancement to region splitting methods – it shares in the good properties of both.

2 Theoretical Framework

We adpot a general modelling framework – the image is completely covered by a collection of non-intersecting connected regions $\{R_i\}$, which contain data D_i with some size measure $N_i = N_i(D_i)$. The region contents are drawn from $\boldsymbol{\alpha}(D_i)$, one of a collection of models indexed by parameter vector $\boldsymbol{\alpha}$.

We assume the existence of an estimator $E(D_i)$ which determines, given a set of data, a 'best' model with vector $\hat{\boldsymbol{\alpha}}_i$ from the collection. We place a commonplace restriction on these models:

Items of data from the modelled regions are viewed as independent, identically distributed (i.i.d.) random variables; i.e. the regions are those with some form of *stationary statistics*. This assumption is required at a point in the development mentioned in Section 2.3.

2.1 Statistical Framework

By design, the system proceeds by deterministic, minimal statistically significant discrete perturbations of the configuration, which at all times conforms with the description given above.

Which perturbations are, or are not, statistically significant, will vary during the course of the algorithm, and different sizes of perturbation will be appropriate according to the contents of the regions.

We will define this size in terms of a function $\delta(D_i, D_j)$, the *discriminability* of two sets of region contents.

The algorithm will be deemed to have converged when

1. No significant perturbations of a boundary between two regions will improve its discriminability.
2. Each boundary considered as a whole is significant.

We require that (i) δ has the dimensions of a log probability, and (ii) for fixed N_1, as $N_2 \to \infty$, $\delta \to \beta$, some $\beta \neq 0, -\infty$. This guides us to form δ using the hypothesis of homogeneity of two neighbouring regions; i.e. $H_\omega : \boldsymbol{\alpha}(D_{12}) = \hat{\boldsymbol{\alpha}}_{12}$, against the alternative hypothesis $H_a : \boldsymbol{\alpha}(D_1) = \hat{\boldsymbol{\alpha}}_1$ and $\boldsymbol{\alpha}(D_2) = \hat{\boldsymbol{\alpha}}_2$, where R_{12} with data D_{12} is the region formed by considering R_1 and R_2 one region. We choose δ to be the log level of confidence at which the null hypothesis H_ω of homogeneity would be rejected, if in fact we were required to make the decision.

This choice now constrains our estimators E to be the *maximum likelihood estimators* (m.l.e.s) of the parameters under the required hypothesis.

We now present the derivation of δ for a simple case; our parameters will be the means μ_i and standard deviations σ_i of univariate Gaussians $G(x; \mu_i, \sigma_i)$, and so $\hat{\boldsymbol{\alpha}}_i = (m_i, s_i)$, where m_i and s_i are the standard (biased) maximum likelihood estimates of the parameters.

2.2 Univariate Gaussians

The log likelihood of data of size N drawn from a population $G(\mu, \sigma)$ under a hypothesis specifying $\mu = m$, $\sigma = s$ is

$$\log\left(\lambda(\mu, \sigma, m, s)\right) = -\frac{N}{2}\left(\log(2\pi\sigma^2) - \frac{s^2}{\sigma^2} - \frac{(\mu - m)^2}{\sigma^2}\right). \quad (1)$$

Consider two such regions, whose parameters have m.l.e.s m_i, s_i under the inhomogeneous hypothesis H_a, and $m_1 = m_2 = m_i^\alpha$, $s_1 = s_2 = s_i^\alpha$ under the homogeneity hypothesis H_α; the forms of these standard estimators are not presented here.

The likelihood ratio test (LRT) statistic for testing H_a against H_ω is obtained by summing three terms of the form of equation (1); it is given by

$$-\log \lambda^\alpha = N \log(s^\alpha)^2 - N_1 \log s_1^2 - N_2 \log s_2^2. \quad (2)$$

This quantity arises through the testing of a hypothesis allowing 2 degrees of freedom (the parameters m_i^α and s_i^α) as against a null hypothesis with 4. Thus, this statistic has an asymptotic (as both N_1 and N_2 become large) χ^2 distribution with 2 degrees of freedom. The reader is referred to Silvey [13], p. 114 for details of the argument, which is general for all tests of this sort.

Although formally this result is only applicable asymptotically, empirically it is highly reliable in this case for extremely small sample sizes.

Thus, the discriminability δ is defined as

$$\delta(D_i, D_j) = \log \left(1 - \int_0^{\log \lambda^\alpha(D_i, D_j)} \chi^2(2, x)\, dx \right) \qquad (3)$$

i.e. the log level of confidence under which the data D_i, D_j would give rise to the rejection of H_ω.

A generalised derivation for multivariate Gaussians is found in [10], p. 140.

2.3 Optimisation Framework

A *border* B_{ij} between regions R_i and R_j is the set of points neighbouring both regions. Neighbouring a border B_{ij} are two sets L_{ij} and L_{ji} of *border elements*; these are (not necessarily non-overlapping) subregions of R_i and R_j, which are considered suitable candidates for exchange with the opposite region.

The minimum size of subregion below which the membership of R_i or R_j cannot significantly be determined is $N^*(D_i, D_j)$, the *magic number* for the border B_{ij}; members of L_{ij} must be larger than this.

To determine N^*, we consider subregions $S_j'^N$ of R_j with size N, assumed to give rise to the same model vector $\hat{\alpha}_j$ (not true in general, depending on the form of E – this is where we need the uniformity assumption mentioned in 2).

$N^*(D_j, D_i)$ is defined to be the largest N such that $\delta(D_j^N, D_i) > t_c$ where t_c is a level of confidence to be defined later.

3 Progress of Optimisation

In overview, the optimisation procedure has a three–phase structure:

Phase A – Seeding – the region configuration must be initialised in a suitable form, aiding efficiency and accuracy of convergence.

Phase B – Creeping – estimated regions are repeatedly deformed until no further significant deformation is possible.

Phase C – Merging – regions with weak (to be defined) boundaries are merged with the regions on the other side of the weak boundary.

Phase A is executed once, and alternating passes of Phases B and C are applied until both produce no further change.

3.1 Phase A

Phase A determines regions of maximum local discriminability from a selection of neighbours, where the maximum is taken through all scales of the image. This aims at allocating (a) at least one seed region to each actual image region (b) minimising seed regions straddling actual region boundaries.

The current system is rather generous in seeding regions, as figure 1(b), a seeding of image 1(a), shows. However, the remaining phases are quite robust; speed could be improved by seeding fewer, more well–positioned regions.

Each seed is converted into a region R_i, and the border configurations L_{ij} described in section 2.3 are initialised.

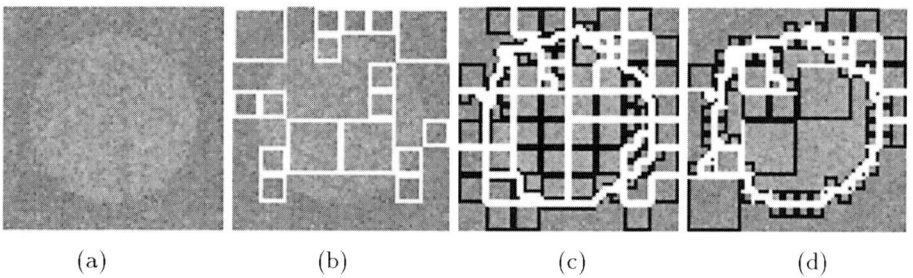

(a)   (b)   (c)   (d)

Fig. 1. (a) Original image; the regions differ in means by 15 units, and have common deviation 9. (b) Configuration after seeding; (c) After one phase of creeping; (d) After one phase of merging. The white boundaries represent regions R_i, black squares are border elements L_{ij}.

3.2 Phase B

Phase B applies operations analagous to the region deformation procedures of region growing/snake/balloon systems — in our system this procedure further generalises to the split–and–merge/link systems; we can decompose the deformation step as follows:

> Choose a border element L_{ij}^n from some border B_{ij}
> Form new regions R_i' and R_j' by exchanging it with R_j
> Evaluate the discriminability $\delta(D_i', D_j')$
> If it has increased, return L_{ij}^n to R_i; else leave with R_j.

Thus, although at a high level, the regions appear to be bounded by deformable contours, at a low level, the system is proceeding by deterministic split–and–merge. Some connectivitiy issues arise here, which there is not space to treat; for more details, see [4].

Figure 1(c) shows the configuration after Phase B is applied to figure 1(b)

3.3 Phase C

Region creeping and region merging proceed from a consistent and unified viewpoint. Creeping is a modification of the boundary configuration variables by significant increments (i.e. border elements). Merging occurs when these configuration variables as a whole are not significant, and *all possible efforts* have been made to improve their significance.

There are thus two criteria by which boundaries may be insignificant variables:

1. The discriminability $\delta(D_i, D_j)$ associated with the boundary as a whole is below the confidence threshold t_c.

OR

2. Due to the geometry of the region, NO border elements $L^n{}_{ij}$ of the requisite size as defined in section 2.3 can be formed.

Figure 1(d) shows the figure 1(c) configuration after Phase C.

3.4 Representation

The algorithm is not tied to a particular choice of representation.

Phase A depends on a hierarchical image partition, but the following operations may use e.g. the polygons or spline chains of section 1.1. However, in this implementation, a quadtree framework is used for all three phases, the best way to handle the rapid changes in connectivity and adjacency that occur in phases B and C. Cells C_i and border elements L_{ij} are therefore single quadtree cells.

3.5 Confidence

The confidence criterion t_c remains to be set; we use a base level of -5, giving a confidence level of 99.6%. A quadtree size N contains $2N/k$ pairs of adjacent cells area k; for a uniform error rate we set $t_c = -12 + \max(7, \log(\min(N_i, N_j)))$.

4 Evaluation

We demonstrate the power of the algorithm to adapt automatically to the extremes of scales/topologies present in images. Figure 2 is the final segmentation of figure 1(d) - the system correctly descends to the appropriate scale, and the circle is accurately localised after 3 iterations of phases B and C.

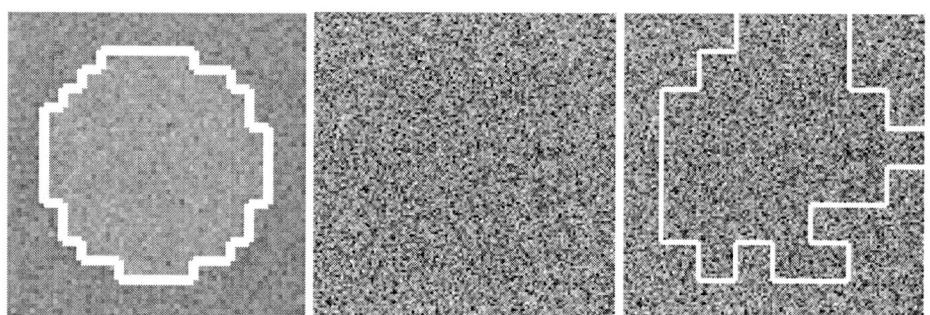

Fig. 2. Final circle image   **Fig. 3.** Original 4/32 image   **Fig. 4.** 4/32 result

4.1 Noise robustness

Figure 3 shows a severe test of the algorithm; a dark circle drawn from $G(124, 32^2)$ is successfully segmented (figure 4) from background $G(128, 32^2)$ (the magic number was artificially lowered one scale, leading to the irregular boundary). This level of noise is too severe for any contour-based technique, and indeed too severe for segmentation by the human visual system. The system continues to detect the circle until the difference in means drops below 2.

A split–and–merge system, however, only succeeds on an image where the difference in means is 5. The run time on 128x128 images of this sort is around 1 second, a factor of 2–3 slower than split–and–merge, which [8] shows to be the fastest of region segmentation schemes.

Figure 5 shows a systematic evaluation of the algorithm's breakdown under noise, using an evaluation criterion and results taken from [8]. The criterion is a weighted sum of terms involving region size and boundary length. Our system (the dark line) is compared against five competing region- and edge-based systems, on varying noise images similar to figure 6.

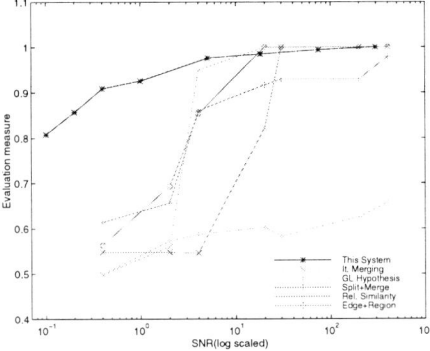

Fig. 5. Performance evaluation

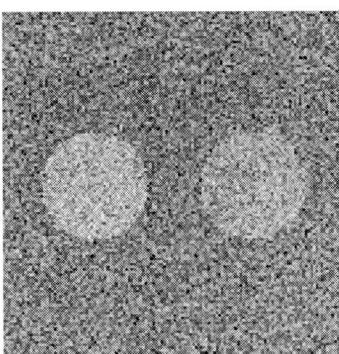

Fig. 6. Typical image from [8]

Our method is considerably superior in high noise, and shows excellent performance in low noise also; the lack of smoothness regularisation leads to misclassification of a few pixels in moderate noise. Such a term could be added as post-processing; however, the score would degrade on images with corners.

We also note that the other systems' results are the result of optimising the criterion with respect to their adjustable parameters, over several dozen runs. Our system also scores heavily over the others in that it has no such parameters.

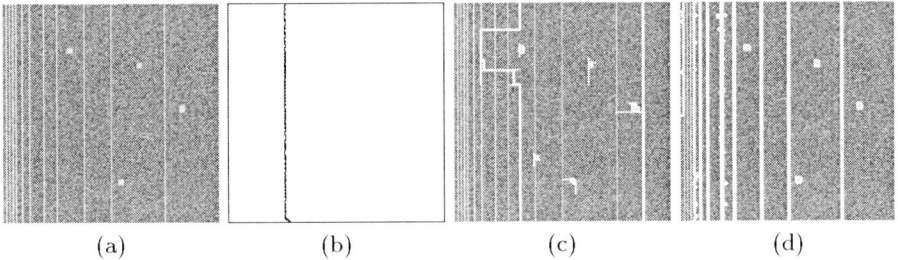

Fig. 7. (a) Original degenerate image; (b) Result from Yuille+Zhu system[15]; (c) Split and merge result; (d) Proposed system.

4.2 Scale adaptation

In the other direction, we show a successful segmentation of figure 7(a), taken from [15]; the algorithm correctly identifies the thin lines and blocks on the right-hand side and centre of the image; the left-hand strip is however still seen to be a single, high-variance region. Contrary to the claim made in [15], this success is made in a unified framework without the addition of any form of edge term. In order to segment the strip with closely-spaced lines, it will be necessary to handle pixels on an individual basis, which is inefficient in the current representation.

Both split-and-merge, and the system of [15], which claims to generalise and improve all contour-based methods, fail to segment this image.

5 Conclusions and Further Work

Our choice of criteria and optimisation procedure have been shown to resolve the problems of regularisation, boundary bias, automation, inflexibility and computational inefficiency from which region-based optimisation schemes often suffer.

Work will now turn to application of the multivariate version of the algorithm to the segmentation of coloured and textured images, and the development of appropriate surface models for the segmentation of images of real scenes.

References

1. P. Andrey and P. Tarroux. 'Unsupervised Texture Segmentation using Selectionist Relaxation'. In *Proc. 4th European Conf. on Computer Vision*, Volume I, pp. 482–491. Springer-Verlag, 1996. LNCS 1064.
2. H.J. Autonisse. Image segmentation in pyramids. *Computer Vision, Graphics and Image Processing*, 19(4):367–383, 1982.
3. J.A. Bangham, R. Harvey, P.D. Ling, and R.V. Aldridge. Nonlinear scale-space from n-dimensional sieves. In *Proc. 4th European Conf. on Computer Vision*, Volume I, pp. 189–198. Springer-Verlag, 1996. LNCS 1064.
4. A.M. Basman, J. Lasenby, and R. Cipolla. The Creep-and-Merge segmentation system. Technical Report CUED/F-INFENG/TR295, University of Cambridge, July 1997.
5. L.D. Cohen. On active contour models and balloons. *Computer Vision, Graphics and Image Processing*, 53(2):211–218, May 1991.
6. J.H. Elder and S.W. Zucker. Local scale control for edge detection and blur estimation. In *Proc. 4th European Conf. on Computer Vision*, Volume II, pp. 58–69. Springer-Verlag, 1996. Lecture Notes in Computer Science 1065.
7. S.A. Hojjatoleslami and J. Kittler. Region growing: A new approach. Technical report, University of Surrey, 1995.
8. H. Jiang, J. Toriwaki, and H. Suzuki. Comparative performance evaluation of segmentation methods based on region growing and division. *Systems and Computers in Japan*, 24(13):28–42, 1993.
9. M. Kass, A.P. Witkin, and D. Terzopoulos. Snakes: Active contour models. *Int. Journal of Computer Vision*, 1(4):321–331, Jan 1998.
10. K.V. Mardia, J.T. Kent, and J.M. Bibby. *Multivariate Analysis*. Academic Press, 1979.
11. T. Pavlidis. *Structural Pattern Recognition*, Chapter 5. Springer-Verlag, 1977.
12. P. Shroeter and J. Bigun. Hierarchical image segmentation by multi-dimensional clustering and orientation-adaptive boundary refinement. *Pattern Recognition*, 28(5):295–709, May 1995.
13. S.D. Silvey. *Statistical Inference*, Chapter 6. Penguin, 1970.
14. F. van der Heijden. *Image Based Measurement Systems*, Chapter 6. John Wiley & Sons, 1994.
15. S.C. Zhu and A. Yuille. 'Region Competition: Unifying Snakes, Region Growing, and Bayes/MDL for Multiband Image Segmentation'. *IEEE Trans. Pattern Analysis and Machine Intell.*, 18(9):884–900, September 1996.

An Automatic Transformation from Bimodal to Pseudo-Binary Images

José M. Iñesta, Pedro J. Sanz, and Ángel P. del Pobil
Departamento de Informática, Universitat Jaume I. Campus Penyeta Roja, E-12071
Castellón, Spain
{inesta,sanzp,pobil}@inf.uji.es

Abstract

In this paper a procedure is proposed to transform bimodal images into enhanced ones with unified properties. These (pseudo-binary) images keep valuable grey level information and can be easily segmented by a global thresholding technique, using always the same threshold in the middle of the grey scale. This transformation is automatically self-adjusted from the statistical characterization of the histogram modes. It can be useful for the automatic segmentation of bimodal images found in controlled illumination environments, dealing with possible uncontrolled light variations. In addition, the grey level values in the pseudo-binary image can be considered as occupation percentage of the pixels, so subpixel reasonings can be easily inferred from the data in this new image.

Key Words : Grey levels, Global thresholding techniques, Segmentation, Histogram Gaussian distributions, Subpixel precision.

1 Introduction

The method for segmenting a digital image is always closely related to the illumination scheme utilized. In certain environments, the digitization conditions (illumination technique, digitization process, etc.) are controlled. These conditions often provide well contrasted images in which segmentation is rather easy by means of global thresholding. These images present dark objects on a bright background or viceversa. But certain problems may arise such as presence of shadows, etc.[1] A good segmentation preserving the object shape is the first step towards a reliable attribute measurement when a high standard of precision is needed, or in any case in which shape is an important issue. Complex schemes have been used[2] for the automation of this process in non trivial images, with success only on specific conditions and with a high computational cost.

The information we "a priori" have about an image is its histogram, $H(z)$. A number of authors have used histogram characterization for image segmentation[3-5] or for avoiding expense computations on the image[6]. This permits a fast processing procedure, because, once the histogram is computed, all operations are $O(N_z)$, where N_z is the number of grey levels. If the distributions of dark and bright pixels can be clearly identified, then the image histogram is said to be bimodal, and each mode may be assigned either to objects of interest or background. As the frequency distributions for the bright and dark models become increasingly overlapped, the choice of a threshold in the valley between both modes becomes more difficult.

The digital nature of the image introduces errors in object feature measurements due to the pixel quantization. Certain magnitudes like areas and centroids are subjected to low precision when they are assessed using only the information contained in binary images. This is an important drawback when reliability in measurements with morphometric significance is needed[7]. Such effects could be minimized using the information contained in the grey levels of the contour pixels and the adjacent ones. This would increase the precision and invariance of descriptors[8].

In this work we propose to translate the global threshold determination problem in bimodal images into a transformation of the original image into an enhanced one that holds unified criteria. This new image is useful both to be easily thresholded always by the same threshold (to ease the segmentation problem) and for keeping grey levels in the transition zones (to deal with the subpixel approach). Due to the transformation definition, the grey level values in the transformed image can be considered as occupation percentage of the pixels, so subpixel reasonings can be inferred from the data in this new image. This transformation is self-adjusted from the analysis of the image histogram and implemented through a LUT. We name this procedure *pseudo-binarization*. The output images are then called *pseudo-binary* images.

This technique is useful in all situations in which the illumination conditions can be controlled enough to provide well contrasted images with objects and background that can be objectively isolated, like in many robotic applications[9], but also if images are well contrasted in any context[7,8]. Obviously, it is of interest when grey levels are wanted to be kept in transition zones.

2 The pseudo binarization procedure

2.1 Histogram analysis

One of the main issues of this procedure is its automatism. This automation comes from the analysis of the image histogram. But this image property is not usually suitable for the application of function analysis procedures, due to the stochastic component of its nature. A low-pass filtering of the histogram previous to its characterization, deletes unwanted "noise" letting the desired "signal" (the modes) arise. Let $H(z)$ the histogram of the original image. This discrete one-dimensional function will be filtered to produce a smooth reference histogram $H'(z)$ by means of a discrete convolution with a gaussian kernel,

$$G(z,w) = \frac{1}{\sqrt{2\pi w}} e^{-\frac{z^2}{2w^2}}$$

so the new histogram is expressed as $H'(z,w) = H(z) \otimes G(z,w)$.

This H' is easier to be analyzed, but is also dependent on the width of the filter w. This parameter is not conditionant because it can be chosen within a wide range of values without important variations in the outcome. The filtering provides a reference histogram for further characterization of both modes, but this process does not affect the image, because H' is just a tool for mode extraction, while the pseudo-binarization transformation is done on the original H. Thus, a reasonable wide gaussian is desirable: this way, only valid information survives. From these considerations, w has been heuristically set to 2.5 with good results in all images tested.

Once we have $H'(z,w)$ is which the two modes are hypothetically distinguishable, the aim is to extract the mode characterization through their approximation with two normal distributions; i.e. the means and standard deviations computed from the frequencies of H:

$$\mu_m = \frac{\sum_i z_i h(z_i)}{\sum_i h(z_i)} \quad \text{and} \quad \sigma_m^2 = \frac{\sum_i z_i h(z_i)}{\sum_i h(z_i)} - \mu_m^2 ,$$

where $h(z_i)$ represent the frequencies for each grey level z_i, $i \in [z_{min}, z_{max}]$, and $m = \{0,1\}$ for our problem (two modes: dark and bright). For isolating both modes, the bottom $h_0 = H'(z_0)$ of the valley between both modes in H' is seeked, and only values of $z \mid h'(z_0) > h_0$ are considered, being z_0 level that separates the modes (see Fig. 1).

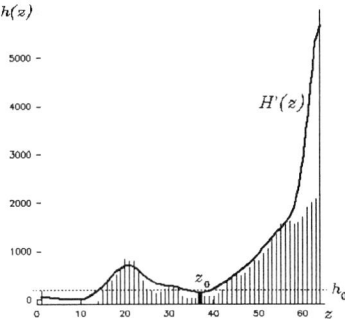

Fig. 1. Smoothed histogram and valley detection for mode characterization.

2.2 Image transformation

The pseudo-binarization method has been designed based on a contrast stretching procedure[9]. Using the above described characterization, a classification of the image pixels into three sets is performed. We assume $z_{min} = 0$.

1. *Pixels clearly belonging to mode 0*: those with their grey levels between the minimum level (z_{min}) and the dark mode mean plus a given number of times, n, its standard deviation ($\mu_0 + n\sigma_0$). In the transformed image the grey level $z' = z_{min}$ will be assigned to these pixels.

2. *Pixels clearly belonging to mode 1*: those with their grey levels between the bright mode mean minus n times its standard deviation ($\mu_1 - n\sigma_1$) and the maximum level (z_{max}). In the transformed image the maximum grey level, $z' = z_{max}$, will be assigned to these pixels.

3. *Intermediate pixels*: those between $\mu_0 + n\sigma_0$ and $\mu_1 - n\sigma_1$. They will take grey level values linearly expanding the histogram to cover the whole grey level range. If the image assumptions are held, these pixels will be adjacent to boundary zones.

The implementation of this classification generates a look up table (LUT) for the transformation, as observed in Fig. 2, that produces a contrast stretching in the histogram. This way, the objects can be isolated from the background and furthermore the grey level values are preserved at the edge points (in which the intermediate values in the grey scale are generated). This transformation can be analytically expressed as follows:

$$z' = T[H(z)] = \begin{cases} 0 & \text{if } z \leq \mu_0 + n\sigma_0 \\ \dfrac{z_{max}}{\mu_1 - n\sigma_1 - (\mu_0 + n\sigma_0)} z + \dfrac{z_{max}(\mu_0 + n\sigma_0)}{\mu_0 + n\sigma_0 - (\mu_1 - n\sigma_1)} & \text{if } \mu_1 - n\sigma_1 > z > \mu_0 + n\sigma_0 \\ z_{max} & \text{if } z \geq \mu_1 - n\sigma_1 \end{cases} \quad (1)$$

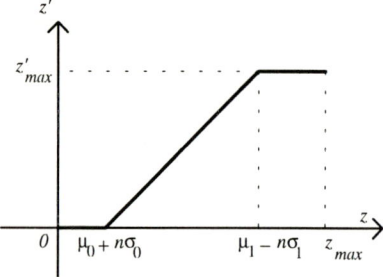

Fig. 2. Plot of the transference function used for the histogram transformation described in Eq. (1).

Only n remains to be selected to define the transformation. This n is also extracted from the characterization of both modes. For this, n is incremented, $n = 1,2,3,...$, and, for each n, $\mu_0 + n\sigma_0$ and $\mu_1 - n\sigma_1$ are computed until $\mu_0 + n\sigma_0 \geq \mu_1 - n\sigma_1$. Then, n is set one unity less in order to make room for the intermediate grey levels.

The pseudo-binary image obtained this way is easily segmentable by means of an edge tracking algorithm that separates pixels over the central value of the grey intensity range ($z' = z_{max}/2$) from those under it. Moreover, since image has not been binarized, the grey levels at the edge-adjacent zones are kept and linearly expanded to a maximum range (see result section), so subpixel precision can be inferred from these values interpreted as percentage of occupation in order to compute precise shape descriptors[7,8].

3 Results and Discussion

The behaviour of the proposed method is shown for images of real applications in different domains such as robotics[9] and medical imaging[7,8]. In Fig. 3, a number of images are presented: tools, CT scan, and a Moiré topography. The differences between dark and light modes have been maximized in the transformed image, and they can be segmented by a threshold = $z_{max}/2$. Note that, in the first example, false contours can be removed through morphological filters, if neccessary. In spite of the trimodal character of the CT image, the transformation is able to separate bone from soft tissue. Working with the resulting image is much easier and morphological filters or edge detection algorithms can help the final segmentation of the objects of interest. The third image shows how the method cope with a wide range of variations in illumination conditions, resolving low contrasted images. Finally, in the Moiré topography, fringes become well contrasted for segmentation.

The histograms for the first and second examples in Fig. 3 show the three diferent regions of the histogram after the application of the transformation: two single level modes at z_{min} and z_{max}, and some few intermediate levels (thick bars for the pseudo-binary image and thin bars for the original one). On the other hand, those of the third and fourth examples show the histograms H and H', and the detection of the separation level, z_0.

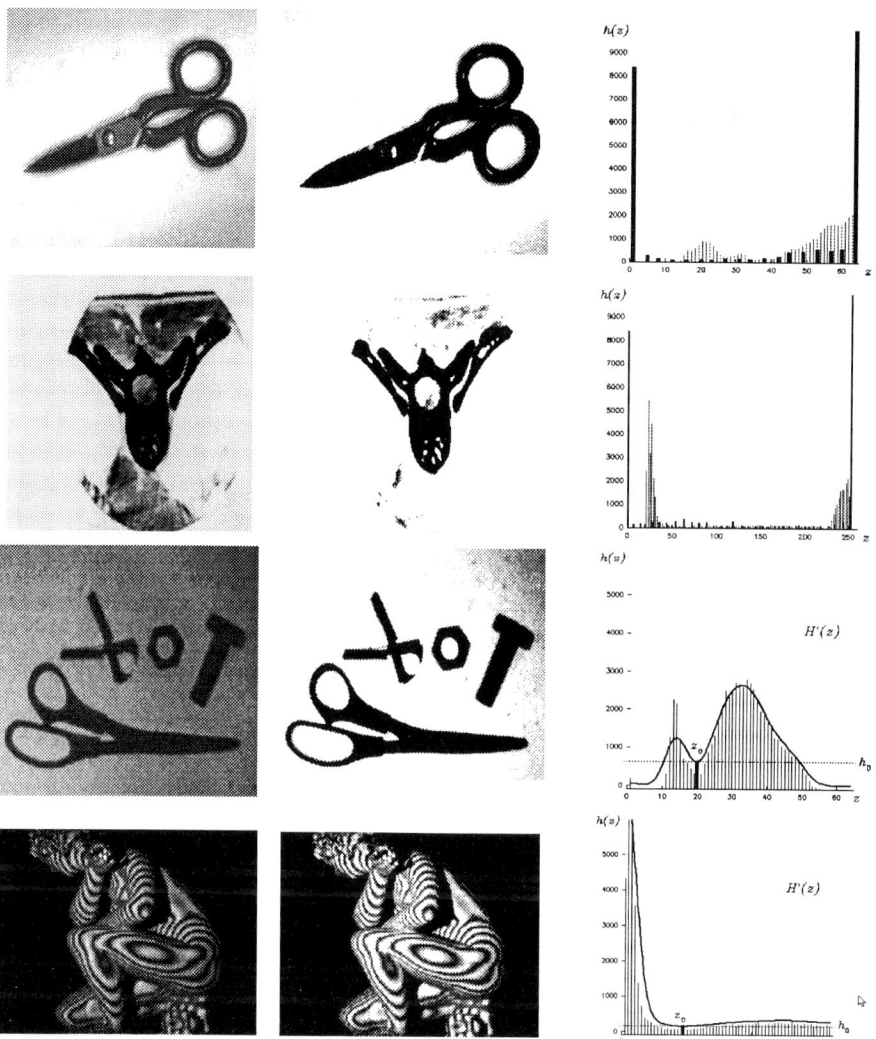

Fig. 3. (Left) Original images; (center) Pseudo-binary images; (right) Comparative histograms for the original (raw and smoothed) and transformed images. First row: scissors; Second row: CT scan of a human vertebra; Third row: low contrast image; Fourth row: Moiré topography.

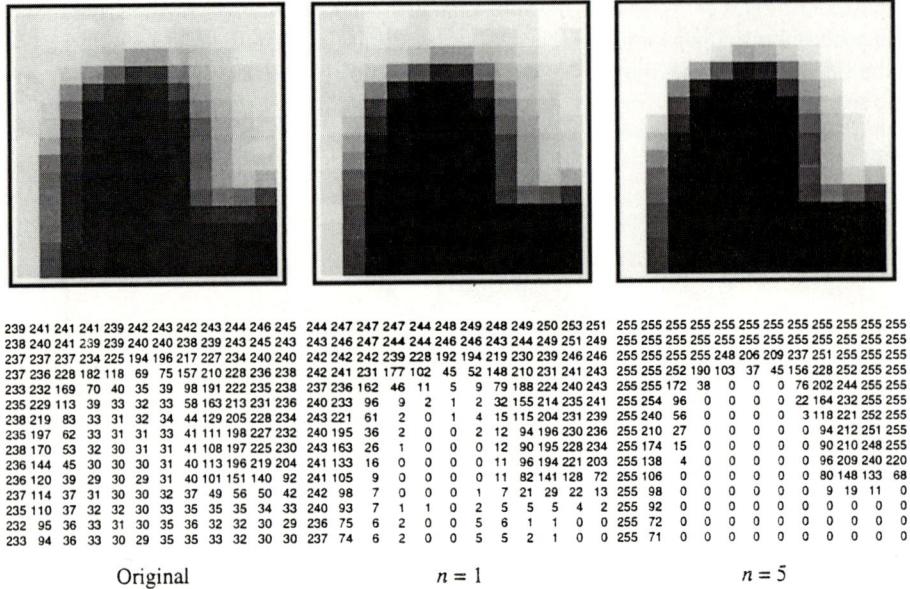

```
239 241 241 241 239 242 243 242 243 244 246 245   244 247 247 247 244 248 249 248 249 250 253 251   255 255 255 255 255 255 255 255 255 255 255 255
238 240 241 239 239 240 240 238 239 243 245 243   243 246 247 244 244 246 246 243 244 249 251 249   255 255 255 255 255 255 255 255 255 255 255 255
237 237 237 234 225 194 196 217 227 234 240 240   242 242 242 239 228 192 194 219 230 239 246 246   255 255 255 255 248 206 209 237 251 255 255 255
237 236 228 182 118  69  75 157 210 228 236 238   242 241 231 177 102  45  52 148 210 231 241 243   255 255 252 190 103  37  45 156 228 252 255 255
233 232 169  70  40  35  39  98 191 222 235 238   237 236 162  46  11   5   9  79 188 224 240 243   255 255 172  38   0   0   0  76 202 244 255 255
235 229 113  39  33  32  33  58 163 213 231 236   240 233  96   9   2   1   2  32 155 214 235 241   255 254  96   0   0   0   0  22 164 232 255 255
238 219  83  33  31  32  34  44 129 205 228 234   243 221  61   2   0   1   4  15 115 204 231 239   255 240  56   0   0   0   0   3 118 221 252 255
235 197  62  33  31  31  33  41 111 198 227 232   240 195  36   2   0   0   2  12  94 196 230 236   255 210  27   0   0   0   0   0  94 212 251 255
238 170  53  32  30  31  31  41 108 197 225 230   243 163  26   1   0   0   0  12  90 195 228 234   255 174  15   0   0   0   0   0  90 210 248 255
236 144  45  30  30  30  31  40 113 196 219 204   241 133  16   0   0   0   0  11  96 194 221 203   255 138   4   0   0   0   0   0  96 209 240 220
236 120  39  29  30  29  31  40 101 151 140  92   241 105   9   0   0   0   0  11  82 141 128  72   255 106   0   0   0   0   0   0  80 148 133  68
237 114  37  31  30  30  32  37  49  56  50  42   242  98   7   0   0   0   1   7  21  29  22  13   255  98   0   0   0   0   0   0   9  19  11   0
235 110  37  32  32  30  33  35  35  35  34  33   240  93   7   1   1   0   2   5   5   5   4   2   255  92   0   0   0   0   0   0   0   0   0   0
232  95  36  33  31  30  35  36  32  32  30  29   236  75   6   2   0   0   5   6   1   1   0   0   255  72   0   0   0   0   0   0   0   0   0   0
233  94  36  33  30  29  35  35  33  32  30  30   237  74   6   2   0   0   5   5   2   1   0   0   255  71   0   0   0   0   0   0   0   0   0   0
```

      Original $n = 1$ $n = 5$

Fig. 4. Numerical and grey values for pixels in an image window subjected to the transformation with different n values: (left) original image; (centre) using one standard deviation; (right) using five standard deviations.

    Fig. 4 shows how the contrast is being improved as n is increased. The linear definition for the central zone of the transformation transference function, permits to map the grey level values of the pseudo-binary image into occupation percentages. This fact implies that a global threshold that preserves shape should be defined like the one that sets the edges at 50% of object occupation, so the enhanced image should be always thresholded by the central value of the grey level scale.

    Once the image has been pseudo-binarized we can extract shape contours. Thus, the contours will be defined as the border points for the different 8-connected sets of pixels with grey values greater or less than $z_{max}/2$. Fig. 4 (left) shows the set of pixels that verify this condition. For locating the edges with subpixel precision, first the direction of the gradient vector at each border point is found. Then, the displacement of the center of the boundary pixel in the direction of the gradient is determined by the following equation:

$$\Delta = 1.5 - \frac{z + z_{+1}}{z_{max}} ; \qquad (2)$$

where z is the grey level at the boundary pixel and z_{+1} that of the 8-neighbour in the gradient direction. See in Fig. 5 (right) the results of applying this method with the values displayed in Fig. 5 (left). This equation works regardless of dark and bright modes respresenting either objects and background or vice versa.

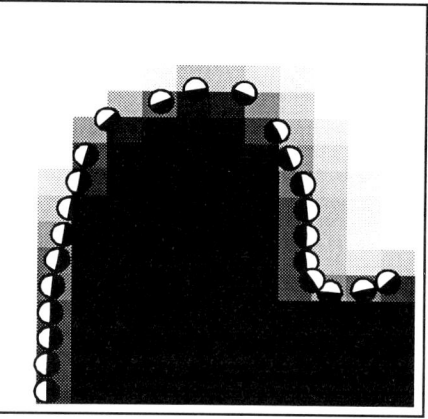

Fig. 5. (left) Boundary points. (right) Edge location with subpixel precision through the use of the grey level values in the pseudo-binary image. The circle represents orientation and position of the edge at each boundary point.

4 Conclusions

An efficient and simple procedure has been proposed to transform bimodal images into an enhanced one with unified criteria. This approach permits to translate the global threshold determination problem into an histogram analysis and characterization problem. In addition, this new image is useful either to be easily thresholded always by $z_{max}/2$ (to ease the segmentation problem) or to keep grey levels in the transition zones (to deal with the subpixel approach). Due to the transformation definition, the grey level values in the transformed image can be considered as occupation percentage of the pixels, so subpixel reasoning can be inferred from the data in this new image. Moreover, due to the linear expansion of the intermediate grey levels, the possible segmentation would be very respectful with the shapes in the image.

This method has been successfully utilized in robotics[10] and in medical imaging[7,8] applications.

Acknowledgements

This work has been funded by the projects: Fundació Caixa-Castelló 6I390; CICYT TAP 95-0710-C-01, Generalitat Valenciana GV-2214/94, Fundació Caixa-Castelló P1A94-22, and ESPRIT ("IOTA").

References

[1]   Davies ER. (1990) "Machine vision: theory, algorithms, practicalities". Academic Press Ltd.

[2]   Haralick RM and Shapiro LG. (1992). "Computer and robot vision". Addison-Wesley Publishing Company, Inc.

[3]   Price K. (1984) "Image Segmentation: a Comment on Studies in Global and Local Histogram-Guided Relaxation Algorithms". IEEE Transactions on Pattern Analysis and Machine Intelligence, vol. 6: 247-249.

[4]   Sang U, Lee K, Chung YS, and Park RH. (1990) "A comparative performance study of several global thresholding techniques for segmentation". Computer Vision, Graphics, and Image Processing, vol. 52(2): 171-190.

[5]   Whatmough RJ. (1991) "Automatic threshold selection from a histogram using the exponential hull". Computer Vision, Graphics, and Image Processing. Graphical Models and Image Processing, vol. 53(6): 592-600.

[6]   Maravall D, Sanandrés JA, and Baumela L. (1995) "Mobile Detection Based on Histogram Difference". Lecture Notes in Computer Science, 970. Springer-Verlag. pp. 286-293.

[7]   Sanz PJ, Iñesta JM, Buendia M, and Sarti MA. (1994) "A Fast and Precise Way for Computation of Moments for Morphometry in Medical Images". Proceedings of the V Int. Symposium on Biomedical Engineering, pp. 99-100. Santiago de Compostela, Spain.

[8]   Iñesta JM. (1994) "Algoritmos de Visión Artificial y de Reconocimiento de Patrones para el Estudio Morfométrico del Eje Raquídeo Humano". Tesis Doctoral. Dpto de Informática, Univ. de Valencia, Spain. (In Spanish).

[9]   Pratt WK. (1991) "Digital Image Processing". John Wiley and Sons, New York.

[10]   Sanz PJ. (1996) "Razonamiento Geométrico Basado en Visión para la Determinación y Ejecución del Agarre en Robots Manipuladores". Tesis Doctoral. Dpto. de Informática, Univ. Jaume I, Spain. (In Spanish).

A New Deformable Model for 3D Image Segmentation

Zixin Zhang, Michael Braun and Phillip Abbott

Department of Applied Physics, University of Technology, Sydney,
P.O.Box 123, Broadway, NSW 2007, Australia

Abstract. A fully 3D active surface model is presented with self–inflation and self–deflation forces. The model makes full use of 3D image information, deforms locally and allows strong deformation. The self–inflation and self–deflation forces enable the active surface to travel a long distance without the help from any external forces. We introduce a method of adapting model parameters, which enables our model to bypass some noise and irrelevant edge points. The model is tested with synthetic and real images. Accurate segmentation results are obtained in the presence of image noise and imperfect image data. Importantly, the model is capable of converging to the correct boundary even if the initial estimate is not close. Computational efficiency of segmentation with our model is addressed.

1 Introduction

Increasing availability of 3D image data, e.g. clinical CT and MR, has spurred growing interest in 3D image segmentation. By making full use of 3D image data, a segmentation algorithm is more accurate and more robust to noise and imperfect data than a 2D slice–by–slice approach.

Active surfaces [11] are a 3D generalization of the 2D active contour models (snakes) [8]. They are generalized splines with piecewise smoothness and have a diverse coverage of region shapes. They segment a 3D image by fitting the surface to desired image features, e.g. edge points. Once initialized in the vicinity of the desired image features, the surface evolves under the influence of internal and external forces until the balance of all forces is reached. The process of segmentation is formulated as the minimization of the energy cost functional of the active surface [11, 3]

$$E = \int\int_\Omega \left[w_1 E_{\text{int}}(s(u,v)) + w_2 E_{\text{ext}}(s(u,v)) \right] du dv, \qquad (1)$$

where $s(u,v) = (x(u,v), y(u,v), z(u,v))$ defines the Ω surface parameterized by u and v; E_{int} and E_{ext} are the internal and external energy, respectively; w_1 and w_2 are the corresponding weights. The internal energy enables the active surface to bridge gaps in edge data and to bypass image noise. The external energy serves to attract the active surface to the desired image features.

Polygonal closed meshes are commonly used to represent an active surface. Triangular and simplex [4] meshes can deform locally and thus are able to fit various shapes accurately. A surface can also be represented by nodes (vertices of mesh elements). The number of surface nodes can be constant or variable. With a constant number of nodes, the inter–nodal distance (or the size of the mesh), and thus the surface resolution will generally change during surface deformation. To maintain adequate surface resolution, it is generally necessary to vary the number of nodes as the surface deforms. Surface nodes are typically restricted to move within a plane. In fully 3D deformable models, surface nodes are free to move in any direction, which enables full use of 3D image data.

Of the several active surface models proposed [12, 3, 6, 4, 1, 13], most are not fully 3D. For example, Cohen & Cohen's model [3] allows surface node to move only within a plane. Whitaker *et al* [13] use 2D active contours but introduce interslice energy generated by extrapolating slice contours across slices. Some 3D models are excessively rigid due to the fixed number of surface nodes [12, 3, 4]. Huang *et al* [6] and Bulpitt & Efford [1] propose fully 3D surface models with triangular meshes and a variable number of surface nodes. Huang's model is designed specifically for the reconstruction of the left ventricle motion. Bulpitt & Efford's model aims at maintaining constant computational cost while setting adequate surface resolution by allocating more nodes to the high curvature parts of the surface.

Current models lack appropriate long distance attraction forces [9]. In particular, gradient forces operate over a short range. Some models overcome the problem by introducing spring forces [8, 4, 6, 1] which attract surface nodes to desired image features, e.g. edges. However, finding a node's corresponding edges is difficult and computationally expensive [6]. In the balloon model [2], a force normal to the surface inflates or deflates the surface to help drive it towards desired image features. It is an artificially added external force which, as implemented, is not adaptive.

Another limitation, crucial to successful implementation, is the shortage of practical methods for adapting model parameters to image data [9]. It is often desired to vary the values of model parameters from one image to another, from one iteration to another, and from one surface node to another. Among other major limitations, active surface models tend to shrink under the influence of internal force. They are also prone to becoming trapped by noise and irrelevant edge points. Though computationally efficient compared with region-based segmentation methods, computational cost for 3D image segmentation is significant due to the large data set.

2 Surface representation

Closed triangular meshes are used to represent the surface in our model (Figure 1). Each edge of a triangle is shared by exactly two triangles. This is the so called conforming triangular mesh [10]. Non–conforming triangular mesh will complicate the calculations of continuity and smoothness energies. The mesh elements are planar. The surface can also be described by surface nodes. Each node is directly connected to its (first order) neighbour nodes, and is shared by its (first order) neighbour triangles. The number of the neighbour nodes equals the number of the neighbour triangles. This number can vary from node to node, enabling the model to deform locally, thus permitting a diverse range of surface shapes.

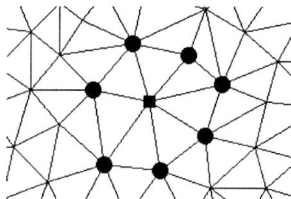

Fig. 1. Triangular meshes. A node j (solid square) is surrounded by its first order neighbour nodes (solid circles). Triangles with node j as a vertex are the first order neighbour triangles.

Our model allowes surface nodes to move in all directions during surface deformation to make full use of the available image information. As a result of node movement, the inter–nodal distances will change. A large inter–nodal distance may lead to important image features being ignored. On the other hand, a small inter–nodal distance requires more surface nodes to represent a surface, thus increasing computational cost. To maintain adequate surface resolution and high computational efficiency, node–insertion and node–removal [5, 7, 1] are introduced. A new node is inserted somewhere between two directly connected nodes if their inter-nodal distance is larger than a threshold. A node is removed if the inter-nodal distance is smaller than another threshold (see [15] for details).

The thresholds for both node-insertion and node-removal are application dependent. Also, they can be made dependent on surface smoothness and time (iteration). Surface smoothness dependency is effected by raising the thresholds where the surface is smooth. This will make dense distribution of nodes on parts of the surface with high curvature and sparse distribution on parts with low curvature. The time dependency is implemented by starting with and maintaining a small number of surface nodes during surface evolution, and adding more nodes in the final stages of the evolution. This control over node density can help speed up the evolution while providing the final segmentation result with adequate surface resolution (see section 5 for more discussion).

3 Energies

In this section, we describe the internal and external energies of our model as well as the self–inflation/deflation forces.

3.1 Surface energy

In our model, only smoothness energy is used for the internal energy, and the external energy simply consists of the edge energy. Kass' continuity energy [8] is redundant because our model maintains inter-nodal distance by node insertion and removal. Furthermore, continuity energy tends to pull surface nodes closer, leading to undesirable surface shrinking. The discrete form of our energy function is expressed as a sum over M surface nodes,

$$E = \sum_{j=1}^{M} \left(w_j^s E_j^s + w_j^e E_j^e \right), \quad \text{where } E_j^e = -|\nabla I_j|^2 ; \qquad (2)$$

at each node j, E_j^s is the smoothness energy, E_j^e the edge energy; w_j^s and w_j^e the corresponding weights, and ∇I_j denotes the gradient at node j of image intensity I. Since the method for calculation of the smoothness energy of a 3D active surface cannot be readily generalized from that of the 2D active contour, we designed our smoothness measures based on the angles formed by the surface normal at a node and the edges connecting the node to its neighbour nodes (see [15] for details).

3.2 Self-inflation and self-deflation forces

The self-inflation/deflation forces arise from the manipulation of the smoothness energy. The inflation/deflation switch and the adjustment of the force strength can be done merely by tuning a model parameter. The self-inflation/deflation forces enable the active surface to travel a long distance without any help from external forces (see [15] for details).

3.3 Energy minimization

Energy minimization is performed iteratively by the greedy algorithm [14]. A search neighbourhood consists of six closest voxels in 3D. A surface node is free to move to one of its neighbour voxel positions, or stay at its current position, depending on the location of the energy minimum. The size of the neighbourhood affects the number of iterations needed for surface evolution. Also, it affects the computational cost for each iteration. The larger the size of the neighbourhood, the higher the computational cost for each iteration.

4 Adaptation of model parameters

Noise and irrelevant edge points give rise to difficulties in image segmentation. Active surface nodes often stick to such points and fail to reach true edges. To reduce these effects, edge maps are commonly thresholded. However, thresholding eliminates only some noise and irrelevant edge points. We propose an adaptation of the edge energy weight w_j^e (Eq. 2) during surface evolution to help the model bypass noise and isolated irrelevant edge points.

The algorithm for adaptation of the edge energy weight is based on the assumption that noise and irrelevant edge points are isolated compared with true edge points. During surface evolution, once a surface node attaches to a true edge point, most of its neighbour nodes are expected to adhere to other true edge points in the next few iterations. If a surface node has stuck on a high gradient point and, after a few iterations, few of its neighbour nodes also stick to some other high gradient points, the point is regarded as an isolated noise or irrelevant edge point. By relaxing w_j^e (Eq. 2) at that surface node, active surface can bypass the noise or isolated irrelevant edge point under the action of the smoothness force. The procedure is as follows:

1. test if the edge strength at a surface node is larger than a certain threshold; if not, test the next surface node;
2. count the neighbour nodes whose edge strength also exceeds the threshold;
3. if the counter is less than, say 30%, in several consecutive iterations, say 5, relax the weight of edge energy at that surface node.

In generally, this algorithm is less effective for 2D active contour models because each contour node has only two neighbour nodes. This manifests one of the advantages of full 3D segmentation over 2D.

5 Computational efficiency

Though active surface models are computationally efficient relative to region-based segmentation methods, computational cost for 3D image segmentation is significant due to the large image matrix. Computational cost of our model with greedy algorithm is proportional to the number of iterations needed for the active surface to converge to region boundaries. For each iteration, the computational cost is proportional to the number of surface nodes and the size of the search neighbourhood for each node.

A large surface mesh size suggests a small number of surface nodes needed for surface representation, reducing the computational cost for surface evolution. It also suggests low surface resolution. To minimize computational cost while obtaining adequate surface resolution, a large surface mesh size is applied at early stages of surface evolution, whereas a small size of surface mesh is applied at final stages. The size of the surface mesh is controlled by the inter-nodal distance thresholds for surface node insertion and removal. Therefore, by manipulating the inter-nodal distance thresholds, computational cost can be reduced.

The active surface model, with the greedy algorithm, lends itself well to a parallel implementation because the measures of both smoothness and edge energies are local. We implement the parallelisation by splitting the linked list, which connects surface nodes, into sections and assigning each section to a different processor. The algorithm runs on a multi-processor architecture (Sun Ultra–2 with two processors). We are investigating the computational gain of this and other possible approaches to parallelisation.

6 Experiments and results

Two 3D synthetic images and a 3D CT image were used. The synthetic images were of size 128 × 128 × 128 with added Gaussian noise of signal to noise ratio of 1. A simple image contains a single ellipsoid and a compound image contains two partly merged ellipsoids. The CT image, from the Visible Human Set at the National Library of Medicine, Bethesda, MD, USA, was interpolated to 1mm interslice spacing. A Sun SPARC–10 workstation was used in our experiments.

Experiment 1: To demonstrate the deformability of the active surface model with the self–inflation/deflation forces, the compound image was segmented with an initial spherical surface placed within the object (Figure 2). The segmentation was successful even though the object splits into disjoint regions in some slices.

Experiment 2: To demonstrate the ability to bypass isolated irrelevant edge points by parameter adaptation, a bar and a cluster of high gradient points were added to the edge map of the simple image (Figure 3). Normally, the active surface sticks to the high gradient points (Figure 3c). With the adaptation of edge energy weight w_j^e (Eq. 2), the active surface bypassed the high gradient points (Figure 3d).

Experiment 3: To investigate the computational efficiency, the simple synthetic image was segmented with and without manipulating the number of nodes during surface evolution. An initial spherical surface was placed within the object. Controlling the number of surface nodes resulted in about 40% reduction of computational time (from 5m 35s to 3m 15s).

Experiment 4: To test the model incorporating the self–inflation/deflation forces and the adaptation of model parameters, a kidney in the 3D CT image was segmented (Figure 4). The initial surface was a sphere totally contained within the kidney. Examination of each cross section of the segmentation result by eye revealed that the segmentation was accurate.

7 Conclusion

We present a fully 3D active surface model with the self–inflation and self–deflation forces. Our surface model has the following features: surface nodes are free to move in all directions, thus enabling the model to make full use of 3D image information; the model is deformable locally, thus providing coverage for a diverse range of surfaces; the model is capable of strong deformations, thus reducing its sensitivity to initial shape and position. The self–inflation/deflation forces permit the surface to travel a long distance without the aid of external forces. They are simple and easy to control. We introduce an adaptation of model parameters. By adapting the weight of edge energy at each surface node, the model is capable of bypassing some noise and isolated irrelevant edge points. By adapting the inter–nodal distance thresholds for node insertion and node removal, the model is capable of adjusting the number of surface nodes during surface evolution so that the computational cost is minimum compatible with adequate surface resolution of segmentation.

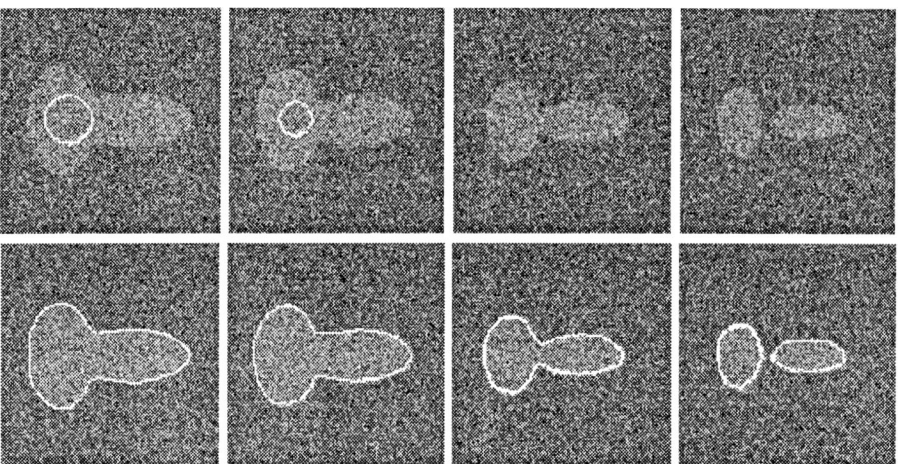

Fig. 2. The representative parallel cross sections of the compound synthetic image showing the initial (top row) and the final surface (bottom row).

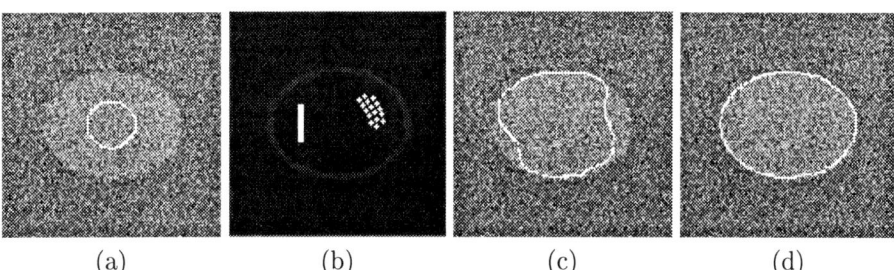

Fig. 3. A central cross section of the simple synthetic image showing the initial surface (a), edge map with added high gradient points (b), and the final surface segmented without (c), and with (d), parameter adaptation.

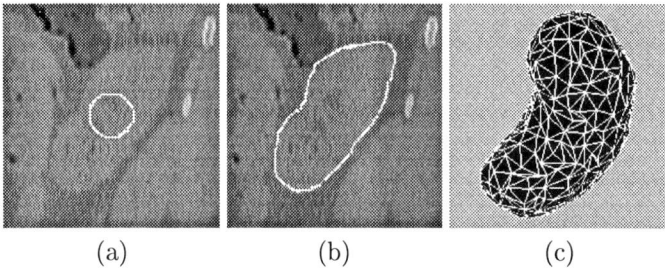

Fig. 4. A central cross section of the CT image with the initial (a) and final (b) surface. A perspective view of the final surface mesh is shown in (c).

The model is tested with two synthetic images and a 3D CT image. Accurate segmentation results were obtained. The full 3D active surface model produced correct segmentation where an object splits into disjoint regions in some slices. By combining the self–inflation/deflation forces and the adaptation of model parameters, our model is computationally efficient, capable of converging to the correct boundary even if the initial estimate is not close, and appears to be robust to noise and imperfect image data.

References

1. Bulpitt A.J. and Efford N.D.: An effective 3d deformable model with a self-optimising mesh. *Image Vision Computing* **Vol.14** pp 573–580, August 1996.
2. Cohen L.D.: On active contour models and balloons. *Computer Graphics and Image Processing: Image Understanding* **Vol.53, No.2**, pp 211–218, 1991
3. Cohen L.D. & Cohen I.: Finite–element methods for active contour models and balloons for 2–D and 3–D images. *IEEE Trans. Pattern Analysis and Machine Intelligence* **Vol.PAMI–15, No.11**, pp 1131–1147, 1993
4. Delingette H.: Adaptive and deformable models based on simplex meshes. In *IEEE Computer Vision and Pattern Recognition* **CVPR'94**, pp 152–157 1994
5. Hoppe H. et al: Piecewise smooth surface reconstruction. *Computer Graphics Proceedings, Annual Conference Series,1994* **SIGGRAPH 94**, pp 295–302 1994
6. Huang W.C. & Goldgof D.B.: Left ventricle motion modeling and analysis by adaptive–size physically–based models. *SPIE—Biomedical Image Processing and Three–Dimensional Microscopy* **Vol.1660**, pp 299–310 1992
7. Gupta A., O'Donnell T. & Singh A.: A 3–D deformable model for segmentation and tracking of anisotropic cine cardiac MR images. *SPIE—Image Processing* **Vol.2167**, pp 77–86, 1994
8. Kass M., Witkin A. & Terzopoulos D.: Snakes: active contour models. *International Journal of Computer Vision* **Vol.1**, pp 321–331, 1988
9. Leymarie F. & Levine M.D.: Tracking deformable objects in the plane using an active contour model. *IEEE Trans. Pattern Analysis and Machine Intelligence* **Vol.PAMI–15, No.6**, pp 617–634, 1993
10. Rivara M.C.: Algorithms for refining triangular grids suitable for adaptive and multigrid techniques. *International Journal for mathematical methods in engineering* **Vol.20**, pp 745–756, 1984
11. Terzopoulos D., Witkin A. & Kass M.: Constraints on deformable models: recovering 3D shape and nonrigid motion. *Artificial Intelligence* **Vol.36**, pp 91–123, 1988
12. Terzopoulos D. & Vasilescu M.: Adaptive surface reconstruction. *SPIE—Sensor fusion III: 3–D perception and reconstruction* **Vol.1383**, pp 257–264, 1990
13. Whitaker J.M. & Braun M.: Three dimensional image segmentation using active contours with interslice energy. *Proc. APRS/CBT Image Segmentation Workshop*, Sydney, Australia pp 47-51, December 1996
14. Williams D.J. & Shah M.: A fast algorithm for active contours and curvature estimation. *Computer Graphics and Image Processing: Image Understanding* **Vol.55, No.1**, pp 14–26, 1992
15. Zhang Z. & Braun M.: Fully 3D active surface models with self-inflation and self-deflation forces. *Proc. IEEE Computer Society Conference on Computer Vision and Pattern Recognition (CVPR'97)* San Juan, Puerto Rico, June 1997

Evolutionary Image Segmentation

Primo Zingaretti, Antonella Carbonaro, Paolo Puliti

Istituto di Informatica, Facoltà di Ingegneria, Università di Ancona
Via Brecce Bianche, I-60131 Ancona (Italy)
e-mail: {zinga, carbonar, puliti}@inform.unian.it

Abstract - We describe an approach to image segmentation based on a two-layer module that is executed until a good segmentation is achieved, providing an evolution of previous segmentation results at each execution. The first layer performs a global segmentation of an image of decreasing area at each evolution by adopting a genetic algorithm learning technique to select segmentation parameters that give better results. The second layer provides the input to the next evolution by selecting the segmented regions that need further optimisation. A main goal of our system is to perform the segmentation without using neither ground-truth information nor human judgement. Thus, edge detection is performed to assess the performance of region segmentation and to guide the evolution of segmentation. Experimental results are consistent with what is observed visually.

1. Introduction

Image segmentation is the process of classifying an image into a set of disjoint regions whose characteristics such as intensity, colour, texture, etc. are similar. Image segmentation is a very important process because it is typically the first task of any automatic image understanding process, and all subsequent steps, such as feature extraction and classification, object detection and recognition, depend heavily on its results. The segmentation process is a complicated task because most of available segmentation techniques use numerous control parameters that must be adjusted to obtain an optimal performance. In practice, we do not know general automatic systems whose parameters provide optimal performance for all images; even the choice of an ad hoc setting based on the actual algorithm or *a priori* domain knowledge may result in an unsatisfactory segmentation. Many different segmentation algorithms have been developed [6, 9]. Most of them are based on thresholding techniques, which usually are further classified in global, local, region or edge based techniques. A performance evaluation of the different algorithms proposed for a given application is inherently difficult to obtain, due to the lack of appropriate measures for judging the quality of segmentation results. None of the segmentation quality measures suggested in the literature [6, 14] has achieved widespread acceptance as a universal measure. At present most segmentation results are evaluated visually and qualitatively.

Image segmentation is formulated in this paper as an optimisation problem. A genetic algorithm (GA) [4] is used to efficiently search, in the hyperspace of segmentation parameters, the set that maximises the segmentation quality criteria. GAs have been recently applied in machine vision problems (for example, image segmentation [2, 3], edge detection [1] and target recognition [8]) mainly because they can locate an approximate global maximum in a search space in a way not dependent on the particular application domain and without using detailed knowledge about the processing technique. The goal of our system is not to build a new segmentation algorithm for a certain image class, but, rather, to develop a most general possible technique that uses neither ground-truth information nor human judgement in the evaluation of segmentation results.

The paper is organised as follows. Section 2 describes the evolutionary image segmentation process. Experimental results are presented in Section 3, while some conclusions and future research plans in Section 4.

2. Image Segmentation Using a Genetic Algorithm

In this Section we describe the architecture and some implementation aspects of the developed system.

2.1. System architecture

The proposed segmentation system is based on a module consisting of two layers: *global segmentation layer* and *single-region evaluation layer*. The module is executed until a good segmentation is achieved, providing an evolution of previous segmentation results at each execution. Our strategy is based on the hypothesis that, by modifying processing parameters, it is possible to obtain a good segmentation at least for some regions. This strategy extends the space search, but the difficulty in finding a good solution among many alternatives should be overcome by the GA. Fig. 1 shows the global architecture and the information flow for the approach described in this paper.

The global segmentation layer adopts a genetic learning technique to produce a new classification of pixels into regions. The GA analyses the characteristics of the input image and uses this information to select an appropriate parameter combination to segment the image. A fitness is then evaluated and the GA cycles until a segmentation result of acceptable quality is reached, that is, the stopping criterion of the GA is based on a fitness threshold T_1 that is independent from the adopted processing parameters. In this way, the choice of the number of thresholds (over-under segmentation) is not a problem any more.

The single-region evaluation layer is mainly based on the local evaluation of fitness criteria for each region resulting from the global segmentation. The aim of this layer is to select those regions that need further optimisation. The set of all these regions will constitute the object-of-interest image (OOI) for the next evolution.

The global segmentation layer performs a segmentation of the whole image at the first application of the module and of the generated OOIs at the successive evolutions. First a subset of individuals (seed population, SP) that has image features more similar to those ones of currently processed OOI is extracted from a long term population (LTP) characterised by N individuals (strings). The similarity measure is

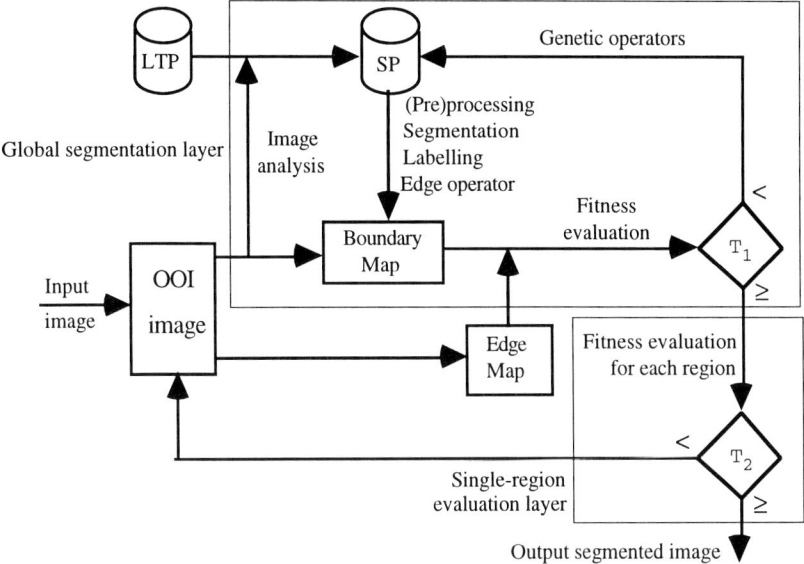

Fig. 1. System architecture.

based on the Euclidean distance normalised with respect to the range of actual feature values, and weighted on the importance of each feature, that is:

$$dist_{ab} = \frac{1}{NF} \sum_{i=1}^{NF} w_i \left| \frac{F_{ia} - F_{ib}}{\max(F_i) - \min(F_i)} \right|$$

where: $dist_{ab}$ is the similarity measure between the two individuals a and b; NF is the number of features represented in the image feature part of the string; w_i is the weight assigned to each feature; F_{ix} is the value of feature i for the individual x; $\max(F_i)$ and $\min(F_i)$ are maximum and minimum values of feature i in the current global population. If the features have the same importance, as in the present application, $w_i = 1$ for all i. Then, image segmentation is performed using the processing parameters of the individuals extracted from the seed population and the genetic operators are applied on the processing parameters of these individuals until the global fitness does not exceed T_1 (see Fig. 1).

Our segmentation system is based on the multithresholding of the intensity histogram [12]. Histogram-based thresholding is computationally simpler than other existing algorithms and does not need a priori information of the image, but using global features it may be not optimal for all objects of the input image. The parameters of the global segmentation layer include the (pre)processing of the input image, which may be based on global features, such as histogram transformations, or may concern local operations, such as spatial filtering and noise removal. Both these operations should be considered as "global processing" because the processing is applied to the whole OOI, even if it enhances some local features of the input image. A real "local processing" is introduced by means of successive evolutions. In fact, at each evolution the second layer should exclude from the input OOI those regions with a fitness value greater than a given threshold T_2. Then in the next evolution the global processing will be performed on the histogram of a smaller OOI, that is, in a more local manner. In this sense we say that our approach uses global techniques in a local context, and then it is able to adapt to varying image characteristics.

The aim of this layer is to reduce the total computational cost of the system by decreasing the number of fitness evaluations, which are computational expensive. As our approach requires the segmentation evaluation of each region of the segmented image, the splitting of the system in two layers avoids to compute a high number of fitness evaluations at each segmentation performed by the GA. In fact, for each segmentation performed by the GA in the first layer, the fitness is evaluated only one time by considering all the regions as a whole, while in the second layer the fitness is computed for each region resulting from the global segmentation. Thus, in our system the total number of fitness computations is given by: $N + R$, where N is the number of segmentations performed by the GA and R is the number of regions in the final segmented image. On the contrary, a system with a unique layer should be compelled to evaluate about $N*R$ fitness evaluations, hypothesising that the regions after each segmentation are R on the average.

2.2. System implementation

In this Section we describe in detail the three parts that constitute the knowledge structure of each individual of the GA population: image features (used to discriminate the images presented to the system), segmentation parameters (processing alternatives) and fitness (used to quantitatively evaluate the segmentation effectiveness).

2.2.1 Image features

The system uses two categories of features: global features of the image and features of the image histogram. We have considered as global features the normalised abscissa and ordinate of the intensity centroid, the ratio between the inertial major axes, the angles of the inertial major axes, and the seven invariant moments proposed by Hu [7]. From the image histogram of grey level intensities we have computed the number of peaks, the grey levels and the normalised frequencies of the maximum peak and the minimum valley, the distance between the two valleys with the greater number of pixels and the normalised maximum frequency of these regions, the amplitude of the image spectrum. In addition, we have considered the mean, standard deviation, skewness, kurtosis, and entropy of the histogram to measure its shape.

2.2.2. Segmentation parameters

The segmentation parameter part of each individual of the GA population is constituted by: (pre)processing operations (such as averaging, sharpening and median filtering, noise removal and introduction of local characteristics in the image intensity histogram by using gradient and laplacian operators [11, 13]), number of histogram averaging (to modify the number of thresholds by which the image will be segmented), object size filter (to reduce the computational cost of the labelling process by excluding small regions), and coefficient of small object removal (to determine the number of erosions and dilations performed to fill small holes inside regions).

2.2.3 Fitness

We perform edge detection to assess the performance of region segmentation and then to guide the evolution of segmentation. In this sense the contribution of our paper may be viewed as a new method for the integration of region-based and edge-based segmentation techniques [5, 10]. In this context, segmentation results suffer from three kinds of errors: 1) *false boundary* - a region boundary is not an edge and there are no edges nearby (for example, if our uniformity criterion is to keep intensity approximately constant over a given area and the light intensity varies linearly within a region then artificial boundaries need to be set even if there is no clear line where a transition occurs); 2) *imprecise boundary* - a region boundary does not coincide with an edge (for example, highlights may displace the real region boundary, at least in some of its parts); 3) *missing boundary* - there exist edges in the image with no region boundaries near them. It should be pointed out that if we reduce errors of the last kind by performing an over-segmentation (for example, by allowing small valleys in the histogram or by gathering thresholds according to different criteria) the probability of errors of the first kind will increase.

We have chosen as fitness criterion the correspondence between boundaries of regions resulting from segmentation (*boundary map*) and edges obtained from the Roberts edge operator (*edge map*). This correspondence is difficult to find out in real images even in presence of a good segmentation, so we have introduced a matching tolerance, dependent on image and region size, to overcome errors in the extraction of significative edges. The achievement of a good segmentation is represented by a fitness higher than T_2. For the efficiency of our two-layer approach the threshold T_1, adopted in the global segmentation layer, should be high enough to permit the existence of at least one region exceeding T_2, but also not too high to avoid a great number of GA iterations. In any case $T_1 < T_2$.

In particular, we compute two types of correspondences: F_1, an exact correspondence between boundaries and edges, and F_2, a fuzzy correspondence. In particular, imprecise boundaries occur whenever F_1 is low and F_2 high. These values are computed using the following two formula:

$$F_1 = \frac{P_1}{\min(N_1, N_2)} \qquad F_2 = \left(P_1 + \sum_{i=0}^{d-1} \frac{hist(i) \cdot (d-i)}{d} - \frac{E_1 + E_2}{2} \right) \Big/ N_1$$

where: P_1 is the number of pixels resulting from a logical *'and'* between boundary and edge maps; N_1 is the total number of pixels in the boundary map; N_2 is the total number of pixels in the image obtained from the binarization of the edge map at the threshold that minimises the difference between N_1 and the pixels resulting from the binarization itself; d is the tolerance in the matching distance; $hist(i)$ is the number of boundary pixels that have a distance i from the nearest edge pixel in the corresponding edge map; E_1 is the number of boundary pixels that have a distance greater than d from any edge pixel in the corresponding edge map; E_2 is the number of edge pixels that have a distance greater than d from any boundary pixel in the corresponding boundary map.

The fitness function, F, also used in the single-region evaluation layer, is obtained by weighting in the same way the two correspondence measures:

$$F = \frac{1}{2}F_1 + \frac{1}{2}F_2$$

A final global fitness F_{tot} is computed as a weighted average of the areas and the fitness values of each resulting region for the objective evaluation of the output segmented image:

$$F_{tot} = \sum_i F_i \cdot \frac{area_i}{area}$$

3. Experimental Results

The work is in progress. Further (pre)processing techniques need to be coded in the segmentation parameter part of an individual and many other images need to be processed to construct an efficient LTP. Due to these limitations the convergence of the GA in some images is not so quickly as it should be.

The system has been tested on real images using an LTP consisting of 100 individuals characterised by processing parameters that yielded visually good segmentation results on a set of sample images. As a processing example, elaborations performed on the standard image "camera", shown in Fig. 2, are described in the following. The processing parameters adopted by the first evolution of the global segmentation layer, consisting of 5 iterations of the GA, were: a noise removal coefficient of 1 as pre-processing operation, 40 histogram averagings, and a minimum object size of 50 pixels. The selected thresholds yielded the 38 regions shown in Fig. 3. The global fitness was 0.806 (F_1 = 0.708 and F_2 = 0.904, T_1 = 0.750). The single-region evaluation layer extracted the coat-region, white region R reproduced in Fig. 4, as the only one region with a fitness value (F = 0.907) greater than T_2. Normally the system continues by considering the complement $\overline{R}$ of R as the OOI for the next evolution. In this case, as R constituted a large part of the initial image, two separate processings of the evolutionary system were executed for each of the two OOIs, R and $\overline{R}$, and a combination (add operation) of the results was performed at the end. After other 9 evolutions for the processing of R (in order they were extracted: tower and near buildings, sky, leftmost grassland, rightmost grassland, etc.) and only one evolution for $\overline{R}$, the final segmented image shown in Fig. 5 was obtained. This image is constituted of 39 regions plus a set of unclassified small regions (black holes representing only 4.38% of the whole image). The global fitness for the objective evaluation of the segmentation result was F_{tot} = 0.829. Comparisons with other segmentation algorithms, as above mentioned, are difficult to perform. Anyway, our final segmentation result was an improvement with respect to the global segmentation of Fig. 3, at least from a visual comparison.

The computational cost of the GA, synthesised by the number of evolutions and of iterations for each evolution, and the segmentation results, in terms of number of regions, percentage of unclassified regions (holes) and global fitness, have been analysed also for other standard grey level images. Different behaviours for these

Fig. 2. The grey level standard image "camera" 256x256x8 pixels.

Fig. 3. The image obtained from the global segmentation layer.

Fig. 4. The OOIs generated by the first evolution of the system.

Fig. 5. Final result of the segmentation process.

parameters have been noticed. In particular, some images required a great number of iterations of the first layer and many holes resulted in some other segmentations. In both these cases better results were obtained by lowering the T_1 threshold. The T_2 threshold is also a critical factor. While a low T_2 impairs the segmentation results, a high T_2 heavily increments the computational cost because in many evolutions no region resulting from the first layer can be considered as well segmented, that is, no real evolution occurs. Finally, in other images, for example in the standard "Lena" image, we have to introduce a compactness factor in the fitness function to avoid the inclusion of regions of high complexity that overcome the object size filter.

4. Conclusions

In this work we have described an automatic two-layer approach to image segmentation that, using an adaptive methodology, globally searches for a good segmentation, and then performs subsequent local segmentation refinements without using neither ground-truth information nor human judgement. We have outlined a GA-based system using many of standard segmentation techniques to select optimal image processing parameters.

Experimental results exhibit a promising performance for the proposed method. In fact, the final segmented images are consistent with what is observed visually. Great improvements appear to be possible without requiring substantial modifications of the structure. The integration of other (pre)processing alternatives, such as histogram modification techniques, need only to be included in the parameter part of the GA string, apart from their implementation. In addition, future works will focus on other evaluation techniques, for example by introducing perceptually motivated features, and on improvements regarding refinements of extracted regions.

References

[1] S.M.Bhandarkar, Y.Zhang, W.D.Potter, An Edge Detection Technique Using GA-Based Optimisation, *Pattern Recognition*, 27(9), 1159-1180, (1994).

[2] B.Bhanu, S.Lee, J.Ming, Adaptive Image Segmentation using a genetic algorithm, *IEEE Trans. on SMC*, 25(12), 1543-1567, (1995).

[3] D.N.Chun, H.S.Yang, Robust image segmentation using genetic algorithm with a fuzzy measure, *Pattern Recognition*, 29(7), 1195-1211, (1996).

[4] D.E.Goldberg, *Genetic Algorithms in Search, Optimisation, and Machine Learning*, Addison-Wesley, Reading, MA, (1989).

[5] J.F.Haddon and J.F.Boyce, Image Segmentation by unifying region and boundary information, *IEEE Trans. on PAMI*, 12(10); 929-948, (1990).

[6] R.M.Haralick, L.G.Shapiro, Image segmentation techniques, *Computer Vision, Graphics, and Image Processing*, 29, 100-132, (1985).

[7] M.K.Hu, Visual problem recognition by moment invariants, *IRE Trans. Inf. Theory*, 8, 179-187, (1962).

[8] A. J. Katz, P. R. Thrift, Generating Image Filters for Target Recognition by Genetic Learning, *IEEE Trans. on PAMI*, 16(9); 906-910, (1994).

[9] N.R.Pal and S.K.Pal, A review on image segmentation techniques, *Pattern Recognition*, 26, 1277-1294, (1993).

[10] J.T.Pavlidis and Y.T.Liow, Integrating region growing and edge detection, *IEEE Trans. on PAMI*, 12(3); 225-233, (1990).

[11] G.Tascini, P.Puliti, P.Zingaretti, Region Detection in grey-level images, in *"Progress in Image Analysis and Processing"*, Cantoni, Cordella, Levialdi, Sanniti di Baja Eds., World Scientific Publ. Co., Singapore, 106-110, (1990).

[12] S.Wang and R.M.Haralick, Automatic multithreshold selection, *Computer Vision, Graphics, and Image Processing*, 25, 46-67, (1984).

[13] J.S. Weszka and A.Rosenfeld, Histogram modification for threshold selection, *IEEE Trans. on SMC*, 9(1), 38-52, (1979).

[14] Y.J.Zhang, A survey on evaluation methods for image segmentation, *Pattern Recognition*, 29(8), 1335-1346, (1996).

Discontinuity Adaptive MRF Model for Synthetic Aperture Radar Image Analysis

P.C. Smits[1], S.G. Dellepiane[1], and G. Vernazza[2]

[1]University of Genoa
Dept. Biophysical and Electronic Eng
Via all'Opera Pia 11A, Genova Italy
{smits,silvana}@dibe.unige.it

[2]University of Cagliari
Dept. Electrical and Electronic Eng
Piazza d'Armi, Cagliari, Italy
vernazza@diee.unica.it

Abstract. In this paper, an approach is presented for the reconstruction and analysis of synthetic aperture radar (SAR) images that preserves better fine structures and borders in the image than classical methods. The method uses the discontinuity adaptive MRF model proposed by Li [1] in combination which an observation model that exploits a gamma distribution. This resulted in a new algorithm that is suited to the analysis of SAR images.

1. Introduction and problem definition

Image restoration, segmentation and classification of images can be formulated as ill-posed problems. Although the quality of many modern imaging sensors is such that these problems are not too ill-posed, this situation changes in cases where one wants to extract types of information for which the sensor has not been build in the first place. In order to tackle an ill-posed problem in such cases, a-priori constraints or other sources of information are important for the regularization of the problem. In this article, we focus on the problem of SAR intensity image restoration for land-cover mapping applications. This data may contain regions that have geometrically difficult shapes, like fine structures and critical borders between classes, that are obscured by speckle noise. Existing literature does address the incorporation of adequate statistical models in segmentation algorithms for this type of data, but fails to pay sufficient attention to the aspect of preserving small structures.

Markov Random Field (MRF) approaches have shown to be useful because of the ability to define the spatial interaction between the pixels in the image. In the literature, the interactions between pixels are fixed for the entire image, and may lead to undesired smoothing. This paper builds on two lines of research reported in the literature. One concerns aspects related to the statistics of SAR intensity data, and the other relates to recent developments in the use of a more precise MRF models. The novelty of this paper is the combination of a discontinuity adaptive (DA-) MRF model that accounts for small structures and discontinuities as proposed in [1,2] and the Gamma distribution as the image model, which appeared to be useful for modeling SAR data [3].

This article is organized as follows. Section 2 gives an overview of the proposed method. Section 3 elucidates the observation model as presented in [3]. Section 4 outlines Li's DA-MRF model. How the observation and the regularization model are merged is explained in section 5. Section 6 shows some experimental results, followed by discussion and conclusions (section 7).

2. Method overview

In the approach proposed in this paper, an MRF model is utilized in which regularization constraints like smoothness are encoded into an energy following a probabilistic route. Basically, the model consists of the sum of two energy terms. One term is the observation model, which defines the relation between the observed intensity data and the image labels (see Fig. 1). The energy returned by this term is a measure for the closeness of the observation to an image class model. The second term is a regularizer, and penalizes the irregularities according to the a-priori smoothness constraint encoded in it (see Fig. 2).

Both energy terms have been object of study: the observation model since it has to cope with images degraded by speckle noise with its typical gamma distribution [3], and the regularizer because in many image analysis applications the assumption of the uniform smoothness everywhere can lead to undesirable, over smoothed, solutions ([2],[4]).

Fig. 1. Data flow diagram of the observation model.

Fig. 2. Data flow diagram of the regularizer.

3. Observation model

In this section, we are concerned with the observation model, i.e., the relation between the intensity data I and our image labels L. In the literature agreement exists on the gamma distribution being one of the most suitable one for SAR data. The distribution for multilook intensities are modeled as [3]:

$$p(I_s / L_s) = \frac{N^N I_s^{N-1}}{\langle I \rangle_l^N \Gamma(N)} \exp\left(-\frac{NI_s}{\langle I \rangle_l}\right), \qquad (1)$$

where s is the index for the location, I is the intensity of the SAR data, L the label, N the number of independent one-look samples used to form each multilook intensity sample I_s and $\Gamma(N) = (N-1)!$. The energy function U_1 is given by

$$U_1(I_s / L_s) = \frac{NI_s}{\langle I \rangle_{L_s}} - (N-1)\log(I_s) + N\log\left(\langle I \rangle_{L_s}\right). \qquad (2)$$

Equation (2) is the class-conditional energy function or image model that will be used in section 5. Basically it says that, given a class L_s it returns a low value (good) if the intensity of a pixel under test s is close to the mean intensity of the suggested class L_s, and a high value (not so good) if otherwise.

4. Discontinuity adaptive MRF model

The MAP labeling with a prior potential is equivalent to the regularization of an ill-posed problem. An important assumption of the *classical* implementation of a regularization constraint is the smoothness. It is incorporated into the energy function whereby the cost of the solution is defined. A regularized solution corresponds to the maximum a-posteriori estimate of an MRF.

In this article, the smoothness constraint is not blindly applied to an entire image, but the introduction of line processes or weak continuity constraints will control the application of the strength and shape of the smoothness. The DA-MRF model for image analysis has proven to be a valuable alternative to the classical MRF model. We follow [2] in the definition of the DA-MRF model, and optimize the model for use in SAR image analysis.

In [3], the prior distribution of the region labels is modeled as

$p(L_s / L_r, r \in N_s) = \frac{1}{Z_2} \exp[-MU_2(L_s / L_r, r \in N_s)]$, where the Gibbs energy function U_2 is

$$U_2(L_s / L_r, r \in N_s) = -\frac{\beta}{M} \sum_r \delta(L_s - L_r). \qquad (3)$$

β is a clustering parameter equal to 1.4, Z_2 is a normalizing constant independent of L_s, and $\delta(.)$ is the Kronecker delta. U_2 returns a low value (good) if all the pixels in the neighborhood N_s have the same label, and high (not so good) if otherwise.

As is clear from (3), the region labeling is modeled as an isotropic MRF process with local dependencies. In [2] it is argued that in many real-world problems this approximation may not be sufficiently accurate, and this holds true especially in the application of land-cover mapping, where often a great level of detail is needed.

In [2] solutions to this problem have been reviewed. Li focuses solely on the smoothness priors, and his analysis of the different solutions proposed in the literature, such as weak string and membrane [5,6], line process [7], resulted in the so-called *discontinuity adaptive* smoothness model, of which the before-mentioned models are special instances.

In what follows, we first explain in detail how Li comes to his DA-MRF prior, after which the integration between the DA-MRF prior and the Gamma-observation model shall be outlined. In its general 1-D form, the smoothness term $U(f)$ is defined as

$$U(f) = \sum_{n=1}^{N} U_n(f) = \sum_{n=1}^{N} \lambda_n \int_a^b g(f^{(n)}(x)) dx, \qquad (4)$$

being f the signal to be restored, $U_n(f)$ the nth order regularizer, N is the highest order to be considered and $\lambda \geq 0$ is a weighting factor. A potential function $g(f^{(n)}(x))$ is the penalty against the irregularity in $f^{(n-1)}(x)$ and corresponds to prior clique potentials in MRF models. Limiting the model to be of the first order, and considering the general string model, from (4) we can derive

$$E(f) = U(f/d) = \int_a^b u(f/d) dx, \qquad (5)$$

where, using for the time being the Gaussian observation model

$$U(d/f) = \int_a^b \chi(x)[f(x) - d(x)]^2 dx,$$

$$u(f/d) = \chi(x)[f(x) - d(x)]^2 + \lambda g(f'(x)), \qquad (6)$$

$\chi(x)$ being an appropriate weight function.

The solutions minimizing $U(f/d)$ must satisfy the associated Euler-Lagrange differential equation

$$u(f, f') - \frac{d}{dx} u'(f, f') = 0,$$

with the boundary conditions $f(a) = f_a$ and $f(b) = f_b$ which are prescribed constants. Writing out the Euler equation for the Gaussian, one dimensional model yields

$$2\chi(x)[f(x) - d(x)] - \lambda \frac{d}{dx} g'(f'(x)) = 0. \qquad (7)$$

A potential function g is chosen to be even ($g(\eta) = g(|\eta|)$) and the derivative of g can be expressed as $g'(\eta) = 2\eta h(\eta)$, where h is called an *interaction function*. The interaction $h(\eta)$ must be small for large $|\eta|$ and approaches to 0 as $|\eta|$ goes to ∞. In [2] different Adaptive Interaction Functions (AIF) are proposed, that all satisfy Definition 1:

Definition 1 ([2]): An adaptive interaction function (AIF) h_γ parameterized by γ (>0) is a function that satisfies: (i) $h_\gamma \in C^1$ (ii) $h_\gamma(\eta) = h_\gamma(-\eta)$ (iii) $h_\gamma(\eta) > 0$ (iv) $h_\gamma^1(\eta) < 0$ ($\forall \eta > 0$) (v) $\lim_{\eta \to \infty} |\eta h_\gamma(\eta)| = C < \infty$. In section 5 more details on the AIF are given.

Extending (7) to 2-D yields

$$\chi[f-d] - \lambda \frac{\partial}{\partial x}[f_x h(f_x)] - \lambda \frac{\partial}{\partial y}[f_y h(f_y)] = 0.$$

The Euler equation can be solved by minimizing the corresponding energy (5). Using the first order backward difference as an approximation of the first derivative, (5) can be approximated by

$$E(f) = \sum \chi_i[f_i - d_i] + \lambda \sum_{i=1}^{m} \sum_{r \in N_i} g(f_i - f_r),$$

where N_i consists of the neighbors of i. From the above, using the gradient-decent method, we obtain the following updating equation:

$$f_i^{(t+1)} \leftarrow f_i^{(t)} - 2\mu \left\{ \chi_i[f_i^{(t)} - d_i] - \lambda \sum_{r \in N_i} (f_r^{(t)} - f_i^{(t)}) h(f_r^{(t)} - f_i^{(t)}) \right\},$$

where μ is the update strength, and is chosen constant during the optimization process. At each iteration, the label of each pixel is updated based on the contextual information in the label image at the previous itaration. The updating process, which is a deterministic relaxation, starts with a certain initial label configuration $f^{(t=0)}$. A suited stop criteria should be defined.

5. DA-MRF model with gamma observation statistics

Since the approach of Rignot and Chellappa deals with classification rather than with segmentation, the values of L_s and $f(x)$ in (2) and (7), respectively, are class indexes and have a semantic value rather than a quantitative one. To solve this problem, and to enable one to convert the non-linear equation into a linear one, it is proposed to substitute the class indexes with the mean class-intensity values:

$$\rangle d \langle_{f_i} = f_i. \tag{8}$$

Using (8), and substituting the Gaussian data distribution with the Gamma distribution

$$U_1(d_{i,j} / f_{i,j}) = \frac{N d_{i,j}}{f_{i,j}} - (N-1)\log(d_{i,j}) + N \log(f_{i,j})$$

allows us to re-write (7) into

$$u(f/d) = \frac{N d_{i,j}}{f_{i,j}} - (N-1)\log(d_{i,j}) + N \log(f_{i,j}) + \lambda g(f_{i,j}'). \tag{9}$$

Writing out the Euler equation for the Gamma case yields

$$-\frac{N d_s}{(f_{x,y})^2} + \frac{N}{f_{x,y}} - \lambda \frac{\partial}{\partial x}(f_x) h(f_x) - \lambda \frac{\partial}{\partial y}(f_y) h(f_y) = 0.$$

The posterior energy can now be estimated by

$$E(f) = \sum_{i=1}^{m} U_1(d_i / f_i) + \lambda \sum_{i=1}^{m} \sum_{r \in N_i} g(f_i - f_r),$$

and using the classical, deterministic gradient based minimization method yields the following updating scheme (10):

$$f_s^{(t+1)} \leftarrow f_s^{(t)} - \mu \left\{ \left[\frac{N(f_s^{(t)} - d_s)}{(f_s^{(t)})^2} \right] - \lambda \sum_{r \in N_s} (f_r^{(t)} - f_s^{(t)}) h(f_r^{(t)} - f_s^{(t)}) \right\}$$

The regularizer takes effect when $\lambda > 0$. Its strength depends on the qualitative and quantitative shape of the AIF, basically determined by h_γ. For the experiments that follow, the AIF was chosen to be $h_\gamma = (1 + |\eta|/\gamma)^{-1}$ in both the x and y directions. The main reason is that it allows bounded but non-zero smoothing at discontinuities, useful properties when analyzing speckle images. Additional advantages are that the resulting energy function is convex and that the algorithm becomes computational efficient.

6. Experimental results

6.1. Parameter settings

Table I should give the reader a feeling about the influence of the various parameters.

Table I. Overview of parameters, their effects, and their suggested values for a 4 look SAR intensity image. The values correspond to those used to produce the results.

Parameter	Function	Effect		Suggested value
		within order of magnitude	between o.o.m.	Filtering
λ	Relation between observation and regularization model	Small	large: strong smoothing. Low: none.	0.001
γ	Choice of regularization strength.	Strong	Strong	0.1-3
μ	Update strength (optimization procedure)	None.	Large: unstable; small: slow convergence	50
K	Number of iterations (optimization procedure)			100

6.2 Data description

In order to assess the usefulness of the proposed method, experiments have been done on portions of a 4-Look SAR intensity image (HH polarization) of a Flevoland (NL) scene, acquired during the Maestro I campaign in 1989. The original test image is shown in Fig. 3a.

6.3. Comments on results

Experimental results are reported in Fig. 3. Fig. 3b shows that, applying a low λ value, the final result tends to convert to the original data. With an increasing λ, clustering of regions gets stronger (Fig. 3c). The strong effect of γ, reported in table I, can be seen in Fig. 3c and d.

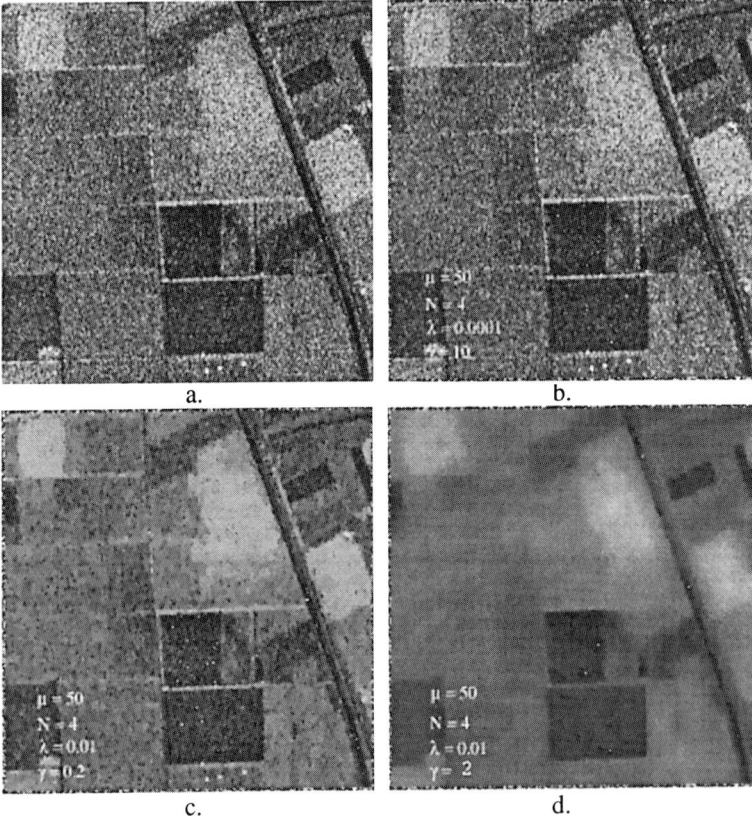

Fig. 3. Results. a) 256x256 Portion of a Flevoland 4 look SAR intensity image (HH polarization); b) Light filtering: the small λ value (0.0001) practically "switches off" the regularizing term; c) And d) show the effect of a small and large γ value, respectively. All results in the image were obtained after 100 iterations.

7. Discussion and conclusions

In this paper, an approach has been presented for the analysis of synthetic aperture radar (SAR) images that preserves better fine structures and borders in the image than classical methods. The method uses the discontinuity adaptive MRF model in which the Gaussian observation model has been replaced by a gamma distribution. This resulted in a new algorithm that is more suited to the segmentation of SAR images if one is interested in preserving details. The clustering effect of regions with homogeneous back scatter signals is principally determined by mainly two parameters whose function is easy and intuitively to understand.

Results on various real-world data sets have shown that the method proposed in this paper is a useful tool for the analysis and interpretation of SAR images. However, it is stressed that at the end of the day it remains the end-user to decide what level of detail is required.

The software of the proposed algorithm for Windows 95 will be made available via URL http://dibe.unige.it/TMR_Smits. Readers are encouraged to try the method on their own imagery and to report their findings.

Acknowledgements

The authors with to thank Dr. A. Freeman of NASA/JPL for providing the SAR data. This work was supported by the European Community program Training and Mobility for Researchers (Marie Curie Fellowship) under contract ERBF MBICT 95257.

References

[1] Li, S.Z. (1995b), Markov Random Field modeling in computer vision, Springer Verlag, New York.
[2] Li, S.Z. (1995), Discontinuity-adaptive MRF prior and robust statistics: a comparative study, IEEE Trans. on Pattern Analysis and Machine Intelligence **17**(6), 576-586.
[3] Rignot, E., Chellappa, R. (1993). Maximum a posteriori classification of multifrequency, multilook, synthetic aperture radar intensity data, J. Opt. Soc. Am. A, Vol. 10, No. 4, pp. 573-582.
[4] Smits P.C. and S. Dellepiane (1996), "Information fusion in a Markov Random Field based image segmentation approach using adaptive neighbourhoods," 13th Int. Conf. on Pattern Recognition, Vienna, August 1996, pp. 570-575.
[5] Blake (1983), The least disturbance principle and weak constraints, Pattern Recognition Letters, Vol. 1, pp. 393-399.
[6] Blake A. and A. Zisserman (1987), Visual reconstruction. Cambridge, MA: MIT Press.
[7] Geman S. and Geman D. (1984), Stochastic relaxation, Gibbs distributions and the Bayesian restoration of images, IEEE Trans. Pattern Anal. Machine Intell., Vol. PAMI-6, nov. 1984, pp. 721-741.

Region Growing Euclidean Distance Transforms

Olivier Cuisenaire

Telecommunication and Remote Sensing Laboratory
Université Catholique de Louvain, Belgium

Abstract

By propagating a vector for each pixel, we show that nearly Euclidean distance maps can be produced quickly by a region growing algorithm using hierarchical queues. Properties of the propagation scheme are used to detect potentially erroneous pixels and correct them by using larger neighbourhoods, without significantly affecting the computation time. Thus, Euclidean distance maps are produced in a time comparable to its commonly used chamfer approximations.

1. Introduction

Distance maps are images where the value of each pixel of the foreground is the distance to the nearest pixel of the background. Generating such maps using a Euclidean distance metric is a complex problem since a direct application of this definition usually requires an excessive computation time.

The development of fast algorithms for producing Distance Transforms (DT), as approximations of the Euclidean distance maps, has allowed their applications in various fields such as chamfer matching, registration of medical images [6], generation of morphological skeletons or active contour models.

Numerous DT algorithms have been proposed, offering various trade-offs between computation time and quality of the approximation of the Euclidean metric. The DT in the literature belong to two categories: the Chamfer DT originally proposed by Borgerfors in [2] and the Vector DT proposed by Danielsson in [1]. These algorithms rely on a number of raster scans over the image, although some parallel implementations have also been proposed [4]. In this paper, we present a region-growing Vector DT where pixels are scanned by increasing value of the distance.

In section 2 we present the principles of existing DT. In section 3, the region growing new algorithm is described. In section 4, the quality of the approximation is increased by using larger neighbourhoods for a subset of potentially erroneous points. In section 5, we analyse the types and values of errors made by our algorithm and compare it with CDT. In section 6 the complexity of the algorithm is considered.

2. A review of Distance Transforms

Chamfer Distance Transforms (CDT) are commonly used to approximate the Euclidean metric. They are based on the assumption that the value of the distance for each pixel can be computed from of the values of its neighbours plus a mask constant.

CDT are usually produced in 2 raster scans over the image, using half of the neighbour pixels as a mask for each scan. The simplest CDT are the chessboard and city-block DT using masks a and b of fig 1. G. Borgerfors introduced better approximations with mask c in [2] and mask d in [3].

There are two sources of errors in CDT. First, lines are approximated by segments following the main directions available in the mask, i.e. every 45° for the CDT 3-4 and every 22.5° for CDT 5-7-11. Secondly, the values of the mask constants are approximated by integers which leads to errors even in the main directions of mask.

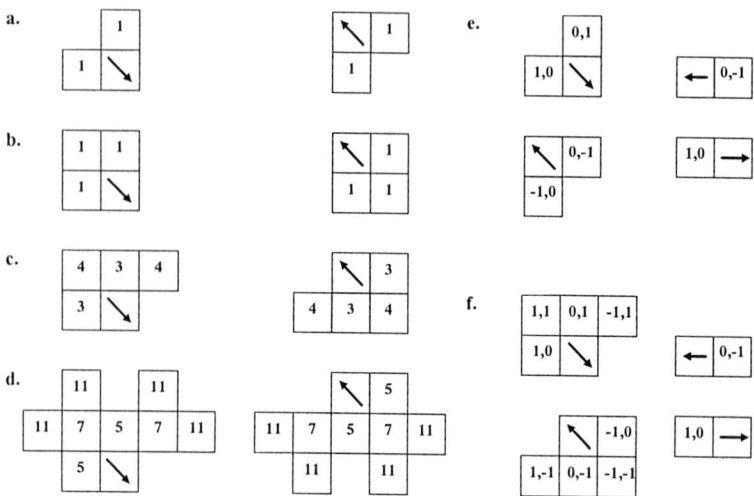

Fig. 1. Masks of the various DT: a) chessboard - b) City-block - c) Chamfer 3-4 - d) Chamfer 5-7-11 - e) 4SSED - f) 8SSED.

To produce better approximations of the Euclidean metric, one should transmit more information from pixels to their neighbours. In [1], Danielsson proposes to propagate a vector localising the nearest background pixel or NBP.

This requires the use four scans. Masks e of fig 1.corresponds to 4SED (4 neighbours Sequential Euclidean Distance) and masks f to 8SED. In 3 dimensions, 6 scans are needed. The increased number of scans and the higher complexity of vector comparisons makes the Vector Distance Transform (VDT) significantly slower than CDT.

Using VDT, most pixels are error-free, but some errors occur for particular background pixel configurations, when the basic assumption - that the closest background point to a pixel is also the closest point to one of its neighbours - is not satisfied.

3. Region Growing Algorithm

Our method uses the 4SED mask, but instead of using the raster scans, pixels are considered by increasing value of the distance. This is implemented with a data structure called hierarchical queues (HQ), illustrated at fig 3. The queue labelled i in the HQ contains the pixels for which i is the square of the distance to their NBP. With a Euclidean metric, this value is an integer. For each pixel in the HQ, its location and its NBP are stored. An image called map is also created, containing the square of the distance at each pixel. In what follows, our aim is to compute this map.

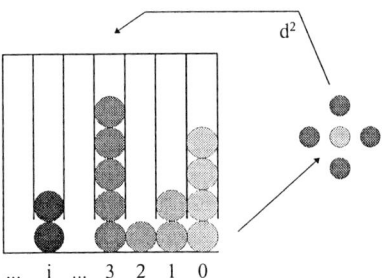

Fig. 2. Hierarchical Queues are made of a collection of FIFO queues. In-going elements enter any of the queues, outgoing elements are taken from the smaller numbered non-empty queue.

The HQ is initialised with queue 0 containing the background pixels and all other queues empty. The map is initialised with zero for background pixels and the maximum integer value everywhere else.

Pixels are then treated in the HQ order. For each pixel, we consider that its neighbours have the same NBP. If this leads to a smaller value of the distance than stored in the map, the map is updated with this value and the neighbour is inserted in the HQ. In the HQ, pixels are treated by increasing value, which on the map corresponds to the border of a growing region centred on the background pixels.

More formally, the algorithm is written

Initialisation: $HQ(0)$ is the list of background pixels
   $HQ(i)$ empty for all $i \neq 0$
   $map[p]$ is maxint for all pixel p
   $map[p]$ is 0 for background pixels

Main: while HQ not empty
   {   get p from HQ
      for each neighbour n of p
         { $d = dist^2(n, NBP(p))$
            if $d < map[n]$
            { $map[n] = d$
               add n to $HQ(d)$ with $NBP(n) = NBP(p)$
            }
         }
   }

Note that this algorithm allows pixels to be first mislabelled with a wrong NBP and later corrected as closer to another. Nonetheless, the HQ scan processes the correction before the initial error since the distance is smaller hence the correction is in a queue of smaller label. Errors are therefore not propagated.

Also, with the HQ scan there is no need of propagating the information back to pixels of a smaller value. Therefore, we only consider the part of the neighbourhood which decreases none of the components of the relative position of the NBP. In most cases, only one quadrant of the mask will need to be considered, which reduces the search to 2 pixels our of 4 for 4SED, 3 out of 8 for 8SED, ...

4. Use of Larger neighbourhoods

As we will see in next section, VDT produces errors only for few pixels, with specific background pixel configurations. With 4SED for instance, the largest relative error is illustrated at fig 7.a, where the information "closer to pixel B" cannot be propagated to pixel x. This could be solved by using the mask of 8SED allowing the information to be transmitted diagonally so that x could be reached from the pixel labelled 2, as suggested by the arrow.

In general, two main strategies can be envisaged to reduce the errors of a DT. First, one can use larger neighbourhoods for the masks. This leads to a significant increase of the computation time, which is proportional to the product of the size of the mask by the number of pixels. The second method, proposed in [5], consists of increasing the amount of information transmitted from pixel to pixel. εVDT(0) stores the list of all NBP instead of only one of them. εVDT(1) also includes nearly NBP in the list. It produces an error-free map, but is orders of magnitude slower than 4SED or 8SED.

We propose to use the first approach and to quicken it significantly. The distance map is first computed quickly but roughly with the smallest possible mask (4SED). Then, larger neighbourhoods are used to correct errors made with this first scan. Fortunately, these corrections only need to be made for a small subset of pixels. Indeed, after using 4SED, most pixels are error-free. And the erroneous pixels share some properties we can use to restrict the set of points to treat with larger masks.

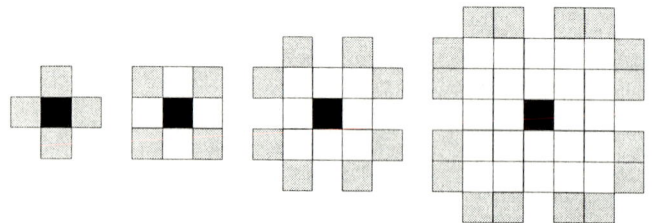

Fig. 3. a sequel of increasing size neighbourhoods: from left to right: 4SED mask, 8SED or 3x3 mask, 5x5 mask, 7x7 mask. Light grey pixels are those added to the smaller mask to create the larger one.

First, we consider neighbourhoods N_0, N_1, ... N_i ... of increasing size, such as illustrated at figure 6. We define an err_i pixel for N_i a pixel in the map for which the values computed using N_i and N_{i+1} are different. An end_i pixel is one that was not propagated while using the algorithm described in section 3 with N_i. end_i pixels can easily be detected while running the algorithm.

Once we have a map created using N_i, we want to correct some errors to make it as good as if it had been created with N_{i+1}. For this, we only need to correct the err_i pixels. We use the following property: Any err_i pixel is a N_{i+1} neighbour of either another err_i pixel or of an end_i pixel. This can easily be proved ab absurdum. Therefore, the N_{i+1} mask only needs to be propagated from end_i pixels and from corrected err_i pixels. This leads to the following algorithm for 2 neighbourhoods, which can be extended immediately for any number of neighbourhoods

```
while HQ1 not empty
{    get p from HQ1
     for each N1 neighbour n of p
     {      d=dist²(n, NBP(p) )
            if d<map[n]
            {      map[n]=d
                   add n to HQ1(d) with NBP(n)=NBP(p)
            }
     }
     if no neighbour n was put in HQ1 (p is endi)
            add p to HQ2(map[p])
}
while HQ2 not empty
{    get p from HQ2
     for each N2 neighbour n of p
     {      d=dist²(n, NBP(p) )
            if d<map[n]
            {      map[n]=d
                   add n to HQ2(d) with NBP(n)=NBP(p)
}    }     }
```

This can further be improved by noticing that each neighbourhood provides a perfect map up to a certain value. There is no need to use larger neighbourhoods for smaller distances. For instance, the smallest possible error with 4SED is illustrated at figure 7.a. The pixel labelled 2 is end_{4sed}. It should used to reach pixel x with the 8SED mask. Hence end_{4sed} pixels labelled 0 or 1 should not be considered while using 8SED to improve a map created with 4SED. The smaller value for which 8SED shall be used is 2.

The thresholds for some masks in 2 and 3 dimensions can be found in tables 1 and 2. In 3D, the background pixels should be included in the HQ2 to take into account the fact that the direct neighbourhood can produce errors even with $d^2=0$.

5. Error Analysis

The error analysis for our DT compared to the Euclidean metric is similar to the analysis made in [1]. In this paper, we consider, for each mask, the smallest values for which a pixel can be mislabelled. It also corresponds to the largest relative error. The configurations of NBP leading to those errors for 4SED and 8SED are illustrated in fig 7 and numerical results found by extensive search are listed in tables 1 and 2.

They should be compared with the relative errors obtained with CDT: 8% for CDT 3-4 and 2% for CDT 5-7-11. This ranks the 4SED mask between the two chamfer methods and 8SED or any larger mask as better than any published CDT.

Furthermore, VDT finds the correct value for most pixels. The number of erroneous pixels is image dependant. In the case of figure 9, less than 300 out of 65500 points, or 0.5% are erroneous when using the 4SED mask. Typically, errors are located in areas where the "propagation front" is shrinking while expansion areas are error-free since no information loss occurs there. In particular, for a single background pixel (fig. 8), the 4SED mask provides a perfect Euclidean map.

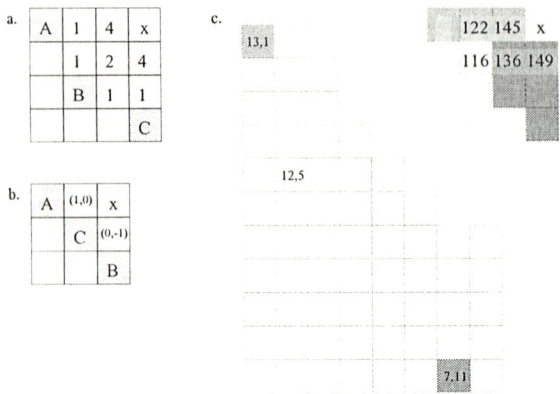

Fig. 4. a) 4SED error: pixel x is assigned a squared distance value of 9 rather than 8 as A or C are mistaken as closer to x than B. **b)** New type of error: pixel x is assigned a value of 4 instead of 2 if pixels are scanned in a specific order A,B, C **c)** using 8SED, pixel x is wrongly assigned $170=7^2+11^2=13^2+1^2$ instead of $169=12^2+5^2$

Mask	d^2 correct	d^2 assigned	d correct	relative error	d^2
4SED	8	9	2.89	6,1%	2
8SED	169	170	13	0,3%	116
5x5	964	965	31.05	0.05%	778
7x7	2404	2405	49.03	0.02%	2089

Table 1: Smallest errors for each mask size. The last column (d^2) gives the value from which the non-propagating pixels should be propagated with a larger mask.

Mask	d^2 correct	d^2 assigned	d correct	relative error	d^2
6SED	3	4	1,73	15,5%	0
3x3x3	49	50	7	1,0%	24
5x5x5	228	229	15,1	0,2%	146
7x7x7	626	627	25,02	0,08%	441

Table 2. Same as table 1 for 3D distance maps.

Other differences between VDT and CDT are illustrated at fig 8. First, the shape of iso-distance curbs: CDT gives polygons of 8 or 16 sides, VDT a perfect circle. This influences applications such as the generation of skeletons [1]. Secondly, CDT only allow 8 or 16 directions for the gradient, VDT a continuous range of directions. This influences convergence domains and speed in gradient-based minimisation as in [6].

Fig. 5. From left to right: CDT 3-4 / CDT 5-7-11 / EDT. Up: distance from a central pixel. Colours are mod. 30 for better visibility. Down: direction of the gradient of those maps.

6. Complexity Analysis

The usual method for assessing the complexity of a DT is to consider the number of comparisons per pixel required by the algorithm. Unfortunately, in order to compare our algorithm to a CDT, this is not a valid method since the type of comparison is different (scalars for CDT, vectors for VDT) and their number is image dependant.

Furthermore, hierarchical queues are dynamic data structures. Hence, a significant part of the time is used for dynamic memory allocation. This makes the comparison of CDT and our algorithm dependant on the relative efficiency of each type of operation, and thus machine-dependant.

We therefore choose a heuristic approach, using CDT and our method on test images several types of workstations. In average, our DT with the 4SED mask is 1,5 slower than CDT 3-4 and similar to CDT 5-7-11. In three dimensions, it is between 1.05 and 3 times slower than CDT 3-4-5, depending on the workstation.

These good results - despite the complexity of the basic operation -are explained by the small amount of comparisons required. On the test images, our method requires an average of 2,1 and 3,35 comparisons per pixel in 2 and 3 dimensions respectively, compared with 8 and 26 for CDT. In d dimensions, CDT requires $o(d^3)$ comparisons while our method requires $o(d)$ comparisons whose complexity grow like $o(d)$. Hence the global complexity growth like $o(d^2)$. Furthermore, the algorithm can be stopped at any step and provide a partial map where all pixels within a certain distance have been computed. This is of interest for the last steps of minimisation methods where all points of interest are close to their targets.

Let us now consider the multi-mask algorithm of section 4. The additional cost for using the larger mask can be estimated from the number of end_{4sed} pixels. This number is image related. For instance, with the 256x256 image of fig. 9, out of the 65536 pixels treated with the 4SED mask, 8894 are considered for the 8SED mask, 3765 for the 5x5 mask and 2282 for the 7x7 mask. This leads to 65536*2,1 + 8894*1 +

3765*2 + 2282*4 =163177 comparisons if, for the larger masks, only one quadrant of the pixels highlighted at fig 3 are used. The additional cost of using the 7x7 mask instead of 4SED is less then 20%, while using this 7x7 mask with the single-mask algorithm of section 3 represents an additional cost of around 400%.

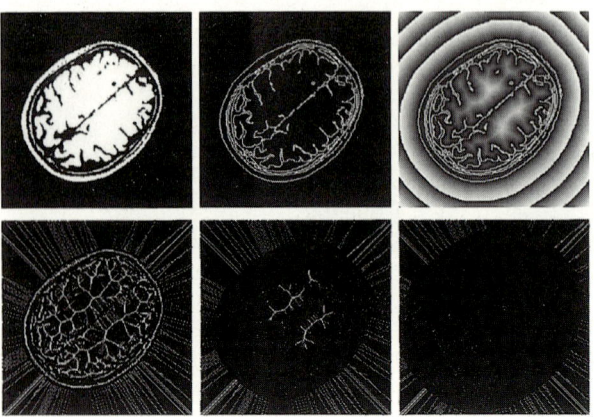

Fig.6. Up: a typical image (left), its edges (centre) and the Euclidean map (right) to these edges. Colours are displayed modulo 25 for better visibility. Down: set of err$_{4SED}$ pixels used for 8SED (left, threshold at $d^2=2$), 5x5 (threshold at 116) and 7x7 masks (threshold at 778).

7. Conclusion

We have developed an algorithm computing Euclidean distance maps in a time similar to the chamfer DT approximations. It is progressive since it can be stopped at any time and provide a sensible result, either by computing only within a certain distance, or improving the map by progressively using larger masks.

Finally, the method seems easily adaptable to other metrics and in particular to non-isotropic grids or higher dimensions. This and the use of the distance maps to generate skeletons of binary objects is the subject of future research.

Acknowledgement

Olivier Cuisenaire's work is funded by Belgium's F.R.I.A

References

1. P.E. Danielsson, Euclidean Distance Mapping, *CGIP* 14, 1980, 227-248.
2. Borgerfors, Distance Transformations in Arbitrary Dimensions, *CVGIP* 27, 1984, 321-345
3. G. Borgerfors, Distance Transformations in Digital Images, CVGIP 34, 1986, 344-371
4. H. Embrechts and D. Roose, A parallel Euclidean Distance Transformation Algorithm, *CVIU* 63, 1996, 15-26
5. J. Mullikin, The vector distance transform in two and three dimensions, *CVGIP* 54(6), 1992, 526-535
6. O. Cuisenaire, J.Ph. Thiran, B.Macq, Ch. Michel, A. De Volder and F. Marques, Automatic Registration of 3D MR Images with a Computerised Brain Atlas, *SPIE Medical Imaging 1996*, SPIE vol. 1710, 438-449 .

COP: A New Method for Extracting Edges and Corners

Sun Cheol Bae, In So Kweon
Dep. of Automation and Design Eng., Korea Advanced of Institute of Science and Technology.
207-43, Cheongryangri-dong, Dongdaemoon-gu, Seoul, KOREA.
Internet Address:iskweon@cais.kaist.ac.kr   TEL. +82-2-958-3465, FAX. +82-2-960-0510

Abstract

This paper presents a new, simple and effective low level processing method to obtain features such as edges and corners. In both edge and corner extraction algorithms, we use two oriented cross operators called COP (Crosses as Oriented Pair). To obtain edges, many conventional edge operators use derivative convolution masks and are followed by the conventional non-maximum suppression algorithm which needs rather complicated edge direction calculation. Moreover, most conventional derivative-based operators suffer from poor connectivity at junctions, sensitivity to noise and two extrema when they are applied to a line. But COP makes it possible to find edge direction very easily, to localize edge position accurately and to obtain connected edges. Second, we propose a new corner detection algorithm using COP. Most conventional corner detectors have shortcomings such as missing junctions, poor localization, sensitivity to noise and high computational cost. With the characteristics of COP and simple rules, we can accomplish a very fast, accurate and robust corner detection than any other corner detector. Performances of two proposed algorithms are described with test results.

1. Introduction

Edge detection has been one of the most active fields in computer vision. Most of previous research on edge detection have been based on discrete approximations to differential operations. Recently, Smith and Brady [1] proposed a new low-level processing "SUSAN", which uses simple masking operations, namely, only counts the number of pixels dissimilar to the center pixel of a circular window. So it is very robust to noise. Most of the edge detectors use conventional non-maximum suppression algorithm to obtain meaningful edges. But it often produces undesirable effects such as thick and broken edges or bad localized edges.

The recent literature on corner detection can be widely divided into two approaches. The first approach is to initially segment the image into regions, represent the object boundary with a chain code, and then search for points having maximum curvature or perform a polygonal approximation on the chains and then search for the line segment intersections. The major drawback of this approach is that the performance of corner detection depends on the preceding image segmentation result. To overcome this, most of recent attempts use the second approach, which mainly uses gradients and surface curvatures [1,2,3,4,5,6,7,8]. This approach, however, also suffers from localization of corners and noise. Smith and Brady [1] use a circular window, composed of 37 pixels, to obtain USAN area whose centroid is then used to suppress false positives. Finally, non-maximum suppression finds corners. The corner localization and noise robustness are good or better than other algorithms. But it takes much computational cost in obtaining USAN area and finding corners due to the size of the window. In some cases, they also produce false positive and negative corners.

2. COP (Crosses as Oriented Pair)

In this section, we describe the characteristics of COP which is composed of two 3 by 3 oriented crosses as shown in Figure 1.

Figure 1. Two crosses of COP.

Each of COP has a very interesting property. The left one responds strongly to ±45° edges, and the right one to the horizontal and vertical edges. Figure 2 shows these characteristics of each of COP with simple images. Each response shows the inverted USAN area, namely, the number of pixels dissimilar to the mask's nucleus. Moreover, each of COP shows peak values near the corners. Another important characteristic is that COP makes it possible to determine orientation in a very simple manner. Briefly speaking, if the inverted USAN output of the left cross is larger than that of the right one, edge direction is +45° or -45°. In this case, if sum of the inverted USAN output of top-left and bottom-right neighbors is larger than that of top-right and bottom-left ones, then edge direction is -45° and vice versa. If the inverted USAN output of the right cross is larger than that of the left one, edge direction is 0° (horizontal) or 90° (vertical). In this case, if sum of the inverted USAN output of left and right neighbors is larger than that of top and bottom ones, then edge direction is 0° and vice versa. Other cases which are not mentioned in the above happen at corners or some places where orientation changes.

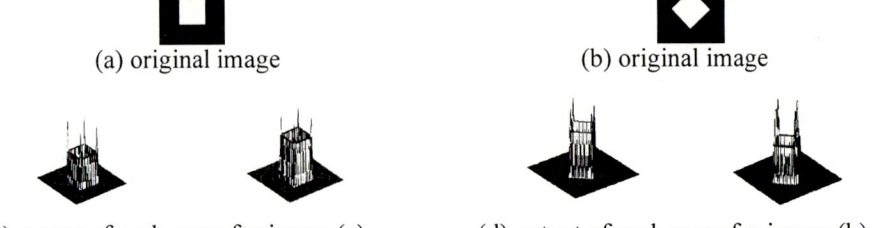

(a) original image   (b) original image

(c) output of each cross for image (a).   (d) output of each cross for image (b).

Figure 2. Each response of COP to simple synthetic images.

3. Edge extraction

3.1 Algorithm
Our edge detection algorithm performs the following five steps.

Step 1. Obtain inverted USAN area, R1 and R2, using COP (R1: output of ✚, R2: output of ✖);
Step 2. Calculate edge direction;
Step 3. Discard the elements having lowest edge response (when R1*R2 < 2);
Step 4. Apply non-maximum suppression; and
Step 5. Edge refinement is added if needed.

3.2 Experimental results

Experiments are carried out for several synthetic and real images. Figure 3 shows the responses of COP to some synthetic images including multi-region junctions ("T", "Y", "X"). All outputs of COP show good performances. Some experiments are carried out on real scenes. COP produces satisfactory results as shown in Figure 4. One complicated outdoor road scene is also used to test COP edge detector. Some problems such as fluctuating edges take place in this case due to noise. But this may be cured if refining process is fulfilled.

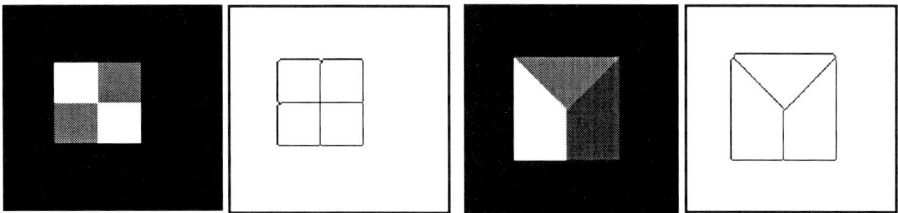

Figure 3. Outputs of COP for synthetic images.

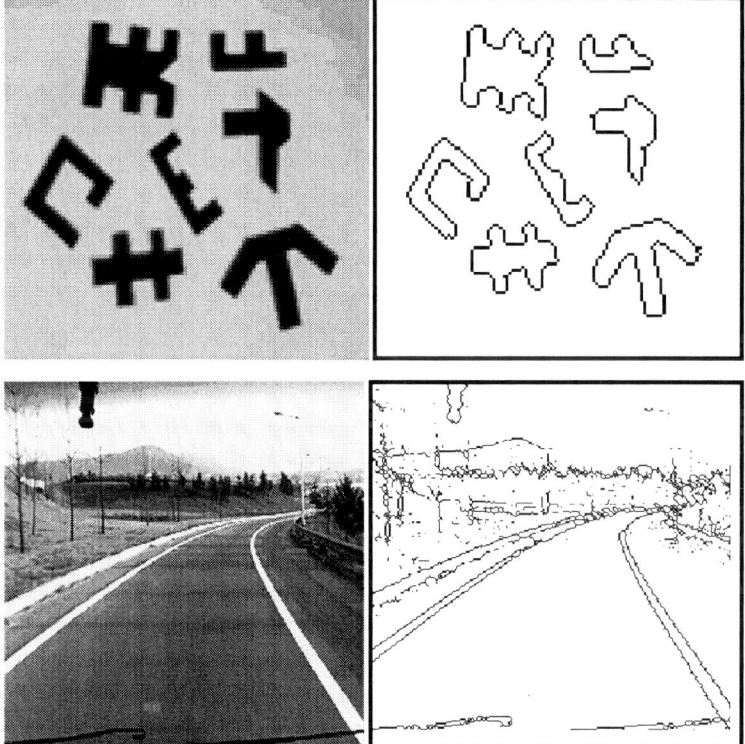

Figure 4. Outputs of COP for real images.

4. Corner extraction

Our corner detection algorithm is mainly divided into the following five steps:
Step 1. Obtain inverted USAN area using COP;
Step 2. Obtain corner candidates;
Step 3. Compare each output of COP to obtain useful directional information;
Step 4. Calculate 2nd order tangential derivatives of original intensity value among the candidates obtained in step 2 except the predetermined corners; and
Step 5. Find local maximum in a search window.

In step 1, each output of COP has values ranging from 0 to 4. In step 2, we perform our algorithm only when $[R1(i, j)+R2(i, j)]>3$ to obtain corner candidates. In step 3, we can simply obtain edge direction information using COP. All we have to do is compare each output of COP. In step 4, we use a second order directional derivative mask, [-2, -1, 6, -1, -2], to obtain tangential derivative of intensity. This mask represents the one dimensional Laplacian of Gaussian and corner checking rules are applied to some of the obtained corner candidates. Finally, in step 5, corners are identified by taking local maxima in the search window.

4.1 Corner candidates and rules

We consider any point in the image as corner candidate if each inverted USAN output of COP, R1 and R2, satisfies the condition that $R1+R2 > 3$ at that point. To overcome the limitation of the size of COP and make full use of the characteristics of COP, we define simple rules as follows.

- Corner checking rules: Only when $R1(i, j) = R2(i, j)(=2$ except rule 6)
  - Rule 1 : $[R1(i, j-1) - R2(i, j-1)] \times [R1(i, j+1) - R2(i, j+1)] < 0$
  - Rule 2 : $[R1(i-1, j) - R2(i-1, j)] \times [R1(i+1, j) - R2(i+1, j)] < 0$
  - Rule 3 : $[R1(i, j+1) = R2(i, j+1) = 2]$ & $[R1(i+1, j) = R2(i+1, j) = 2]$ & $[R1(i+1, j) = R2(i+1, j) = 2]$
  - Rule 4 : { $[R1(i, j+1) = R2(i, j+1) = 2]$ or $[R1(i+1, j) = R2(i+1, j) = 2]$}
    &{[Any neighbor of R1 on diagonal] > [corresponding one of R2]}
  - Rule 5 : { $[R1(i, j-1) = R2(i, j-1) = 2]$ & $[R1(i, j+1) = R2(i, j+1) = 2]$
    & $[R1(i+1, j-2) = R1(i+1, j+2) = 2]$ or $[R1(i-1, j-2) = R1(i-1, j+2) = 2]$}
    { $[R1(i-1, j) = R2(i-1, j) = 2]$ & $[R1(i+1, j) = R2(i+1, j) = 2]$
    & $[R1(i-2, j-1) = R1(i+2, j-1) = 2]$ or $[R1(i-2, j+1) = R1(i+2, j+1) = 2]$}
  - Rule 6 : $R1(i,j) = R2(i,j) = 3$

If a pixel satisfies any of six rules, we select that point as corner. Rules 1 and 2 mean that there exist two edges whose directions are changing in the horizontal and the vertical sides of a pixel, respectively. Rules 3 and 4 are similar to rules 1 and 2, but used to consider more smooth corners. Junctions are also considered in rule 5. Rule 6 is clear because this happens when sharp corner exists.

4.2 Rule analysis

There exist only 7 patterns satisfying the case that R1 and R2 are all equal to 2, as shown in Figure 5. We avoid the duplicated cases such as the axis-symmetric and rotationally symmetric ones. Especially, the first pattern is important and should be well analyzed because this case has the highest possibility of occurrence. In Figure 6, we analyze this case in detail. In Pattern A, R1 is larger than R2 at left neighbor, which means that there may be +45° or -45° edge locally. In this case, as you know,

+45° edge exists. And R2 is larger than R1 at right neighbor, which means that there may be locally horizontal or vertical edge. In this case, horizontal edge exists. Anyway the center point is regarded as corner point by rule 1. But rule 1 is also applied to the case like pattern D. By rules, pattern B is not considered as corner. Look at the pattern C, in this case the center pixel and its right neighbor may be regarded as noises on flat edge, or may be corners if significant orientation change happens in the right part of this pattern. We also consider this case by rule 5. There are many other patterns satisfying the case that R1 and R2 are all equal to 2. Such patterns are also checked whether they are corners or not by proposed corner checking rules.

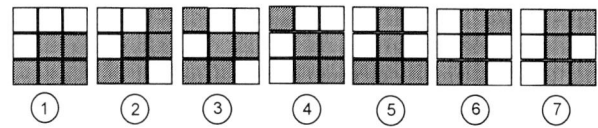

Figure 5. Typical seven patterns when R1 and R2 are all equal to 2.

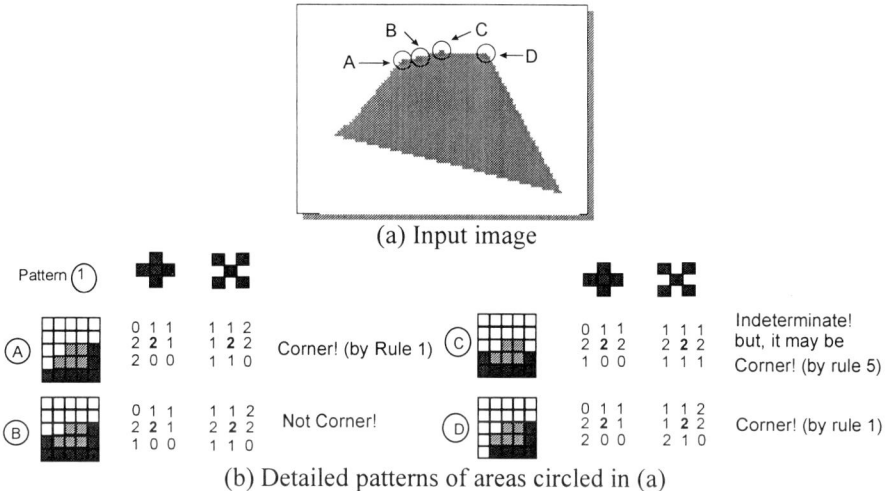

Figure 6. Rule analysis for a synthetic image.

4.3 Experimental results

The results of testing this corner algorithm are shown and discussed for several synthetic and real images. Some images are used to test its sensitivity to noise. Finally, one example is added to test how flexible our corner finder is, as threshold varies. All images are tested to compare COP's performance with SUSAN. For fair comparison, we use original SUSAN algorithm supported by Internet. Our algorithm produces satisfactory results for all cases. Results for real images are shown in Figures 7~8. Figure 9 shows the output of two corner finders for one real image as the threshold value varies from 30 to 60 in 10 increments. If the threshold value is lower than some optimal threshold, false positives show up. In the opposite case, false negatives show up. As you see, the performance of the COP corner detector doesn't degrade so much as the threshold value varies. With respect to speed, COP only needs small 3 by 3 masking operations. Therefore, time complexity is at least 4 times lower than SUSAN. Simplicity in orientation calculation is another factor to shorten the computation time. For example, it took at most 0.2~0.3 seconds to process 256 by 256 real scene on a 586 PC.

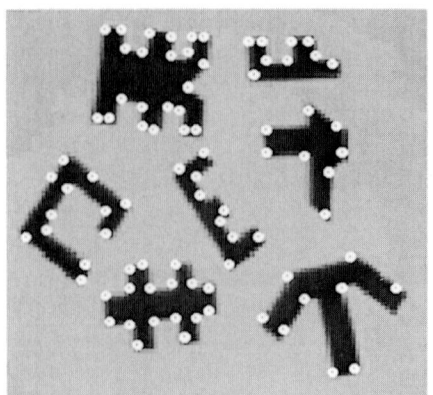 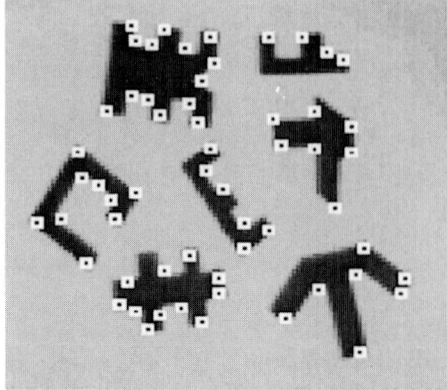

Figure 7. Outputs of COP (left) and SUSAN (right).

Figure 8. Outputs of COP (left) and SUSAN (right).

4.4 Comparison of each corner detector

In this section, we briefly summarize the performance of each detector for some images only in view of detectability. Figure 10 summarizes the results of each corner detector to a varing Gaussian smoothing factor σ whose value is 0.5, 1.0 and 1.5, respectively. COP has no false positives and less false-alarms. In Table 1, all numbers denoting false-positives and negatives are expressed relatively to the number of real corners. Though more false-positives than SUSAN, COP has less false-negatives, which means good detection. In this test, we use a five by five search window to find local maxima in each corner detector.

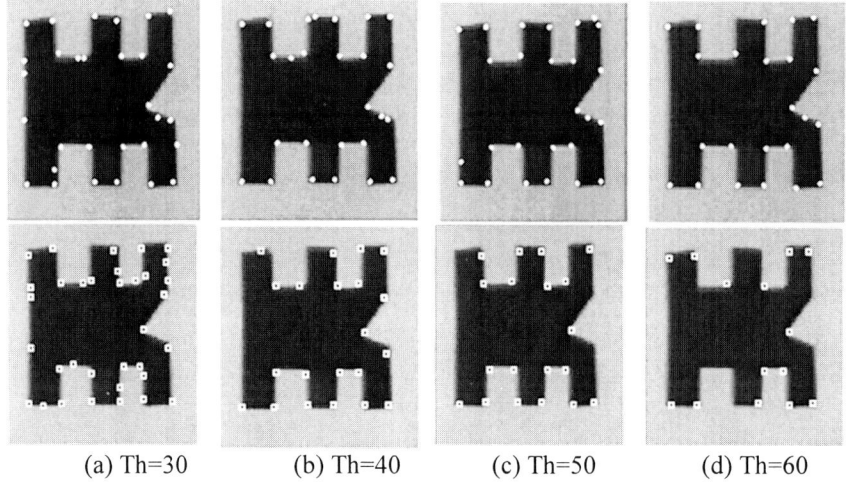

(a) Th=30 (b) Th=40 (c) Th=50 (d) Th=60

Figure 9. The threshold flexibility test of COP (top row) and SUSAN (bottom row).

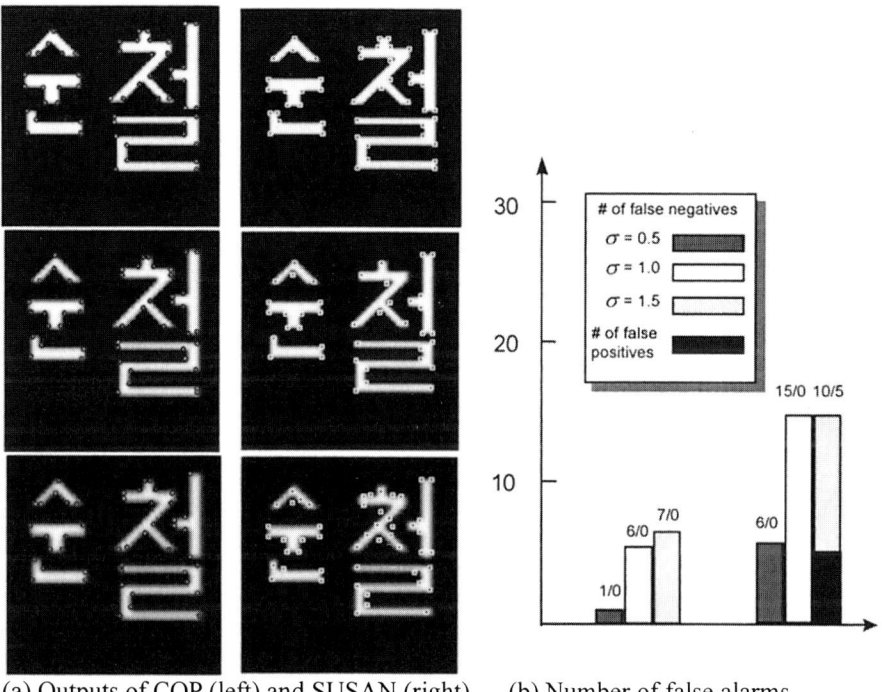

(a) Outputs of COP (left) and SUSAN (right)   (b) Number of false alarms

Figure 10. Detectability comparison of each corner detector with a varying Gaussian smoothing factor, σ ranging from 0.5 to 1.5 in 0.5 increments.

Table 1. Detectability comparison of each corner detector.

Test image	detector / false alarm	COP	SUSAN
순철	+	0/55	0/55
	−	0/55	0/55
ㅅㅐ	+	0/33	0/33
	−	0/33	5/33
(image)	+	3/95	1/95
	−	12/95	38/95
(image)	+	0/70	0/70
	−	0/70	1/70
total	+	3/253	1/253
	−	12/253	44/253

5. Conclusions

In this paper, we proposed two oriented cross operators, COP, which provide a useful information to extract low-level features due to its characteristics, preference for different edges and simple orientation determination. With the help of COP, we proposed two new feature detection algorithms, for edges and corners. In edge detection algorithm, we don't use the conventional non-maximum suppression algorithm but use a new non-maximum suppression rules. In complicated outdoor scene, there remains some problems such as fluctuating and broken edges like other edge extraction algorithms though it gives good results for synthetic and some real images. However, COP "corner detector" provides satisfactory results for all synthetic and noisy real scenes. Its small size lowers computational cost and increases localization accuracy. Moreover, its simple orientation selection and noise robustness make it possible to produce very effective and good performance.

References

[1] S. M. Smith and J.M. Brady, "A new Approach to Low Level Image Processing", IJCV, 1996. In publication.
[2] P. R. Beaudet, "Rotational invariant image operators", ICPR pp579-583,1978.
[3] L. Kitchen and A. Rosenfeld, "Gray-level corner detection", P. Recognition Letters,1, 1982.
[4] L. Dreschler and H.-H. Nagel, "Volumetric Model and 3D Trajectory of a Moving Car Derved from Monocular TV-frame Sequence of a Street Scene", CVGIP, 20(3):pp199- 228, 1981.
[5] C.G. Harris and M. Stephens, "A combined corner and edge detector", In 4th Alvey Vision Conference, pp147-151,1988.
[6] R. Deriche and G. Giraudon, "A computational Approach for Corners and Vertex Detection ", IJCV, 10(2):pp101-124, 1993.
[7] H. Wang and M. Brady, "Real-time corner detection algorithm for motion estimation", 13(9): pp 695-703, 1995.
[8] C. H. Chen, J. S. Lee and Y. N. Sun, "Wavelet transformation for gray-level corner detection", 28(6), pp853-861, Pattern Recognition, 1995.
[9] J. F. Canny, "Finding edges and lines in images", Master's thesis, MIT,Cambridge, USA, 1983.

An Integrated Approach for Segmentation and Representation of Range Images

Olga R. P. Bellon
UFPR-Depto. de Informática
81513-970 Curitiba-PR-Brazil
olga@inf.ufpr.br

Clesio L. Tozzi
UNICAMP-FEEC-DCA
13081-970 Campinas-SP-Brazil
clesio@dca.fee.unicamp.br

Abstract: This paper presents an integrated approach for segmentation and representation of objects in range images. The segmentation is based on the association of edge detection with clustering techniques, and it produces a set of labeled regions plus the coefficients of the plane fitted to each region. With this information, the reconstruction error for the whole image is estimated, and the initial number of regions supplied to the clustering algorithm can be updated, based on this error. Once the image reconstruction error is smaller than the desired, the next step is to create a polyhedral representation to the surfaces of the image. For each segmented region, the representation process yields its 3D vertices' coordinates, ordered to form a polygon. The final representation is robust enough to guarantee that there are no cracks on the reconstructed surfaces. The main contributions of this work are: (1) A new approach for determining the optimal number of image regions; (2) A suitable representation that can be applied to both polyhedral and non-polyhedral objects.

1. Introduction

In image analysis, it is usually convenient to transform the results of the segmentation into a more suitable form, for instance, to perform interpretation and recognition of objects. This representation must be robust enough to allow the object reconstruction with a small error. Since the representation is tightly related to the segmentation results, they should have a combined treatment.

This paper presents an integrated method for segmentation and representation of objects in range images. The segmentation is based on the *K-Means* clustering technique [4] which is combined with edge detection information, yielding a set of segments well approximated by planes. The representation makes use of these segments to build a suitable polyhedral description

The main contributions of this work are: (1) A new approach for determining the optimal number of image regions; (2) A suitable representation that can be applied to both polyhedral and non-polyhedral objects.

This work is organized as follows. Section 2 presents a brief discussion on the related work. In Section 3, the segmentation/representation process is presented. Finally, in Section 4, the experimental results are shown, followed by our conclusions.

2. Related Works

Clustering techniques show great independence from rigid threshold values, as compared with other segmentation techniques like region growing [1] and split-and-merge [5],[12]. However, in clustering algorithms, it is necessary to determine the ideal number of image regions. There possibilities to find this number are: (a) Splitting/merging analysis after the clustering process; (b) Generating various sets of clusters, one for each possible number of regions, and to develop a methodology to determine what set best represents the image. Possibility (b) is inefficient in many applications. Possibility (a) suffers from the same problem related to other techniques: the threshold determination.

Two relevant research lines [8],[9] use possibility (b). They analyze the segmentation results to determine the optimal number of clusters to the image based on *within-cluster* and *between-clusters* measures. However, these two methods are computationally costly, and none of them guarantee the correct determination of the searched number. Bhandarkar and Siebert [2] use a clustering algorithm to segment range images of polyhedral objects. The number of regions is adopted as the maximum expected in the image, and the clustering result is analyzed to identify regions that can be splitted/merged, based on experimental thresholds.

After the segmentation is done, the next step in the image analysis is to create a robust representation to the segmented regions. This representation must allow image reconstruction with small error, as well as be suitable to image interpretation. The representation is tightly related to the segmentation results but such association, from the topological viewpoint, has been poorly contemplated in the literature.

Some researchers deal with image segmentation and representation [1],[8],[12]. However, they do not treat the representation problem in terms of the topology of the segmented regions. Bhandarkar and Siebert [2] create a topological representation of segmented objects in range images, but their work is limited only to polyhedral objects. Faugueras and Herbert [5] present an approach to solve the representation problem of segmented images by using a triangulation processes, but the obtained representation is computationally costly.

The segmentation/representation method developed is presented in the next section.

3. The Method of Segmentation/Representation

To solve the problem of determining the optimal number of image regions, we propose to start the clustering process by using an initial estimate of this number, and to update it based on the clustering results without rigid threshold values. This estimate is supplied by performing an edge detection process in the range image. We also propose a simple and efficient method to create an approximate, topologically driven, polyhedral representation for both polyhedral objects and non-polyhedral objects in segmented range images. The method is based on the assumption that the object segments are well approximated by planes, in accordance with the segmentation process used. For each segment (region), the method yields the 3D vertices' coordinates, ordered to form a polygon, and the region's fitted plane equation. This representation is easily built from the range information, and it yields a fast image reconstruction, as shown below.

3.1. The Segmentation Process
3.1.1. Edge Detection and Normal Computation
First of all, the range image undergoes an edge detection process to identify jump and roof edges. In depth edges there is a significant discontinuity of depth values between neighbouring pixels. In roof edges there is a significant variation in the normal orientation of the adjacent surfaces.

To identify the jump edge pixels, a common border operator can be used. We used the DRF (Direct Recursive Filter) operator, developed by Chen and Castan [5]. The threshold T_1 used to define the edge pixels is obtained based on [7]. It is equal to the average T_m added by 1.5 plus the standard deviation σ of all the resulting values from the DRF application.

The roof edges are detected by a more complex process. In this work, we compute the normal parameters to each pixel and, from this information, identify the roof edges by the variation of the normal orientation angles. The normal (a,b,c) to each pixel (i,j) is computed by a plane fitting process, using the Least-Mean-Square method [10], in a 3x3 window centered in pixel p. In this computation, the jump edge pixels already detected are not taken into account to reduce spurious effects. The angle that measures the normal change between a pixel and its neighbours is given by the greater value of its normal variation related to its neighbours. These values are calculated from the normal vectors' inner product. When this angle is greater than a threshold T_2, which is set experimentally to 10 degrees, the pixel is considered to belong to a roof edge.

Once the jump edges and the roof edges are detected, the edge map can be created. From the edge map, after it suffers an edge closing process, it is possible to estimate the initial number K of regions by counting the number of its closed regions. This number is then supplied to the clustering algorithm.

3.1.2. Approximation Error Computation

The computation of the normals to each pixel is based on a 3x3 window, and it contains an error related to the process of fitting a plane to the window. This error, named approximation error ε_a, can be estimated to each image pixel p by:

$$\varepsilon_a^2(p) = \frac{1}{n_e} \sum_{i \in v} \left(d_p - a_p x_i - b_p y_i - c_p z_i \right)^2 \quad (3.2)$$

where a_p, b_p, c_p, d_p are the plane coefficients at pixel p, (x_i, y_i, z_i) are the 3D coordinates of the pixels belonging to the neighbourhood v of pixel p, and n_e is the number of pixels in the neighbourhood v that contribute to its normal computation. The global approximation error is the mean of all approximation errors, and can be computed by:

$$\varepsilon_a = \frac{1}{n - n_b} \sum \varepsilon_a^2(p) \quad (3.3)$$

where n_b is the total number of image edge pixels. The global approximation error will be used as a parameter to estimate the quality of the next phase of the process.

3.1.3. Clustering

The basic concepts of the clustering algorithm adopted [6] are described next.

Let the i-th pattern, $i=1,2,...,n$, from the data set under observation be written as $x_i = (x_{i1}, x_{i2}, ..., x_{iN})^T$, where $(x_{i1}, x_{i2}, ..., x_{iN})^T$ is its feature vector and N is the number of features. The number of patterns, n, is assumed to be significantly larger than N. A clustering is a partition $[C_1, C_2, ..., C_K]$ of the integers $[1,2,...,n]$ that assigns each pattern a single cluster label. The patterns corresponding to the integers in C_k form the k-th cluster, whose center is: $c_k = (c_{k1}, c_{k2}, ..., c_{kN})^T$, where

$$c_{kj} = \frac{1}{M_k} \sum_{i \in C_K} x_{ij} \qquad (3.4)$$

and M_k is the number of patterns in cluster k. The squared error for cluster k is:

$$e_k^2 = \sum_{i \in C_K} (x_i - c_k)^T (x_i - c_k) \qquad (3.5)$$

and the squared error for the clustering is:

$$E_K^2 = \sum_{k=1}^{K} e_k^2 \qquad (3.6)$$

The objectives of such a clustering process are to define, for a given K, a clustering that minimizes E_K^2 and to find a suitable K, much smaller than n.

The clustering algorithm applied in the method described herein has the normal vector components as input features. The initial number of regions, K, is also provided, based on the edge map. After the clustering algorithm is applied, the obtained result is a segmented image, for which each pixel has a label associated to a connected region. We wish that each segmented region has a small error when approximated by a plane.

3.1.4. Plane Equations Computation

The plane fitting process follows the same idea applied to estimate the normals. In this case, all the pixels that belong to each region, without taking into account the edge pixels, are used to compute the plane equation coefficients (a_k, b_k, c_k, d_k) most suitable to each region K.

3.1.5. Reconstruction Error Computation

The reconstruction error ε_r is computed to each non-edge pixel, based on the region plane equations, by Equation 3.7. Based on this equation, it is possible to compute the global reconstruction error ε_{gr}, and the reconstruction error to each region K, given respectively by Equations 3.8 and 3.9, where n_b is the total number of image pixels that belong to an edge, and n_k is the number of effective pixels, without taking into account the edge pixels, which belong to each region K. These equations are similar to those presented by Besl and Jain [1] but equations 3.7-3.9 are adapted to planar approximations.

$$\varepsilon_r^2(p) = (d_k - a_k x_i - b_k y_i - c_k z_i)^2 \qquad (3.7)$$

$$\varepsilon^2_{gr} = \frac{1}{n - n_b} \sum \varepsilon_r^2(p) \qquad (3.8)$$

$$\varepsilon_K^2 = \frac{1}{n_k} \sum_{i \in K} (d_k - a_k x_i - b_k y_i - c_k z_i)^2 \qquad (3.9)$$

If the global reconstruction error obtained is greater than desired, as based on the approximation error, the number of regions, K, can be increased and the clustering process is reapplied. Otherwise, the next step is to build the polyhedral representation.

3.2. The Representation Process

The input information to the representation process is a labeled segmented image. These segments are approximated by planes, and must be grouped to compose one or more objects. To create a robust polyhedral representation from this information, this representation must satisfy the polyhedral surface conditions [11]. The surface must be divided in planar faces that have edges as its intersections. The segments will be approximated by polygons, and their intersections will be line segments. For these conditions to be sufficient, it is necessary to consider the intersections between adjacent segments. The representation process steps are presented below.

3.2.1. Vertices and Edges Computation

In the context of this work, a vertex is defined by a pixel p which has three or four different regions on its neighbourhood, including its own region, using 4-neighbourhood and a 3x3 window centered at the pixel p. When a vertex is defined, a set of edges is also defined, with one edge corresponding to each pair of vertices neighbour segments. The first step to build the desired representation is to identify the vertices'neighbour regions. This can be done by using a scan image process, analyzing the neighbouhood of the pixels which have three or four different neighbour regions.

Because of the planar approximation, the 3D coordinates that best represent each vertex may not be exactly those related to the pixel identified. The suitable coordinates must be computed by considering the intersections between neighbour planes in the vertex. Three situations must be considered to compute the best 3D coordinates to each representation vertex: (a) A vertex is formed by the intersection of three-neighbour segments; (b) A vertex is formed by the intersection of four-neighbour segments; and (c) A vertex belongs to a depth edge.

In the case (a), the 3D vertices' coordinates can be computed by solving a simple linear equation system, using the plane equations of the three neighbour segments. Then, the computed vertex must be inserted in the vertices list and the three edges must be inserted in the edges list. When the same idea is used to solve case (b), the linear equation system may be undetermined, because the number of plane equations is greater than the number of variables. Thus, a vertex composed by four segments must be transformed in two vertices linked by an edge, where each vertex is associated to three different segments. In this case, the two vertices must be inserted in the vertices list, after the computation of their 3D coordinates, and their corresponding edges must be inserted in the edges list. To compute the vertices in case (c), the discontinuity between their neighbour segments must be considered. The detected vertex must be inserted in the vertices list without the refinement of its 3D coordinates, because it is not a result of the intersection of their neighbour segments, and the corresponding edges must be inserted in the edges list.

Note that, in accordance to the segmentation process applied to the image, both the depth edges and the roof edges will always keep their original shape. Then, if these edges are not straight lines, they are unsuitable to be directly approximated by our representation of edges, since much information may be lost along the borders. The solution adopted is to perform polygonalization on the edges identified in the edge detection process. This polygonalization produces a set of edges and its two corresponding vertices. After the polygonalization, the new vertices and edges are inserted in the vertices list and in the edges list, respectively.

The main steps of the algorithm to construct the polyhedral representation are:
1. *Perform the polygonalization on the edge map;*
2. *For each polygon edge do:*
   2.1. For each edge vertex do:
      2.1.1. If the vertex is not inserted in the vertices list do:
         2.1.1.1. Insert the vertex in the vertices list;
   2.2. Insert the edge in edges list;
3. *Scan the segmented image to detect vertex candidates;*
   3.1. For each detected pixel do:
      3.1.1. If the pixel is not inserted in the vertices list:
         3.1.1.1. If the pixel has four different neighbour regions:
            3.1.1.1.1. Split the pixel in two vertices, as shown in Figure 3.1;
         3.1.1.2. For each vertex do:
            3.1.1.2.1. If its neighbour regions have any depth discontinuities do:
               3.1.1.2.1.1. Compute its new 3D coordinates;
            3.1.1.2.2. Insert its 3D coordinates in the vertices list and the edge in the edges list;
4. *Evaluate the vertices list:*
   4.1. If there are two or more vertices too close do:
      4.1.1. To merge those vertices;
      4.1.2. To reorganize the vertices list and the edges list.
5. *Build the regions list, based in the vertices list and in the edges list.*

3.2.3. Representation Quality Measure

Once an image representation is created, its quality can be measured based on the corresponding initial range image. To compute this measure, the representation obtained must be placed in spatial correspondence with the input range image. This matching can be performed by a calibration process [6], since the correspondence between the 3D points and the image points is known.

After the calibration, the 3D vertices' coordinates of the obtained representation are projected and the corresponding edges are built, to create a 2-D wire-frame image. This image can be used as a visual measure of the representation quality. A quantitative measure of the representation quality can be computed based on the difference between the original 3D vertices' coordinates of the range image and the 3D vertices' coordinates of the obtained representation.

In the next section, we present our experimental results.

4. Experimental Results

To illustrate the experimental results, we present a set of images and one Table.

(a) Segmented image K=15.   (b) Segmented image K=20.   (c) Segmented image K=30.
Figure 4.1: Segmentation results (sphere without noise).

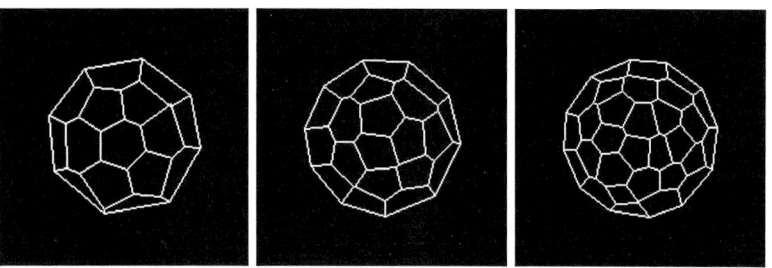

(a) Wire-frame image K=15.   (b) Wire-frame image K=20.   (d) Wire-frame image K=30.
Figure 4.2: Wire-frame results (sphere without noise).

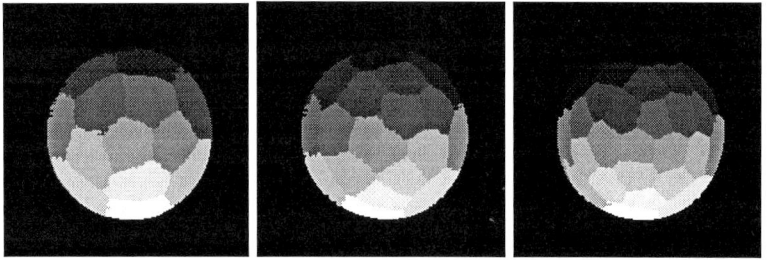

(a) Segmented image K=15.   (b) Segmented image K=20.   (c) Segmented image K=30.
Figure 4.3: Wire-frame results (sphere added by Gaussian noise).

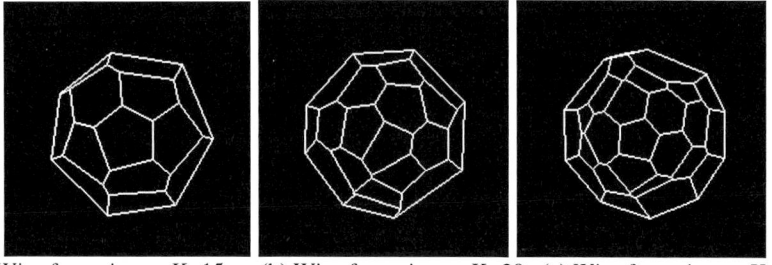

(a) Wire-frame image K=15.   (b) Wire-frame image K=20.   (c) Wire-frame image K=30.
Figure 4.4: Wire-frame results (sphere added by Gaussian noise).

Table 1: Comparison of measure errors

Image	ε_{ga}	ε_{gr}	ε_v	ε_v/n_v
Sphere K=15	0.001574	0.455996	120.63270	4.79
Sphere K=20	0.001574	0.324371	97.755966	2.57
Sphere K=30	0.001574	0.225245	76.374603	1.32
Noisy Sphere K=15	0.006114	1.516631	400.29523	16.68
Noisy Sphere K=20	0.006114	1.208808	262.17148	7.28
Noisy Sphere K=30	0.006114	0.978072	272.94876	5.45

Figure 4.1(a)-(c) presents the segmentation results for a range image of a sphere. Figure 4.2(a)-(c) presents the wire-frame results for Figure 4.1(a)-(c). In Figure 4.3 (a)-(c), the segmentation results to the original range image added by the effect of Gaussian noise (mean=0 and variance=0.01) are presented. Finally, Figure 4.4(a)-(c) shows the obtained wire-frame representations to Figure 4.3(a)-(c). Table 1 shows a

comparison of measure errors where εv is the 3D vertices' coordinates error, and nv is the number of vertices. Surprisingly, the vertices' coordinates error εv for noisy sphere with $K=20$ is greater than the one for the sphere with $K=30$. This is due to the stability of the plane equations' computation.

5. Conclusions

This work presents a new integrated segmentation/representation process to range images. The main contributions of this work are: (1) A new approach for determining the optimal number of image regions; (2) A suitable representation that can be applied to both polyhedral and non-polyhedral objects. The advantages of the process described here are: (1) It is loosely dependent on threshold values; (2) It integrates image segmentation and image representation in a cooperative manner; (3) The final representation is robust; (4) It allows the computation of a concrete process quality measure; (5) It can be applied to polyhedral and non-polyhedral objects. There are three future directions to improve on the results of this work. The first one is to use the region reconstruction error to guide the clustering process to reduce the computational cost. The second research line is to compute a more significant representation quality measure based on the difference between the reconstructed surface and the original one. The last direction is to create a full 3-D object representation, by matching different views of the same object. These ideas are currently under study.

6. References

[1] P. Besl and R. Jain, 1988. "Segmentation through variable-order surface fitting". *IEEE T-PAMI*, Vol.10, No.2, pp.167-192.

[2] S. Bhandarkar and A. Siebert, 1992. "Integrating edge and surface information for range image segmentation". *Pattern Recognition*, Vol.25, No.9, pp.947-962.

[3] J. Chen, S. Castan, 1986. "An optimal linear operator for edge detection". *Proc. of CVPR'86*, Miami.

[4] R. Dubes and R.C. Jain, 1976. "Clustering techniques: the user's dilemma". *Pattern Recognition Letters*, Vol.8, pp.247-260.

[5] O.D. Faugeras and M. Herbert, 1987. "The representation, recognition, and positioning of 3D shapes from range data". *Techniques for 3D Machine Perception*, Ed. North Holland, Netherlands, pp.13-52.

[6] R.C. Gonzalez, R.E. Woods, 1992. *Digital Image Processing*. Addison-Wesley.

[7] J.F. Haddon, 1988. "Generalized threshold selection for edge detection", *Pattern Recognition*, Vol.3, pp.195-203.

[8] R. Hoffman and A. Jain, 1987. "Segmentation and classification of range images". *IEEE T-PAMI*, Vol.9, No.5, pp.608-620.

[9] R. Krishnapuram and A Munshi, 1991. "Cluster-based segmentation of range images using differential-geometric features". *Optical Engineering*, Vol.30, No.10, pp.1468-1478.

[10] B. Noble and J.W. Daniel, 1988. *Applied Linear Algebra*. Prentice-Hall Int.

[11] F.P. Preparata and M.I. Shamos, 1985. *Computational Geometry: An Introduction*. Springer-Verlag, NY-USA, 1985.

[12] N. Yokoya and M. Levine, 1989. "Range image segmentation based on differential geometry: a hybrid approach". *IEEE T-PAMI*, Vol.11, No.5, pp.643-649.

Two-Dimensional Fractal Segmentation of Natural Images

Vo Anh[1], Junji Maeda[2], Tohru Ishizaka[2], Yukinori Suzuki[2], and Quang Tieng[1]

[1] Centre In Statistical Science And Industrial Mathematics, Queensland University of Technology, GPO Box 2434, Brisbane, Q.4001, Australia
[2] Department of Computer Science And Systems Engineering, Muroran Institute of Technology, 27-1 Mizumotocho, Muroran 050 Japan

Abstract. This paper presents a method that integrates fractal dimension and edge information into a region growing algorithm for segmentation of natural images. The local fractal dimension is estimated by two methods, one based on the 1-D Fourier-wavelet transform and the other on the 2-D Fourier transform. The algorithm also consists of a technique which stores edge informaiton not on a pixel itself but on a boundary between pixels in the region-edge integrating algorithm in order to use the edge information more effectively and to simplify the algorithm.Experimental results are presented to evaluate the performance of the methods.

1 Introduction

Image segmentation is an important task in image analysis and computer vision. In particular, it is an essential component in an algorithm for pattern recognition and classification (an example being recognition of road signs for use in a geographic information system). Pentland (1984) showed that natural images are commonly isotropic fractals over a range of scales, and fractional Brownian motion (fBm) can be used to model textured and shaded image regions. The defining characteristic of a fractal surface is its fractal dimension (FD). This characteristic furnishes a scale-invariant description of a natural image and corresponds closely to the intuitive notion of surface roughness.

Because of these desirable properties, the FD is an important feature of a natural image that can be used in an algorithm for its segmentation (see Keller et al. (1989) and Keller and Seo (1990)). However, it is known that the FD alone does not perform a satisfactory segmentation due to its high resolution, hence still yielding a large number of details (see Maeda et al. (1996)). On the other hand, a procedure based on both region and edge information is known to be effective for image segmentation (Pavlidis and Liow (1990)).

Maeda et al. (1996) proposed a method that integrates the FD information and the edge information into a region growing algorithm for natural image segmentation. In this method, the FD is used as a local feature that contains the region information, while the edge information is used to control the region growing algorithm. Maeda et al. (1996) used a Fourier-wavelet method to estimate the FD of a local region of the image. This FD is related to the exponent in the power-law form of the spectral density of the random field generating that image (Anh et al. (1996)).

The method of Maeda et al. (1996) in estimating this exponent, and hence the FD, is one-dimensional; that is, each local region is scanned horizontally (or vertically or diagonally) to obtain a time series for the estimation. This 1-D approach is fast to implement, but may not be as efficient as a 2-D approach, which uses information in the entire neighbourhood of each point. This paper proposes a 2-D method to estimate the FD of a local region. The key element is the 2-D isotropic version of fBm, whose spectral density also involves an exponent relating to the FD. Our periodogram-based method to estimate this exponent works well for a local region (even as small as 5×5) because it only requires that portion of the periodogram in a small neighbourhood of frequency 0. We will compare the performance of this 2-D method with the 1-D Fourier-wavelet method in this paper.

Another component of our segmentation algorithm is based on the technique of edge representation by boundary pixels, called the boundary edge (Maeda et al. (1996)). This technique improves the control ability of the edge information in the region growing algorithm and also simplifies the algorithm.

The local spectral density and the Fourier-wavelet method for the general case of random fields in $\Re^n$ are described in Section 2, while Section 3 details the 2-D spectral method of estimating the FD. The segmentation algorithm is outlined in Section 4. The experimental results are reported in Section 5. Finally, some conclusions are drawn in Section 6.

2 Local spectral density

In this section, we define the local spectral density and local periodogram of a random field defined on $\Re^n$. We will consider the wavelet functions ψ satisfying the usual admissibility condition (Koornwinder (1993), p. 30) and denote

$$\psi_{a,b}(t) = \frac{1}{a^{\frac{n}{2}}} \psi\left(\frac{t-b}{a}\right), a > 0, b, t \in \Re^n. \tag{1}$$

Then the Fourier transform of $\psi_{a,b}(t)$ is

$$\hat{\psi}_{a,b}(\omega) = a^{\frac{n}{2}} \hat{\psi}(a\omega) e^{-i(b,\omega)} \tag{2}$$

The Fourier-wavelet transform of a random field $X(t)$ is defined as

$$\hat{X}_{\psi_{a,b}}(\omega) = \int_{\Re^n} X(t) e^{-i(t,\omega)} \overline{\psi_{a,b}(t)} dt, \omega \in \Re^n \tag{3}$$

if it exists. In order to motivate a concept of local spectral density, we assume for the time being that $X(t)$ is stationary with covariance function $R(t)$ and spectral density $f(\omega)$. Then the following spectral representation holds:

$$E(X(t)X(s)) = R(t-s) = \int_{\Re^n} e^{i(t-s,\omega)} f(\omega) d\omega. \tag{4}$$

It now follows that, using (4),

$$E\left|\hat{X}_{\psi_{a,b}}(\omega)\right|^2$$
$$= \int_{\Re^n}\int_{\Re^n} E(X(t)X(s)) e^{-i(t-s,\omega)} \psi_{a,b}(t) \overline{\psi_{a,b}(s)} dt ds$$
$$= \int_{\Re^n}\int_{\Re^n} \left(\int_{\Re^n} e^{i(t-s,\lambda)} f(\lambda) d\lambda\right) e^{-i(t-s,\omega)} \psi_{a,b}(t) \overline{\psi_{a,b}(s)} dt ds$$
$$= \int_{\Re^n} f(\lambda) \left[\left(\int_{\Re^n} e^{i(t,\lambda-\omega)} \psi_{a,b}(t) dt\right) \left(\int_{\Re^n} e^{-i(s,\lambda-\omega)} \overline{\psi_{a,b}(s)} ds\right)\right] d\lambda$$

$$= \int_{\Re^n} f(\lambda) \left|\hat{\psi}_{a,b}(\lambda-\omega)\right|^2 d\lambda. \tag{5}$$

In view of (2) and (5), the quantity $\frac{1}{a^n} E\left|\hat{X}_{\psi_{a,b}}(\omega)\right|^2$ may be considered as the scale-space energy density of $X(t)$. As a result, we define the local spectral density of $X(t)$ as

$$\lim_{l\to\infty} \frac{1}{l} \int_{0+}^{l} E\left|\hat{X}_{\psi_{a,b}}(\omega)\right|^2 \frac{da}{a^{n+1}} \tag{6}$$

if the limit exists. Now, by choosing the wavelet function ψ so that

$$\lim_{l\to\infty} \frac{1}{l} \int_{0+}^{l} \left|\hat{\psi}_{a,b}(\lambda-\omega)\right|^2 \frac{da}{a^{n+1}} = C_\psi \delta(\lambda-\omega), \tag{7}$$

where C_ψ is a constant dependent only on ψ, and δ is the Dirac delta function (Koornwinder (1993), p. 30), we obtain from (5) and (7) that

$$\lim_{l\to\infty} \frac{1}{l} \int_{0+}^{l} E\left|\hat{X}_{\psi_{a,b}}(\omega)\right|^2 \frac{da}{a^{n+1}} = C_\psi f(\omega) \tag{8}$$

using Fubini's theorem. The result (8) suggests a definition of the local periodogram as

$$I_\psi(\omega) = \frac{1}{C_\psi} \frac{1}{T^n l} \int_{[0,T]^n} \int_{0+}^{l} \left|\hat{X}_{\psi_{a,b}}(\omega)\right|^2 \frac{da}{a^{n+1}} db. \tag{9}$$

Remark. For an example of a wavelet function which satisfies condition (7), we may take the Marr wavelet (second derivative of the 1-D Gaussian):

$$\psi(t) = (1-t^2) e^{-\frac{t^2}{2}}, \hat{\psi}(\omega) = \sqrt{2\pi}\omega^2 e^{-\frac{\omega^2}{2}}.$$

Then,

$$\lim_{l\to\infty} \frac{1}{l}\int_{0+}^{l} |\hat{\psi}_{a,b}(\lambda-\omega)|^2 \frac{da}{a} = \lim_{l\to\infty} \frac{1}{l}\int_{0+}^{l} |\hat{\psi}(a(\lambda-\omega))|^2 da$$

$$= \lim_{l\to\infty} \frac{2\pi}{l}\int_{0+}^{l} a^4 (\lambda-\omega)^4 e^{-a^2(\lambda-\omega)^2} da$$

$$= \lim_{l\to\infty} \frac{3\pi}{2}\frac{1}{l}\int_{0+}^{l} e^{-a^2(\lambda-\omega)^2} da$$

$$= \frac{3\pi}{2}\delta(\lambda-\omega).$$

3 Local fractal dimension

As suggested by Pentland (1984), we will consider a fractal random field X with spectral density of the form

$$f(\omega_1,\omega_2) = \frac{\sigma^2}{(\omega_1^2+\omega_2^2)^{2\beta-\frac{1}{2}}}, \beta > \frac{1}{4}, \sigma, \omega_1, \omega_2 \in \Re \tag{10}$$

as $\omega_1^2+\omega_2^2 \to 0$ to represent natural image data. As shown in Anh et al. (1996), (10) has the approximation

$$f_D(\omega_1,\omega_2) \sim \frac{\sigma^2}{4\pi^2}(\omega_1^2+\omega_2^2)^{\frac{1}{2}-2\beta}, \omega_1, \omega_2 \in [-\pi,\pi] \tag{11}$$

as $\omega_1^2+\omega_2^2 \to 0$. Noting that

$$|4-(e^{i\omega_1}+e^{-i\omega_1}+e^{i\omega_2}+e^{-i\omega_2})|^2 = \left[4\sin^2\frac{\omega_1}{2}+4\sin^2\frac{\omega_2}{2}\right]^2 \to (\omega_1^2+\omega_2^2)^2$$

near the origin, the approximation (11) suggests the model

$$f_D(\omega_1,\omega_2) = \frac{\sigma^2}{4\pi^2}\frac{1}{|4-(e^{i\omega_1}+e^{-i\omega_1}+e^{i\omega_2}+e^{-i\omega_2})|^{2d}} \tag{12}$$

$$d = \beta - \frac{1}{4}$$

as a discrete approximation to (10) near the origin. As a result, Proposition 2 of Anh et al. (1996) implies that the dimension of the random field X with spectral density (10) can be inferred as

$$\dim(graph(X)) = 3 - 2d. \tag{13}$$

The problem now is to estimate d from the data, hence yielding the fractal dimension of $X(t)$ via (13). It is seen from (12) that this can be done via the regression

$$\ln f_D(\omega_{1j},\omega_{2j}) = a - d\ln|4-(e^{i\omega_{1j}}+e^{-i\omega_{1j}}+e^{i\omega_{2j}}+e^{-i\omega_{2j}})|^2$$
$$+u(\omega_{1j},\omega_{2j}) \tag{14}$$

where a is the constant term, $\omega_{ij} = \frac{2\pi j}{T}, 1 \leq j < T, T$ being the sample size (e.g. $T = 5$), $i = 1, 2$, and u is planar white noise. For an estimate of f_D, Maeda et al. (1996) used the 1-D version of the local periodogram (9) based on a B-spline wavelet, which is a good approximation to the 1-D Marr wavelet. It is noted that a fast algorithm for a 2-D discrete wavelet transform is not available (unless the separability condition is assumed, in which case this is performed as a composition of two 1-D wavelet transforms). Hence, in this paper, in addition to the 1-D version of the local periodogram (9), we also estimate f_D by the usual periodogram, which is computed via the 2-D FFT. The result will then be compared with the 1-D Fourier-wavelet approach.

4 Segmentation procedure

Our segmentation procedure integrates the region information in the form of the FD map and the edge information in the form of the edge map on boundary pixels obtained from an original image by boundary Sobel operator. The edge information in the region growing algorithm plays an important role to decide whether the two adjacent regions should be merged or not. If there is no edge between the current region and the adjacent neighbouring region, the neighbouring region is merged into the current region. If there is an edge between them, the growth toward that direction is terminated. When the growth towards all possible directions is completed, one segmented region including the current region is formed. These operations are repeatedly performed for the complete image.

The steps of the algorithm are as follows:

1. The FD map and the edge map are computed from an original image. The FD map is calculated from a 3×3 local area for each pixel of the original image in the 1-D Fourier-wavelet method and from a 5×5 local area for the 2-D Fourier method. The edge map is obtained by the boundary Sobel operator and used after thresholding.
2. The original image is divided into initial kernel regions according to the region growing algorithm based on the grey levels of the original image. The neighbouring pixel is merged into the initial kernel region R_0 if

$$|g_{ave}(R_0) - g(i,j)| \leq T_{GL}$$

is satisfied, where $g_{ave}(R_0)$ is the average grey level within the kernel region R_0, $g(i,j)$ is the grey value of the neighbouring pixel at (i,j), and T_{GL} is a predetermined threshold.
3. Starting from the initial kernel regions, the region growing algorithm is carried out based on the FD information and the edge information. A neighbouring pixel is merged into the intial kernel region R_0 if both of the following two conditions are satisfied:
   - There is no edge between the initial kernel region R_0 and the neighbouring pixel at (i,j) that is located in the direction of the growth.

- $|d(R_0) - d(i,j)| \leq T_{FD}$, where $d(R_0)$ is the average FD value within the initial kernel region R_0, $d(i,j)$ is the FD of the neighbouring pixel at (i,j), and T_{FD} is a predetermined threshold.

   The above region growing algorithm is executed for all neighbouring pixels around the initial kernel regions, and a newly merged region is created.
4. The region growing algorithm is carried out by repeatedly performing Step 3 until there is no remaining region that is to be merged in the whole image.
5. The smaller regions that remain after the above steps are removed as noise under the following conditions: Calculate the differences between the average FD values within the small region under consideration and the average FD values within all neighbouring regions around the small region. Select one region from several neighbouring regions that have the smallest value of the above differences. If the smallest difference is less than the threshold value T_{FD}, the small region is merged into the selected neighbouring region. In other words, the small regions are removed.

5 Experimental results

Figure 1 shows an original natural image that has 400 × 400 pixels and 256 grey levels. The local FD maps calculated from the original image using the 1-D Fourier-wavelet method and the 2-D Fourier method are shown in Figures 2 and 3 respectively. These figures indicate that the FD maps from both methods are still very detailed, but invariant to the lighting condition (e.g. shading); the wood texture of the two walls of the house is captured clearly although one wall is in the shade in the original image. This is an advantage of using the FD as a key feature in a segmentation scheme.

The segmented images by the region-edge integrating algorithm based on the FD maps of Figures 2 and 3 are shown in Figures 4 and 5 respectively. Both methods provide good segmentation and are quite comparable to each other. The circular signs on the two walls show clearer details in the 2-D Fourier method. This will be advantageous in applications such as road sign recognition, where the only relevant details come from the road sign which must be recognised in a natural scene. It seems that a segmentation with less detail but higher accuracy can be achieved if proper image processing is applied to the FD map of the 2-D method prior to executing the segmentation algorithm, since this FD map (Fig. 3) possesses sufficient information for this purpose. Furthermore, the 2-D FFT is relatively fast, while the Fourier-wavelet technique requires more processing time, since a discrete wavelet transform must by performed for each frequency used in the regression for estimating the local FD (see Eq. (12) of Maeda *et al.* (1996)).

6 Conclusions

This paper has put forward a method for the segmentation of natural images that integrates the FD information and the edge information into a region growing algorithm. We have proposed two highly-accurate methods to estimate the local FD based on a 1-D Fourier-wavelet transform and a 2-D Fourier transform. We have also investigated the boundary edge in order to improve the control ability in the region-edge integrating algorithm, and to simplify the segmentation algorithm. A comparison has been made between the 1-D Fourier-wavelet method and the 2-D Fourier method. Experimental results indicate that both methods provide promising results in the segmentation of natural images.

References

1. Anh, V.V., Gras, F. and Tsui, H.T., Multifractal description of natural scenes, Fractals 4 (1996) 35-43.
2. Keller, J.M., Chen, S. and Crownover, R., Texture description and segmentation through fractal geometry, Computer Vision, Graphics and Image Processing 45 (1989) 150-166.
3. Keller, J.M. and Seo, Y.B., Local fractal geometric features for image segmentation, Int. J. Imaging Systems and Technology 2 (1990) 267-284.
4. Koornwinder, T.H., Wavelets: An Elementary Treatment of Theory and Applications, World Scientific, Singapore, 1993.
5. Maeda, J., Anh, V.V., Ishizaka, T. and Suzuki, Y., Integration of local fractal dimension and boundary edge in segmenting natural images, Proc. IEEE Int. Conf. on Image Processing, P. Delogne (Ed.), Ceuterick, Leuven, 1996, Vol. 1, 845-848.
6. Pavlidis, T. and Liow, Y.T., Integrating region growing and edge detection, IEEE Trans. PAMI 12 (1990) 225-233.
7. Pentland, A.P., Fractal-based description of natural scenes, IEEE Trans. PAMI 6 (1984) 661-674.

Fig. 1. The original image.

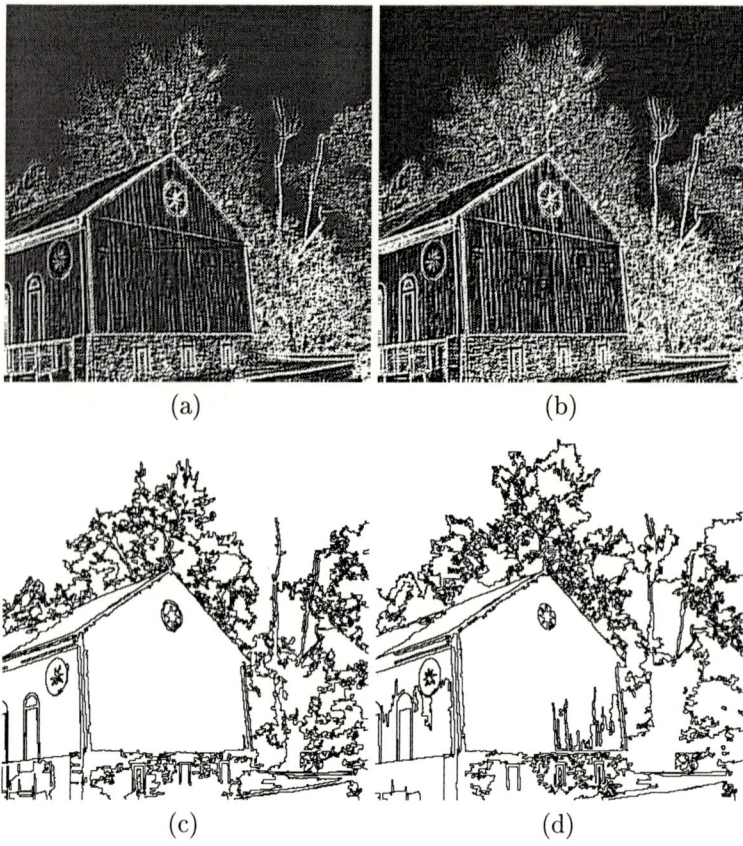

Fig. 2. (a)The FD map using the 1-D Fourier-wavelet method with 3×3 local blocks; (b)The FD map using the 2-D Fourier method with 5×5 local blocks; (c)The segmented image based on the FD map of (a); (d)The segmented image based on the FD map of (b).

Fast Segmentation of Range Images

Michal Haindl[1,2] and Pavel Žid[2]

[1] Center for Machine Perception, Czech Technical University,
[2] Institute of Information Theory and Automation of the Czech Academy of Sciences,
Praha, Czech Republic
E-mail: {haindl,zid}@utia.cas.cz

Abstract. A new type of range image segmentation method is introduced. The image segmentation is based on a recursive adaptive regression model prediction for detecting range image step discontinuities which are present at object face borders. Border pixels are detected in two perpendicular directions and detection results are combined together. Two predictors in each direction use identical contextual information from the pixel's neighbourhood and they mutually compete for the most optimal discontinuity detection. The method suggested can be successfully applied also to other image segmentation applications, e.g. panchromatic or multispectral image data, etc.

1 Introduction

Segmentation is a fundamental process affecting the overall performance of a machine vision system. Range image segmentation is a crucial part of autonomous navigation systems and as such it has been an active research area for past fifteen years. Segmentation of a range image should partition this image into meaningful patches representing single objects faces, but the task is complicated with outliers resulting from a sensing operation and corrupting boundary between distinct shapes. Optimal segmentation algorithm should be stable, accurate and numerically efficient.

There are many segmentation algorithms published in computer vision literature and a number of good survey articles [1],[10] is available. However their mutual comparison is very difficult because of lack of sound experimental evaluation results. A rare exception is results published in [7] together with experimental data available on their Internet server. These data and results are used also for our algorithm evaluation. The common approaches to solve the range image segmentation problem are region growing based algorithms [7],[2] where single regions are formed by iteratively growing from seed regions, split-and-merge, clustering (e.g. clustering in a fitted planes parameter space [4]), and edge based techniques [12]. Region growing and split-and-merge algorithms present the problem that they have to deal with different threshold values that are difficult to obtain (see [7]) and depend on an application. Clustering algorithms suffer usually less influence from this kind of problem, although other problems exist. Current segmentation algorithms most often miss small regions and perform poorly when the required region precision is high. Experiments in

[7] indicate that missed and false detected regions occur more frequently than over or under segmentation.

The present paper is organised as follows. In Section 2, a proposed method general concept under a Bayesian framework is introduced. Section 3 completes the algorithm with a locally optimal model selection rule design. Section 4 deals with numerical realisation problems while Section 5 contains evaluation results obtained on range data test set [7].

2 The Adaptive Regression Model

The regression method uses high spatial correlation (and spectral in the case of multispectral segmentation applications) between neighbours of a predicted pixel. We assume the mono-spectral line to be modelled as:

$$Y_{\acute{t}} = P^T Z_{\acute{t}} + E_{\acute{t}} \qquad (1)$$

where $P^T = [a_1, \ldots, a_\beta]$ is the $1 \times \beta$ unknown parameter vector $\beta = cardI_{\acute{t}}$. We denote the $\beta \times 1$ data vector

$$Z_{\acute{t}} = [Y_{\acute{t}-i} : \forall i \in I_{\acute{t}}]^T \qquad (2)$$

with a multi-index $\acute{t} = (m, n, d)$; $Y_{\acute{t}}$ is a predicted mono-spectral pixel value, m is the row number, n the column number, $d(d \geq 1)$ denotes the number of spectral bands, $E_{\acute{t}}$ is the white noise component. $I_{\acute{t}}$ is some neighbour index shift set.

Note that although the model predicts a mono-spectral pixel component, the model can use information from all other spectral bands in the case of a multi-spectral image ($d \geq 1$) as well. For range images $d = 1$.

Let us denote another multiindex $t = (m, n, d)$ and choose a direction of movement on the image plane e.g. $t - 1 = (m, n - 1, d), t - 2 = (m, n - 2, d), \ldots$. The white noise component E_t has zero mean and constant but unknown dispersion Ω. We assume that the probability density of E_t has a normal distribution independent of previous data and is the same for every time t. The task consists in finding the conditional prediction density $p(Y_t|Y^{(t-1)})$ given the known process history $Y^{(t-1)} = \{Y_{t-1}, Y_{t-2}, \ldots, Y_1, Z_t, Z_{t-1}, \ldots, Z_1\}$ and taking its conditional mean estimation $\tilde{Y}$ for the predicted data. If a prediction error is greater than an adaptive threshold the algorithm assumes an object face edge pixel. We have chosen the conditional mean estimator for data prediction, because of its optimal properties ([3]):

$$\tilde{Y}_t = E[Y_t|Y^{(t-1)}] \qquad (3)$$

Assuming normality of the white noise component E_t, conditional independence between pixels and an a priori probability density for the unknown model parameters chosen in the form (this normal form of a priori probability results in analytically manageable form of a posteriori probability density)

$$p(P, \Omega^{-1}|Y^{(0)}) = (2\pi)^{-\frac{\gamma(0)}{2}} |\Omega|^{-\frac{\gamma(0)}{2}} \exp\left\{-\frac{1}{2}tr\{\Omega^{-1}\begin{pmatrix}-I\\P\end{pmatrix}^T V_0 \begin{pmatrix}-I\\P\end{pmatrix}\}\right\},$$
(4)

where V_0 is a positive definite $(\beta+1)*(\beta+1)$ matrix and $\gamma(0) > d$, we have shown ([5]) that the conditional mean value is:

$$\tilde{Y}_t = \hat{P}_{t-1}^T Z_t .$$
(5)

The following notation is used in (4) and (5):

$$\hat{P}_{t-1} = V_{zz(t-1)}^{-1} V_{zy(t-1)} ,$$
(6)

$$V_{t-1} = \tilde{V}_{t-1} + V_0 ,$$
(7)

$$\tilde{V}_{t-1} = \begin{pmatrix} \tilde{V}_{yy(t-1)} & \tilde{V}_{zy(t-1)}^T \\ \tilde{V}_{zy(t-1)} & \tilde{V}_{zz(t-1)} \end{pmatrix} .$$
(8)

$$\tilde{V}_{xw(t-1)} = \alpha \tilde{V}_{xw(t-2)} + X_{t-1} W_{t-1}^T$$

It is easy to check (see [5]) also the validity of recursive (9). We assume slowly changing parameters, consequently these equations were modified using a constant exponential "forgetting factor" α to allow parameter adaptation.

$$\hat{P}_t = \hat{P}_{t-1} + (\alpha^2 + Z_t^T V_{zz(t-1)}^{-1} Z_t)^{-1} V_{zz(t-1)}^{-1} Z_t (Y_t - \hat{P}_{t-1}^T Z_t)^T$$
(9)

If the prediction error is larger than the adaptive threshold

$$|\tilde{Y}_t - Y_t| > \frac{2.5}{l} \sum_{i=1}^{l} |\tilde{Y}_{t-i} - Y_{t-i}|$$
(10)

then the pixel t is classified as an object edge pixel. The adaptive threshold is proportional to the local mean prediction error estimation.

3 Locally Optimal Model Selection

Let us assume two regression models (1) M_1 and M_2 with the same number of unknown parameters ($\beta_1 = \beta_2 = \beta$) and an identical neighbour index shift sets $I_{\tilde{t}}$ they differ only in their forgetting factors $\alpha_1 > \alpha_2$. The model $M_1, \alpha_1 \approx 1$ represents homogeneous image areas while the second model better represents new information coming from crossing some face borders because it allows quicker adaptation to this new information. According to the Bayesian theory, the optimal decision rule for minimizing the average probability of decision error chooses the maximum a posteriori probability model, i.e. a model whose conditional probability given the past data is the highest one. Predictors used in the presented algorithm can be therefore completed as in (11):

$$\tilde{Y}_t = \begin{cases} \hat{P}_{1,t-1}^T Z_t & \text{if } p(M_1|Y^{(t-1)}) > p(M_2|Y^{(t-1)}) \\ \hat{P}_{2,t-1}^T Z_t & \text{otherwise} \end{cases} \quad (11)$$

where Z_t is a data vector identical to both models. Following the Bayesian framework used in our paper and choosing uniform a priori model in the absence of contrary information, $p(M_i|Y^{(t-1)}) \sim p(Y^{(t-1)}|M_i)$, the simultaneous conditional probability density can be evaluated from

$$p(Y^{(t-1)}|M_i) = \int\int p(Y^{(t-1)}|P,\Omega^{-1})p(P,\Omega^{-1}|M_i)dPd\Omega^{-1} \ . \quad (12)$$

Under the already assumed conditional pixel independence, the analytical solution has the form

$$p(M_i|Y^{(t-1)}) = k\ \Gamma\left(\frac{\gamma(t-1)-\beta+2}{2}\right)|V_{i,zz(t-1)}|^{-\frac{1}{2}}\ \lambda_{i,t-1}^{-\frac{\gamma(t-1)-\beta+2}{2}}, \quad (13)$$

where k is a common constant. All statistics related to a model M_1 (7)-(9), (13) are computed using the exponential forgetting constant α_1 while symmetrical statistics of the model M_2 are computed using the second constant α_2. The solution of (13) uses the following notations:

$$\gamma(t) = \alpha_i^2 \gamma(t-1) + 1 \ , \quad (14)$$

$$\lambda_{t-1} = V_{yy(t-1)} - V_{zy(t-1)}^T V_{zz(t-1)}^{-1} V_{zy(t-1)} \ . \quad (15)$$

The determinant $|V_{zz(t)}|$ as well as λ_t can be evaluated recursively ([5]):

$$|V_{zz(t)}| = |V_{zz(t-1)}|\ \alpha_i^{2\beta}\ (1+Z_t^T V_{zz(t-1)}^{-1} Z_t) \ , \quad (16)$$

$$\lambda_t = \lambda_{t-1}\ \alpha_i^2\ (1+(Y_t - \hat{P}_{t-1}^T Z_t)^T \lambda_{t-1}^{-1}\ (Y_t - \hat{P}_{t-1}^T Z_t)(\alpha_i^2 + Z_t^T V_{zz(t-1)}^{-1} Z_t)^{-1}) \ . \quad (17)$$

4   Numerical Realization

The predictors in (11) can be evaluated using updating of matrices $V_{i,t}$ (7) and their following inversion. Another possibility is the direct updating of $\hat{P}_{i,t}$ (9). To ensure the numerical stability of the solution, it is advantageous to calculate $\hat{P}_{i,t}$ (9) using a square-root filter, which guarantees the positivity of matrix (7). The filter updates directly the Cholesky square root of matrices $V_{i,t}^{-1}$.

Alternatively it is possible to use the UDU filter (a factorization into two triangular and one diagonal matrices) for this purpose. Initialisation of recursive

Fig. 1. ABW test image and intensity image #0.

(9) and (17) must keep the condition of positive definiteness of matrices $V_{i,0}$ (4). We implemented in our algorithm the uniform a priori start $V_{i,0} = I$. This solution not only conforms with the initial lack of information at the start of algorithm, but also simplifies the calculation of the integral (12). Another possibility could be for example a local condition start, which ensures a quicker adaptation .

5 Results

In this section we present segmentation results of the proposed method and compare them with methods surveyed in [7]. Our goal was to properly detect single object face borders, while the performance evaluation in [7] is based on region comparison. Thus we were forced to use different performance criteria and to apply them also on test images, ground truth, and segmentation results from [7] to objectively compare the proposed segmentation algorithm performance with the state of art in the range image segmentation field. A version of our algorithm complemented with a region growing step to produce single object faces will be reported elsewhere. The performance of the methods is compared on two types real range images - the laser range finder and structured light scanner images. Objects present on these images are restricted to the objects with planar surfaces. The reason for this were difficulties of direct comparison of segmented curved surface patches. The first 512×512 test image shown on Fig. 1 was acquired from an ABW structured light scanner [11] and contains five polyhedral objects placed in a variety of angles and with different interobject spacing. Single objects have from 10 to 16 faces. The second test image (Fig. 2) with four more complex polyhedral objects (16 to 20 faces) was acquired from a Perceptron laser range finder [9]. The ABW range camera recovers depth from triangulation using structured light while the Perceptron computes the phase shift between an outgoing laser beam and its returned signal. Structured light scanner images have some shadow areas, where the sensor cannot make a

Fig. 2. Perceptron test image and intensity image #26.

range measurement. Laser images on the other hand are more noisy than the ABW data and their segmentation results are reported in [7] to be worse in all performance metrics used.

Ground truth for both images were hand made by a human operator outlining the boundary of each apparent surface patch in each image and subsequently corrected by another human operator. We took over these sets for our results evaluation too. Comparison of range image segmentations to the ground truth is done using the criteria of probability of finding a correct border pixel

$$P_c = \frac{card\{\mathcal{T} \cap \mathcal{G}\}}{card\{\mathcal{G}\}} \quad , \qquad P_w = \frac{card\{\mathcal{T} \cap (\mathcal{I} - \mathcal{G})\}}{card\{\mathcal{I}\}}$$

and a probability of wrong border pixel detection where $\mathcal{G}, \mathcal{T}, \mathcal{I}$ are the ground truth set, the segmentation result set, and the set of all image pixels, respectively. Pixels corresponding to the contextual support set I_i are denoted $*$ and the predicted pixel $\circ$, respectively. The segmentation results of both examples (ABW # 0, Perceptron # 26) are shown in Fig.3. The optimal regression models M_i for the ABW image # 0 were found to be: $M_i \quad * \quad \circ \quad *$ The second tested image (Perceptron # 26) optimal support set was found: $M_i \quad * \quad \circ$

Performance metrics used to compare segmentation algorithms in [7] were based on comparison of segmented regions. Our algorithm (A) is based on border detection between regions hence the comparison was made on region borders extracted from result images obtained using segmentation methods USF [2], WSU [6], UB [8], and UE [7]. Results in Table 1. show good performance of our method in comparison with previously published range segmenters, while being much faster than all these methods. Although we could not directly compare processing times of all evaluated methods because average processing times in [7] were measured on different platforms, we can estimate our processing time to be comparable to the quickest method UB and two orders of magnitude lower than the slowest method USF. The presented method clearly outperforms the tested methods in the overall number of wrongly detected border pixels, but in

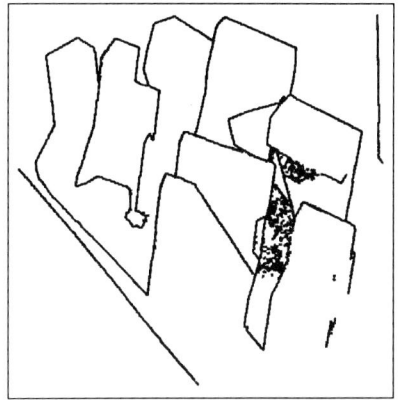 

Fig. 3. ABW segmentation and Perceptron segmentation results.

Table 1. Segmentation performance criteria

	ABW # 0		Perceptron # 26	
method	P_c	P_w	P_c	P_w
USF	0.69	0.02	0.69	0.07
WSU	0.68	0.009	-	-
UB	0.77	0.009	0.63	0.02
UE	0.86	0.011	0.62	0.02
A	0.64	0.004	0.52	0.002

the same time misses some correct border pixels. Missing pixels are usually from inner object border faces. Visual comparion of results demonstrates superior quality of detected borders using our algorithm. Detected borders are clean and accurately located. Compared segmenters generally produce worse quality face borders (double edges, noisy borders, wrong face overlapping, etc.).

6 Conclusion

We proposed the novel efficient and robust method based on a range profile prediction modelling. A range profile is modelled using an adaptive regression model. The adaptive predictor uses spatial correlation from neighbouring data what results in improved robustness of the algorithm over rigid schemes, which are affected with outliers often present at the boundary of distinct shapes. The proposed algorithm is recursive and therefore numerically effective. A parallel implementation of the algorithm is straightforward, every image row and column can be processed independently by its dedicated processor. The numerical stability is guaranteed using the Cholesky factorization of data gathering matrices. Usual handicap of segmentation methods is their lot of application dependent parameters to be experimentally estimated. Some methods need nearly a dozen adjustable parameters. Our method on the other hand requires only a contextual neighbourhood selection, which can be done using Bayesian statistics of section three. The algorithm performance is demonstrated on the set of test range images used for evaluation of some other range image segmentation methods in literature [7]. Our algorithm demonstrates encouraging segmentation quality while being of an order of magnitude faster than these techniques.

The test results of the algorithm are encouraging. The proposed method was always able to find all objects present in all our experimental scenes with excellent border localization precision. Detected borders can guide a subsequent image region recognition step.

The proposed method is fully adaptive, numerically robust and still with moderate computation complexity so it can be used in an on-line virtual reality acquisition system, robot navigation system or some other image acquisition systems.

References

1. Besl, P.J., Jain, R.C.: Three-dimensional object recognition. ACM Computing Surveys **17** (1985) no.1 75–145
2. Besl, P.J., Jain, R.C.: Segmentation Through Variable-Order Surface Fitting. IEEE Trans. PAMI **10** (1988) no.2 167–192
3. Broemeling,L.D.: Bayesian Analysis of Linear Models. New York, Dekker 1985
4. Flynn, P.J. Jain,A.K.: BONSAI: 3D Object Recognition Using Constrained Search. IEEE Trans. PAMI **13** (1991) no.10 1066–1075
5. Haindl,M., Šimberová, S.: A Multispectral Image Line Reconstruction method. In: Theory & Applications of Image Analysis. P. Johansen, S. Olsen Eds., World Scientific Publishing Co., Singapore, 1992
6. Hoffman, R.L., Jain, A.K.: Segmentation and Classification of Range Images. IEEE Trans. PAMI **9** (1987) no.5 608–620
7. Hoover, A., Jean-Baptiste, G., Jiang, X., Flynn, P.J., Bunke, H., Goldof, D.B., Bowyer, K., Eggert, D.W., Fitzgibbon, A., Fisher, R.B.: An Experimental Comparison of Range Image Segmentation Algorithms. IEEE Trans. PAMI **18** (1996) no.7 673–689
8. Jiang, X.Y., Bunke, H.: Fast Segmentation of Range Images into Planar Regions by Scan Line Grouping. Machine Vision and App. **7** (1994) no. 2 115–122
9. Perceptron Inc., LASAR Hardware Manual, 23855 Research Drive, Farmington Hills, Michigan, 1993
10. Sinha, S.S., Jain, R.: Handbook of Pattern Recognition and Image Processing. Wiley, New York, 1994
11. Stahs, T.G., Wahl, F.M.: Fast and Robust Data Acquisition in a Low-Cost Environment., SPIE no.1395: Close-Range Photogrammetry Meets Machine Vision (1990) 496–503, Zurich
12. Zhang, X.; Zhao, D.: Range image segmentation via edges and critical points. Proc. SPIE **2501** (1995) no. 3 1626–1637

Image Compression Based on Centipede Model [*]

Binnur Kurt[1], Muhittin Gökmen[1], Anil K. Jain[2]

[1] Istanbul Technical University,
Faculty of Elecrical and Electronics, Department of Computer Science,
Maslak, ISTANBUL 80626 - TURKEY
kurt@cs.itu.edu.tr, gokmen@cs.itu.edu.tr
[2] Michigan State University, Department of Computer Science,
East Lansing, MICHIGAN 48824 - U.S.A.
jain@cps.msu.edu

Abstract. We present an efficient contour based image coding scheme based on Centipede Model. Unlike previous contour based models which presents discontinuities with various scales as a step edge of constant scale, the centipede model allows us to utilize the actual scales of discontinuities as well as location and contrast across them. The use of the actual scale of edges together with other properties enables us to reconstruct a better replica of the original image as compared to the algorithm lacking this feature. In this model, there is a centipede for each edge segment which lies along the segment and the gray level variation across an edge point is represented by the difference between footholds and distance between left and right feet of the centipede. We obtain edges by using the recently introduced Generalized Edge Detector (GED) [1] which controls the scale and shape of the filter, providing edges suitable to the application in hand. The detected edge segments are ranked based on the weighted sum of the length of the segment, mean contrast and standard deviation of gray values on the segment. In our scheme, the compression ratio is controlled by retaining the most significant segments and by adjusting the distance between the successive foot pairs. The original image is reconstructed from this sparse information by minimizing a hybrid energy functional which spans a space called $\lambda\tau$-space. Since the GED filters are derived from this energy functional, we utilized the same process for detecting the edges and reconstructing the surface from them. The proposed model and the algorithm have been tested on both real and synthetic images. Compression ratio reaches to 180:1 for synthetic images while it ranges from 25:1 to 100:1 for real images. We have experimentally shown that the proposed model preserves perceptually important features even at the high compression ratios.

1 INTRODUCTION

The model based image compression has been attracted great interest as an attempt to exceed the boundaries of classical coding methods such as predictive

---

[*] This study is partially supported by TUBİTAK grant EEEAG-154 and NSF grant INT-9500652.

coding, information theory based methods, transform coding and vector quantization. One general image model often used characterizes the image in terms of contours and regions surrounded by them. In this framework, one class of algorithm requires an accurate partitioning of image into homogeneous closed regions [2, 3] (region based), whereas the other class of algorithms attempt to reconstruct the image from edge segments (edge based) and their neighborhood [2]. The former approach utilizes the uniformity of the each segmented region while the latter utilizes the differences between two regions. Even though much of the research efforts devoted to the region based approaches, it is more practical and efficient to extract edges by using high performance edge detectors instead of costly region segmentation. Gökmen et al. [4] developed a model based coding of fingerprint images based on ridges and valleys. Acar and Gökmen [5] utilized the weak membrane model of image in an attempt to carry out the edge detection and surface reconstruction by using the same process. This unification enables to use the same minimization process for both edge detection and surface reconstruction, and simplifies the coding scheme. However, the reconstructed images look somewhat artificial due to treating each edge as a step function by disregarding its actual scale. As known, the boundaries of physical structures in the world give rise blurred transitions in image intensities corrupted by noise instead of step discontinuities. Elder and Zucker [6] suggested a method to estimate the scale of edge by modeling it as a step function blurred by a Gaussian kernel.

In the following section, we introduce a coding scheme which combines all these features by utilizing a novel model called centipede model. In section 3, we describe how to encode the model parameters, together with ranking the edge segments and modeling the gray level variation along a segment by fitting polynomial curves. Also included in this section is the reconstruction of original image from encoded model parameters by minimizing a hybrid energy functional. In the last section, the performance of the algorithm on synthetic and real images are quantitatively and qualitatively analyzed.

2 CENTIPEDE MODEL

In our coding scheme, each connected edge segment as well as intensity variations around segment is described by a model called Centipede Model. A centipede, shown in Fig. 1, consists of a backbone along the edge segment and a number of legs approximately parallel to each other and normal to the backbone at the intersection points. Our centipede model consists of six parameters. When a centipede is placed over an edge segment, the transition of gray values can be captured by these parameters. Representing the position of a centipede parametrically by $V(s) = (x(s), y(s))$, the model can be described by the pentuple $(V(s), I(s), C_L(s), C_R(s), W_L(s), W_R(s))$ defined as

$$C_L(s) = I(V_L(s)) - I(V(s)), \quad C_R(s) = I(V_R(s)) - I(V(s))$$
$$W_L(s) = | V(s) - \inf_{V_L}\{| V - V_L |: | \frac{\partial I(x,y)}{\partial n} |_{V=V_L} < T_L\} |$$

$$W_R(s) = \mid V(s) - \inf_{V_R}\{\mid V - V_R \mid : \mid \frac{\partial I(x,y)}{\partial n} \mid_{V=V_R} < T_R\} \mid \qquad (1)$$

where $I(x,y)$ is the image intensity at (x,y), $\frac{\partial I(x,y)}{\partial n} = n.\bigtriangledown I$ is intensity gradient in the direction n, normal to $V(s)$, T_L and T_R are very small threshold values for the intensity gradients.

Thus, the model consists of geometry of edge segment $V(s)$, intensity value on edge $I(s)$, the lengths of left and right legs W_L and W_R capturing the scale of the edge, and left and right footholds, C_L and C_R corresponding to the contrast across this edge, respectively. The length of a leg, W_L, corresponding to the scale of an edge, is determined as the distance from this edge point to the closest point for which the gradient magnitude along the direction normal to the edge direction is less than a predetermined threshold value T_L. The entire image is modeled as a family of centipedes placed on edge segments as shown in In Fig. 1. The image

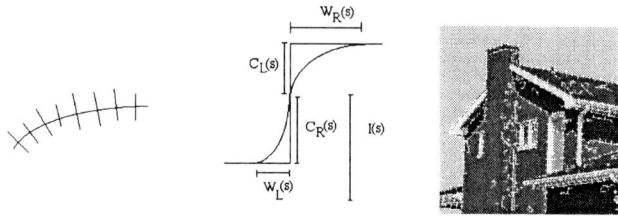

Fig. 1. Centipede Model, Parameters, Centipede Model overlied on the House image (From left to right)

is compressed by efficiently encoding the parameters of the centipede, and the original image is formed by reconstructing a surface from these parameters by minimizing a hybrid energy functional.

3 ENCODING AND DECODING IMAGES

The first part of the algorithm is to detect the edges by using the Generalized Edge Detector which encompasses most of the widely used edge detectors such as Canny's, Deriche's, Sarkar and Boyer's, Shen and Kastan's edge detectors. Edges are then traced to detect the distinct contours. These contours are coded by differential chain coding. Edge following algorithm forces the edge segments being as smooth as possible so that the gain of the compression of binary image with differential chain code followed by Huffman coding is maximized.

The second part of the compression algorithm is to select the perceptually most important edge segments among these contours obtained by tracing the edge segments. This is achieved by first assigning a priority to each edge segment by simply calculating the weighted sum of normalized set of contour length, average contrast along normal direction and average curvature evaluated directly from the differential chain code representation of the curve. Priority assigned to the contour C_i is given by

$$Priority(C_i) = w_l \cdot Length(C_i) + w_c \cdot Contrast(C_i) + w_u \cdot Curvature(C_i). \quad (2)$$

In the previous section, a method is introduced to extract the model parameters. The parameters along a segment are modeled by polynomials, and coefficients of each polynomial are coded. At the decoding part, decoder constructs two images, a binary edge map and intensity image obtained from model parameters which can be constructed simply by evaluating the polynomial at each edge point. Since both images are sparse, we use hybrid energy functional to span a surface through these points. For this purpose, the surface reconstruction problem is set as finding a function $f(x,y)$ which minimizes

$$E(f; \lambda, \tau) = \int\int_\Omega \beta(x,y)(f(x,y) - d(x,y))^2 + \lambda(1-\tau)(f_x^2(x,y) + f_y^2(x,y))$$
$$+ \lambda\tau(f_{xx}^2(x,y) + 2f_{xy}^2(x,y) + f_{yy}^2(x,y))\,dx\,dy \quad (3)$$

where λ controls the smoothness of the surface and τ controls the continuity of the surface. Properties of the hybrid model is explained in [1]. Minimization of functional given by (3) is obtained iteratively by Successive Over-Relaxation.
In order to eliminate a possible blurring across discontinuities, we defined the centipede foots, i.e. C_L and C_R, as crease points and vanish the last term in (3) including the second derivatives at these points. As described in [1], the Generalized Edge Detector is derived from the functional in (3), the scale and the shape of the GED is controlled by the λ and τ parameters in the hybrid functional. Thus we utilize the same process for both detecting edges and reconstructing images from them, unlike unrelated processes such as Canny edge detection and surface interpolation.

4 RESULTS AND CONCLUSIONS

The proposed model-based image coding has been applied to various synthetic and real images. We first consider the effect of ranking the edge segments and selecting only most significant segments on the reconstructed image quality. Fig.2(a) shows the original Lenna and House images. Fig.2(b,c,d,e) show the selected edges from a complete edge map obtained by the GED (with $\lambda = 0.5$ and $\tau = 0.5$). From these edge maps, 100%, 75%, 50% and 25% of edges are retained and these edges together with the reconstructed images are shown in Fig.2(b),(c),(d) and (e), respectively. The quantitative results for this test are also shown in this figure. The qualitative and quantitative evaluation of results indicates that the selection scheme works quite successfully. Even if very large portion of the edge segments are removed, the reconstructed image still contains most of the perceptually pertinent features. When we code the centipede parameters, we divided the each edge segment into blocks and approximate the parameters $I(s), C_L(s), C_R(s), W_L(s)$, and $W_R(s)$ over a block by fitting curves of order n. Thus in our coding scheme, the compression ratio and the quality of reconstructed image can be controlled by the the percentage of selected edge segments, the block size and the orders of approximations for the intensity, the contrast, and the edge width over the block.

Fig.3 shows the detected edges, V(s), the intensities and estimated scales of edges and the centipede model superimposed on the original House and Lenna images in Fig.2(a). To reveal the coding performance of the proposed model, we considered various synthtetic and real images. For the synthetic checkerboard and bar images[5], compression ratios of 127:1 and 157:1 are achieved, respectively. Fig.4 shows the original image, edge map and reconstructed images, from left to right, for House and Cameraman images. Compression ratios are 44 : 1 and 29 : 1, respectively. Fig.5 shows how the reconstructed image is degraded as the compression ratio increases. As seen from these results, relatively high compression ratios can be achieved by the proposed scheme. One of the advantageous of this contour based approach as compared to transform based coding is that this scheme does not cause excessively blurred image or blocking artifacts as the compression ratio increases. It retains the most important features even for the very high compression ratios.

In conclusion, we presented a new model for contour based image compression. This model enables us to utilize the scale, brightness and contrast of edges. We developed a ranking scheme for edge segments so that the most significant edge segments can be kept after the removal of the edge segments to increase the compression ratio. We utilized an efficient way of encoding the model parameters by means of curve fitting, differential chain coding and Huffman coding. We used the similar process controlled by the same parameters, λ and τ, for both detecting edges and reconstructing original image from edges. All these combined features make the proposed centipede model very attractive alternative to the existing model based schemes.

References

1. M. Gökmen, A. K. Jain, "$\lambda\tau$-space representation of images and generalized edge detector," to appear in *IEEE Trans. on PAMI*, June 1997.
2. M. Kunt, M. Benard, R. Leonardi, "Recent results in high compression image coding," *IEEE Transactions on Circuits and Systems*, Vol. CAS-34, No. 11, pp. 1306-1336, November 1987.
3. H. Sanderson, G. Crebbin, "Region-based image coding using polynomial intensity functions," *IEE Proc. on Vision and Image Proc.*, No.1, pp.15-23, Feb. 1996.
4. M. Gökmen, İ. Ersoy, A. K. Jain, "Compression of fingerprint images using hybrid image model," *Proc. IEEE Int'l Conf. On Image Proc.*, Vol.III, pp. 395-398, 1996.
5. T. Acar, M. Gökmen, "Image coding using weak membrane model," *Proc. of Visual Comm. and Image Processing'94*, pp. 1221-1230, Chicago, Illinois, 1994.
6. J.H. Elder, S.W. Zucker, "Scale space localization, Blur and Contour based image coding," *Proc. of Int. Conf. on Image Processing*, pp. 1221-1230, 1996.

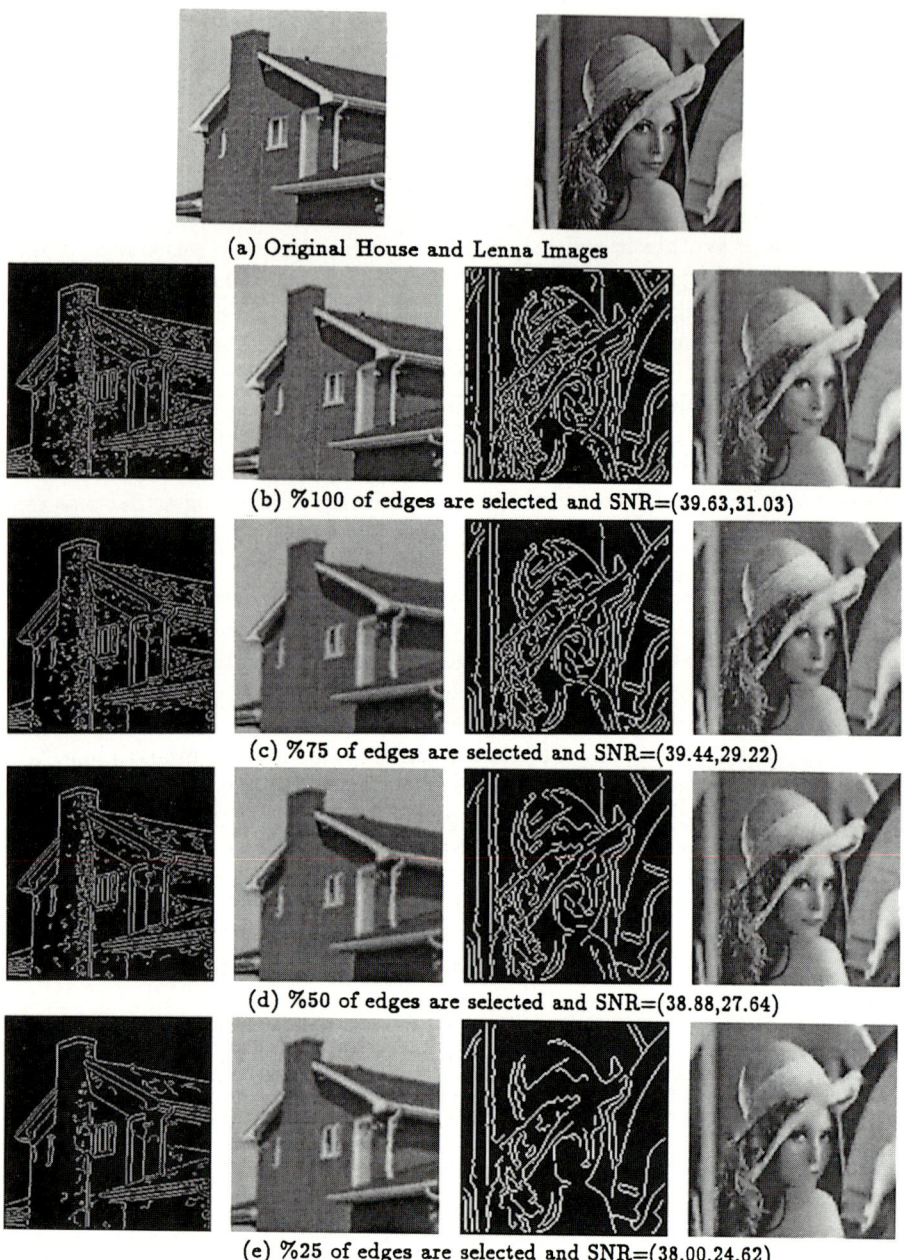

(a) Original House and Lenna Images

(b) %100 of edges are selected and SNR=(39.63,31.03)

(c) %75 of edges are selected and SNR=(39.44,29.22)

(d) %50 of edges are selected and SNR=(38.88,27.64)

(e) %25 of edges are selected and SNR=(38.00,24.62)

Fig. 2. The selected edges and reconstructed images for House and Lenna images and corresponding SNR(dB) values. From top to bottom, 100, 75, 50 and 25% of edges are retained.

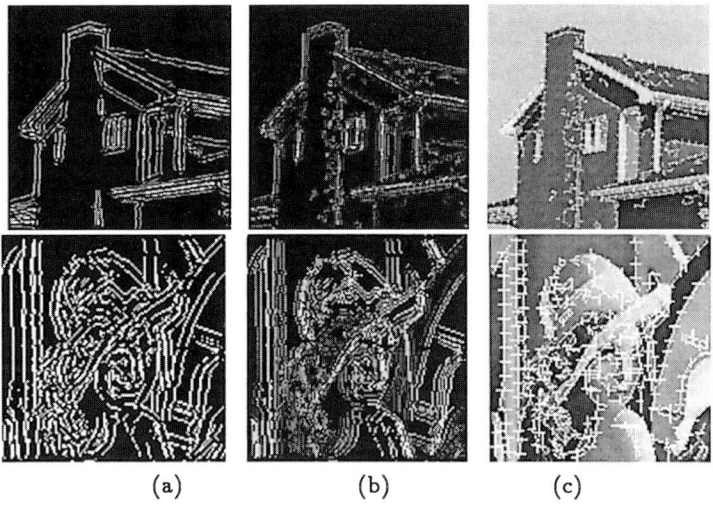

Fig. 3. Representation of detected edge scales.(a) Centipede Footholds, (b) Intensities on edges and footholds, (c) Centipede Model superimposed on the original House and Lenna images.

Compression Ratio : 44 : 1

Compression Ratio : 29 : 1

Fig. 4. Original image, Edge Map and Reconstructed image for House and Cameraman images.

Fig. 5. The reconstructed House and Cameraman images for the specified Compression Ratios (CR).

Unsupervised Texture Segmentation Using Feature Distributions

Timo Ojala and Matti Pietikäinen
Machine Vision and Media Processing Group, Infotech Oulu
University of Oulu, FIN-90570 Oulu, Finland
ojala@ee.oulu.fi, mkp@ee.oulu.fi

Abstract
This paper presents an unsupervised texture segmentation method, which uses distributions of local binary patterns and pattern contrasts for measuring the similarity of adjacent image regions during the segmentation process. Nonparametric log-likelihood test, the G statistic, is engaged as a pseudo-metric for comparing feature distributions. A region-based algorithm is developed for coarse image segmentation and a pixelwise classification scheme for improving localization of region boundaries. The performance of the method is evaluated with various types of test images. The same set of parameter values is used in all the experiments with texture mosaics in order to demonstrate the robustness of our approach.

1 Introduction

Segmentation of an image into differently textured regions is a difficult problem. Usually one does not know *a priori* what types of textures exist in an image, how many textures there are, and what regions have which textures [1]. In order to distinguish reliably between two textures relatively large samples of them must be examined, i.e., relatively large blocks of the image. But a large block is unlikely to be entirely contained in a homogeneously textured region and it becomes difficult to correctly determine the boundaries between regions.

Many different approaches to texture segmentation have been proposed. Segmentation methods are usually classified as region-based, boundary-based or as a hybrid of the two. For surveys of image and texture segmentation techniques, see [2,3]. The segmentation can be supervised or unsupervised. In unsupervised segmentation no *a priori* information about the textures present in the image is available. This makes it is a very challenging research problem in which only limited success has been achieved so far. Examples of different approaches to unsupervised segmentation are presented, e.g., in [4-10].

Our recent studies show that excellent texture discrimination can be obtained with local texture operators and nonparametric statistical discrimination of sample and prototype distributions. Texture classification results obtained by using distributions of local binary patterns (LBP) or gray scale differences have been better than those obtained with the existing methods [11-14]. Our method can be easily generalized to utilize multiple texture features, multiscale information, color features and combinations of multiple features using the new multichannel approach presented in [14].

This paper presents an efficient method for unsupervised texture segmentation based on texture description with feature distributions. A region-based algorithm is developed for coarse image segmentation and a pixelwise classification scheme for improving the localization of region boundaries.

2 Texture Description

The texture contents of an image region are characterized by the joint distribution of Local Binary Pattern (LBP) and Contrast (C) features [12]. The original 3x3 neighborhood (Fig. 1a) is thresholded by the value of the center pixel. The values of the pixels in the thresholded neighborhood (Fig. 1b) are multiplied by the binomial weights given to the corresponding pixels (Fig. 1c) and obtained values (Fig. 1d) are summed for the LBP number (169) of this texture unit. By definition LBP is invariant to any monotonic gray scale transformation. LBP describes the spatial structure of the local texture, but it does not address the contrast of the texture. For this purpose we combine LBP with a simple contrast measure C, which is the difference between the average gray level of those pixels which have value 1 and those which have value 0 (Fig. 1b).

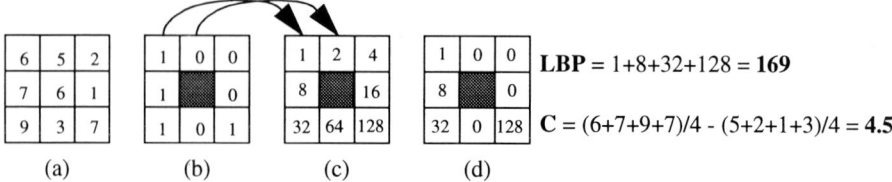

(a)　(b)　(c)　(d)

Fig. 1. Computation of Local Binary Pattern (LBP) and contrast measure C.

The LBP/C distribution is approximated by a discrete two dimensional histogram of size 256xb, where b is the number of bins for C. A log-likelihood-ratio, the G statistic [15], is used as a pseudo-metric for comparing LBP/C distributions. The value of the G statistic indicates the probability that the two sample distributions come from the same population: the higher the value, the lower the probability that the two samples are from the same population. We measured the similarity of two histograms with a two-way test of independence:

$$G = 2\left(\left[\sum_{s,m}\sum_{i=1}^{n} f_i \log f_i\right] - \left[\sum_{s,m}\left(\sum_{i=1}^{n} f_i\right)\log\left(\sum_{i=1}^{n} f_i\right)\right] - \left[\sum_{i=1}^{n}\left(\sum_{s,m} f_i\right)\log\left(\sum_{s,m} f_i\right)\right] + \left[\left(\sum_{s,m}\sum_{i=1}^{n} f_i\right)\log\left(\sum_{s,m}\sum_{i=1}^{n} f_i\right)\right]\right) \quad (1)$$

where s, m are the two sample histograms, n is the number of bins and f_i is the frequency at bin i. See [15] for a detailed derivation of the formula.

3 Segmentation Algorithm

The segmentation method consists of three phases: hierarchical splitting, agglomerative merging and pixelwise classification. First, hierarchical splitting is used to divide the image into regions of roughly uniform texture. Then, agglomerative merging procedure merges similar adjacent regions until a stopping criterion is met. At this point we have obtained rough estimates of the different textured regions present in the image and complete the analysis by a pixelwise classification to improve the localization. Fig. 2 illustrates the progress of the segmentation algorithm on a 512x512 mosaic containing five different Brodatz [16] textures.

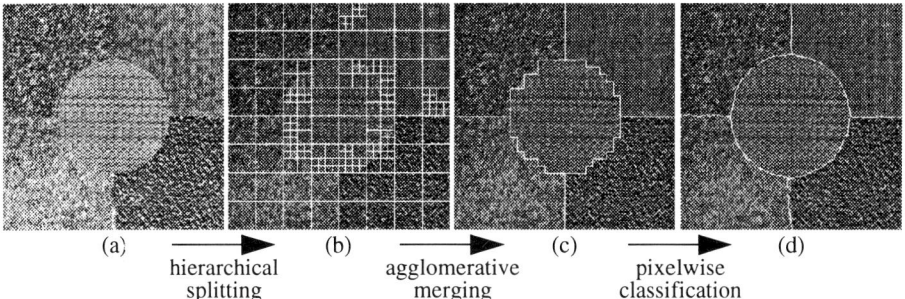

Fig. 2. Texture mosaic #1; the main sequence of the proposed segmentation algorithm. $MIR_{stop}=9.5$, $MIR_{hi}=1.2$, $ERR_a=1.4\%$, $ERR_p=1.7\%$, sweeps=16.

3.1 Hierarchical Splitting

A necessary prerequisite for the agglomerative merging to be successful is that the individual image regions are uniform in texture. For this purpose we apply the hierarchical splitting algorithm, which recursively splits the original image into square blocks of varying size. The decision whether a block is split to four subblocks is based on a uniformity test. We measure the six pairwise G distances between the LBP/C histograms of the four subblocks. If we denote the largest of the six G values by G_{max} and the smallest by G_{min}, the block is found to be nonuniform and is thus split further into four subblocks, if a measure of relative dissimilarity within region is greater than a threshold

$$R = \frac{G_{max}}{G_{min}} > X \quad (2)$$

Regarding the proper choice of X, one should rather choose a too small value for X instead of a too large one. It is better to split too much than too little, for the following agglomerative merging procedure is able to correct errors, where an uniform block of a single texture has been needlessly split. But error recovery is not possible, if segments containing several textures are assumed to be uniform.

To begin with, we divide the image into rectangular blocks of size S_{max}. If we applied the uniformity test on arbitrarily large image segments, we could fail to detect small texture patches and end up treating regions containing several textures as uniform. The next step is to use the uniformity test. If a block does not satisfy the test, it is divided into four subblocks. This procedure is repeated recursively on each subblock until a predetermined minimum block size S_{min} is reached. It is necessary to set a minimum limit for the block size, for the block has to contain a sufficient number of pixels for the LBP/C histogram to be reliable.

Fig. 2b illustrates the result of the hierarchical splitting algorithm with $X=1.2$, $S_{max}=64$ and $S_{min}=16$. As expected, the splitting goes deepest around the texture boundaries.

Note that the hierarchical splitting phase is not mandatory, but we could skip it by dividing the input image directly to blocks of size S_{min} and the successive agglomerative merging phase would still succeed. This is particularly true for easier problems of homogeneous and clearly distinct textures. However, our experiments have shown that finding larger areas of uniform texture with the hierarchical splitting method improves the convergence of the agglomerative merging algorithm.

3.2 Agglomerative Merging

Once the image has been split into blocks of roughly uniform texture, we apply an agglomerative merging procedure, which merges similar adjacent regions until a stopping criterion is satisfied. At a particular stage of the merging, we merge that pair of adjacent segments, which has the smallest Merger Importance (*MI*) value. *MI* is defined as

$$MI = p \times G \qquad (3)$$

where p is the number of pixels in the smaller of the two regions and G is the distance measure defined in Eq. 1. In other words, at each step the procedure chooses that merger of all possible mergers, which introduces the smallest change in the segmented image. Once the pair of adjacent segments with the smallest *MI* value has been found, the regions are merged and the two respective LBP/C histograms are summed to be the histogram of the new image region. Before moving to the next merger we compute the G distances between the new region and all adjacent regions to it. Merging is allowed to proceed until the stopping rule

$$MIR = \frac{MI_{cur}}{MI_{max}} > Y \qquad (4)$$

triggers. Merging is halted if *MIR*, the ratio of MI_{cur}, Merger Importance for the current best merge, and MI_{max}, the largest Merger Importance of all preceding mergers, exceeds a preset threshold *Y*. In theory, it is possible that the very first merges have a zero *MI* value (i.e. there are adjacent regions with identical LBP/C histograms), which would lead to a premature termination of the agglomerative merging phase. To prevent this the stopping rule is not evaluated for the first 10% of all possible merges.

Fig. 2c shows the result of the agglomerative merging phase after 174 merges. The *MIR* of the 175th merge (MIR_{stop}) is 9.5 and the merging is halted. The highest *MIR* value up to that point (MIR_{hi}) had been 1.2. The relationship between MIR_{stop}, MIR_{hi} and threshold *Y* reflects the reliability of the result of the agglomerative merging phase. The very large value of MIR_{stop} and very small value of MIR_{hi} underline the easiness with which the rough estimate of the texture regions is obtained for mosaic #1. Note that the segmentation error of 1.4% after the agglomerative clustering phase (ERR_a) is a somewhat biased in this problem, for the horizontal and vertical texture boundaries are accidentally aligned with the initial blocks.

3.3 Pixelwise Classification

To improve the localization of the boundaries a simple pixelwise classification algorithm is used. If the hierarchical splitting and agglomerative merging phases have succeeded, we have obtained quite reliable estimates of the different textured regions present in the image. Treating the LBP/C histograms of the image segments as our texture models we switch into a texture classification mode. If an image pixel is on the boundary of at least two distinct textures (i.e. the pixel is 4-connected to at least one pixel with a different label), we place a discrete disc with radius r on the pixel and compute the LBP/C histogram over the disc. We compute the G distances between the histogram of the disc and the models of those regions, which are 4-connected to the pixel in question. We relabel the pixel, if the label of the nearest model is different from the current label of the pixel and there is at least one 4-connected adjacent pixel with the tentative new label. The latter condition improves smooth adaption of texture boundaries and decreases the probability of small holes occurring inside the regions. If the pixel is relabeled, i.e. it is moved from an image segment to the adjacent segment, we update the corresponding texture models accordingly, hence the texture models become more accurate during the process. Only those pixels at which the disc is entirely inside the image are examined, hence the final segmentation result will contain a border of r pixels wide.

In the next scan over the image we only check the neighborhoods of those pixels, which were relabeled in the previous sweep. The process of pixelwise classification continues until no pixels are relabeled or maximum number of sweeps is reached. This is set to be two times S_{min}, based on the reasoning that the boundary estimate of the agglomerative merging phase can be at most this far away from the 'true' texture boundary. Setting an upper limit for the number of iterations ensures that the process will not wander around endlessly, if the disc is not able to capture enough information of the local texture to be stable. According to our experiments the algorithm generally converges quickly with homogeneous textures, whereas with locally stochastic natural scenes maximum number of sweeps may be consumed. We did not apply any post-processing method to improve the final segmentation result, e.g. by smoothing the texture boundaries or removing small regions as many existing algorithms do.

Fig. 2d demonstrates the final segmentation result after the pixelwise classification phase. Disc with radius of 11 pixels was used and 16 sweeps were needed. The final segmentation error (ERR_p), computed over the area processed by the disc which excludes the border of r pixels, is 1.7%.

4 Experimental Results

Next, we present some quantitative results obtained with the method. The segmentation results for two additional texture mosaics and a natural scene are presented. The same set of parameter values was used for all texture mosaics to demonstrate the robustness of the approach: $b=8$, $S_{max}=64$, $S_{min}=16$, $X=1.2$, $Y=2.0$, and $r=11$. See [17] for results for additional images and for a detailed discussion on parameter selection. For each mosaic we provide the original image, the rough segmentation result after the agglomerative merging phase and the final segmentation result after the pixelwise classification phase. The segmentation results are superpositioned on the original image.

Mosaic #2 (Fig. 3a) is a 512x512 image containing four textures made by a GMRF process and a circle of painted surface in the middle [18]. The more difficult nature of this problem shows in the values of MIR_{stop} (5.2) and MIR_{hi} (1.6), which are clearly closer to threshold Y than what was the case with mosaic #1. Nevertheless, the rough segmentation result (Fig. 3b) with segmentation error of 4.2% is quite decent. The final segmentation result (Fig. 3c) after 23 sweeps with segmentation error of 1.2% is excellent.

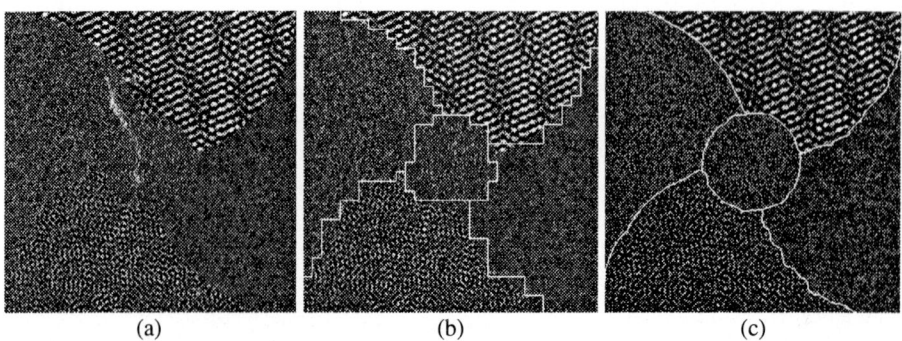

Fig. 3. Texture mosaic #2. MIR_{stop}=5.2, MIR_{hi}=1.6, ERR_a=4.2%, ERR_p=1.2%, sweeps=23.

Mosaic #3 (Fig. 4a) is 384x384 pixels in size and it is composed of textures of outdoor scenes [10]. In their study Jain and Karu tackled the problem of texture segmentation with a neural network generalization of the traditional multichannel filtering method, using various filter banks for feature extraction. For this mosaic Jain and Karu reported labeling error of 6% with Laws' filters in supervised mode. Our unsupervised method gives a clearly better segmentation result of 2.1%. Note that the pixelwise classification clearly improves the result of the agglomerative merging phase (7.8%). The difference between MIR_{stop} (2.8) and MIR_{hi} (1.2) is still noticeable, but by far the smallest in the three cases, reflecting the inherent difficulty of this problem.

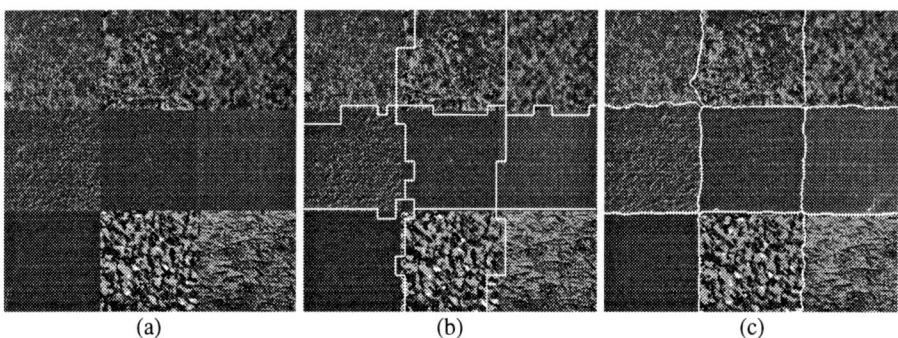

Fig. 4. Texture mosaic #3. MIR_{stop}=2.8, MIR_{hi}=1.2, ERR_a=7.8%, ERR_p=2.1%, sweeps=24.

We also applied the texture segmentation method to natural scenes. The scenes were originally in RGB format [9], but we converted them to gray level intensity images. As an example, scene #1 (Fig. 5a) is a 384 x 384 image of rocks in the sea. As we can observe from the image, the textures of natural scenes are generally more non-uniform than the homogeneous textures of the test mosaics. Also, in natural scenes adjacent textured regions are not necessarily separated by well-defined boundaries, but the spatial pattern smoothly changes from one texture to another. Further, we have to observe the infinite scale of texture differences present in natural scenes; choosing the right scale is a very subjective matter. For these reasons there is often no 'correct' segmentation for a natural scene, as is the case with texture mosaics.

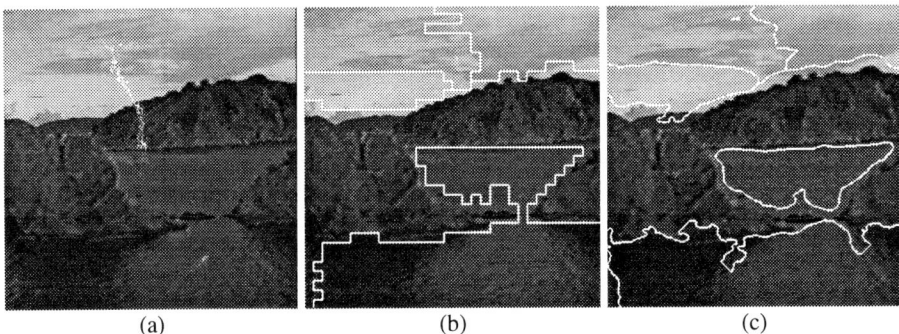

Fig. 5. Natural scene #1.

The parameters X and Y primarily control the scale of texture differences that will be detected. With values $X=1.1$ and $Y=1.5$ the rough segmentation result after the agglomerative merging phase is presented in Fig. 5b and the final segmentation result is shown in Fig. 5c. If we decreased Y further, the segmentation result would contain an increasing number of regions. The invariance of the LBP/C transform to average gray level shows in the bottom part of the image, where the sea is interpreted as a single region despite the shadows. The obtained result is very satisfactory, considering that important color or gray scale information is not utilized in the segmentation.

5 Conclusion

We proposed a solution to unsupervised texture segmentation, in which a method based on comparison of feature distributions is used to find homogeneously textured image regions and to localize boundaries between regions. Texture information is measured with a method based on local binary patterns and contrast (LBP/C) that we have recently developed. A region-based algorithm is developed for coarse image segmentation and a pixelwise classification scheme for improving the localization of region boundaries.

The method performed very well in experiments. It is not sensitive to the selection of parameter values, does not require any prior knowledge about the number of textures or regions in the image, and seems to provide significantly better results than existing unsupervised texture segmentation approaches. The method can be easily generalized, e.g., to utilize other texture features, multiscale information, color features, and combinations of multiple features [14].

Acknowledgements
The financial support provided by the Academy of Finland, the Technology Devel-

opment Center of Finland, and the Graduate School in Electronics, Telecommunications and Automation is gratefully acknowledged. The authors also wish to thank following persons for providing images used in this study; Richard C. Dubes, Anil K. Jain, John Lees and Kalle Karu from the Michigan State University and Glenn Healey and David Slater from the University of California at Irvine.

References

[1] M. Tuceryan and A.K. Jain, "Texture analysis," Chapter 2.1 in *Handbook of Pattern Recognition and Computer Vision*, C.H. Chen, L.F. Pau, P.S.P. Wang (eds.), World Scientific, Singapore, pp. 235-276, 1993.
[2] R.M. Haralick and L.G. Shapiro, "Image segmentation techniques," *Computer Vision, Graphics, and Image Processing*, vol. 29, pp. 100-132, 1985.
[3] T.R. Reed and J.M.H. Du Buf, "A review of recent texture segmentation and feature extraction techniques," *CVGIP Image Understanding*, vol. 57, pp. 359-372, 1993.
[4] M. Pietikäinen and A. Rosenfeld, "Image segmentation by texture using pyramid node linking," *IEEE Transactions on Systems, Man, and Cybernetics*, vol. 11, pp. 822-825, 1981.
[5] A.K. Jain and F. Farrokhnia, "Unsupervised texture segmentation using Gabor filters," *Pattern Recognition*, vol. 24, pp. 1167-1186, 1991.
[6] B.S. Manjunath and R. Chellappa, "Unsupervised texture segmentation using Markov random field models," *IEEE Transactions on Pattern Analysis and Machine Intelligence*, vol. 13, pp. 478-482, 1991.
[7] F.S. Cohen and Z. Fan, "Maximum likelihood unsupervised textured image segmentation," *CVGIP: Graphical Models and Image Processing*, vol. 54, pp. 239-251, 1992.
[8] M. Unser, "Texture classification and segmentation using wavelet frames," *IEEE Transactions on Image Processing*, vol. 4, pp. 1549-1560, 1995.
[9] D.K. Panjwani and G. Healey, "Markov random field models for unsupervised segmentation of textured color images," *IEEE Transactions on Pattern Analysis and Machine Intelligence*, vol. 17, pp. 939-954, 1995.
[10] A.K. Jain and K. Karu, "Learning texture discrimination masks," *IEEE Transactions on Pattern Analysis and Machine Intelligence*, vol. 18, pp. 195-205, 1996.
[11] D. Harwood, T. Ojala, M. Pietikäinen, S. Kelman and L.S. Davis, "Texture classification by center-symmetric auto-correlation, using Kullback discrimination of distributions," *Pattern Recognition Letters*, vol. 16, pp. 1-10, 1995.
[12] T. Ojala, M. Pietikäinen and D. Harwood, "A comparative study of texture measures with classification based on feature distributions," *Pattern Recognition*, vol. 29, pp. 51-59, 1996.
[13] T. Ojala, M. Pietikäinen and J. Nisula, "Determining composition of grain mixtures by texture classification based on feature distributions," *International Journal of Pattern Recognition and Artificial Intelligence*, vol. 10, pp. 73-82, 1996.
[14] T. Ojala, "Multichannel approach to texture description with feature distributions", Technical Report CAR-TR-846, Center for Automation Research, University of Maryland, 1996.
[15] R.R. Sokal and F.J. Rohlf, *Introduction to Biostatistics*, 2nd ed. W.H. Freeman and Co, New York, 1987.
[16] P. Brodatz, *Textures: A Photographic Album for Artists and Designers.* Dover Publications, New York, 1966.
[17] T. Ojala and M. Pietikäinen, "Unsupervised texture segmentation using feature distributions", Technical Report CAR-TR-837, Center for Automation Research, University of Maryland, 1996.
[18] P.P. Ohanian and R.C. Dubes, "Performance evaluation for four classes of textural features," *Pattern Recognition*, vol. 25, pp. 819-833, 1992.

Color Based Object Recognition

T. Gevers and A.W.M. Smeulders

Faculty of WINS, University of Amsterdam, The Netherlands
email: gevers@wins.uva.nl
Images and recognition scheme can be experienced on-line at
http://www.wins.uva.nl/research/isis/zomax/

abstract

Assuming white illumination and dichromatic reflectance, we propose new color models $c_1c_2c_3$ and $l_1l_2l_3$ invariant to the viewing direction, object geometry and shading. Further, it is shown that $l_1l_2l_3$ is also invariant to highlights. Further, a change in spectral power distribution of the illumination is considered to propose a new photometric color invariant $m_1m_2m_3$ for matte objects.

To evaluate photometric color invariant object recognition in practice, experiments have been carried out on a database consisting of 500 images taken from 3-D multicolored man-made objects.

On the basis of the reported theory and experimental results, it is shown that high object recognition accuracy is achieved by $l_1l_2l_3$ and hue H followed by $c_1c_2c_3$ and normalized colors rgb under the constraint of white illumination. Finally, it is shown that solely $m_1m_2m_3$ is invariant to a change in illumination color.

1 Introduction

Color provides powerful information for object recognition. A simple and effective recognition scheme is to represent and match images on the basis of RGB histograms as proposed by Swain and Ballard [6]. This color-based recognition method has been extended by Funt and Finlayson [2] to become illumination independent by indexing on an illumination-invariant set of color descriptors. Furthermore, Healey and Slater [4] use illumination invariant moment descriptors for object recognition. The method fails, however, when objects are occluded as the moments are defined as an integral property on the (region) object as one.

Our aim is to analyze and evaluate various color models to be used for the purpose of object recognition by color-metric histogram matching according to the following criteria: 1. Robustness to a change in viewpoint; 2. Robustness to a change in object orientation; 3. Robustness to a change in the intensity and the direction of the illumination; 4. Robustness to a change in the color of the illumination; 5. High discriminative power; 6. Robustness to noise; 7. Robustness to object occlusion and cluttering.

The general application is considered of recognition of 3-D multicolored objects from 2-D color images.

This paper is organized as follows. In Section 2, the dichromatic reflectance under "white" reflection is introduced and new photometric invariant color features are proposed. The performance of object recognition by histogram matching differentiated for the various color models is evaluated and compared on an image database of 500 reference images in Section 3.

2 Photometric Color Invariance

In this paper, we concentrate on the following standard, *essentially different*, color features derived from RGB: intensity $I(R,G,B) = R + B + G$, RGB, normalized colors $r(R,G,B) = \frac{R}{R+G+B}$, $g(R,G,B) = \frac{G}{R+G+B}$, $b(R,G,B) = \frac{B}{R+G+B}$, hue $H(R,G,B) = \arctan(\frac{\sqrt{3}(G-B)}{(R-G)+(R-B)})$ and saturation $S(R,G,B) = 1 - \frac{\min(R,G,B)}{R+G+B}$.

2.1 The Reflection Model

Consider an image of an infinitesimal surface patch. Using the red, green and blue sensors with spectral sensitivities given by $f_R(\lambda)$, $f_G(\lambda)$ and $f_B(\lambda)$ respectively, to obtain an image of the surface patch illuminated by a SPD of the incident light denoted by $e(\lambda)$, the measured sensor values will be given by Shafer [5]:

$$C = m_b(\mathbf{n},\mathbf{s}) \int_\lambda f_C(\lambda) e(\lambda) c_b(\lambda) d\lambda + m_s(\mathbf{n},\mathbf{s},\mathbf{v}) \int_\lambda f_C(\lambda) e(\lambda) c_s(\lambda) d\lambda \quad (1)$$

for $C = \{R, G, B\}$ giving the Cth sensor response. Further, $c_b(\lambda)$ and $c_s(\lambda)$ are the albedo and Fresnel reflectance respectively. λ denotes the wavelength, $\mathbf{n}$ is the surface patch normal, $\mathbf{s}$ is the direction of the illumination source, and $\mathbf{v}$ is the direction of the viewer. Geometric terms m_b and m_s denote the geometric dependencies on the body and surface reflection respectively.

Considering the neutral interface reflection (NIR) model (assuming that $c_s(\lambda)$ has a constant value independent of the wavelength) and "white" illumination, then $e(\lambda) = e$ and $c_s(\lambda) = c_s$. Then, we propose that the measured sensor values are given by:

$$C_w = em_b(\mathbf{n},\mathbf{s}) k_C + em_s(\mathbf{n},\mathbf{s},\mathbf{v}) c_s \int_\lambda f_C(\lambda) d\lambda \quad (2)$$

for $C_w \in \{R_w, G_w, B_w\}$ giving the red, green and blue sensor response under the assumption of a white light source. $k_C = \int_\lambda f_C(\lambda) c_b(\lambda) d\lambda$ is a compact formulation depending on the sensors and the surface albedo.

If the integrated white condition holds (as we assume throughout the paper) $\int_\lambda f_R(\lambda) d\lambda = \int_\lambda f_G(\lambda) d\lambda = \int_\lambda f_B(\lambda) d\lambda = f$, we have:

$$C_w = C_b + C_s = em_b(\mathbf{n},\mathbf{s}) k_C + em_s(\mathbf{n},\mathbf{s},\mathbf{v}) c_s f \quad (3)$$

2.2 Reflection with White Illumination

Photometric Invariant Color Features for Matte, Dull Surfaces According to the body reflection term of eq. (3), $C_b = em_b(\mathbf{n},\mathbf{s})k_C$, a uniformly painted surface (i.e. with fixed k_C) may give rise to a broad variance of RGB values due to the varying circumstances induced by the image-forming process. The same argument holds for intensity I.

In contrast, normalized color rgb is insensitive to surface orientation, illumination direction and intensity as can be seen from:

$$r(R_b, G_b, B_b) = \frac{em_b(\mathbf{n},\mathbf{s})k_R}{em_b(\mathbf{n},\mathbf{s})(k_R + k_G + k_B)} = \frac{k_R}{k_R + k_G + k_B} \quad (4)$$

only dependent on the sensors and the surface albedo. Equal arguments hold for g and b.

Saturation S is an invariant for the set of matte, dull surfaces illuminated by white SPD mathematically specified by:

$$S(R_b, G_b, B_b) = 1 - \frac{\min(em_b(\mathbf{n},\mathbf{s})k_R, em_b(\mathbf{n},\mathbf{s})k_G, em_b(\mathbf{n},\mathbf{s})k_B)}{em_b(\mathbf{n},\mathbf{s})(k_R + k_G + k_B)} = 1 - \frac{\min(k_R, k_G, k_B)}{(k_R + k_G + k_B)} \quad (5)$$

Similarly, hue H is an invariant for matte, dull surfaces:

$$H(R_b, G_b, B_b) = \arctan\left(\frac{\sqrt{3}em_b(\mathbf{n},\mathbf{s})(k_G - k_B)}{em_b(\mathbf{n},\mathbf{s})((k_R - k_G) + (k_R - k_B))}\right) = \arctan\left(\frac{\sqrt{3}(k_G - k_B)}{(k_R - k_G) + (k_R - k_B)}\right) \quad (6)$$

In fact, any expression defining colors on the same linear color cluster formed by the body reflection vector in RGB-space are photometric color invariants for dichromatic reflectance for matte surfaces under white light. To that end, we propose the following photometric color invariant model:

$$c_1 = \arctan(\frac{R}{\max\{G, B\}}), c_2 = \arctan(\frac{G}{\max\{R, B\}}), c_3 = \arctan(\frac{B}{\max\{R, G\}}) \quad (7)$$

denoting the angles of the body reflection vector and consequently being invariants for matte, dull objects:

$$c_1(R_b, G_b, B_b) = \arctan(\frac{em_b(\mathbf{n},\mathbf{s})k_R}{\max\{em_b(\mathbf{n},\mathbf{s})k_G, em_b(\mathbf{n},\mathbf{s})k_B\}}) = \arctan(\frac{k_R}{\max\{k_G, k_B\}}) \quad (8)$$

only dependent on the sensors and the surface albedo. Equal arguments hold for c_2 and c_3.

Photometric Invariant Color Features for Both Matte and Shiny Surfaces Note that under the given conditions (the NIR model), the color of highlights is not related to the color of the surface on which they appear, but only on the color of the light source. Thus for the white light source, the surface reflection color cluster is on the diagonal grey axis of the basic RGB-color space corresponding to intensity I. For a given point on a shiny surface, the contribution of the body reflection component and surface reflection component are added together. Hence, the observed colors of the surface must be inside the triangular color cluster in the RGB-space formed by the two reflection components.

Because H is a function of the angle between the reference line and color point, all possible colors of the same shiny uniformly colored surface have to be of the same hue mathematically specified as:

$$H(R_w, G_w, G_w) = \arctan\left(\frac{\sqrt{3}(k_G - k_B)}{(k_R - k_G) + (k_R - k_B)}\right) \tag{9}$$

Only dependent on the sensors and the surface albedo. Obviously other color features depend on the contribution of the surface reflection component and hence are sensitive to highlights.

In fact, any expression defining colors on the same linear triangular color cluster, formed by the two reflection components in RGB-space, are photometric color invariants for the dichromatic reflectance under white light.

To that end, a new photometric color invariant model $l_1 l_2 l_3$ is proposed uniquely determining the direction of the linear triangular color cluster: $l_1 = \frac{(R-G)^2}{(R-G)^2+(R-B)^2+(G-B)^2}, l_2 = \frac{(R-B)^2}{(R-G)^2+(R-B)^2+(G-B)^2}, l_3 = \frac{(G-B)^2}{(R-G)^2+(R-B)^2+(G-B)^2}$ the set of normalized color differences which is, similar to H, an invariant for the set of matte and shiny surfaces.

2.3 Reflection with Colored Illumination

The Reflection Model We consider the body reflection term of the dichromatic reflection model:

$$C_c = m_b(\mathbf{n}, \mathbf{s}) \int_\lambda f_C(\lambda) e(\lambda) c_b(\lambda) d\lambda \tag{10}$$

for $C = \{R, G, B\}$ where $C_c = \{R_c, G_c, B_c\}$ gives the red, green and blue sensor response of a matte infinitesimal surface patch under unknown spectral power distribution of the illumination.

Suppose that the sensor sensitivities of the color camera are narrowband with spectral response be approximated by delta functions $f_C(\lambda) = \delta(\lambda - \lambda_C)$, then we have:

$$C_c = m_b(\mathbf{n}, \mathbf{s}) e(\lambda_C) c_b(\lambda_C) \tag{11}$$

By simply filling in C_c in the color feature equations, we can see that all color feature values change with a change in illumination color.

Color Constant Feature for Matte, Dull Surfaces Funt and Finlayson [2] proposes a simple and effective illumination-independent color feature for the purpose of object recognition. The method runs short, however, when images are contaminated by shading and highlights. To that end, we propose a color constant feature not only independent of the illumination color but also discounting shading cues:

$$m(C_1^{\mathbf{x}_1}, C_1^{\mathbf{x}_2}, C_2^{\mathbf{x}_1}, C_2^{\mathbf{x}_2}) = \frac{C_1^{\mathbf{x}_1} C_2^{\mathbf{x}_2}}{C_1^{\mathbf{x}_2} C_2^{\mathbf{x}_1}}, C_1 \neq C_2 \qquad (12)$$

expressing the color ratio between two neighboring image locations, for $C_1, C_2 \in \{R, G, B\}$ where $\mathbf{x}_1$ and $\mathbf{x}_2$ denote the image locations of the two neighboring pixels.

Having three color components of two locations, color ratios obtained from a RGB-color image are:

$$m_1 = \frac{R^{\mathbf{x}_1} G^{\mathbf{x}_2}}{R^{\mathbf{x}_2} G^{\mathbf{x}_1}}, m_2 = \frac{R^{\mathbf{x}_1} B^{\mathbf{x}_2}}{R^{\mathbf{x}_2} B^{\mathbf{x}_1}}, m_3 = \frac{G^{\mathbf{x}_1} B^{\mathbf{x}_2}}{G^{\mathbf{x}_2} B^{\mathbf{x}_1}} \qquad (13)$$

For the ease of exposition, we concentrate on m_1 based on the RG-color bands in the following discussion. Without loss of generality, all results derived for m_1 will also hold for m_2 and m_3.

If we assume that the color of the illumination is locally constant (at least over the two neighboring locations from which ratio is computed), the color ratio is independent of the illumination color, and also a change in viewpoint, the surface geometry, and illumination intensity as follows from:

$$m_1 = \frac{(m_b^{\mathbf{y}_1}(\mathbf{n},\mathbf{s})e^{\mathbf{y}_1}(\lambda_R)c_b^{\mathbf{y}_1}(\lambda_R))(m_b^{\mathbf{y}_2}(\mathbf{n},\mathbf{s})e^{\mathbf{y}_2}(\lambda_G)c_b^{\mathbf{y}_2}(\lambda_G))}{(m_b^{\mathbf{y}_2}(\mathbf{n},\mathbf{s})e^{\mathbf{y}_2}(\lambda_R)c_b^{\mathbf{y}_2}(\lambda_R))(m_b^{\mathbf{y}_1}(\mathbf{n},\mathbf{s})e^{\mathbf{y}_1}(\lambda_G)c_b^{\mathbf{y}_1}(\lambda_G))} = \frac{c_b^{\mathbf{y}_1}(\lambda_R)c_b^{\mathbf{y}_2}(\lambda_G)}{c_b^{\mathbf{y}_2}(\lambda_R)c_b^{\mathbf{y}_1}(\lambda_G)}$$
(14)

only dependent on the surface albedo, where $\mathbf{y}_1$ and $\mathbf{y}_2$ are two neighboring locations on the object's surface not necessarily of the same orientation.

Taking logarithms of both sides of equation 12 results for m_1 in:

$$\ln m_1(R^{\mathbf{x}_1}, R^{\mathbf{x}_2}, G^{\mathbf{x}_1}, G^{\mathbf{x}_2}) = \ln R^{\mathbf{x}_1} + \ln G^{\mathbf{x}_2} - \ln R^{\mathbf{x}_2} - \ln G^{\mathbf{x}_1} \qquad (15)$$

The color ratios can be seen as differences at two neighboring locations $\mathbf{x}_1$ and $\mathbf{x}_2$ in the image domain:

$$d_{m_1}(\mathbf{x}_1, \mathbf{x}_2) = \ln R^{\mathbf{x}_1} + \ln G^{\mathbf{x}_2} - \ln R^{\mathbf{x}_2} - \ln G^{\mathbf{x}_1} \qquad (16)$$

When these differences are taken between neighboring pixels in a particular direction, they correspond to finite-difference differentiation.

The results obtained so far for m_1 hold also for m_2 and m_3, yielding a 3-tuple $(\mathcal{G}_{m_1}(\mathbf{x}), \mathcal{G}_{m_2}(\mathbf{x}), \mathcal{G}_{m_3}(\mathbf{x}))$ denoting the gradient magnitude for every neighborhood centered at $\mathbf{x}$ in the image.

For pixels on a uniformly painted region, in theory, all three components will be zero whereas at least one the three components will be non-zero for pixels on locations where two regions of distinct color meet.

3 Color Based Object Recognition: Experiments

The database consists of $N_1 = 500$ reference images of multicolored domestic objects, tools, toys, etc.. Objects were recorded in isolation (one per image) with the aid of the SONY XC-003P CCD color camera (3 chips) and the Matrox Magic Color frame grabber. Objects were recorded against a white cardboard background. Two light sources of average day-light color are used to illuminate the objects in the scene. A second, independent set (the test set) of recordings was made of randomly chosen objects already in the database. These objects, $N_2 = 70$ in number, were recorded again (one per image) with a new, arbitrary position and orientation with respect to the camera (some recorded upside down, some rotated, some at different distances (different scale)).

Histograms are constructed on the basis of different color features representing the distribution of discrete color feature values in a n-dimensional color feature space, where $n = 3$ for RGB, rgb, $l_1 l_2 l_3$, $c_1 c_2 c_3$ and $m_1 m_2 m_3$, and $n = 1$ for I, S and H. During histogram construction, all pixels in a color image are discarded with a local saturation and intensity smaller then 5 percent of the total range. Consequently, the white cardboard background as well as the grey, white, dark or nearly colorless parts of objects as recorded in the color image will not be considered in the matching process. For comparison reasons in the literature, in this paper, the histogram similarity function is expressed by histogram intersection [6].

For a measure of match quality, let rank r^{Q_i} denote the position of the correct match for test image Q_i, $i = 1, ..., N_2$, in the ordered list of N_1 match values. The rank r^{Q_i} ranges from $r = 1$ from a perfect match to $r = N_1$ for the worst possible match. Then, for one experiment, the average ranking percentile is defined by $\bar{r} = (\frac{1}{N_2} \sum_{i=1}^{N_2} \frac{N_1 - r^{Q_i}}{N_1 - 1}) 100\%$. The cumulative percentile of test images producing a rank smaller or equal to j is defined as $\mathcal{X}(j) = (\frac{1}{N_2} \sum_{k=1}^{j} \eta(r^{Q_i} == k)) 100\%$, where η reads as the number of test images having rank k.

For more information see [3]. The image database and the performance of the recognition scheme can be experienced within the ZOMAX system on-line at
http://www.wins.uva.nl/research/isis/zomax/.

4 Results

In this subsection, we report on the recognition accuracy of the matching process for $N_2 = 70$ test images and $N_1 = 500$ reference images for the various color features. As stated, white lighting is used during the recording of the reference images in the image database and the independent test set. However, the objects were recorded with a new, arbitrary position and orientation with respect to camera. In Fig. 1 accumulated ranking percentile is shown for the various color features.

From the results of Fig. 1 we can observe that the discriminative power of $l_1 l_2 l_3$, H followed by $c_1 c_2 c_3$ and rgb is higher then the other color features achiev-

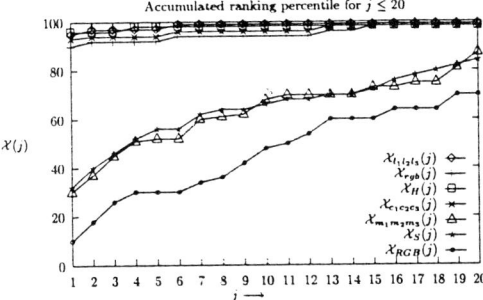

Fig. 1. The discriminative power of the histogram matching process differentiated for the various color features plotted against the ranking j. The cumulative percentile $\mathcal{X}$ for H, $l_1 l_2 l_3$, $c_1 c_2 c_3$, rgb, S, $m_1 m_2 m_3$ and RGB is given by $\mathcal{X}_H$, $\mathcal{X}_{l_1 l_2 l_3}$, $\mathcal{X}_{c_1 c_2 c_3}$, $\mathcal{X}_{rgb}$, $\mathcal{X}_S$, $\mathcal{X}_{m_1 m_2 m_3}$ and $\mathcal{X}_{RGB}$ respectively.

ing a probability of respectively 99, 98, 94 and 92 perfect matches out of 100. Saturation S and color ratio $m_1 m_2 m_3$ provides slightly worse recognition accuracy. As expected, the discrimination power of RGB has the worst performance due to its sensitivity to varying imaging conditions.

4.1 The Effect of a Change in the Illumination Intensity

The effect of a change in the illumination intensity is approximated by a multiplication of each RGB-color by a uniform scalar factor α. To measure the sensitivity of different color feature in practice, RGB-images of the test set are multiplied by a constant factor varying over $\alpha \in \{\ 0.5, 0.7, 0.8, 0.9,\ 1.0, 1.1, 1.2, 1.3, 1.5\}$. The discrimination power of the histogram matching process differentiated for the various color features plotted against illumination intensity is shown in Fig. 2. As expected, RGB and I-color features depend on the illumination intensity.

4.2 The Effect of a Change in the Illumination Color

Based on the coefficient rule or von Kries model, the change in the illumination color is approximated by a 3x3 diagonal matrix among the sensor bands and is equal to the multiplication of each RGB-color band by an independent scalar factor [1]. Note that the diagonal model of illumination change holds in the case of narrowband sensors. To measure the sensitivity of the various color feature in practice with respect to a change in the color of the illumination, the R, G and B-color bands of each image of the test set are multiplied by a factor $\beta_1 = \beta$, $\beta_2 = 1$ and $\beta_3 = 2 - \beta$ respectively (i.e. $\beta_1 R$, $\beta_2 G$ and $\beta_3 B$) by varying β over $\{0.5, 0.7, 0.8, 0.9, 1.0, 1.1, 1.2, 1.3, 1.5\}$. The discrimination power of the histogram matching process differentiated for the various color features plotted against the illumination color is shown in Fig. 3. For $\beta < 1$ the color is reddish whereas bluish for $\beta > 1$.

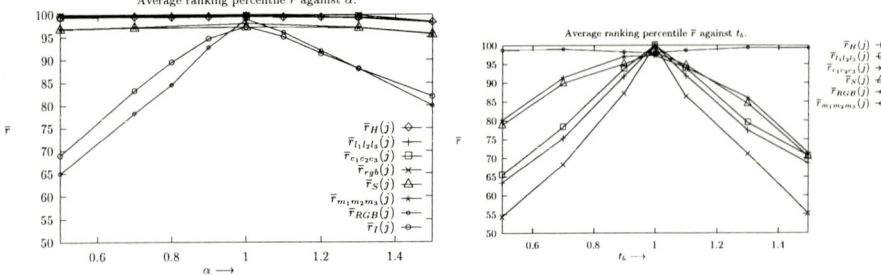

Fig. 2. The discriminative power plotted against the illumination intensity represented by variation as expressed by the factor α.

Fig. 3. The discriminative power plotted against the change β in the color composition of the illumination spectrum.

As expected, only the color ratio $m_1 m_2 m_3$ is insensitive to a change in illumination color. From Fig. 3 we can observe that color features H, $l_1 l_2 l_3$ and $c_1 c_2 c_3$, which achieved best recognition accuracy under white illumination, see Figures 1 and 2, are highly sensitive to a change in illumination color.

5 Conclusion

On the basis of the above reported theory and experiments, it is concluded that the proposed invariant $l_1 l_2 l_3$ followed by H are most appropriate to be used for photometric color invariant object recognition by color-metric histogram matching under the constraint of a white illumination source. When no constraints are imposed on the imaging conditions (i.e. the most general case), the newly proposed color ratio $m_1 m_2 m_3$ is most appropriate.

References

1. Finlayson, G. D., Drew M. S., and Funt B. V., *Spectral Sharpening: Sensor Transformations for improved Color Constancy* J. Opt. Soc. Am., 11(5), pp. 1553-1563, 1994.
2. Funt, B. V. and Finlayson, G. D., *Color Constant Color Indexing*, IEEE PAMI, 17(5), pp. 522-529, 1995.
3. Gevers, T., *Color Image Invariant Segmentation and Retrieval*, PhD Thesis, ISBN 90-74795-51-X, University of Amsterdam, The Netherlands, 1996.
4. Healey, G. and Slater D, *Global Color Constancy: Recognition of Objects by Use of Illumination Invariant Properties of Color Distributions*, J. Opt. Soc. Am. A, Vol. 11, No. 11, pp. 3003-3010, Nov 1995.
5. Shafer, S. A., *Using Color to Separate Reflection Components*, COLOR Res. Appl., 10(4), pp 210-218, 1985.
6. Swain, M. J. and Ballard, D. H., *Color Indexing*, Int. Journal of CV, Vol. 7, No. 1, pp. 11-32, 1991.

Color Texture Classification by Wavelet Energy Correlation Signatures

G. Van de Wouwer*, S. Livens, P. Scheunders, D. Van Dyck

Vision Lab, Department of Physics, University of Antwerp, Groenenborgerlaan 171, 2020 Antwerpen, Belgium

Abstract. In the last decade, multiscale techniques for gray-level texture analysis have been intensively studied. In this paper, we aim on extending these techniques to color images. We introduce wavelet energy-correlation signatures and we derive the transformation of these signatures upon linear color space transformations. Classification experiments demonstrate that the wavelet correlation features contain more information than the intensity or the energy features of each color plane separately. The influence of image representation in color space is evaluated.

1 Introduction

For image analysis, color and texture are two of the most important properties, especially when one is dealing with real world images. Classical image analysis schemes only take into account the pixel gray-levels, which represents the total amount of visible light at the pixels position. The performance of such schemes can be improved by adding color information [1]. The color of a pixel is typically represented with the RGB tristimulus values, corresponding to the Red, Green and Blue frequency bands of the visible light spectrum. Color is then a feature in the 3-dimensional RGB color space, which contains information regarding the spectral distribution of light complementary to the gray-level information.

An important topic when processing color images is their representation. The RGB representation is frequently being transformed into other color spaces [2] [3]. The performance of an image analysis system can strongly depend on the choice of the color representation [4] [5]. However, there does not appear to be a systematic means of determining an optimum color-coordinate system for a particular task.

In the analysis of color images, the description of image regions has mainly been performed using color histograms [3] [6]. However, they no longer suffice when local spatial correlations are important to characterize a region. The extra information needed to adequately describe the image regions is commonly known as "texture". Texture has been studied extensively and many texture analysis schemes have been proposed [7]. The fundamental property which they all have in common is that they exploit local spatial interactions between pixels.

---

* corresponding author: email: wouwer@ruca.ua.ac.be

A rather limited number of systems use combined information of color and texture, and even when they do, both aspects are mostly dealt with using separate methods [8] [9]. It is only recently that attempts are being made to combine both aspects in a single method, by extending gray-level texture analysis methods to color images [10] [11]. This combination can be made more formal by defining "color-texture" as "the set of local statistical properties of the colors of image regions". Efficient characterization of color texture requires the exploitation of spatial correlations as well as correlations between color bands.

The importance of a joint color-texture characterization is expected to grow rapidly in the near future, e.g. for indexing image databases. At present time, the color extensions of several major texture analysis methods are still unexplored. We will investigate one of them, based on multiresolution decomposition. These techniques give rise to an interesting class of texture analysis methods. Strong arguments for their use can be found in psychovisual research, which offers evidence that the human visual system processes images in a multiscale way [12]. Wavelets provide a convenient way to obtain a multiresolution representation [13], from which texture features are easily extracted [14] [15] [16].

We propose a scheme for the characterization of colored texture images. Feature extraction using wavelet decomposition is described. Wavelet correlation signatures are defined which contain the energies of each color plane and the cross-correlation between different planes. While the first have already successfully been used for texture characterization, the latter represent the coupling between texture and color. We will show that these features transform linearly upon linear color space transformation. The experiments will demonstrate the usefulness of correlation signatures as texture features. The influence of the choice of color space representation on classification performance will be investigated.

2 Wavelet Signatures

The (continuous) wavelet transform of a 1-D signal $f(x)$ is defined as

$$(W_a f)(b) = \int f(x) \psi_{a,b}^\star(x) dx \quad \text{with} \quad \psi_{a,b}(x) = \frac{1}{\sqrt{a}} \psi(\frac{x-b}{a}) \tag{1}$$

The *mother wavelet* ψ has to satisfy the admissibility criterion to ensure that it is a localized zero-mean function. (1) can be discretized by restraining a and b to a discrete lattice ($a = 2^n$, $b \in \mathcal{Z}$). Typically some more constraints are imposed on ψ to ensure that the transform is non-redundant, complete and constitutes a multiresolution representation of the original signal. This has led to an efficient real-space implementation of the transform using quadrature mirror filters.

The extension to the 2-D case is usually performed by applying a separable filter bank to the image:

$$L_n(\mathbf{b}) = [H_x * [H_y * L_{n-1}]_{\downarrow 2,1}]_{\downarrow 1,2} (\mathbf{b}) \tag{2}$$

$$D_{n1}(\mathbf{b}) = [H_x * [G_y * L_{n-1}]_{\downarrow 2,1}]_{\downarrow 1,2} (\mathbf{b}) \tag{3}$$

$$D_{n2}(\mathbf{b}) = [G_x * [H_y * L_{n-1}]_{\downarrow 2,1}]_{\downarrow 1,2}(\mathbf{b}) \tag{4}$$

$$D_{n3}(\mathbf{b}) = [G_x * [G_y * L_{n-1}]_{\downarrow 2,1}]_{\downarrow 1,2}(\mathbf{b}) \tag{5}$$

where $\mathbf{b} \in \mathrm{R}^2$, $*$ denotes the convolution operator, $\downarrow 2,1$ ($\downarrow 1,2$) sub-sampling along the rows (columns) and $L_0 = I(\mathbf{x})$ is the original image. H and G are a low and bandpass filter respectively. L_n is obtained by low pass filtering and is therefore referred to as the low resolution image at scale n. The D_{ni} are obtained by bandpass filtering in a specific direction and thus contain directional detail information at scale n; they are referred to as the *detail images*. The original image I is thus represented by a set of subimages at several scales; $\{L_d, D_{ni}\}_{i=1,2,3}^{n=0,\ldots,d-1}$ which is a *multiscale representation of depth d* of the image I.

The energy of a subimage D_{ni} is defined as

$$E_{ni} = \int (D_{ni}(\mathbf{b}))^2 d\mathbf{b} \tag{6}$$

The *wavelet energy signatures* $\{E_{ni}\}_{n=0,\ldots,d-1, i=1,2,3}$ reflect the distribution of energy along the frequency axis over scale and orientation and have proven to be very useful for gray-level texture characterization. Since most relevant texture information has been removed by iterative low pass filtering, the energy of the low resolution image L_d is generally not considered a texture feature.

The most straightforward extension of the wavelet energy signatures to color images is to transform each color plane separately and extract the energies of each transformed plane; i.e. replace I by the R,G and B-plane consecutively in (2)-(6). We denote such an energy by $E_{ni}^{X_j}$ where the X_j indicates the color plane. This triples the amount of features w.r.t. the gray-level case.

Let us define

$$C_{ni}^{X_j X_k} = \int D_{ni}^{X_j}(\mathbf{b}) D_{ni}^{X_k}(\mathbf{b}) d\mathbf{b} \tag{7}$$

and call the set $\{C_{ni}^{X_j X_k}\}_{n=0,\ldots,d-1, i=1,2,3}^{j,k=1,2,3,\, j \leq k}$ the *wavelet covariance signatures*. They include the energies for $j = k$; the others represent the covariance between different color planes and consequently the coupling between the color and texture properties of the image.

The covariance signatures, however, are by definition proportional to the energies. They are normalized to remove this redundant information:

$$\tilde{C}_{ni}^{X_j X_k} = \begin{cases} E_{ni}^{X_j} & j = k \\ \dfrac{C_{ni}^{X_j X_k}}{E_{ni}^{X_j} E_{ni}^{X_k}} & j \neq k \end{cases} \tag{8}$$

The features $\{\tilde{C}_{ni}^{X_j X_k}\}_{n=0,\ldots,d-1, i=1,2,3}^{j,k=1,2,3,\, j \leq k}$ are the *wavelet correlation signatures*.

3 Color Space Transforms

For compression purposes, transformations to different color spaces are often employed to achieve image bandwidth reduction without significantly degrading image quality. However, since our goal is to efficiently characterize texture, the choice of color space should enable extraction of useful features rather than visual image representation. Non-linear transforms are mainly employed to obtain a color space in which the 3 coordinates have an intuitive meaning (mostly a luminance, a saturation and a hue component) [17]. They typically introduce some non-removable singularities, which is very impractical for further processing. We will therefore limit ourselves to linear color space transforms, i.e.

$$X' = MX \qquad (9)$$

where $X = (X_1(\mathbf{x})\ X_2(\mathbf{x})\ X_3(\mathbf{x}))^\tau$ contains the original components of the signal ($^\tau$ means transpose), M is a 3 by 3 invertible transformation matrix and X' contains the transformed signal.

Three particular color space transforms (for which $X = (R\ G\ B)^\tau$) are:

$$M_1 = \begin{pmatrix} 0.405 & 0.116 & 0.133 \\ 0.299 & 0.587 & 0.114 \\ 0.145 & 0.827 & 0.627 \end{pmatrix} \quad M_2 = \begin{pmatrix} 0.299 & 0.587 & 0.114 \\ 0.596 & -0.274 & -0.322 \\ 0.211 & -0.253 & 0.312 \end{pmatrix} \quad M_3 = \begin{pmatrix} 0.333 & 0.333 & 0.333 \\ 0.500 & 0.000 & -0.500 \\ -0.500 & 1.000 & -0.500 \end{pmatrix}$$

The first transforms RGB to the UVW-space (V=Y=luminance). This is a "perceptually uniform" space constructed so that equal changes in the space are experienced as equal changes in color by human perception. M_2 represents the YIQ-space. The Y signal is the image luminance and the I and Q signals carry the chrominance information. The last one (M_3) represents the K-L space (Karhunen-Loève transform), which transforms an image to an orthogonal basis in which the axes are statistically uncorrelated, and in that sense decorrelates the information present in RGB space.

Effect of linear color transform on the wavelet signatures
We now investigate how the wavelet covariance signatures transform under a linear color space transform. Let us fix n and i and rewrite (1) in vector notation for a color image $X = (X_1(\mathbf{x})\ X_2(\mathbf{x})\ X_3(\mathbf{x}))^\tau$:

$$(W_{2^n,i}X)(\mathbf{b}) = \int X(\mathbf{x})\varphi_{2^n,\mathbf{b}}^{i\,\star}(\mathbf{x})d\mathbf{x} \qquad (10)$$

Define C_{ni} as a (symmtric) matrix with the wavelet covariance signatures as elements:

$$C_{ni} = \int (W_{2^n,i}X)(\mathbf{b})\left((W_{2^n,i}X)(\mathbf{b})\right)^\tau d\mathbf{b} \qquad (11)$$

After a color space transformation $X' = MX$ the covariance signatures become:

$$C'_{ni} = \int (W_{2^n,i}X')(\mathbf{b})\,((W_{2^n,i}X')(\mathbf{b}))^T\,d\mathbf{b}$$
$$= \int \left(\int X'(\mathbf{x})\varphi^{i\,\star}_{2^n,\mathbf{b}}(\mathbf{x})d\mathbf{x}\right)\left(\int X'(\mathbf{x})\varphi^{i\,\star}_{2^n,\mathbf{b}}(\mathbf{x})d\mathbf{x}\right)^T d\mathbf{b}$$
$$= \int (\int MX(\mathbf{x})\varphi^{i\,\star}_{2^n,\mathbf{b}}(\mathbf{x})d\mathbf{x})(\int MX(\mathbf{x})\varphi^{i\,\star}_{2^n,\mathbf{b}}(\mathbf{x})d\mathbf{x})^T d\mathbf{b}$$
$$= MC_{ni}M^T \qquad (12)$$

Or, explicitly:
$$C_{ni}^{X'_j,X'_k} = \sum_{r,s=1}^{3} m_{jr}m_{ks}C_{ni}^{X_r X_s} \qquad (13)$$

For the energies this means (taking the RGB-space for X):

$$E_{ni}^{X'_j} = C_{ni}^{X'_j X'_j} = \sum_{r,s=1}^{3} m_{jr}m_{js}C_{ni}^{X_r X_s}$$
$$= m_{j1}^2 E_{ni}^R + m_{j2}^2 E_{ni}^G + m_{j3}^2 E_{ni}^B + 2m_{j1}m_{j2}C_{ni}^{RG}$$
$$+ 2m_{j1}m_{j3}C_{ni}^{RB} + 2m_{j2}m_{j3}C_{ni}^{GB} \qquad (14)$$

These formulas offer an interesting insight in the effect of linear color space transform on the wavelet signatures. (13) shows that a linear color space transform implies a linear transform of the covariance signatures. However, from (14) it follows that this is not true for the energy signatures. The first 3 terms reveal that the "new" energy features are linearly obtained from the "old" ones; the next 3 terms however depend on the covariances between the R, G and B planes for the same subimage. There is no clear connection between the energies in the original and transformed color space; to compute the latter the wavelet covariance signatures are required.

(14) also shows that performing a simple linear transform from RGB space to another color space results in a clearly different feature set. Hence, the quality of the features (i.e. their ability to characterize and discriminate between color textures) shall be heavily dependent on the choice of color space. This shall be demonstrated in the experimental section.

When one experiments using several color transforms, a practical advantage of the relation (13) comes into play. It is sufficient to perform the wavelet transform once (for the R,G and B planes) and to compute the covariance signatures. The new wavelet signatures are then obtained using (13) without the need of performing several wavelet transforms.

For the correlation signatures, the simple relation (13) does not hold. To transform correlation signatures into other color spaces, it is therefore convenient to transform the covariance signatures first and to normalize them afterwards.

4 Classification Methods

A k-nearest neighbor classifier (k-nn) [18] is used to estimate recognition performance. Since the emphasis in this work is on the feature extraction stage, k-nn provides an efficient and robust classification scheme for evaluation of recognition rates and comparison of feature sets.

Recognition rate is estimated by the leave-one-out method. This method sequentially picks each available data sample and classifies it (by the k-nn rule) using the remaining samples. Each available sample is thus employed once as a test sample. The *recognition rate* is estimated by counting the total number of samples classified correctly.

Well known in pattern recognition literature is the curse of dimensionality phenomenon, which dictates that classification performance not necessarily increases with an increasing number of features (given a fixed amount of data samples). Therefore, given a feature extraction scheme and a finite number of training instances, there exists an optimal number of features for a particular task. This becomes inherently important when dealing with colored images, since the number of extracted features is much larger compared to the gray-level case. Therefore, it is crucial to adopt a feature selection (or extraction) scheme to find a (sub-)optimal set of features. In this work the Floating Forward Feature Selection scheme (FFFS) [19] is adopted. This algorithm is initialized by taking the best feature ("best" is defined here as giving the best recognition performance). The selection then continues by iteratively adding (or deleting) a feature in each step to obtain a subset of all available features which gives the highest classification performance.

5 Experiments and Conclusion

30 real-world (512x512) RGB color-images from different natural scenes [20] were selected: Bark0, Bark4, Bark6, Bark8, Bark9, Brick1, Brick4, Brick5, Fabric0, Fabric4, Fabric7, Fabric9, Fabric11, Fabric13, Fabric16, Fabric17, Fabric18, Food0, Food2, Food5, Food8, Grass1, Sand0, Stone4, Tile1, Tile3, Tile7, Water6, Wood1 and Wood2. A database of 1920 color image regions of 30 texture classes was constructed by dividing each image into 64 non-overlapping 64x64 subimages. The following classification experiments were conducted:

1. Intensity (gray-level) images were generated by computing the luminance, hereby discarding color information. A wavelet transform of depth 4 was performed and energy signatures were computed for each of the 12 detail images. (total: 12 features)
2. Each R, G and B component was wavelet transformed (depth 4) and energy signatures were computed from each detail image. (total: 36 features)
3. Each R, G and B component was wavelet transformed (depth 4) and correlation signatures were computed from each detail image. (total: 72 features)
4. 72 correlation signatures were computed using (13) for the 3 color spaces mentioned in section 3: a) UVW space, b) YIQ space, c) K-L space

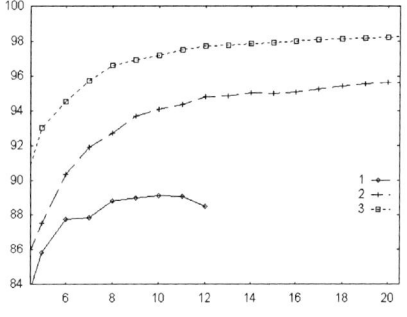 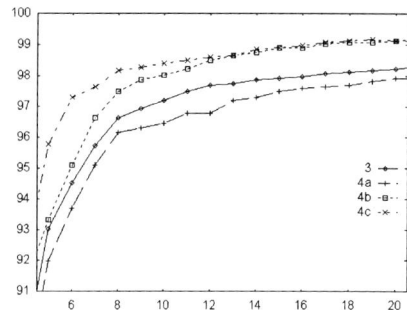

Fig. 1. recognition rate (%) versus feature set dimensionality graphs. a) 1. intensity 2. energy RGB 3. correlation RGB. b) correlation signatures in different color spaces: RGB (3), UVW (4a), YIQ (4b), K-L(4c).

Classification results are depicted in Fig. 1. One observes that the recognition rates saturate about a dimensionality of 10, at which point the error rates for each texture class were investigated. We found that recognition for Fabric0-7-17-18, Food5, Sand0, Tile7 and Wood2 was 100% for all classifiers. Recognition rates for Fabric9-16 and Water6 were 97-100% and did not differ much between classifiers. This shows that intensity alone contains sufficient information to characterize some textures, but fails to do so on others. Fig. 1,a) shows that adding color information does significantly increase recognition performance. Comparing curves 2 and 3 in Fig. 1,a) shows that the correlation signatures offer a clear advantage over the energies.

Fig. 1,b) compares the performance of the wavelet correlation signatures in different colour spaces, which demonstrates that recognition performace is color space dependent. It is apparent that the recognition rate for the UVW space is lower than in the K-L and YIQ spaces and also lower than for the RGB space. Recognition performance is thus indeed color space dependent. Overall, the best results are obtained with the K-L transform.

The conducted experiments demonstrate that color texture can adequately be described by the wavelet correlation signatures. These features are not only suited for image classification, but can easily be employed for other color texture analysis tasks. For instance, for segmentation the wavelet signatures are computed over a (small) local window centered on each pixel of the image, resulting in one feature vector per pixel. Each pixel is then assigned to a particular image region, e.g. by (unsupervised) clustering techniques in the space of feature vectors.

References

1. G.J. Klinker, S.A. Shafer, and T. Kanade. A physical approach to color image understanding. *Int. J. Comput. Vision*, 4:7–38, 1990.
2. G. Wyszecki and W. S. Stiles. *Color Science, Concepts and Methods, Quantitative Data and Formulas, 2nd ed.* J. Wiley and Sons, New York, 1982.
3. Q-T Luong. Color in computer vision. In C.H. Chen, L.F. Pau, and P.S.P. Wang, editors, *Handbook of Pattern Recognition & Computer Vision*, chapter 2.3, pages 311–368. World Scientific, Singapore, 1993.
4. Y.I. Ohta, T. Kanade, and T. Sakai. Color information for region segmentation. *Computer graphics and image processing*, 13:222–241, 1980.
5. W.K. Pratt. Spatial transform coding of color images. *IEEE Trans. Comm. Tech.*, 19(6):980–992, 1971.
6. C.L. Novak and S.A. Shafer. Methods for estimating scene parameters from color histograms. *J. Opt. Soc. Am. A.*, 11:3020–3036, 1994.
7. T.R. Reed and J.M.H. du Buf. A review of recent texture segmentation and feature extraction techniques. *CVGIP: Image Understanding*, 57(3):359–372, 1993.
8. D. Lee, R. Barber, W. Niblack, M. Flickner, J. Hafner, and D. Petkovic. Indexing for complex queries on a query-by-content image database. In *Proc. of the 12nd IAPR Int. Conf. on Pattern Recognition*, volume I, pages 142–146, Jerusalem, Israel, 1994.
9. J. R. Smith and S. Chang. Local color and texture extraction and spatial query. In *IEEE Proc. Int. Conf. on Im. Proc.*, volume III, pages 1011–1014, 1996.
10. T. Caelli and D. Reye. On the classification of image regions by colour texture and shape. *Pattern Recognition*, 26(4):461–470, 1993.
11. D.K. Panjwani and G. Healey. Markov random field models for unsupervised segmentation of textured color images. *IEEE Trans. Pattern Anal. Machine Intell.*, 17(10):939–954, 1995.
12. T.S. Lee. Image representation using 2d gabor wavelets. *IEEE Trans. Patt. Anal. Machine Intell.*, 18(10):959–971, 1996.
13. S. Mallat. A theory for multiresolution signal decomposition: the wavelet representation. *IEEE Trans. Patt. Anal. Machine Intell.*, 11(7):674–693, 1989.
14. T. Chang and C.-C.J. Kuo. Texture analysis and classification with tree-structured wavelet transform. *IEEE Trans. Im. Proc.P*, 2(4), 1993.
15. G. Van de Wouwer, P. Scheunders, D. Van Dyck, M. De Bodt, F. Wuyts, and P.H. Van de Heyning. Wavelet-filvq classifier for speech analysis. In *Proc. Int. Conf. Patt. Rec., Vienna*, pages 214–218, 1996.
16. O. Pichler, A. Teuner, and B.J. Hosticka. A comparison of texture feature extraction using adaptive gabor filtering, pyramidal and tree structured wavelet transforms. *Pattern Recognition*, 29(5):733–742, 1996.
17. H. Levkowitz and G. T. Herman. Glhs: A generalized lightness, hue and saturation model. *CVGIP*, 11(11):3011–3019, 1994.
18. P.A. Devijver and J. Kittler. *Pattern recognition: a statistical approach.* Prentice/Hall, Englewood Cliffs, New Jersey, 1982.
19. P. Pudil, J. Novovicova, and J. Kittler. Floating search methods in feature selection. *Pattern Recognition Letters*, 15:1119–1125, 1994.
20. VisTex. Color image database. http://www-white.media.mit.edu/vismod/-imagery/VisionTexture, 1995. MIT Media Lab.

Cross-Media Color Matching Using Neural Networks

E. Boldrin, R. Schettini
Institute of Multimedia Information Technologies (ITIM)
National Research Council (CNR)
Via Ampere 56, 20131, Milano, Italy
E-mail: Centaura@ITIM.MI.CNR.IT

Abstract

Cross-media color reproduction is receiving a great deal of attention as a result of the increasing availability of color devices. A practical approach to accurate color reproduction, integrating colorimetric and interactive methods by means of feed-forward neural networks trained by back-propagation, is proposed. Experimental results confirming the feasibility of this approach are reported.

Introduction

Cross-media color reproduction demands device- and viewing condition- independent color description. While most color management systems already incorporate device profiles that make colorimetric, i.e. device independent, color reproduction possible, they often neglect color appearance modeling. At the moment there is not even any general agreement as to the choice and use of a color appearance model, which also requires expensive measuring devices, tightly controlled viewing conditions, and rigorous device modeling and calibration [1]. Thus, the most effective and common approach to the matching of colors on different devices or supports is to have the user define the color combinations interactively. Taking this state of affairs as our point of departure, we propose a practical method, that integrates colorimetric and interactive methods by means of neural networks, for the accurate reproduction of surface color ranges and images on displays. Experimental results are encouraging.

Cross-media color matching by back-propagation

An early had verified the feasibility of using a feed-forward neural network trained by back-propagation on a small, predefined set of colors to support the entry of color data bases in the context of textile CAD [15]. Color ranges were reproduced colorimetrically on the display, and the user was given editing tools with which to match the colors of the training set. The neural network learned the correct response, using the device-independent color descriptions of the mapping interactively defined by a user. The implicit mapping coded in the neural network was then applied to correct all the colors of the color range to be reproduced on the screen. Although the users considered the matches obtained interactively good, the small, predefined, training set (less than 50 elements) was not sufficient to train the network. It was able to learn the training set correctly, but the corrections proposed for the other samples, while congruent with those of the user, were still not enough to dispense with further editing. Larger and more complete training sets appeared to be needed. The procedure presented here has been designed to deal with this drawback, minimizing user efforts.

1. *Color ranges are measured using a spectrophotometer or a colorimeter.*

2. *Color ranges are reproduced colorimetrically on the display.*

This involves a colorimetric characterization of the device that defines an appropriate mapping between the device-dependent RGB and standard CIE tristimulus values [15]. Since original and reproduced colors are normally seen in quite different contexts and lighting conditions, a chromatic adaptation model is applied. We have chosen here the simplest and most widely used: the CIELAB color appearance model [19]. The following procedure can, however, be applied to correct colors reproduced using any appearance model for a first rendering on the display (the Hunt color appearance model can be used to improve the match [3,6]). Unreproducible colors are approximated by projecting them onto the surface of the monitor gamut.

3. *The user is given editing tools based on visual interaction with which to match the colors of a suitable training set on the display [14].*

4. *A neural network learns to correct the colors to be displayed using the CIELAB device-independent color description of the mapping interactively defined by the user.*

A multilayer feed-forward neural network, consisting of three input and three output units and three hidden layers of 7 units each is used. This architecture was determined empirically [3]. The first, or input layer serves as a holding site for the values to be processed by the network (CIELAB coordinates of the color to be corrected). The last, or output layer is the point at which the final state of the network can be read (CIELAB coordinates of the matching color). Links connect each unit of one layer only to units in the next layer. The activation function of the nodes of the input layer is linear, while that of the hidden layers is sigmoidal. The back-propagation algorithm gives the prescription for changing the weights of any feed-forward network so it can learn the training set constituted by a set of input-output data. This algorithm is designed to reduce the error between the current and the desired output of the network, using gradient descent. A detailed derivation of the back-propagation algorithm can be found in references [4,12]. The learning rate and momentum constant are adaptively modified during the iterative learning process, while the training set is randomized to produce an unbiased mapping. The training is continued for as many epochs as are necessary to reduce the overall error of the training set to an acceptably low value. The network is initialized by training it on a set of 24 color samples that cover a wide range of chromaticity.

5. *The implicit mapping coded in the neural network is applied to correct subsequent colors of the color range to be reproduced on the screen.*

6. *The user is allowed to modify the colors proposed and run the learning phase again.*

The system memorizes the mappings corrected by the user during data entry; when a significant number of new mappings have been defined, the user reruns the learning phase. This last step can be repeated during data entry until no further corrections are necessary.

Cross-media color image matching

Some processing steps must be modified to extend the procedure described above to pictorial images in cross-media matching. This procedure is graphically depicted in Figure 1. When dealing with images, colors cannot be acquired by a spectrophotometer or colorimeter (Step 1), but must be acquired by scanning. Each device has its own color space, defined by the relationship between the input colors and the corresponding RGB codes used to represent them [11]. Consequently, waiving device calibration, which converts the native color space into a standard device-independent one, will result in unmatched colors throughout the system. Moreover, images acquired with different devices can not be reliably compared and stored. Several authors have shown that a 3x3 matrix transformation can be used to relate RGB values to CIE tristimulus values XYZ *if and only if* the digitizer channel sensitivities are a linear transformation of the color matching function defining the CIE colorimetric observer e.g. [5]. Relatively few of the scanners and cameras that can be found on the market at present have been designed to meet this constraint, motivating several different approaches to the problem. Strachan et al. [18] describe a method to determine the linear transformation that "best" relates RGB and the CIE XYZ values. While effective, this method may generate unreliable results when the actual filter responses are far from being a linear transform of the human cone responses. To overcome this drawback Kang and Anderson [8,9] have experimented both multiple linear regression and cascade correlation neural networks for scanner calibration. Although the latter have produced good results on the training sets adopted, the generalization results obtained are rather poor. We have applied a feed-forward neural network trained by back-propagation to relate the output of our non-colorimetric scanner (a Sharp JP600) with CIELAB standard coordinates [16]. The training set the algorithm uses is the ANSI IT8 7.2 color target, designed specifically for scanner calibration [10]. Any set of colors not in the training set may be used to test the network's capability of generalization numerically. The Macbeth ColorChecker [16] was chosen for out experimental application so, that the method we propose could be compared with others that have used the same chart to evaluate calibration accuracy. Despite its simplicity, the calibration accuracy of the method is superior to that reported by other authors in similar experiments [8,9,18]. A second important difference is the color editing tool (Step 4). Since we can not modify the image appearance by editing single colors, and global modification of the image's appearance often does not result in correct matching, a color cluster editor is needed. We have applied the original Kanamori and Kotera idea of selective color editing [7] to develop a soft color cluster editor which allows the user to correct or modify the image colors as they appear on the display, in a friendly and effective way, until satisfactory visual matching with the original is obtained [17].

A color cluster is composed of colors that are "similar" to a selected color centroid; the farther a color lies from the centroid, the less it will be changed in editing. In order to effectively define the cluster we have exploited sighting, the best medium for color communication, and the fact that computer-driven displays allow the user to select and view colors forming composite images on the screen in real time. Visual interaction allows the user to select color centroids without considering their internal representation, physical qualities, or names. When only small precise color adjustments are required, this can be done by clicking on the color of the image (zoom facility is provided), or browsing the image look-up table, or using a color editor [14]. Our experience, however, suggests that taking the color centroid averaging the colors of an image region interactively selected by the user is more suitable for broader color adjustments. Consequently this option has also been made available. As said above, a cluster is composed of colors "similar" to the centroid. Similarity is a typically fuzzy concept. In previous studies we defined three primary fuzzy sets corresponding to similarity in the lightness (L^*), hue (h^*) and chroma (C^*) color dimensions, and used them to permit color image retrieval by pictorial example [2]. These fuzzy sets are used here as the default in the cluster editor.

Since for larger color adjustments the user generally chooses a region of the image to define the cluster, color feature membership functions are defined using the statistical data of the region selected. More precisely, for both lightness and hue a trapezoidal membership function is defined, with a major bases four times the feature's standard deviation and a minor bases twice the standard deviation. For chroma the major bases is set at six times its standard deviation (these numbers were defined in preliminary elicitation experiments).

The membership functions defined, $\mu_{h^*}(\Delta h^*)$, $\mu_{C^*}(\Delta C^*)$, and $\mu_{L^*}(\Delta L^*)$, associate a degree of similarity of a given color feature (hue, chroma, or lightness) with respect to the same feature of the color centroid. The product of these three membership functions is used to define the cluster membership function [17], i.e. the degree of similarity of a given color to the cluster centroid:

$$\mu_{h^*C^*L^*}(\Delta h^*, \Delta C^*, \Delta L^*) = \mu_{h^*}(\Delta h^*)\mu_{C^*}(\Delta C^*)\mu_{L^*}(\Delta L^*)$$

When $\mu_{h^*C^*L^*}(\Delta h^*, \Delta C^*, \Delta L^*) = 1$, that is $\mu_{h^*}(\Delta h^*) = \mu_{L^*}(\Delta L^*) = \mu_{C^*}(\Delta C^*) = 1$, the color considered will be modified in the described manner. When $\mu_{h^*C^*L^*}(\Delta h^*, \Delta C^*, \Delta L^*) = 0$, that is, when at least one of the membership functions is zero, the color will not be changed in editing, because it does not belong to the cluster. When $0 < \mu_{h^*C^*L^*}(\Delta h^*, \Delta C^*, \Delta L^*) < 1$, the color will be modified in proportion to its similarity to the centroid.

Since similarity is a subjective, context-sensitive property, there are several situations in which default, or statistically defined membership functions do not coincide with the user's idea of "similarity". Rather than model all the possible situations, we have again taken advantage of the possibility of having a visual feed-back of the user's choices: the system displays a gray-level image B(x,y) in which the darkness of a pixel corresponds to its degree of similarity to the centroid considered. This picture is easily obtained as follows:

$$B(x,y) = \left[(L*_{bg} - L*_0)(1 - \mu_{h*C*L*}(\Delta h*, \Delta C*, \Delta L*))\right](x,y)$$

where $L*_{bg}$, and $L*_0$ are the background and black-point lightness respectively.

Figure 2 illustrates color centroid selection using the color editor we have developed. The default membership functions are given in the lower window of the Figure, while the gray-level picture on the right shows the pixels that would be modified in editing (the darker they are, the more similar they are to the centroid). The user can move the sliders provided to define the shape of the feature membership functions and have an immediate visual feed-back of the derived cluster membership function (the feature membership functions thus defined may have a rectangular shape: this corresponds to a crisp definition of the cluster boundaries).

The appearance of the cluster defined can be modified by changing one color feature at a time. Altogether the user can modify seven interindependent features: hue (h*), lightness (L*), chroma (C*), redness (r*), greenness (g*), yellowness (y*), and blueness (b*) (the last four variables correspond to positive and negative variations of the u* and v* features). Given any possible color changes selected by the user, the new color coordinates of the cluster colors are modified proportionally to their degree of similarity to the cluster centroid.

The user has an immediate visual feed-back of the editing activity. If a part of a cluster is moved outside the display gamut during editing, the system gives an acoustic warning and automatically performs an approximation of the out-of-gamut colors by desaturating them. Once the cluster has been modified to the user's satisfaction, the whole procedure can be repeated for different colors. Colors already edited can be protected so that they will not be inadvertently altered in successive operations.

The same network topology and learning strategy used in the first experiment have been adopted for the color appearance mapping of the image to be reproduced (the learning rate and momentum constant are adaptively modified during the iterative learning process, while the training set is randomized to produce an unbiased mapping). To initialize the network a picture of the IT8 7.2 is reproduced on the screen and the user is allowed to modify its colors with the soft color cluster editor until he feels it matches the appearance of the original picture. The initial training set is composed of all the colors thus modified.

Preliminary results

The method proposed here has been proved effective when tested with the assistance of a team of three users familiar with color manipulation, on real data within the framework of textile CAD. The data entry of color ranges in a CAD system is a sequential process which is usually performed by a trained user through a trial and

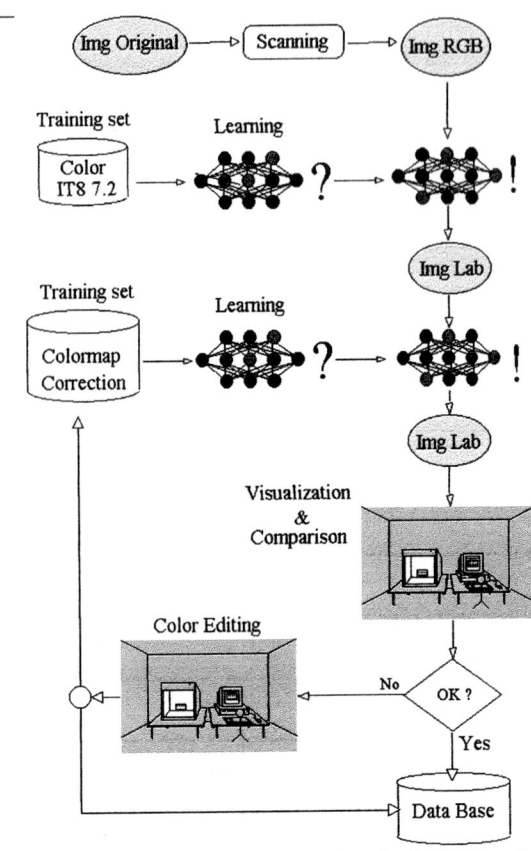

Fig. 1: The strategy developed strategy for cross-media color image matching.

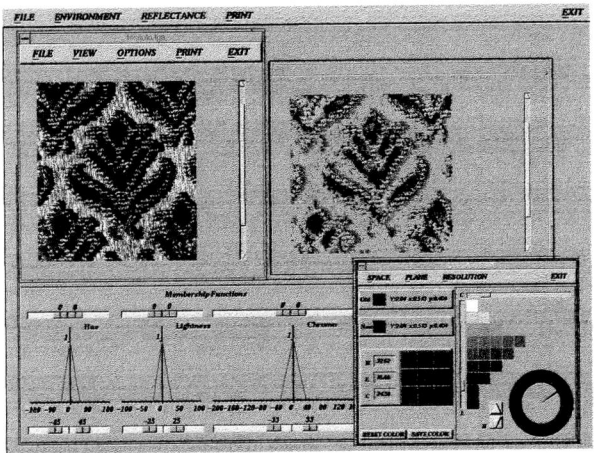

Fig.2: Color centroid selection using the color editor. The similarity picture and default membership functions are also shown

error approach. Using our approach the entry of a color range of over 800 samples in silk was begun by applying the color correction proposed by the network trained on a set of 24 color samples randomly chosen. Each user then "personalized" the training set, modifying some of the color mappings proposed by the network during data entry. Subsequent colors were entered at random (although within the CAD system the users could organize and browse them in several different ways), selecting only colors that were likely to be used together with those colors already reproduced. When at least 25 mappings had been redefined during color entry, the user ran the learning phase again.

We found that the mapping proposed by the neural network after the first training stage reduced the average correction of new colors by about half for all three users. Further personalization of the training set produced less regular results for different users; however, with a training set of about 100 elements the network learned to model each user well enough to require only user supervision in subsequent color data entry.

The procedure has been also been extended successfully to pictorial cross-media image in trial experiments regarding the entry of more than 200 images in a multimedia catalogue of ancient textiles. The mapping proposed by the neural network after the first training stage always improved the quality of the image matching. Subsequent color corrections generally reduced the average color distance between the CIELAB coordinates of the image colors proposed by the network and those of the matching colors obtained by the users' editing. Sometimes, however, the color harmony of the image was modified unexpectedly, making further interactive corrections necessary. The method has still to be tested on different image databases and its effectiveness evaluated in paired-comparison with more standard color correction techniques [1]. We must also solve a "technical problem" related to the color correction network design. During the acquisition of the image database many thousands of colors are "more or less" modified. If the colors in the training set are over four times the number of network links, the network may no longer be able to generalize the desired mapping. To overcome this drawback a maximum must be set for the size of the training set, or else the network must be allowed to evolve dynamically as the training set grows [4]. In the present version we add to the original training set (the color corrections of the IT8 7.2 color set) only centroids of the color clusters that have been modified. We are, however, exploring other hypotheses.

Several experiments are currently in progress to increase the effectiveness of application of our procedure [3] and evaluate them according to CIE guidelines [1].

References

1. P.J. Alessi "CIE guidelines for coordinate research on evaluation of colour appearance models for reflection print and self-luminous display image comparison" Color Research and Application Vol. 19, pp. 48-58, 1994.

2. E Binaghi, I Gagliardi, R. Schettini "Image retrieval using fuzzy evaluation of color similarity" International Journal of Pattern Recognition and Artificial Intelligence, Vol. 8, pp. 945-968, 1994.
3. E. Boldrin, P. Campadelli, R. Schettini "Effective and efficient mapping of color appearance" Color Research and Application, 1997 (in print).
4 J. Hertz, A. Krogh, R.G. Palmer "Introduction to the theory of neural computing" Addison-Wesley, New York, Vol. 1, pp. 115-162, 1991.
5 B.K.P. Horn (1984) Exact Reproduction of Colored Images, Computer Vision, Graphics and Image Processing, Vol. 26, pp. 135-167.
6 R.G.W. Hunt "Revised colour-appearance model for related and unrelated colours" Color Research and Application, Vol. 16, pp. 146-165, 1991.
7. K. Kanamori, H. Kotera "A Method for Selective Color Control In Perceptual Color Space" Journal of Imaging Technologies, Vol. 35(5), pp. 307-316, 1991.
8 H.R. Kang P.G. Anderson "Neural network application to the color scanner and printer calibrations" Journal of Electronic Imaging, Vol. 1, pp. 25-135, 1992.
9 H. Kang "Color Scanner Calibration. Journal of Imaging and technology", Vol. 36, 162-170, 1992.
10 N. Otha "Development of color targets for input scanner calibration" Proc. 7th Congress of the International Color Association "Colour 93", Technical University of Budapest, Budapest (Hungary), p. 130, 1993.
11 Proc. 3nd IS&T and SID's Second Color Imaging Conference: Color Science, Systems and Applications, 1995.
12 D.E. Rumelhart, G.E. Hinton, R.J. Williams "Learning internal representations by error propagation" In D.E. Rumelhart, J.L. McClelland Eds, Parallel distributed processing,, MIT Press, Cambridge (MA), Vol. 1 pp. 145-168, 1986.
13. H. Saarelma, P. Oittinen "Automatic Picture Reproduction" Graphics Art in Finland, Vol. 22(1), pp. 3-11, 1993.
14. R. Schettini, A. Della Ventura, M.T. Artese "Color Specification by Visual Interaction" The Visual Computer, Vol 9(6), pp. 143-150, 1992.
15 R. Schettini "The faithful rendition of color ranges on display, Proc. IS&T and SID's Color Imaging Conference: Transforms & Transportability of Color, Scottsdale, Arziona, pp. 160-163, 1993.
16 R. Schettini, B. Barolo, E. Boldrin "Colorimetric calibration of color scanners by back propagation" Pattern Recognition Letters, Vol. 16(10), pp. 1051-1056, 1995.
17 R. Schettini, B. Barolo, E. Boldrin "A soft color cluster editor" Image Processing and Communications, Vol. 1(1), pp. 17-32, 1995.
18 N.J.C. Strachan, P. Nesvadba e A.R.Allen "Calibration of video camera digitising system in the L*u*v* colour space" Pattern Recognition Letters, 11, 771-777, 1990.
19 G. Wyszecki e W.S. Stiles "Color science: concepts and methods, quantitative data and formulae", Wiley, New York, 1982.

Object Recognition and Performance Bounds*

J. K. Aggarwal and Shishir Shah

Computer and Vision Research Center
Department of Electrical and Computer Engineering, ENS 522
The University of Texas at Austin
Austin, TX 78712-1084, U.S.A.

Abstract. Object recognition is the classification of objects into one of many *a priori* known object classes. In addition, it may involve the estimation of the pose of the object and/or the track of the object in a sequence of images. Bayesian statistical pattern recognition, neural networks and rule based systems have been used to address the object recognition problem. In the case of statistical pattern recognition it is assumed that the *a priori* probability density functions are known or that they can be estimated from the given samples. For neural networks the samples may be used to train a network and the coefficients for the network function may be estimated. Whereas, in the case of the rule based system, rules may be given by an expert or they may be estimated from the samples. However, Bayesian framework provides a methodology for the estimation of error bounds on the performance of the recognition system. The paper discusses the Bayesian paradigm and contrasts its ability to provide performance bounds as compared to neural networks and rule based systems. Future direction of results on object recognition and performance bounds will also be discussed.

1 Introduction

Humans recognize objects and understand complex scenes with multiple objects, noise, clutter, occlusion, and camouflage with great ease. Humans are able to recognize as many as 10,000 distinct objects [Bie85] under varying viewing conditions, while a state-of-the-art object recognition system can recognize relatively few objects. We know very little about the physiological mechanisms with which the human visual system solves and uses solutions to lower-level processes such as depth and shape in the task of object recognition [CJR93]. Modeling human object recognition systems in terms of evidence-based systems accounts for the issues of view-independence, partial occlusions, variation between objects within object classes, and novel exemplar of object classes. As long as an object has enough similarity to the other objects in its class, the same set of evidence is accumulated, which helps in its recognition as a member of that object class. The evidence-based approach is also able to account for both perceptual and

---

* This work was supported by the Army Research Office Contracts DAAH-94-G-0417 and DAAH 049510494.

semantic considerations with explanatory efficiency. Due to the lack of working knowledge of the human visual system, there are no algorithmic descriptions for the human or other biological object recognition systems (ORS). Machine ORS have been driven to duplicate this diversity and remarkable performance. It is safe to say that machine ORS have progressed significantly in the past decade. A number of machine vision systems are now available in the marketplace for applications in inspection, target recognition, robotic manipulation, etc.

The dominant paradigm for object recognition in machine vision research is inverse optics, pioneered by Marr [Mar82]. Inverse optics is a bottom-up process where edges, surfaces, depth cues, etc. are identified before object recognition. While no precise definition of object recognition has been accepted, it is usually considered as the description of the three-dimensional object/scene that accounts for the two-dimensional imagery. It is perceived as a high-level task in computer vision, relating semantic knowledge in terms of a configuration of known objects [Ros84]. Object recognition is then achieved by comparing descriptions of *a priori* known object models, which are generalized descriptors that define object classes. In contrast to the bottom-up process, model-driven or top-down approaches to object recognition employ object models to predict image features and seek to find these features in the image or in a transformed feature space. In both approaches, the task of object recognition involves processing at all levels of computer vision. Typically, the input to the process is an image or a set of images from a sensor or multiple sensors. Some preprocessing is performed on the data and relevant information is extracted from the processed data and associated with a known description of the object. Therefore, object recognition involves lower-level vision, as with edge detection and image segmentation; mid-level vision, as with representation and description of pattern shape, and feature extraction; and higher-level vision, as with pattern category assignment or classification with an *a priori* known object descriptor.

In order to build a system that can achieve success in a realistic environment, certain simplifications and assumptions about the environment and the problem being tackled are generally made. This process of simplification introduces uncertainties into a problem that may create inaccuracies or difficulties in the system's reasoning abilities if these uncertainties are not represented and handled in a suitable manner. Some ways of dealing with uncertainty are by using: (1) methods that employ nonnumerical techniques, primarily nonmonotonic logic, (2) methods that are based on traditional probability theory, (3) methods that use neo-calculi techniques such as fuzzy logic, confidence factors and Dempster-Shafer calculus to represent uncertainties, and (4) approaches that are based on heuristic methods, where the uncertainties are not given explicit notations but are instead embedded in domain-specific procedures and data structures.

It is not the intent of the authors to present another review of object recognition systems. A number of good reviews of various paradigms and techniques have appeared in the past [AA93b, BJ85, CD86, SFH92]. The purpose of this paper is to look at the fundamental problems and discuss various ways of formulation in practical object recognition systems. This paper is organized into the following sections: Section 2 briefly reviews the object recognition problem and some of the solutions proposed using both the model-based and bottom-up approaches. Next, a review of classification methods which allows for the incorporation of uncertainties into the system and provides a theoretical foundation for the inaccuracies in the reasoning ability of object recognition systems is presented in section 3. The classification paradigms considered are statistical or Bayesian, neural network based, and rule-based. We present a coherent comparison of the methods and discuss the ability of each in measuring the performance of the object recognition process by incorporating a degree of uncertainty. Finally, section 4 summarizes the trends of object recognition and discusses future directions of research.

2 Object Recognition

A wide range of approaches have been proposed and applied with limited success to the machine recognition of objects. Recognizing 3-dimensional (3D) objects from 2-dimensional (2D) images is an important part of computer vision [MA77]. The success of most computer vision applications (robotics, automatic target recognition, surveillance, etc.) is closely tied to reliable recognition of 3D objects or surfaces. The study of object recognition and the development of experimental object recognition systems has had a significant impact on the direction and content of computer vision research. Although a plethora of paradigms, algorithms and systems has been proposed over the past two decades, a versatile solution has not yet been developed; thus far, only partial solutions and limited success in constrained environments has been achieved. Practical implementation of an ORS can be viewed as a multi-stage process, as illustrated in Figure 1. Ideally, all objects of interest pass through each step and are included in the output list. As the data moves through the stages, the processing algorithms become more object specific and the number of data items processed and the number of false alarms decrease. The bottom-up approach mentioned briefly earlier has been successfully applied in a number of application. Here minimal amount of *a priori* information about the objects is used in the earlier part of the recognition process.

In model-based recognition, a 3D model(s) of the object(s) to be recognized is available. The 3D model contains concise and complete information about the object in terms of shape descriptions [VMA86], object parts information, relationship between object parts, etc. The 3D structure of an object is frequently represented by CAD models [AA93a], where volume-based representations of the object are built using primitives such as generalized cones, generalized cylinders and spheres. A method that uses a rectangular parallelepiped as the primitive volume element to represent objects was developed in [KA86]. Octrees [CA84] have also been used for the volumetric representation of objects. Typically, recognition involves extracting 3D information from the image and comparing it with the model features [AA93a], or deriving a 2D description from the image and then comparing it with 2D projections of the model. In using the former method,

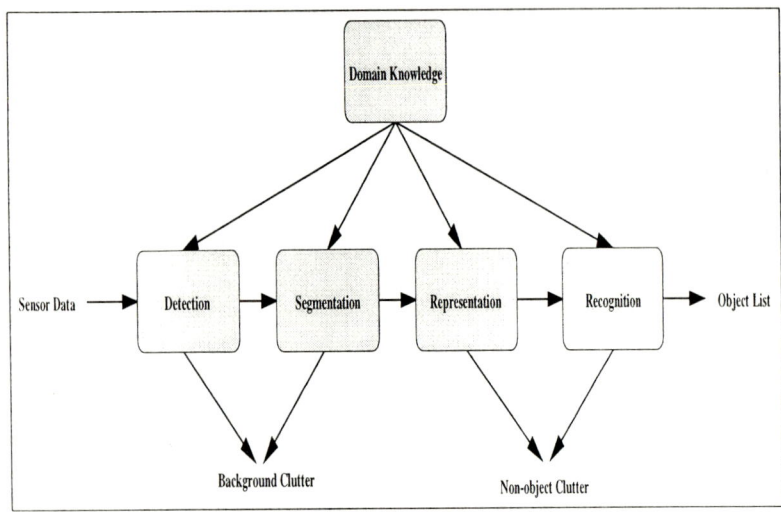

Fig. 1. Conceptual data flow in object recognition systems.

the sensing device should be able to provide 3D information in some form (such as range data or depth information using a stereo setup) which can then be compared with the model. In the latter case, the task is more difficult because (1) the effects of self-occlusions and perspective must be considered, and (2) the projection direction needs to be determined. In [WMA84], the 3D structure of an object is constructed using an observed sequence of silhouettes. During matching, the 3D structure of the unknown object is constructed from different image views, and more views are added to the construction process until features extracted from the object match one of the object models. A comprehensive survey of model-based vision systems using dense-range images is presented in [AA93b], and a recent survey is found in [Pop94].

View-based object recognition is often referred to as *viewer-centered* or *2D object recognition,* because direct information about the 3D structure of the object (such as a 3D model) is not available; the only *a priori* information is in the form of representations of the object viewed at different angles (aspects) and distances. Each representation (or characteristic view) describes the object from a single viewpoint, or from a range of viewpoints yielding similar views. Evidence shows that object recognition in human vision is viewer-centered rather than object-centered [KvD79]. The characteristic views may be obtained by building a database of images of the object or may be rendered from a 3D model of the object [PC93], [ZSB93]. Matching, in this case, is simpler than in model-based recognition because it involves only a 2D/2D comparison. However, considerable storage space is required to represent all of the characteristic views of an object. The number of model features to search among also increases, because each characteristic view can be considered to be a model. Methods have been developed to reduce the search space by grouping similar views [Pop94] [BR92], [PPK92].

Broadly speaking, there are two ways to approach this problem. The first is based on matching salient information, (e.g., corner points, lines, contours etc.,) that has been extracted from the image to the information obtained from the image database [MA77], [CJ93]. Based on the best match, the object is recognized and its pose estimated. The second approach extracts translation, rotation and scale invariant features (such as moment invariants [Hu62], Zernike moments [KH90] or Fourier descriptors [CH91]) from each image and compares them to the features that have been extracted from sample images of all the objects. The comparison is usually done in the form of a classification operation [DBM77].

Motivated by the human visual system, which strongly suggests a hierarchical approach to recognition, machine vision systems have been developed which attempt to mimic this process. Psychologists suggest that recognition of objects is guided by *perceptual organization* in the visual cortex. The principles of perceptual organization are the grouping of low-level, generic features to detect symmetry, collinearity, and parallelism from an input image. These principles have been shown to be useful in machine ORS, especially when no prior information of the image content is available [LA92]. Perceptual organization has been used to segment images into visible object surfaces [MN89]. Detection and recognition of various manmade objects in complex scenes has been accomplished using these principles. Most work in this area has concentrated on extracting groups of features, recognition of objects with exact models, and using additional sensing information.

In some sense, every object recognition algorithm is model based because every algorithm makes and uses *a priori* assumptions about the image and object characteristics. It would be difficult indeed to find an object about which we know nothing! With clutter, noise, occlusions, varying environmental conditions, and imperfect sensor information, these assumptions about the objects play an important role in the overall process. It becomes critical to incorporate a measure of uncertainty into the assumptions and algorithms we develop to evaluate the performance of the developed system. The final step for all recognition systems involves the classification of detected features to an *a priori* model. The success of this classification depends heavily on two main issues: (a) identifying the type of features to use in the matching, and (b) determining the best procedure to establish the correspondence between image and model features. The reliability and efficiency of an object recognition system directly depends on how carefully these issues are addressed.

3 Recognition Paradigms

A multitude of paradigms have been used to achieve success in constrained object recognition systems. Figure 2 shows the main technologies applied to this problem. Bayesian statistical pattern recognition, neural networks and rule based systems have been used extensively and successfully in addressing the object recognition problem. In this section we provide an overview of each of these methods and discuss their abilities to provide a performance measure.

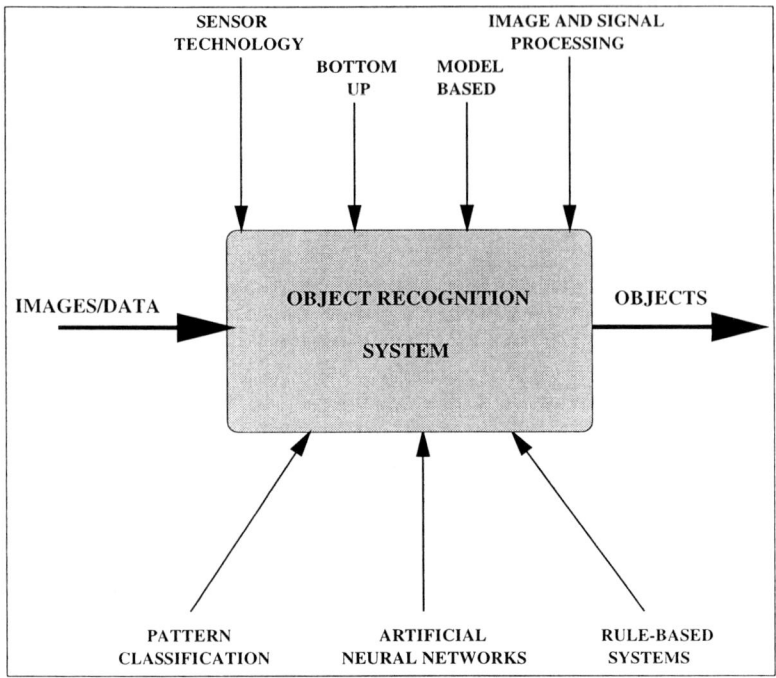

Fig. 2. Object recognition paradigms.

3.1 Bayesian Formulations

Bayesian methods provide a formal means to reason about partial beliefs under conditions of uncertainty [Pea88] [DH73]. Bayesian statistics have been used at various stages of the object recognition process to provide a firm theoretical footing as well as to improve performance and incorporate error estimates for the overall process. The biggest advantage of a Bayesian (or probabilistic) framework is in its ability to incorporate uncertainty elegantly into a process. Bayesian approaches also provide error estimates with their decisions, which give another perspective for analyzing systems. Bayesian statistics have been used in the object recognition paradigm for indexing, model matching and incorporating neighborhood relations under different contexts with some degree of success. In order to apply Bayes' theorem, one needs to have an estimate of the *prior* probabilities and also the underlying *likelihood* distributions. Depending on the application, different methods are used to determine these factors. *Prior* probabilities are usually estimated as the percentage of occurrence of the proposition over a period of time. The *likelihoods* are often estimated by making an assumption that simplifies the relationship between the hypothesis and the evidence. A commonly used assumption is that the evidence and the hypothesis are related by a normal (Gaussian) distribution.

Let us Consider a simple example. Suppose we are to recognize two objects, A and B, where the prior information is such that the object A occurs 70%

of the time and object B, 30% of the time. This provides the estimate of a priori probabilities, $P(A) = 0.7$, and $P(B) = 0.3$. Now consider that given the object data, we are able to extract a relevant feature for recognition, X. Thus the recognition problem can be posed as the identification of object A or B, given only the feature X. From a set of training samples, we can compute the parametrized density function that represents each of the objects. Assuming a normal distribution,

$$p(X|O) = \frac{1}{\sqrt{2\pi}\sigma_O} \exp \frac{-1}{2}(\frac{X - \mu_O}{\sigma_O})^2 \qquad (1)$$

where O may be object A or B, μ_O and σ_O are the mean and variance for the respective object feature distribution. Given the prior probability and the likelihood, the posterior probability of recognizing the objects is given by the inversion formula,

$$P(O|X) = \frac{P(X|O)P(O)}{P(X)}, \qquad (2)$$

The denominator $P(X)$, given by $P(X|A)P(A) + P(X|B)P(B))$, is a normalizing constant. Thus the recognition is based on deciding object A if $P(A|X) > P(B|X)$ and vice versa. In most practical formulations, the classification rule does not lead to perfect classification. One reason for this is that features are common to two or more classes and the regions for supports, or likelihoods, overlap. The Bayesian framework provides an estimate of the probability of classification error associated with each decision. These may take into account the significance of a classification error in addition to the probability of an error. The simple, two-object problem described above, which led to an intuitively appealing classification rule, can be extended to consider the probability of a classification error as a function of the measured feature X. We incur an error if we choose object B and the true class is object A or if we choose A and the true class is B. The error corresponding to this decision can be formulated as:

$$P(error|X) = P(A|X) \; if \; we \; decide \; B \qquad (3)$$
$$= P(B|X) \; if \; we \; decide \; A$$

The total error of classification can be expanded as:

$$\begin{aligned} P(error) &= P(error|A)P(A) + P(error|B)P(B) \qquad (4)\\ &= P(X \in R_B|A)P(A) + P(X \in R_A|B)P(B) \\ &= P(X < \psi|B)P(B) + P(X > \psi|A)P(A) \\ &= P(B)\int_{-\infty}^{\psi} p(X|B)dX + P(A)\int_{\psi}^{\infty} p(X|A)dX \end{aligned}$$

The Bayesian formulation can easily be extended to n-classes, thus n different objects can be represented by parametrized density functions. The types and parameters of these functions can vary between different objects. To realize a Bayesian object recognition system, three main steps have to be followed:

1. Training, where the parameters θ_α, $1 \leq \alpha \leq n$, of the model density functions have to be estimated from a sample set of objects, A and B in our simple example.
2. Localization, where the image information is processed to estimate data that is most relevant to learned object models. This marks the use of relevant features X.
3. Recognition, where the localized image features are matched to the object model to determine the object class number α, by evaluating the discriminant function derived.

Generalizing for classification error in the n class decision problem, the expected risk or error is given by application of the total probability theorem [DH73]:

$$R[\psi(X)] = \int R[\psi(X)|X]p(X)dX \qquad (5)$$

where $\psi(X)$ is the set of decision rules which maps the observed feature, X to its respective class.

For indexing formulation in object recognition, a feature set(s) (index vector) is identified that maps each unique object model (or part of a model) into a distinct point in the index space. This point is stored in a table with a pointer back to the object model. At runtime, the same type of feature set(s) are obtained from the image to form an index vector, which is then used to quickly access nearby pre-stored points. Thus a set of possible matches is found through correspondence of all possible image/model pairs. The distributions of the entries in the table could be organized based on similarities between object features or could be organized hierarchically such that the object classes are represented by a prototype table entry and further indexing is done to match the particular type of object within a class. Indexing using three points can be achieved using a probabilistic indexing scheme, which is based on the *probabilistic peaking effect* [BA90]. Alignment [HU90] and geometric hashing [GG92] are related techniques that are used for recognizing 3D object from 2D scenes. Both of these methods use a small number of points to find a transformation between the model space and the image space. Recognition then consists of finding evidence for instances of the models in the data, either by transforming the image into the model space and voting for an object's pose or by hypothesizing a pose and then transforming it into image space to guide the search. In [Wel93], a two-stage statistical formulation is used for feature-based object recognition. This work clearly shows how the Bayesian theory can be applied to model matching both in the *correspondence* space and the *transformation* space. A more detailed review of Bayesian techniques can be found in [AGNT96].

3.2 Neural Networks

Artificial neural networks (ANN) are motivated by biological systems which implement pattern recognition computations via interconnections of physical cells, called neurons. The idea that the computations underlying the emulation of intelligent behavior may be accomplished by interactions of a large number of simple processing units is explored using ANNs. ANNs are highly parallel networks of simple computational elements (nodes) [JMM96], where each node performs operations such as summing the weighted inputs coming into it and then amplifying / thresholding the sum. The properties of the nodes, their interconnection topology (number of layers and number of nodes per layer), the connection strengths between pairs of nodes (weights) and the method used to update these weights (learning rule) characterize a neural network. Figure 3 shows a typical two-layer structure for an ANN. Neural networks are data-driven, and modifying patterns of internode connectivity as a function of the training data is the learning approach. In other words, the knowledge is stored in the form of network weights. Neural networks are trained so that subsequent associative behavior would recognize new patterns that are similar to the learned patterns. Learning in a neural network is usually performed using two distinct techniques: supervised and unsupervised. In supervised learning, the network is presented with both the input and the desired output for each input, and learning takes place to determine the weight structure that best realizes this input/output relationship. In unsupervised learning, the network is presented only with the input data and the network uses statistical regularities in the data to group it into categories.

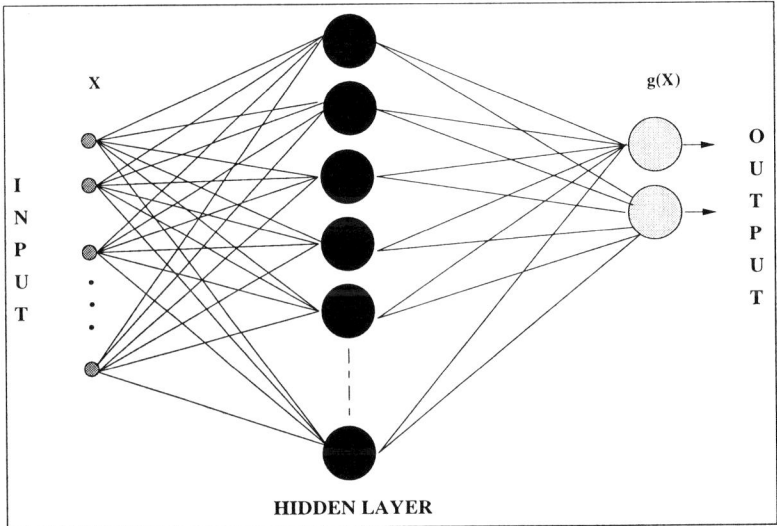

Fig. 3. A two-layer neural network.

Several types of neural networks can serve as adaptive classifiers that learn through examples and thus do not require a good *a priori* mathematical model for the underlying physical characteristics. These include feed-forward networks such as the Multi-Layer Perceptron (MLP), as well as kernel-based classifiers such as those employing Radial Basis Functions (RBFs). A second group of neural-like schemes such as the Learning Vector Quantization (LVQ) have also received considerable attention. These are adaptive, exemplar-based classifiers that are closer in spirit to the classical K-nearest neighbor method. The strength of both groups of classifiers lies in their applicability to problems involving arbitrary distributions. Most neural network classifiers do not require the simultaneous availability of all training data and frequently yield error rates comparable to Bayesian methods without needing *a priori* information. Techniques such as fuzzy logic can be incorporated into a neural network classifier for applications with little training data. A good review of probabilistic, hyperplane, kernel and exemplar-based classifiers that discusses the relative merit of various schemes within each category is available in [NL91]. Although neural networks do not require geometric models, they do require that the set of examples used for training should come from the same (possibly unknown) distribution as the set used for testing the networks, in order to provide valid generalization and good performance on classifying unknown signals [GT94]. To obtain valid results, the number of training examples must be adequate and comparable to the number of effective parameters in the neural network. A deeper understanding of the properties of feed-forward neural networks has emerged recently that can relate their properties to Bayesian decision making and to information theoretic results [Bis95]. To compare the ANN structure to the Bayesian approach, consider the problem of recognizing 1 of C objects in an environment. The evidence of an observed object is given by the feature vector $X \in R^N$. Using the Bayesian decision, a network can be constructed where the hidden unit outputs represent the posterior estimates for each of the C classes and the final output unit performs the *max* operation. This is just the Bayes decision rule, as the probabilities are learned prior to their use in the network.

The common approach to learning in any feed-forward neural network is to perform gradient descent on a criterion function. For a simple two-class problem, the input features to the network are represented by $Z = (-1, X_1, X_2, \ldots, X_K)$ where -1 is provided as a bias term. The weights to H hidden units are initialized to be $\Lambda = (\theta, W_1, W_2, \ldots, W_J)$. The activation or output due to a single feature X is given as:

$$g(X) = W^t X + \theta \qquad (6)$$

The weights to the network are learned over the entire training set, so the final decision function is:

$$g(Z) = \Lambda^t Z \qquad (7)$$

and the class label is assigned based on the output sign. In order to learn the input-output relationship and update the weights, the criterion function to be minimized is chosen as the mean squared error. Thus, if the true output or class is d_i and the network output is $g_i(X)$, over M training patterns, the expected error cost is:

$$\Delta = E \sum_{i=1}^{M} [g_i(X) - d_i]^2 \qquad (8)$$

If we consider a two-layer network with linear output units and hidden units with the logistic sigmoid activation, solving for a C class problem, the standard error function for each pattern to be minimized in an iterative manner is

$$E^m = \frac{1}{2} \sum_{k=1}^{C} (g_k - d_k)^2 \qquad (9)$$

Using gradient descent, the derivative of the error is obtained by differentiating the error function with respect to the weights. As the output unit is linear, the error for each unit is simply given by

$$\delta_k = g_k - d_k \qquad (10)$$

while for units in the hidden layer, the errors are

$$\delta_j = z_j(1 - z_j) \sum_{k=1}^{C} w_{kj} \delta_k \qquad (11)$$

where z_j is the activation for the jth hidden unit, w_{kj} are the weight connections between the hidden and output layer, and the sum runs over the output units. Thus the weight updates are given by

$$\Delta w_{kj} = -\eta \delta_k z_j \qquad (12)$$
$$\Delta w_{ji} = -\eta \delta_j x_i$$

for the output layer and hidden layer respectively, where η is the learning rate.

After learning the network weights, the resultant error indicates the performance of the network. Networks which can provide an estimate of probabilities associated with each decision can be used to determine the recognition/classification performance. Taking $P(X, C_j)$ to be the joint probability of input and the corresponding class label, and since $P(X, C_j) = P(C_j|X)P(X)$, the expected error cost can be evaluated as:

$$\Delta = \int \sum_{j=1}^{C} \sum_{i=1}^{M} [g_i(X) - d_i]^2 P(X, C_j) dX \qquad (13)$$

$$= \int \sum_{j=1}^{C} \sum_{i=1}^{M} [g_i(X) - d_i]^2 P(C_j|X) P(X) dX$$

$$= \int \sum_{k=1}^{C} [\sum_{j=1}^{C} \sum_{i=1}^{M} [[g_i(X) - d_i]^2 P(C_j|X)] P(X, C_k) dX$$

$$= E[\sum_{j=1}^{C} \sum_{i=1}^{M} [[g_i(X) - d_i]^2 P(C_j|X)]]$$

$$= E[\sum_{j=1}^{C} \sum_{i=1}^{M} [g_i(X)^2 P(C_j|X) - 2g_i(x) d_i P(C_j|X) + d_i^2 P(C_j|X)]]$$

Simplifying using the expectation of true class given input features:

$$\Delta = E[\sum_{i=1}^{M}[g_i(X) - Ed_i|X]^2] + \underbrace{E[\sum_{i=1}^{M} Var[d_i|X]]}_{Independent\ of\ network} \quad (14)$$

The first term on the right is simply the mean squared error between the network outputs and the conditional expectation of the desired outputs. Thus, when the network parameters are chosen to minimize a squared error cost function, outputs estimate the conditional expectation. For a 1 of C classification, d_i equals 1 if the input X belongs to class C_i and 0 otherwise. Therefore,

$$E[d_i|X] = \sum_{j=1}^{M} d_i P(C_j|X) \quad (15)$$
$$= P(C_j|X)$$

which are nothing but posterior probabilities. It has been demonstrated that classifiers provide outputs which accurately estimate known Bayesian probabilities and the outputs sum to one even though they are not explicitly constrained during training. More details regarding estimating probabilities in other networks and a survey of neural network approaches to machine inspection can be found in [Gho94].

3.3 Rule Based Approaches

Artificial Intelligence (AI) techniques have proven to fit well in high-level tasks that require reasoning capabilities and prior domain knowledge representation. A typical AI system has two main components: (1) *A knowledge-base* component which includes general facts about the application domain as well as task specific knowledge, and (2) *a control strategy* such as an *inference engine* which controls the reasoning or search process. The knowledge-base component of an AI system can be represented either as a set of procedures or in a declarative (i.e., non-procedural) fashion. Propositional logic, predicate calculus, decision trees, production rules, semantic nets, frames and slots, fuzzy logic and probabilistic logic are some of the commonly used knowledge representation techniques in the AI field. Although top-down (or goal-driven) and bottom-up (or data-driven) are the most commonly used control strategies, many successful AI systems use a hybrid top-down and bottom-up control strategy. Rule-based approaches have commonly been used in relation to object recognition systems due to their emergence from inductive learning and explanation-based learning. Reasoning under uncertainty is how humans perform object recognition, and it is rarely done with 100% certainty. Evidence supporting or refuting each particular decision is collected, examined, and weighed against all evidence supporting or refuting other possible conclusions. Similarly, in real world complex problems such as machine ORS, some type of probabilistic or uncertain reasoning is required [SP90]. Consider a hypothetical rule, expressed in standard logical notation

$$a \wedge b \wedge c \wedge d \to O_1 \qquad (16)$$

for recognizing an object O_1. An expert considering the same decision may choose the object despite lack of evidence for d if sufficient evidence exists for $a \wedge b \wedge c$. Without the knowledge of certainty in each of the evidences considered, it is hard to incorporate this notion into a rule-based ORS.

Rule-based paradigms provide a logical and understandable manner for using symbolic knowledge or domain knowledge in performing complex and heuristic tasks. Many object recognition systems have been developed based on these principles [Won87, Tou87, DMPA93, RH92]. In the overall structure of the object recognition paradigm, rule-based systems provide added advantages by increasing the system abstraction level, system maintainability, and uncertainty handling, providing reasoning and explanation capability, providing a built-in control strategy, and adding learning capabilities. Due to their use of symbolic representation, knowledge-based systems can be utilized to abstract many segmentation and labeling details. In rule-based ORS, the knowledge base and the matching criterion are two separate modules. Therefore, they both can be updated with little effort and time. Rule-based systems can handle uncertain decisions by attaching a measure of belief to each of their output decisions. In real object recognition applications it is important, if not essential, to have an explanation modality to clarify why and how a specific decision has been chosen over other decisions and to subsequently tune up the reasoning process. Reasoning and explanation capabilities are two unique features of rule-based systems. A rule-based system provides a built-in inference engine that can be used in a bottom-up, top-down or hybrid top-down and bottom-up fashion. A bottom-up control strategy can be used in a system when the noise level in the row data is low or when the search span in the solution space is large and hard to prune. In other cases, when there is a lot of interaction between data in the lower level tasks, a top-bottom or a goal-driven control strategy is more appropriate. However, in both cases, having a built-in control strategy with heuristic search criteria helps to reduce object recognition system complexity and implementation effort.

A system for multisensor image interpretation using a rule-based approach was developed by Chu and Aggarwal [CA95]. The AIMS (Automatic Interpretation system using Multiple Sensors) system has three main building blocks: (1) a segmentation module that integrates segmentation information from thermal, range, intensity, and velocity images and combines them into an integrated segmentation map; (2) a representation module, in which the outcome of the segmentation module is represented in a structural knowledge-based format that can be utilized by the KEE package; and (3) an interpretation module that uses KEE and supplementary LISP procedures, in a bottom-up manner to recognize different objects in an image. AIMS' reasoning process depends on knowledge in the form of rule-bases that are based on: (K1) knowledge of the imaging geometry and device parameters, which are independent of the imaged scene; (K2) information on the segmented image regions, such as size, average temperature within the region, average distance, etc.; (K3) neighborhood relationships between the image regions; (K4) features and models of objects; and (K5) other general heuristics that are derived from known facts about the application domain and common sense. Using the above knowledge, a forward-chaining reasoning approach is adopted to recognize the objects that appear in an image, using six main consequent types of rules to: (R1) handle the difference between individual segmentation maps and the integrated segmentation map. These rules are also used to compute low-level attributes and place them in the corresponding knowledge structure. (R2) distinguish between man-made objects and background (MMO/BG). One such example is:

If (Segment A is relatively hot) AND
(Segment A has a compact contour)
Then (Segment A is a MMO, Confidence=Func(temperature, shape)),

(R3) to group similar segments (regions) into objects based on neighborhood relationships and other similarity measures. (R4) to classify back-ground (BG) into SKY, TREE, and GROUND types, and (R5) to classify man-made objects (MMO) into different types such as BULLETIN-BOARD, TANK, JEEP, APC, or TRUCK based on shape and size analysis. One such rule is:

IF (Segment A is of type MMO) AND
(Segment AS has a cool sub-region located at its lower-half) AND
(Segment A is about 2.0-2.5m high) AND
(Segment A has a trapezoidal contour) AND
THEN (Segment A is an APC with confidence of 0.8)

and finally, (R6) to verify the interpretation of an object and its surrounding objects. As example, a region recognized as a SKY cannot be surrounded by a region classified as GROUND. Any conflicting interpretations lead to reduced certainty factor recognitions.

Several algorithms have been developed for learning the domain knowledge from a set of learning examples in the form of a rule set. Sequential covering algorithms learn one rule at a time, subtracting out the covered examples and repeating the process on the remaining examples. In contrast, decision tree algorithms learn an entire set of disjuncts simultaneously as part of the single search for an acceptable decision tree. The main difference in the two approaches is in the partitions of data that they generate. Decision tree algorithms make fewer independent choices in selecting the precondition of each rule. Rule-based systems have been able to perform 3D shape recovery and orientation from a single view [SSY92]. The system uses some geometric regularity assumptions about perceived objects and image formation to recognize the objects from the 2D images. The system uses the expert system paradigms to perform some geometric reasoning from a given 2D image and form a set of possible 3D views and orientations that correspond to the given 2D object view. The reasoning process is done in a forward-chaining fashion using OPS5, a production system language. The outcome of the reasoning process may result in more than one interpretation, each with an attached certainty factor that quantifies the system measure of belief in the recovered 3D object from the given prospective view.

Overall, current rule-based systems are limited in their ability to interpret typical knowledge bases in object recognition. Better outlier or exception dealing capabilities need to be explored along with retrieval or associative knowledge. Further, their capacity to evaluate error in the decision capability is limited and the authors are not aware of any means for characterizing the performance which can have a bound as provided through neural-network and Bayesian systems. Above all, better techniques to connect them with other kinds of representations need to be addressed, so that we can use rule-based approaches in object recognition systems in conjunction with different kinds of models and search procedures.

4 Future Directions

A number of distinct paradigms have been applied in our continuing attempts at development of machine object recognition systems. Most of them have been successful at least partially in constrained environments. A general purpose object recognition system is not in sight as yet. The object recognition problem is like an "elephant" being examined by a number of "visionless" persons. Each of the visionless persons gives a self-consistent and accurate description of the elephant. However, it would be difficult if not impossible to discover a complete description of the elephant from these partial "visionless" descriptions. Bayesian methodology is driven by probability theory. ANN methodology is motivated by the presumed behavior of a collection of biological neurons. Rule-based systems tend to emulate the presumed behavior of a human expert. An ideal ORS may be a combination of all three methodologies unified in a "visionary" fashion. Systematic methods and formalisms need to be developed for the design of hybrid systems consisting of the basic paradigms, performing characteristic tasks while simultaneously interacting with other modules. Such interaction would allow for the flow of information and decisions with competition and cooperation, all in the context of a global constraint, while minimizing the error in object recognition. The Bayesian paradigm in its formulation provides error estimates in the statistical formulation and yields similar estimates in the case of ANNs and possibly rule-based paradigms, especially if the relevant features have a Gaussian distribution.

References

[AA93a]   F. Arman and J. K. Aggarwal. CAD-based vision: Object recognition in cluttered range images using recognition strategies. *Computer Vision, Graphics, and Image Procesing*, 58(1):33–47, 1993.

[AA93b]   F. Arman and J. K. Aggarwal. Model-based object recognition in dense depth images - a review. *ACM Computing Surveys*, 25(1):5–43, 1993.

[AGNT96] J. K. Aggarwal, J. Ghosh, D. Nair, and I. Taha. A comparative study of three paradigms for object recognition: Bayesian, neural network and expert systems. In K. Bowyer and N. Ahuja, editors, *Advances in Image Understanding: A Festschrift to Azriel Rosenfeld*, chapter 15, pages 300–324. Springer-Verlag, 1996.

[BA90]   J Ben-Arie. The probabilistic peaking effect of viewed angles and distances with application to 3D object recognition. *IEEE Transactions on Pattern Analysis and Machine Intelligence*, 12(8):760–774, 1990.

[Bie85]   I. Biederman. Human image understanding: Recent research and a theory. *Computer Vision, Graphics and Image Processing*, 32:29–73, 1985.

[Bis95]   C. M. Bishop. *Neural Networks for Pattern Recognition*. Oxford University Press, New York, 1995.

[BJ85]   P.J. Besl and R.C. Jain. Three-dimensional object recognition. *ACM Computing Surveys*, 17(1):75–145, March 1985.

[BR92]   J. B. Burns and E. M. Riseman. Matching complex images to multiple 3D objects using view description networks. In *Proceedings of IEEE Conference on Computer Vision and Pattern Recognition*, pages 328–334, 1992.

[CA84]   C. H. Chien and J. K. Aggarwal. A volume/surface octree representation. In *7th International Conference on Pattern Recognition*, pages 817–820, 1984.

[CA95]   C.C. Chu and J.K. Aggarwal. The interpretation of a laser rader images by a knowledge-based system. *Machine Vision and Applications*, 4:145–163, 1995.

[CD86]   R.T. Chin and C.R. Dyer. Model-based recognition in robot vision. *ACM Computing Surveys*, 18(1):67–108, March 1986.

[CH91]   Z. Chen and S. Ho. Computer vision for robust 3D aircraft recognition with fast library search. *Pattern Recognition*, 24(5):375–390, 1991.

[CJ93]   S. Chen and A. K. Jain. Strategies of multi-view multi-matching for 3d object recognition. *Computer Vision and Image Processing*, 57(1):121–130, 1993.

[CJR93]   T. Caelli, M. Johnston, and T. Robinson. 3d object recognition: Inspiration and lessons from biological vision. In A. K. Jain and P. J. Flynn, editors, *Three-Dimensional Object Recognition Systems*, pages 1–16. Elsevier Science Publishers, 1993.

[DBM77]   S.A. Dudani, K.J. Breeding, and R.B. McGhee. Aircraft identification by moment invariants. *IEEE Transactions on Computers*, C-26:39–46, 1977.

[DH73]   R. O. Duda and P. E. Hart. *Pattern Classification and Scene Analysis*. A Wiley-Interscience Publication, 1973.

[DMPA93] M. De Mathelin, C. Perneel, and M. Acheroy. IRES: an expert system for automatic target recognition from short-distance infrared images. In L.E. Garn and L.L. Graceffo, editors, *SPIE, Architecture, Hardware, and Forward-Looking Infrared Issues in Automatic Object Recognition*, volume 1957, pages 68–84, 1993.

[GG92]   D. Gavrila and F. Greon. 3D object recognition from 2D image using geometric hashing. *Pattern Recognition Letters*, 13(4):263–278, 1992.

[Gho94]   J. Ghosh. Vision based inspection. In C. H. Dagli, editor, *Artificial Neural Networks for Intelligent Manufacturing*, pages 265–297. Chapman and Hall, London, 1994.

[GT94] J. Ghosh and K. Tumer. Structural adaptation and generalization in supervised feedforward networks. *Journal of Artificial Neural Networks*, 1(4):431–458, 1994.

[Hu62] M. Hu. Visual pattern recognition by moment invariants. *IRE Transactions on Information Theory*, February:179–187, 1962.

[HU90] D. P. Huttenlocher and S. Ullman. Recognizing solid objects by alignment with the image. *International Journal on Computer Vision*, 5(2):195–212, 1990.

[JMM96] A. Jain, J. Mao, and K. M. Mohiuddin. Artificial neural networks: A tutorial. In *Computer*, pages 31–44, March 1996.

[KA86] Y. C. Kim and J. K. Aggarwal. Rectangular parallepiped coding: A volumetric representation of three-dimensional objects. *IEEE Transactions on Robotics and Automation*, 2(3):127–134, 1986.

[KH90] A. Khotanzad and Y.H. Hong. Invariant image recognition by Zernike moments. *IEEE Transactions on Pattern Analysis and Machine Intelligence*, 12:489–497, 1990.

[KvD79] J. Koenderink and A. van Doorn. The internal representation of solid shape with respect to vision. *Biological Cybernetics*, 32:211–216, 1979.

[LA92] H. Q. Lu and J. K. Aggarwal. Applying perceptual organization to the detection of man-made objects in non-urban scenes. *Pattern Recognition*, 25(8):835–853, 1992.

[MA77] J. W. McKee and J. K. Aggarwal. Computer recognition of partial views of curved objects. *IEEE Transactions on Computers*, C-26(8):790–800, 1977.

[Mar82] D. Marr. *Vision*. W. H. Freeman, 1982.

[MN89] R. Mohan and R. Nevatia. Using perceptual organization to extract 3-d structures. *PAMI*, 11(11):1121–1139, November 1989.

[NL91] K. Ng and R.P. Lippmann. Practical characteristics of neural network and conventional pattern classifiers. In J.E. Moody R.P. Lippmann and D.S. Touretzky, editors, *Neural Information Processing Systems*, pages 970–976, 1991.

[PC93] A. Pathak and O. I. Camps. Bayesian view class determination. *IEEE Conference on Computer Vision and Pattern Recognition*, pages 407–412, 1993.

[Pea88] J. Pearl. *Probabilistic Reasoning in Intelligent Systems: Networks of Plausible Inference*. Morgan Kaufmann Publishers, Inc. San Mateo, California, 1988.

[Pop94] A. Pope. Model-based object recognition-a survey of recent research. *Technical Report*, TR-94-04, 1994.

[PPK92] S. Petitjean, S. Ponce, and D. J. Kriegman. Computing exact aspect graphs of curved objects: Algebraic surfaces. *International Journal on Computer Vision*, 9(3):231–255, 1992.

[RH92] E.M. Riseman and A.R. Hanson. A methodology for the development of general knowledge-based vision system. In C. Torras, editor, *Computer Vision: Theory and Industrial Applications*, pages 293–336. Springer Verlag, 1992.

[Ros84] A. Rosenfeld. Image analysis: Problems, progress and prospects. *Pattern Recognition*, 17(1):3–12, January 1984.

[SFH92] P. Suetens, P. Fua, and A.J. Hanson. Some computational strategies for object recognition. *ACM Computing Surveys*, 24(1):5–62, March 1992.

[SP90]   G. Shafer and J. Pearl, editors. *Readings in Uncertain Reasoning*. Morgan Kauffman, Inc., 1990.

[SSY92] W.J. Shomar, G. Seetharaman, and T.Y. Young. An expert system for recovering 3D shape and orientation from a single view. In L. Shapiro and A. Rosenfield, editors, *Computer Vision and Image Processing*, pages 459–516. Academic Press, 1992.

[Tou87] J. T. Tou. Knowledge-based systems for robotic application. In A. Wong and A. Pugh, editors, *Machine Intelligence and Knowledge Engineering for Robotics Applications, Proc. NATO/ASI Workshop*, pages 145–189. Springer Verlag, 1987.

[VMA86] B. Vemuri, A. Mitiche, and J. K. Aggarwal. Curvature-based representation of objects from range data. *Image and Vision Computing*, 4(2):107–114, 1986.

[Wel93] W. M. Wells. *Statistical Object Recognition*. PhD thesis, Cambridge, MIT, November 1993.

[WMA84] Y. F. Wang, M. J. Magee, and J. K. Aggarwal. Matching three-dimensional objects using silhouettes. *IEEE Transactions on Pattern Analysis and Machine Intelligence*, 6(4):513–518, 1984.

[Won87] A. Wong. Knowledge representation for robot vision and path planning using attributed graphs and hypergraphs. In A. Wong and A. Pugh, editors, *Machine Intelligence and Knowledge Engineering for Robotics Applications, Proc. NATO/ASI Workshop*, pages 113–143. Springer Verlag, 1987.

[ZSB93] S. Zhang, G. Sullivan, and K. Baker. The automatic construction of a view-independent relational model for 3D object recognition. *IEEE Transactions on Pattern Analysis and Machine Intelligence*, 15(6):778–786, 1993.

Relating Image Warping to 3D Geometrical Deformations

A.L. Yuille[1], Mario Ferraro[2], and Tony Zhang[3]

[1] Smith-Kettlewell Eye Research Institute, San Francisco, CA 94115.
[2] Dipartimento di Fisica Sperimentale, Universita' di Torino, via Giuria 1, 10125 Torino, Italy.
[3] Division of Applied Sciences, Harvard University, Cambridge MA, 02138

Abstract. We demonstrate that, for a large class of reflectance functions, there is a direct relationship between image warps and the corresponding geometric deformations of the underlying three-dimensional objects. This helps explain the hidden geometrical assumptions in object recognition schemes which involve two-dimensional image warping computed by matching image intensity. In addition, it allows us to propose a novel variant of shape from shading which we call shape from image warping. The idea is that the three-dimensional shape of an object is estimated by determining how much the image of the object is warped with respect to the image of a known prototype shape. Therefore detecting the image warp relative to a prototype of known shape allows us to reconstruct the shape of the imaged object. We derive properties of these shape warps and illustrate the results by recovering the shapes of faces.

1 Introduction

Recent work on object recognition [6] uses two-dimensional warping of intensity images to allow for the changing three-dimensional geometry of the objects. For example, the image of a viewed object will change with the angle of view. For small changes of angle, or small deformations of shape, the change can be modeled as a spatial warp of the intensity image. The approach makes the implicit assumption that we can model geometrical changes in three-dimensions by warps in the two-dimensional image plane. When is this assumption valid?

In this paper we show that, for a large class of reflectance functions, there is a direct relationship between image warps and geometrical changes of the underlying three-dimensional objects. We also demonstrate that not all warps are physically reasonable and determine constraints that physical warps must satisfy.

Our analysis explains the hidden assumptions used by Hallinan and other workers on object recognition [6]. It allows us to understand the relationship to other object recognition theories based on three-dimensional geometry.

In addition, it offers a novel approach to shape from shading [7], [8], [11]. We call our method *shape from warping*. By contrast to standard techniques, our approach works by assuming prototype models of shape and by estimating the warp between the input image and the image of the prototype. From these image warps we show how to deduce the three-dimensional shape. This method allows us to introduce object specific knowledge into the shape estimation. We will prove that this approach works without needing to know the precise reflectance function of the object.

Our approach has some similarities to recent work [1], [3], which shows that shape can be recovered from an intensity image provided prior assumptions are made about the shape class. The techniques used in this paper, however, are very different.

This paper is organized as follows. Section (2) demonstrates the basic relationship between two-dimensional image warping and three-dimensional geometrical variations. In Section (3) we describe how the surface integrability condition put constraints on the class of warps by requiring that they generate consistent surfaces. In Section (4) we discuss the underlying assumptions of object recognition theories which involve two-dimensional image warping. Section (5) illustrates shape from warping be recovering the shape of faces.

2 Relating Image and Geometrical Warps

Suppose we have a surface, or object, of the form $z(\mathbf{x}) = f(\mathbf{x};\ )$. (More precisely, we consider a *Monge patch*, which is assumed to be smooth, C^∞, or at least C^2.) This surface corresponds to an intensity image $I(\mathbf{x})$. The image is related to the surface by a reflectance function, see [7]. For example, it is common to assume that:

$$I(\mathbf{x}) = R(\mathbf{n}(\mathbf{x}), \mathbf{k}, \mathbf{s}), \tag{1}$$

where R is the reflectance function, $\mathbf{n}(\mathbf{x})$ are the surface normals, $\mathbf{k}$ is the viewer direction, and $\mathbf{s}$ is the light source direction. (This is the most general class of reflectance function in common use and includes, for example, the Phong model as a special case provided the albedo is constant).

Our key result is that *warping the image corresponds to warping the surface normals of the underlying surface*. More precisely, suppose we apply a warp $\phi(.)$ to the image by the mapping:

$$\phi : \mathbf{x} \to \phi(\mathbf{x}), \tag{2}$$

then this induces a warp:

$$I(\mathbf{x}) \mapsto I(\phi(\mathbf{x})), \tag{3}$$

to the image and, by equation (1), a warp:

$$\mathbf{n}(\mathbf{x}) \mapsto \mathbf{n}(\phi(\mathbf{x})), \tag{4}$$

to the *surface normals* of the surface.

This result can be used in two ways. Firstly, it can used to explain the hidden geometrical assumptions of theories of object recognition which make use of two dimensional image warps [6], see section (4) for more details. Secondly, the result can be exploited to develop a novel approach to shape from shading which we call *shape from warping.* If the warping assumption is valid then we can use the image warps to recover three-dimensional shape assuming that we have a known three-dimensional prototype with surface normals $\{\mathbf{n}_0(\mathbf{x})\}$ and a corresponding image $I_0(\mathbf{x})$. For a given image $I(\mathbf{x})$ we find the warp $\phi(\mathbf{x})$ such that $I(\mathbf{x}) = I_0(\phi(\mathbf{x}))$ and hence determine the shape to be $\mathbf{n}(\mathbf{x}) = \mathbf{n}_0(\phi(\mathbf{x}))$.

It should be emphasized that the approach assumes that the reflectance function is of form $I(\mathbf{x}; \alpha) = R(\mathbf{n}(\mathbf{x}; \alpha), \mathbf{k}, \mathbf{s})$. This will not be true if the object has significant albedo changes but may be a reasonable approximation for many objects. (We note that almost all shape from shading algorithms assume constant albedo). This seems to be true for our experiments on faces, see section (5).

But what warps are allowable? From equation (4) we see that they warp the surface normals of a prototype shape, and not all such warps will form consistent surfaces. This is investigated in the next section.

3 Normal Consistency

The preceeding section showed that image warps often corresponded to warps of surface normals. In what situations will this give a consistent surface? Warps transform a given set of surface normals $\mathcal{N}_t$ into a different set say $\mathcal{N}$, but this does not guarantee the existence of a consistent surface with these normals. To ensure the existence of a such a surface a further condition must be satisfied, that is referred to as the integrability condition [12]. Let us derive the *surface integrability condition* (see, for example, [8]).

Suppose we have a surface $z = f(x, y)$. Then its surface normals are defined by:

$$\mathbf{n}(x, y) = \frac{1}{\{1 + \nabla f \cdot \nabla f\}^{1/2}}(-f_x, -f_y, 1), \qquad (5)$$

where $\nabla f = (f_x, f_y)$. We can write $\mathbf{n} = (n_1, n_2, n_3)$ and observe that:

$$\frac{n_1}{n_3} = -f_x, \quad \frac{n_2}{n_3} = -f_y. \qquad (6)$$

Hence we derive the *surface integrability condition*:

$$\frac{\partial}{\partial y}\left(\frac{n_1}{n_3}\right) = \frac{\partial}{\partial x}\left(\frac{n_2}{n_3}\right). \qquad (7)$$

We have shown that this is a necessary condition for the surface to be consistent. To see that it is sufficient we observe that, by elementary vector calculus [4], equation (7) implies that there exists a function $\psi(x, y)$ such that:

$$\left(\frac{n_1}{n_3}\right) = \frac{\partial \psi}{\partial x}, \quad \left(\frac{n_2}{n_3}\right) = \frac{\partial \psi}{\partial y}. \qquad (8)$$

We can solve equation (8) for n_1, n_2, n_3, using the normalization condition $n_1^2 + n_2^2 + n_3^2 = 1$, and obtain the surface expression (5) after identifying ψ with $-f$. Thus we see that equation (7) is also a sufficient condition.

We must now tackle the harder task of putting consistency conditions on the warp so that the warped normals are consistent. Suppose $\mathbf{n}(\mathbf{x})$ are the surface normals of the prototype surface, and hence obey the surface integrability condition. Let us apply a warp $\phi(\mathbf{x}) = (\phi_1(\mathbf{x}), \phi_2(\mathbf{x}))$ to the prototype surface. The resulting surface $\mathbf{n}(\phi(\mathbf{x}))$ is consistent provided:

$$\frac{\partial}{\partial y}\left(\frac{n_1(\phi(\mathbf{x}))}{n_3(\phi(\mathbf{x}))}\right) = \frac{\partial}{\partial x}\left(\frac{n_2(\phi(\mathbf{x}))}{n_3(\phi(\mathbf{x}))}\right). \tag{9}$$

By equation (6) we have:

$$\frac{n_1(\phi(\mathbf{x}))}{n_3(\phi(\mathbf{x}))} = -\frac{\partial f}{\partial x}_{(\phi(\mathbf{x}))},$$
$$\frac{n_2(\phi(\mathbf{x}))}{n_3(\phi(\mathbf{x}))} = -\frac{\partial f}{\partial y}_{(\phi(\mathbf{x}))}, \tag{10}$$

where the derivatives on the right hand sides are evaluated at $\phi(\mathbf{x})$ (a example is given below for readers unfamiliar with this notation). We observe that

$$\frac{\partial}{\partial y}\left(\frac{\partial f}{\partial x}_{(\phi(\mathbf{x}))}\right) = \frac{\partial^2 f}{\partial x^2}_{(\phi(\mathbf{x}))}\frac{\partial \phi_1}{\partial y} + \frac{\partial^2 f}{\partial x \partial y}_{(\phi(\mathbf{x}))}\frac{\partial \phi_2}{\partial y}. \tag{11}$$

A similar result holds if we replace x by y.

Substituting from (10) into (9) and using (11) yields the result:

$$\phi_{1,y}\{\frac{\partial^2 f}{\partial x^2}\}_{\phi(\mathbf{x})} + \phi_{2,y}\{\frac{\partial^2 f}{\partial x \partial y}\}_{\phi(\mathbf{x})} =$$
$$\phi_{1,x}\{\frac{\partial^2 f}{\partial x \partial y}\}_{\phi(\mathbf{x})} + \phi_{2,x}\{\frac{\partial^2 f}{\partial y^2}\}_{\phi(\mathbf{x})}. \tag{12}$$

This gives a relationship between the Hessian of the surface (terms $\partial^2 f/\partial x^2$, etc.) and the *stress tensor* of the warps which is defined to have components $\phi_{1,x}, \phi_{1,y}, \phi_{2,x}, \phi_{2,y}$.

We can get some clarification of the notation used in this derivation by considering the one-dimensional case where the surface $f(x)$ can be expressed as a power series:

$$f(x) = \sum_{r=0}^{N} a_r x^r, \tag{13}$$

and warping it by $\phi(x)$ gives:

$$f(\phi(x)) = \sum_{r=0}^{N} a_r \{\phi(x)\}^r. \tag{14}$$

In this case we find that:

$$\frac{\partial f}{\partial x}_{\phi(x)} = \sum_{r=0}^{N} a_r r \{\phi(x)\}^{r-1} = \frac{\partial}{\partial \phi(x)} f(\phi(x)). \tag{15}$$

Some intuition can be obtained by considering surfaces of revolution. These surfaces are essentially one-dimensional and it can be shown [13] that warps which preserve the revolution property will satisfy this equation.

4 Underlying Geometric Assumptions of Image Warping

The Hallinan model for face recognition [6] assumes that the image of a given face can be written as[4]:

$$I(\mathbf{x}) = \sum_{i=1}^{5} \alpha_i B_i(\phi(\mathbf{x})), \qquad (16)$$

where the $\{B_i(.)\}$ are lighting basis functions [5] – which model the appearance of the face under different lighting conditions – and $\phi(.)$ is a geometric warp which is required to be monotonic. The $\{\alpha_i\}$ are coefficients corresponding to the specific lighting conditions.

The Hallinan model therefore assumes that any image of a face can be obtained by monotonically warping a prototype face model. Now the albedo is approximately constant over most of the face and, in particular, it is constant over those regions of the face where most of the shape changes occur – the cheeks, forehead, and nose. Thus we argue that faces approximately satisfy the assumptions of shape from warping.

Therefore the image warps of the Hallinan model should correspond directly to warps of the surface normals of faces[5]. If so, the monotonicity assumption of Hallinan's warps can be shown to imply that no surface extrema can be created in the images objects [13]. In other words, the model given by equation (16) will break down if some of the objects have a different number of surface extrema than the prototype. This will not happen for faces but it will be a problem if the model is applied to other object classes.

It is interest to constrast Hallinan's face model with the one recently proposed by Atick et al. [1]. This model represents the face in terms of a three-dimensional model which is generated by doing principal component analysis of a set of three-dimensional range data of faces. A face can therefore be represented by its coefficients of expansion in a basis of *eigenheads*. By contrast, Hallinan has computed a class of two-dimensional eigenwarps. Our paper therefore suggests that Hallinan's image eigenwarps correspond to eigenwarps of the surface normals and hence are closely related to Atick's eigenheads.

5 Examples of Shape from Warping

We illustrate our theory on real images of faces obtained from [6]. Our prototype model consists of a face image $I_p(\mathbf{x})$ and its corresponding surface shape $z = f_p(\mathbf{x})$ obtained from laser-range data. From this surface we compute the surface normals $\mathbf{n}_p(\mathbf{x})$.

For given input images $\{I_\alpha(\mathbf{x})\}$ we compute the spatial warps $\phi_\alpha(\mathbf{x})$ relative to the prototype image by using the algorithm, and code, described in [6], see

---

[4] The full model also includes a global affine warp which we will neglect for simplicity.
[5] This will be verified by the experiments described in the following section.

figure (1). These warps are calculated to minimize a matching energy function:

$$E[\phi] = \frac{1}{|D|} \int_D \psi\{I(\phi(\mathbf{x})) - I_p(\mathbf{x})\} \, d\mathbf{x}$$
$$+ \int_D (tr J(\mathbf{x})^T J(\mathbf{x})\{1 + \frac{1}{\det J(\mathbf{x})^2}\} - 4) \, d\mathbf{x}, \qquad (17)$$

where J is the Jacobian matrix of the warp field $\phi(\mathbf{x})$ and ψ is a robust norm [9].

More sophisticated versions of the algorithm in [6] will obtain the spatial warps even if the lighting conditions are unknown. But we will not deal with this case in this paper.

The code described in [6] does not put any restrictions on the class of warps other than requiring monotonicity. It would be interesting to impose the consistency condition, see equation (12), while calculating the warps but this is difficult because the warps appear in the arguments of the surface derivative terms.

Instead we use the algorithm to compute the image warps $\phi(\mathbf{x})$, calculate the warps normal fields $\mathbf{n}(\mathbf{x}) = \mathbf{n}_p(\phi(\mathbf{x}))$ and then determine the closest consistent surface by minimizing the cost function:

$$E[z] = \int \{(\frac{\partial z}{\partial x} - \frac{n_1(\mathbf{x})}{n_3(\mathbf{x})})^2 + (\frac{\partial z}{\partial y} - \frac{n_2(\mathbf{x})}{n_3(\mathbf{x})})^2 \, d\mathbf{x}. \qquad (18)$$

The resulting warps and reconstructed faces are shown in figures (1, 2). The warps are generally well calculated by Hallinan's algorithm though close inspection, see [13], shows some errors, particularly on the right boundary of object number 3. These errors become apparent in the reconstructions, see figure (2), and we see that errors arise on the edge of object 3.

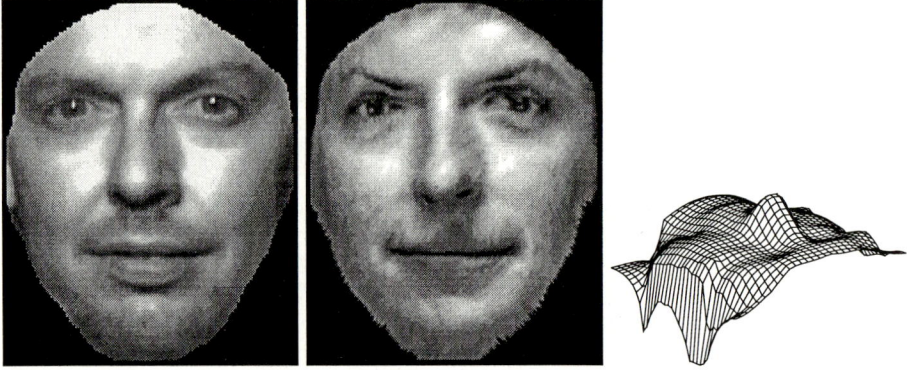

Fig. 1. Applying our algorithm to the first object. Left the intensity of the first object, center warping the prototype to match the object, right the estimated shape of the first object.

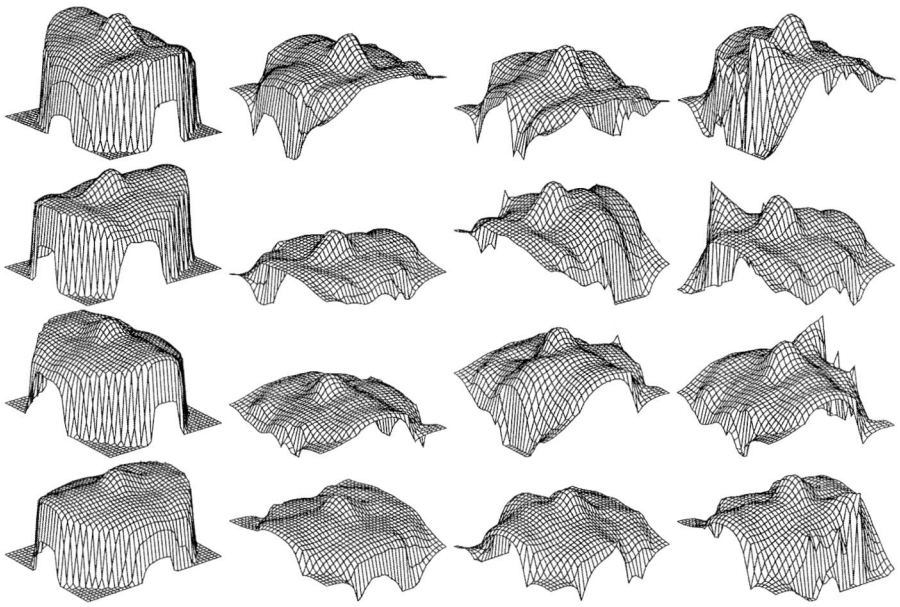

Fig. 2. Showing the estimated face for the objects, from different viewpoints, and comparing them to the prototype shape. The four rows correspond to different viewpoints. The first column shows the prototype shape and the next three columns show the three objects. Observe the errors which sometimes occur at the object boundaries. These are side effects from the warping algorithm.

It seems, therefore, that the errors in reconstruction are due to partial failures in the warping algorithm. Thus, since these are nontrivial images, we regard this as proof of concept for shape from warping.

6 Summary

This paper has demonstrated a fundamental relationship between two-dimensional image warps and three-dimensional geometrical transformations.

Our analysis shows the underlying geometrical assumptions about three-dimensional shape used in models of object recognition which rely on dense two-dimensional warps computed by matching image intensity [6].

We also described *shape from warping*, a variant of shape from shading, which allows shape to be determined for a class of surfaces without needing to know the exact reflectance function. We investigated the class of warps that are consistent and what types of surfaces could be reconstructed in this way. The theory was illustrated by applying it to recovering the shape of faces.

Acknowledgements

We would like to thank Peter Belhumeur, Gaile Gordon, and particularly Peter Hallinan for access to their computer code. Support was provided by NSF Grant IRI 93-17670 and ARPA/ONR Contract N00014-95-1-1022.

References

1. J.J. Atick, P.A. Griffin, and A.N. Redlich. "Statistical Approach to Shape from Shading: Reconstruction of 3D Face Surfaces from Single 2D Images". Preprint. The Computational Neuroscience Laboratory. The Rockefeller University. New York, NY. 1995.
2. R. Epstein, P.W. Hallinan and A.L. Yuille. "5 ± Eigenimages Suffice: An Empirical Investigation of Low-Dimensional Lighting Models". In *Proceedings of IEEE WORKSHOP ON PHYSICS-BASED MODELING IN COMPUTER VISION.* 1995.
3. R. Epstein, A.L. Yuille, and P.N. Belhumeur. "Learning object representations from lighting variations".In **Object Representation in Computer Vision II**. Eds. J. Ponce, A. Zisserman, and M. Hebert. Springer Lecture Notes in Computer Science 1144. 1996.
4. M.D. Greenberg. *Foundations of Applied Mathematics.* Prentice-Hall Inc. Englewood Cliffs, NJ. 1978.
5. P.W. Hallinan. "A low-dimensional lighting representation of human faces for arbitrary lighting conditions". In. *Proc. IEEE Conf. on Comp. Vision and Patt. Recog.*, pp 995-999. 1994.
6. P.W. Hallinan. *A Deformable Model for Face Recognition under Arbitrary Lighting Conditions.* PhD Thesis. Division of Applied Sciences. Harvard University. 1995.
7. B.K.P. Horn. **Computer Vision**. MIT Press, Cambridge, Mass. 1986.
8. B.K.P. Horn and M. J. Brooks, Eds. **Shape from Shading**. Cambridge MA, MIT Press, 1989.
9. P.J. Huber. *Robust Statistics.* John Wiley and Sons. New York. 1981.
10. S.K. Nayar, K. Ikeuchi and T. Kanade "Surface reflections: physical and geometric perspectives" *IEEE trans. on Pattern Analysis and Machine Intelligence*, vol 13 p611-634. 1991.
11. J. Oliensis. "Shape from shading as a partially well-constrained problem". *Computer Vision, Graphics, and Image Processing: Image Understanding.* 54, pp 163-183. 1991.
12. C. Von Westenholz. **Differential forms in Mathematical Physics.** North-Holland publishing company. Amsterdam. 1981.
13. A.L. Yuille, M. Ferraro, and T. Zhang. "Surface Shape from Warping and Object Recognition". Smith-Kettlewell Eye Research Institute Preprint. 1996.

Using Top-Down and Bottom-Up Analysis for a Multi-Scale Skeleton Hierarchy

Gunilla Borgefors[1], Giuliana Ramella[2], Gabriella Sanniti di Baja[2]

[1]Centre for Image Analysis, Swedish University of Agricultural Sciences, Uppsala, Sweden
Fax: +46 18 553447; email: gunilla@cb.uu.se
[2]Istituto di Cibernetica, Italian National Research Council, Arco Felice, Naples, Italy
Fax: + 39 81 5267654; email: gr, gsdb@imagm.na.cnr.it

Abstract Multi-scale skeletons can be conveniently employed in the matching phase of a recognition task. The multi-scale skeletons are here obtained by first computing the skeleton at all levels of a resolution structure and then establishing a hierarchy among skeleton components at different scales, using a parent-child relationship. Although subsets of the skeleton expected to represent given pattern subsets may consist of different number of components at different scales, a component preserving decomposition is obtained that produces a hierarchy in accordance with human intuition.

1 Introduction

Pattern recognition can be based on pattern decomposition and description. In this respect, the skeleton is a convenient tool to facilitate the matching phase of a recognition process, in the case of patterns whose shape can be perceived as the superposition of elongated regions [1,2]. The skeleton is a linear subset of the pattern, centred within the pattern, and is characterised by the same topological and geometrical structure. Skeleton branches are in correspondence with the elongated regions constituting the pattern. Thus, the spatial relationships among pattern subsets can be easily derived, e.g., while tracing the corresponding skeleton branches. This would not happen if the morphological skeleton , e.g., [3], is used, since it does not generally reflect the topological properties of the pattern.

If skeleton components are hierarchically ranked, a better pattern description becomes available and recognition is facilitated [4]. Moreover, the complexity of the matching phase can be reduced by using multi-scale skeletons, e.g., [5,6]; in fact, one can initially match only lower scale skeletons, which represent the most significant pattern subsets, and thus reduce the number of comparisons among higher scale skeletons, necessary to achieve an exact match.

In this paper, we use the multi-scale skeletons obtained by simultaneously extracting the skeleton at all levels of a resolution pyramid; moreover, a hierarchical skeleton decomposition is obtained at all resolution levels by identifying and ranking skeleton subsets, based on their permanence in the skeleton at the various scales.

A first step towards a hierarchical decomposition of the multi-scale skeletons has been taken recently [7]. Analogously to [7], the resolution structure we use here to obtain the multi-scale skeletons is the AND-pyramid. The AND-pyramid is easy to implement but, as the resolution decreases, the pattern is shrunk and narrow regions of the initial pattern may either completely vanish or become disconnected (see Figure 1, where the AND-pyramid of a test pattern is shown). Since skeletonization is a

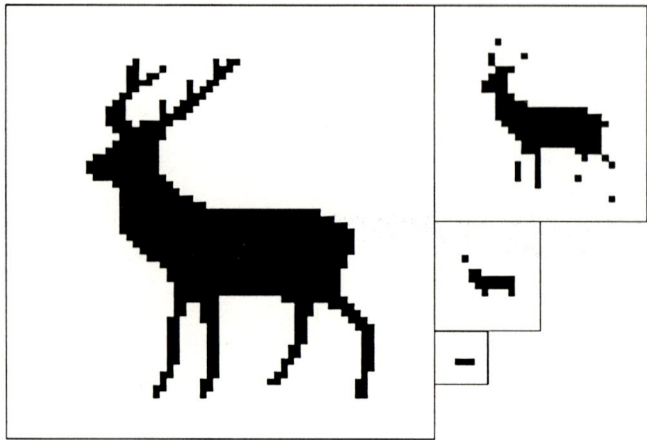

Figure 1. The four significant levels of the AND-pyramid of a test $2^6 \times 2^6$ image.

topology preserving process, the multi-scale skeletons computed on the AND-pyramid will have the same fate. Thus, subsets of the skeleton expected to represent given pattern subsets may consist of different number of components at different scales. We here address and solve the problem of identifying a set of non-connected skeleton fragments as belonging to the same component by inferring connectedness information from the higher scale skeletons onto the lower scale ones during a bottom-up analysis of the pyramid. In this way we can establish a correspondence among skeleton subsets at various scales, even when the subsets do not consist of the same number of connected components. Another novel feature of the current skeleton decomposition is that small noise components are absorbed by adjacent more significant components. The identification of disjunct components and the removal of noise components both contribute to reduce the number of components in the skeleton decomposition to the most significant ones and, accordingly, simplify the skeleton hierarchy.

2 Skeleton Hierarchy

Let P be a $2^n \times 2^n$ binary picture, where black and white pixels respectively constitute the pattern and its complement. We assume that all the pixels on the border of P are white and store P in the highest resolution level (also called the first level, or the bottom level) of an AND-pyramid. The next, lower, resolution level of the pyramid is built from the first level. All pixels with four black *children* in the first level are set to black. Similarly, the third level is built from the second, and so on until all (n+1) resolution levels are obtained. Indeed, resolution levels sized $2^2 \times 2^2$, $2^1 \times 2^1$ and $2^0 \times 2^0$ are not meaningful for skeletonization purposes and we will consider the $2^3 \times 2^3$ pixel image as the last pyramid level (also called the top level). The AND-pyramid is easy to compute, but it is *not* shape preserving and, when the resolution decreases, some of the subsets present in the initial pattern either completely vanish or become disconnected, [8]. Thick regions appear at all resolution levels and constitute the most significant pattern components.

Skeletonization is accomplished simultaneously at all resolution levels. Any skeletonization algorithm can be used. We favour algorithms requiring two distinct phases, respectively tailored to the identification of an at most two-pixel wide set of skeletal pixels, and to the reduction of this set to unit width, see, e.g., [9]. Using this approach, we can postpone the second skeletonization phase, till after the desired hierarchy has been built at all levels of the pyramid. This is preferable to guarantee that the parent-child relationship among skeleton components at different scales can be correctly established. Even though we are aware that the term skeleton should be used only for the set resulting after both skeletonization phases have been accomplished, to avoid lengthy periphrases in the following we will refer also to the original sets of skeletal pixels as the skeletons.

2.1 Top-Down Approach

The main idea that in [7] guided the construction of the skeleton hierarchy by means of a top-down process was the observation that skeleton components present at lower resolution levels were definitely also present at higher levels. The process used in that work can be summarised as follows. The connected components of the skeleton are identified at the lowest resolution level ($2^3 \times 2^3$ level). Each component is *parent* of a *child* component at the next level ($2^4 \times 2^4$), *grandparent* of a *grandchild* at level $2^5 \times 2^5$, and so on; a suitable process allows one to establish the parent-child relationship and to identify all the descendants. As the structure of the skeleton is generally more and more complex as soon as the resolution increases, not all skeletal pixels at the successive levels are assigned to descendants of components in the top level. Pixels not assigned to child components at level $2^4 \times 2^4$ are grouped into connected components that are interpreted as new parent components, directly originating at level $2^4 \times 2^4$. Their children and grandchildren can be found in the higher levels. Similarly, new parent components and their descendants can be found at all levels. During the process, skeleton components at each single pyramid level are assigned a *permanence number*, counting the number of levels up to the most remote corresponding ancestor component. At each level, the most significant components are those with the largest permanence. The maximal permanence is equal to the number of levels of the resolution pyramid.

Three sub-processes are done to establish the parent-child relationship. In fact, due to the discrete nature of the AND-pyramid and the skeleton, a child component might have some of its pixels slightly shifted compared to their "expected" positions. The black children of the pixels in a parent component do not exhaust the black pixels expected to constitute the child component. The first sub-process is termed *projection*. Every skeletal pixel p in a (parent) skeleton component projects, at the immediately higher resolution level, over a 2×2 set *(quadruplet)*, generally including both white pixels and skeletal (black) pixels; skeletal pixels in the quadruplet Q associated to p are assigned to the corresponding child component. The second sub-process is termed *expansion* and is active whenever Q includes only white pixels. Expansion interprets as belonging to the current child component the skeletal pixels possibly found in the four quadruplets placed North, East, South and West of Q, provided that these pixels have not already been marked as belonging to any other child component. Either projection or expansion is accomplished. Both processes involve pixels placed in a pair of successive levels. The third sub-process, termed *propagation*, is accomplished after projection (or expansion) has been performed from all pixels at a given level onto the successive higher resolution level. Propagation involves only pixels on the latter level. For every pixel p ascribed to a child component by projection or expansion, propagation assigns also the black neighbours of p to the same component, provided that they have not already been assigned to other components.

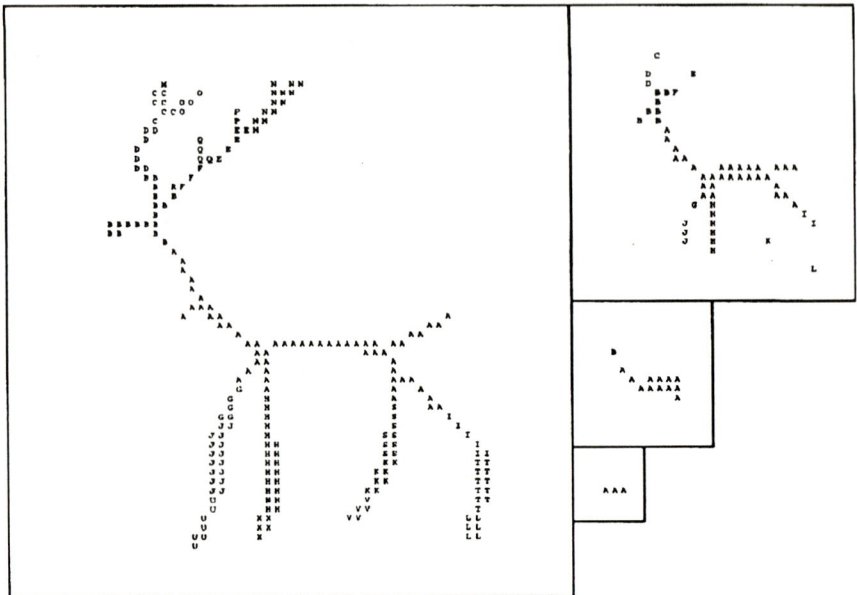

Figure 2. Skeleton hierarchy obtained by using the "old" top-down process [7].

At any pyramid level, the skeleton components with the highest permanence are processed first to identify the corresponding child components. Components with smaller and smaller permanence are then processed and assign to their child components only pixels that have not yet been included in other child components with larger permanence. The criteria used to establish the parent-child correspondence favours skeleton components that are more stable, that is components that are present at many pyramid levels.

In Figure 2, the hierarchy produced on the skeleton of the test pattern by the top-down process introduced in [7] is shown. The same letter is used to denote skeletal pixels belonging to components enjoying the parent-child relationship. At each level, letter A denotes the skeleton component with the highest permanence, letters B, C-L and M-W are components with smaller and smaller permanence. We note that a number of components larger than the intuitively expected one characterises the hierarchy. Skeletal pixels that are grouped into a unique connected component onto a given level, correspond to pixels constituting distinct components onto the immediately smaller resolution level, due to unavoidable pattern disconnections occurred while building the AND-pyramid. Since the hierarchy is built starting from the lower resolution levels where disconnections are likely to occur, at higher levels subsets of the skeleton representing regions of the pattern perceived as a whole are segmented into a number of components with different permanence that, hence, are assigned to different hierarchy levels (see, for example, the skeleton subsets in correspondence with the legs of the deer in Figure 2). The top-down process used to establish the parent-child relationship could not recover information on connectedness, lost when storing the pattern in the AND-pyramid.

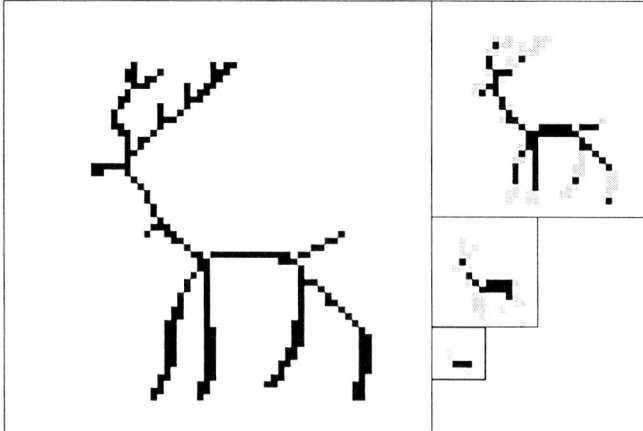

Figure 3. Gray squares denote the extra pixels added to the skeletal pixels (black squares) by the bottom-up OR-projection process.

2.2 Bottom-Up Component Restoring Process

To recover the information on how skeleton fragments should be grouped into components, the pyramid has to be checked bottom-up, so that skeleton subsets that are connected at a given level can transfer connectedness information onto the successive smaller resolution levels. The OR logic operation is used for this purpose.

An *OR-projection* process is accomplished simultaneously for all pairs of successive levels, $2^k \times 2^k$ and $2^{k-1} \times 2^{k-1}$, $3 < k \leq n$. Any level $2^k \times 2^k$ is partitioned into 2×2 blocks of pixels (quadruplets) and, for every quadruplet with at least a skeletal pixel, its parent pixel on level $2^{k-1} \times 2^{k-1}$ is changed to black, (if it was white).

As an effect of the OR-projection, a number of extra pixels are added to the skeletal pixels found during skeletonization (see Figure 3). The extra pixels modify both the number of connected components and the structure of the set of the skeletal pixels on level $2^{k-1} \times 2^{k-1}$, making it resemble more closely the set of skeletal pixel on level $2^k \times 2^k$. We distinguish two types of extra pixels. Extra pixels of type 1 link skeletal pixels on level $2^{k-1} \times 2^{k-1}$, that would otherwise been grouped into a number of distinct components. Extra pixels of type 1 should be kept if we like a more intuitive skeleton decomposition, that is able to identify a subset of the skeleton as a unit even when, due to resolution problems, the represented region appears as disconnected at that level of the AND-pyramid. Extra pixels of type 2 correspond to regions of the pattern that are totally absent at level $2^{k-1} \times 2^{k-1}$. Extra pixels of type 2 should be removed. To this aim, topology preserving removal operations can be repeatedly applied to the extra pixels. For any of them, the connectivity number C_8, as defined in [10], is used to count the number of components of black pixels (skeletal and extra pixels) in its neighbourhood. Extra pixels having $C_8 \leq 1$ are sequentially removed. These pixels are, in fact, not necessary for connectedness maintenance. Removal is iterated as far as removable extra pixels are found.

The above sketched removal process removes all type 2 extra pixels, but also reduces to unit width the set of extra pixels of type 1. This might prevent the identification of all pixels expected to constitute the child components, when

Figure 4. Gray squares denote the extra pixels restoring skeleton connectedness.

establishing the parent-child relationship. Indeed, the process sketched in Section 2.1 uses the two-pixel wide set of skeletal pixels, rather than the unit wide skeletons, to build the hierarchy. Thus, the set of type 1 extra pixels should not be reduced to unit width. We proceed as follows. Rather than actually removing the extra pixels for which it is $C_8 \leq 1$, we *mark* them as pixels candidates for removal. (Note that, when computing C_8, marked pixels in the neighbourhood of any extra pixel are interpreted as if they were already white pixels.) Marking is iterated as long as extra pixels that can be marked are found. Then, the marker is simultaneously removed from all pixels having a horizontal/vertical neighbour that is a non-marked extra pixel. Only pixels that are still marked after this process are type 2 pixels and are finally removed. In Figure 4, the effect of the removal process is shown.

The process described in Section 2.1 is then applied to build the hierarchy on the skeleton modified to restore connectedness. The result are shown in Figure 5, where the extra pixels have all been removed to facilitate the comparison of the performance of the new algorithm with that of [7], shown in Figure 2. In the bottom level, letters A, B, C-H and I-Q denote components with permanence 4, 3, 2 and 1, respectively. The number of components has significantly been reduced with respect to Figure 2.

A few noise components, defined as components consisting of single pixels, still affect the hierarchy. They can be found at the periphery of other components (e.g., at level $2^6 \times 2^6$ the tips of the horns of the deer labelled I or M, both adjacent to components with permanence 2) or in between components whose parents were indeed adjacent at the immediately lower resolution level (e.g., at level $2^6 \times 2^6$ pixel labelled N in between component B, with permanence 3, and D, with permanence 2, in the horns of the deer). Noise components can be originated at any level and are all characterised by permanence equal to 1. They should be absorbed by adjacent components to simplify the structure of the resulting decomposition. This process is performed at each level of the pyramid, before identifying the connected components directly originating at that level, i.e., the components with permanence equal to 1. Each skeletal pixel not already assigned to any component that has *all* its skeletal neighbours already assigned to some component is assigned to the neighbouring component having the highest permanence. Finally, reduction to unit width is performed on all pyramid levels by means of an iterative thinning, based on topology

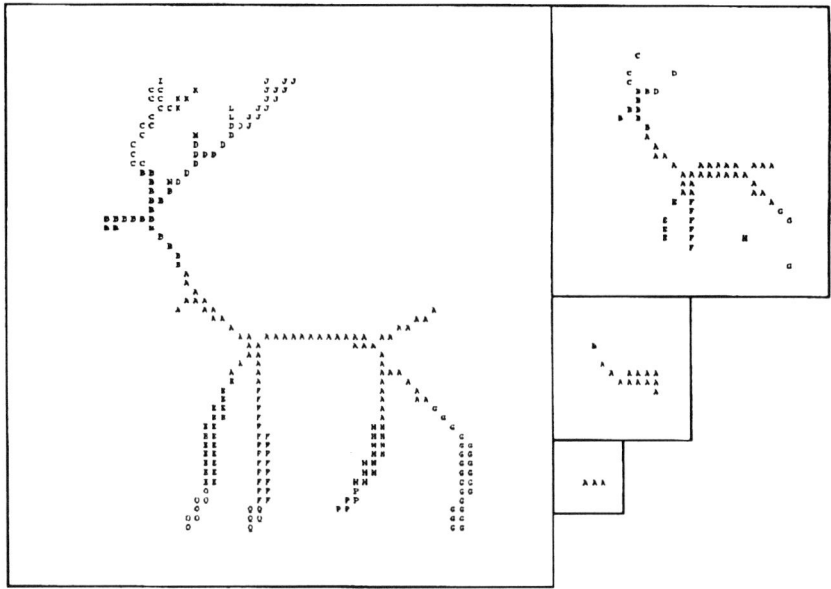

Figure 5. Skeleton hierarchy obtained by applying the top-down decomposition after using a bottom-up connectedness restoring process.

preserving removal operations. At each iteration, removal is active only on pixels having a given permanence, starting from the lowest permanence. In this way, removal of pixels belonging to components with lower permanence (and hence having smaller significance) is favoured. The number of iterations required is equal to the maximum permanence in the hierarchy, i.e., is equal to the number (n-2) of pyramid levels.

The resulting hierarchy for the unit wide skeleton can be seen in Figure 6, where again the extra pixels, that are only used to establish correspondences between components, are not shown. Letters A, B, C-H and I-N denote components with decreasing permanence (4, 3, 2 and 1 in the bottom level, respectively). A smaller number of components, all significant, has been obtained and the hierarchical decomposition is more in accordance with human intuition.

3 Conclusion

A method to hierarchically rank components of multi-scale skeletons computed at all resolution levels of an AND-pyramid has been presented. The proposed method uses both top-down and bottom-up processes to identify skeleton components. Subsets of the skeleton expected to represent given pattern subsets are taken as a whole even if they consist of different number of connected sets of skeleton pixels at different scales. This is achieved by inferring connectedness information from the higher resolution scales, using a bottom-up analysis of the pyramid, in addition to the more intuitive top-down process used to build the hierarchy. Moreover, small noise components are absorbed by adjacent more significant components. The decomposition of the skeleton obtained in this way is in accordance with human intuition.

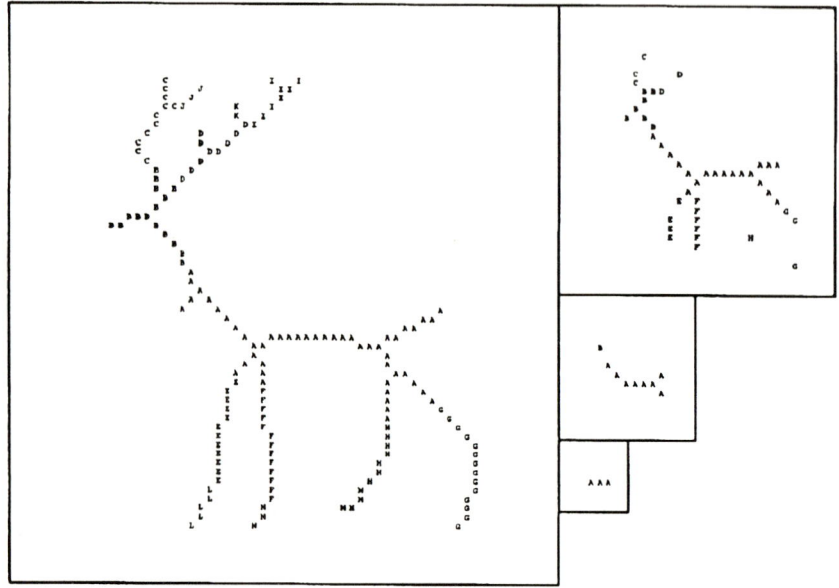

Figure 6.   The resulting skeleton hierarchy.

References
1. H.Blum, R.N.Nagel, "Shape description using weighted symmetric axis features", *Pattern Recognition*, **10**, 167-180 (1978).
2. G. Sanniti di Baja, E. Thiel, "(3,4)-weighted skeleton decomposition for pattern representation and description", *Pattern Recognition*, **27**, 1039-1049 (1994).
3. A.Toet, "A hierarchical morphological image decomposition", *Pattern Recognition Letters*, **11**, 267-274 (1990).
4. H. Rom, G.Medioni, "Hierarchical decomposition and axial shape description", *IEEE Trans. on PAMI*, **15**, 973-981 (1993).
5. A.R. Dill, M.D. Levine, P.B. Noble, "Multiple resolution skeletons," *IEEE Trans. on PAMI* **9**, 495-504 (1987).
6. S.M.Pizer, W.R.Oliver, S.H.Bloomberg, "Hierarchical shape description via the multiresolution symmetric axis transform," *IEEE Trans. on PAMI* **9**, 505-511 (1987).
7. G.Borgefors, G.Ramella, G.Sanniti di Baja, "Multi-scale skeletons from binary pyramids", *Proc. 3rd Int.Workshop on Visual Form*, May 97, Capri, Italy.
8. G.Borgefors, G.Ramella, G.Sanniti di Baja, "Multiresolution representation of shape in binary images", in *Discrete Geometry for Computer Imagery*, S.Miguet, A.Montanvert and S.Ubéda (Eds.), LNCS **1176**, Springer-Verlag, 51-58 (1996).
9. C.Arcelli, G.Sanniti di Baja, "A one-pass two-operations process to detect the skeletal pixels on the 4-distance transform," *IEEE Trans. on PAMI* **11**, 411-414, 1989.
10. S.Yokoi, J.I.Toriwaki, T.Fukumura, "An analysis of topological properties of digitized binary pictures using local features", *CGIP*, **4**, 63-73 (1975).

A New Algorithm for 3D Profilometry Based on Phase Measurement

Luigi Di Stefano[1] and Frank Boland[2]

[1] DEIS, University of Bologna
Viale Risorgimento 2, 40136 Bologna, Italy
[2] EEE, University of Dublin
Trinity College, Dublin 2, Ireland

Abstract. This paper describes a new phase extraction algorithm for phase profilometry. The algorithm uses a square wave to demodulate phase and moving averages and comb-shaped filters to extract the phase information from low-frequency. The proposed algorithm is compared with the two major profilometry techniques, namely Fourier domain profilometry and signal domain profilometry based on FIR low-pass filtering.

1 Introduction

Automated, non-contact 3D optical profilometry gathers ever-increasing attention in industrial applications such as on-line inspection, quality control, gauging of manufactured parts [1]. In such a context active methods based on the projection of structured light are very popular and commercially successful [2]. In this paper we address structured light techniques based on the projection of a 2D regular grating and phase measurement, i.e phase profilometry techniques, which are particularly attractive among active methods due to the fast acquisition time, very simple optical arrangement and low cost.

In phase profilometry a periodic structured light pattern is projected onto the object and viewed by a camera from an angularly offset position. The typical pattern is a sequence of equally spaced horizontal or vertical dark lines, called "fringes", which are generated projecting a square wave grating. Since the imaged pattern is phase-modulated according to the topography of the object, the extraction of phase information from the image allows reconstruction of the 3D shape. Of the two major phase extraction techniques, one, called Fourier Transform Profilometry (FTP), relies on processing the image in the frequency domain while the other is based on demodulation and convolution operations executed in the real signal domain (we will refer to this technique as Signal Domain Profilometry, SDP). In this paper we present a new signal-domain phase extraction technique which uses a fast demodulation scheme and relies entirely on very simple operators such as moving averages and comb filters. We also compare this new technique with FTP and SDP, addressing adaptiveness to patterns of different frequencies, ability to deal with surfaces of nonuniform reflectivity and computational complexity.

2 Phase Profilometry Techniques

It can be shown that, once a reference plane is selected, the image of the pattern used in phase profilometry can be expressed as

$$g(x,y) = r(x,y) \sum_{n=-\infty}^{+\infty} A_n e^{j(2\pi n f_0 x + n\phi(x,y))} \qquad (1)$$

while the reference image, i.e. the image obtained projecting the pattern on the reference plane, is given by

$$g_0(x,y) = \sum_{n=-\infty}^{+\infty} A_n e^{j(2\pi n f_0 x + n\phi_0(x))} \qquad (2)$$

with $\phi(x,y) - \phi_0(x)$, referred to as "phase deviation", being a known function of the object's height distribution with respect to the reference plane. In (1) and (2) x and y are the horizontal and vertical axis, the pattern lines are vertical, f_0 is the pattern frequency, A_n are the Fourier coefficients of the pattern and $r(x,y)$ represents the reflectivity of object's surface. Hence, image rows are expressed as a summation of phase-amplitude modulated harmonics, amplitude modulation due to non-uniform reflectivity and phase modulation embodying the object's 3D shape. It is worth noting that, since the grating is a square wave, the spectrum of a row contains only the odd harmonics and that the low frequency term is due to $r(x,y)$.

Based on this principle, two major techniques aimed at extracting phase from the imaged pattern have been devised. In FTP [3], [4] the phase profile associated with each image row is evaluated as the angle of the analytic representation of the modulated first harmonic. This is obtained by computing the FFT of the row, filtering out all frequencies outside a narrow band centered at f_0 and computing the IFFT. In SDP [5] each image row is multiplied by $e^{-j2\pi f_0 x}$ so as to shift the spectrum towards the left by f_0. Thus, the phase signal associated with the first harmonic moves to low-frequency and is extracted through a low-pass FIR filter. Phase is evaluated as the angle of the complex signal extracted by the filter. With both techniques all the processing steps are applied also to the reference image in order to evaluate phase deviation. SDP yields a substantial reduction of computing time with respect to FTP [5].

3 A New Phase Extraction Algorithm

We have developed a new phase extraction algorithm based on moving the phase information to low-frequency and extracting it by means of signal-domain filters. Let us consider the two periodic real functions $c_1(x)$ and $c_2(x)$:

$$c_1(x) = \sum_{n=-\infty}^{+\infty} C_n e^{j2\pi n f_0 x} \qquad (3)$$

$$c_2(x) = c_1(x - p_0/4) = \sum_{n=-\infty}^{+\infty} C_n e^{j2\pi n f_0 x} e^{-jn\pi/2} = \sum_{n=-\infty}^{+\infty} \tilde{C}_n e^{j2\pi n f_0 x} \qquad (4)$$

and multiply the row $y = \overline{y}$ of the deformed pattern image by $c_1(x)$ and $c_2(x)$:

$$q_r(x,\overline{y}) = r(x,\overline{y}) \left(\sum_{n=-\infty}^{+\infty} A_n e^{j(2\pi n f_0 x + n\phi(x,\overline{y}))} \right) \left(\sum_{n=-\infty}^{+\infty} C_n e^{j2\pi n f_0 x} \right) \qquad (5)$$

$$q_{im}(x,\overline{y}) = r(x,\overline{y}) \left(\sum_{n=-\infty}^{+\infty} A_n e^{j(2\pi n f_0 x + n\phi(x,\overline{y}))} \right) \left(\sum_{n=-\infty}^{+\infty} \tilde{C}_n e^{j2\pi n f_0 x} \right) \qquad (6)$$

If the mean value of $c_1(x)$ is zero, the low-frequency components of q_r and q_{im}, indicated as $\overline{q}_r$ and $\overline{q}_{im}$, can be expressed as

$$\overline{q}_r(x,\overline{y}) = 2r(x,\overline{y}) \sum_{n=-\infty}^{+\infty} |A_n||C_n| \cos(\alpha_n + n\phi(x,\overline{y}) - \gamma_n) \qquad (7)$$

$$\overline{q}_{im}(x,\overline{y}) = 2r(x,\overline{y}) \sum_{n=-\infty}^{+\infty} |A_n||C_n| \cos(\alpha_n + n(\phi(x,\overline{y}) - \pi/2) - \gamma_n) \qquad (8)$$

where α_n and γ_n are the angles of the complex coefficients A_n and C_n. If a simple smoothing operator such as a moving average is applied to the image prior (5) and (6) in order to attenuate the higher harmonics in $g(x,\overline{y})$, the dominant term in $\overline{q}_r$ and $\overline{q}_{im}$ is that associated with the main harmonic:

$$\overline{q}_r(x,\overline{y}) \cong 2r(x,\overline{y})|A_1||C_1| \cos(\alpha_1 + \phi(x,\overline{y}) - \gamma_1) \qquad (9)$$

$$\overline{q}_{im}(x,\overline{y}) \cong -2r(x,\overline{y})|A_n||C_n| \sin(\alpha_1 + \phi(x,\overline{y}) - \gamma_1) \qquad (10)$$

Hence, phase information is embedded into the angle of the low-frequency complex signal

$$\overline{q}(x,\overline{y}) = \overline{q}_r(x,\overline{y}) - j\overline{q}_{im}(x,\overline{y}) \cong 2r(x,\overline{y})|A_1||C_1|e^{\alpha_1 + \phi(x,\overline{y}) - \gamma_1} \ . \qquad (11)$$

Since execution of the same procedure on the reference image yields

$$\overline{q}_0(x,\overline{y}) \cong 2r(x,\overline{y})|A_1||C_1|e^{\alpha_1 + \phi_0(x) - \gamma_1} \qquad (12)$$

the phase deviation can be extracted by subtracting the angle of $\overline{q}_0(x,\overline{y})$ from the angle of $\overline{q}(x,\overline{y})$.

Once shown that the phase information can be moved to low-frequency by multiplying the image by two generic periodic functions, a suitable choice for these functions is a square wave swinging between -1 and +1. Thus, demodulation requires no more than proper manipulation of pixel signs. The step following demodulation is extraction of the low-frequency spectrum, the main task being rejection of the unwanted spectra which are generated in the demodulation process. Since these are centered at specific harmonics of the grating frequency, we propose to apply first a comb-shaped filter in order to reject selectively unwanted harmonics and then a moving average to obtain the required low-pass behavior. The proposed technique is potentially faster than SDP since the demodulation cost is practically negligible and only very fast filters such as comb or modified comb filters [6] are used.

4 Simulation Analysis

We use simulations to discuss adaptiveness of phase extraction techniques with respect to changes of the fringe frequency. Adaptiveness is a key requirement since coarse fringe patterns are effective for measuring coarse surface variations while fine patterns are suited to the measurement of fine surface details [7]. In our analysis we will take into account speed of design, accuracy and computational complexity. Simulation methodology and parameters are as follows. Line size is set to 256 pixel. Assuming uniform reflectivity, the image of a one-dimensional fringe pattern is phase-modulated based on a trapezoidal phase profile with 0.15 pixel slope. We simulate projection of a fine fringe pattern ($f_0 = 0.125\,\text{pixel}^{-1}$, 32 fringes per line) and of a coarse one ($f_0 = 0.625\,\text{pixel}^{-1}$, 16 fringes per line).

First we consider the fine pattern and the new technique. According to previous section, the line is smoothed using a 3-point moving average. Since reflectivity is uniform, we subtract the mean value before demodulation in order to get rid easily of the DC component. Then, demodulation is carried out by proper manipulation of pixel signs. Since after demodulation the main unwanted spectra are centered at frequencies $\mp 2f_0$, we apply the comb filter capable of eliminating these frequencies, given by $y_k = x_k + x_{k-2}$. As a result, the low-frequency signal can be easily extracted by a 9-points moving average. The left plot of Fig. 1 provides a proof of the feasibility of the method: the extracted phase profile (solid line) closely approximates the target phase profile (dashed line); the mean absolute phase error is 31 mrad. As far as FTP is concerned, we implemented the frequency domain filter by hanning window in order to reduce ripple. With regard to SDP, we use the 9 taps FIR filter proposed in [5]. Phase profiles extracted by FTP and SDP, not shown here, can be considered equivalent to that in the left plot of Fig. 1, phase errors being 29 mrad for FTP and 35 mrad for SDP.

We consider now the coarse fringe pattern. Adapting our technique to this new pattern consists in tuning the comb-shaped filter so as to eliminate frequencies $\mp 2f_0, \mp 4f_0, \mp 6f_0$ from the demodulated signals. Since the filter $y_k = x_k + x_{k-2}$ eliminates frequencies $\mp 4f_0$ while the filter $y_k = x_k + x_{k-4}$ eliminates frequencies $\mp 2f_0, \mp 6f_0$, the cascade of the two gives the modified comb filter suited to the coarse fringe pattern: $y_k = x_k + x_{k-2} + x_{k-4} + x_{k-6}$. Thus, the design process is quite simple and fast. Moreover, computational complexity remains practically as low as in the fine pattern case. In fact the new filter requires just one more subtraction per point since it can be split into two independent filters, running on even and odd pixels, computed through the recursive relation $y_k = y_{k-2} + x_k - x_{k-8}$. The phase profile extracted with the coarse pattern is

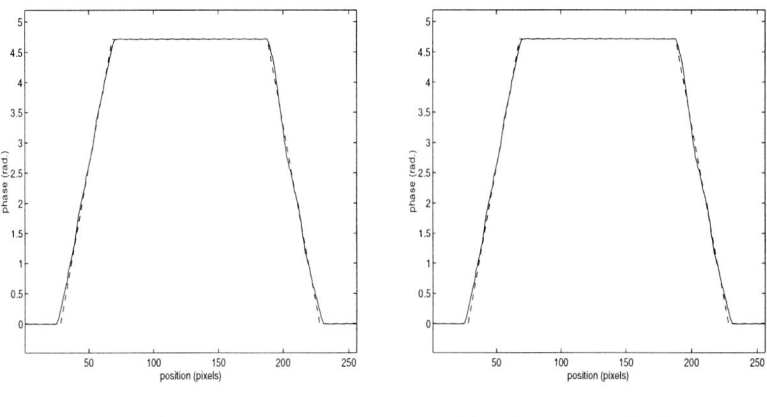

Fig. 1. Phase profiles extracted by the new method

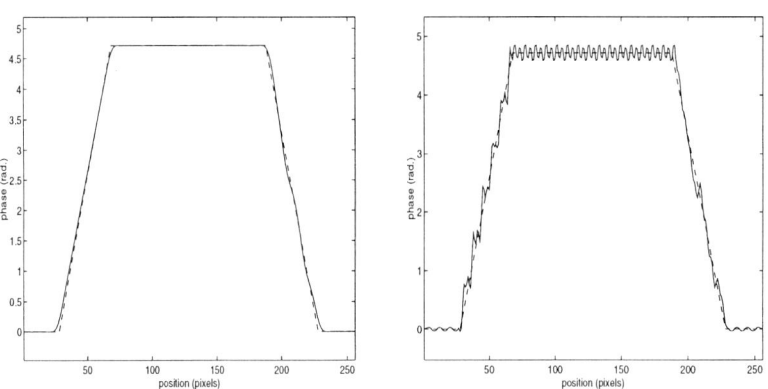

Fig. 2. Coarse pattern: phase profiles extracted by FTP (left) and SDP (right)

shown in the right plot of Fig. 1, the mean absolute phase error being 40 mrad. Adapting FTP to the new pattern requires just halving the size of the window and centering it at the new frequency. Since computation time is largely dominated by the FFT-IFFT pair, complexity is nearly independent of the pattern frequency. On the other hand, a suitable FIR filter must be designed from scratch to adapt SDP to a new pattern frequency. Since in this case the main unwanted spectra are close to DC, a quite sharp transition from passband to stopband is needed to extract the low-frequency spectrum. This makes the design extremely problematic. We have discussed this issue in detail in [8] addressing 9 taps minimax FIR filters; here we use the best 9 taps filter shown in [8]. Profiles extracted by FTP and SDP are plotted in Fig. 2, phase error being 45 mrad with FTP and 96 mrad with SDP.

Thus, with both FTP and our technique one can easily adapt the procedure to a different fringe pattern, no practical variation in the computational complexity is associated with the frequency change, and the accuracy with a coarse pattern remains rather good compared to that of the fine pattern case. Conversely, adapting SDP implies the carrying out of a FIR filter design process which turns out to be very critical in case of a coarse pattern. Moreover, if in this case the order of the filter is low, SDP is significantly less accurate. The only way to improve SDP accuracy with a coarse pattern is to increase the order of the filter. We had to increase the number of taps up to 21 to push the error within 50 mrad (i.e. 45 mrad). Therefore, even though with a coarse pattern SDP may provide an accuracy level similar to that of the other two techniques, this implies a substantial increase of the computational load.

We have assumed uniform reflectivity, but this is not the case of most real images. This would introduce errors with every phase extraction technique since the low-frequency reflectivity spectrum may overlap with the spectrum to be extracted. In addition, with SDP the filter design becomes even more complicated since a much sharper transition is required to attenuate the reflectivity spectrum, which is shifted to $-f_0$ by demodulation. In the following section we address this issue in the context of 3D PCB inspection.

5 Experimental Results

We have investigated the use of phase-profilometry for 3D PCB inspection [8], [9], that consist in reconstructing the 3D shape of the solder paste printed on SMT (Surface Mount Technology) component pads before component placement. The left image in Fig. 3 shows a circular pad on the PCB with the projected fringe pattern. Image size is 256*256 pixels and fringe frequency is $0.0625\,\text{pixel}^{-1}$.

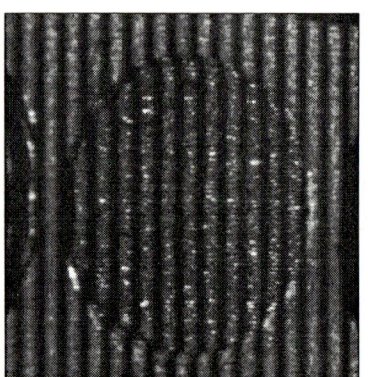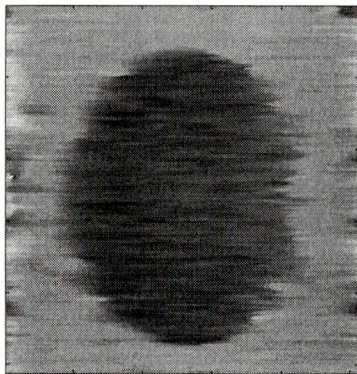

Fig. 3. Original image and phase map extracted by FTP

With FTP we deal with the potential overlap between spectra by reducing the size of the filter. Although this requires some tuning, it is still an easy and rapid process. The phase map extracted from the pad image using FTP is shown in the right image of Fig. 3, darker points representing higher phase deviations

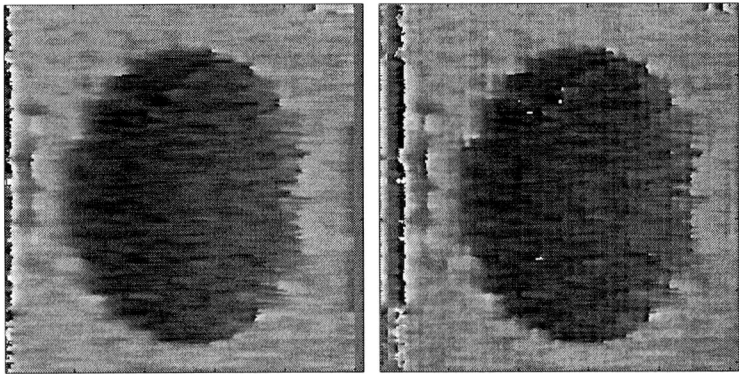

Fig. 4. Phase map extracted by our technique and by SDP

(i.e. heights). As far as the proposed technique is concerned, the comb filter should now attenuate all the fringe frequency harmonics. Thus, we cascade the filter $y_k = x_k + x_{k-8}$ to the filter for the coarse pattern used in the simulations; this yields the modified comb filter: $y_k = x_k + x_{k-2} + x_{k-4} + x_{k-6} + x_{k-8} + x_{k-10} + x_{k-12} + x_{k-14}$. Since the filter can be computed through the relation $y_k = y_{k-2} + x_k - x_{k-16}$, the computational load is the same as that of the filter used in the simulations. The phase map extracted using our technique is shown in the left image of Fig. 4.

Due to unavailability of a suitable high-accuracy metrology equipment, we cannot assess quantitatively these results. However, qualitative judgment provides useful hints with respect to comparative analysis of the addressed technique. In this context, the clear contrast between the flat background area and the circular shaped object region shows that both techniques allow neat detection of the elevation with respect to PCB surface associated with solder paste on pads. Use of SDP for PCB inspection requires the FIR filter to be redesigned, with a much sharper transition required to deal with nonuniform reflectivity. As a result, the filter order must be increased to 29 to obtain results qualitatively similar to those of the other two techniques. The phase map extracted by the 29 taps filter is shown in the right image of Fig. 4.

Eventually, we have measured computation times (90 MHz Pentium PC, image size: 256*256) in order to compare complexity of the addressed techniques. We have considered the time required to extract the complex signal embodying phase deviation (T_1), which is peculiar of a given technique, and the total time for obtaining the phase map (T), obtained by adding to T_1 the constant time ($T_2 = 0.4441$ secs) required to evaluate the angle and subtract the reference. As already observed, with both FTP and the new technique the computational load is nearly constant. Hence, using the procedures developed for PCB inspection, we have measured $T_1 = 1.1156$ secs for FTP and $T_1 = 0.0537$ secs for our technique. Conversely, with SDP we have evaluated T_1 using the 9, 21 and 29 taps filters (suited respectively to the fine pattern, coarse pattern, coarse pattern and nonuniform reflectivity), measuring respectively 0.1992 secs, 0.4173 secs and 0.5390 secs. Hence, our algorithm is, roughly, 20 times faster than FTP and, taking the average of the times for the 3 filters, 7 times faster than SDP. However, due to the substantial value of T_2 the actual speedup is lower. Nonetheless, evaluation of T shows that our algorithm renders profilometry 3 times faster than using FTP, while the speedup with respect to SDP can range from 30%, as in the case of fine patterns, to almost 100%, as in the case of coarse patterns and nonuniform reflectivity.

6 Conclusion

We have presented a new phase extraction algorithm for phase profilometry. The algorithm relies on a fast demodulation scheme and very simple DSP operators such as comb filter and moving averages. We have also compared it with FTP and SDP. The new technique's advantage with respect to FTP is speed since it renders profilometry roughly 3 times faster. With respect to SDP, the main advantage is adaptiveness. In fact, while one can easily tune a comb-shaped filter to reject different frequencies, SDP requires complete re-design of the low-pass FIR filter. This task becomes very critical when the fringes frequency is low and/or reflectivity is nonuniform, with satisfactory results attainable only at the expense of a significant increase of the filter complexity. Therefore, the proposed technique is also notably faster than SDP in the case of coarse patterns and/or nonuniform reflectivity.

References

1. D. Poussart and D. Laurendau. 3-D sensing for industrial computer vision. in J. Sanz, editor, *Advances in Machine Vision*, Springer-Verlag, 1989.
2. P. Besl. Active, optical range imaging sensors. *Machine Vision and Applications*, Vol.1, 1988.
3. M. Takeda, H. Ina and S. Kobayashi. Fourier-transform method of fringe-pattern analysis for computer-based topography and interferometry. *Journal of the Optical Society of America*, Vol. 72, No. 1, 1982.
4. M. Takeda and K. Mutoh. Fourier transform profilometry for the automatic measurement of 3-D object shapes. *Applied Optics*, Vol. 22, No. 24, 1983.
5. S. Tang and Y. Hung. Fast profilometer for the automatic measurement of 3-D object shapes. *Applied Optics*, Vol. 29, No. 20, 1990.
6. E. Cunningham. *Digital Filtering: an introduction*. Houghton Mifflin, 1992.
7. G. Sansoni, L. Biancardi, U. Minoni and F. Docchio. A Novel Adaptive System for 3-D Optical Profilometry Using a Liquid Crystal Light Projector. *IEEE Transactions on Instrumentation and Measurement*, Vol. 43, No. 4, 1994.
8. L. Di Stefano and F. Boland. Three-Dimensional Inspection of Printed Circuit Boards Using Phase Profilometry. In *Proceedings of EUSIPCO-96*, Trieste, 10-13 September, 1996.
9. L. Di Stefano and F. Boland. Solder paste inspection by structured light methods based on phase measurement. In *SPIE Proceedings*, Vol. 2899, 1996.

Surface Modeling and Display from Range and Color Data

Kari Pulli[1], Michael Cohen[2], Tom Duchamp[1], Hugues Hoppe[2], John McDonald[1], Linda Shapiro[1], and Werner Stuetzle[1]

[1] University of Washington, Seattle WA, USA
[2] Microsoft Research, Redmond WA, USA

Abstract. Two approaches for modeling surfaces from a collection of range maps and associated color images are covered. The first approach presents a method that robustly obtains a closed mesh that approximates the object geometry. The mesh can then be simplified and texture mapped for display. The second approach does not attempt to create a single object model. Instead, a set of models is constructed, one model for each view of the object. several of the view-based models are rendered separately, and their information is combined in a view-dependent manner for display.

1 Introduction

In this paper we propose two methods for modeling and displaying real objects using a set of range maps with associated color images as input. The first method follows the traditional surface reconstruction approach, where we attempt to create a single surface model that accurately describes the scanned object. Specifically, our algorithm [9] emphasizes robust recovery of the object topology from the input data. Holes due to missing data, i.e., unobserved surface regions, are automatically filled so that the model remains consistent with the input data. This approximate model can then be fitted more accurately to the input data, textured from the color images, and displayed. Our second method shows that it is not necessary to integrate the input data into a single surface model for display purposes. Instead, one can model each view separately and integrate the separate views at display time in the image space. We call this approach *view-based rendering* [8].

Section 2 describes our robust method for modeling object surfaces. Section 3 presents the view-based rendering method. Section 4 concludes the paper.

2 Robust approximate meshes

Overview. Our algorithm processes a cubical volume surrounding all the input data in a hierarchical fashion. For each cube-shaped partition, it checks whether the cube can be shown to be entirely inside or outside of the object. If neither, the cube is subdivided and the same test is recursively applied to the resulting smaller cubes. Our surface approximation is the closed boundary between cubes that lie entirely outside of the object and all the other cubes.

Assumptions. We assume that the range data is expressed as a set of range maps called views. Each view is an image in which each pixel stores a 3D point instead of a color value. Further, we assume that the calibration parameters of the sensor are known so that we can project any 3D point to the image plane of the sensor. We also assume that the line segments between the sensor and each measured point lie entirely outside of the object we are modeling. Finally, we assume that all range views have been registered to a common coordinate system.

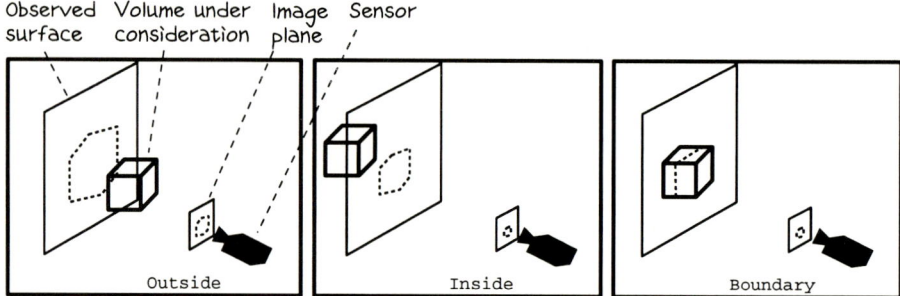

Fig. 1. The three cases of the algorithm. In case 1 the cube is in front of the range data, in case 2 it is entirely behind the surface (with respect to the sensor), while in case 3 the cube intersects the range data.

2.1 Processing a single range view

The initial volume is an axis-aligned cube that fully surrounds all the range data. We use interval analysis to evaluate the volumetric function on the cube. If the cube is neither completely inside nor completely outside the object, we recursively subdivide it into eight smaller cubes, which are added to the octree as children of the current cube. For each of these cubes, we classify their location with respect to the sensor and the range data. A cube can be classified in three ways (see Figure 1):

- In case 1 the cube lies between the range data and the sensor. The cube is assumed to lie outside of the object. It is not processed any further.
- In case 2 the whole cube is behind the range data. As far as this sensor is concerned, the cube will be assumed to lie inside of the object. It will not be further subdivided.
- In case 3 the cube intersects the range map. In this case we subdivide the cube into its eight children and recursively apply the algorithm up to a prespecified maximum subdivision level. A case 3 cube at the finest level is assumed to be at the boundary of the object.

The cubes are classified as follows. We project the eight corners of the cube to the sensor's image plane, where the convex hull of the points forms a hexagon. The rays from the sensor to the hexagon form a cone, which we truncate so that it just encloses the cube. If all the data points projecting onto the hexagon are behind the truncated cone (i.e., are farther than the farthest corner of the cube from the sensor), the cube is outside. If all those points are closer than the closest cube corner, the cube is inside. Otherwise, it is the boundary case. Possible missing data is treated as points that are very close to the sensor. If we can label parts of the depth map as background, data at those locations are treated as being infinitely far away.

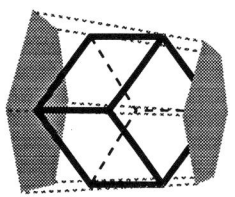

Fig. 2: An octree cube and its truncated cone.

Our labeling method is simple, but conservative. For instance, if all the points projecting onto the hexagon are actually behind the cube, but some of them are inside the truncated cone, we would erroneously label the cube to intersect the surface. This, however, is not a problem because the next subdivision level will most likely determine that all the children of the cube lie in the exterior of the object and therefore remove them.

2.2 Generalizations to multiple views

If multiple views are available, we have a choice of two processing orders. We can traverse the whole octree once and use the consensus of all the views to determine the labeling for each cube (simultaneous processing), or we can process one view at a time, building on the results of the previously processed views (sequential processing).

In simultaneous processing, we traverse the octree as we did in the case of a single view. However, the cube labeling process changes slightly.

- A cube is labeled inside the object only if it would be labeled inside with respect to each single view.
- A cube is labeled outside if it would be labeled outside with respect to any single view.
- Otherwise, the cube is labeled boundary and is further subdivided, unless the maximum subdivision level has been reached.

We use sequential processing if we later obtain a new view that we want to integrate into a previously processed octree. We recursively descend the octree and perform the occlusion test for each cube that has not been determined to lie outside of the object. If the new view determines that a cube is outside, it is relabeled and the subtrees below it are removed. Similarly, a boundary label overrides a previous inside label, in which case the cube's descendants must be recursively tested, potentially up to the maximum subdivision level.

Although both processing orders produce the same result, the simultaneous processing order is in general faster [10]. In sequential processing the silhouette of the object often creates a visual cone (centered at the sensor) that separates volumes known to be outside from those speculated to be inside. The algorithm would have to recurse up to the finest subdivision level to accurately determine this boundary. In simultaneous processing, however, another view could determine at a rather coarse level of subdivision that at least part of that boundary is actually outside of the object, and the finer levels of the octree for that subvolume need never be processed.

2.3 Mesh extraction

The labeling in the octree divides the space into two sets: the cubes known to lie outside of the object and the cubes that are assumed to be part of the object. Our surface estimate will be the closed boundary between these sets. This definition allows us to create a plausible surface even at locations where we failed to obtain data [1]. The boundary is represented as a collection of vertices and triangles that can be easily combined to form a mesh.

The octree generated by the algorithm has the following structure: outside cubes and inside cubes do not have any children, while the boundary cubes have a sequence of descendants down to the finest subdivision level. We traverse the octree starting from the root. At an outside cube we do nothing. At a boundary cube that is not at the finest level, we descend to the children. If we reach the maximum subdivision level and the cube is either at the boundary or inside we check the labeling of the six neighbors. If a neighboring cube is an outside cube, we create two triangles for the face they share. In an inside cube that is not at the maximum subdivision level, we check whether it abuts with an outside cube, and in such case create enough triangles (of same size as the ones created at the finest level) to cover the shared part of the face. The triangles are combined into a closed triangle mesh.

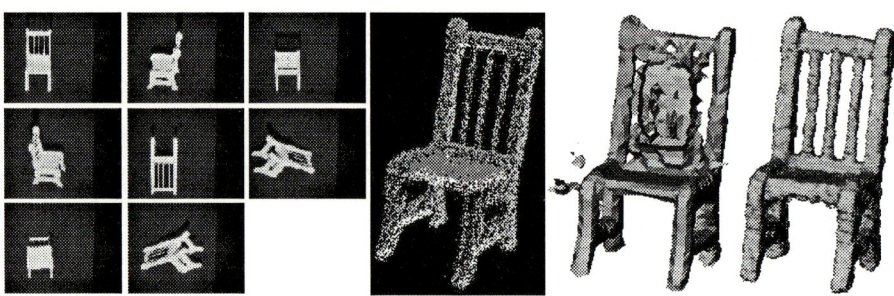

Fig. 3. Eight views of a chair data set, registered points, result from using a previous method, result from our method.

2.4 Reconstruction results

We have tested our method with both real and synthetic data. One of the real data sets consisting of eight views of a miniature chair is shown in Fig. 3, along with the data points, and failed surface reconstruction from another algorithm [5] that was our original motivation for this method. Even though we have cleaned the data and removed most of the outliers, some noisy measurements close to the surface remain, especially between the spokes of the back support of the chair. The algorithm from [5] works with an unorganized point cloud and does not use any extra knowledge (such as viewing directions, etc.) in addition to the points. It works quite nicely if the data does not contain outliers and uniformly samples the underlying surface. Unfortunately real data, such as this data set, often violate both of those requirements. In contrast, our method is able to correctly recover the topology of the chair as shown in the rightmost image.

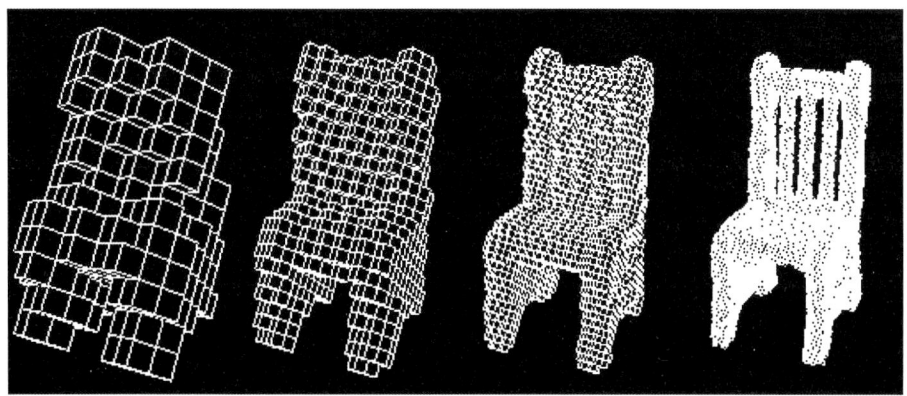

Fig. 4. Chair after 4, 5, 6, and 7 subdivisions.

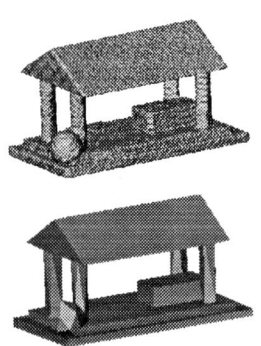

Figure 4 shows intermediate results of our method using the chair data set, displaying the octree after 4, 5, 6, and 7 subdivisions. The final mesh in Fig. 3 was obtained from the level 7 octree. We smooth the mesh before displaying using Taubin's method [11]. Notice that the spokes and the holes between them have been robustly recovered despite the large number of outliers and some missing data. This is partially due to the implicit removal of outliers that the algorithm performs: in most cases, there is a view that determines that the cube containing an outlier lies outside of the object. Fig. 5 also shows the results from a synthetic data set. The smoothed result appears above, while the result after applying Hoppe's mesh optimization algorithm [6] appears below.

Fig. 5: Simulated temple data set, our results and a simplified mesh.

2.5 Discussion

The simplicity of the algorithm leads to a fast implementation. This again allows interactive selection of the subdivision level. For example, with the chair data set one could first subdivide the octree six times. Within ten seconds or so the user can see the current approximation, which is not good enough since all the holes in the object haven't been opened yet. An extra subdivision step (25 sec. later) shows that now the topology of the current approximation agrees with that of the real object.

It is important to notice that holes often can be correctly detected and modeled using only indirect evidence, i.e., using the fact that the scanner can see through the hole. Curless and Levoy suggest using a backdrop [1], placing planes behind the holes that can be detected. However, all range scanning methods based on optical triangulation have limits on how narrow cavities can be scanned. Better results can be obtained if a color (or even intensity) image can be obtained with the range scan (typically easily available with scanners using optical triangulation), the background may be segmented out using image segmentation methods, or even user interaction.

Some methods that employ signed surface distance functions are unable to correctly reconstruct thin objects. Curless and Levoy [1], for example, build a distance function by storing positive distances to voxels in front of the surface and negative distances to voxels behind the surface. In addition to the distance, weights are stored to facilitate combining data from different views. With this method, views from opposite sides of a thin object interfere and may cancel each other, preventing reliable extraction of the zero set of the signed distance function. Our volumetric method carves away the space between the sensor and the object and does not construct a signed distance function. In case of a thin sheet of paper, our algorithm would construct a thin layer of octree cubes (voxels) straddling the paper.

3 View-based rendering

Rendering statically textured surface models produces images that are much less realistic than photographs, which can capture intricate geometric texture and global illumination effects with ease. This is one of the main reasons why image-based rendering algorithms have become popular. This section proposes a new rendering method that does not require creating a full 3D object model. Rather, we create independent models for the depth map observed from each viewpoint, a much simpler task. Instead of having to gather and manipulate a set of images dense enough for purely image-based rendering [4,7], our method only requires images from the typically small set of viewpoints from which the range data were captured. A request for an image of the object from a specified viewpoint is satisfied using the color and geometry in the stored views. This section describes our new *view-based rendering* algorithm and shows results on non-trivial real objects.

The input to our view-based rendering system is a set of color images of the objects. Along with each color image we obtain a range map for the part of the object surface that is visible in the image. Registering the range maps into a common coordinate system gives us the relative camera locations and orientations of the color images with respect to the object. We replace the dense range maps by sparse triangle meshes that closely approximate them. We then texture map each triangle mesh using the associated color image. To synthesize an image of the object from a fixed viewpoint, we individually render the meshes constructed from the three nearest viewpoints and blend them together with a pixel-based weighting algorithm using soft z-buffering.

3.1 A simple approach

To better understand the virtues of our approach, it is helpful to consider a more simple algorithm. If we want to view the object from any of the stored viewpoints, we can place a virtual camera at one of them and render the associated textured mesh. We can even move the virtual camera around the stored viewpoint by rendering the mesh from the new viewpoint. But as the viewpoint changes, parts of the surface not seen from the original viewpoint may become visible, opening holes in the rendered image. If, however, the missing surface parts are seen from one or more other stored viewpoints, we can fill the holes by simultaneously rendering the textured meshes associated with the additional viewpoints. The resulting image is a collage of several individual images. Because individual meshes are likely to overlap, the compound errors from the actual range measurements, view registration, and polygonal approximation make arbitrary which surface is closest to the camera and therefore rendered. Also, the alignment of the color information is not perfect, and there may be additional slight changes in the lighting conditions between the views. These errors cause the unnatural features visible in Figure 6(a).

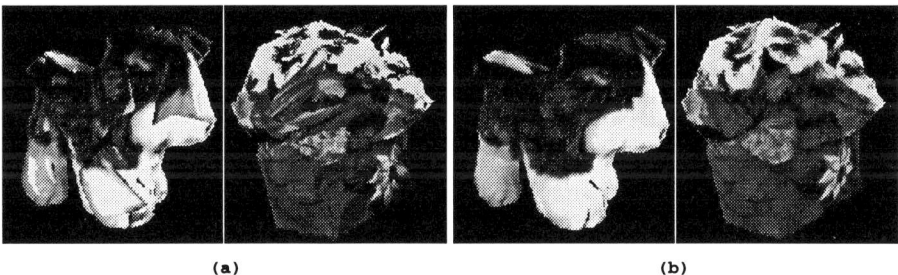

Fig. 6. (a) The result of combining three views by repeatedly rendering the view-based meshes from the viewpoint of the virtual camera. (b) Using the weights and soft z-buffering produces a much better result.

We can improve on this by giving different weights to the views, with the viewpoint closest to the viewpoint of the virtual camera receiving higher weight than the others. The effect of self-occlusion can be minimized by using z-buffering and back-face culling when rendering the individual views. Even with these improvements, several problems remain. The pixels where only some of the views contribute appear darker than others. Even if we normalize the colors by dividing the color values by the sum of the weights of the contributing views, changes in lighting and registration errors create visible artifacts at mesh boundaries. There are also problems with self-occlusion. Without z-buffering the color information from surfaces that should be hidden by other surfaces is blended with the color of the visible surfaces, causing parts of the front-most surface to appear partially transparent. A third problem is related to the uniform weighting of the images generated by the meshes. The color and surface geometry is sampled much more densely at surface locations that are perpendicular to the sensor than at tilted surfaces. Additionally, the range information is usually less reliable at tilted surfaces.

In the next section we describe how we can produce much better images (see Fig. 6(b)) using a more sophisticated approach.

3.2 Three weights and soft z-buffering

To synthesize an image of the object from a fixed viewpoint, we first select n stored views whose viewing directions roughly agree with the direction from the viewpoint to the object. Each selected textured mesh is individually rendered from this viewpoint to obtain n separate images. The images are blended into a single image by the following weighting scheme. Consider a single pixel. Let r be the red channel value (green and blue are processed in the same manner) associated to it. We set

$$r = \frac{\sum_{i=1}^{n} w_i r_i}{\sum_{i=1}^{n} w_i}$$

where r_i is the color value associated to that pixel in the ith image and w_i is a weight designed to overcome the difficulties encountered in the naive implementation described above. The weight w_i is the product of three weights $w_i = w_{\theta,i} \cdot w_{\phi,i} \cdot w_{\gamma,i}$, whose definition is illustrated in Figs. 7 and 8. Self-occlusions are handled by using soft z-buffering to combine the images pixel by pixel.

The first weight, w_θ, measures the proximity of the stored view to the current viewpoint, and therefore changes dynamically as the virtual camera moves. Both the appearance of minute geometric surface details and the surface reflectance change with the viewing direction; the weight w_θ is designed to favor views with viewing directions similar to that of the virtual camera. Figure 7 illustrates how w_θ is calculated. All the views are placed on a unit sphere, and the sphere is triangulated. The current viewpoint is placed onto that sphere, and we determine which triangle contains it. Only the views corresponding to the corners of that triangle will get a nonzero w_θ. Specifically, we calculate the barycentric coordinates for the current view point within the triangle. Each of the three components of the barycentric coordinates corresponds to a corner of the triangle, and the w_θ of the corner views is the corresponding component. The three w_θ's are all in the range $[0.0, 1.0]$ and they sum up to 1.0.

The second weight, w_φ, is a static measure of surface sampling density. As a surface perpendicular to the camera is rotated by an angle ϕ, the surface area projecting to a pixel increases by $1/\cos\phi$ and the surface sampling density

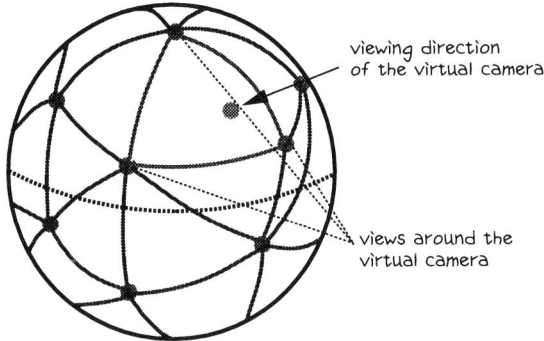

Fig. 7. The views are placed on a unit sphere based on their viewing directions, and the sphere is triangulated using the views as triangle vertices. The views corresponding to the corners of the triangle surrounding the direction of the virtual camera are used, and their weights w_θ are the barycentric coordinates the virtual camera within that triangle.

decreases by $\cos\phi$. In our system, a weight $w_\varphi = \mathbf{n} \cdot \mathbf{d}$ is applied to each mesh triangle, where $\mathbf{n}$ is the external unit normal of the triangle and $\mathbf{d}$ is a unit vector pointing from the centroid of the triangle to the sensor. Figure 8(b) shows w_φ with gray level encoding: the lighter the triangle, the higher w_φ. The scanning geometry ensures that this value is in the range $(0.0, 1.0]$.

The third weight w_γ which we call the *blend weight*, is designed to smoothly blend the meshes at their boundaries. As illustrated by Figure 8 (c), the blend weight linearly increases with distance from the mesh boundary. Like w_φ, the weight w_γ does not depend on the viewing direction of the virtual camera. A similar weight was used by Debevec *et al.* [2].

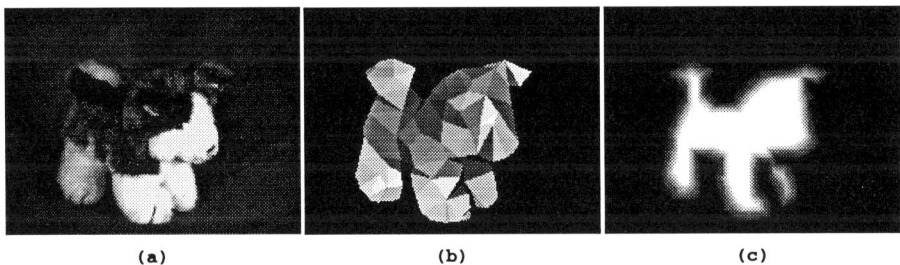

Fig. 8. (a) An image of a toy dog. (b) Weight w_φ is applied to each face of the triangle mesh. (c) Weight w_γ smoothly decreases the influence of the view towards the mesh boundaries.

Most self-occlusions are handled during rendering of individual views using backface culling and z-buffering. When combining the view-based partial models, part of one view's model may occlude part of another view's model. Unless the surfaces are relatively close to each other, the occluded pixel must be excluded from contributing to the pixel color. This is done by performing "soft" z-buffering, in software. First, we consult the z-buffer information of each separately rendered view and search for the smallest value. Views with z-values within a threshold from the closest are included in the composition, others are excluded. The threshold is chosen to slightly exceed an upper estimate of the combination of the sampling, registration, and polygonal approximation errors.

Figure 9 illustrates a potential problem. In the picture the view-based surface approximation of the rightmost camera has failed to notice a step edge due to self-occlusion in the data, and has wrongly connected two surface regions. When performing the soft z-buffering for the pixel corresponding to the dashed line, the wrongly connected step edge would be so much closer than the contribution from the other view that the soft z-buffering would throw away the correct sample. However,

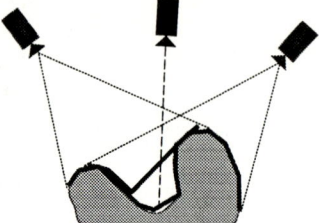

Fig. 9: Problems with undetected step edges.

while doing the soft z-buffering we can treat the weights as confidence measures. If a pixel with a very low confidence value covers a pixel with a high confidence value, we ignore the low confidence pixel altogether.

3.3 Implementation

Triangle mesh creation. We currently create the triangle meshes manually. For each view, the user marks the boundaries of the object by inserting points into the color image, while the software incrementally updates a Delaunay triangulation of the vertices. When the user adds a vertex, the system optimizes the z-coordinates of all the vertices so that the least squares error of the range data approximation is minimized. Triangles that are almost parallel to the viewing direction are discarded since they are likely to be step edges, not a good approximation of the object surface. Triangles outside of the object are discarded as well.

We have begun to automate the mesh creation phase. The segmentation of the image into object and background is facilitated by placing a blue cloth to the background and scanning the empty scene. Points whose position and color match the data scanned from an empty scene can be classified as background. The adding of vertices is easily automated. For example, Garland and Heckbert [3] add vertices to image coordinates where the current approximation is worst. The drawback of this approach is that if the data contains step edges due to self-occlusions, the mesh is likely to become unnecessarily dense before a good approximation is achieved. For this reason we will perform a mesh simplification step using the mesh optimization methods by Hoppe *et al.* [6].

Rendering. We have built an interactive viewer for viewing the reconstructed images (see Figure 10). For each frame, we calculate the dot product of the camera viewing directions for the stored views and the viewing direction of the virtual camera. The three views with highest dot product values (the weight

w_θ) are then rendered separately from the viewpoint of the virtual camera as textured triangle meshes.

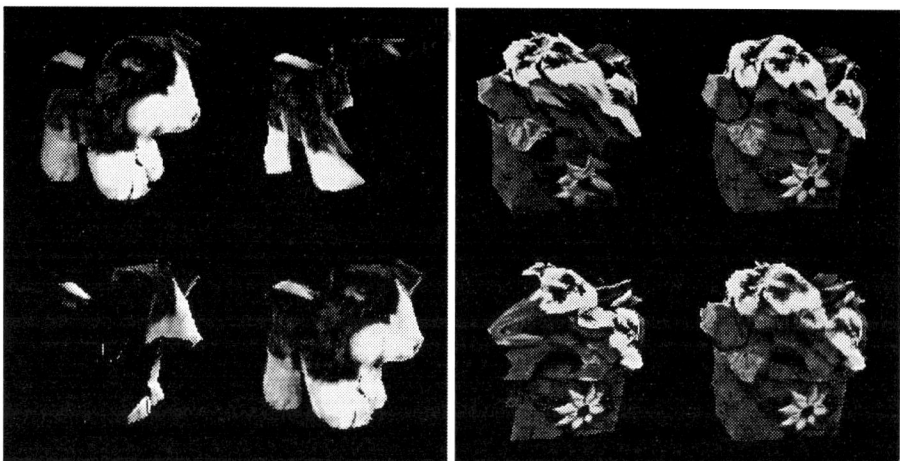

Fig. 10. Our viewer shows the three view-based models rendered from the viewpoint of the virtual camera. The final image is on the bottom right.

Two of the weights, w_φ and w_γ are static for each view, they do not depend on the viewing direction of the virtual camera. We can apply both of these weights offline. w_φ is the weight used to decrease the importance of triangles that are tilted with respect to the scanner. It is applied by assigning the RGBA color $(1, 1, 1, w_\varphi)$ to each triangle. w_γ is the weight used to hide artifacts at the mesh boundary of a view. It is directly applied to the alpha channel of the texture map that stores the color information. We calculate the weights for each pixel by first projecting the triangle mesh onto the color image and painting it white on a black background. We then calculate the distance d for each white pixel to the closest black pixel. The pixels with distances of at least n get alpha value 1, all other pixels get the value $\frac{d}{n}$.

Figure 11 presents pseudo code for the view composition algorithm. The function min_reliable_z() returns the minimum z for a given pixel, unless the closest pixel is a low-confidence (weight) point that would occlude a high-confidence point, in which case the z for the minimum high-confidence point is returned.

When we render a triangle mesh with the described colors and texture maps, the hardware calculates the correct weights for us. The alpha value in each pixel

```
FOR EACH pixel
  zmin         := min_reliable_z( pixel )
  pixel_color  := (0,0,0)
  pixel_weight := 0
  FOR EACH view
    IF zmin <= z[view,pixel] <= zmin+thrsoft_z THEN
      weight       := wθ * wφ * wγ
      pixel_color  += weight * color[view,pixel]
      pixel_weight += weight
    ENDIF
  END
  color[pixel] := pixel_color / pixel_weight
END
```

Fig. 11: Pseudo code for color blending.

is $w_\varphi \cdot w_\gamma$. It is also possible to apply the remaining weight, w_θ, using graphics hardware. After we render the views, we have to read in the information from the frame buffer. Many graphics libraries, such as OpenGL, allow scaling each pixel while reading the frame buffer into memory. If we scale the alpha channel by w_θ, the resulting alpha value contains the final weight $w_\theta \cdot w_\varphi \cdot w_\gamma$.

3.4 Discussion

View-based rendering takes a step from model-based rendering towards image-based rendering. The advantage as compared to model-based rendering is that making view-based models is much easier than making full models for a large class of objects, such as the flower basket illustrated in Figs. 6 and 10. Additionally, it provides view-dependant texturing of the object, which enables one to create believable impressions of fine geometric detail through texturing without actually having to model the fine geometry. Using static texturing the illusion breaks once the object is rotated to a different angle from where the fine detail was originally seen. The advantage of view-based rendering as compared to image-based rendering methods [4,7] is that a much sparser sample set of images has to be obtained and less data has to be stored, since the geometric information enables us to view the same data from a continuous set of viewpoints close to the one where the view was really taken.

We have demonstrated interactive viewing of our reconstructed models from arbitrary viewpoints at speeds of up to eight frames per second.

4 Conclusion

We have presented two alternative approaches for surface reconstruction and display using range and color data. One of the methods involves creating a single object model that can then be viewed from an arbitrary viewpoint, the other method creates a set of view-based models that are texture mapped using color images, and composites a few of the models in image space to a final image. We are currently working on view-based texturing of complete surface models, so we can better compare the relative advantages of these two competing approaches.

References

1. B. Curless and M. Levoy. A volumetric method for building complex models from range images. In *Proceedings of SIGGRAPH '96*, pages 303–312, August 1996.
2. P. E. Debevec, C. J. Taylor, and J. Malik. Modeling and rendering architecture from photographs: A hybrid geometry- and image-based approach. In *SIGGRAPH 96 Conference Proceedings*, pages 11–20. ACM SIGGRAPH, Addison Wesley, August 1996.
3. M. Garland and P. Heckbert. Fast polygonal approximation of terrains and height fields. Technical Report CMU-CS-95-181, Dept. of Computer Science, Carnegie Mellon University, Pittsburgh, PA, 1995.

4. S. J. Gortler, R. Grzeszczuk, R. Szeliski, and M. F. Cohen. The lumigraph. In *SIGGRAPH 96 Conference Proceedings*, pages 43–54. ACM SIGGRAPH, Addison Wesley, August 1996.
5. H. Hoppe, T. DeRose, T. Duchamp, J. McDonald, and W. Stuetzle. Surface reconstruction from unorganized points. In *Proceedings of SIGGRAPH '92*, pages 71–78, July 1992.
6. H. Hoppe, T. DeRose, T. Duchamp, J. McDonald, and W. Stuetzle. Mesh optimization. In *Computer Graphics (SIGGRAPH '93 Proceedings)*, volume 27, pages 19–26, August 1993.
7. M. Levoy and P. Hanrahan. Light field rendering. In *SIGGRAPH 96 Conference Proceedings*, pages 31–42. ACM SIGGRAPH, Addison Wesley, August 1996.
8. K. Pulli, M. Cohen, T. Duchamp, H. Hoppe, L. Shapiro, and W. Stuetzle. View-based rendering: Visualizing real objects from scanned range and color data. Technical Report UW-CSE-97-04-01, Univ. of Washington, Seattle WA 98105, 1997. Available through ftp://ftp.cs.washington.edu/tr/1997/04/UW-CSE-97-04-01.d.
9. K. Pulli, T. Duchamp, H. Hoppe, John McDonald, L. Shapiro, and W. Stuetzle. Robust meshes from multiple range maps. In *Proc. IEEE Int. Conf. on 3-D Imaging and Modeling*, May 1997.
10. R. Szeliski. Rapid octree construction from image sequences. *CVGIP: Image Understanding*, 58(1):23–32, July 1993.
11. G. Taubin. A signal processing approach to fair surface design. In *Proceedings of SIGGRAPH '95*, pages 351–358, August 1995.

An Improved Active Shape Model: Handling Occlusion and Outliers

Nicolae Duta[1], Milan Sonka[2]

[1] Department of Computer Science, Michigan State University
East Lansing, MI 48824
[2] Department of Electrical and Computer Engineering, The University of Iowa
Iowa City, IA 52242

Abstract. An improvement of the *Active Shape* procedure identifying new examples of previously learned shapes using the point distribution model is presented. The novel segmentation and interpretation approach incorporates *a priori* knowledge about the objects of interest and their specific structural relationships to provide robust segmentation and labeling.
The method was utilized to successfully identify 10 neuroanatomic structures in 19 individual MR images and 2 car classes (left-right and right-left oriented) in 400 perspective images of street scenes.

1 Introduction

There have been numerous attempts to build models describing shape and appearance of non-rigid objects, and employ them for automated object identification in the analyzed images [1–5]. Among them, the Point Distribution Model (PDM) representing the variation of a set of shapes around the average that was designed by Cootes and Taylor has many favorable shape representation properties [6, 7].

We report an improvement of the *Active Shape* procedure [7] designed to find new examples of previously learned shapes using the point distribution model. This approach is particularly useful if variations in shape and appearance are difficult to model as is often the case with non-rigid objects or when point-of-view perspective is involved. The method presented below is generally applicable to virtually any task involving deformable shape analysis.

2 Object Occlusion and Shape Outliers in PDM-based Image Interpretation: A New Approach

In many areas of image segmentation and interpretation, reliable a priori knowledge is available to help guide the image analysis process. In many cases, approximate positions of individual objects can be determined from context. Knowledge about object sizes, shapes, gray level appearance, etc. can be acquired from a training set of examples.

2.1 Knowledge-Based Point Distribution Model

In order to take advantage of the available a priori knowledge, three additional features were included in the model: Gray-level appearance, border strength, and average position. In the implementation described below, we also used the implicit knowledge about object context representing inter-relationships of several objects.

Gray-level appearance is calculated in neighborhoods around each of the shape model points. It is determined for every shape model point j of each training image along a profile g_j of a constant length, centered at the point j. Since the profiles vary with gray level scaling, derivatives of the gray levels along each profile are determined and normalized.

Border strength is determined for each border segment of the model. Every two consecutive model points that lie on the object boundary define a border segment. To compute its strength, a local filtering is applied to each clique on that border segment. The filter is based on a pair of close parallel profiles.

Average position of each shape model point that is calculated in the image coordinates is also incorporated in the model.

Our **knowledge-based shape model** combines generally applicable parameters of the point distribution model and the knowledge-specific parameters appropriate for the image segmentation task in question. As such, the complete model is composed of:

1. The eigenvectors corresponding to the largest eigenvalues of the covariance matrix describing the *Allowable Shape Domain* [6,7].
2. The average gray level appearance values for each point of the model.
3. The average border strength for corresponding border segments and the parameters of the mask (width, length) for which the strength was computed.
4. The average position of the points of the average shape.
5. Connectivity information (the number of shapes, point ordering along contours).

2.2 Searching for Objects: Model Fitting

The searching procedure developed for our PDM approach to image segmentation and interpretation is based on a model fitting strategy that substantially differs from the *Active Shape Procedure* of Cootes and Taylor. The difference is twofold. First, our search is entirely model driven meaning that segmentation hypotheses are not influenced by possibly misleading image data and do not use any preprocessing. At each step of the fitting process, several model location hypotheses are considered and evaluated. Second, an outlier detection and replacement procedure has been developed to detect misplaced points and infer their new positions. The outlier detection improves robustness and accuracy of the shape model fitting process.

The searching procedure consists of the following steps: 1) Model fitting using linear transforms, 2) model fitting using piecewise linear transforms, 3) outlier removal, 4) final point adjustment, 5) final outlier removal.

Model Fitting Function As a result of the hypotheses generation processes, shape model locations are sequentially hypothesized. In order to evaluate the model location hypotheses, a *fitness function* is needed to assess the agreement between the image data and the particular model instance. We have designed a fitness function $F = F_B/(F_{GA})^2$ that consists of two components:

1. *Fitness of the gray level appearance* F_{GA} is determined as the average squared Euclidean distance between the actual gray level appearance and the mean gray level profile incorporated in the shape model.
2. *Fitness of the border* F_B is calculated as the ratio between the aggregate response of all four point cliques along the contour and the maximum possible response (twice the number of cliques).

Model fitting using linear transforms Shape instance hypotheses specify the locations of all model points within the analyzed image. The hypotheses are generated using affine transformations and are applied to the model average position. The parameters of the linear transforms that contribute to the hypotheses generation are application dependent and reflect the a priori knowledge one has about the scale and pose of the objects of interest. For the two applications reported here the parameters were: scaling in the range [0.9, 1.1], step 0.1 (both applications); rotation [-8°, 8°], step 4° (brain model) and [-4°, 4°] (car model); and translation [-4, 4] pixels, step 1 pixel in both x, y directions (brain model) respectively [-128, 128] pixels (car model). Each hypothesis represents a rigid transform of the average shape, and all generated hypotheses are sequentially evaluated using the model fitting function and the best fit is determined. If the prior knowledge includes the fact that the objects of interest are always present in the images processed (as is the case of the brain images) no thresholding is subsequently done, otherwise the best fit value is thresholded and no object reported if its value is low.

Model fitting using piecewise linear transforms Since non-rigid objects or object with inter-subject variability are discussed here, rigid linear transforms do not account for any potential deformations of the expected shape. Therefore, linear transforms (translation, rotation, and scaling) with a small range of parameters are applied to subsets of consecutive model points. Each model point is taken in turn as the center of such a subset. After the best position of the subset is obtained, a thresholding of the fitting value is performed and the center point of the set is considered to be occluded and discarded if the value is low. Otherwise, only the position of the center point is kept and the remaining points are discarded. The number of consecutive points that are considered for this transform is application dependent and is a function of local border strength and length. To preserve robustness, the center point is not moved if the two border segments adjacent to it are very weak.

Outlier removal: A new approach Under unfavorable circumstances, the previous step may introduce incorrectly determined vertices – **outliers**. This may happen if a subshape fitted by the previous step exhibits weak edges or if there exists another border of similar properties in the neighborhood. In the existing literature dealing with point distribution models, no outlier detection has been introduced. Typically, when using PDM's, shapes that do not correspond to the allowed shape at any stage of the detection process are rejected (Fig. 1). In other words, the model parameters b_j (Eq. 2) must not exceed some maximum values. Cootes and Taylor propose to derive limits for b_j by examining the distributions of the parameter values required to generate the training set [7]. They recommend that b_j's be chosen such that

$$-3\sqrt{\lambda_j} \leq b_j \leq 3\sqrt{\lambda_j} \qquad (1)$$

since most of the population lies within three standard deviations of the mean. This approach can introduce two kinds of errors: 1) The shape/location hypothesis is completely rejected because one or two points are misplaced, such situation is documented in Fig. 1; or 2) the hypothesis is accepted even though one or two points are misplaced.

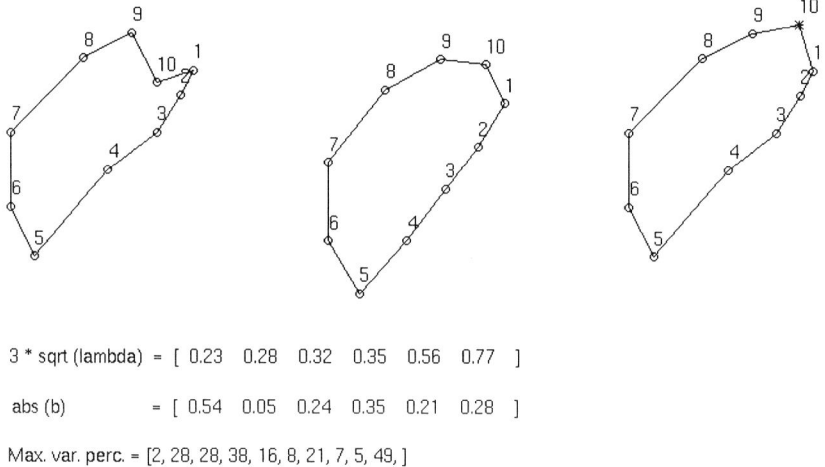

3 * sqrt (lambda) = [0.23 0.28 0.32 0.35 0.56 0.77]

abs (b)   = [0.54 0.05 0.24 0.35 0.21 0.28]

Max. var. perc. = [2, 28, 28, 38, 16, 8, 21, 7, 5, 49,]

Fig. 1. Example of a shape hypothesis rejected by the Cootes' Active Shape procedure (left). Note an outlier responsible for rejection (marked by 10). The same shape hypothesis after our outlier removal and adjustment steps (right, adjusted vertex marked by *). The average shape is shown in the middle. The values given below the figure are specified by Eq. 1, 2, and 6.

To treat the problem of outliers in a systematic fashion, we have developed a new approach to outlier detection and position adjustment. The misplaced points are identified using the information about the relative positions of the shape model vertices that are implicitly included in the shape model. Let $z = (x', y')^T$ be the model point positions after the piecewise linear transforms were applied and the resulting shape was aligned with the shape average. According to the shape model, the hypothesized shape should satisfy

$$\mathbf{z} = \bar{\mathbf{x}} + P\mathbf{b} \qquad (2)$$

where $\bar{\mathbf{x}}$ is the average shape, P is the matrix of the first t eigenvectors, and $\mathbf{b}$ is a vector of weights. Therefore,

$$b_j = \sum_{i=1}^{n} P_{i,j}(z_i - \bar{\mathbf{x}}_i) = \sum_{i=1}^{n} v_{i,j} \qquad (3)$$

where

$$|v_{i,j}| = |P_{i,j}(z_i - \bar{\mathbf{x}}_i)| \qquad (4)$$

is the absolute variation induced by point i in parameter b_j. Let the percentage of variation induced by point i in parameter b_j be defined as

$$V_{i,j} = \frac{|v_{i,j}|}{\sum_{i=1}^{n}|v_{i,j}|} 100 \qquad (5)$$

and let the maximum percentage of variation induced by point i in any of the parameters b_j be defined as

$$u_i = \max_{j=1..t} V_{i,j} \qquad (6)$$

If all the points were to generate an equal amount of variation, then all the percentages u_i were approximately $100/n$, n being the number of object points. However, since outliers may be present, larger variation may be associated with some points – the outliers. A point is considered to be an outlier if the percentage of variance generated by its position in any of the parameters b_j of the model is more than 4× greater than the average amount of variation. If several outliers are present, the variance is distributed among them and (perhaps) the well placed points. As a result, it is difficult to identify outliers if more than a few occur simultaneously. Once such misplaced points are detected, they must be moved to a new position that can be inferred from the alignment of the rest of the shape instance with the average shape.

Final point adjustment Some of the shape model points may have been declared outliers in the previous step. Consequently, their position may have been adjusted solely considering the average shape appearance and not considering the image data. Therefore, they must be subject to the position optimization step to better correspond with the image data.

Final outlier removal Resulting from the previous steps, or newly introduced during the final point adjustment, outliers may remain present in the shape model. Following the same outlier detection procedure as applied in the first outlier removal step, the outliers are identified and removed, no adjustment is attempted in this final step of model fitting.

3 Results

The method presented above was employed to design two PDM shape models. 1) Magnetic resonance images of human brain were segmented into neuroanatomic

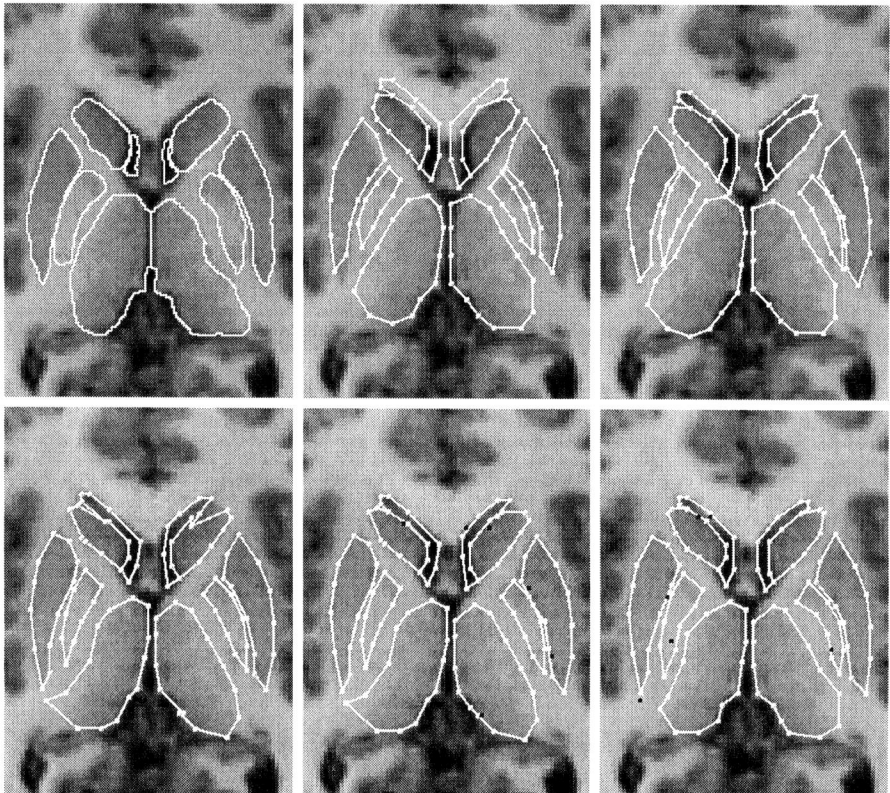

Fig. 2. Example of automated brain image segmentation and interpretation. Upper row from left to right: Manual tracings. Initial average position of the shape model. Optimal shape model position after linear transform step. Bottom row from left to right: Optimal shape model position after piecewise linear transform step. Outlier detection – outliers marked by dark dots). Final outlier detection – marked by dark dots and removed from consideration in the shape model.

brain structures. 2) Perspective images of moving cars on a street were used. Training image sets served for construction of shape models, and the method's performance was quantitatively assessed in separate testing image sets.

First, the PDM approach was trained in 8 images and tested in 19 MR brain images. Ten neuroanatomic structures were successfully segmented and interpreted in all images from the test set. Fig. 2 shows the observer-traced and computer-detected contours together with the intermediate results. In the test set, the neuroanatomic structures were identified with the labeling error of $7 \pm 3\%$, and the average border positioning error was 0.8 ± 0.1 pixels. The automatically-determined borders of the ten neuroanatomic structures exhibit a high level of accuracy and substantial speedup when compared to a previously reported genetic image interpretation approach [8].

Fig. 3. Example of automated car segmentation. a) Original image. b) Optimal left-to-right car model position. c) Optimal right-to-left car model position. d) Second best right-to-left car model position.

Second, the PDM approach was employed in identification of two car classes (left-right and right-left orientation) in perspective images of street scenes. After training in 10 images, our approach correctly detected the cars present in 400 images from the test set. Fig. 3 shows an example of the computer-detected contours for the two car models. Note that, although the model was trained only on sedans, it was also able to accurately detect a hatchback car (except for the one contour point with an opposite convexity). Also note the good detection of the occluded cars in the background. As a matter of fact, all cars exhibiting less than 40% occlusion were successfully detected. The car segmentation results represent a substantial improvement over a recently published study [5].

4 Discussion

We expect further improvement of the method's performance by incorporating better structure positioning information. At this moment, no explicit statistical information is used concerning the relative positions of the objects. The only constraint that is enforced after each step is that no two objects overlap that belong to the same multi-object model. In case of object overlap, the faulting objects have to be grouped together. If an object is misplaced but does not overlap with another object, the hypothesis is accepted and a wrong detection may occur. While we did not experience such a behavior, the possibility remains present. Furthermore, no statistical information is used concerning the model points variation. Incorporation of such information may be useful for the outlier detection procedure.

5 Conclusion

A new fully automated segmentation and interpretation method has been presented. The method was utilized to identify 10 neuroanatomic structures in individual MR images and 2 car classes (left-right and right-left oriented) in perspective images of moving cars. In all cases, the method was independently trained and tested in separate images sets. The method correctly segmented and interpreted images from the two very dissimilar application areas.

Acknowledgments

This work was supported in part by the NSF grant IRI 96-16747.

References

1. M Kass, A Witkin, and D Terzopoulos. Snakes: Active contour models. In *Proceedings, First International Conference on Computer Vision, London, England*, pages 259–268, Piscataway, NJ, 1987. IEEE.
2. L H Staib and J S Duncan. Boundary finding with parametrically deformable models. *IEEE Trans. Pattern Anal. and Machine Intelligence*, 14(11):1061–1075, 1992.
3. A Jain, Y Zhong, and S Lakshmanan. Object matching using deformable templates. *IEEE Trans. Pattern Anal. and Machine Intelligence*, 18(3):267–277, 1996.
4. M Sonka, V Hlavac, and R Boyle. *Image Processing, Analysis, and Machine Vision*. Chapman and Hall, London, New York, 1993.
5. A Jain, M P Dubuisson, and S Lakshmanan. Vehicle segmentation using deformable templates. *IEEE Trans. Pattern Anal. and Machine Intelligence*, 18(3):293–308, 1996.
6. T F Cootes, A Hill, C J Taylor, and J Haslam. Use of active shape models for locating structures in medical images. *Image & Vision Computing*, 12(6):355–366, 1994.
7. T F Cootes, C J Taylor, D H Cooper, and J Graham. Active shape models – their training and application. *Computer Vision and Image Understanding*, 61:38–59, 1995.
8. M Sonka, S K Tadikonda, and S M Collins. Knowledge-based interpretation of MR brain images. *IEEE Trans. Med. Imaging*, 15:443–452, 1996.

Perspective Matching Using the EM Algorithm

A.D.J. Cross and E.R. Hancock

Department of Computer Science,
University of York,
York, Y01 5DD, UK.
erh@minster.york.ac.uk

Abstract. This paper describes a new approach to perspective matching which simultaneously exploits both rigidity-structure and point distribution information. The structural component of the model is represented by a Delaunay triangulation of the point-set. The point-distribution model is represented by a perspective deformation of the point-set. Model-matching is realised using a variant of the EM algorithm. This involves coupling the correspondence matching of the Delaunay triangulation to the recovery of the point deformation parameters. We use a Bayesian consistency measure to gauge the relational structure of the point correspondences. Maximum-likelihood point deformation parameters are estimated using a mixture-model defined over the point error-residuals. In effect, the Bayesian consistency measure is used to weight the contributions to a mean-squares error-criterion. The method is evaluated on matching 2D objects under varying pose.

1 Introduction

Rigidity constraints play an important role in correspondence matching [6,11]. It was Ullman [11] who first suggested the use of point-rigidity as a way of imposing proximity constraints on point-sets. Several authors have drawn inspiration from Ullman's ideas in developing general purpose correspondence matching algorithms using the the Gaussian weighted proximity matrix. There are two contrasting uses of the proximity-matrix which deserve special mention. Scott and Lonquet-Higgins [7] locate correspondences by finding a singular value decomposition of the inter-image proximity matrix. Shapiro of Brady [9], on the other hand, match by comparing the modal eignestructure of the intra-image proximity matrix. In fact, these techniques provide the basic ground-work on which the deformable shape models of Cootes *et al* [3] and Sclaroff and Pentland [8] build. More recently, McReynolds and Lowe [6] have shown how rigidity constraints can be incorporated into the recovery of perspective correspondence matches. This algorithm uses Levenberq-Marquardt optimisation to locate least-squares correspondence matches.

Our aim in this paper is to present a statistical framework that allows both the rigidity structure and deformation properties of point-sets to be utilised in perspective matching. These two aspects of the matching process have distinct representations. Point-set deformation is modelled using a Gaussian mixture-model for the co-ordinate error-residuals under a perspective transformation. Rigidity-structure is represented by a Delaunay triangulation [1] over the point sets being matched. We recover the correspondence matches using a variant of the EM algorithm. This is an iterative process which alternates between transformation parameter estimation in the maximisation step and updating matching probabilities in the expectation step.

Maximisation returns three pieces of information. The first of these is a set maximum-likelihood perspective parameters which are used to project the model-points onto the view-space of the data. The second piece of information is the variance-covariance matrix for the co-ordinate error residuals. Finally, we estimate a set of *maximum a posteriori probability* point correspondence matches between the nodes of the Delaunay triangulations representing the structure of the point-sets. In other words, the M-step models the co-ordinate projection of the deformed shape into a view space where measurements are taken. Expectation involves updating probability distributions which describe the current state of the matching process. Here we dichotomise between the point-distribution and structural aspects of our model.

The outline of this paper is as follows. In Section 2, we detail the elements of our point-matching process. Section 3 describes how we establish Delaunay graphs and exploit graph structure in finding correspondence matches. The two-step iterative matching process is detailed in Section 4. Section 5 presents some experiments concerned with matching different 3D poses of planar shapes. Finally, Section 6 offers some conclusions and describes our future plans.

2 Point Deformation

In the work described here we are interested in matching 2D shapes under different perspective viewing conditions. Projective deformation of a point-set involves translation, rotation, affine shear and foreshortening. Our basic aim is to recover the parameters of the perspective transformation which brings a set of model or fiducial points into correspondence with their counterparts in a set of image data. Each point in the image data is represented by an augmented vector of co-ordinates $\underline{w}_i = (x_i, y_i, 1)^T$ where i is the point index. The available set of image points is denoted by $\mathbf{w} = \{\underline{w}_i, \forall i \in \mathcal{D}\}$ where $\mathcal{D}$ is the point index-set. The fiducial points constituting the model are similarly represented by the set $\mathbf{z} = \{\underline{z}_j, \forall j \in \mathcal{M}\}$. Here $\mathcal{M}$ is the index-set for the model feature-points and the $\underline{z}_j$ represent the corresponding image co-ordinates. The aim of our matching algorithm is to iteratively recover a parameter-matrix $\Phi^{(n)}$ which describes the perspective co-ordinate transformation that brings the feature points in image and model into registration with one-another. The transformed co-ordinate vectors are computed in the following way

$$\underline{z}_j^{(n)} = \frac{1}{\underline{z}_j^T . \underline{\Psi}^{(n)}} \Phi^{(n)} \underline{z}_j \qquad (1)$$

In the above equation the matrix $\Phi^{(n)}$ models the perspective transformation of co-ordinates. In the work reported here we are not concerned with explicitly recovering the geometry of translation, rotation, affine shear and foreshortening. Instead, we adopt the following parameterisation of the perspective deformation of the point-set. The required transformation matrix has nine free parameters and is of the form

$$\Phi^{(n)} = \begin{pmatrix} \phi_{1,1}^{(n)} & \phi_{1,2}^{(n)} & \phi_{1,3}^{(n)} \\ \phi_{2,1}^{(n)} & \phi_{2,2}^{(n)} & \phi_{2,3}^{(n)} \\ \phi_{3,1}^{(n)} & \phi_{3,2}^{(n)} & \phi_{3,3}^{(n)} \end{pmatrix} \quad (2)$$

The superscript on the matrix indicates that the parameters are taken at n^{th} iteration of our algorithm. The vector $\Psi^{(n)} = (\phi_{3,1}^{(n)}, \phi_{3,2}^{(n)}, 1)^T$ is formed from the elements of the bottom row of the perspective parameter matrix. Expressed in this form, the transformation matrix subsumes the simpler Euclidean and affine cases which require four and six parameters respectively.

3 Structural Deformation

One of our goals in this paper is to exploit structural constraints to improve the recovery of perspective parameters from sets of feature points. Because of its well documented robustness to noise and change of viewpoint, we adopt the Delaunay graph as our basic representation of image structure [1]. We establish Delaunay graph representations of data and model, by seeding Voronoi tessellations from the feature-points. Our aim is to exploit the structure of the Delaunay graphs to impose constraints on the recovery of perspective parameters. The overall parameter estimation strategy is based on the expectation-maximisation algorithm. We incorporate the structural information into the parameter estimation process by borrowing an idea from the hierarchical mixture of experts algorithm of Jordan and Jacobs [5]. We use the probabilities of structural matches to gate the expected likelihood function. These gating probabilities are computed using the framework for relational graph matching recently reported by Wilson and Hancock [13,14,12].

The process of Delaunay triangulation generates relational graphs from the two sets of point-features. More formally, the point-sets are the nodes of the a data graph $G_D = \{\mathcal{D}, E_D\}$ and a model graph $G_M = \{\mathcal{M}, E_M\}$. Key to our perspective deformation process is the idea of using the structure of Delaunay graphs to find correspondences between the two point-sets. This correspondence matching is denoted by the function $f : \mathcal{M} \to \mathcal{D}$ from the nodes of the data-graph to those of the model graph. According to this notation the statements $(i,j) \in f^{(n)}$ or equivalently $f^{(n)}(i) = j$ indicate that there is a match between the node $i \in \mathcal{D}$ of the model-graph to the node $j \in \mathcal{M}$ of the model graph at iteration n of the algorithm.

We exploit the structure of the Delaunay graphs to compute the consistency of match using the Bayesian framework for relational graph-matching recently reported by Wilson and Hancock [12–14]. Details of the method are outside the scope of this paper. Suffice to say that consistency of a configuration of matches residing on the neighbourhood $R_i = i \cup \{k \; ; \; (i,k) \in E_D\}$ of the node i in the data-graph and its counterpart $S_j = j \cup \{l \; ; \; (j,l) \in E_m\}$ for the node j in the model-graph is gauged by Hamming distance. The Hamming distance $H(i,j)$ counts the number of matches on the data-graph neighbourhood R_i that are inconsistently matched onto the model-graph neighbourhood S_j. According to Wilson and Hancock [12–14] the probability that the data-graph node i matches to the model-graph node j at iteration n of the algorithm is given by

$$\zeta_{i,j}^{(n)} = \frac{\exp\left[-\beta H(i,j)\right]}{\sum_{j \in \mathcal{M}} \exp\left[-\beta H(i,j)\right]} \tag{3}$$

In the above expression, the Hamming distance is given by

$$H(i,j) = \sum_{(u_k, v_k) \in R_i \bullet S_j} \left(1 - \delta_{f^{(n)}(u_k), v_k}\right) \tag{4}$$

where the symbol $\bullet$ denotes the composition of the data-graph relation R_i and the model-graph relation S_j. The exponential constant $\beta = \ln \frac{1-P_e}{P_e}$ is related to the uniform probability of structural matching errors P_e. This probability is set to reflect the overlap of the two point-sets In the work reported here we set P_e as follows

$$P_e = \frac{2||\mathcal{M}| - |\mathcal{D}||}{||\mathcal{M}| + |\mathcal{D}||} \tag{5}$$

4 The EM Algorithm

In this section we describe the main stages of our iterative matching scheme. Our aim is to extract both maximum likelihood perspective parameters and maximum *a posteriori* matching probabilities using coupled update operations. In the spirit of Dempster, Laird and Rubin's EM algorithm [4], we aim to condition the updated parameter estimates (i.e. $\Phi^{(n+1)}$) on the most recently available correspondence matches (i.e. $f^{(n)}$). In other words, the maximum-likelihood parameters satisfy the following condition

$$\Phi^{(n+1)} = \arg\max_{\Phi} p(\Phi | \mathbf{w}, f^{(n)}) \tag{6}$$

In a similar way, the *maximum a posteriori* matches are conditioned upon the most recently available parameter-estimates. The matching configuration therefore satisfies the following condition

$$f^{(n+1)} = \arg\max_{f} P(f | \mathbf{w}, \Phi^{(n)}) \tag{7}$$

4.1 Expectation

We have recently shown how coupled updates of this form can be realised through the optimisation of single integrated expected likelihood function. Details of the formal development are outside the scope of this paper and can found in the recent account of Turner and Hancock [10]. Suffice to say that the parameters and the correspondence matches may be sought through joint optimisation of the quantity

$$Q(\Phi^{(n+1)}|\Phi^{(n)}) = \sum_{(i,j) \in f^{(n)}} P(\underline{z}_j|\underline{w}_i, \Phi^{(n)}) \zeta_{i,j}^{(n)} \ln p(\underline{w}_i|\underline{z}_j, \Phi^{(n+1)}) \qquad (8)$$

The structure of this expected log-likelihood function requires further comment. The measurement densities $p(\underline{w}_i|\underline{z}_j, \Phi^{(n+1)})$ model the distribution of error-residuals between the observed model-point position $\underline{w}_i$ and the predicted position of the model point $\underline{z}_j$ under the current set of transformation parameter $\Phi^{(n+1)}$. The log-likelihood contributions at iteration $n+1$ are weighted by the *a posteriori* measurement probabilities $P(\underline{z}_j|\underline{w}_i, \Phi^{(n)})$ computed at iteration n of the algorithm. Following Jordan and Jacobs [5] we gate the individual expected-likelihood contributions using the the structural matching probabilities $\zeta_{i,j}^{(n)}$. Finally, the summation extends over the set of correspondence matches $(i,j) \in f^{(n)}$ available at iteration n.

Using the Bayes rule, we can re-write the *a posteriori* measurement probabilities in terms of the of the conditional measurement densities

$$P(\underline{z}_j|\underline{w}_i, \Phi^{(n)}) = \frac{\alpha_j^{(n)} p(\underline{w}_i|\underline{z}_j, \Phi^{(n)})}{\sum_{j' \in \mathcal{M}} \alpha_{j'}^{(n)} p(\underline{w}_i|\underline{z}_{j'}, \Phi^{(n)})} \qquad (9)$$

The mixing proportions are computed by averaging the *a posteriori* probabilities over the set of data-points, i.e.

$$\alpha_j^{(n+1)} = \frac{1}{|\mathcal{D}|} \sum_{i \in \mathcal{D}} P(\underline{z}_j|\underline{w}_i, \Phi^{(n)}) \qquad (10)$$

4.2 Gaussian Error Model

In order to proceed with the development of a point registration process we require a model for the conditional measurement densities, i.e. $p(\underline{w}_i|\underline{z}_j, \Phi^{(n)})$. Here we assume that the required model can be specified in terms of a multivariate Gaussian distribution. The random variables appearing in these distributions are the error residuals for the position predictions of the jth model point delivered by the current estimated transformation parameters. Accordingly we write

$$p(\underline{w}_i|\underline{z}_j, \Phi^{(n)}) = \frac{1}{(2\pi)^{\frac{3}{2}} \sqrt{|\Sigma|}} \exp\left[-\frac{1}{2} \epsilon_{i,j}(\Phi^{(n)})^T \Sigma^{-1} \epsilon_{i,j}(\Phi^{(n)})\right] \qquad (11)$$

In the above expression Σ is the variance-covariance matrix for the vector of error-residuals $\epsilon_{i,j}(\Phi^{(n)}) = \underline{w}_i - \underline{z}_j^{(n)}$ between the components of the predicted measurement vectors $\underline{z}_j^{(n)}$ and their counterparts in the data, i.e. $\underline{w}_i$. Formally, the matrix is related to the expectation of the outer-product of the error-residuals i.e. $\Sigma = E[\epsilon_{i,j}(\Phi^{(n)})\epsilon_{i,j}(\Phi^{(n)})^T]$.

With these ingredients, the expectation step of the EM algorithm simply reduces to computing the weighted squared error criterion

$$Q'(\Phi^{(n+1)}|\Phi^{(n)}) = \sum_{(i,j)\in f^{(n)}} P(\underline{z}_j|\underline{w}_i, \Phi^{(n)})\zeta_{i,j}^{(n)} \epsilon_{i,j}(\Phi^{(n+1)})^T \Sigma^{-1} \epsilon_{i,j}(\Phi^{(n+1)}) \quad (12)$$

In other words, the *a posteriori* probabilities $P(\underline{z}_j|\underline{w}_i, \Phi^{(n)})$ and the structural matching probabilities $\zeta_{i,j}^{(n)}$ effectively regulate the contributions to the likelihood function. Matches for which there is little evidence contribute insignificantly, while those which are in good registration dominate.

4.3 Maximisation

The maximisation step of our perspective deformation algorithm is based on dual coupled update processes. The first of these aims to locate maximum *a posteriori* probability correspondence matches. The second update operation is concerned with locating maximum likelihood perspective parameters. We effect the coupling by allowing information flow between the two processes.

Maximum *a posteriori* probability matches Point correspondences are sought so as to maximise the *a posteriori* probability of structural match. Individual point-correspondences should be updated in the following manner

$$f^{(n+1)}(i) = \arg\max_{j\in\mathcal{M}} P(\underline{z}_j|\underline{w}_i, \Phi^{(n)})\zeta_{i,j}^{(n)} \quad (13)$$

Once this update equation has been applied, the unmatched model-graph nodes are identified for removal from the triangulation. At this point the edited set of model feature-points is re-triangulated along the lines suggested in Section 3 The updated structural matching probabilities $\zeta_{i,j}^{(n+1)}$ are also updated using equations (3) and (4) as outlined in Section 2.

Maximum likelihood parameters Maximising the expected log-likelihood function is equivalent to minimising the following weighted squared error criterion

$$\Phi^{(n+1)} = \arg\min_{\Phi} \sum_{(i,j)\in f^{(n)}} P(\underline{z}_j|\underline{w}_i, \Phi^{(n)})\zeta_{i,j}^{(n)} \epsilon_{i,j}(\Phi)^T \Sigma^{-1} \epsilon_{i,j}(\Phi) \quad (14)$$

5 Experiments

The real-world evaluation of our matching method is concerned with recognising planer objects in different 3D poses. The object used in this study is a 3.5 inch floppy disk which is placed on a desktop. The scene is viewed with a low-quality SGI IndyCam. The feature points used to triangulate the object are corners. Since the imaging process is not accurately modelled by a perspective transformation under pin-hole optics, the example provides a challenging test of our matching process.

Our experiments are illustrated in Figure 1. The first two columns show the views under match. In the first example (the upper row of Figure 1) we are concerned with matching when there is a significant difference in perspective forshortening. In the example shown in the second row of Figure 1, there is a rotation of the object in addition to the foreshortening. The final two rows show the matching of a fronto-parallel view against rotated and forshortened ones. The images in the third column are the initial matching configurations. Here the perspective parameter matrix has been selected at random. The fourth column in Figure 1 shows the final matching configuration after the EM algorithm has converged. In all four cases the final registration is accurate. The algorithm appears to be capable of recovering good matches even when the initial pose estimate is poor.

6 Conclusions

Our main contribution in this paper has been to develop a new algorithm for matching under perspective deformation. The method integrates both relational point-structure and point-distribution information into the deformation process. Matching is realised using the EM algorithm. We have illustrated the effectiveness of the resulting perspective deformation process in the matching of 3D poses of planar objects.

References

1. N Ahuja, "Dot Pattern Processing using Voronoi Neighbourhoods", *IEEE PAMI*, **4**, pp 336–343, 1982.
2. J-D. Boissonnat, "Geometric Structures for Three-Dimensional Shape Representation", *ACM Transactions on Graphics*, **3**, pp. 266-286, 1984.
3. T.F Cootes, C.J. Taylor, D.H. Cooper and J. Graham, "Active Shape Models - Their Training and Application", *Computer Vision, Graphics and Image Understanding*, **61**, pp. 38–59, 1995.
4. A.P. Dempster, Rubin N.M. and Rubin D.B., "Maximum-likelihood from incomplete data via the EM algorithm", J. Royal Statistical Soc. Ser. B (methodological),**39**, pp 1-38, 1977.
5. M.I. Jordan and R.A. Jacobs, "Hierarchical Mixtures of Experts and the EM Algorithm", *Neural Computation*, **6**, pp. 181-214, 1994.

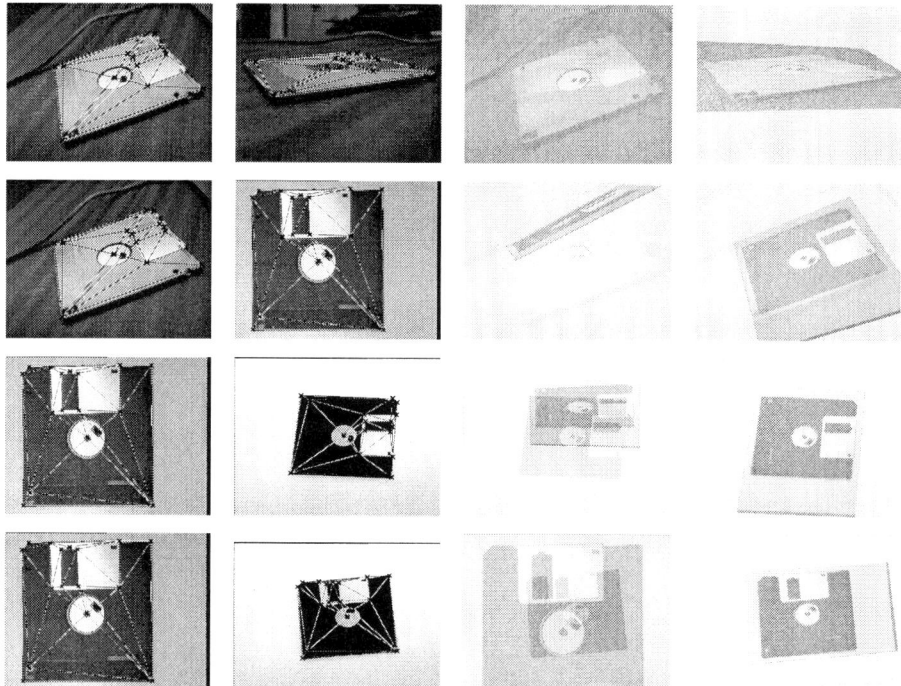

Fig. 1. Left to right : model, data, initial configurations and final configuration for the matching process.

6. D.P. McReynolds and D.G. Lowe, "Rigidity Checking of 3D Point Correspondences under Perspective Projection", *IEEE PAMI*, **18**, pp. 1174–1185, 1996.
7. G. L. Scott and H.C. Longuet-Higgins, "An Algorithm for Associating the Features of 2 Images", *Proceedings of the Royal Society of London Series B-Biological*, **244**, pp. 21–26, 1991.
8. S. Sclaroff and A.P. Pentland, "Modal Matching for Correspondence and Recognition", *IEEE PAMI*, **17**, pp. 545–661, 1995.
9. L.S. Shapiro and J. M. Brady, "Feature-based Correspondence - An Eigenvector Approach", *Image and Vision Computing*, **10**, pp. 283–288, 1992.
10. M. Turner and E.R. Hancock, "An EM-like Relaxation Operator", *Proceedings of the Thirteenth International Conference on Pattern Recognition*, Volume II, pp. 166–170, 1996.
11. S. Ullman, "The Interpretation of Visual Motion", *MIT Press*, 1979.
12. R.C. Wilson, A.N. Evans and E.R. Hancock, "Relational Matching by Discrete Relaxation", *Image and Vision Computing*, **13**, pp. 411–421, 1995.
13. R.C. Wilson and E.R. Hancock, "Relational Matching with Dynamic Graph Structures", *Proceedings of the Fifth International Conference on Computer Vision*, pp. 450-456, 19
14. R.C. Wilson and E.R. Hancock, "Gauging Relational Consistency and Correcting Structural Errors", *IEEE Computer Society Computer Vision and Pattern Recognition Conference*, pp. 54–62, 1996.

Identifying Human Face Profiles with Semi-Local Integral Invariants

Jun Sato and Roberto Cipolla

Department of Engineering, University of Cambridge, Cambridge CB2 1PZ, England

Abstract. In this paper, we propose a method for identifying human face profiles by using invariant representations of image curves. In particular, we show that semi-local invariants are very useful for matching profile curves and identifying human faces. A method for finding face profiles from the changes in apparent contours of faces is also considered. The results of some experiments with real images of human faces show the power of the proposed method.

1 Introduction

Identifying human faces is very important for many applications in computer vision, such as security systems. From ancient times, artists often used human profiles to depict human faces. In fact human face profiles provide important cues to identify an individual person as shown in Fig. 1.

Although human faces are curved surfaces, we can assume that the face profiles are the projections of planar curves, if the viewing axis is perpendicular to the plane of symmetry, and if the distance to the face from the viewer is large enough. Under this assumption, the distortions of the profile caused by viewer motions are described by similarity transformations. Thus, if the similarity invariants are available from the profile curves, we can use these invariants for identifying individual faces.

The question is what sort of invariants are available from the profile curves. Unfortunately, existing invariants [1,6,9-11] in computer vision suffer from problems due to the occlusion of objects, the existence of image noise and the requirement of point or line correspondences. That is, if the invariants are defined locally they are sensitive to noise, and if the invariants are defined globally they suffer from the occlusion and the correspondence problems. The existence and importance of semi-local invariants has recently been shown [7,8]. In this paper, we show that semi-local invariants under the similarity group can be applied successfully for identifying individual human face profiles.

We first consider invariant representations of human face profiles. We next investigate how to find the canonical profile of a face under rotational motions of the face. The preliminary results of face identification experiments are shown. The distinguishability of the proposed method is also investigated.

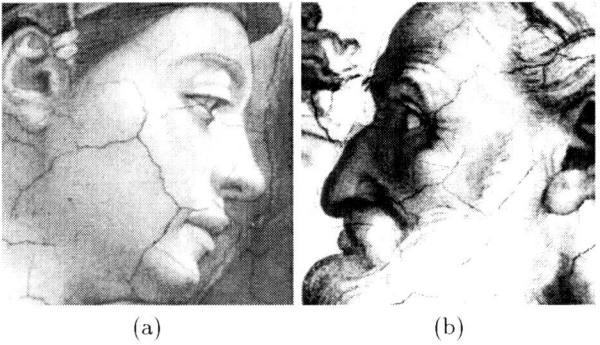

Fig. 1. Face profiles. (a) and (b) are example pictures of human face profiles painted by Michelangelo in the Sistina Chapel in Rome. Profiles have often been used to depict human faces.

2 Invariant Representation of Face Profiles

Consider a face to have 3D bilateral symmetry. The direction of the face and the longitudinal axis define a plane of symmetry (see Fig. 2). The ray from the camera center touches the curved surface of the face. The contour generator is defined on the curved surface as a set of tangent points of the ray to the surface, and separates the visible and occluded parts on the surface. The projection of the contour generator onto the image plane is an apparent contour [4]. In general, the contour generator slips over the surface as the relative position and orientation between the camera and the surface changes.

If the viewing axis is perpendicular to the plane of symmetry as shown in Fig. 2, we call the apparent contour a *canonical profile* of the face. If the distance from the viewer to the face is large compared to the depth of the face, then the contour generator on the face can be considered a planar curve (or approximately a planar curve) which lies on the plane of symmetry. Since the plane of symmetry is perpendicular to the viewing axis, the distortions of the profile curve of a face caused by the changes in viewpoint are described by planar similarity transformations. Thus, if we can extract invariants under the similarity group, then we could use these invariants for identifying individual human face profiles. The problem is what sort of similarity invariants can be applied for this task. To recognise faces reliably the invariants should not be sensitive to noise, should not suffer under partial occlusions and should have enough distinguishability. However, in general, local invariants such as differential invariants [11] are sensitive to noise and are poor in distinguishability, while global invariants such as moment invariants [5,9] suffer from the occlusion problem. To cope with these problems we next introduce semi-local integral invariants, which are less sensitive to noise than classical differential invariants, and unlike moment invariants do not suffer from the occlusion problem.

The semi-local integral invariants were introduced by Sato and Cipolla [7] and Bruckstein et al. [2], and are used for matching image curves and extracting symmetry axes of planar and 3D bilateral symmetry [8]. The idea of semi-local

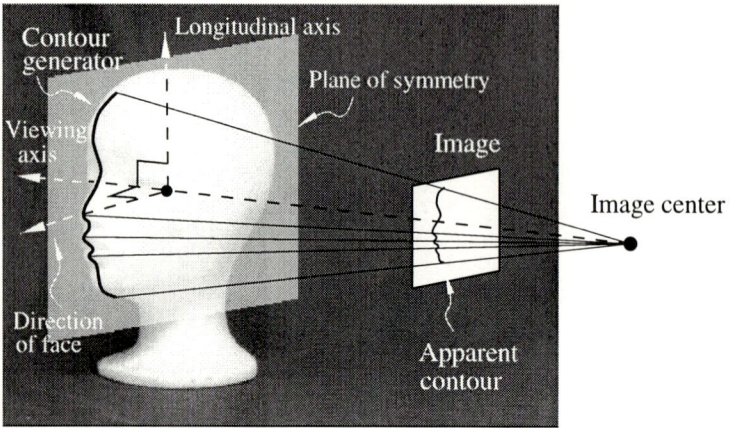

Fig. 2. Apparent contours under the weak perspective assumption. If the viewing axis is perpendicular to the direction of face and the longitudinal axis under weak perspective, the distortion of the apparent contour can be described by similarity transformations.

integral invariants is to define integral invariants based on invariant parameterisations of the group.

Since as we have seen similarity invariants are required for identifying human face profiles, we next consider the semi-local integral invariants under the similarity group. It is known that the invariant parameterisation, s, under similarity transformations can be described by the Euclidean curvature, κ, and the Euclidean arc-length, v, as follows:

$$ds = \kappa dv \qquad (1)$$

where ds and dv are the differentials of s and v respectively. By integrating an invariant function, F, with respect to s along the curve with interval of $[-\Delta s, \Delta s]$, we have a semi-local invariant under similarity transformations:

$$I(s_1) = \int_{s_1 - \Delta s}^{s_1 + \Delta s} F ds \qquad (2)$$

If we choose the function F carefully, the integral formula (2) can be solved analytically, and the resulting invariants have simple forms. For example, if we substitute $F(s) = \langle C(s_1 + \Delta s) - C(s_1 - \Delta s), C_s(s) \rangle$ into (2), we have the following invariant:

$$I_1(s_1) = \langle C(s_1 + \Delta s) - C(s_1 - \Delta s), C(s_1 + \Delta s) - C(s_1 - \Delta s) \rangle \qquad (3)$$

where C_s denotes the derivative of C with respect to s, and $\langle , \rangle$ denotes the scalar product of two vectors. Since I_1 is a relative invariant of weight 1, an absolute invariant can be computed by taking a ratio of two semi-local invariants as follows:

$$I_2(s_1) = \frac{\langle C(s_1 + \Delta s_1) - C(s_1 - \Delta s_1), C(s_1 + \Delta s_1) - C(s_1 - \Delta s_1) \rangle}{\langle C(s_1 + \Delta s_2) - C(s_1 - \Delta s_2), C(s_1 + \Delta s_2) - C(s_1 - \Delta s_2) \rangle} \qquad (4)$$

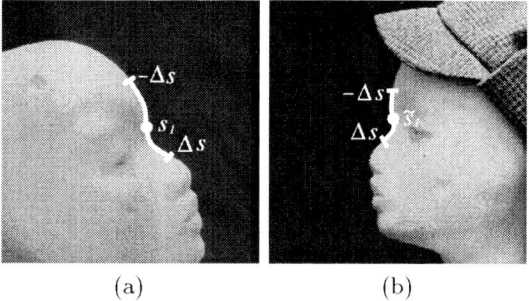

(a) (b)

Fig. 3. Semi-local invariant. The interval of integration can be identified uniquely from invariant arc-length, s. That is, if s_1 and $\tilde{s}_1$ correspond to each other, the interval $[-\Delta s, \Delta s]$ with respect to s corresponds to the same interval $[-\Delta s, \Delta s]$ with respect to $\tilde{s}$ in the second image. Even though the curve is partially occluded (second image), the semi-local integral invariants can be defined on the remaining parts of the curve.

We now summarise the properties of semi-local integral invariants:
1. **Occlusion:**
   Even if a part of the curve is occluded, the semi-local invariants can be defined on the remaining parts (visible parts) of the curve. Thus, these invariants do not suffer from the occlusion problem.
2. **Noise Sensitivity:**
   Since the semi-local invariants enable us to reduce the order of derivatives from that of group curvatures to that of group arc-length, they require lower order derivatives than differential invariants. The semi-local integral invariants are therefore less sensitive to noise than classical differential invariants.
3. **Distinguishability:**
   Since the new invariants are computed from a certain interval of a curve (not from a point), their distinguishability is much higher than local invariants which are computed at a single point on the curve.

3 Finding Profiles from Apparent Contours

Up to now, we have shown that if we have face profiles, we can identify them by using semi-local integral invariants under similarity transformations. In general, however, we do not know whether the extracted apparent contour in an image is of the canonical profile or not. In this section, we describe how to find canonical profiles from the projected apparent contours of faces.

We assume that we do not have any rotation around the direction of face. The distortions of an apparent contour are therefore due to the rotations around the viewing axis and the longitudinal axis. As it is shown in [4], the change in position, $\mathbf{r}_t$, of a point on a contour generator is described by the curvature of the surface in the direction of the viewing ray, κ, the normal to the surface, $\mathbf{n}$, and the change in ray direction, $\mathbf{p}_t$ as follows:

$$|\mathbf{r}_t| = \frac{\langle \mathbf{p}_t, \mathbf{n} \rangle}{\kappa} \qquad (5)$$

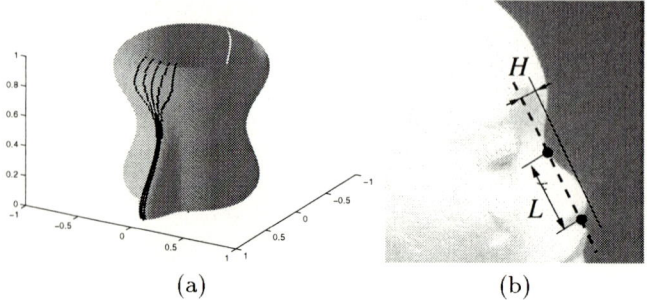

(a)   (b)

Fig. 4. Apparent contours and Face profiles. (a) shows an an artificial face model and the changes in contour generator (black lines) caused by the rotation of the face around the longitudinal axis. The contour generator on the nose is almost fixed on the surface. (b) shows an image of a model face. The dashed line is a bitangent line to the contour curve of the face. L and H are the length and the height of the nose in the image.

From (5) we have the following observations:
1. Under rotations around the viewing axis, $\langle \mathbf{p}_t, \mathbf{n} \rangle = 0$ and $|\mathbf{r}_t| = 0$, so the contour generator does not slip on the surface of a face.
2. Since κ of a nose is large, the changes in contour generator of a nose part are negligible under rotational motions around the longitudinal axis as shown in Fig. 4 (a).

Thus, the contour generator of a nose part can be approximately considered as a fixed planar curve in the space. This property of contour generator is useful for finding the canonical profiles of face. Let us consider a bitangent line to the bottoms of the nose of a face in the image as shown in Fig. 4 (b). We define the length, L, of a nose as a distance between two bitangent points, and the height, H of the nose as a distance between the bitangent line and the top of the nose. If we rotate the face around the viewing axis and the longitudinal axis, the ratio, $R = \frac{H}{L}$, between the height, H, and the length, L, changes. Since the contour generator of a nose part is considered as a planar curve, the apparent contour is of the canonical profile, if the ratio, R, takes a maximum.

4 Experiments

4.1 Procedure for Identifying Face Profiles

We now summarise the procedure for identifying human face profiles.
1. B-spline curves are fitted [3] to the apparent contour of a face.
2. The profile curve of the face is found by rotating the face maximising R.
3. The similarity arc-length and the similarity semi-local integral invariant (4) with an arbitrary but constant Δs are computed at all points on the curves, and subsequently plotted on an invariant graph with similarity arc-length as horizontal axis and the semi-local integral invariant as vertical axis. The derived curve on the graph is an invariant signature up to a horizontal shift.

4. To match curves we simply shift one invariant signature horizontally minimising the total difference between the two signatures.
5. Face profiles are identical if the two signatures match each other.

4.2 Preliminary Results

We next show the results of some experiments. Fig. 5 (a) and (b) show images of a model face observed from two different viewpoints. The extracted contour curves are shown by the white lines in each image. In Fig. 5 (b), we have not only distortion but also occlusions of image curves. The semi-local invariants are computed at every point on the curves, and plotted on the graph as shown in Fig. 5 (c). In this and the following experiments, we chose $\Delta s_1 = 0.15$ and $\Delta s_2 = 1.2$ for the intervals of integration. In this graph, the horizontal and the vertical axes are the arc-length and the semi-local invariants under similarity transformations. The solid and dashed lines are the invariant signatures from Fig. 5 (a) and (b) respectively. As we can see in this graph, the semi-local invariants are quite useful for identifying human face profiles even under occlusions. Fig. 6 (a) and (b) show the profile images of a real human face, and (c) shows the extracted invariant signatures. Again the computed signatures are very stable and useful for identifying the human face. In Fig. 6 (d), (e) and (f), we show the results from another example. Even if the face expressions are different in (d) and (e), the computed invariant signatures are identical as shown in (f), and are thus useful. This is because, the changes in shape of foreheads and noses are small under the changes in expression.

4.3 Distinguishability

We next show the distinguishability of the proposed signatures. Fig. 7 (a) and (b) show the two different profiles, and (c) shows their invariant signatures. As we can see, the signatures are very different, and it is quite easy to distinguish these two faces, while if we use classical differential invariants, they are difficult to distinguish because of the high sensitivity to image noise.

5 Conclusions

In this paper, we have proposed an efficient method for identifying human face profiles from an invariant representation of image curves. Semi-local invariants under similarity transformations are exploited for matching image curves and identifying profiles. Since the semi-local invariants are less sensitive to noise and have higher distinguishability than traditional differential invariants, the proposed method is useful for identifying human faces reliably. The method is implemented and tested on real face images. Even if the face expressions are different the extracted invariant signatures are stable, and useful for identifying faces.

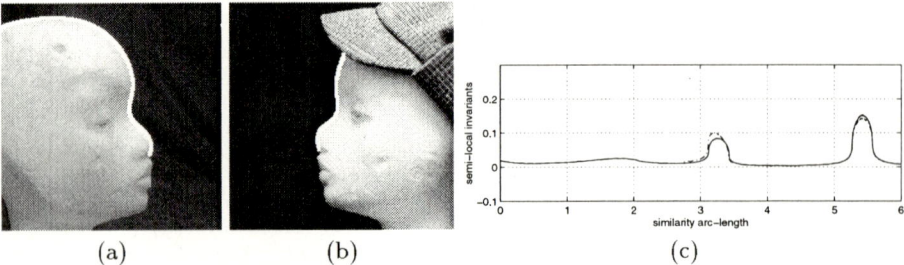

Fig. 5. Results from a model face. (a) and (b) show images of a model face observed from two different viewpoints. The extracted contour curves are shown by the white lines in each image. In Fig. 5 (b), we have not only distortion but also occlusions of image curves. The solid and dashed lines in (c) show invariant signatures computed from the curves in (a) and (b) by using the semi-local invariants. The horizontal and the vertical axes are the similarity arc-length and the semi-local invariants under similarity transformations.

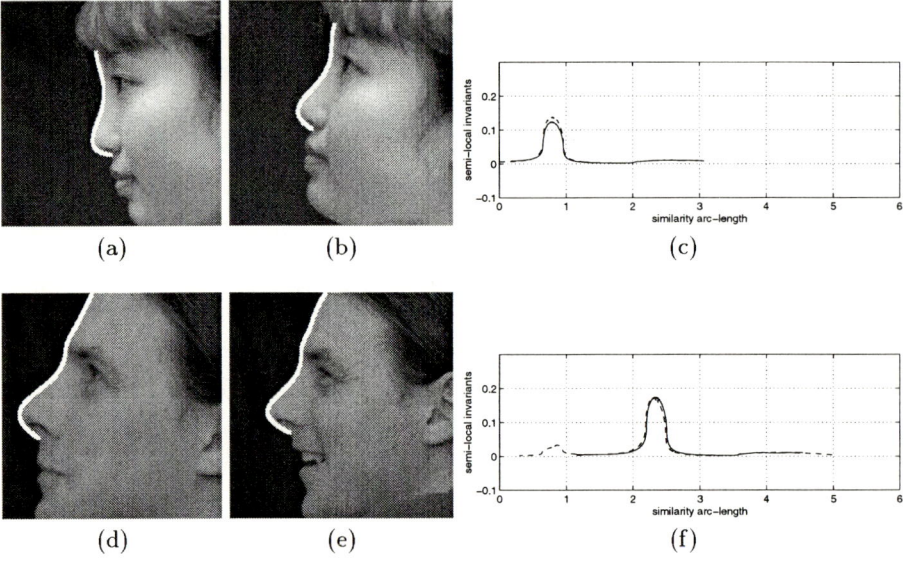

Fig. 6. Results from a real face. (a) and (b) show images of a human face observed from two different viewpoints. The extracted contour curves are shown by the white lines in each image. The solid and dashed lines in (c) show invariant signatures computed from the curves in (a) and (b) by using the semi-local invariants. The horizontal and the vertical axes are the similarity arc-length and the semi-local invariants under similarity transformations. (d), (e) and (f) show the result from another example. Note the face expressions in (d) and (e) are different.

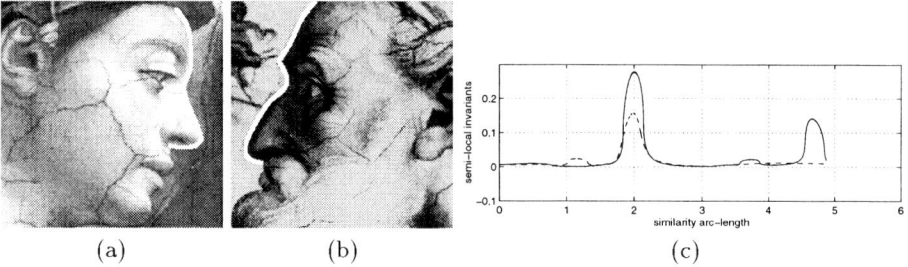

(a) (b) (c)

Fig. 7. Distinguishability of semi-local invariants. (a) and (b) show images of two different profiles. The white lines show the extracted contour curves of the profiles. The solid and the dashed lines in (c) show invariant signatures of these contour curves. The signatures are very different from each other.

References

1. E.B. Barrett and P.M. Payton. General methods for determining projective invariants in imagery. *Computer Vision, Graphics and Image Processing*, 53(1):46–65, 1991.
2. A.M. Bruckstein, R.J. Holt, A.N. Netravali, and T.J. Richardson. Invariant signatures for planar shape recognition under partial occlusion. *Computer Vision, Graphics and Image Processing*, 58(1):49–65, 1993.
3. T.J. Cham and R. Cipolla. Automated B-spline curve representation with MDL-based active contours. In *Proc. British Machine Vision Conference*, volume 2, pages 363–372, Edinburgh, September 1996.
4. R. Cipolla and A. Blake. Surface shape from the deformation of apparent contours. *International Journal of Computer Vision*, 9(2):83–112, 1992.
5. M. Hu. Visual pattern recognition by moment invariants. *IRE Transaction on Information Theory*, IT-8:179–187, February 1962.
6. J.L. Mundy and A. Zisserman. *Geometric Invariance in Computer Vision*. MIT Press, Cambridge, USA, 1992.
7. J. Sato and R. Cipolla. Affine integral invariants and matching of curves. In *Proc. International Conference on Pattern Recognition*, volume 1, pages 915–919, Vienna, Austria, August 1996.
8. J. Sato and R. Cipolla. Affine integral invariants for extracting symmetry axes. In *Proc. British Machine Vision Conference*, volume 1, pages 63–72, Edinburgh, September 1996.
9. G. Taubin and D.B. Cooper. Object recognition based on moment (or algebraic) invariants. In J.L. Mundy and A. Zisserman, editors, *Geometric Invariance in Computer Vision*, pages 375–397. MIT Press, 1992.
10. L.J. Van Gool, T. Moons, E. Pauwels, and A. Oosterlinck. Semi-differential invariants. In J.L. Mundy and A. Zisserman, editors, *Geometric Invariance in Computer Vision*, pages 157–192. MIT Press, 1992.
11. I. Weiss. Projective invariants of shapes. In *Proc. Image Understanding workshop*, volume 2, pages 1125–1134, 1988.

Adaptive Fovea Structures for Space-Variant Sensors

Pelegrín Camacho, Fabián Arrebola and Francisco Sandoval
Dpto. Tecnología Electrónica - E.T.S. Ingenieros de Telecomunicación
Universidad de Málaga - Campus de Teatinos, 29071- Málaga -SPAIN
E-mail : pelegrin@dte.uma.es

Abstract

In this paper we describe the architecture and data structure of space-variant sensors with reconfigurable cartesian geometries. The ability of these sensors to change the position and size of their high resolution regions or electronic foveas, makes them suitable to compensate the limited performance or coarse fixation characteristics of the mechanical systems utilized for gaze tasks in active vision applications where size, weight or cost could be conditioning factors to the performance or feasibility of the whole system. An alternative to the implementation of these sensors is based on off-the-shelf CCD cameras and devices with reconfiguration capabilities, such as FPGAs. In this way, besides the multiresolution data output, sensor reconfiguration systems let generate additional data adapted to the functions of the higher level modules of the active vision systems. As a result of this computing capability at the sensor level, it is possible to unload the processing stages of certain tasks without penalty in time or significant addition of hardware. An approach to selective foveation tasks and motion detection is presented.

1. Introduction

There are technical applications based on active vision systems requiring wide field of view -FOV- and high resolution capability. Facing the implementation of such a systems, several research groups focused their efforts to the development of spatially variant sensors having structures with non-uniform resolution across polar geometries, resembling the retinotopology of the human eye [1][2][3]. An approach based on cartesian topologies was proposed by Bandera and Scott [4], having fovea centered lattices surrounded by sets of concentric rings. The advantages of multiresolution sensing -selective data reduction in space, resolution and time- are obtained at the expense of complex gaze fixation mechanisms to place the sensor fovea onto the regions of interest, considering both the position and size of the objects. Thus, in relation to the development of multiresolution sensors two aspects must be considered: firstly the number of elements, their size and placement on the sensor structure, since they will determine the number and the acuity of the different resolution cells, the levels of the sensor and its data compression and, secondly, the dimensional ratio of the resolution levels or resolution profile because, depending on the size and singularities of the objects, multiple foveations could be required for recognition, leading to complex gaze control procedures. Within the architectures of vision systems [5], our approach has been oriented to space-variant sensors with reconfigurable structures, looking for adaptability to the size and position of the objects of interest, in emulation of saccadic eye movements with an electronic fovea.

2. Rectangular Shifted-fovea Structures for Space-variant Sensors

Cartesian multiresolution geometries [4] consist of concentric structures with a fovea surrounded by rings with decreasing resolution. They are characterized with the following parameters:

m : Number of concentric rings surrounding the central fovea.

h, v : Horizontal and vertical subdivision factors, giving the number of resolution cells -*rexels*- or subrings found in the directions of the cartesian axes within any of the m resolution rings. Fovea size is fixed to $4h \times 4v$ pixels.

Using off-the-shelf CCD cameras, we have applied reconfigurability of non-concentric structures [6], to obtain rexels by pixel averaging on rectangular FOVs sized H and V pixels. Eccentricity is fixed by shifting all rings s_h, s_v rexels, as seen in Fig. 2.a. The minimum jump, J_N, for any ring N, corresponding to $|s_h|$ or $|s_v| = 1$, is the shift of that ring across the FOV, caused by the simultaneous and relative shifts of all rings within the rings surrounding them. The jump, in pixels, is given by

$$J_N = 2^{N+1} + 2^{N+2} + \ldots + 2^m = \sum_{N}^{m-1} 2^{i+1} \qquad (1)$$

valid for all rings, except the peripheral one delimiting the FOV. The effects can be seen on the foveal image of Fig.1.b obtained with a structure having m = 3 and whose fovea was shifted onto the car at the right in the uniresolution image of Fig.1.a. Resolution levels associated to the foveal image for m = 2 are shown in Fig. 1.c. Shifted fovea sensors allow data structures with an equal number of elements on all resolution levels. The size of the fovea and upper levels is fixed, as seen in Fig. 1.c., because once the number of rings has been defined, the values of h and v, their ratio and the fovea size -$4h \times 4v$- are uniquely determined by the expressions

$$h = \frac{H}{2^{m+2}} \quad ; \quad v = \frac{V}{2^{m+2}} \qquad (2)$$

The fact of having a fovea with fixed dimensions implies that targets with a size greater than fovea could only be recognized by refoveating onto the targets, unless the details leading to recognition could be identified in a higher level because of its wider solid angle, and in spite of its lower resolution. In case that neither of those options would be feasible on the system, the alternative would be to reduce m, i.e. to increase the fovea dimensions. However, the convenience of reducing the number of resolution levels must be considered in relation to the increase of data involved: Structures with low m, i.e. having a smooth resolution profile on a fixed FOV, have a reduced compression factor which is equivalent to have a foveal sensor with greater number of elements. Considering the expressions in (2) and the relative sizes of fovea and rings, the compression factor F, defined as the ratio of the number of pixels on the FOV to the number of elements in sensor structures as that of Fig.2.a, will be given by

$$F = \frac{H \cdot V}{4h \cdot 4v \cdot (1 + \frac{3m}{4})} = \frac{h \cdot 2^{m+2} \cdot v \cdot 2^{m+2}}{hv \cdot (16 + 12m)} = \frac{2^{2m+4}}{16 + 12m} \qquad (3)$$

where it can be seen that the higher the value of m, the higher the compression factor.

a) Original image 640x480 pixels   b) Foveal image ($m=3$ $h,v,s_h,s_v = 20,15,11,1$)

c) Prism resolution levels : Fovea, L1, L2, for $m=2$.   (Expanded ≈ 3x3)

Fig. 1 Foveal image and resolution levels of a shifted fovea structure.

3. Rectangular Lattices with Adaptive Data Structure

To avoid the inconveniences of a low m, without loosing the advantages of greater fovea, the approach here proposed has the double effect of expanding the fovea and the reconfigurability of shifted fovea geometries, without increasing the data volume involved, as will be shown later. The procedure consists in making different resolution profiles at each side of the fovea and combining them horizontal and vertically. To do that, as seen in Fig.2.b and 2.c, instead of parameters h, v, s_h, s_v, we introduce a new set of four parameters l, r, t, b to define subdivision factors at each side of fovea. Resolution profiles increase or decrease inversely with lateral subdivision factors. Thus, in Fig. 2.b, at the right of fovea, the gradient is twice the one at the left. Departing from structure in Fig.2.b, the resolution gradients at the left and bottom regions could be increased in one step, leading to the structure of Fig. 2.c, with a greater fovea, which is placed according to the set of parameters l, r, t, b.

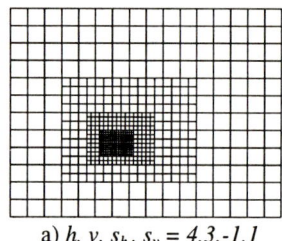 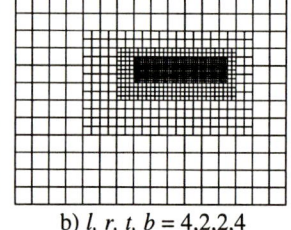 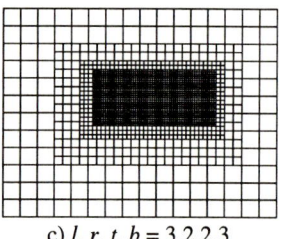

a) $h, v, s_h, s_v = 4,3,-1,1$ b) $l, r, t, b = 4,2,2,4$ c) $l, r, t, b = 3,2,2,3$

Fig. 2 Rectangular structures with $m = 3$. a) Shifted fovea, b) and c) Adaptive fovea.

The variation of lateral resolution profiles is shown in Fig.3, where it can be seen a tilted and truncated inner pyramid sharing its top level with the coarsest one of the outer pyramid, containing the compressed FOV. Projection of that top level onto the FOV, performed through the truncated pyramid, will determine the size and location of fovea in relation to the projecting angles at each side of inner pyramid. Parallel projections of the outer pyramid levels onto the FOV, will determine lateral resolution gradients and the hierarchical data structure of the sensor. At the limit, both pyramids become one, when adaptive fovea is expanded up to cover the whole FOV.

The sensors so obtained will have asymmetric structures whose highest resolution levels can be sized to cover the shape and position of any region of interest or target on the FOV. Fovea adaptation improves greatly the performance of the vision system by eliminating the refoveation processes required when fovea size is smaller than target and, therefore, optimizing data handling and processing time for recognition purposes. Thus, structures of Fig.2 have the same number of resolution levels, but changing the lateral gradients in a regular manner, we are able to fix the asymmetry of the structure and the ratio of each level N to the N+1. That ratio and the data structure associated is regularly maintained in an homothetic way which resembles the outer pyramid data compression.

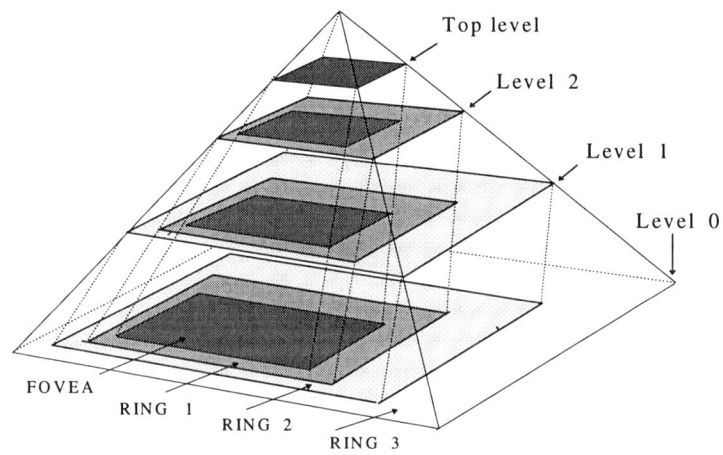

Fig. 3 Merged pyramids structure for adaptive fovea lattices with $m = 3$.

To configure the structure to the size and location of the target, it is assumed that target has an associated bounding box whose coordinates X_{min}, X_{max}, Y_{min}, Y_{max} were previously determined after applying the appropriate algorithms for objects detection. According to these set of coordinates it is inferred the minimum fovea dimensions as well as its location on the FOV. The remaining steps to apply to the algorithm determining the structure consist on the determination of the maximum subdivision factors to have on each side of the fovea, without overskipping the bounding box coordinates. Thus, lateral subdivision factor will be obtained as:

Left subdivision factor $\quad l = INT\left[\dfrac{X_{min}}{J_0}\right]$ (4)

Right subdivision factor $\quad r = INT\left[\dfrac{H - X_{max}}{J_0}\right]$

Top subdivision factor $\quad t = INT\left[\dfrac{Y_{min}}{J_0}\right]$

Bottom subdivision factor $\quad b = INT\left[\dfrac{V - Y_{max}}{J_0}\right]$

where J_0, as defined in (1), refers to the minimum jump to apply on fovea sides to alter the resolution gradient one step. After obtaining l and t, all inner rings and fovea can be referred to the origin or upper left corner of the FOV. The origin, O, and the opposite corner, Z, of any level N are thus related to the jump, J_N, and their coordinates given by

$$O_{xN} = J_N.l \qquad O_{yN} = J_N.t \qquad Z_{xN} = H - J_N.r \qquad Z_{yN} = V - J_N.b \qquad (5)$$

After applying reconfiguration to image in Fig.1.a, resolution levels are obtained as shown in Fig.4, with lateral dimensions given in number of elements, fovea in pixels and upper levels, L_i, in their rexels equivalent to 2^i pixels. It can be seen the relative eccentricity of levels, their different sizes as well as the homothecy relating them.

Fovea -L_0-: 150 x 74 $\qquad\qquad L_1$: 110 x 66 $\qquad\qquad L_2$: 90 x 62 $\qquad L_3$: 80 x 60

Fig. 4 Adaptive fovea levels obtained with $m = 3$, $l,r,t,b = 32, 3, 18, 11$.

To evaluate the effectiveness of adaptive fovea structures -AFSs- the number of cells in these lattices is compared to that in shifted fovea structures. That number is obtained by adding the expressions giving the cells in fovea and rings to derive the expression for the whole structure, considering that the number of subrings on each side of fovea is determined by l, r, t, b. Fovea and rings sizes are given by the differences between the lengths of the FOV and the respective widths of the rings to each side of the fovea or ring considered, therefore, the magnitudes, in pixels, are

Horizontal width of level N $\qquad W_{hN} = H - J_N.(l+r) = H - J_N.x$ (6)
Vertical width of level N $\qquad W_{vN} = V - J_N.(t+b) = V - J_N.y$
Pixels related to L_N or Size of level N $\qquad S_N = (H - J_N.x).(V - J_N.y)$

where x and y refer to total resolution steps horizontal and vertically. Therefore, the number of elements -*rexels*- on ring N, R_N, will be given by

$$R_N = \frac{\text{Size of level } N - \text{Size of level } (N-1)}{\text{Size of rexel in ring } N} \quad (7)$$

The size, in pixels, of one rexel in ring N is given by the exponential 2^{2N}. From (6) and (7), the following expressions are obtained

Elements in Fovea $\quad R_0 = HV - J_0 \cdot (x+y) + x \cdot y \cdot J_0^2 \quad (8)$

Elements in Ring N $\quad R_N = \dfrac{2^N \cdot (Hy + Vx) + x \cdot y \cdot (J_N^2 - J_{N-1}^2)}{2^{2N}}$

Elements in Ring m $\quad R_m = \dfrac{2^m \cdot (Hy + Vx) - x \cdot y \cdot J_{m-1}^2}{2^{2m}}$

and the total number of elements in any adaptive sensor lattice will be given by

$$R_T = \sum_0^m R_i = H.V - (H.y + V.x) \cdot \sum_1^m \left(2^i - \frac{1}{2^i}\right) + \frac{3.x.y}{4} \cdot \left[J_0^2 + \sum_1^{m-1}(J_0 - J_i)^2\right] \quad (9)$$

from which it can be obtained, as a particular case for $x = 2h$ and $y = 2v$, the expression for the number of elements in fixed size fovea lattices, given by

$$R_t = h v (16 + 12m) \quad (10)$$

Table I shows the number of elements in AFSs for some pairs of x and y in lattices with $m = 3$. Sensor elements are compared with those of shifted fovea structures having $m = 2$ and related to the number of pixels on their respective foveas.

TABLE 1: Comparison of foveal structures (H, V = 640 x 480 pixels)

Structure	m	x	y	Sensor elements	% elements	Fovea dimensions	Fovea pixels	% pixels
Shifted	2	80	60	48000	100,0	160 x 120	19200	100
adaptive 1	3	32	24	39936	83,2	192 x 144	27648	144
adaptive 2	3	30	23	47130	98,2	220 x 158	34760	181
adaptive 3	3	31	22	47814	99,6	206 x 172	35432	184
adaptive 4	3	30	22	50220	104,6	220 x 172	37840	197

It is worth to note that adaptive structures 1, 2 and 3 above, in spite of having a lower number of elements, which implies a lower data handling on the system, have foveas with higher size than the one of shifted structure with $m = 2$. Thus, wider regions of the FOV are covered with the highest resolution, eliminating the need to refoveate when target singularities and size require examination with fovea acuity. All structures cover the same FOV, but adaptive fovea enhances high resolution levels, looking for better performance on recognition tasks, at the cost of reducing acuity of coarse resolution levels, which are mainly used for object detection.

4. Motion Detection with Adaptive Fovea Structures

Application of adaptive fovea sensors for motion detection is based on the temporal differences, $D_N(x,y,t)$, between corresponding elements of consecutive images of the $I_N(x, y, i)$ input sequence :

$$D_N = |I_N(x, y, t+1) - I_N(x, y, t)| \qquad (11)$$

Temporal differences are measured for all sensor cells, x and y being their coordinates within the resolution level N containing the cell. Differences are compared to preset thresholds and transformed to binary masks determining a rough estimate of the shape and position of moving objects. Differences on the coarsest resolution level, because of the cell averaging performed, can be considered as the result of applying a low-pass filter to the original images and, therefore, binary masks obtained at this level are used as bounding box delimiters, indicating dynamic regions within the static FOV.

From sequences as in Fig. 5, binary masks sets are obtained as shown in Fig. 6 and 7.

Fig. 5 Motion sequence.

Fig. 6. Binary masks set of first two images of sequence, applied to both images.

Fig. 7. Binary masks set of last two images of sequence, applied to last image.

Fig. 8. AND of Binary masks in Figs. 6 and 7, applied to second image.

Both masks are ANDed to form the second level binary mask. Application of this mask to the image in common to both first level masks, lets extraction of the moving objects on any resolution level, as seen in Fig. 8. Foveations to track the object of interest can be based on predictions obtained from bounding boxes of coarsest levels, which are continuously corrected after obtaining parameters l, r, t, b as given in (4), to extract the segmented regions covering moving objects at the coarsest level.

5. Conclusions

We have described a new methodology to obtain adaptive multiresolution structures for visual sensors. Their main advantage is the possibility to adapt the size and position of their foveas to areas of interest, minimizing the overall data volume related to the resolution levels as well as the number of foveations required to examine a region and, therefore, the time involved on those tasks. Pixel averaging, to obtain rexels from the uniresolution camera data, as well as rexel or pixel additions and subtractions, are easily performed on FPGAs, applying their capability to perform basic arithmetic operations. Similarly, the In-circuit reconfiguration feature of these devices is well suited to select the areas of the FOV where resolution must be changed or brightness values of the image sequence subtracted for motion detection. Adaptive fovea structures allow coverage of wide FOVs, where several moving objects could appear but, at the same time, selected areas of image sequences can be observed in a closer detail, emulating saccadic eye movements for tracking tasks on low cost active vision systems. Mechanical camera orientations, as coarse saccadic head movements, are used to change the FOV whenever the targets get out of scenes being examined, but the stringent mechanical requirements associated to non-reconfigurable sensors are eliminated with the fast and low cost electronic foveation system.

Acknowledgements
The present work has been partially supported by the Spanish Comisión Interministerial de Ciencia y Tecnología , (CICYT), Project No. TIC095 - 0589.

References
[1] J.Van der Spiegel, G.Kreider, C.Claeis, I.Debusschere, G.Sandini, F. Fantini, G.Soncini, "A foveated retina-like sensor using CCD technology" in *Analog VLSI implementation of Neural Systems*, C.Mead, M.Ismail, Eds.,KluwerAcad. Publ, 1989.
[2] M.Tistarelli, G.Sandini," Estimation of depth from motion using anthropomorphic visual sensor", *Image and Vision Computing,* vol.8, No. 4, pp. 271- 278, 1990.
[3] A.S.Rojer, E.L.Schwartz,"Design considerations for a space-variant visual sensor with complex logarithmic geometry", in *Proceedings of the 10th. International Conference on Pattern Recognition,* pp.278 - 285, 1990.
[4] C.Bandera, P. Scott,.:"Foveal Machine Vision Systems" in *IEEE International Conf. on Systs., Man and Cybernetics*, Cambridge, MA, pp.596-599. November 1989.
[5] E. Schwartz, D.N Greve, G.Bonmassar, "Space-variant Active Vision: Definition, Overview and Examples". *Neural Networks,* Vol.8, No. 7/8, pp.1297 - 1308, 1995.
[6] P.Camacho, F.Arrebola, F.Sandoval, "Shifted Fovea Multiresolution Geometries" in *Proceedings of the IEEE International Conference on Image Processing*, Vol.I pp.307- 310. Lausanne, Switzerland. 1996.

Structural Characterisation
of Image Processing Operators

P. Bottoni*, L. Cinque*, S. Levialdi*, P. Mussio+, B. Nebbia*
* - DSI, Università di Roma, Via Salaria 113, 00198 Roma, Italy
+- DEA, Università di Brescia, Via Branze 38, 25231 Brescia

Abstract

We present a methodology to specify the action of image processing operators in terms of transformations induced on structures present in an image. The methodology is based on an analysis of the characteristics of the operator with respect to the adopted segmentation technique and is illustrated through an example of characterisation of the Gaussian filter.

1 Introduction

Automatic generation of image interpretation strategies is facilitated if descriptions of the behaviour of available operators exist specifying which structures of an image are affected and which remain unaltered by the operator application. It is thus possible to decide which operators to choose, based on descriptions of the image at hand, of the quality of results one wants to obtain and of the known features of available operators.

In particular, one is interested in using operators preserving relevant structural features. In this case, the usual characterisation of operator behaviours based on their action on numerical features of the image signal, such as noise or texture [1], is insufficient. For example, in restoring line drawings, one wants to sieve out triangular structures from the borders of a binary image, while preserving significant corners [2]. This method is here generalised to characterise image processing operators (IPOs) via the transformations they induce on structures defined in terms of their shape. Structures are first described as strings of attributed symbols identifying pixels in which the contour changes direction [3]. Shapes are then characterised by grouping these attributed symbols into more abstract shape descriptions, again expressed by attributed symbols [2]. An IPO effect is characterised by defining how descriptions of structures present in the input and output images are transformed.

The diagram of Fig.1 summarises this view. Given an image I, an IPO Tr maps it into a new image I'. The analysis algorithm anI [3] derives a description D of image I and a description D' of I'. The problem is: find a description transformation ? such that the diagram commutes (i.e. identical results are obtained by combining Tr and anI or anI and ?). We characterise Tr by specifying this transformation.

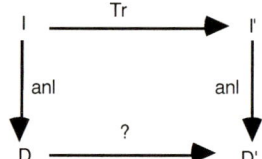

Fig. 1. Image and description transformations

Description transformations are here characterised by rewriting rules. We define an analytical methodology which extracts rules defining the transformations from numerical specification of the given operators. The main contribution of the paper is the proposal of a method for the characterisation of operators through explicit description of the shape transformations they perform, and the consequent definition of notations, based on run codes, to describe symmetrical structures, used for the experimental part. The methodology is designed to solve the problem of defining the behaviour of image operators for automatic generation of image processing and interpretation strategies; in particular, to assess which structures types are invariant for a transformation, and to characterise families of structures with respect to transformations under which they are invariant. The paper develops as follows. In Section 2 descriptions are introduced as strings of attributed symbols. Two attributes characterise structures based on a run-length encoding. Section 3 illustrates the method to characterise shape transformations, and Section 4 draws conclusions.

2. Shape Characterisation

A structural description of an image enumerates the structures recognised in it, with each structure described by a name - identifying its type and denoted by a *symbol* - and a vector of *properties* - the values assumed by a set of variables, called *attributes*. The n-tuple with the symbol and its properties is an *attributed symbol*. Structures can be decomposed into simpler ones. *Primitive* structures are not further decomposable.

Following [3], primitive structures for binary image description are black pixels, called *multiple elements*, in which the contour changes its direction locally. They are associated with the symbol ME with attributes: code, summarising the pixel's 8-neighbourhood, as in

p_1	p_2	p_3
p_0	p	p_4
p_7	p_6	p_5

Fig. 2, and coordinates; code is: $code(p) = \sum_{i=0}^{7} u_i \cdot 2^i$,

Fig. 2. The ordering of 8 neighbourhoods

where u_i is 1 if the pixel p_i is white, 0 if it is black.

Hence, each multiple element is described by an attributed symbol <ME, code, coordinates> and the whole image by the string of symbols describing its multiple elements. In [3] it is shown that such a description is sufficient to reconstruct any original black and white image. For example, Fig. 3 is an image in which white pixels are background, while textured and black pixels constitute the image. Textured pixels are the multiple elements. The whole image is described by the string of attributed symbols describing the seven MEs, Im: <ME,143,5,1> <ME,6,5,5> <ME,31,2,8> <Me,12,8,14> <Me,62,8,16> <ME,248,9,16> <ME,227,9,1>. The MEs marked with A, B and C describe a particular structure called, *r-triangle* which can be described by an attributed symbol <tri,base,length,fstpt>. A is associated with the code 31, B with the code 6 and C with code 12. A *Conditional Attributed L-System with Interaction* (CAIL) [2] can be defined (see Fig. 6), which: 1) identifies the substring Tr: <ME,6,5,5> <ME,31,2,8> <Me,12,8,14> in the string describing the image, 2) recognises that the three elements are vertices of a rectangle triangle, 3) evaluates the triangle attributes and 4) generates the new attributed symbol.

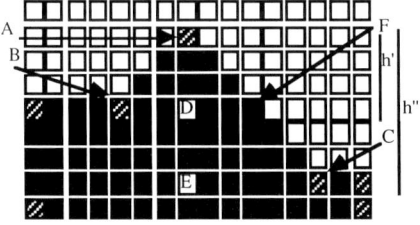

Fig.3. An upward triangle defined by the sequence of codes (6,31,12).

R-triangles are defined in geometrical terms and constitute a well defined set of contour structures in b/w images. Analogously, the set of substrings describing these structures is a language on ME specified by a CAIL. In general, the set V of types of structures is assumed as alphabet; each single structure is described by an attributed symbol, formed by a symbol in V, and by its property vector. An image description is a string of attributed symbols. Structure description based on MEs is formalised by CAILs. CAILs extend the grammars traditionally used in syntactic pattern recognition [4] in two main ways typical of L-systems [5]. First, they use one single alphabet, not partitioned in terminals and non-terminals. Each symbol denotes a structure, not a possible derivation of it. Second, direct generation is parallel.

A *CAIL* is a rewriting system RW=(V,P,$\Rightarrow$) and a semantic domain D=(Ω, Φ, Γ). V is a finite set of symbols, the alphabet. P is a finite set of productions of the form $<<\omega_1, \delta, \omega_2>, \varepsilon>$, meaning that for δ to be rewritten as ε, it must be embedded in the context represented by ω_1, $\omega_2 \in V^*$ $\Omega = \{D_1,...,D_k\}$ is a finite set of domains. Φ is a set of functions. Γ is a set of predicates, also comprising the constant predicates *true* and *false*. For each $p \in P$ $\exists \gamma \in \Gamma$, called the *condition* to be satisfied for p to apply, and $\exists r_p \subseteq \Phi$, a finite set of functions computing attributes of the consequent as a function of those of the antecedent. The rewriting relation $\Rightarrow$ specifies that rules in P are applied in parallel to directly generate a string Z from a string W (in symbols W $\Rightarrow$ Z).

Rules are applied to strings of attributed symbols. Strings are cyclical, so that two consecutive symbols describe two consecutive multiple elements in the contour. In the image, structures are set of pixels described by substrings of multiple elements.

A *side* is a set of contour pixels connecting two consecutive multiple elements. The side *length* is length(p_1,p_2)=max$(|x_1-x_2|, |y_1-y_2|)$. Geometrical entities are also used as attributes to describe structures. For example, the *symmetry axis* of a structure is a line in the image splitting the structure into two isomorphic parts. A *run* is a sequence of adjacent black pixels orthogonal to a symmetry axis. In the following, we exploit possible partial symmetries of structures emerging from the contour of components to derive transformation laws of their descriptions. We consider contour structures for which an intuitive geometrical notion exists and provide their formal definition in terms of strings of attributed symbols, through suitable CAILs. CAILs are also used to specify how structures are modified by application of IPOs.

2.1 Characterisation of Trapezes

An *i-trapeze*, as shown in Fig. 4a, is the digital version of an isosceles trapeze appearing on the contour of an image. The trapeze is delimited by the multiple elements marked with A,B,C,D. The upper base (upbase) is the sequence of pixels from the second to the third multiple element, while the lower base (lowbase) is the sequence of pixels from the first to the fourth, including these two. The *height* of an i-trapeze is the sequence of pixels from the second multiple element to the nearest pixel in the base. Similar trapezes can appear oriented in the eight directions and with different lengths of their upper base and of their height. However, for each direction and for each size, the four multiple elements delimiting the trapeze present a specific sequence of codes. For example, the one shown in Fig. 4a is characterised by the code sequence (6,15,30,12), and size changes only affect the coordinates of the multiple elements in the description of the particular image. Hence, there is a finite number of possible sequences of codes for the multiple elements which can delimit an i-trapeze.

The set of descriptions of possible structures of type i-trapeze, denoted by the symbol itrpz, is defined by the rule in Fig. 5, where TRPZ= {(6,15,30,12), (3,135,195,129), (192,255,240,96), (48,120,60,24), (12,30,15,6), (129,195,135,3), (96,240,255,192), (24,60,120,48)}. This rule allows the recognition of description of i-trapezes and the substitution of the sequence of four elements denoting the delimited i-trapeze with a more synthetic attributed symbol itrpz, whose attributes lowbase, upbase and height and fstpt are computed by the semantic rules and allow the reconstruction of the trapeze.

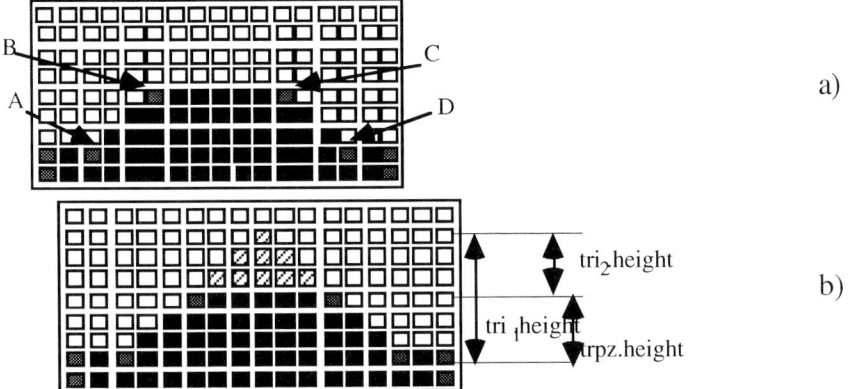

Fig. 4. a) An i-trapeze on the contour; b) i-trapeze as difference of two isosceles triangles.

Tuples in TRPZ define the clockwise sequences of codes for extremal points of an i-trapeze. The construction in Fig. 4b shows the relation among the attributes. In particular, note that a trapeze results from the difference of two triangles tri_1 and tri_2. Its height is thus defined by the law $trpz.h = tri_1.h - tri_2.h$. Since for a triangular structure $h=(b+1)/2$, we obtain $tri_1.height = (trpz.lowbase+1)/2$ and $tri_2.height = [(trpz.base-2)+1]/2$. Therefore: $trpz.height = [(trpz.lowbase +1)/2] - \{[(trpz.upbase-2)+1]/2\} = (trpz.lowbase-trpz.upbase+2)/2$.

Syntactic part: $<<, ME_1°ME_2° ME_3 ° ME_4,>$, itrpz>

Semantic part:

γ: $(ME_1.code, ME_2.code, ME_3.code, ME_4.code) \in TRPZ)$ && $(length(ME_1,ME_2) = length(ME_3,ME_4))$

r_p: itrpz.lowbase=$|ME_1.x-ME_4.x|+1$; itrpz.upbase=$|ME_2.x-ME_3.x|+1$; itrpz.height=$|ME_2.y-ME_3.y|+1$, itrpz.fstpt = $<ME_1.x, ME_1.y>$

Fig. 5. The rule for defining a trapeze.

2.2. Characterisation of Rectangle Triangles

In a similar way, an *r-triangle*, as the one shown in Fig. 3, is the digital version of a rectangular triangle. The *base* of an r-triangle is the sequence of pixels from the first to the third multiple element, including these two. The *height* of an r-triangle is the sequence of pixels from the second multiple element to the nearest pixel in the base. Again, the fstpt attribute defines the initial coordinates of the triangle.

> Syntactic part: <<,$ME_1°ME_2° ME_3$,>, tri>
>
> Semantic part
> γ: (ME_1.code,ME_2.code,ME_3.code)∈ TRI)
> r_p:tri.base=|ME_1.x-ME_3.x|+1, tri.height=|ME_2.y-ME_1.y|+1, tri.fstpt=<ME_1.x,ME_1.y>

Fig. 6. The rule for defining an r-triangle

The set of descriptions of possible structures of type r-triangle, denoted by the symbol tri, is defined by the rule in Fig. 6, where TRI={(6,31,12), (3,199,129), (192,241,96), (48,124,24), (12,31,6), (129,199,3), (96,241,192), (24,124,48)}. Triples in TRI define the clockwise sequences of codes for extremal points of an r-triangle. Such a triangle can be seen as the superposition of a symmetrical triangle on a rectangle trapeze. For example, in Fig. 3, the pixel marked with A is the vertex of the symmetrical triangle which extends between vertices B and F, while the trapeze has bases DF and EC respectively. h' and h" are the heights of these triangles. We call *symmetrical component* the substructure delimited by the pixels B, A and F, and *residual component* the substructure delimited by pixels F and C.

The relation holds tri.base = h'+h"-1. For Fig. 3, tri.base=10, h'=4, h"=7. The semantic rule r_p is used in the case of upwards and downwards triangles. The case for leftwards and rightwards triangles is derived by exchanging the x and y coordinates. Constraints exist on base and height of an r-triangle and on their relations. In particular, the minimum height is 2 and the minimum base is 3. Due to the structure partial symmetry, for each pixel added to the height h', the base is increased of two pixels, for each pixel added to the height h"-h', the base is increased of one pixel.

2.3 Description of runs

Structures of interest are characterised through a description of the runs composing them. The study of the variations of the run-lengths of structures helps to derive rules defining the transformation of a structure into another, by considering the independent transformations of symmetrical and non-symmetrical components of structures, based on a run-length encoding. Indeed, each structure in which the sides maintain their orientation can be seen as the composition of a symmetrical structure with an asymmetrical one. We call such structures *oriented structures*. Each component is formed by a sequence of parallel runs. With each run a number is associated representing its length. Hence an oriented structure is characterised by the sequence of lengths of its runs. We call such a sequence a *length-sequence*, and it becomes the value of an attribute lenseq for symbols denoting oriented structures.

Theorem 1: For each type T of oriented structure there is a language L(T) of strings of natural numbers, coding the sequences of lengths, each string uniquely describing a structure of type T up to localisation.

The proof is based on the following lemma.

Lemma 1 A symmetrical structure is composed only of runs of the same parity.

Based on these definitions, a function is derived for computing the attribute lenseq associated with an oriented structure. For example, for r-triangles the function trilen depends on the two heights h' and h" of the composing triangles and is inductively defined as follows: trilen (1,h")=$\Delta 1 \Diamond 2°...°h"$, trilen (h',h")= Δ sym(trilen(h'-1,h"))°(2h'-1) $\Diamond$ 2h'°...°(h'+h"-1), where the function sym extracts the subsequence between the two separators Δ and $\Diamond$. The symbol '°' denotes concatenation, '$\Diamond$' is a separator between the symmetrical and the residual components. Note that, following the contour clockwise, the description of residual component comes after the separator ◊. Had this component been before the symmetrical part, its description would have preceded the separator Δ. In the following, we develop the argument only for structures where the residual component follows the symmetrical component. The inverse case is easily derived.

As an example, the r-triangle of Fig. 3 is described by the string of naturals, tri(4,7) = Δ1°3°5°7 ◊ 8°9°10, where each number represents the length of a run and 1 denotes the vertex of the r-triangle, considered as a run of length one; 3 and 5 are the lengths of the following runs, 7 is the length of the base of the symmetrical component, and the sequence 8°9°10 describes the length of the runs of the residual components.

Another useful notation associates an oriented structure with the *variation-sequence* in the length of its runs. This sequence is the value for an attribute varseq.

Theorem 2: For each type T of oriented structure there is a language L(T) of strings of natural numbers, coding the variations in length of the runs, each string uniquely describing a structure of type T up to the localisation and length of the top run.

As an example: tri(4,7) = Δ1°3°5°7 ◊ 8°9°10 ≡ Δ 2;2;2 ◊ 1;1;1 = Δ $2^3 ◊ 1^3$, where ';' is now used to denote concatenation. In general, for an r-triangle of heights h' and h" the attribute varseq has value computed by the function trivar(h',h")=Δ$2^{h'-1}$◊$1^{h''-h'}$, while for an i-trapeze of height h, the attribute is calculated as trpzvar(h)=Δ2^{h-1}◊. Note that symbols in the strings associated with symmetrical components are even numbers.

3. Characterising Shape Transformations

In this Section we illustrate the methodology for analytical extraction of rewriting rules, using as example the transformations induced by a Gaussian filter [6], in the case that the segmentation operator is a thresholding.

We use a Gaussian filter with kernel of size 5, and we consider the same size for the neighbourhood of pixels. The convolution kernel is symmetrical and generated from the vector 1/20×[1,5,8,5,1], through the law: w(i,j)=w(i)×w(j). Due to the symmetry characteristics of both the filter and the considered structures, it is possible to identify some minimum values for the size of the structures, in order for the transformation to be characterised by a regular, formalisable behaviour. In a binary image, there are four cases for the centre pixel of a neighbourhood: 1) internal pixel, 2) pixel close to background, 3) pixel close to foreground, and 4) background pixel.

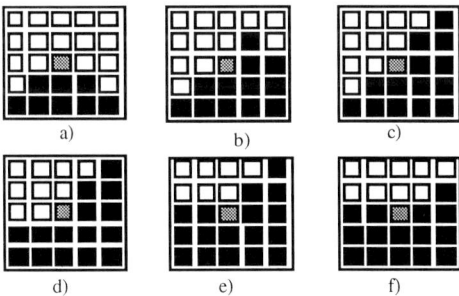

Fig. 7. Some configurations of neighbourhoods on the contour of an r-triangle.

We refer to grey level images with values in a set {0,...,255}. The segmentation algorithm is a simple thresholding. Here we adopt as threshold the value 200; a pixel of coordinates (n,m) is transformed into a pixel with grey value 255 if and only if formula 1 is satisfied, where w(i,j) denotes the weight of the element of position (i,j) in the convolution kernel and [g(n+i,m+j) = 255] denotes the elements with value 255.

$$W = \left(\sum_{i,j \in \{-2,+2\}} w(i,j) \cdot [g(n+i,m+j) = 255] \right) \geq \frac{200}{255} = 0.7843 \quad (1)$$

To define the effect of the filtering on the segmented image, neighbourhood configurations are analysed to identify those allowing the central pixel to be transformed into 255. Fig. 7 shows neighbourhoods on the contour of an r-triangle.

3.1. Transformation of an r-triangle

The method described above is here used for the case of the r-triangle. In Fig. 8a, A indicates from which pixel all pixels maintain their value while descending toward the base. In Fig. 8b, B indicates until which pixel the value is maintained.

For the neighbourhoods considered in Fig. 7, the central pixel is transformed into 0. All the other pixels in the r-triangle which are close to the background are instead transformed into 255. This result is extended to neighbourhoods resulting from reflection and rotation of the neighbourhoods above. There exists a minimal height, viz. 4 pixels, above which a general law can be identified. Below this height, the methodology can describe the behaviour of the transformation for the single cases.

Theorem 3 (r-triangle). The variation-sequence associated with an r-triangle of heights h'≥4, h"≥h'+2 is transformed by a Gaussian filter in a variation-sequence, by the law: $\Delta 2^{h'-1} \Diamond \, 1^{h''-h'} \rightarrow \Delta 2^{h'-2} \Diamond \, 1^{h''-h'}$.

<<segm$_1$, r-tri, segm$_2$> , trpz>
γ: (tri.height≥4) && (segm$_1$.length≥3) && (segm$_2$.length≥3))
rp: trpz.height'=tri.height'-1; trpz.height"=tri.height"-1; trpz.varseq=g(tri.varseq);

The overall structure transformation erases the r-triangle vertex, while leaving the same base length. Hence, an r-triangle symmetrical part is transformed into a trapeze, while the residual component maintains its size, but with a lower base run. The structural transformation induced by the Gaussian filter is summarised by the following rule, with g defined by Theorem 3, and trpz a symbol denoting a trapezoidal structure with the same sequence of codes as an i-trapeze, but different side lengths. As an example, for the case of Fig. 9 we have: G[(tri(5,7))]=$\Delta 2^4 \Diamond 1^2$.

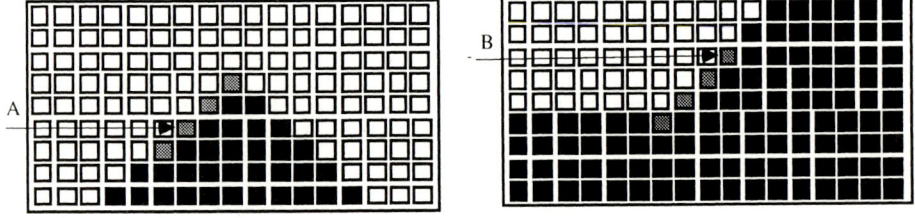

Fig. 8. Initial (a) and final (b) configurations on the contour of an r-triangle

The transformation law for variation sequences in i-trapezes is analogous to that for

triangles, but the structure size does not change, since the overall effect is that of lowering it. Hence i-trapezes are invariant under iteration of Gaussian filtering (until there are sufficient pixels in the whole object).
Theorem 4 (i-trapeze). The variation-sequence associated with an i-trapeze of height h≥4 and upbase>1 is transformed by a Gaussian filter as: $2^{h-1} \to 2^{h-1}$

The CAIL for trapeze transformation derives, with the semantic rule including $trpz_1.lowbase=trpz_2.lowbase$; $trpz_1.upbase=trpz_2.upbase$; $trpz_1.height= trpz_2.height$.

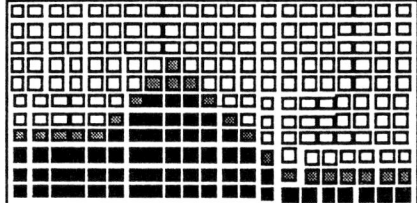 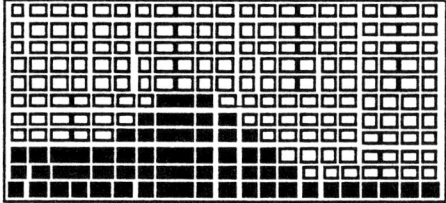

Fig. 9. Gaussian transformation of an r-triangle.

4. Conclusions

We have extended the structural approach to image description via rewriting rules to the problem of describing transformations of structures in an image, under application of global image transformations. The methodology is based on an analysis of the characteristics of the operator with respect to the adopted segmentation technique.

The conditions for its application are that structures are composed of a symmetrical part and of a residual, non-symmetrical, but regular -i.e. maintaining its orientation, component. Both components must consist of contiguous pixels along a contour. Moreover, the structure size must be sufficient to avoid high-frequency effects.

This research aims at providing symbolic descriptions of the characteristics of IPOs, in order to facilitate their automatic selection during the generation of an image interpretation strategy. This will require defining a language for operator description and a metalevel interpreter able to select suitable operators based on such descriptions.

5. References

[1] C.Spinu, C.Garbay, J.M.Chassery, "A cooperative and adaptive approach to medical image segmentation", in P.Barahona, M.Stefanelli, J.Wyatt eds., *Artificial Intelligence in Medicine*, Springer, 1995, 379-390.
[2] P.Bottoni, P.Mussio, P.Protti, R.Schettini, "Knowledge-based contextual recognition and sieving of digital images", *PRL.*, **10**(2):101-110, 1989
[3] P.Mussio, D.Merelli, M.Padula, "An approach to Definition, Description and Extraction of structures in Binary Digital Images", *CVG&IP*, **31**:19-49, 1985.
[4] K.S.Fu, *Syntactic Pattern Recognition and Applications*, Prentice-Hall, 1982.
[5] G.T.Herman, H.Rozenberg, *Developmental Systems and Languages*, North-Holland, 1975.
[6] P.J.Burt, "Smart sensing within a pyramid vision machine", *Proc. of the IEEE*, **76**(8):1006-1015, 1988.

Easy Calibration of Pan/Tilt Camera Heads and Online Computation of the Epipolar Correspondences

Stephan Spiess[1] and Mengxiang Li[2]

[1] Institute of Flexible Automation (INFA)
Vienna University of Technology
A-1040 Wien, Austria
sp@flexaut.tuwien.ac.at

[2] Computational Vision and Active Perception Laboratory (CVAP)
Department of Numerical Analysis and Computing Science
Royal Institute of Technology (KTH), S-100 44 Stockholm, Sweden

Abstract. In this article we describe a calibration procedure of a binocular camera head with two independent pan and tilt axes. For the calibration procedure itself no separate rotations of the respective axes and no fixation is required. To get reliable calibration data just a plane surface with known 2D coordinates of distinguishable target points is needed. This publication describes facts and techniques that are known to the robotic people but mostly unknown to the computer vision society.

1 Introduction

A binocular camera system usually comprises a pair of cameras mounted on a platform or a robot arm. Depending on the design of the system, the degrees of freedom (DOF) may be different. Normally, the head (neck) has two DOF: pan and tilt. Each eye (camera) may have one or two DOF: pan and tilt. In addition eyes may have the freedom of zooming, focusing and aperture control. With such a system one can manipulate its visual parameters in a controlled manner in order to extract useful information about the scene in time and space.

We will focus on the problem of calibrating the relative position and orientation of the two camera coordinate frames which depend on the commanded angular values, the geometry of the system and initially unknown errors of this description. If these errors are identified they can be included in the geometric model and together with the commanded position we obtain an exact kinematic description of the system. We can assume for simplicity that the intrinsic parameters of the cameras are known, although they may vary with different zoom or focus settings. The interdependence of these parameters can be recorded in look-up-tables [7], thus our assumption does not restrict the general case. The knowledge of the intrinsic parameters, gained by camera calibration, and the extrinsic parameters, gained by a kinematic calibration, enables the fast and reliable computation of the essential and fundamental matrix without the need of point matching.

The presented work is related in the same way to the field of robot calibration as it is to the field of camera calibration. In both disciplines the visual system or the robot is calibrated in a world coordinate system with measurement points with accurately known position. Besides a world-base transform (the extrinsic parameters in the case of camera calibration) a least squares estimation yields the parameters of the object model such that an error function is minimized. From robot calibration it is known that the results of this estimation are only reliable if a wide range of the working space is covered in the calibration procedure [12]. For a camera with known intrinsic parameters the recovery of the pose with respect to a world coordinate system requires that at least four distinguishable non-collinear points with known 3D coordinates are in the view. The recovery gets more reliable if the points cover a larger part of the image. For the calibration of a "robot" with two axes and limited link length, like it is the case with a pan-tilt camera device, we encounter a trade-off dilemma that is described in the following. If the calibration object is very far away, then the axes of the device can move considerably and still have the object in view. But in that case we can not rely on the camera pose recovery, because the image of the object will be small. If, on the other hand, the object is very close the pose recovery gets reliable but we can only cover little of the working space of the axes of rotation, the object has to be kept in view (see figures 1, 2). A way out of this dilemma would be the use of a large calibration device with high accuracy, which is very expensive.

Li, Brady, and Wiles [5] attack the problem of binocular calibration by using point matchings at certain camera poses. At those poses the fundamental matrix is computed by an eight points algorithm and afterwards corrections are applied. Their method is only applicable to mechanisms with common elevation, that means with a common tilt axis.

Young et al. [9] describe a method where the 3D motion between different robot positions is recovered by analysing the image contents. This motion is compared to a nominal motion of separately moved axes of the robot. For a binocular camera head this approach is trapped in the described dilemma.

Davidson, Reid, and Murray [1] fixate points of known 2D coordinates to recover a plane-plane homography and calibrate their binocular head. However fixation means closed loop control of the axes of the platform and that means avoidable effort. Furthermore their approach requires that the axes of rotation are perpendicular to each other and aligned with the camera axes.

Since Zhuang, Wang, and Roth [16] use minimal models for robot and camera, parameter identification can be done simultaneously. No separate motion of axes is necessary. Every measured image point simply contributes two equations to a large system of nonlinear equations. The use of a calibration board of 0.002mm accuracy makes it difficult and expensive to confirm their results. The main drawback of their publication is the unnecessarily complicated mathematics.

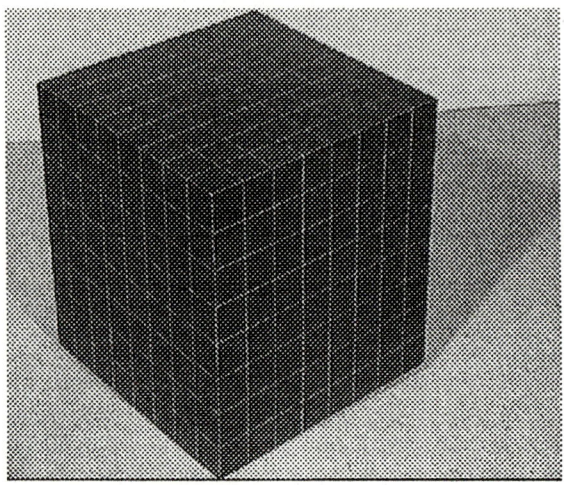

Fig. 1. A calibration object too close. A very small range of action of the cameras decreases the reliability of the kinematic calibration.

Many authors solve the problem of hand-eye calibration by solving homogeneous matrix equations of the form $AX = XB$, see [3] for a brief overview. These approaches need also the 3D recovery and suffer therefore from the mentioned dilemma. We think that solving the problem in Euclidean space is generally a bad idea, because then we have to cope with the inaccuracies introduced by pose estimation.

The subsequently presented approach is similarly to Zhuang's method capable of simultaneous calibration of camera and platform, no fixation is needed, no separate movements of axes are necessary. The mathematics used is easily understandable, the MATLAB-source code is provided [10]. Experimental results show that for the recovery of the epipolar geometry satisfactory results can be achieved without any sophisticated calibration device.

2 Problem Formulation

2.1 A Kinematic Model

Since we are interested in the calibration of the relative transformation of the two cameras, the neck does not influence our calibration result and is therefore not considered here. For the sake of simplicity we place the base coordinate system of the camera head such that its origin lies in the middle of the nominal intersections of the respective pan and tilt axes, the z axis is vertical and the y axis points towards the scene. Let A'_w denote the 6 DOF world-base transform

Fig. 2. A large distance enables a larger range of action but the pose recovery gets unreliable.

relating the coordinate system of a calibration plane to the base of the binocular head of the form

$$\boldsymbol{A'}_w = Transl(x, y, z)Rot(x, \alpha)Rot(y, \beta)Rot(z, \theta). \tag{1}$$

In the two kinematic chains first we rotate about the pan axes (vergence) then about the tilt axes (elevation). Enhancing $\boldsymbol{A'}_w$ with (initially unknown) error parameters and translations to the respective pan axes yields for the left base coordinate system

$$\boldsymbol{A}_{w,l} = \boldsymbol{A'}_w Transl(e/2 + \Delta x_l, \Delta y_l, \Delta z_l)Rot(x, \Delta\alpha_l)Rot(y, \Delta\beta_l)Rot(z, \Delta\theta_l). \tag{2}$$

Where e denotes the eye to eye distance of the camera head. For the right base coordinate system replace l by r and e by $-e$. Now the z axis of $\boldsymbol{A}_{w,l}$ is aligned with the left pan axis and similarly the z axis of $\boldsymbol{A}_{w,r}$ is aligned with the right pan axis. We follow the two different kinematic chains to the respective cameras. Using the Denavit-Hartenberg model [8,14] we obtain

$$\boldsymbol{A}_{pan,l} = Rot(z, \phi_l + \pi/2)Transl(\Delta a_l, 0, 0)Rotx(\Delta\alpha_l - \pi/2) \tag{3}$$

as the transformation relating the pan and tilt axis. A similar equation is obtained for the right kinematic chain. Here ϕ_l denotes the variable pan angle. In the model we assume that zero pan angles result in a straight forward look with approximately parallel optical axes. This is no restriction to the general case, because adding a appropriate constant offsets to the angular values will always lead to such a system. Generally the Denavit-Hartenberg model has four (error) parameters, but here only two, the normal distance Δa_l of the two axes and the twist angle $\Delta \alpha_l$ appear. That is because the unused parameters are redundant with Δz_l and $\Delta \theta_l$.

The rotations about the tilt axes and the transformations into the camera frames have again 6 DOF. Again left and right transformations are the same.

$$A_{tilt,l} = Rot(z, \psi_l + \pi)Rot(y, \pi/2)Rot(z, \Delta\theta_l)Rot(y, \Delta\beta_l) \\ Rot(x, \Delta\alpha_l)Transl(\Delta x, \Delta y, \Delta y) \quad (4)$$

With ψ_l we denoted the left tilt angle.

All together we have 28 error parameters, which corresponds to the rule that a serial manipulator with rotational axes consisting of rigid links has $4N + 6$ parameters [11,16], where N is the number of degrees of freedom and our head is considered as two 2 DOF robots.

2.2 Camera Model

The left camera frame is described by the homogeneous transformation

$$C_l(\phi_l, \psi_l, param) = A_{w,l} A_{pan,l} A_{tilt,l}$$

$$C_l = \begin{pmatrix} r_{11} & r_{12} & r_{13} & X_0 \\ r_{21} & r_{22} & r_{23} & Y_0 \\ r_{31} & r_{32} & r_{33} & Z_0 \\ 0 & 0 & 0 & 1 \end{pmatrix} \quad (5)$$

In our case the positive z axis of the camera frame directs away from the scene. In this case we derive the left image coordinates $(x_{mod}, y_{mod})^T$ as the projection of a 3D point $(X, Y, Z)^T$ by equations 6, 7 [4].

$$x_{mod} = x_0 + (x - x_0)d_r - f_x \frac{r_{11}\hat{X} + r_{21}\hat{Y} + r_{31}\hat{Z}}{r_{13}\hat{X} + r_{23}\hat{Y} + r_{33}\hat{Z}} \quad (6)$$

$$y_{mod} = y_0 + (y - y_0)d_r - f_y \frac{r_{12}\hat{X} + r_{22}\hat{Y} + r_{32}\hat{Z}}{r_{13}\hat{X} + r_{23}\hat{Y} + r_{33}\hat{Z}} \quad (7)$$

where $f_x = f/s_x$, $f_y = f/s_y$, f is the focal length and s_x, s_y are the pixel lengths in x and y directions. We use the abbreviations $\hat{X} = X - X_0, \hat{Y} = Y - Y_0, \hat{Z} = Z - Z_0$. The coordinates of the principal point are (x_0, y_0) and d_r denotes the radial distortion of the lens with

$$d_r = a_1(\frac{r^2}{R_0^2} - 1) + a_2(\frac{r^4}{R_0^4} - 1) \quad (8)$$

$$r = \sqrt{(x - x_0)^2 + (y - y_0)^2}. \quad (9)$$

The constant R_0 is set to half of the image border length. This computation of the distortion exhibits better numerical stability than a conventional model [6,15]. The intrinsic parameters $f_x, f_y, x_0, y_0, a_1, a_2$ are not accurately known and therefore for each camera the number of parameters is now 14+6=20.

A similar equation holds for the right image coordinates.

2.3 Cost Function

Every measured target point with known 3D coordinates and viewed with the left camera produces an image point $(x_{meas}, y_{meas})^T$. Together with the currently commanded angular variables ϕ_l, ψ_l this point contributes two equations

$$x_{mod} = x_{meas} \tag{10}$$
$$y_{mod} = y_{meas} \tag{11}$$

to a system of nonlinear equations. The equations depend on the unknown parameters and the known joint variables and image coordinates. We can solve for the unknown parameters by a least squares estimation if we have at least 10 different points recorded with different commanded poses.

In practice we have to use an overdetermined system to gain reliable results. Due to measurement noise and limits of the model, equations 10, 11 will never be fullfilled accurately. Thus the cost function for m measurement points can be written as

$$\sum_{i=1}^{m} \left((x_{mod,i} - x_{meas,i})^2 + (y_{mod,i} - y_{meas,i})^2 \right). \tag{12}$$

The minimization can be done with the Gauss-Newton method which converges very fast in the neighbourhood of a solution. This method is the iterated application of a locally linearised balanced adjustment [4]. The iterations are stopped when the amount of adjustment on each of the parameters is below a prescribed threshold. The Jacobian can be approximated using finite differences. A start estimation for the intrinsic camera parameters can be obtained from the camera specifications. The initial error parameters of the pan/tilt unit can be set to zero.

The world-to-base transformation (eq. 1) has to be measured manually with measurement tape to obtain a feasible start value for the parameter estimation. Since the accuracy of this measurement is mostly far worse than the initial estimation of the other parameters, an estimation procedure for the world-to-base parameters (eq. 2) with the other parameters fixed to their default values should be done before doing the adjustment with all parameters. After obtaining the final parameters the parameters of the world-to-base transform are worthless for the epipolar problem.

Although like [16] the described technique is capable of simultaneous calibration of all parameters we recommend in the case of imperfect calibration data the splitting into two stages. Stage one being the camera calibration, thus the identification of the intrinsic parameters, and stage two being the kinematic calibration with fixed intrinsic parameters.

3 Epipolar Geometry

Let $D = C_l^{-1} C_r$ with elements d_{ij} be the transformation from the left camera frame to the right. Then the *essential matrix* [13] is given by

$$E = T D_{3\times 3} \tag{13}$$

where $D_{3\times 3}$ is the upper left 3×3 submatrix of D and

$$T = \begin{pmatrix} 0 & d_{34} & -d_{24} \\ -d_{34} & 0 & d_{14} \\ d_{24} & -d_{14} & 0 \end{pmatrix}. \tag{14}$$

If we neglect distortion the transformation of a point $(x, y, 1)^T$ given in homogeneous pixel coordinates into a metric representation is given by a multiplication with a matrix [2,7]

$$P^{-1} = \begin{pmatrix} -\frac{1}{f_x} & 0 & \frac{x_0}{f_x} \\ 0 & -\frac{1}{f_y} & \frac{y_0}{f_y} \\ 0 & 0 & 1 \end{pmatrix}. \tag{15}$$

Now the *fundamental matrix* can be derived as

$$F = (P_l^{-1})^T E P_r^{-1}. \tag{16}$$

For any point $(x_l, y_l, 1)^T$ in the left image given in homogeneous pixel coordinates F yields the epipolar line $Ax + By + C = 0$ in the right image as follows

$$(A, B, C) = (x_l, y_l, 1) F. \tag{17}$$

Note that x and y are given in pixel coordinates. For the reverse direction we have to use F^T. Further note that this nice property is only valid for a distortion-free lens. That means in practice we have to correct the distortion of a measured image point $(x, y)^T$ by an appropriate shift about $(-d_r(x - x_0), -d_r(y - y_0))^T$.

4 Experimental Results

In all simulations and experiments we first calibrated the cameras with the calibration cube shown in figure 1 and fixed the intrinsic parameters for the kinematic calibration. This was done because the accuracy of the used calibration object for the camera calibration was several times higher than the accuracy of the 3D coordinates of the target points used for kinematic calibration. The calibration worked well, despite of the fact that we measured the 2D points of measurement targets manually, details can be found in [17] and can be extracted from [10].

5 Summary

We presented a minimal, but complete offset model for the kinematics of a binocular head with separated eye pan and tilt and a, for the calibration procedure, fixed neck. We demonstrated that the online calculation of the essential and fundamental matrix can be done to a, for the sake of computer vision, satisfactory precision with easy mathematics and without expensive calibration mechanisms. The accuracy of the obtained results is in the range of that of competing methods but accomplished with very little effort.

References

1. A. Davidson, I. Reid, D. Murray: "The active camera as a projective pointing device." *Proc. of the British Machine Vision Conf. '95*, Birmingham 1995, pp. 453 – 462.
2. O. Faugeras: *Three-Dimensional Computer Vision. A Geometric Viewpoint.* Cambrige, MA, MIT Press, 1993.
3. R. Horaud, F. Dornaika: "Hand-Eye Calibration." *The Int. Journal of Robotics Research*, Vol. 14, No. 3, June 1995, pp. 195 – 210.
4. K. Kraus: *Photogrammetry.* Dümmler Verlag Bonn, 1993.
5. F. Li, M. Brady, C. Wiles: "Fast Computation of the Fundamental Matrix for an Active Stereo Vision System." *Computer Vision – ECCV'96.* Lecture Notes in Computer Science, Vol. 1064, ed. by B. Buxton and R. Cipolla, Springer-Verlag, 1996.
6. M. Li: "Camera Calibration of the KTH Head-Eye System." Technical Report ISRN KTH/NA/P-9407-SE, Dep. of Numerical Analysis and Computing Science, Royal Institute of Technology, Stockholm 1996.
7. M. Li: "Camera Calibration of a Head-Eye System for Active Vision." *Computer Vision – ECCV'94*, Lecture Notes in Computer Science, Vol. 800, ed. by J.-O. Eklundh, Springer-Verlag 1994, pp. 543 – 554.
8. R. Paul: *Robot Manipulators: Mathematics, Programming, and Control.* Cambridge, MA, MIT Press, 1981.
9. G.-S. Young, T.-H. Hong, M. Herman, J. Yang: "Kinematic Calibration of an Active Camera System." *Proc. of IEEE Conf. CV '92*, pp. 748 –751.
10. MATLAB source code for the calibration; anonymous ftp to antigone.infa.tuwien.ac.at. File /pub/BinoCal.tar.Z
11. B. Mooring, Z. Roth, M. Driels: *Fundamentals of Manipulator Calibration.* New York, Wiley & Sons, 1991.
12. K. Schröer: "Theory of kinematic modelling and numerical procedures for robot calibration." *Robot Calibration*, edited by R. Bernhardt and S. Albright, Chapman & Hall, London 1993.
13. R. Tsai, T. Huang: "Uniqueness and Estimation of Three-Dimensional Motion Parameters of Rigid Objects with Curved Surfaces." *IEEE PAMI*, Vol. PAMI-6, No. 1, Jan. 1984.
14. M. Vincze, K.M. Filz, H. Gander, J.P. Prenninger, G. Zeichen: "A Systematic Approach to Model Arbitrary Non Geometric Kinematic Errors." *Advances in Robot Kinematics and Computational Geometry*, edited by J. Lenarcic and B. Ravani, Kluwer Academic Publishers, 1994, pp. 129 – 138.
15. J. Weng, P. Cohen, M. Herniou: "Camera Calibration with Distortion Models and Accuracy Evaluation." *IEEE Transactions on Pattern Analysis and Machine Intelligence*, Vol. 14, No. 10, Oct. 1992.
16. H. Zhuang, K. Wang, Z. Roth: "Simultaneous Calibration of a Robot and a Hand-Mounted Camera". *IEEE Transactions on Robotics and Automation*, Vol. 11, No. 5, Oct. 1995, pp. 649 – 660.
17. S. Spiess, M. Li: "Kinematic Calibration of an Active Binocular Head for Online Computation of the Epipolar Geomatry", *Technical Report of the Computer Vision and Active Perception Laboratory*, Royal Institute of Technology, Stockolm, Sweden. See www.bion.kth.se/abstracts/cvap205.html

Integration of Spatio-Temporal Information for Motion Detection by Means of Fuzzy Reasoning

M. Barni, F. Bartolini, V. Cappellini, F. Lambardi

Dipartimento di Ingegneria Elettronica, Università di Firenze,
Via S. Marta 3, 50139, Firenze
e-mail barto@cosimo.die.unifi.it

Abstract. In this paper a motion detection system based on fuzzy reasoning is presented. Each pixel of a frame of the sequence is attempted to be classified as belonging to one of four classes (moving, still, uncovered background, covered background). The classes are treated as fuzzy sets, and as such, they are characterized by membership functions. After an initialization step, the degree of membership of each pixel to each class is refined by the application of a reasoning module driven by a set of fuzzy rules. Such fuzzy rules are designed so that the spatio-temporal correlation of image sequences is exploited by integrating information extracted from a small spatio-temporal neighborhood. The proposed system results to be flexible, thanks to the use of the reasoning approach (rules can be easily changed), and robust, thanks to the use of fuzzy logic.

1 Introduction

Motion detection techniques play an important role in the field of image sequence processing. Sensing the objects of the scene that are moving is useful, for example, to reduce the area of the images to be compressed in video coding systems, to produce alarms when an intrusion is occurring in restricted areas, or to count vehicles for traffic control applications.

Motion detection techniques are aimed to detect if (and possibly where) something is moving in the imaged scene.

When the goal is simply the surveillance of restricted areas it is not usually needed to detect where moving objects are localized but only whether they are present or not, this allows to deal with image noise by integrating the information obtained over the whole image [2,5].

For more complex tasks, as for example traffic monitoring, it is, on the contrary, required to segment the image in moving objects and static areas. For this goal two classes of algorithms can be identified. The algorithms of the first class analyze the differences between the current image and a reference static background. In order to deal with global illumination changes, the static background has to be adaptively updated [6] or edge images (less affected by the variation of the ambient lighting) have to be used [7]; nevertheless, the sensitivity of the algorithms of this class to sudden illumination changes remains a major drawback.

The algorithms of the second class, on the other side, analyze the differences between couples of successive frames of the sequence. For changing pixels detection, likelihood ratio tests are usually used [8]. Furthermore to enhance robustness it is important to exploit the a priori knowledge about motion spatio-temporal coherence, in that spatial and temporal neighboring pixels usually move in a similar way. Spatio-temporal integration of motion information can be, for example, achieved by modeling the image sequence as a 3D gaussian markov random field and by estimating the segmentation with a maximum a posteriori criterion [9]. This method is quite robust, but it is difficult to be driven by a priori heuristic knowledge about the image sequence. Furthermore, it does not allow to distinguish covered and uncovered areas from moving objects. Another approach for integrating spatio-temporal information is based on a set of heuristic rules that allow to assign each image pixel to a class identifying the behavior of the corresponding point in the 3D real world [10].

In the system described in this paper integration is achieved by imposing some heuristic rules to hold in a spatio-temporal neighborhood of each pixel. The adopted rules allow to label each pixel as belonging to a moving object (class M), to an area of covered (class C) or uncovered (class U) background, or to the still background (class S). The use of 4 classes instead of 2 only (still/moving) permits to better shape the mask of moving objects given that only class M and C pixels are assigned to it. We also propose to cast the heuristic rules in the framework of fuzzy reasoning. Fuzzy reasoning is, in fact, very suitable for modeling human knowledge about real world and also for producing very robust algorithms [4].

2 Fuzzy reasoning

Fuzzy or approximate reasoning is the core of any fuzzy system, since it is in charge of deriving actions or taking decisions by starting from imprecise or vague data. According to the original approach proposed by Zadeh [11], and further developed in [12] and [13], the fuzzy reasoning module is primarily characterized by a set of fuzzy rules, by means of which raw data are processed and decisions are taken. In the case of a multi-input-single-output (MISO) system, fuzzy rules have the form

R_1: if x_1 is A_{11} and ... x_n is A_{1n} then y is B_1
...
R_m: if x_1 is A_{m1} and ... x_n is A_{mn} then y is B_m

where x_i are the input linguistic variables, y is the output linguistic variable and A_{ji}, B_j are fuzzy sets. The problem of finding the value of the output variable y (i.e. the fuzzy set associated to it) by starting from the values assumed by the variables x_i is called the *fuzzy inference* or *fuzzy reasoning process*. Such a process involves three main steps:

1. determination of the matching between the variables x_i and the sets A_{jj};

2. computation of the output of each single rule R_j;
3. aggregation of the outputs of the single rules.

A comprehensive survey of the possible approaches that can be followed to perform each of the above steps is outside the scope of this note, in the following only the strategy used throughout the motion detection system will be described. Interested readers may find more details in [12,13,4].

Let λ_j be the degree of activation of the j-th rule. To evaluate λ_j's, let us consider first the quantities

$$\rho_{ji} = \max_{u \in U}\{\min\{\mu_{x_i}(u), A_j(u)\}\}, \qquad (1)$$

where $\mu_{x_i}(u)$ is the membership function of the fuzzy set assumed by the linguistic variable x_i, and ρ_j represents the degree of matching between the actual value of the variable x_i and the fuzzy set A_j. Once ρ_j have been calculated the activation degrees λ_j are computed as

$$\lambda_j = \min_{i=1,n}\{\rho_{ji}\}. \qquad (2)$$

Steps 2 and 3 of the fuzzy inference process, i.e. computation of the output of each rule and combination of such outputs, is accomplished by means of the formula

$$\mu_y(u) = \max_{j=1,m}\{\min\{\lambda_j, \mu_{B_j}(u)\}\}. \qquad (3)$$

It should be noted that being y a linguistic variable, the output of the first three steps of the inference process is a fuzzy set, whose memberships function is defined by means of equations 1 through 3. The defuzzification of y can be carried out in many ways [13]. In this work the following simple rule is applied to produce a crisp version Y of the linguistic variable y

$$Y = \max_{u \in U}\{\mu_y(u)\}. \qquad (4)$$

A generalization of the scheme of fuzzy system given above comprises the possibility of including an else statement [11,13].

3 Overview of the System

As outlined in Sections 1 and 2, the system aims at classifying pixels as being in one of 4 possible states: moving (M), still (S), covering (C) and uncovering (U). In particular, for each frame, pixels are assigned a degree of membership to one of the above 4 classes.

As a first step pixels classification is initialized by relying on frame differences. Then a former set of fuzzy rules is applied in order to refine the rough classification produced by the initializer. The status of a pixel in position (i,j) is determined by reasoning upon the states of the pixels lying in a 3×3 window centered in (i,j) and the states of the pixels in position (i,j) in the previous and the subsequent frame.

In order to understand how the fuzzy reasoning module operates, some definitions must be given first. Six linguistic variables are taken into account by the system: 2 of them, x_{-1} and x_{+1}, refer to the status of pixels (i,j) in the previous and subsequent frame respectively, whereas the other 4 are related to the global status of the pixels inside the 3 × 3 working window in the current frame. Let us call the variables belonging to this second group x_m, x_s, x_c and x_u.

Fuzzy sets assigned to x_{-1} and x_{+1} are defined in a space U' consisting of 4 elements: let us call these elements m (moving), s (still), c (covering) and u (uncovering). Such fuzzy sets are built by specifying for each element in U' the possibility that the previous/subsequent pixel is a moving, still, covering or uncovering pixel (in Table 1 the fuzzy sets used to define the rules the reasoning module consists of are depicted).

Fuzzy set	Elements of U'			
	m	s	c	u
moving	1	0	0.1	0.1
covering	0.2	0	1	0.1
uncovering	0	0.2	0.1	1
still	0	1	0.1	0.1

Table 1. Values of the membership functions of the four fuzzy sets.

With regard to x_m, x_s, x_c and x_u, they account for the global status of the pixels in the current frame. To build the space U_M fuzzy sets referring to x_m belong to, the degrees of membership to the class M of pixels inside the working window are summed. Let us call such a sum M-sum; the possible values that the M-sum can assume form the U_M space. A fuzzy set defined in U_M gives the possibility, for each value in U_M, that the $M - sum$ takes that value. Let us consider for example the statement *most of the pixels in the working window are moving (belong to class M)*; such a statement can be cast in the framework of approximate reasoning by introducing a fuzzy set whose shape is illustrated in Figure 1, the fuzzy set meaning that for a window in which *most* of the pixels are known to belong to the M-class the possibility that the M-sum is, for example, 2.5 equals 0.4. Of course, similar considerations hold for the linguistic variables x_s, x_c and x_u.

By means of fuzzy inference, the inputs to the reasoning modules are processed and pixel classification significantly improved. Following the first reasoning module a second step is carried out to further refine pixel classification. For this second step a different set of rules is applied since different configurations are likely to occur with regard to the output of the initializer.

The fuzzy reasoning module produces four memberships with the possibility that the pixels belong to the M, S, C and U classes. At this point, defuzzification

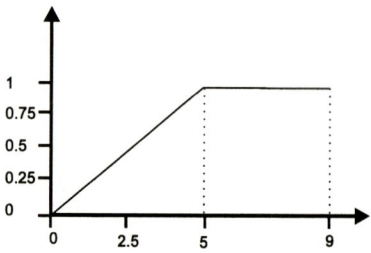

Fig. 1. The function describing the *most of the pixels* fuzzy set.

is performed to definitely classify pixels: more specifically, each pixel is assigned to the class it exhibits the maximum membership to.

4 Initialization and set of rules

With regard to the initialization of memberships, a procedure derived from [9] is adopted. A likelihood-ratio-like parameter L is evaluated inside a 5×5 window W centered at the current pixel position:

$$L = \frac{1}{n^2} \sum_{(i,j) \in W} \delta(i,j)^2 +$$

$$\frac{1}{\sum_{(i,j) \in W} \Delta x^2(i,j)} \left\{ \left[\sum_{(i,j) \in W} \Delta x(i,j) \delta(i,j)^2 \right]^2 + \left[\sum_{(i,j) \in W} \Delta y(i,j) \delta(i,j)^2 \right]^2 \right\}$$

(5)

where $\delta(i,j)$ is the frame difference computed for pixels in position (i,j) and $\Delta x(i,j)$ and $\Delta y(i,j)$ are the horizontal and vertical distances of pixel in position (i,j) from the center of the window. The value of this parameter it is used as input to the 4 memberships initialization functions (see Figure 2).

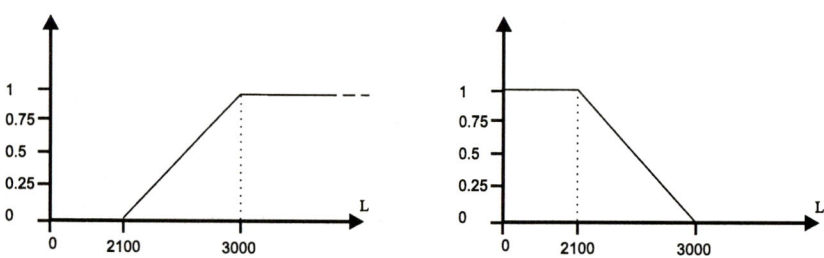

Fig. 2. Initialization functions used for the Moving, Covering and Uncovering fuzzy sets (*Left*) and for the Still fuzzy set (*right*).

In this section the complete list of fuzzy rules the system consists of is given. Let us begin with the first reasoning step, the following rules are applied to the output of the initializer:

R_1: if x_p was *moving* and
   x_s is *still* and
   at least some pixels in the current frame are *uncovering*,
   then the current pixel is *certainly uncovering*;

R_2: if x_p was *moving* or *covering* and
   x_s is *moving* or *uncovering* and
   at least some pixels in the current frame are *moving*,
   then the current pixel is *certainly moving*;

R_3: if x_p was *still* and
   x_s is *moving* and
   at least some pixels in the current frame are *covering*,
   then the current pixel is *certainly covering*;

R_4: if x_p was *still* or *uncovering* and
   x_s is *covering* or *still* and
   at least some pixels in the current frame are *still*,
   then the current pixel is *certainly still*;

else

G_1: if *most* of the pixels in the current frame are *still*
   then the current pixel is *certainly still*;

G_2: if *most* of the pixels in the current frame are *moving*
   then the current pixel is *certainly moving*;

else

H_1: if *most* of the pixels in the current frame are *covering*
   then the current pixel is *certainly covering*;

H_2: if *most* of the pixels in the current frame are *uncovering*
   then the current pixel is *certainly uncovering*;

The above rules have been designed by taking into account the situations that are most likely to occur in the analysis of image sequences; when an unusual pixel configuration is encountered which is not encompassed by these rules, the fuzzy engine produces a vague output to be refined during subsequent steps. The second set of rules is very similar to the first one, the only difference being that rules H_1 and H_2 are at the same level as rules G_1 and G_2. The rationale for such a minor modification is that, due to the particular initialization adopted, the first module is likely to reason upon images in which the degree of membership to the M-class is equal to those of the C and U-classes.

5 Results and Conclusions

The proposed system has been tested on synthetic and real world images. The results obtained by processing a sequence representing vehicles on a highway lane are here described.

In Figure 3 a frame of the processed sequence is shown: in the field of view both the lanes are included. In Figure 4 the values computed by the algorithm for the four membership functions at each pixel location are visualized (the images labeled M, S, C and U display, respectively, the Moving, the Still, the Covering and the Uncovering membership functions). In Figure 5 the results of

Fig. 3. A frame of the sequence representing vehicles on a highway.

the defuzzification process are shown. In each map the pixels belonging to the respective class are highlighted. From the analysis of the maps it is evident that the algorithm allows the estimation of the direction of motion, even if it does not estimate speed. For vehicles approaching the camera, covered areas are, in fact, behind and uncovered are on the back. The converse it is true for vehicles moving in the opposite direction. Furthermore the classification into four classes instead of only two (i.e. moving/still) helps in managing ambiguous situation and increases, then, the robustness of the algorithm.

The presented fuzzy reasoning approach to motion detection allows to combine the flexibility of the systems based on reasoning algorithms and the robustness which is characteristic of fuzzy methodologies.

6 Acknowledgment

This work was partially funded by the *Progetto Finalizzato Trasporti II* of CNR (Italian National Research Council) under grant nr. 96.00027.PF74 .

References

1. S.B. Chae, J.S. Kim and R.H. Park "Video coding by segmenting motion vectors and frame differences," vol. 32, no. 4, Apr. 1993, pp. 870–876.
2. I. Distein, "A new technique for visual motion alarm," *Pattern Recognition Letter*, vol. 8, no. 5, Dec. 1988, pp. 347–351.
3. P.G. Michalopoulos, "Vehicle detection video through image processing: the Autoscope system," *IEEE Transactions on Vehicular Technology*, vol. 40, no. 1, Feb. 1991, pp. 21–28.

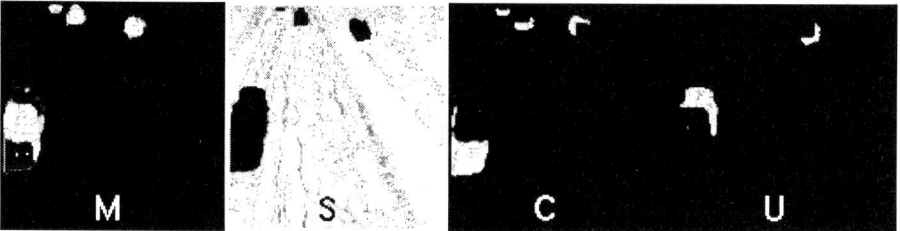

Fig. 4. Representation of the values computed by the system for the four membership functions at each pixel postion (values are normalized in the range 0–255).

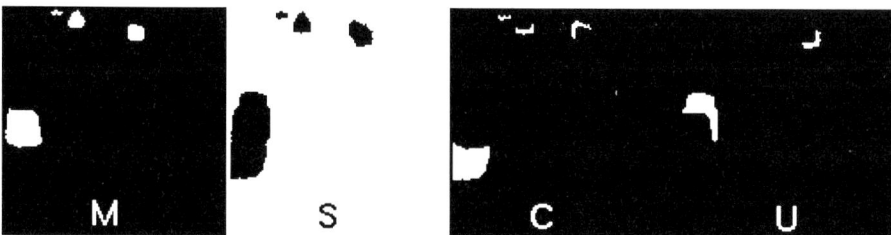

Fig. 5. Maps of the pixels belonging to the four classes (Moving, Still, Covering, Uncovering).

4. T. Terano, K. Asai and M. Sugeno, *Fuzzy Systems Theory and its Applications*, Academic Press Inc. 1992.
5. P. Iliev and L. Tsekov, "Motion detection using image histogram sequence analysis," *Signal Processing*, vol. 30, no. 3, Feb. 1993, pp. 373–384.
6. K.P. Karmann, A. Von Brandt and R. Gerl, "Moving object segmentation based on adaptive reference images," *Proc. EUSIPCO-90.*, Barcelon, Sept. 18–21, 1990.
7. M.A. Vicencio, B. Qiu and M.G. Hartley, "Algorithms and architectures for the analysis of road-traffic movements," *Proc. 3rd Int. Conf. Image Proc. and Appl.*, Warwich, UK, July 1989.
8. R. Jain and H.H. Nagel, "On the analysis of accumulative difference pictures from image sequences of real world scenes," *IEEE Transactions on Pattern Analysis and Machine Intelligence*, vol. 1, no. 2, Apr. 1979, pp. 206–214.
9. P. Bouthemy and P. Lalande, "Recovery of moving object masks in an image sequence using local spatio-temporal contextual information," *Optical Engineering*, vol. 32, no. 6, June 1993, pp. 1205–1212.
10. R. Jain, "Extraction of motion information from peripheral processes," *IEEE Transactions on Pattern Analysis and Machine Intelligence*, vol. 3, no. 5, Sept. 1981, pp. 489–503.
11. L. A. Zadeh, "Outline of a new approach to the analysis of complex systems and decision processes", *IEEE Trans. Syst. Man Cybern.*, vol.SMC–3, pp. 28–44, 1973.
12. D. Dubois and H. Prade, *Possibility Theory: an Approach to Computerized Processing of Uncertainty*, New York, Plenum Press, 1988.
13. C. C. Lee, "Fuzzy logic in control systems: fuzzy logic controller - Part I and II", *IEEE Trans. Syst. Man Cibern.*, vol.SMC–20, no.2, pp. 404–435, Mar./Apr. 1990.

Adaptive Motion Estimation and Video Vector Quantization Based on Spatiotemporal Non-linearities of Human Perception

J. Malo, F. Ferri*, J. Albert* & J.M. Artigas
*Departament d' Òptica, *Institut de Robòtica,
Facultat de Física, Universitat de València,
C/ Dr. Moliner 50, 46100, Burjassot, València (Spain)
e-mail: Jesus.Malo@uv.es*

Abstract

The two main tasks of a video coding system are motion estimation and vector quantization of the signal. In this work a new splitting criterion to control the adaptive decomposition for the non-uniform optical flow estimation is exposed. Also, a novel bit allocation procedure is proposed for the quantization of the DCT transform of the video signal. These new approaches are founded on a perception model that reproduce the relative importance given by the human visual system to any location in the *spatial frequency, temporal frequency and amplitude* domain of the DCT transform. The experiments show that the proposed procedures behave better than their equivalent (fixed-block-size motion estimation and fixed-step-size quantization of the spatial DCT) used by MPEG-2.

Key Words: Video Coding, Motion Estimation, Perceptual oriented Quantization.

1 Introduction

The objective of video coding is to find a lossy non-redundant expression of the signal to minimize the transmission or storage requirements in such a way that the retrieved sequence and the original one satisfy some criterion of similarity. In natural video sequences two kinds of redundacies can be indentified: objective (related to spatio-temporal correlations) and subjective (related to the human visual system characteristics). These strong redundancies will give rise to strong compression ratios if they are removed. In practical applications judged by a human observer the compression procedure has to be designed to minimize the *subjective* distortions of the reconstructed sequence. In those cases, perceptual properties of the human viewer must be included in the coder design. The redundancy removal in today widely accepted video compressors (H.261, MPEG-1, MPEG-2 [1,2]) is based in the temporal predictability of the signal [3]. In this encoding scheme a motion estimation block extracts the motion information, and a predictor block uses this to make a prediction of the next frame. In general, the motion estimation/prediction process cannot exactly reproduce the next frame, so an error signal has to be transmitted to the decoder. In this error sequence there still exists a certain degree of redundancy which must be reduced through a quantization module. Research effort in the field has been focused in the motion estimation and error quantization modules considered in an *isolated way*. In this sense, better motion estimation techniques [4,5] try to give better predictions and intrinsically lower complexity error signals. On the other hand, better quantization

---

Acknowledgements: This work has been supported by CICYT projet TIC95-676-C02-01 and IVEI (Generalitat Valenciana) project Nº96/003-035.

techniques [6] are intended to distribute the quantization noise in such a way that the required quality is preserved. In order to achieve a lower complexity error signal, Dufaux et al. [4] have proposed a multiresolution motion estimation algorithm based on minimizing the spatial entropy of the corrections and displacement field, but this approach does not take into account the frequency-dependent transform coding that follows motion estimation. On the other hand, most common implementations of perceptual quantizers are based on the human visual system still-threshold spatial frequency response [1,2,7]. 3D transform coders have been recently proposed [8,9], but in all these approaches, the spatio-temporal suprathreshold characteristics of human perception have not been deeply exploited [6].

Following this, a new video coding scheme is introduced in this work to improve in several ways the temporal and spatial redundancy removal used in MPEG-2. First, a new adaptive-block-size matching algorithm for motion estimation is presented. In the proposed scheme, the splitting criterion is based on minimizing the entropy *in the frequency domain*, taking into account the frequency selective quantization made after the motion estimation stage. In this way, the motion estimation is modified according to the frequency selective encoding requirements. And second, a novel non-linear bit allocation procedure is used for the quantization of the DCT transform of the video signal. This new approach is founded on a perception model that reproduce the relative importance given by the human visual system to any location in the 4-dimensional *spatial frequency, temporal frequency and amplitude* domain of the DCT [10,11].

2 Improvements and Alternatives to the Standard Coding Schemes

In the present standards (up to MPEG-2) the motion estimation module is just an optical flow estimator [3,4,5]. There are several methods to compute the image flow (eg. *differential methods* [4,5], *frequency domain* methods [5]), but standard video coders estimate the displacement field through a *matching technique* [3,4,5]. The usual MPEG implementation of this general matching idea consists of: using rectangular neighbourhoods (blocks), searching for displacement candidates in a restricted set, and using standard correlation as a measure of similarity [1,2,4]. This is called fixed-block-size Block Matching Algorithm (BMA). Regarding to the quality of the predicted frame, the most important parameter in the BMA is the block size. Decreasing the block size leads to a more accurate prediction and to a lower complexity error signal, but it implies a more exhaustive optical flow computation increasing superfluous motion information and reducing the robustness of the displacement estimate.

Standard still image compressors [7] and video coders [1,2] introduce human visual system characteristics at the quantization stage. A transform of the error signal to a frequency domain (DCT) is done because human visual sensitivity is highly uneven in these domains. The bit allocation procedure is founded in human visual system models that consider the perception as a linear filtering process in the Fourier domain. The quantization matrix is inspired in the threshold frequency response of the human viewer [7] and then a variable number of quantization levels per coefficient is allowed. These levels are uniformly distributed in the amplitude range [7,10]. The threshold linear filter model reproduces the relative importance given by the human visual system to the different frequencies, but its validity is restricted to near-threshold contrasts. Neither amplitude sensitivity non-linearities, nor temporal frequency selectivity are taken into account in such a quantization scheme.

2.1 Adaptive-Block-Size BMA with Frequency Domain Splitting Criterion

It has been shown [4,12,13] that an adaptive block size algorithm reduces the number of necessary vector displacement computations concentrating the effort in highly moving areas and improves the quality of moving objects contours in the predicted frames. The adaptive-block-size BMA starts with a motion estimation at a coarse resolution level (big block size). Then, subsequent local refinements of this estimation are made, increasing the resolution in certain areas (splitting selected blocks). The key parameter in this adaptive-block-size BMA is the *splitting criterion*, which is usually based in some *objective* difference measurement related to the energy of the error [12,13]. Taking into account that compression is the main objective of a video coder, Dufaux et al. [4] proposed a splitting criterion for adaptive-block-size motion estimation based on minimizing the entropy of the corrections and displacement field. A coarse block was divided if the entropy of the divided block error signal and its associated flow field was lower than the entropy of the full block error signal and its displacement field:

if $H_{\text{error (split)}} + H_{\text{displacements (split)}} < H_{\text{error (nosplit)}} + H_{\text{displacements (nosplit)}} \Rightarrow$ split   (1)

In this scheme the entropy of the error signal is calculated in the spatial domain before the quantization stage. It means that this *spatially measured entropy* splitting criterion is independent of the quantization process. Splittings that reduce the error signal complexity can be superfluous depending on the type of quantizer used. Following this, and assuming a certain quantization scheme Q in the frequency DCT domain, a novel splitting criterion is proposed:

if $H_{Q[\text{error (split)}]} + H_{\text{displac.(split)}} < H_{Q[\text{error (nosplit)}]} + H_{\text{displac.(nosplit)}} \Rightarrow$ split   (2)

With this *spectral entropy* criterion, motion estimation takes into account the particular quantizer used and some theoretical adventages of this new criterion can be outlined:

1) The *spectral entropy* criterion should be more restrictive than the *spatial entropy* criterion: if the quantization is severe, the action of the quantizer over the non-split block significantly reduce its complexity, so no benefit will be obtained from the splitting. The practical consequence of this is that a simpler quadtree decomposition (fewer blocks) is obtained and usually, a more robust motion estimate is associated with this decomposition.

2) The splitting criterion is matched with the quantizer requirements. The motion estimation is refined only if it improves the error signal in the areas which are significant to the quantizer. If the quantizer is designed with a certain objective, the *spectral entropy* criterion controls the motion estimation to optimize the performance in the same direction. In this case, the *spatial entropy* criterion will minimize the entropy measured in the spatial domain, but the final entropy (the entropy of the transform coefficients) should be minimized by the *spectral entropy* criterion.

2.2 The Information Allocation Function Perception Model

Non-linear response of the human visual system to the amplitude of sinusoidal patterns has lead to question the threshold linear filter based models. The existence of perceptual tolerances to changes in the amplitude and frequency of still gratings implies that the visual system maps the continuous spatial frequency/amplitude range into a finite set of non-uniformly distributed discrete perceptions. This fact has been used to propose a new suprathreshold and non-linear perception model [10,11]. In this model, low level perception is considered as an information removal process characterized by an Information Allocation Function (*IAF*) giving the amount of information used by the system to encode each area of the spatial frequency/amplitude domain. This perceptual bit allocation function has been succesfully used to design a non-uniform step size quantizer for image compression improving the performance of the JPEG-like uniform quantization [10]. The *IAF* model has been used to define a subjective quality measure which accurately reproduce the opinion of the observers under a variety of noise conditions [11]. In this work we generalize this non-linear vector quantizer model to non-zero temporal frequencies, modifiying the threshold spatio-temporal filter [14] with suprathreshold contributions as in the 2-dimensional case, to postulate a spatio-temporal *IAF* of the human observer. Figure 1 show different views of this function.

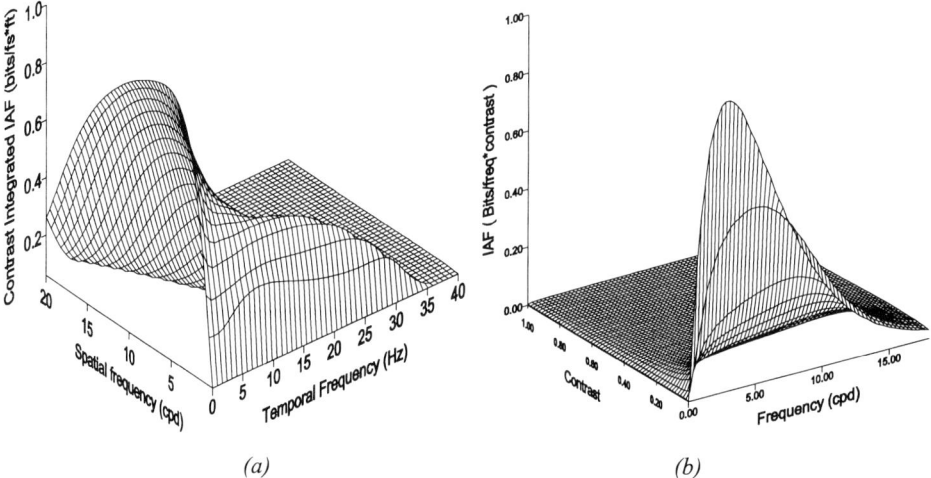

(a)   *(b)*

Fig. 1. (a) Contrast integrated *IAF* (relative amount of information used by the human visual system to encode each spatio-temporal component of an input sequence). (b) Spatial frequency and amplitude dependency of the *IAF* at a particular temporal frequency (non-uniform amplitude bit allocation for different spatial coefficients).

2.3 New Bit Allocation in the DCT Transform Domain

Assuming the *IAF* model, the non-uniformities of perceptual sensitivity in the spatial frequency, temporal frequency and amplitude domain allows us to concentrate the encoding effort in the middle spatial frequencies, low amplitudes and low temporal frequencies. In this work the spatio-temporal *IAF* has been used in two ways. First in a frame-by-frame way, without considering temporal characteristics of the human visual system. We have applied the *IAF* at zero temporal frequency to quantize each frame of the error sequence as does MPEG with its Q matrix. And second, considering both spatial and temporal characteristics of the human viewer. In this later case a quantization of the coefficients of the 3D DCT of the error sequence is performed. In both cases, the codebook design is as follows [10,11]: in the 2D case, one starts deciding the number of quantization levels per coefficient. This is done by means of contrast integrated IAF at zero temporal frequency. Then, the quantization steps and the decision boundaries are non-uniformly distributed in amplitude for each coefficient by a process similar to non-uniform random number generation.

In the 3D case, the first step is deciding how much information is used to encode each temporal frequency frame of the 3D DCT of the error signal. This is done through the contrast and spatial frequency integrated *IAF*. After the temporal frequency dependent bit allocation, there is a spatial frequency dependent and amplitude dependent bit allocation process, analogous to the 2D case for every temporal frequency frame.

3 Results

Both classical and proposed video coding schemes have been applied to different natural sequences. The flow fields and the decoded frames are compared using the exposed techniques at similar bit rates (below 0.5 bpp) on the *Taxi* test sequence [5]. The size of the estimation blocks is 8*8 in the fixed-block-size BMA while this size ranges from 64*64 to 4*4 in the adaptive cases (resolution level of the quadtree decomposition ranging from 2 to 6). The size of the quantization blocks is 16*16 in every case. The splitting criterion is applied in such a way that no interaction between blocks or resolutions is considered [4]. Only forward prediction is used for motion compensation.

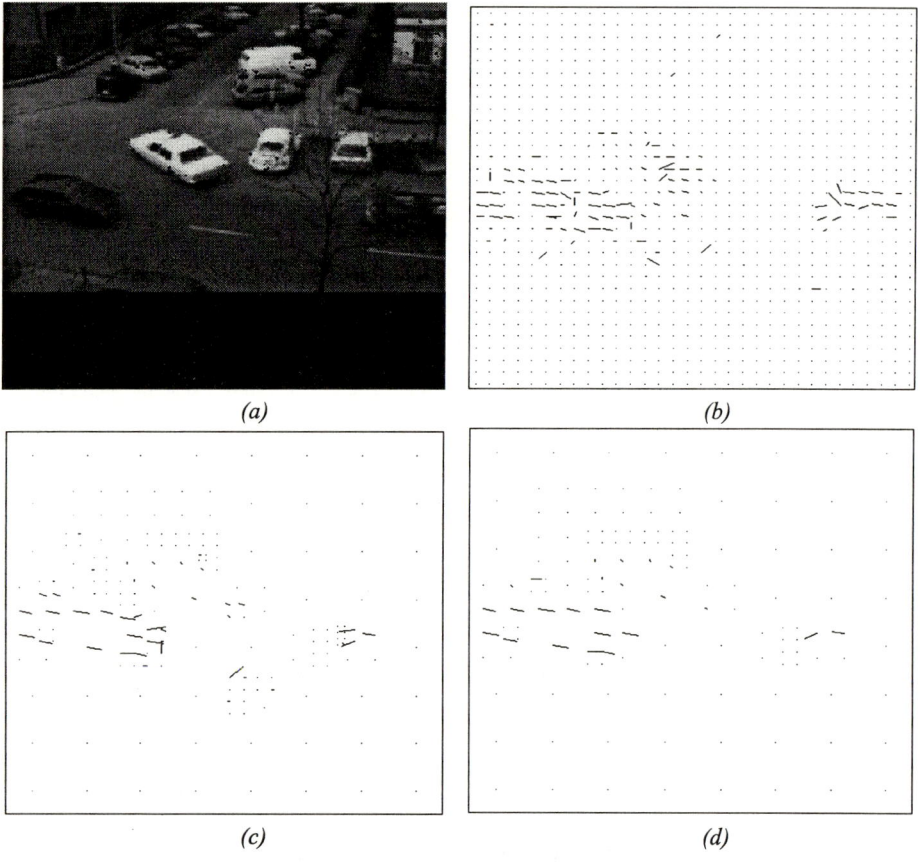

Fig. 2. (a) Original frame 2 of the test sequence. (b) Fixed-block-size (8*8) BMA flow field between frames 2 and 3. (c) Adaptive-block-size (64*64 to 4*4) BMA flow field with spatially measured entropy criterion. (d) The same with spectral entropy criterion.

3.1 Motion Estimation

A number of interesting remarks arise from the experimental results that confirm the statements made in previous sections.

Fixed versus adaptive schemes: Figure 2 shows how the adaptive algorithm reduces the motion estimation effort and improves it only in the complex motion areas (moving cars). The fixed-block-size algorithm is more sensitive to noise (see the false alarms at the bottom, a static area). Little moving structures are not resolved in any case.

	Blocks		Displacement estimations		Entropy computations	
Initial resol. Level	2	3	2	3	2	3
Spatial criterion	97.3	161.2	484.8	795.0	234.5	365.7
Spectral criterion	68.5	137.5	352.0	700.1	168.0	318.4

Table 1. Average number (per frame) of blocks -size of the vector field-, displacement estimations and entropy computations as a function of the splitting criterion and the initial resolution level.

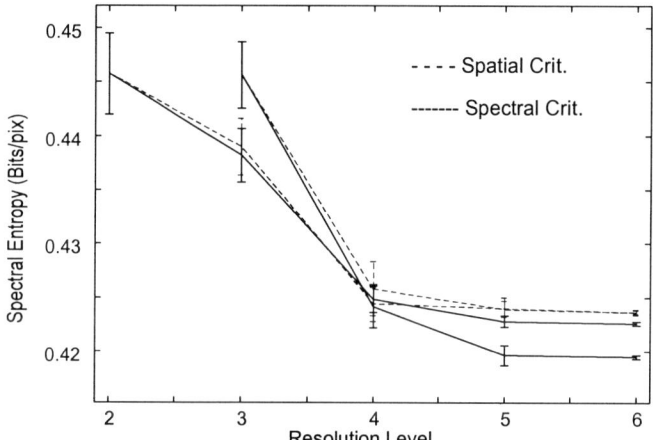

Fig. 3. Average spectral entropy (and standard deviations) of the error signal as a function of the resolution level, starting from level 2 and level 3.

Spatial entropy criterion versus spectral entropy criterion: The spectral entropy criterion is more restrictive than the spatial one. Figure 2 shows how the spatial criterion give rise to further block splitting and a more complex motion information. Note how the spatial criterion splits some blocks in static areas. The data of table 1 show that spectral criterion reduces the number of split blocks and hence the final vector field size and the number of displacement estimations and entropy computations. As the motion estimation blocks are kept bigger, the spectral criterion gives a more robust motion estimation. Note (figure 2) that in the spatial case when splitting occurs in a static area the block matching search gives rise to the same false alarms as in the exhaustive computation. The relation between motion estimation and vector quantization in the spectral criterion case implies a better reduction of the entropy of the encoded error. In figure 3, the decrease of spectral entropy of the error signal with different inial resolution levels and different splitting criteria is compared. This figure shows that spectral criterion always achieves a final lower entropy. It means that minimizing the entropy in the spatial domain does not ensure that it will be minimized after the transform coding.

Effect of the initial resolution level: It is clear that increasing the initial resolution level increases the complexity: if a lower resolution is taken, both criterion keep some blocks unsplit, so the complexity is reduced. (See the values of the table 1). Figure 3 shows that a more exhaustive initial level achieve better entropy values of the error signal (at a cost of bigger motion information complexity).

3.2 Vector quantization

Figure 4 shows the reconstructed frame 5 of the sequence by the different schemes. Distorted contours of moving objects are due to errors of the motion estimation algorithms. The high frequency textured noise is due to the frequency shaped quantization. The adaptive motion estimation (fig. 4b,c) gives clearly better motion compensated images. It is apparent that the quantization errors are *less visible* in the perceptually improved cases.

Fig. 4. Detail of reconstructed frame 5 of the TAXI sequence. (a) MPEG-like scheme. Fixed-block-size BMA, amplitude uniform bit allocation (Linear filter model). (b) Adaptive-size BMA, Frame-by-frame non-linear quantization. 2-d *IAF* model. (c) Adaptive-size BMA, spatio-temporal non-linear quantization. 3-d *IAF* model.

4 Concluding Remarks

Some specific improvements of a tipical video scheme have been presented. First, a perceptually based vector quantization has been designed using a recently developed perception model. Furthermore, an adaptive BMA with an alternative splitting criterion specially related to the quantization block has been proposed. This criterion gives rise to simpler and more robust description of the scene motion. The joint application of the proposed methods substantially improves the subjective quality of the reconstructed signal at a given compression ratio. Future research, and more exhaustive experimentation, is still needed to quantify which are the relative importance of each proposed alternative with regard to the final performance of the system.

References

[1] D. LeGall. MPEG: A video compression standard for multimedia applications. Comm. ACM Vol. 34, N° 4, 47-58. (1991)

[2] ISO/IEC 13818 Draft International Standard: Generic coding of moving pictures and associated audio, Part 2: Video. (1993)

[3] H.G. Musmann, P. Pirsh & H.J. Grallert. Advances in video coding. Proc. IEEE Vol.73, N°4, 523-548. (1985).

[4] F. Dufaux & F. Moscheni. Motion estimation techniques for digital TV: A review and new contribution. Proc IEEE Vol.83, N°6, 858-876. (1995)

[5] J.L. Barron, D.J. Fleet & S.S. Bauchemin. Performance of optical flow techniques. IJCV, Vol.12, N°1, 43-77. (1994)

[6] N. Jayant, J. Johnston & R. Safranek. Signal compression based on models of human perception. Proc. IEEE, Vol.81, N°10, 1385-1422. (1993)

[7] G.K. Wallace. The JPEG still picture compression standard. Comm. ACM. Vol.34, N°4, 31-43. (1991)

[8] F. Bosveld, R.L. Lagendijk & J. Biemond. Compatible spatio-temporal subband encoding of HDTV. Signal Proc. Vol.28, 271-290 (1992)

[9] J. Luo, C.W. Chen, K.J. Parker & T.S. Huang. Three dimensional subband video analysis and synthesis with adaptive clustering in high-frequency subbands. Proc IEEE Int. Conf. Im. Proc. Austin TX (1994)

[10] J. Malo, A.M. Pons & J.M. Artigas. Bit allocation algorithm for codebook design in vector quantization fully based on human visual system non-linearities for suprathreshold contrasts. Electr. Lett. Vol.24, 1229-1231. (1995)

[11] J. Malo, A.M. Pons & J.M. Artigas. Subjective image fidelity metric based on bit allocation of the human visual system in the DCT domain. (Accepted in Image Vis. Comp.)

[12] M.H. Chan, Y.B. Yu, & A.G. Constantinides. Variable size block matching motion compensation with applications to video coding. Proc. IEE, Vol.137, N°4, 205-212 (1990)

[13] F. Duffaux & M. Kunt. Multigrid block matching motion estimation with an adaptive local mesh refinement. SPIE Proc. Visual Commun. and Image Process 92', Vol. 1818. (1992)

[14] D.H. Kelly. Motion and vision II: Stabilized spatio-temporal threshold surface. JOSA, Vol.69, N°10, 1340-1349. (1979)

Integral Based Approach for Determining Motion Vector Fields

Atsushi Nomura

Department of Cultural and International Studies, Yamaguchi Prefectural University
Sakurabatake 3-2-1, Yamaguchi, 753 Japan

Abstract. This paper describes a method determining a motion vector field from an image sequence. In this method, a couple of a basic constraint equations in integral form and an additional constraint equation of spatio-temporal local constancy of a motion vector field is solved by the over-deterministic approach such as the least squares method. The method is applied to artificially synthesized image sequences and a real image one. As the results, robustness of the proposed method to noisy image sequences is confirmed. In addition, it is confirmed that the method requires less calculation cost than a conventional spatio-temporal optimization method.

1 Introduction

A motion vector field represents an apparent velocity vector field of brightness patterns in an image sequence. When a TV camera is translated in a static real world, we can obtain 3-dimensional structure of the world from the motion vector field [1]. Since recovering the 3-dimensional structure is useful for object recognition, the reliable method determining the motion vector field is highly required in computer vision research. In the research field of scientific measurements, visualization techniques of scientific phenomena have been developed recently. For example, fluid flow can be observed by tracer particles and a slit light laser illumination system. Besides this, dynamics of a human heart is visualized by magnetic resonance imaging (MRI). Thus, determining velocity vector fields from image sequences is commonly required for understanding the phenomena.

Many methods determining the motion vector fields have been proposed. Most of the methods are divided into the matching based approach and the gradient based one. The matching based approach traces a brightness pattern or a feature point (spot) between two successive images by using a mutual correlation function [2]. Then the displacement vector field is obtained. On the other hand, the gradient based approach utilizes a basic constraint equation relating brightness distribution of an image sequence $f(x, y, t)$ to two motion vector components (v_x, v_y). A typical basic constraint equation [3] is,

$$\frac{\partial f}{\partial t} + v_x \frac{\partial f}{\partial x} + v_y \frac{\partial f}{\partial y} = 0. \tag{1}$$

A motion vector is determined by solving a set of the basic constraint equation and additional constraints representing nature of the vector field, such as smoothness [3], spatial constancy [4] and temporal stationariness [5]. In consequence, appropriate constraint equations should be applied to an image sequence for the recovery of reliable motion vector fields. For instance, if an image sequence has much noise in brightness level, constraint equations having the effect of reducing the noise should be applied to that.

In addition to the above constraint equations, some researchers proposed new basic and additional constraint equations. For example, Aisbett proposed the extended basic constraint equation including a divergence term of a motion vector field and a rate of brightness generation which represents the brightness change of a pattern under illumination change [6]. Bimbo et al. analyzed the divergence term and confirmed its usefulness [7]. Some researchers proposed additional constraint equations on the rate of brightness generation [8, 9].

In this study, a basic constraint equation in integral form is proposed for the determination of an accurate motion vector field. In particular, the determination of the motion vector field from a noisy image sequence is focused on. Furthermore, the basic constraint equation is coupled with the simple additional constraints which have been already proposed [10]. The set of the equations are solved by the least squares method. The proposed method is applied to artificially synthesized image sequences and a real image one. Through the analysis of the image sequences, the usefulness of the proposed basic constraint equation is confirmed.

2 Theory

The conventional gradient based approach utilizes the differential form of basic constraint equations, say eq.(1). Since the equations have differential operations: $\partial f/\partial x, \partial f/\partial y, \partial f/\partial t$, they are not so robust to noise. In contrast to this, integral operations are expected to be robust to noise. Therefore, in this paper, a basic constraint equation having integral operations is introduced,

$$\frac{d}{dt}\int_{\delta S} f \, ds = -\oint_{\delta C} f \boldsymbol{v} \cdot \boldsymbol{n} \, dc, \qquad (2)$$

where δS is a spatial domain surrounded by δC, $\boldsymbol{v}$ is a motion vector to be determined and $\boldsymbol{n}$ is a unit length vector normal to δC and pointing to its outside (Fig.1). The equation shows that the total brightness change observed in δS is represented by movement of brightness pattern into δS. (This is known as Euler's observation method in fluid physics.)

Now we propose a new method utilizing the integral based equation as basic constraint and spatio-temporal local constancy of a motion vector field as additional constraint [10]. Let us consider a square region having $(2 \times L+1) \times (2 \times L+1)$ (pixel^2) as δS (Fig.2). Then eq.(2) is rewritten into the following discrete form,

$$v_x I_x(x,y,t) + v_y I_y(x,y,t) + I_t(x,y,t) = 0, \qquad (3)$$

where,

$$I_x(x,y,t) = \sum_{m=-L}^{L} \{f(x+L,y+m,t) + f(x+L+1,y+m,t)\}/2$$
$$- \sum_{m=-L}^{L} \{f(x-L,y+m,t) + f(x-L-1,y+m,t)\}/2, \quad (4)$$

$$I_y(x,y,t) = \sum_{l=-L}^{L} \{f(x+l,y+L,t) + f(x+l,y+L+1,t)\}/2$$
$$- \sum_{l=-L}^{L} \{f(x+l,y-L,t) + f(x+l,y-L-1,t)\}/2, \quad (5)$$

$$I_t(x,y,t) = \sum_{l=-L}^{L}\sum_{m=-L}^{L} \{f(x+l,y+m,t+1) - f(x+l,y+m,t-1)\}/2. \quad (6)$$

In conversion from the continuous form eq.(2) into discrete form eq.(3), a spatial local constancy of the motion vector field is assumed. Equation (3) is also obtained by a summation of discrete versions of basic constraint equations (1) over δS. This indicates that the terms: I_x, I_y, I_t with large L are insensitive to random noise in contrast to the partial derivatives: $\partial f/\partial x, \partial f/\partial y, \partial f/\partial t$. Since we have only one constraint equation (3), one more additional constraint equations are required to obtain the set of solutions v_x and v_y. Here we also introduce a temporal constancy of the motion vector field as the additional constraint. From this constancy, we obtain a set of basic constraint equations (3) sharing the same v_x and v_y along a time coordinate system. Thus we can estimate them by minimizing the following error function,

$$E = \sum_{n=0}^{N-1} \{v_x I_x(x,y,t+n) + v_y I_y(x,y,t+n) + I_t(x,y,t+n)\}^2, \quad (7)$$

where N represents frame number utilized for the minimization. Note that N should be greater than 2 for over-deterministic optimization.

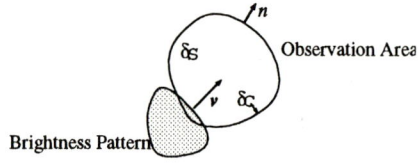

Fig. 1. Graphical representation of the integral based equation (2). Movement v of a brightness pattern causes total brightness change observed in the fixed observation area δS surrounded by δC. The unit length vector n is perpendicular to δC.

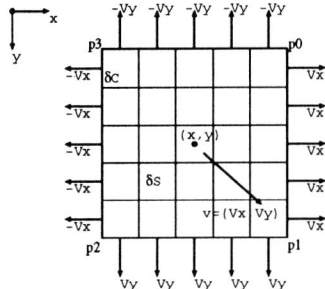

Fig. 2. Graphical representation of the discrete form of the integral based equations (2) and (3). The motion vector components (v_x, v_y) are assumed to be constant in the local region δS. An inner product of n and v becomes v_x along the vertical line p0-p1. On the other hand, along the line p2 - p3, the product becomes $-v_x$. In the same way, the product along p1-p2 and the one along p3-p0 become v_y and $-v_y$, respectively.

3 Experimental Results

3.1 Analysis of Artificially Synthesized Image Sequences

In this section, we confirm usefulness of the proposed method. For this purpose, we apply the proposed method and a conventional gradient based one to artificially synthesized image sequences and compare them in accuracy. The conventional gradient based method [10] utilizes the basic constraint equation (1) combined with spatio-temporal local constancy of a motion vector field as additional constraint. That is, a motion vector is estimated by minimizing the following error function in the conventional method,

$$E = \sum_{l=-L}^{L} \sum_{m=-L}^{L} \sum_{n=0}^{N-1} \{v_x f_x(x+l, y+m, t+n) + v_y f_y(x+l, y+m, t+n)$$
$$+ f_t(x+l, y+m, t+n)\}^2, \qquad (8)$$

where $f_x(x, y, t)$, $f_y(x, y, t)$ and $f_t(x, y, t)$ are spatial and temporal partial derivatives in discrete form and $(2 \times L + 1) \times (2 \times L + 1)(\text{pixel}^2)$ represents size of a square local region for the minimization. The least squares method is carried out in the minimization. We tentatively call the conventional gradient based method as "conventional Spatio-Temporal Optimization Method (STOM)" and the proposed method as "Proposed Integral Based Method (PIBM)". Both of them assume spatial constancy of a motion vector field within the square region $(2 \times L + 1) \times (2 \times L + 1)(\text{pixel}^2)$ and temporal stationariness of the field during time N (frame).

On the other hand, image sequences are artificially synthesized by,

$$f(x, y, t) = (1 - r) \cdot f_0(x, y, t) + r \cdot n(), \qquad (9)$$

where $f_0(x,y,t)$ represents an image sequence obtained by rotating a real static scene around its center, the function $n()$ generates uniformly distributed random number, r is a ratio between $f_0()$ and $n()$. The angular velocity of the rotation is set to 1.0(degree/frame). For the overall field, a velocity component and a directional one in the rotational motion field distribute from 0 to 1.3 (pixel/frame) and from 0 to 2π (radian), respectively. Thus, the image sequence having the rotational motion is useful to the test of accuracy in wide ranges of velocity and direction components. Moreover, varying the ratio r from 0 to 0.5 at intervals 0.1, we obtain 6 kinds of image sequences. All sequences consists of spatial resolution 150×150 (pixel^2).

We define a mean error rate of a determined motion vector field to evaluate accuracy of the determined motion vector field by the following equation,

$$E = \frac{1}{M} \sum_x \sum_y \frac{|v_t(x,y) - v_e(x,y)|}{|v_t(x,y)|} \times 100(\%), \qquad (10)$$

where M represents the number of grid points (150×150), v_t is the true motion vector and v_e is the evaluated motion vector at each point.

We compare accuracy of the two methods by varying the parameters N, L and r. We show their results in Tables 1, 2 and 3. From these tables, when we can not utilize so many grid points, say 1(pixel) $\times$ 8(frame), for determination of a motion vector field, we had better use STOM. Otherwise, PIBM is superior than STOM in accuracy. Furthermore, to evaluate calculation cost we measure their calculation times. As the results, PIBM is about 7 times faster than STOM in calculation time. From these simulation experiments, we understand that PIBM is reliable and it requires less calculation cost than STOM.

3.2 Analysis of a Real Image Sequence

Now the proposed method is applied to a real image sequence to test its performance for real scene. The image sequence is generated by a TV camera attached to a passenger seat of a running car in a dark environment in the night. The optical axis of the camera is parallel to the running direction of the car and the camera lens faces toward the front of it. One of the captured image frames is shown in Fig.3. Because of dark environment, its brightness level is very low and thus noise level is relatively high. The two methods: STOM, PIBM are applied to the image sequence. Motion vector fields determined by STOM and by PIBM are shown in Figs.4 (a) and (b), respectively.

From comparison between the two fields, we understand that the field determined by STOM is completely averaged. On the other hand, around the right bottom region in Fig.4(b) we find that motion vectors are in the horizontally right direction which is consistent with the velocity vector field caused by relative motion between the car and the environment.

Table 1. Dependence of accuracy on the parameter N which represents frame number utilized for the determination of a motion vector field. The parameter L is fixed to be 3(pixel). The methods STOM and PIBM are applied to a rotational image sequence synthesized with $r = 0$.

N	8	16	32	64	128
STOM	18.1	13.9	10.3	9.15	8.27
PIBM	25.9	11.7	3.94	2.12	1.74

Table 2. Dependence of accuracy on the parameter L which represents local size utilized for the determination of a motion vector field. The parameter N is fixed to be 32(frame). The methods STOM and PIBM are applied to a rotational image sequence synthesized with $r = 0$.

L	0	1	2	3	4	5
STOM	14.3	11.6	10.7	10.3	9.72	9.36
PIBM	25.4	9.30	5.30	3.94	3.72	3.96

Table 3. Robustness of the two methods STOM and PIBM against to a random noise component. The parameter r represents the ratio between the noise component and rotational image sequence $f_0(x, y, t)$. The parameters N and L are fixed to be $N = 32$(frame) and $L = 3$(pixel).

r	0.0	0.1	0.2	0.3	0.4	0.5
STOM	10.3	15.5	28.5	42.7	51.3	55.3
PIBM	3.94	9.27	18.6	29.6	41.2	53.0

Fig. 3. Real road image sequence acquired from a running car in a dark environment. Sampling frequency is 30 Hz. Brightness is quantized into 256 levels. Right upper square portion is analyzed.

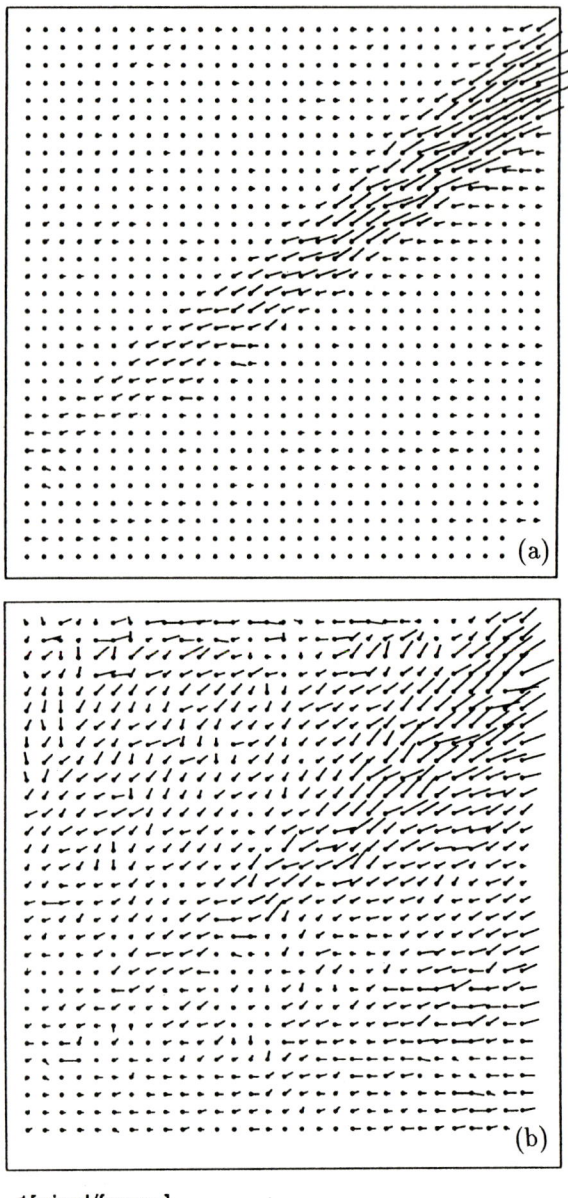

_1[pixel/frame]

Fig. 4. Motion vector fields obtained by (a)STOM and (b)PIBM from the real image sequence(Fig.3). Both methods assume spatial constancy in square portion with $7 \times 7(\text{pixel}^2)$ ($L = 3$) and temporal stationariness during 69(frame) ($N = 69$) corresponding to 2.3 second.

4 Conclusion

In this study, a basic constraint equation in integral form was proposed for the determination of a motion vector field from a noisy image sequence. Moreover, a method employing the proposed basic constraint equation and spatio-temporal local constancy of a motion vector field as an additional constraint was proposed for determination of the field. The proposed method and a conventional gradient based method employing the same additional constraint were applied to artificially synthesized image sequences and a real image one. As the results, the usefulness of the proposed method was confirmed quantitatively for the synthesized image sequence and qualitatively for the real one. In particular, the proposed method was more accurate and it requires less calculation cost.

Acknowledgements

The author would like to thank Prof. Hidetoshi Miike (Yamaguchi University) for helpful comments and stimulating discussions.

References

1. S. Maybank: Theory of reconstruction from image motion. Springer-Verlag (1993)
2. P. Anandan: Computing optical flow from two frames of an image sequence. *COINS Technical Report* 86-16 (1986)
3. B. K. P. Horn and B. G. Schunck: Determining optical flow. *Artificial Intelligence* **17** (1981) 185–203
4. J. K. Kearney, W. B. Thompson and D. L. Boley: Optical flow estimation: an error analysis of gradient-based methods with local optimization. *IEEE Transactions on Pattern Analysis and Machine Intelligence* **PAMI-9** (1987) 229–244
5. A. Nomura, H. Miike and K. Koga: Field theory approach for determining optical flow. *Pattern Recognition Letters* **12** (1991) 183–190
6. J. Aisbett: Optical flow with an intensity-weighted smoothing. *IEEE Transactions on Pattern Analysis and Machine Intelligence* **11** (1989) 512–522
7. A. D. Bimbo, P. Nesi and J. L. C. Sanz: Analysis of optical flow constraints. *IEEE Transactions on Image Processing* **4** (1995) 460–469
8. N. Mukawa: Estimation of shape, reflection coefficients and illuminant direction from image sequences. *Proceedings of the 3rd International Conference on Computer Vision* (Osaka, Japan, 1990) 507–512
9. A. Nomura, H. Miike and K. Koga: Determining motion fields under non-uniform illumination. *Pattern Recognition Letters* **16** (1995) 285–296
10. K. Nakajima, A. Osa, T. Maekawa and H. Miike: Evaluation of body motion by optical flow analysis. *Japanese Journal of Applied Physics* **36** (1997) 2929–2937

A Practical Algorithm for Structure and Motion Recovery from Long Sequence of Images

Miroslav Trajković and Mark Hedley
Department of Electrical Engineering, Sydney University, NSW 2006, Australia
E-mail: {miroslav, hedley}@ee.usyd.edu.au

Abstract. In this paper we present an algorithm for structure and motion (SM) recovery under affine projection from video sequences that is suitable for real time applications. The algorithm tracks the motion of a single structure, be it an object or the entire scene itself, allowing for any type of camera motion. This could be used for example to track the motion of a vehicle in a warehouse (single object, static camera) or for visual navigation from a moving platform (track scene from moving camera). The algorithm requires a set of features to be detected in each frame, and that at least four features are correctly matched between each three consecutive frames. Compared to previous algorithms, this novel algorithm has a lower computational cost, dynamically detects outliers and allows for previously lost features to reappear in the sequence. The algorithm has been tested on real image sequences, and compared to other algorithms we have found that our algorithm has both a smaller error and a lower computational time.

1. Introduction

In any machine vision task the objects of most interest, and the most demanding to handle, are those that are moving. The reason for this is that once we have information about static objects, by virtue of their being static, we do not need to update information about them at the video rate. Moving objects on the other hand should be tracked frame by frame, and these are the objects that the machine vision system will usually need to respond to (for example collision avoidance). A fundamental machine vision task then is to identify and track moving objects in a video sequence, and to determine their structure, which is assumed to be rigid. In this paper we will approach the problem of determining the motion and structure of a single rigid object moving in the field of view of a static camera. The same algorithm can be used if there is a moving camera in a static environment to determine the motion of the camera and the structure of the environment.

One of the most widely used approaches to structure and motion (SM) recovery is the factorization algorithm proposed by Tomasi and Kanade [8]. Image feature points are tracked over several frames and the SM parameters are computed using a singular value decomposition (SVD). This algorithm cannot be used for real time applications as this is a batch method and can only compute the parameters once all measurements have been made. The algorithm is also computationally very expensive (due to the use of the SVD). Morita and Kanade [5] developed a sequential version that has almost the same accuracy but has a much lower computational burden. The issues not addressed in these papers were how to prevent outliers incorrectly biasing the result, and how to handle a set of feature points that changes dynamically.

Held [1] partially extended the Morita and Kanade algorithm to include new features. The video sequence was divided into subsequence, and the features tracked over the subsequence. To track between the subsequences there needed to be common tracked features between consecutive subsequences. The algorithm was verified using synthetic measurements only, with added Gaussian noise. McLauchlan *et al.* [2,3] used the variable state-dimension filter (VSDF) to handle the problem of missing and new features. They posed the structure from motion problem as a parameter estimation problem and solved it by using the Extended Kalman Filter (EKF). They updated the structure and *only the last* motion estimate at each time step and therefore achieved a low computational cost $O(k)$ compared to when the complete motion matrix has been updated. The authors mentioned outlier rejection, but they gave neither a method nor a reference to handle this problem. Another limitation of their algorithm is that features that have been obscured and reappear are treated as new feature points, and are not recognised as a previously tracked feature.

In this paper we present an algorithm that iteratively computes the SM parameters at each time step. It has the same computational cost $O(n+k)$ as [2] but it will update both the structure and complete motion parameters, so it converges to the true solution. It is made possible by noting that structure should remain constant over time (rigid body assumption) and by updating it at each time step. Motivated by work of Reid and Murray [6] we have extended their algorithm so that it can easily include new features and recover the complete structure and motion providing that for each three consecutive frames there are at least four correctly matched feature points. Furthermore, our algorithm detects outliers at each time step. Finally we developed a procedure that checks whether newly appeared features are genuinely new, or old features that have reappeared. The correct classification of these new features both decreases the computational cost of the algorithm and improves its accuracy.

2. Problem Background

In this paper we assume an affine camera model which has the form:

$$p = MS + t \qquad (1)$$

where S is 3×1 is the world coordinate of a feature point, p is 2×1 image projection of S, M is 2×3 projection matrix and t is 2×1 translation vector. This model is a generalisation of the orthographic camera model [5,8] and is a good approximation of the perspective projection when the change in depth is small compared to the average distance from the optical center. This condition is almost always satisfied for independently moving objects. If the object of interest undergoes rigid motion the image projections will not change if we fix the world coordinates of the object of interest and change camera parameters M and t accordingly. Hence, without loss of generality, we may assume that scene is static and that the camera is moving.

Given k projections (images) of n scene points we want to recover the camera motion parameters $M(j)$, $t(j)$, $j = 1, 2, \cdots, k$ and scene structure parameters $S_i, i = 1, 2, \cdots, n$ (world vectors of tracked scene points). These parameters are related by the measurement equation:

$$p_i(j) = M(j)S_i + t(j) \qquad (2)$$

where $p_i(j)$ is the projection of ith point onto the jth image, and i and j vary from 1...n and 1...k respectively. To compute the the SM parameters using the measurement equations, we must minimise the cost function:

$$C(k) = \sum_{j=1}^{k}\sum_{i=1}^{n} \|p_i(j) - M(j)S_i - t(j)\|^2 \qquad (3)$$

For real time implementation the parameters must be computed following the acquisition of each frame. It may be shown that the translational component $t(j)$ at each frame is given by the centroid of the feature locations $t(j) = \overline{p_i(j)}$ and equation (2) can now be rewritten as $w_i(j) = M(j)S_i$ where $w_i(j) = p_i(j) - t(j)$, or in matrix form:

$$W = MS \qquad (4)$$

where $W = [w_i(j)]_{k \times n}$, $M = [M(1)^T \cdots M(k)^T]^T$ and $S = [S_1 \cdots S_n]$, and this is the equation obtained (although in different way) by Tomasi and Kanade [8]. To solve it, they employed singular value decomposition of measurement matrix and showed that $M = U_3 \Sigma_3^{1/2}$ and $S = \Sigma_3^{1/2} V_3^T$, where U_3, Σ_3 and V_3 are submatrices of U, Σ and V corresponding to the three largest singular values of W. However the structure and motion parameters can be determined without computing the full SVD, since

$$W^T W = V \Sigma^2 V^T \text{ and } WW^T = U \Sigma^2 U^T \qquad (5)$$

hence the structure and motion parameters are given by the eigenvector decomposition of $W^T W$ and WW^T respectively.

Note that equation (4) does not have unique solution. In fact, if A is an arbitrary invertible 3×3 matrix the matrices MA and $A^{-1}S$ are valid solutions [8]. To ensure a unique solution, the first two rows of M are fixed, and now in equation (3) j will range from 2...k [3].

3. Algorithm Details

3.1 Structure and Motion Recovery

Unlike Morita and Kanade who have computed motion and structure by iteratively updating equation (4) and McLauchlan who applied EKF to the measurement equation (2), we employ direct minimisation of the cost function (3). For some initial number of frames (usually 3) we initialise the motion and structure matrices by solving equation (4). Once we have the matched features at time $k+1$, and have computed values M, t and S at time k, we wish to find the updated parameters that minimise

$$C(k+1) = C = \sum_{j=2}^{k+1}\sum_{i=1}^{n} w_{ij} \|p_i(j) - M(j)S_i - t(j)\|^2 = \sum_{j=2}^{k+1}\sum_{i=1}^{n} w_{ij} \|v_i(j)\|^2 \qquad (6)$$

where w_{ij} is a binary weight which is unity if feature i is present in frame j, and zero otherwise.

It is convenient to group the motion parameters M and t into a single vector defined as $m = [M_{11} M_{12} M_{13} t_1 M_{21} M_{22} M_{23} t_2]^T$. To determine the structure and motion parameters that minimise the cost function we must solve:

$$\frac{\partial C}{\partial m(j)} = \sum_{i=1}^{n} w_{ij} D_i^T v_i(j) = 0, \; j = 2,3,\ldots,k+1$$

$$\frac{\partial C}{\partial S_i} = \sum_{j=2}^{k+1} w_{ij} E^T(j) v_i(j) = 0, \; i = 1,2,\ldots,n \quad (7)$$

where $D_i = -\partial v_i/\partial m$ is a simple function of S_i, and $E = -\partial v_i/\partial S_i = M$. For simplicity the time parameter j has been dropped. Equation (7) forms a non-linear set of equations to be solved. There are three possible approaches to doing this:

1. Use the previously determined values of M and S_i to calculate D_i and E, then linearise the equations using

$$v_i(j) = \hat{v}_{ij} + \frac{\partial v_i}{\partial m}\Delta m(j) + \frac{\partial v_i}{\partial S_i}\Delta S_i = \hat{v}_{ij} - D_i\Delta m(j) - E^T(j)\Delta S_i \quad (8)$$

From this we can rewrite the system as a linear set of equations to be solved for $\Delta m(j)$ and ΔS_i. This is the set of equations obtained in [3], and the algorithm has linear convergence.

2. Linearise all terms in equation (7), rather than assuming that the derivatives are constant. From this a different set of linear equations is determined that are similar to those in the previous technique, but this has quadratic convergence.

3. The simplest method is to solve for $m(j)$ using the previous values for the derivatives and S_i. Then use the updated parameters to solve for S_i. This is alternated until convergence. This method has linear convergence, but the lowest computational cost.

We used the second method because it has quadratic converge, and usually found three to five iterations sufficient. For the initial estimates of the parameters at each stage of the algorithm we used the parameters determined in the previous stage.

3.2 Matching and Outlier Rejection

Before we can recover the structure and motion parameters we need to have matched feature points (here corners) between consecutive images. This usually involves matching two sets of corners using a cross correlation of a window about each corner [9]. This procedure will often give some small number of incorrect matches (outliers), which can significantly bias the structure and motion estimates. These outliers should be detected so they can be ignored. For this we developed a technique based on robust statistics, using the Least Median of Squares (LMedS) estimator [4], which is described in this section.

Matching between two images.

Having matched the corners between two consecutive images, we randomly select four matched corners to obtain a sample matrix $W_S = [\mathbf{x}_1 \ \mathbf{y}_1 \ \mathbf{x}_2 \ \mathbf{y}_2]'$ where $\mathbf{x}_i$ is a vector containing the x coordinates of the four selected corners in image i, and $\mathbf{y}_i$ is likewise defined. We can decompose this 4×4 matrix using the SVD:

$$W_S = M_S S_S + t_S.$$

where M_S and t_S can be found using equation (5), then we find the structure vector S_R for the remaining set of matched corner points C_R by solving the equation

$$W_R = M_S S_R + t_S,$$

where W_R is the measurement matrix corresponding to C_R. The position error from the model (M_S, t_S) is given by

$$D = W_R - (M_S S_R + t_S)$$

The standard variation of the error vector $\varepsilon = [|d_i|]$ is computed using the equation [4]

$$\sigma^0 = 1.4826 \left(1 + \frac{5}{n-p}\right) \operatorname*{median}_i (\varepsilon_i)$$

where n is number of points in C_R and p is the number of sample points (4 in our case). The number of samples of four sets of matched corners which have to be taken to assure that at least one consists only of inliers with probability higher than γ is given by

$$N_S = \frac{\log(1-\gamma)}{\log(1-(1-\pi)^p)}$$

where π is expected percentage of outliers in the sample. We find the sample with the lowest standard variance, and use the corresponding vector ε.

The next step is outlier rejection. For each point we determine the initial weight w_i.

$$w_i = \begin{cases} 1 & \text{if } \varepsilon_i \leq 2.5\sigma^0 \\ 0 & \text{if } \varepsilon_i > 2.5\sigma^0 \end{cases}$$

Then we compute the *robust* standard deviation estimate as

$$\sigma = \sqrt{\sum_i w_i \varepsilon_i^2 \bigg/ \left(\sum_i w_i - p\right)}$$

Finally, outliers are found as points whose residual error is outside the confidence interval 2.5σ.

Matching between three images

By matching feature points between three images we can make use of the rigidity constraint. For each set of three frames, the corners are matched between the consecutive frames and the outliers rejected. Then, given the position of a feature in any two images, and the motion parameters M and t, its position in the third image is uniquely determined. Hence, by using this procedure, we can detect those outliers which are "correctly" matched over each pair of consecutive images (satisfy the epipolar constraint) but do not satisfy the rigidity constraint over three frames.

3.3 New Feature Points

New feature points will be included in the measurement matrix if they appear and are matched in three consecutive frames. Consider frames n-2, n-1, n and n+1, and let C_1 denote a set of features which are matched over frames n-2 to n, while C_2 denotes the set of features which have been matched *only* over frames n-1 to n+1. Let P be an arbitrary feature from C_2 and let $p^{(n)}$ denote its image projection in the nth frame. Since the motion (and structure) parameters are estimated over the first n frames, the structure parameter S of this point is found by solving the set of equations,

$$\begin{bmatrix} p^{(n-1)} \\ p^{(n)} \end{bmatrix} = \begin{bmatrix} M^{(n-1)} \\ M^{(n)} \end{bmatrix} S + \begin{bmatrix} t^{(n-1)} \\ t^{(n)} \end{bmatrix}.$$

Hence the initial parameters for the algorithm in Section 3.1 have been found.

Before we add the point from C_2 to the measurement matrix, and perform an update, we first check whether this point is genuinely new or whether it is an old feature that has reappeared. Let P_i be a previous feature that had disappeared and P_l be an arbitrary feature from C_2. We can say that P_i and P_l represent the same feature if the world distance between them is small enough, *i.e.* if

$$d_{il} = \|\Delta S\| = \|S_i - S_l\| < \alpha \qquad (9)$$

where α is an unknown threshold to be determined.

While we do not know what is *small* in the structure space (especially because it is only unique up to an affine transformation), we can say what is small in the image space (typically three to five pixels). If the features are close in structure space then the distance between their projections in the image plane has to be small but the reverse need not be true. We use the criteria that the two points are the same one if the distance between their image projections is less than threshold β for all camera locations. Mathematically, this condition can be expressed as

$$\max_{\Delta S, j} \left(\| p_i(j) - p_l(j) \| \right) \le \beta . \qquad (10)$$

The difference between image projections is given by $\Delta p = p_i - p_l = M\Delta S$ where the index j has been dropped for clarity. By applying the Euclidean norm to both sides we obtain $\|\Delta p\|_E \le \|M\|_S \|\Delta S\|_E$ where the index S denotes the spectral norm, which is defined as a square root of the maximum eigenvalue of the matrix $M^T M$, *i.e.* $\|M\|_S = \sqrt{\max \lambda(M^T M)}$. Since the spectral norm is subordinate to Euclidean norm there exists at least one ΔS such that equality holds, and from equation (10) we get $\|\Delta p\|_E = \|M\|_S \|\Delta S\|_E \le \beta$. If $\lambda_j = \|M(j)\|_S$, then by combining (9) and (10) the threshold $\alpha = \beta/\lambda$ where λ is the largest value of λ_j.

4. Experimental Results

4.1 POV image sequence

This artificially generated sequence consists of 90 frames of an object rotating around the fixed axes (see Fig. 1 a–c). The rotation is a constant four degrees per frame. The corners are detected and matched using an algorithm similar to [9]. This is a difficult sequence, because no single corner appears in all the frames, and because the number of points is small – it varies from 6 to 13 per frame. For this reason, both the [8] and [5] are not applicable to this sequence as the number of features is not constant. The VSDF algorithm may be used, but in its original form (without the reappearance of features) this will have a growing number of features as old features reappear, this will lead to slower computation and greater errors in the calculated parameters.

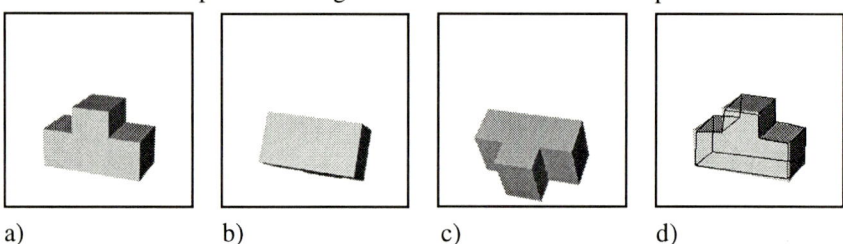

a) b) c) d)

Fig. 1. POV sequence: 1st (a), 30th (b) and 60th (c) image in the sequence of 90 images. d) consistent features and recovered structure (the lines are shown for visualisation only).

Using our algorithm we obtained a total of 22 features, and by disregarding all of them which appeared in less than 20 frames we were left with total of 17 features which is quite close to the 16 corners on the object. Corners that have been effectively tracked over all the frames are shown on Fig 1d. As we can see they are all close to the true locations, including those which are currently occluded or not detected. Some distortion is present due to imperfection of the affine model, but the structure has been correctly recovered.

4.2 Hotel Sequence

This sequence consists of 197 points tracked over 181 frames. Corners were detected in the first image and their location in consequent images were determined using an optical flow technique [7] (see Fig. 2 a,b).The SM parameters were calculated using the SVD algorithm and our algorithm, which produced the same results (without outlier rejection), but the former took about an order of magnitude longer to compute. As in [3], we have found that updating motion parameters over all previous frames in each step is unnecessary. A very important step here is the outlier detection. Since we have matches over all frames, the rigidity constraint was checked not over last three frames only, but over the whole sequence. The comparison of mean squared error with, and without outlier removal is given in Fig. 2c.

5. Conclusion

This paper addresses the problem of structure and motion recovery. The first problem is speed and we presented a fast recursive algorithm that updates the *complete* motion and structure after each time step and have shown that it gives the same results as in

[8]. Furthermore, we have developed an outlier detection technique that decreased the error even further. The paper is also one of the first that deals the problem of feature dropout and reappearance. We have developed a procedure that that allows previously lost features to reappear in the sequence and demonstrated that it works in practice.

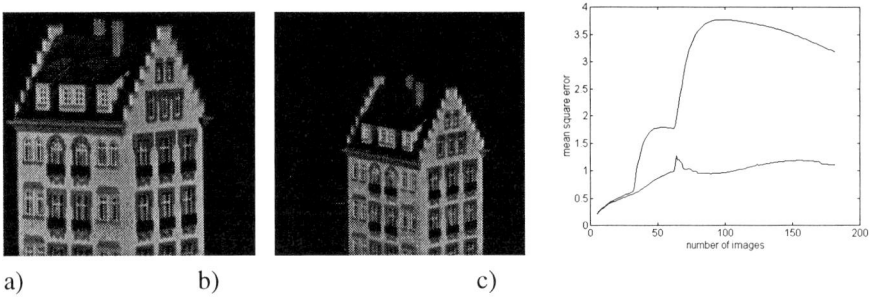

a)   b)   c)

Fig.2 1st (a) and 181st (b) image from the hotel image sequence; c) Plots of the mean square error (cost function) with (lower curve) and without (upper curve) outlier rejection

Acknowledgments

Financial support for this project was provided by the Australian Research Council. The hotel image sequence was provided by the "Modeling by Videotaping group" in the Robotics Institute, Carnegie Mellon University.

REFERENCES

[1] Held A., "Piecewise Shape Reconstruction by incremental factorization", *Proc.7^{th}British Machine Vision Conference, vol I*, UK, 1996, pp. 333-342.

[2] McLauchlan P. *et al.* "Recursive affine structure and motion from image sequences" *Proc. 3rd ECCV, vol. I*, Stockholm, pp. 217-224, 1994.

[3] McLauchlan P. and Murray D.. "A unifying framework for structure and motion recovery from image sequences" *Proc. 5th ICCV*, MIT, pp. 314-320, 1995.

[4] Meer P. *et al.* "Robust regression methods for computer vision: a review", *Intl. Journal Computer Vision* vol. 6., pp. 59-70, 1991.

[5] Morita T. and Kanade T., "A sequential factorization method for recovering shape and motion from image streams", *Proc. 1994 ARPA Image Understanding Workshop, vol. II*, pp. 1177-1188, Monterey, CA1994.

[6] Reid I. D. and Murray D. W., "Active Tracking of Foveated Feature Clusters Using Affine Structure", *Intl. J. Computer Vision,* vol. 18, pp. 41-60, 1996.

[7] Shi J. and Tomasi C., "Good Features to Track". *IEEE Conference on Computer Vision and Pattern Recognition*, June 1994, pp. 593-600.

[8] Tomasi C. and Kanade T., "Shape and motion from image streams under orthography: A factorization approach", *Intl. J. Comp. Vision,* vol. 9, pp. 137-154, 1991.

[9] Trajković M. and Hedley M., "Fast feature detection and matching for machine vision", *Proc. 7th British Machine Vision Conference, vol. I*, Edinburgh, UK, 1996, pp. 91-100.

Object Pose by Affine Iterations

Fadi Dornaika and Christophe Garcia

Institute For System Design Technology
GMD – German National Research Center for Information Technology
Sankt Augustin, Germany, E-mail: Christophe.Garcia@gmd.de

Abstract. The problem of a real-time pose estimation between a 3D scene and a camera is a fundamental task in most 3D computer vision and robotics applications such as object tracking, visual servoing, and virtual reality. In this paper we present a fast method for estimating the 3D pose using 2D to 3D point and line correspondences. This method is inspired by DeMenthon's method (1995) which consists of determining the pose from point correspondences. In this method the pose is iteratively improved with a weak perspective camera model, at convergence the computed pose corresponds to the perspective camera model. Our method is based on the iterative use of a paraperspective camera model which is a first order approximation of perspective. Experiments involving synthetic data as well as real range data indicate the feasibility and robustness of this method.

1 Introduction and motivation

The problem of object pose from 2D to 3D correspondences has received a lot of attention both in the photogrammetry and computer vision literatures. Various approaches to the object pose (or external camera parameters) problem fall into 2 distinct categories: closed-form solutions and non-linear solutions. Closed-form solutions may be applied only to a limited number of correspondences [2], [5]. Whenever the number of correspondences is larger than 4 then closed-form solutions are not efficient and iterative non-linear solutions are necessary [7]. The latter approaches have two drawbacks: i) they need a good initial estimate of the true solution and ii) they are time consuming. Therefore, such approaches can not be used in tasks that require high speed performance (visual servoing, object tracking, ...) [3]. To our knowledge, the method proposed by DeMenthon & Davis [1] is among the first attempts to use linear techniques, associated with the weak perspective camera model in order to obtain the pose that is associated with the perspective camera model. The method starts with computing the object pose using weak perspective model and after a few iterations converges towards a pose estimated under perspective.

In this paper we establish a link between paraperspective model and perspective model in order to estimate the pose using both points and straight lines. It has been argued that since features like straight lines are determined by a large number of pixels, the redundancy makes it possible to locate them accurately in the image. Furthermore, lines can be extracted even if they are partially occulted.

2 Background and notations

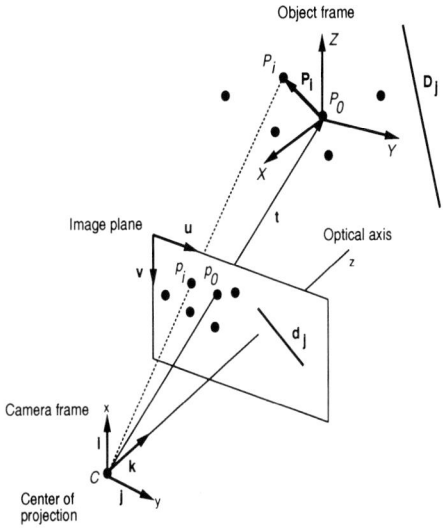

Fig. 1. The pin-hole camera model.

We denote by P_i a 3D point with coordinates (X_i, Y_i, Z_i) in a frame that is attached to the object - the object frame. The origin of this frame is the object point P_0. We denote by $\mathcal{D}_j$ a 3D line that is described parametrically by its direction $\mathbf{V}_j$ and by a point vector $\mathbf{W}_j$. We suppose that the observed scene contains n points $(P_1, ..., P_n)$ (in addition to the reference point P_0) and m straight lines $(\mathcal{D}_1, ..., \mathcal{D}_m)$. These points and lines are expressed in the object frame (see Figure 1).

An object point P_i projects onto the image in p_i with normalized camera coordinates x_i and y_i. An object line $\mathcal{D}_j$ projects onto the image in d_j with normalized coefficient (a_j, b_j, c_j). We denote by $\mathbf{P}_i$ the vector from point P_0 to point P_i. The normalized camera coordinates of p_i are given by:

$$x_i = \frac{X_{ci}}{Z_{ci}} = \frac{\mathbf{i} \cdot \mathbf{P}_i + t_x}{\mathbf{k} \cdot \mathbf{P}_i + t_z} \qquad (1)$$

$$y_i = \frac{Y_{ci}}{Z_{ci}} = \frac{\mathbf{j} \cdot \mathbf{P}_i + t_y}{\mathbf{k} \cdot \mathbf{P}_i + t_z} \qquad (2)$$

These equations describe the classical perspective camera model where the rigid transformation from the object frame to the camera frame is:

$$T = \begin{bmatrix} \mathbf{i}^T & t_x \\ \mathbf{j}^T & t_y \\ \mathbf{k}^T & t_z \\ 0\ 0\ 0 & 1 \end{bmatrix} = \begin{bmatrix} R & \mathbf{t} \\ 0\ 0\ 0 & 1 \end{bmatrix}$$

The relationship between the normalized camera coordinates and the image coordinates may be obtained by introducing the intrinsic camera parameters:

$$u_i = \alpha_u\, x_i + u_c$$
$$v_i = \alpha_v\, y_i + v_c$$

In these equations α_u and α_v are the horizontal and vertical scale factors and u_c and v_c are the image coordinates of the intersection of the optical axis with the image plane.

Similarly one can express the normalized perspective projection of the straight line $\mathcal{D}_j$ as:

$$d_j :\ a_j\, x + b_j\, y + c_j = 0 \tag{3}$$

Where x and y are related to the pose parameters by these 2 equations:

$$x = \frac{\mathbf{i} \cdot \mathbf{P}_j + t_x}{\mathbf{k} \cdot \mathbf{P}_j + t_z} \tag{4}$$

$$y = \frac{\mathbf{j} \cdot \mathbf{P}_j + t_y}{\mathbf{k} \cdot \mathbf{P}_j + t_z} \tag{5}$$

with $\mathbf{P}_j$ being a point on the line $\mathcal{D}_j$.

We divide both the numerator and the denominator of eqs. (1), (2), (4), and (5) by t_z. We introduce the following notations:

- $\mathbf{I} = \mathbf{i}/t_z$ is the first row of the rotation matrix scaled by the z-component of the translation vector;
- $\mathbf{J} = \mathbf{j}/t_z$ is the second row of the rotation matrix scaled by the z-component of the translation vector;
- $x_0 = t_x/t_z$ and $y_0 = t_y/t_z$ are the normalized camera coordinates of p_0 which is the projection of P_0 (the origin of the object frame);
- $\epsilon_i = \mathbf{k} \cdot \mathbf{P}_i/t_z$.

One can notice that $\mathbf{I}$ and $\mathbf{J}$ encapsulate the pose parameters (R and $\mathbf{t}$). We now rewrite the perspective equations (1), (2), and (3) as:

$$x_i = \frac{\mathbf{I} \cdot \mathbf{P}_i + x_0}{1 + \epsilon_i} \tag{6}$$

$$y_i = \frac{\mathbf{J} \cdot \mathbf{P}_i + y_0}{1 + \epsilon_i} \tag{7}$$

$$a_j\,(\mathbf{I} \cdot \mathbf{P}_j + x_0) + b_j\,(\mathbf{J} \cdot \mathbf{P}_j + y_0) + c_j\,(1 + \mathbf{k} \cdot \mathbf{P}_j/t_z) = 0 \tag{8}$$

Each line $\mathcal{D}_j$ is described parametrically by its direction $\mathbf{V}_j$ and by a point vector $\mathbf{W}_j$. Thus, we can write:

$$\mathbf{P}_j = \mathbf{W}_j + \lambda_j\, \mathbf{V}_j \quad (\lambda_j \in \mathbb{R})$$

By substituting this expression into eq. (8) and considering that this equation holds true for all λ_j, we obtain the following two constraints:

$$a_j\ \mathbf{W}_j \cdot \mathbf{I} + b_j\ \mathbf{W}_j \cdot \mathbf{J} + a_j\ x_0 + b_j\ y_0 + c_j\ (1 + \eta_j) = 0 \qquad (9)$$
$$a_j\ \mathbf{V}_j \cdot \mathbf{I} + b_j\ \mathbf{V}_j \cdot \mathbf{J} + c_j\ \xi_j = 0 \qquad (10)$$

where η_j and ξ_j are given by:

$$\eta_j = \mathbf{k} \cdot \mathbf{W}_j / t_z \quad \text{and} \quad \xi_j = \mathbf{k} \cdot \mathbf{V}_j / t_z$$

3 Pose by paraperspective iterations

3.1 Definition and equations

The notion of paraperspective projection was introduced by Ohta et al.[6]. Paraperspective may be viewed as a first-order approximation of perspective:

$$\frac{1}{1 + \epsilon_i} \approx 1 - \epsilon_i \quad \forall i,\ i \in \{1...n\}$$

By using this approximation in eqs. (6) and (7) we obtain the paraperspective projection of P_i:

$$\begin{aligned} x_i^p &= (\mathbf{I} \cdot \mathbf{P}_i + x_0)(1 - \epsilon_i) \\ &\approx \mathbf{I} \cdot \mathbf{P}_i + x_0 - x_0 \epsilon_i \\ &= \frac{\mathbf{i} \cdot \mathbf{P}_i}{t_z} + x_0 - x_0 \frac{\mathbf{k} \cdot \mathbf{P}_i}{t_z} \end{aligned}$$

where the term $1/t_z^2$ was neglected. There is a similar expression for y_i^p. By identification with eqs. (6) and (7) we obtain the relationship between the paraperspective and the perspective projections of P_i:

$$x_i^p = x_i\ (1 + \epsilon_i) - x_0 \epsilon_i \qquad (11)$$
$$y_i^p = y_i\ (1 + \epsilon_i) - y_0 \epsilon_i \qquad (12)$$

The paraperspective coordinates are related to the pose parameters by:

$$x_i^p - x_0 = \frac{\mathbf{i} - x_0\ \mathbf{k}}{t_z} \cdot \mathbf{P}_i \qquad (13)$$
$$y_i^p - y_0 = \frac{\mathbf{j} - y_0\ \mathbf{k}}{t_z} \cdot \mathbf{P}_i \qquad (14)$$

By substituting eqs. (11) and (12) in eqs. (13) and (14), we obtain:

$$\mathbf{P}_i \cdot \mathbf{I}_p = (x_i - x_0)(1 + \epsilon_i) \qquad (15)$$
$$\mathbf{P}_i \cdot \mathbf{J}_p = (y_i - y_0)(1 + \epsilon_i) \qquad (16)$$

with:

$$\mathbf{I}_p = \frac{\mathbf{i} - x_0\ \mathbf{k}}{t_z} \quad \text{and} \quad \mathbf{J}_p = \frac{\mathbf{j} - y_0\ \mathbf{k}}{t_z} \qquad (17)$$

By using these relationships between vectors ($\mathbf{I} = \mathbf{i}/t_z$, $\mathbf{J} = \mathbf{j}/t_z$) and vectors ($\mathbf{I}_p$, $\mathbf{J}_p$) in eqs. (9) and (10), these ones become:

$$a_j\,\mathbf{W}_j \cdot \mathbf{I}_p + b_j\,\mathbf{W}_j \cdot \mathbf{J}_p + (a_j\,x_0 + b_j\,y_0 + c_j)\,(1 + \eta_j) = 0 \qquad (18)$$
$$a_j\,\mathbf{V}_j \cdot \mathbf{I}_p + b_j\,\mathbf{V}_j \cdot \mathbf{J}_p + (a_j\,x_0 + b_j\,y_0 + c_j)\,\xi_j = 0 \qquad (19)$$

In brief, each point correspondence provides the 2 constraints (15) and (16), and each line correspondence provides the 2 constraints (18) and (19). In matrix form these equations can be written as:

$$\underbrace{G}_{(2n+2m)\times 6} \begin{bmatrix} \mathbf{I}_p \\ \mathbf{J}_p \end{bmatrix} = \underbrace{\mathbf{z}_p}_{(2n+2m)\times 1} \qquad (20)$$

where G and $\mathbf{z}_p$ are a $(2n+2m)\times 6$ matrix and a $(2n+2m)$ vector respectively:

$$G = \begin{bmatrix} \vdots & \vdots \\ \mathbf{P}_i^T & \mathbf{0}^T \\ \vdots & \vdots \\ \mathbf{0}^T & \mathbf{P}_i^T \\ \vdots & \vdots \\ a_j\,\mathbf{W}_j^T & b_j\,\mathbf{W}_j^T \\ \vdots & \vdots \\ a_j\,\mathbf{V}_j^T & b_j\,\mathbf{V}_j^T \\ \vdots & \vdots \end{bmatrix} \qquad \mathbf{z}_p = \begin{bmatrix} \vdots \\ (x_i - x_0)(1 + \epsilon_i) \\ \vdots \\ (y_i - y_0)(1 + \epsilon_i) \\ \vdots \\ -(a_j\,x_0 + b_j\,y_0 + c_j)\,(1 + \eta_j) \\ \vdots \\ -(a_j\,x_0 + b_j\,y_0 + c_j)\,\xi_j \\ \vdots \end{bmatrix}$$

3.2 Pose by successive approximations

One may notice if ϵ_i, η_j, and ξ_j are set to zero then (i) equation (20) becomes linear in $\mathbf{I}_p$ and $\mathbf{J}_p$ and (ii) the image features are supposed to be obtained with a paraperspective camera model (see eqs. (11) and (12)). Therefore, it is possible to solve this equation by successive linear approximations. In the following we show how the pose parameters can be computed from $\mathbf{I}_p$ and $\mathbf{J}_p$.

Pose parameters The pose parameters (R and $\mathbf{t}$) can be derived from $\mathbf{I}_p$ and $\mathbf{J}_p$ as follows.

First, one may notice that:

$$\|\mathbf{I}_p\|^2 = \frac{(\mathbf{i} - x_0\,\mathbf{k}) \cdot (\mathbf{i} - x_0\,\mathbf{k})}{t_z^2} = \frac{1 + x_0^2}{t_z^2}$$
$$\|\mathbf{J}_p\|^2 = \frac{1 + y_0^2}{t_z^2}$$

We obtain:
$$t_z = \frac{1}{2}\left(\frac{\sqrt{1+x_0^2}}{\|\mathbf{I}_p\|} + \frac{\sqrt{1+y_0^2}}{\|\mathbf{J}_p\|}\right); \quad t_x = x_0\, t_z; \quad t_y = y_0\, t_z$$

Second, we derive the three orthogonal unit vectors **i**, **j**, and **k**. From (17) we can write:
$$\mathbf{i} = t_z\, \mathbf{I}_p + x_0\, \mathbf{k} \tag{21}$$
$$\mathbf{j} = t_z\, \mathbf{J}_p + y_0\, \mathbf{k} \tag{22}$$

The third vector, **k** is the cross-product of these two vectors:
$$\mathbf{k} = \mathbf{i} \times \mathbf{j}$$
$$= t_z^2\, \mathbf{I}_p \times \mathbf{J}_p + t_z y_0\, \mathbf{I}_p \times \mathbf{k} - t_z x_0\, \mathbf{J}_p \times \mathbf{k}$$

Let $\Omega(\mathbf{a})$ be the skew-symmetric matrix associated with a 3-vector **a** and $I_{3\times 3}$ the identity matrix. The previous expression can now be written as follows:
$$(I_{3\times 3} - t_z y_0\, \Omega(\mathbf{I}_p) + t_z x_0\, \Omega(\mathbf{J}_p))\, \mathbf{k} = t_z^2\, \mathbf{I}_p \times \mathbf{J}_p \tag{23}$$

This equation allows us to compute **k** since it has full rank. Therefore, one can easily determine **k** using eq. (23) and **i** and **j** using eqs. (21) and (22).

Pose by successive approximations The algorithm can be written as follows.

1. For all i and j, $i \in \{1...n\}$, $j \in \{1...m\}$, $(n+m) \geq 3$, $\epsilon_i = 0$, $\eta_j = 0$, $\xi_j = 0$.
2. Solve the overconstrained linear system (20) which provides an estimation of vectors $\mathbf{I}_p$ and $\mathbf{J}_p$:
$$\begin{bmatrix}\mathbf{I}_p \\ \mathbf{J}_p\end{bmatrix} = (G^T G)^{-1} G^T\, \mathbf{z}_p$$

3. Compute the pose parameters, i.e. the position $(t_x, t_y, \text{and } t_z)$ and orientation (**i**, **j**, and **k**) as explained above;
4. For all i and j, compute:
$$\epsilon_i = \frac{\mathbf{k} \cdot \mathbf{P}_i}{t_z}, \quad \eta_j = \frac{\mathbf{k} \cdot \mathbf{W}_j}{t_z}, \quad \xi_j = \frac{\mathbf{k} \cdot \mathbf{V}_j}{t_z}$$

If the changes in ϵ_i, η_j, and ξ_j in two consecutive iterations are below a fixed threshold then stop the procedure, otherwise go to step 2.

The matrix G has full rank since it is assumed that the observed scene is non coplanar. One may notice that the pseudo-inverse of G (i.e. $(G^T G)^{-1} G^T$) can be computed once for all and hence it can be computed independently of the loop presented above. Therefore, the estimation of $\mathbf{I}_p$ and $\mathbf{J}_p$ is particularly efficient.

4 Experiments

Figure 2 shows an example of convergence of the paraperspective algorithm when it is applied to compute the pose of a cube. The first iteration of the algorithm found a paraperspective pose (left). After only six iterations the algorithm correctly determined the pose of the cube (right). This computation takes 6 iterations (3.1 ms on an Ultra-Sparc). Figure 3 illustrates the pose estimation of a cube (its size is 7 cm) and a gripper by the paraperspective algorithm. The gripper is identified by 5 vertices and 5 edges, the cube is identified by 6 vertices and 7 edges. By combining the 2 obtained poses one can obtain the relative position and orientation of the gripper with respect to the cube. For example, the relative position which is given by the translation vector gripper-cube has been found to be : $(20cm, 1.9cm, -5.9cm)^T$. The origins of the 2 coordinate systems are shown by big crosses (see Figure 3 (right)). Therefore, by tracking the gripper location in the image, one can apply visual servoing approaches in order to guide the gripper such that it can grasp the cube [4]. Table 1 gives the residual errors in the image plane between the true features and the projected 3D model associated with the 2 computed poses (gripper and cube).

5 Conclusion

In this paper we focused on the problem of pose computation from 2D to 3D point and line correspondences. We propose a fast method which establishes a link between paraperspective and perspective. The resulting method is very elegant, very fast, and quite accurate. It can be included in real-time vision and robotics applications. The iterative paraperspective method has better convergence properties than the iterative weak perspective method.

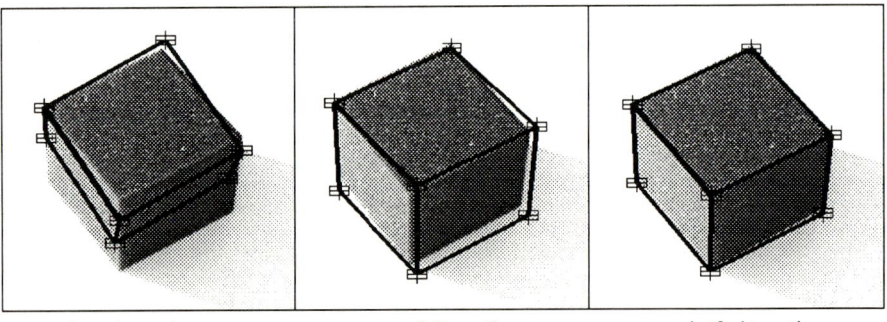

first iteration   second iteration   sixth iteration

Fig. 2. An example of applying the iterative paraperspective algorithm to a cube using 7 vertices and 6 edges (peripheral). This computation takes 6 iterations (3.1 ms on an Ultra-Sparc).

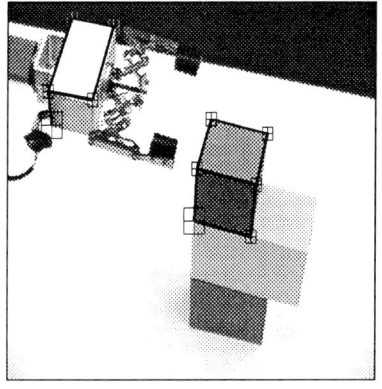 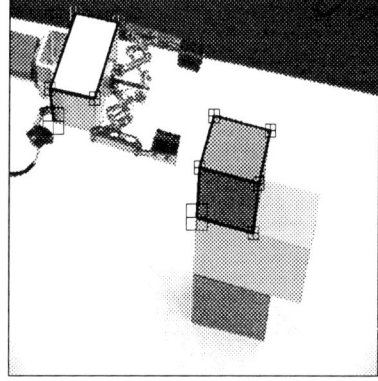

first iteration — third iteration

Fig. 3. An example of applying the paraperspective algorithm to both a gripper and a cube. The 2 obtained poses allow one to compute the relative position and orientation between them.

Residual error (image space)	gripper	cube
Vertices locations (pixels)	0.8	0.9
Edges orientations (deg.)	0.38	1.4
Edges locations (pixels)	0.72	6.0
Number of iterations	3	3
CPU time (ms)	2.1	2.5

Table 1. Pose estimation of both the gripper and the cube using the paraperspective algorithm, the computer being used is an Ultra-Sparc.

References

1. D. DeMenthon and L. Davis. Model-based object pose in 25 lines of code. *International Journal of Computer Vision*, 15:123–141, June 1995.
2. M. Dhome, M. Richetin, J.T. Lapreste, and G. Rives. Determinition of the attitude of 3d objects from a single perspective view. *IEEE Trans. on Pattern Analysis and Machine Intelligence*, 11(12):1265–1278, 1989.
3. B. Espiau, F. Chaumette, and P. Rives. A new approach to visual servoing in robotics. *IEEE Transactions on Robotics and Automation*, 8(3):313–326, June 1992.
4. N. Hollinghurst and R. Cipolla. Uncalibrated stereo hand-eye coordination. In *Proceedings of the Fourth British Machine vision Conference (BMVC 93)*, 1993.
5. R. Horaud, B. Conio, O. Leboulleux, and B. Lacolle. An analytic solution for the perspective 4-point problem. *Computer Vision, Graphics, and Image Processing*, 47(1):33–44, July 1989.
6. Y. Ohta, K. Maenobu, and T. Sakai. Obtaining surface orientation from texels under perspective projection. In *Proceedings of the 7th IJCAI*, 1981.
7. J. Yuan. A general photogrammetric method for determining object position and orientation. *IEEE Transactions on Robotics and Automation*, 5(2):129–142, 1989.

Robust Motion Estimation Using Chrominance Information in Colour Image Sequences

Julian Magarey, Anil Kokaram, and Nick Kingsbury

Signal Processing and Communications Laboratory
Cambridge University Engineering Department
Cambridge CB2 1PZ, United Kingdom

Abstract. This paper describes a method for incorporating the chrominance information when estimating motion in a colour image sequence. It is based on a Maximum-Likelihood (ML) formulation of the motion estimation problem which assumes homogeneous additive Gaussian noise in each colour component, with known inter-field correlation statistics. It defines a noise-decorrelating colour space transform which provides a simple implementation of the ML formulation. Results for noisy synthesised colour sequences with known motion and noise statistics demonstrate the superiority of the exact ML formulation over straightforward, unweighted three-component estimation, most noticeably in high noise conditions.

1 Introduction

Colour video signals consist of both luminance (intensity) and chrominance information. The chrominance has two degrees of freedom, so a full colour signal consists of three *fields* for every frame. There are a variety of representations of these three fields, most commonly defined as some linear transformation of the red-green-blue (RGB) basis which is rooted in the human visual system for analysing colour. For example, the YUV colour space is defined as [2, page 67]

$$\begin{bmatrix} y \\ u \\ v \end{bmatrix} = \begin{bmatrix} 0.3 & 0.6 & 0.1 \\ -0.15 & -0.3 & 0.45 \\ 0.4375 & -0.375 & -0.0625 \end{bmatrix} \begin{bmatrix} r \\ g \\ b \end{bmatrix} \quad (1)$$

where y is the luminance component and the u and v fields contain chrominance information (the matrix entries vary somewhat depending on the visual constants used). Most algorithms for motion estimation of colour sequences have worked in YUV-space and ignored the chrominance components.

It was suggested by Mitiche et al. [7] that chrominance could be used as an additional cue in a multiconstraint method for estimating motion between video frames, though they reported no results which incorporated chrominance. Konrad and Dubois [4] extended their original maximum likelihood (ML) motion estimation framework for scalar signals to encompass three component fields. To do this, they defined a quantity called the *vector* displaced pel difference (DPD) by analogy with the scalar DPD of monochrome estimation. They demonstrated improvements in the quality of the colour-based motion field over the purely luminance-based field for synthetic and real colour sequences. Their method assumed that the colour fields contribute equally and independently to a single motion estimate field. This is implicitly based in turn on the assumption that the noise in each field is additive Gaussian, uncorrelated and equivariant. In their paper on the use of colour in video resolution enhancement, Tom and Katsaggelos [8] used the same implicit assumption.

However, this cannot be relied on in practice. For example, if noise is uncorrelated in YUV-space, it will be far from uncorrelated in RGB-space. If there is no knowledge of the original image-gathering system and its noise properties, unweighted three-component estimation could go seriously wrong. This paper describes the true ML estimator for vector image sequences which allows for correlated additive noise in the three colour components. The ML formulation may be applied to extend the standard region-based matching and gradient-based algorithms. We then define a noise-decorrelating transform, akin to the Karhunen-Loeve transform, into an "optimal" colour space. If this transform is applied to the three original colour fields, the true ML estimate may be found using the common implicit assumption of equal and independent contributions by the three transformed components.

Our test results, obtained on colour sequences with synthesised motion and correlated additive noise, demonstrate the superiority of the optimal method over estimation based on equal and independent contributions from the each of the red, green, and blue fields. The improvement becomes more noticeable as the amount of noise increases. Our results also suggest that the most efficient strategy would be adaptive, based primarily on luminance and only incorporating chrominance appropriately where required.

2 The ML Motion Estimator for Colour Signals

We follow the formulation of Konrad and Dubois [4] in setting up the ML estimator for a three-component sequence. The vector $\mathbf{u} = [u_1, u_2, u_3]^T$ represents the true or underlying image, with its three colour components, which would be obtained by a noise-free optical system. The sequence $\{\mathbf{u}_n, n \in \mathcal{Z}\}$ represents the sequence of images sampled at integer time instants. We use the common assumption of intensity conservation, which requires that intensity *in each component* is constant along the motion trajectory defined by the displacement $\hat{\mathbf{d}}(\mathbf{x})$ at pel $\mathbf{x}$:

$$\mathbf{u}_n(\mathbf{x}) = \mathbf{u}_{n-1}(\mathbf{x} - \hat{\mathbf{d}}(\mathbf{x})) \qquad (2)$$

Taking into account observation noise, we can rewrite (2) using the *observed* signal sequence $\{\mathbf{g}_n, n \in \mathcal{Z}\}$ as

$$\mathbf{g}_n(\mathbf{x}) - \mathbf{g}_{n-1}(\mathbf{x} - \hat{\mathbf{d}}(\mathbf{x})) = \mathbf{e}_n(\mathbf{x}) \qquad (3)$$

where $\mathbf{e}_n$ is the differential noise vector at frame n, which has twice the variance of the individual frame noise. Equation (3) may be rewritten as

$$\mathbf{DPD}(\mathbf{x}, \hat{\mathbf{d}}(\mathbf{x})) = \mathbf{e}_n(\mathbf{x}) \qquad (4)$$

having replaced its left hand side with the *vector displaced pel difference*.

Our aim is to estimate the translation model parameter $\hat{\mathbf{d}}$ based on the observations $\mathbf{g}_{n-1}$ and $\mathbf{g}_n$. The Maximum Likelihood (ML) estimate is defined as

$$\hat{\mathbf{d}}(\mathbf{x}) = \arg \max \ \{p(\mathbf{g}_n(\mathbf{x})|\mathbf{d}, \mathbf{g}_n(\mathbf{x}))\} \tag{5}$$

If we assume zero-mean Gaussian noise statistics, we can write the joint probability density function (pdf) of $\mathbf{e}(\mathbf{x})$ as

$$p(\mathbf{e}(\mathbf{x})) \propto \exp\left(-\frac{1}{2}\mathbf{e}^T(\mathbf{x})R^{-1}(\mathbf{x})\mathbf{e}(\mathbf{x})\right) \tag{6}$$

where $R(\mathbf{x})$ is the 3-by-3 covariance matrix characterising the interaction of the three noise components at pel $\mathbf{x}$. The likelihood is given by the vector noise joint pdf. Combining this fact with (4), we can write

$$p(\mathbf{g}_n(\mathbf{x})|\mathbf{g}_{n-1}(\mathbf{x}), \mathbf{d}) \propto \exp\left(-\frac{1}{2}\mathbf{DPD}^T(\mathbf{x}, \mathbf{d})R^{-1}(\mathbf{x})\mathbf{DPD}(\mathbf{x}, \mathbf{d})\right) \tag{7}$$

so the ML estimator becomes

$$\hat{\mathbf{d}}(\mathbf{x}) = \arg \min \ \{\mathbf{DPD}^T(\mathbf{x}, \mathbf{d})R^{-1}(\mathbf{x})\mathbf{DPD}(\mathbf{x}, \mathbf{d})\} \tag{8}$$

2.1 ML Estimation Using Region-based Matching

To obtain a more robust estimate of $\hat{\mathbf{d}}$, an assumption of *constant local flow* over a region of pels $\Omega = \{\mathbf{x}_i, i = 1, \ldots, N\}$ is commonly invoked. This assumption is approximately valid if the motion field is continuous and the regions are not too large. If we define the $3N$-element *displaced region difference* vector as

$$\mathbf{DRD}(\Omega, \mathbf{d}) = \left[\mathbf{DPD}(\mathbf{x}_1, \mathbf{d}) \ldots \mathbf{DPD}(\mathbf{x}_N, \mathbf{d})\right]^T \tag{9}$$

we can estimate $\hat{\mathbf{d}}$ over the region Ω as

$$\hat{\mathbf{d}}(\Omega) = \arg \min \ \{\mathbf{DRD}^T(\Omega, \mathbf{d})R_\Omega^{-1}\mathbf{DRD}(\Omega, \mathbf{d})\} \tag{10}$$

where R_Ω is the $3N$-by-$3N$ noise component covariance matrix over the region Ω.

Equation (10) may be simplified by the assumption that there is no noise correlation between different pels in the region Ω. This gives R_Ω a block diagonal structure, where each block is the 3-by-3 matrix R of component noise covariance of (6) (the $\mathbf{x}$ argument may be dropped if we further assume homogeneity, i.e. position-independence). When R_Ω has this structure, so too does its inverse, with each block equal to R^{-1}. The ML estimator becomes

$$\hat{\mathbf{d}}(\Omega) = \arg \min \ \left\{\sum_{i=1}^N \mathbf{DPD}^T(\mathbf{x}_i, \mathbf{d})R^{-1}\mathbf{DPD}(\mathbf{x}_i, \mathbf{d})\right\} \tag{11}$$

Equation (11) shows how to find $\hat{\mathbf{d}}$ by an exhaustive search over a set of $\mathbf{d}$ candidates. This is the optimal *region-matching* strategy in the presence of correlated component noise.

If, furthermore, the noise in each component is uncorrelated, R becomes diagonal:

$$R = \text{diag }(\sigma_k^2, k = 1, 2, 3) \tag{12}$$

In this case (11) becomes

$$\hat{\mathbf{d}}(\Omega) = \arg\min \left\{ \sum_{i=1}^{N} \sum_{k=1}^{3} \frac{1}{\sigma_k^2} DPD_k^2(\mathbf{x}_i, \mathbf{d}) \right\} \tag{13}$$

which now involves minimising the sum, over the region, of the squared (scalar) DPDs of each component, weighted by the inverse of the noise variance. This is the formulation obtained by Konrad and Dubois [4]. In effect the contribution of each component to the estimate is weighted by the SNR of the corresponding difference image.

2.2 Gradient-based ML Estimation

An approximate solution to the region-based vector ML estimator (10) may be found by expanding $\mathbf{g}_{n-1}(\mathbf{x}_i - \mathbf{d})$ around $\mathbf{x}_i$ using a first-order Taylor series:

$$\mathbf{g}_{n-1}(\mathbf{x}_i - \mathbf{d}) \approx \mathbf{g}_{n-1}(\mathbf{x}_i) - (\nabla \mathbf{g}_{n-1}(\mathbf{x}_i))^T \mathbf{d} \tag{14}$$

where

$$\nabla \mathbf{g}_{n-1}(\mathbf{x}_i) = \begin{bmatrix} \frac{\partial}{\partial x} g_{1,n-1}(\mathbf{x}_i) & \frac{\partial}{\partial y} g_{1,n-1}(\mathbf{x}_i) \\ \frac{\partial}{\partial x} g_{2,n-1}(\mathbf{x}_i) & \frac{\partial}{\partial y} g_{2,n-1}(\mathbf{x}_i) \\ \frac{\partial}{\partial x} g_{3,n-1}(\mathbf{x}_i) & \frac{\partial}{\partial y} g_{3,n-1}(\mathbf{x}_i) \end{bmatrix}^T \tag{15}$$

This approximation allows a closed-form least-squares solution to be found for $\mathbf{d}$, called the *gradient-based* ML estimator:

$$\hat{\mathbf{d}}(\Omega) = (G^T R_\Omega^{-1} G)^{-1} G^T R_\Omega^{-1} \mathbf{z} \tag{16}$$

where

$$G = \begin{bmatrix} \nabla \mathbf{g}_{n-1}(\mathbf{x}_1) \ldots \nabla \mathbf{g}_{n-1}(\mathbf{x}_N) \end{bmatrix}^T \tag{17}$$

$$\text{and } \mathbf{z} = \begin{bmatrix} \mathbf{g}_{n-1}(\mathbf{x}_1) - \mathbf{g}_n(\mathbf{x}_1) \ldots \mathbf{g}_{n-1}(\mathbf{x}_N) - \mathbf{g}_n(\mathbf{x}_N) \end{bmatrix}^T \tag{18}$$

As with region-based matching, the assumption that noise at different pels is uncorrelated and homogeneous simplifies the computation in (16). In practice this method is severely limited in its measurement range because of the neglect of higher order terms in (14). The range may be increased by using an iterative approach which uses an equation similar to (16) to compute updates to an initial estimate of $\hat{\mathbf{d}}$ [3].

2.3 A Decorrelating Transform

Equation (1) defines the transform from the RGB colour space to the YUV space. A general linear colour space transform on RGB space may be written as

$$\mathbf{g} = C \begin{bmatrix} r \\ g \\ b \end{bmatrix} \quad (19)$$

for an n-by-3 matrix C. In the new colour space ("C-space"), the inter-component noise correlation matrix becomes

$$R_C = C R_{rgb} C^T \quad (20)$$

Clearly if we can find a matrix C such that $R_C = \sigma^2 I$ for some σ, the vector ML gradient-based estimator (16) and region-based matching estimator (11) revert to straightforward formulations in which the colour components make equal and independent contributions to the final estimate, as postulated by Konrad and Dubois [4].

To find such a *noise-decorrelating* transform, given R_{rgb} (or R_{yuv}, in which case we apply the transform to the YUV fields), we can use singular value decomposition (SVD). Because R_{rgb} is square, symmetric and non-negative definite [2, page 33], it is orthogonally diagonalisable with non-negative eigenvalues:

$$R_{rgb} = V D V^T \quad (21)$$

where D is diagonal with non-negative entries, and V is orthogonal. The number of non-zero eigenvalues is the rank n of R_{rgb}. The case $n < 3$ results when at least one component is a linear combination of the others. In this (exceptional) case, when R_{rgb} is non-invertible, (11) and (16) may not be used. The SVD-based method will identify this case and project to a colour space of appropriately reduced dimensionality, thus saving computation. This is done by extracting the invertible n-by-n portion D' of D and the corresponding rows V' of V. Setting

$$C = (\sqrt{D'})^{-1} V'^T \quad (22)$$

guarantees that $C R_{rgb} C^T = I_n$ as desired. This procedure is similar to the Karhunen-Loeve Transform for compressing images [2, page 163], except it is carried out using noise rather than signal statistics.

3 Tests on Synthesised Sequences

The synthesised test sequences were obtained by applying motion fields of three distinct kinds—uniform translation, rotation, and divergence—to the 128-by-128 pel central portions of frame 1 of the "carphone", "foreman", and "suzie" colour sequences respectively.

To add correlated noise, we first found three 128-by-128 uncorrelated, equivariant white Gaussian noise images with variance σ^2. An invertible 3-by-3 matrix

M was used to transform the noise fields into M^{-1}-space; the transformed noise was added to the RGB signal fields. This procedure gave

$$R_{rgb} = \sigma^2 M^{-1}(M^{-1})^T$$
$$= \sigma^2 \begin{bmatrix} 1.7393 & 0.1871 & -0.1886 \\ 0.1871 & 0.1318 & -0.0742 \\ -0.1886 & -0.0742 & 0.3654 \end{bmatrix} \quad (23)$$

This particular M was chosen to give a clearly non-diagonal R_{rgb} in order to best illustrate the potential improvement from the proposed approach over unweighted RGB-space estimation.

A modified version of the iterative gradient-based algorithm was used. The algorithm includes a stabilising term in the matrix-inversion step of the update (16) to give better convergence performance and robustness during iteration. To further increase the measurement range, the algorithm was implemented *hierarchically*, based on a 4-level Gaussian pyramid decomposition—see [3] for details. The final full-density field was obtained by bilinear interpolation from the field of region vectors.

To measure the accuracy of the full-density motion field, we used Fleet and Jepson's angular measure of error [1], averaged over the field excluding a strip of width 16 pels around the boundary of the image. This is is akin to a relative measure of error, except it does not give undue weighting to errors in very small motion vectors as relative error would.

For each of the three test sequences, three sets of results were obtained: those using unweighted RGB-space ME; those from luminance-only estimation; and those from vector ML estimation, obtained by first transforming from RGB to the optimal colour space. Figure 1 shows these results, plotted as mean error angle against σ. In each case the optimal strategy is clearly the most robust to noise. The difference between the RGB and optimal strategies is illustrated in Fig. 2, which shows the lower right portion of the motion fields at $\sigma = 36$ for the rotation sequence, superimposed on images of error angle (darker means greater error.) The improvement using the optimal strategy under conditions of high, correlated noise is clear.

The results also show that at low noise levels, there is little to be gained by using the optimal approach as opposed to RGB-space ME. Furthermore, luminance-only estimation loses little by comparison with the optimal strategy at the lower noise levels. These results have been repeated for the complex-wavelet-domain ME algorithm of Magarey and Kingsbury [5, 6].

Luminance-only ME requires only slightly more than a third of the amount of computation required for full-colour ME. Our results suggest that this strategy provides near-optimal performance except where noise overwhelms luminance contrast. In such cases, chrominance information may be incorporated (according to the ML formulation) to increase the robustness of the estimates. An adaptive strategy, in which luminance is the primary quantity for estimation, with some criterion to indicate where the chrominance information should be incorporated, would provide the best tradeoff between accuracy and efficiency. Our on-going

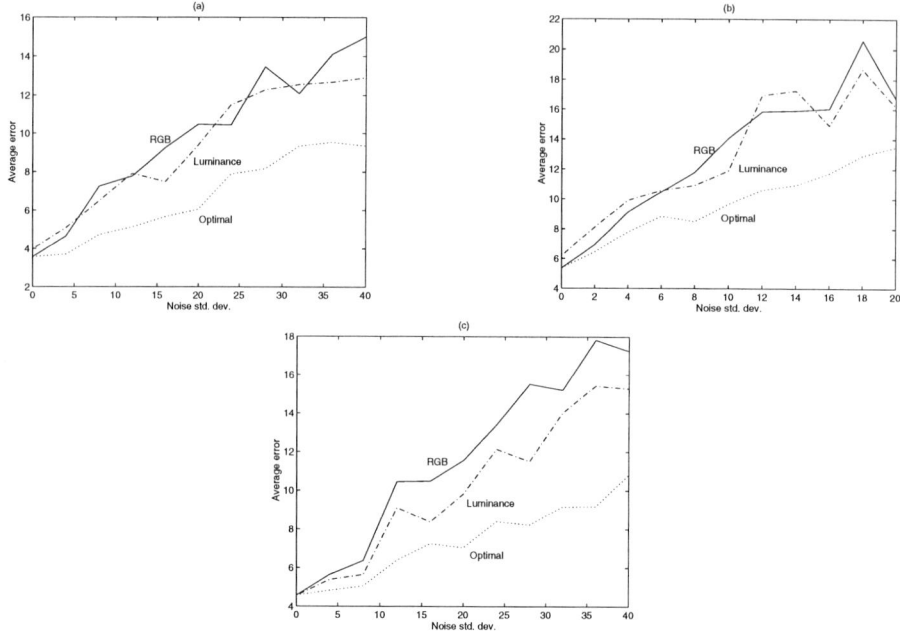

Fig. 1. Mean motion field error vs σ for RGB, luminance-only, and optimal estimation. (a) Translation sequence. (b) Divergence sequence. (c) Rotation sequence.

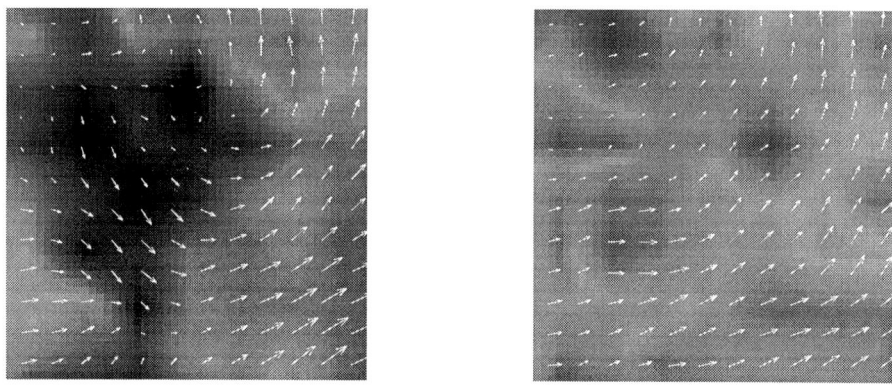

Fig. 2. Lower right portion of motion fields superimposed on error angle images (darker means greater error.) Sequence: rotation, with correlated noise $\sigma = 36$. (Left) RGB-estimated field (one estimate per 4 by 4 pels). (Right) Optimally-estimated field (same resolution).

work is aimed at finding a general technique for characterising the noise from typical video sources, and finding a criterion for incorporation of chrominance information.

4 Conclusion

In this paper we have shown how to formulate the ML motion estimator for colour image sequences in the presence of correlated, homogeneous Gaussian noise in the three component fields. Vector ML formulations for region-matching and gradient-based approaches were given in terms of the inter-component noise covariance matrix. If the covariance matrix is diagonal, each component contributes independently to the ML estimate. We have shown how to define a linear transformation into a new colour space in which the noise covariance is the identity matrix. In the new, "optimal" colour space, the components may be treated as equal and independent contributors to the ML estimate. The effectiveness of the optimal colour space transformation was demonstrated for a modified gradient-based algorithm applied to three synthesised test sequences containing additive noise with deliberately induced covariance. Our tests also showed that luminance-only estimation performs reasonably well by comparison with the more expensive full-colour approach, particularly at low noise levels. This suggests that the best strategy for a general colour sequence would be adaptive.

References

1. D.J. Fleet and A.D. Jepson. Computation of component image velocity from local phase information. *Intern. J. Comput. Vis.*, 5:77–104, 1990.
2. A.K. Jain. *Fundamentals of Digital Image Processing*. Prentice-Hall International, 1989.
3. A. C. Kokaram and S. J. Godsill. A system for reconstruction of missing data in image sequences using sampled 3D AR models and MRF motion priors. In *Computer Vision - ECCV '96*, volume II, pages 613–624. Springer Lecture Notes in Computer Science, April 1996.
4. J. Konrad and E. Dubois. Use of colour information in Bayesian estimation of 2-D motion. In *Proc. ICASSP*, pages 2205–2208. IEEE, 1990.
5. J.F.A. Magarey. *Motion estimation using complex wavelets*. PhD thesis, Cambridge University Department of Engineering, 1997.
6. J.F.A. Magarey, A.C. Kokaram, and N.G. Kingsbury. Optimal schemes for motion estimation using colour image sequences. In *Proc. Int. Conf. On Image Processing*, October 1997. *(To appear)*.
7. A. Mitiche, Y.F. Wang, and J.K. Aggarwal. Experiments in computing optical flow with the gradient-based, multiconstraint method. *Pattern Recognition*, 20:173–179, 1987.
8. B.C. Tom and A.K. Katsaggelos. Resolution enhancement of colour video. In *Proc. EUSIPCO-96*, pages 145–148, September 1996.

Temporal Prediction of Video Sequences Using an Image Warping Technique Based on Color Segmentation

N. Herodotou, and A.N. Venetsanopoulos
Digital Signal & Image Processing Laboratory
Department of Electrical and Computer Engineering
University of Toronto
10 King's College Road
Toronto, Ontario
M5S 3G4
CANADA
E-mail: nicos@dsp.toronto.edu
E-mail: anv@dsp.toronto.edu
URL: http://www.comm.toronto.edu/~dsp/dsp.html

Abstract. An image warping technique based on segmented regions is introduced for the temporal prediction of videophone-type sequences. At the encoder, a set of control points are determined from the previous frame and their corresponding best matched points are determined from the current frame. The selection process of these points is achieved by segmenting the previous frame into different regions using a color segmentation technique based on a recursive histogramming approach and the control points are subsequently chosen along the region boundaries. The spatial offset of these points between the previous and current frame are represented as motion vectors. The facial area is intraframe encoded to avoid distortions of the facial features which convey critical information. At the decoder, the same segmentation and control point selection algorithm is used along with the motion vectors in order to find the region boundaries of the predicted frame. The facial area is decoded and an affine transformation is finally used to determine the remaining regions and form the predicted frame. This technique produces results that are free from blocking artifacts as in the conventional block matching method, with only a moderate increase in computational complexity.

1 Introduction

A number of multimedia applications have newly emerged due to the recent advances in the area of mobile communications and the tremendous growth of the *Internet*. Applications such as portable communicators, video email, and video databases, to name a few, have placed even greater demands for more effective video coding schemes. However, future coding techniques must focus on providing better ways to represent and exchange visual information in addition to efficient compression methods. These efforts aim to provide the user with greater flexibility for *content-based* access and manipulation of multimedia data as in the proposed MPEG 4 and future MPEG 7 standards [1].

In conventional coding methods such as H.261 and MPEG 1 and 2, video sequences are compressed using an information-theoretic-based approach, that is, by exploiting the stochastic properties of the signals. Recently, however, greater attention has been paid to a newer generation of coding schemes which are *object-based* [2, 3]. These methods rely on the techniques of image analysis and computer graphics to represent the image signals using their structural features such as contours and regions. In this latter approach, an input video sequence is first segmented into an appropriate set of arbitrarily shaped regions. The features of each region such as shape, motion, and texture/color information are subsequently used in the encoding process. Thus, the success of an object-based method depends largely on the segmentation of the scene based on its image contents.

Motion compensation is typically used to remove the temporal correlation that exists between frames in an image sequence. Conventional motion compensated prediction methods rely on standard block matching approaches where displacement vectors are estimated over rectangular blocks of the image. This approach is favorable due to its straightforward approach, however, it fails to adequately model object motion which is non-translatory (i.e. object rotation, deformation, or change of scale). This scheme also suffers from annoying blocking artifacts when components of an image feature are assigned different motion vectors. In order to alleviate these problems, several approaches based on digital image warping have been introduced in the past [4, 5, 6, 7]. In these schemes, the predicted frames (also referred to as the current frames) are formed by geometrically transforming or *warping* the previous frames. These methods also fit well within the framework of the object-based coders described earlier.

In this paper, we focus our attention on an object-based approach to motion compensated prediction for videophone-type applications. The techniques of image analysis and digital image warping are utilized to form the predicted frame. This is achieved by segmenting the previous frame into a facial region and a number of arbitrarily shaped regions based on the color information. Each of these regions (represented by a suitable set of control points) are subsequently transformed to form the predicted image. Thus, the motion compensated warping prediction scheme described here consists of three stages: i) Face localization and image segmentation, ii) Control point selection and motion vector assignment, and, iii) Image warping of the previous frame.

2 Face Localization and Image Segmentation

The first step in the coding system described above is the segmentation of the scene into an appropriate set of *regions* or *objects*. Once again, we focus our attention on the specific application of a head-and-shoulders videophone-type sequence. We have found that warping the facial region of the previous frame to obtain the predicted facial region yields unsatisfactory results due to the deformations of the facial features. The opening/closing of the eyes and mouth are failure areas of the warping model due to the covered and uncovered areas created by these deformable features. Thus, we opt to extract the facial area and code this arbitrarily shaped region using a conventional intraframe technique (i.e. Discrete Cosine Transform, Karhunen Loeve Transform, etc.). This will alleviate any coding artifacts within the facial region and provide a more intelligible and perceptually pleasing image to a human observer.

The identification and location of the facial region is determined by utilizing the apriori knowledge of the skin-tone distributions in the perceptual HSV color space. It has been found that skin-colored clusters form within a rather well defined region in the HSV hexcone model for a variety of different skin types [8]. Pixels that fall within a defined polyhedron [8] are classified as skin regions. A set of binary operations which include median filtering, and region filling/removal are used to refine the extracted facial area.

Once the facial area has been identified, then the remaining pixels in the scene must be segmented into a set of distinct regions. A recursive histogramming approach is followed in order to achieve this [9]. Color space conversion is first performed on the RGB image sequences to obtain their corresponding HSV representation. Singularities in the Hue component (i.e. where R,G, and B values are equal) are grouped into different regions according to brightness (i.e. Value) prior to any histogram processing. A histogram of the Hue values is then formed and the most prominent peak is selected. In order to extract these prominent peaks, then the histograms above must be smoothed to remove any meaningless local extrema. For this purpose, we apply the well-known scale space filter [10], where the Hue histogram is convolved with a Gaussian function of zero mean. The peaks and valleys can be determined by examining the first and second derivatives of the convolved result. This process is repeated recursively until no dominant peaks remain in the Hue histogram. After the extraction of each region above, a binary median filter is applied as a post-processing step to smoothen and eliminate small holes within each region, and also to remove any misclassified pixels. A final step in the segmentation process is to classify the remaining pixels into their appropriate regions and merge any regions with a similar hue. A Euclidean distance measure is used in this latter merging step.

3 Control Point Selection & Motion Vector Assignment

Having segmented the image into a set of distinct regions, then a sufficient number of control points must be appropriately selected to represent each region. The partitioning of the image in this way allows the set of regions in the previous frame to be *warped* into their corresponding regions in the current frame. Both, the previous and current frames are available at the encoder while only the former of the two is available at the decoder side. The prediction of the current frame at the decoder is determined by utilizing the previous frame along with the received *warping* instructions (i.e. side information) as follows. The spatial offset (i.e. motion vectors) of the selected control points is determined at the encoder by using the previous and current frames. The encoder finally transmits the previous frame, the computed motion vectors, and the intraframe encoded facial region. The decoder then receives the previous frame along with the necessary side information. The same segmentation and control point selection algorithm as in the encoder are then used to partition the previous frame. As a result, the same control points determined at the encoder are also found at the decoder. These points are spatially shifted to their appropriate positions in the predicted frame according to the received motion vectors. The warping algorithm can finally be applied along with the decoded facial region to obtain the predicted frame. The overhead in the form of side information in this approach consists of the transmitted motion vectors along with the encoded facial area. In [4], the above scheme is known as a forward matching technique and has the advantage that the control points can be selected based on the contents of the image.

A simple way of selecting the control points is by forming a rectangular mesh that partitions the previous frames into a uniform set of non-overlapping blocks. However, a uniform spacing of the selected control points can lead to inaccurate motion vectors which can cause geometric distortions. In order to prevent this from happening, the selection algorithm we use here chooses control points that reside on the edges of the segmented regions determined earlier. In order to simplify the selection algorithm we choose every 10th pixel on each region border and also place a control point at the center of mass of each region. This selection allows each region to be broken up into triangular patches which are formed by the edge points and the center of mass. In this way, each of these patches can be individually *warped* via transformation equations. Control points are also selected at all corners and midpoints of each side of the image frame. These points, however, are stationary and are not spatially offset in the predicted frame. For a better representation of the shape of each region, we also choose control points at pixel boundaries which border three or more distinct regions.

The assignment of motion vectors for each selected control point once again is determined at the encoder where both, the previous and the current frames are available. A modified block matching technique, similar to the one in [4] is also used here, where a 21×21 window size is employed in which the central pixels within this block structure are more heavily weighted. Unlike the scheme in [4], however, the square block used in the matching process here, is not subsampled. Further to this, the mean squared error (MSE) criterion using the Euclidean distance measure is utilized due to the color information. The best match of each selected control point is determined by finding the minimum MSE value within a search space of + or − 15 pixels. These motion vectors are transmitted to the decoder as overhead information along with the encoded facial region.

4 Image Warping

When the decoder receives the previous frame along with the transmitted motion vectors it must predict the current or missing frames. This is accomplished by *warping* the triangles in each of the regions of the previous frame to the corresponding triangles in the predicted frame. The vertices of the triangles in the predicted frame are found by using the same segmentation and control point selection algorithm as in the encoder and spatially shifting these appropriately as mentioned earlier using the motion vector information. Once these are found, then a triangle to triangle mapping follows using an affine transformation [11]. As a result of this, the points within each triangle are geometrically transformed to their corresponding positions. Bilinear interpolation is used when non-integer positions are found. The facial region is finally added to the predicted frame by using the intraframe encoded information. Thus, the current or predicted frame is reconstructed by a number of geometrically transformed patches plus the encoded facial region.

5 Results

The performance of this scheme was evaluated using Frames 100 and 110 of the Claire CIF sequence. A comparison of the warping prediction was made with the conventional block matching motion compensator. In Figures 1a) and 1b) we present Frames 100 and 110 of the original Claire sequence. In Figure 2a), the results of the segmentation and control point selection are shown while Figure 2b) illustrates the performance of the point matching process. As we can see, the extraction of the facial area and segmentation of the remaining regions leads to an appropriate representation of the image contents. Furthermore, we note that the selected control points are tracked quite accurately as shown in 2 b). The standard block matching results are shown in Figure 3a) where 18×18 size blocks were used. The blocking artifacts in this figure are clearly evident in this case where abrupt head rotations are encountered. The transmission overhead required for this 360×288 pixel image amounts to 320 motion vectors. The predicted image using the object-based warping approach is finally shown in Figure 3b) which does not suffer from the annoying blocking artifacts, as in the conventional case. In our object-based approach, a lossless technique has been used to encode the facial region. Only a few minor degradations can be observed in the results of Figure 3 b). The left collar region, and the hair to the right of the facial area are slightly distorted, however, these areas are not visually as important as the critical information conveyed by the facial features. In addition to the improved subjective quality, only 240 motion vectors are required to be transmitted as side information along with the encoded facial area (approximately 500 bits for lossless compression). Future improvements to this technique must focus on a more robust estimation of the motion vectors for the cases of covered and uncovered regions.

6 Conclusions

A object-based warping technique was examined for the temporal prediction of video sequences. In this scheme, a set of control points were determined at the encoder to represent the image contents within a video sequence. These points were selected along the edges of a set of regions that were obtained by a color segmentation method. The facial region was separately encoded to avoid distortions of the facial features. A modified block matching method was then employed at the encoder in order to assign the appropriate motion vectors to the selected points. At the decoder, the same segmentation and control point selection algorithm were used to obtain the control points in the received current frame. These points were subsequently spatially shifted according to the motion vector information. A warping algorithm using an affine transformation was finally used to form the predicted frame. A significant subjective improvement was found in the predicted image using this technique when compared to the conventional block matching approach. Furthermore, the improved visual quality was achieved with only a moderate increase in the transmission overhead and computational complexity of the coding scheme.

References

1. L. Chiariglione, 'MPEG and Multimedia Communications', *IEEE Transactions on Circuits and Systems for Video Technology*, Vol. 7, No. 1, pp. 5-18, February 1997.
2. H.G. Musmann, M. Hotter, J. Ostermann, 'Object-oriented analysis-synthesis coding of moving objects', *Signal Processing: Image Communication*, Vol. 1, No. 2, pp. 117-138, October 1989.
3. M. Hotter, 'Object-oriented analysis-synthesis coding based on moving two-dimensional objects', *Signal Processing: Image Communication*, Vol. 2, No. 4, pp. 409-428, December 1990.
4. J. Nieweglowski, J. Campbell, P. Haavisto, 'A novel video coding scheme based on temporal prediction using digital image warping', *IEEE Trans. on Consumer Electronics*, Vol. 39, no.3, pp. 141-150, 1993.
5. G. Sullivan, 'Motion Compensation for video compression using control grid interpolation', *IEEE Int. Conf. on ASSP*, pp. 2713-2716, 1991.
6. V. Seferidis, M. Chanbari, 'Generalized block matching motion estimation', *Visual Communications and Image Processing*, SPIE Vol. 1818, pp. 110-119, 1992.
7. J. Nieweglowski, T. Moisala, P. Haavisto, 'Motion compensated video sequence interpolation using digital image warping', *IEEE Int. Conf. on ASSP*, Vol. 5, pp. 205-208, 1994.
8. N. Herodotou, A.N. Venetsanopoulos, 'Image Segmentation for Facial Image Coding of Videophone Sequences', *13th International Conference on Digital Signal Processing*, Santorini, Greece, July 1997.
9. R. Ohlander, K. Price, D.R. Reddy 'Picture Segmentation Using a Recursive Region Splitting Method', *Computer Graphics and Image Processing* Vol. 8, pp. 313-333, 1978.
10. A. Witkin, 'Scale-space Filtering', *Proceedings IJCAI-83*, pp. 1019-1022, Aug 1983.
11. G. Wolberg 'Digital image warping', *IEEE Computer Society Press*, Los Alamitos, California, 1990.

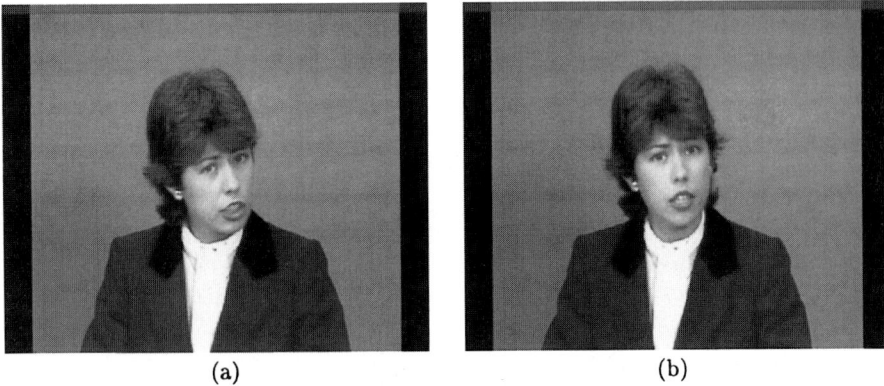

Fig. 1. a) Original Claire, Frame 100, b) Original Claire, Frame 110.

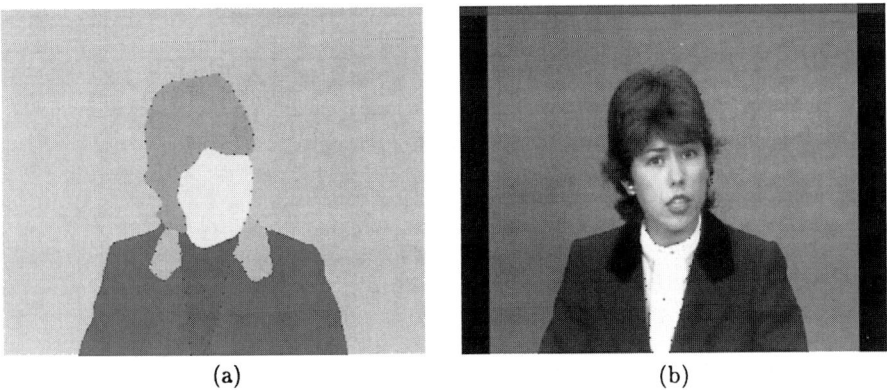

Fig. 2. a) Segmentation and control point selection, b) Matching of control points.

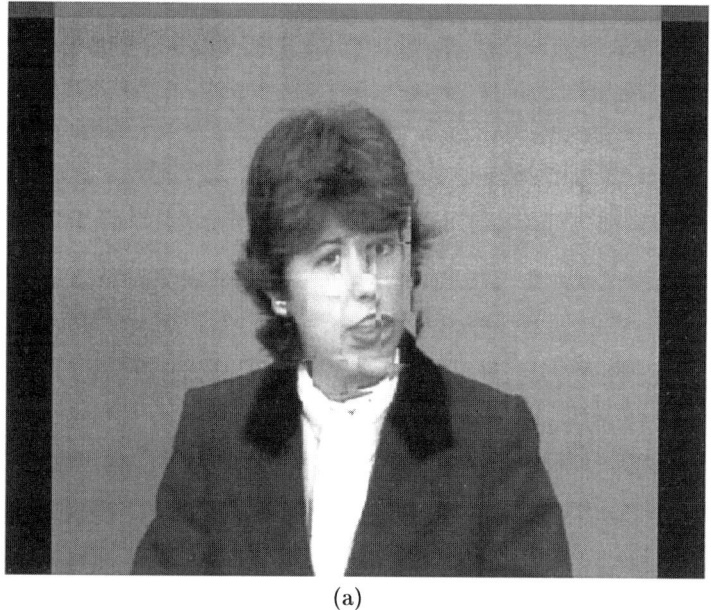

(a)

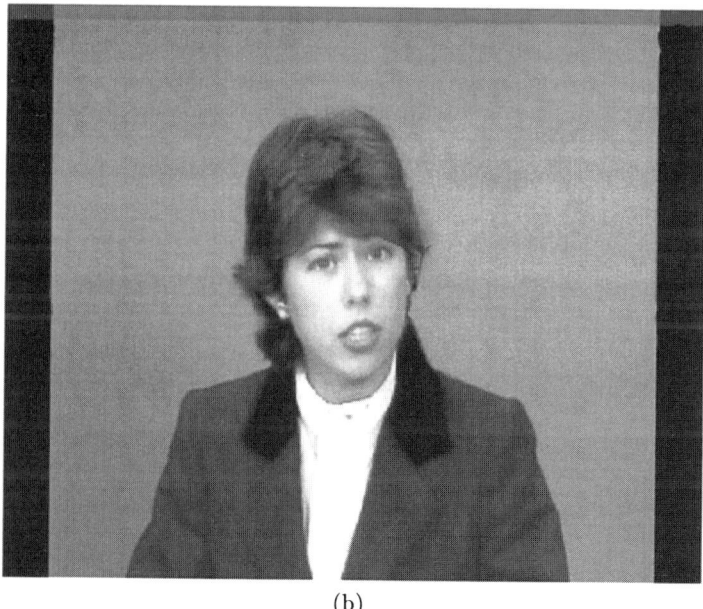

(b)

Fig. 3. a) Conventional block matching prediction of Frame 110, b) Object-based warping prediction of Frame 110.

Motion and Intensity-Based Segmentation and Its Application to Traffice Monitoring*

Jorge Badenas[1], Miroslaw Bober[2], Filiberto Pla[1]

[1] Dept. Informática, Universitat Jaume I, Castellón 12071 (SPAIN)
[2] Dept. Elec. & Elec. Eng., University of Surrey, Guildford, Surrey GU2 5XH,(U.K.)

Abstract. This paper is concerned with an efficient estimation and segmentation of 2-D motion from image sequences, with the focus on traffic monitoring applications. In order to reduce the computational load and facilitate real-time implementation, the proposed approach makes use of simplifying assumptions that the camera is stationary and that the projection of vehicles motion on the image plane can be approximated by translation. We show that a good performance can be achieved even under such apparently restrictive assumptions. To further reduce processing time, we perform gray-level based segmentation that extracts regions of uniform intensity. Subsequently, we estimate motion for the regions. Regions moving with the coherent motion are allowed to merge. The use of 2D motion analysis and the pre-segmentation stage significantly reduces the computational load, and the region-based estimator gives robustness to noise and changes of illumination.

1 Introduction

Motion estimation and segmentation is one of the fundamental problems in image sequence processing. In recent years, as a result of advances in information technology both in terms of computational power and cost, it has become possible to use computer vision techniques to solving many everyday problems. A good example of such a task is traffic monitoring. Traffic monitoring system should be able to carry out many operations including estimation of traffic mean velocity, counting the number of moving vehicles, tracking individual vehicles and detecting whether traffic is moving freely or not.

One of the crucial components of the traffic monitoring system is the motion analysis algorithm. The task is difficult for several reasons: there are multiple moving objects, the objects of interest are usually small (in the image plane) and poorly textured, and the camera may vibrate due to the weather conditions. The illumination conditions may be poor and may change rapidly. Multiple occlusions are likely and the environment may be cluttered. But probably the most challenging is the requirement of real-time or close to real-time performance on relatively cheap hardware. These specific difficulties and constraints require that

---

* The present work was supported in part by the projects ESPRIT PROJECT EP-21007 IOTA and CICYT TIC95-0676-C02-01

a standard of-the-shelf algorithm cannot usually be applied and a dedicated algorithm must be designed. In this paper we present a novel algorithm, which modifies and combines several known approaches to image segmentation and motion estimation. Before we explain the main premises behind our approach, we very briefly review the state of the art in motion estimation.

Motion analysis methods are usually divided into two groups: gradient-based methods and feature-based methods. *Gradient-based methods* exploit the relationship between the spatial and temporal gradients of intensity through the well-known *optic flow equation* [6]. These methods only work satisfactorily in regions where there are sufficient spatial intensity variations (texture) and where the motion between frames is relatively small. Unfortunately, in the case of traffic monitoring, objects exhibit little intensity variations and may be quite small, consequently ruling-out this class of approaches. *Feature-based methods* use features extracted from the frames such as interest points, corners, lines, zero crossing points or regions to determine a motion field by analysing the change of position of features through several frames. Here again, due to small size of the vehicle in the image plane, grey-level edges and corners are short and can be barely detected. Even if features are detected, a confusion during the feature-matching stage is quite likely as the environment is highly cluttered and there are many occlusions and disocclusions. Recently, Bober and Kittler [3] developed a *region-based* motion analysis technique called RHT. They combine the Hough Transform and Robust Statistical kernels. It has been shown that RHT can extract motion parameters (such as displacement or parameters of affine transformation) from two frames with excellent accuracy. Moreover, the technique offers simultaneous motion estimation and segmentation and is computationally efficient.

Great deal of work has been done in the area of motion estimation and segmentation to traffic monitoring. We shall very briefly mention several contributions that show the current trend. Dubuisson and Jain [4] describe a technique that combines motion and colour segmentation. In the paper [5] a system for traffic monitoring is described which fit 3-D wireframe models of generic vehicles to the vehicles projected onto the image plane. A similar approach is shown in [7] where the model of a generic car is projected from the 3-D scene onto the 2-D image. Methods for tracking of 2-D contours have been proposed in [2] and [1]. Weber at al [8], presented other method which represents contours by curves. It employs a contour tracker based on intensity and motion boundaries which uses two Kalman Filters.

The main premise behind our approach are that for the application at hand, an efficient and robust region-based motion analysis should be used. It is expected that region-based estimation is more robust that feature-based approaches, and is also likely to outperform the contour based trackers, which rely solely upon the outer contour of the tracked object. To further aid motion segmentation and reduce computational cost, we propose to apply a grey-level based segmentation as the pre-processing stage. Such approach removes motion-based ambiguity in regions of uniform grey-level and speeds-up motion segmentation by reducing the number of passes in multi-pass stage.

The paper has the following structure. In next section, we present an outline of the algorithm. Sections 3, 4 and 5 describe in greater detail three stages of processing, namely pre-segmentation, motion analysis and post-processing. Section 6 shows the experimental results on real-word sequences and, finally, conclusions are drawn is section 7.

2 Outline of the Algorithm

The algorithm consists of three stages: pre-processing involving gray-level based segmentation, motion analysis (including further segmentation if needed) and post-processing stage, where regions are allowed to merge.

The segmentation of the reference image is designed to group pixels of similar gray-levels. Since the road surface and the cars are usually poorly textured, the gray-level segmentation is likely to group pixels belonging to objects (cars) or background (road).

The first step of motion analysis stage attempts to reduce the number of regions in order to simplify the subsequent operations. This reduction is based on difference images analysis, and, at least, it allows merging the majority of the static regions.

Motion estimation is applied to the rest of regions. The motion estimator uses a translational motion model. Although more complicated motion models are probably more appropriate for the road-traffic sequences, the translational model is computationally less expensive and can still cope when the scaling effect is small compared to the translation.

The postprocessing stage, uses the spatial neighbourhood relations between regions to improve the final segmentation. If two neighbouring regions have similar motion parameters they are likely to belong to the same moving object.

To this end we obtain a 2D segmentation map with the large background region and smaller foreground regions (cars). An accurate estimate of the translation is given to each region.

3 Segmentation of the Reference Frame

The purpose of the first stage is to obtain groups of pixels with similar intensity. Motion will be estimated for these regions, so it is very important that the pixels which are grouped are likely to belong to a single object.

The clustering algorithm used is a modification of the technique developed by Kottle and Sun and described in [9]. This technique is an adaptation of the classical k-means algorithm, employing a three-dimensional space of features: the two image coordinates and the pixels intensity. Pixels are assigned to one of the clusters in an iterative process. At each iteration a pixel i is assigned to a cluster j which minimises the following criterion:

$$E^{ij} = (\overline{p}^i - \overline{m}^j) W^j (\overline{p}^i - \overline{m}^j) \qquad \text{for } j \in \{1, 2, ...k\} \qquad (1)$$

where $\overline{p}^i$ is a vector composed of the coordinates and the intensity of the pixel i, $\overline{m}^j$ is the vector which contains the mean coordinates and mean intensity of the cluster j, k is the number of clusters in the image and W^j is a weight matrix that makes the algorithm to adapt itself to the image.

The original algorithm has an important drawback, namely it requires that the number of regions (clusters) is provided as an input parameter. It is very difficult to predict reliably how many regions are needed, because it depends not only on the number of moving objects, but also on their size. This is because the clustering tends to favour a uniform distribution and the average size of the clusters depends on their number. If too small number of regions is present, the clustering may group pixels from different objects, but too large a parameter used will cause over-segmentation and increase in computational load.

Our approach removes this problem by introducing a multistage segmentation where the original image is segmented initially into a relatively small number of clusters, and each cluster in turn is considered for further segmentation. Since in traffic scenes the road and cars do not exhibit much texture, the decision if a cluster should be further divided is based on the cluster intensity variance σ_j. In addition, since it is very difficult to estimate motion for a very small region reliably, a minimum region-size μ is used to prevent over-segmentation. On the other hand, large regions with a small intensity variance are also not desirable, since they may contain a small region of different intensity. Therefore, a maximum region size η is also restricted. The cluster is divided if the following condition is fulfilled: $(N_j \geq \eta$ or $(N_j \geq \mu$ and $sigma_j^2 \geq \sigma_t))$ where N_j is the number of pixels in cluster j and σ_t is the variance threshold. The following values were used in experiments: $\sigma_t = 12$, $\mu = 1500$ and $\eta = 150$ (image size 192x144 pixels). The selection of the parameter values is somewhat arbitrary, and will depend on the image size, and the minimum size of the object (on the image plane) that should be detected. The parameters are constant for a given system (eg fixed image resolution and camera location).

The proposed modification makes the technique more adaptable to the content of the images, and the final number of clusters is no longer fixed. We still have to specify the initial number of clusters and the number of divisions per iteration, but the results are not sensitive to the value of these parameters. Furthermore, the minimum size of the cluster is now restricted, preventing from extreme over-segmentation. We initially create 6 to 10 clusters and every cluster that passes the division test is again subdivided into 4 clusters. When the size of a cluster is smaller than two times η, then the cluster is only divided into 2 clusters. The technique does not guarantee that all pixels are spatially-connected and we need to perform connected component analysis.

4 Motion Estimation

Since motion estimation is computationally heavy, one does not want to apply it to stationary regions. Therefore, in the first stage we perform a simple and crude test on each of the pre-segmented regions to determine if they are stationary or not. The test is based on a simple observation that if there is an intensity edge between two regions and at least one of them is moving, then the frame difference for some pixels on the boundary is large. (namely for the fractions of edges which are perpendicular to the direction of motion). We follow pixels along the boundary and calculate what proportion of them has large frame difference.

In this step, static regions are merged and those regions that were divided by the clustering algorithm but in fact form a single region without a substantial discrepancy in the intensity of their pixels.

The region-based motion estimation involves finding the parameters of translation that minimises the sum of displaced frame differences (DFD), transformed by a robust kernel ρ. The summation is over all pixels from the reference region. In fact, we are minimising an error measure E defined as follows:

$$E_i(dx, dy) = \frac{1}{N_j} \sum_{(x,y) \in Cluster_j} \rho(I_1(x,y) - I_2(x+dx, y+dy), \alpha, \lambda) \quad (2)$$

where $I_1(x,y), I_2(x,y)$ are the pixel intensity values at location (x,y) in the reference and consecutive frames respectively. $\rho()$ is the robust redescending function and α, λ are the function parameters. When multiple motions are present within a region, the pixels that are not consistent with the dominant motion may bias the estimate. These pixels are referred to as outliers. Application of the robust kernel ρ reduces the influence of outliers so that they will not affect the value of motion estimate. We have used the following kernel due to its low cost:

$$\rho(x, \alpha, \lambda) = \begin{cases} \lambda |x| & if \ |x| < \frac{\alpha}{\lambda} \\ \alpha & otherwise \end{cases}$$

Our approach is a variant of the steepest descent algorithm but it requires less computations. At each iteration, the value of the error function $E_i(dx, dy)$ is compared to the values of E_i computed for eight modified displacements: $(dx + k*r, dy + l*r), k, l \in \{-1, 0, 1\}$ $kl \neq 0$, where r is the current resolution. The procedure terminates when the value of E_i cannot be further improved by modifications to (dx, dy).

To avoid the local minima of the E_i function and to accelerate the process, we use a multiresolution approach. A coarse value of the motion parameters is calculated from the image sequence at coarse resolution. This value is then used as a starting point for iterations at finer resolution. At the finest resolution we obtain the parameters of the translation to subpixel accuracy (0.1 pixel). Bilinear interpolation is used to approximate intensity value at inter-pixel locations.

5 Final Motion Segmentation

This final stage attempts to merge regions moving with coherent motion. This stage is needed, since the initial greylevel based segmentation may split object into smaller regions. All pairs of adjacent regions are considered as candidates

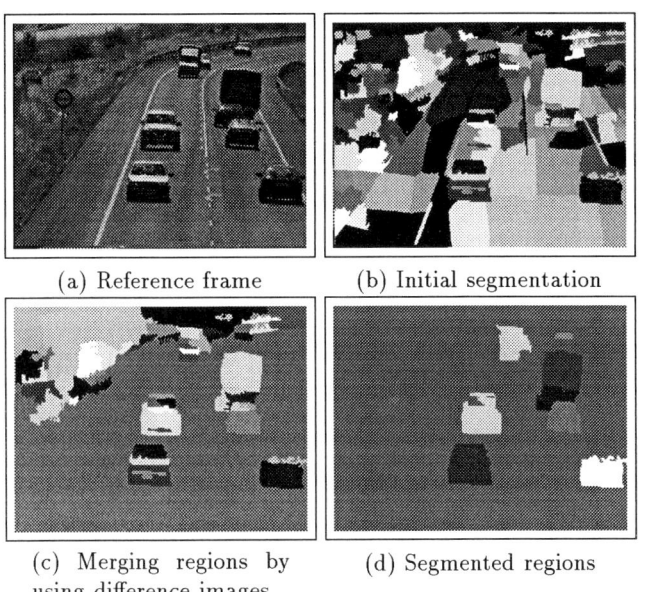

(a) Reference frame (b) Initial segmentation

(c) Merging regions by using difference images (d) Segmented regions

Fig. 1. Segmentation process of a sequence with nine vehicles.

to be merged. The two regions, say A and B, are merged if at least one of the following conditions is satisfied:

$$(E_{AB} \leq Q_1) AND (E_{AB} \leq E_{AA} + Q_2) \quad (3)$$
$$(E_{BA} \leq Q_1) AND (E_{BA} \leq E_{BB} + Q_2) \quad (4)$$

where E_{XY} is the error function for the region X displaced with motion parameters calculated for region Y, and Q_1 and Q_2 are two positive constants. Values assigned to Q_1 and Q_2 are: $5 \leq Q_1 < 9$ and $0 \leq Q_2 \leq 1$. This process is repeated for all pairs of adjacent regions until no pair can be merged.

6 Results

Figure 1 presents the segmentation for a sequence in which 9 vehicles are moving in both directions. Subfigure a is the reference frame, subfigure b is the result of the gray-level based segmentation, subfigure c is the result of the reduction of regions step, and subfigure d is the final motion segmentation.

For this sequence, it can be seen that eight vehicles, and not nine, are detected. There is a car that is moving onto the lane on the left that is united to the van that is hiding part of it. Due to the small relative velocity difference that exits between the velocities of both vehicles and their distant position in relation to the camera, we believe that this can not be considered as a defect of the algorithm. In any case, the algorithm separates both vehicles after a few frames.

For the rest of vehicles shown in figure 1 we can see that they are correctly segmented, in spite of some of them hide parts of other vehicles.

In both sequences the camera was vibrating slightly, provoked by the wind. However, the algorithm can cope with these situations. Note that the method has successfully segmented the moving vehicles.

In subfigure 1.c the importance of the reduction of number of regions step can be seen. This step allows grouping 60-80% of the static pixels of the image into one region. Thus, the subsequent operations are carried out more easily and using less time.

In the first stage, the parameters initial number of regions, σ, μ and η were chosen as 6, 12.0, 300, and 4000, respectively. In the motion segmentation stage, the parameters Q_1 and Q_2 received the values 8.0 and 1.0, respectively.

Table 1 represents the computational costs of the different parts of the algorithm when using diferent configurations for this sequence. These times have been measured on a Hewlett Packard workstation Apollo Model 725/75 with a processor PA-RISC 7100 (75 MHz).

The first row of the table shows the costs when the size of the image is 384x288. We can see that the computational cost of the two first stages is large in comparison with the third stage. The total cost is 47.8 seconds per frame. The algorithm can be speeded up by calculating an 'Activity Map' [10] that marks the pixels that are always static. These pixels are discarded since they never will be occupied by a vehicle. The time of computation of the algorithm when the 'Activity Map' is used is shown in the second row. Now, it has been reduced to 26.63 seconds. We can see that the cost of the *clustering stage is significatively reduced. However, the reduction in the* Motion Estimation stage is not so large, because the cost of the static pixels is small compared with the cost of estimating the motion for the moving pixels.

	Clustering	Motion estimation	Motion segmentation	Total
384x288	15.87	21.69	9.88	47.8
384x288 and A.M.	5.84	14.54	6.25	26.63
192x144 and A.M.	1.55	5.45	1.03	8.03

Table 1. Computational cost (seconds) of the algorithm.

Since 384x288 can be considered as a big image size, and the algorithm does not require so much resolution, we can reduce the computational cost of the algorithm by reducing the size of the images. The third row of the table 1 shows the computational costs of the algorithm when using a 192x144 image size and an 'Activity Map' (A.M.). Now, the total time of computation is 8.03 seconds.

7 Conclusions

We have presented an approach for motion analysis in traffic scenes, which segments moving vehicles and estimates their velocities in the image plane. Our approach is a novel combination of several existing techniques and algorithms.

The clustering algorithm is an improved variant of the method presented in [9] in which we have carried out modifications. With these modifications is not necessary to indicate the final number of regions. We have also developed a method based on difference images which reduce the number of regions by uniting almost all the static regions of the image into a stationary background. The motion estimation process is based on finding the motion that minimizes the gray level difference between the motion-compensated pixels in the original frame and corresponding pixels in the consecutive frame. The estimation is performed with sub-pixel accuracy and uses a multi-resolution approach that allows to avoid the local minima of the *Displaced Frame Difference* function and speeds up the computational process. In order to achieve a reliable motion estimation, we use a *redescending robust kernel*. The final motion segmentation is achieved by merging regions moving with coherent motion.

We have demonstrated that a motion analysis based on a translational model is sufficient for the segmentation purpose when scaling effect of the motion is small compared with translation, and therefore it is not necessary to employ a more complex model such as affine or perspective.

The proposed method proved to be robust to camera vibrations and it copes well with multiple moving objects. Unlike some other techniques it does not use feature points and it consequently works well in a cluttered environment.

References

1. A. Blake, R. Curwen, and A. Zisserman. Affine-invariant contour tracking with automatic control of spatiotemporal scale. In *Proceedings of the Fourth International Conference on Computer Vision, Berlin, May 1993*, pages 66–75, 1993.
2. A. Blake, R. Curwen, and A. Zisserman. A framework for spatio-temporal control in the tracking of visual contours. pages 127–45, 1993.
3. M. Bober and J. Kittler. Estimation of general multimodal motion: an approach based on robust statistics and Hough transform. *Image and Vision Computing*, 12(12):661–668, 1994.
4. M.P. Dubuisson and A.K. Jain. Contour extraction of moving objects in complex outdoor scenes. *International Journal of Computer Vision*, 14:83–105, 1995.
5. J.M. Ferryman, A. D. Worrall, G.D. Sullivan, and K.D. Baker. A generic deformable model for vehicle recognition. In *BMVC95*, pages 128–136, 1995.
6. B.K.P. Horn and B.G. Schunck. Determining optical flow. *Artificial Intelligence*, 17:185–203, 1981.
7. D. Koller, K. Daniilidis, and H.H. Nagel. Model-based object tracking in monocular image sequences of road traffic scenes. *I.J.Computer Vision*, 10:257–281, 1993.
8. D. Koller, J. Weber, and J. Malik. Robust multiple car tracking with occlusion reasoning. In Jan-Olof Eklundh, editor, *Proceedings, 5th European Conference on Computer Vision, (Berlin, 1994)*, pages 189–196. Springer-Verlag, 1994.
9. Kottle and Sun. Motion estimation via cluster matching. *PAMI*, 16, 1994.
10. B.D. Steward, I. Reading, M.S. Thomson, T.D. Binnie, K.W. Dickinson, and C.L. Wan. Adaptative lane finding in road traffic image analysis. pages 133–136. IEEE Conference Publications, 1994.

A Geometrically Deformable Contour Model

Ahmed Raji, Eric Petit, Jacques Lemoine[1]
and Salim Djeziri[2]

[1] LERISS Laboratory, EEA Department, University of Paris12, av Gal De Gaulle, 94010 Creteil, France (e-mail: raji@univ-paris12.fr)
[2] Trois Rivieres University, Departement Mathematiques-Informatique, C.P.500, Trois Rivieres, Quebec, Canada

Abstract. In this paper, we present a discrete dynamic contour model which simulates a liquid expansion on homogeneous flat surface and the outlining of encountered obstacles. It is implemented as a radial expansion of an initial closed curve which determines a reference region. The curve is locally stopped when it attempts to expand in regions which do not exhibit the same features as the reference region. The main interesting properties of this model are : it does not require an initialization close to the solution, it is able to detect several objects of an image just by performing the initialization on the background of the image and it allows short processing times compared to methods based on optimization procedures. Nevertheless, this approach involves some delicate algorithmic problems that are exposed and solved. Experimental results on real and synthetic images are presented.

keywords: deformable contours, edge detection, image segmentation, region growing.

1 Introduction

Image segmentation refers to the decomposition of a scene into its components usually called objects. It is one of the most significant problems of image analysis because segmentation is a necessary step before carrying out scene interpretation or recognition. Image segmentation can be performed by detecting object contours that are locally characterized by image attribute variations. However, most of edge detection methods are based on local derivative operators which are very sensitive to high frequency noise.
Kass et al. [1] [2] have introduced deformable contour models as a new approach to find object contours. The main idea consists in the deformation of an initial closed curve by a variational computation that minimizes an energy function related to internal and external forces acting on the deformable curve. Many definitions of the deformation constraints have been proposed leading to various deformable models [3] [4] [5] [6] [7].
In the present paper, we propose a geometrically deformable contour model that

simulates a liquid expansion and leads to the outlining of the objects of the image. The model is implemented as a deformation of a discrete closed curve which is iteratively expanded on regions presenting the same features as the reference region determined by the initial curve.

2 Deformable contour model

In image segmentation it is wished to represent a contour of an object by a closed curve. In a continuous 2D space, this curve is made of an infinite number of points, and it divides the plane into two regions (internal and external). In a bounded discrete 2D space, a closed curve is described by a linked list made of a finite number of points, with the last point connected to the first one. In the following sections, the curve points are called vertices.

2.1 Liquid expansion simulation

To initialize the process, the user only need to point at the object to segment. At this place, we put a basic discrete curve constituted of three equidistant vertices. The pixels belonging to this triangle determine a reference region. Then, an iterative deformation of the curve is carried out by moving the position of its vertices in a radial direction. Vertices get a blocked status when they attempt to expand on regions which do not present the same features as the reference region. The mobility of each vertex also depends on the status of its neighbours in order to control the curvature. The deformation process is completed when all the vertices are blocked.

When the process is initiated on the background of the image (rather than inside an object) the expansion allows to simultaneously delineate all the objects of the scene. In such a case, the algorithm simulates the expansion of a liquid on a flat surface and the outlining of objects placed on this surface. Implementing this geometrical algorithm on a discrete 2D space involves two main difficulties. On the one hand, the deformed curve can generate intersection with itself. On the other hand, the resolution of the curve decreases during the expansion process. Therefore, after each iteration of the deformation process, the intersections of the curve have to be detected and suppressed by mean of a reconstruction process while its resolution is updated by a resampling process. All of these issues are explained in the next sections.

2.2 Deformation process

In a cartesian coordinate system, the position of a vertex at iteration k of the deformation process is represented by the vector $\vec{P}_{i,k} = (x_{i,k}, y_{i,k})$. The deformable contour can then be defined by :

$$(C_k = \{\vec{P}_{i,k}; i = 1, ..., N_k\}; k = 1, ..., M)$$

where N_k represents the number of vertices at iteration k, and M the number of iterations of the deformation process.

The deformation is performed by iterative radial displacements of the vertices of the curve. The new position of a vertex at the next iteration is obtained by a geometrical transformation of its curent position expressed as follows :

$$\vec{P}_{i,k+1} = T(\vec{P}_{i,k}, \alpha_{i,k}, \vec{\eta}_{i,k})$$

where $\alpha_{i,k}$ represents the magnitude of the displacement of the vertex $\vec{P}_{i,k}$, and $\vec{\eta}_{i,k}$ the local radial unitary vector of the curve at $\vec{P}_{i,k}$.

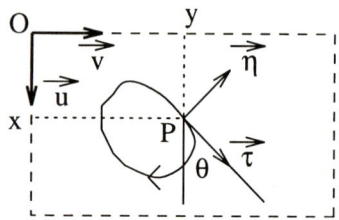

Fig. 1. Local radial unitary vector of a continuous curve

As illustrated in fig.1, in a continuous 2D space $(O, \vec{u}, \vec{v})$, the local tangent unitary vector of a closed continuous curve can be expressed at $P = (x, y)$ in the local coordinate system $(P, \vec{u}, \vec{v})$, with regard to the orientation of the curve, by :

$$\vec{\tau} = (\cos\theta, \sin\theta)$$

with:

$$\theta = \arctan \left.\frac{dy}{dx}\right\|_P$$

The normal unitary vector of the curve can be derived from $\vec{\tau}$ by a rotation of $-\frac{\pi}{2}$ radians :

$$\vec{\eta} = \begin{pmatrix} 0 & -1 \\ 1 & 0 \end{pmatrix} \vec{\tau} = (-\sin\theta, \cos\theta)$$

In the discrete case, the derivative dy/dx can be numerically approximated in the neighbourhood $\{\vec{P}_{i-1,k}, \vec{P}_{i,k}, \vec{P}_{i+1,k}\}$ by:

$$\left.\frac{dy}{dx}\right\|_{P_{i,k}} = \frac{y_{i+1,k} - y_{i-1,k}}{x_{i+1,k} - x_{i-1,k}}$$

The radial displacement of the vertex $\vec{P}_{i,k}$ leads to the new position given by :

$$\vec{P}_{i,k+1} = \vec{P}_{i,k} + \alpha_{i,k} \vec{\eta}_{i,k}$$

or :

$$\begin{cases} x_{i,k+1} = x_{i,k} - \alpha_{i,k} \sin \theta \\ y_{i,k+1} = y_{i,k} + \alpha_{i,k} \cos \theta \end{cases}$$

Some examples of such vertex translations are graphically represented in fig.2. At the end of the deformation process, the final result which is the contour of the object is truncated and reported on the discrete bounded space represented by the image.

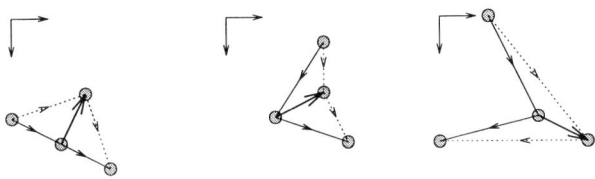

Fig. 2. Examples of radial vertices translations ($\alpha=2$)

2.3 Deformable contour reconstruction process

In the proposed model, the contour of the object is found by iterative deformations of an initial closed curve. This model is valid as long as the deformable curve does not intersect itself. Such situations may occur when the deformable curve has to outline a small size region (representing either a subregion of the object or an artefactual region due to the presence of noise -fig.3-).

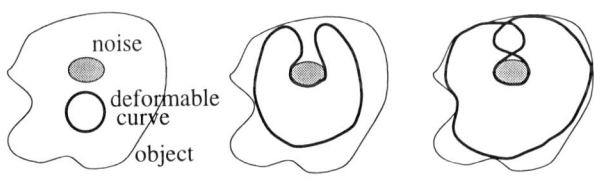

Fig. 3. Generation of secondary closed curves

Intersections are detected and processed by a reconstruction process after each iteration of the curve expansion. Let us consider the previous example reproduced in details in fig.4. The curve deformation leads to the intersection of the segment $[P_i, P_{i+1}]$ with $[P_{i+m}, P_{i+m+1}]$. The reconstruction process consists in changing the vertex connexions by interverting P_{i+1} and P_{i+m+1}. Thus, when all the curve intersections are processed in this way, we obtain two closed curves with positive orientation and one closed curve with negative orientation.

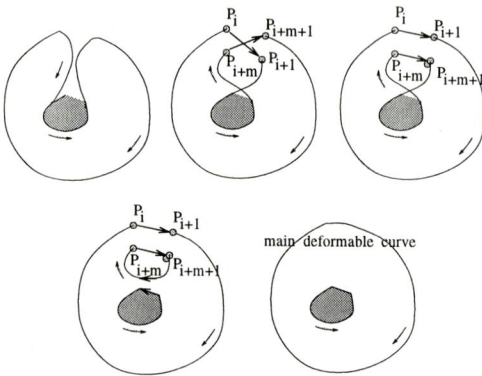

Fig. 4. Reconstruction process

Closed curves having negative orientation correspond to secondary curves generated by the main curve expansion. Furtheremore, since the direction of a vertex displacement is determined with regard to the orientation of the curve, negative orientation means that the vertices will now be displaced toward the interior of the closed curve. Thus, the next iterations of the expansion process will produce a contraction of the deformable curve which leads to the detection of the contour of the outlined object.

All the positively orientated closed curves are artefactual and must be suppressed except the external one denoted C_k that corresponds to the main deformable curve.

The orientation of secondary curves is determined by the sign of $\overrightarrow{AB} \wedge \overrightarrow{BC}$ where the vertices A, B and C are randomly chosen from the linked list describing the considered curve.

2.4 Model resolution

The length of the line segment $[P_i, P_{i+1}]$, denoted d_i, represents the local contour resolution. If d_i is large, the model will not be able to outline accurately objects having irregular frontiers. In addition, the length d_i is iteratively modified by the curve deformation. This involves variation in the local model resolution. To keep this resolution close to a user specified value, the lengths of the edge segments are evaluated at each step of the contour deformation, and updated by a resampling process which consists in inserting additional vertices in the deformable contour.

In fact, the model resolution can be characterized by the maximum distance d allowed between neighbouring vertices. The resampling process consists in checking after each deformation step the entire contour for segments longer than the maximum length d. As illustrated in fig.5, such an edge segment is divided into two shorter ones of equal length by inserting a new vertex at the middle of this segment.

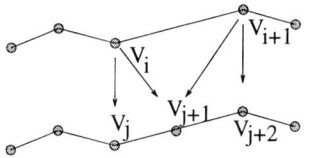

Fig. 5. Model resolution updating

2.5 Controlling the expansion by a region growing strategy

To control the expansion of the curves, we assess the similarity between the features of the reference region and those of the adjacent regions to each vertex. The adjacent region to the vertex $P_{i,k}$ is determined by the set of pixels $\{P_{i-1,k}, P_{i-1,k+1}, P_{i,k+1}, P_{i+1,k}, P_{i,k}\}$ (fig.6-1).

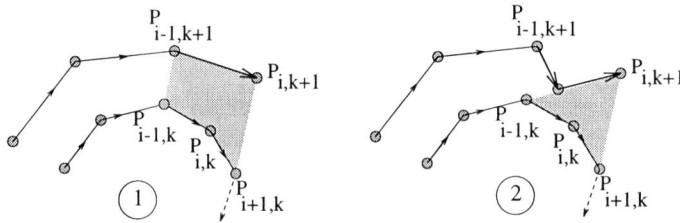

Fig. 6. Controlling the expansion by assessing the similarity between the reference region and the adjacent regions to the deformable contour

When the similarity criterion is satisfied, the vertex $P_{i,k}$ is displaced to $P_{i,k+1}$. Otherwise, a smaller region defined by $\{P_{i-1,k}, P_{i,k+1}, P_{i+1,k}, P_{i,k}\}$ (fig.6-2) is considered. If the similarity criterion is satisfied in this case, in addition to displacing the vertex $P_{i,k}$ to $P_{i,k+1}$, a new vertex is inserted at the intersection between the edge segments $[P_{i-1,k+1}, P_{i,k}]$ and $[P_{i-1,k}, P_{i,k+1}]$ to include only the tested region. If the similarity criterion is still not verified, the vertex $P_{i,k}$ is blocked after a dichotomic adjustment of its position on the object contour.

2.6 Interaction between vertices

Most of active contour models consider a deformation constraint deduced from the local contour curvature. In our model, this parameter is introduced through a mobilty coefficient which depends, for each vertex, on the status of its neighbours. Thus, we avoid to derive strong curvature around the blocked vertices of the deformable contour.

Given the initial deformation magnitude α of the model, each vertex $P_{i,k}$ is affected a mobility coefficient $\mu_{i,k} \in [0, 1]$ so that the magnitude of the displacement of $P_{i,k}$ is given by :

$$\alpha_{i,k} = \mu_{i,k}\, \alpha$$

At the first step of the deformation process, the mobility coefficient is set to 1 for all the vertices. During the next iterations of the curve deformation, the mobility coefficient $\mu_{i,k}$ remains constant as long as $P_{i,k}$ has no blocked vertices in its vicinity. When a vertex is blocked, on the one hand its mobility becomes null, and on the other hand it influences the neighbouring vertices in such a way that it slows down their displacements by reducing their mobility coefficients. This interaction can be expressed by the following relationship :

$$\mu_{i,k} = \mu_{i,k-1} F_{i,k}$$

where $F_{i,k}$ is a decreasing function which depends on the number of blocked vertices in the neigbourhood of $P_{i,k}$. $F_{i,k}$ is defined as follows :

$$F_{i,k} = 1 - \sum_{j, P_j blocked} e^{-\rho s_j}$$

where ρ is a positive constant which determines the size of the neighbourhood of $P_{i,k}$, and s_j is the length of the arc linking $P_{i,k}$ to the blocked vertex P_j.

3 Results

The examples of fig.7 show a free deformation (with no constraints). The vertices of the deformable contour are affected a constant mobility coefficient $\mu = 1$ during the deformation process.

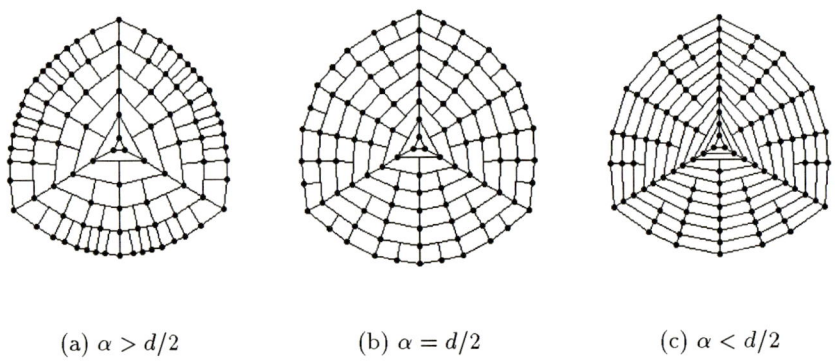

(a) $\alpha > d/2$ (b) $\alpha = d/2$ (c) $\alpha < d/2$

Fig. 7. Free expansion of the model

We note that the curve deformation is quasi-uniform. The shape and dimension of the elementary regions merged by the contour deformation depend on the model resolution d and on the deformation magnitude α.

Fig.8 shows the results obtained by the deformable contour model on a leaver image where it is wished to detect nodules. The similarity criterion is evaluated as follows : let V_r represent the grey level average of the reference region and $V_{i,k}$ that of the adjacent region to the vertex $P_{i,k}$; the similarity criterion is verified if $|V_r - V_{i,k}| \leq th$ (where th is a threshold level interactively set by the user).

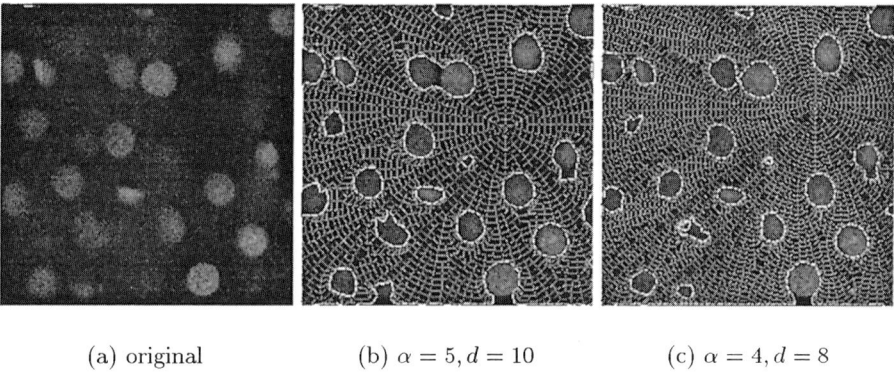

(a) original (b) $\alpha = 5, d = 10$ (c) $\alpha = 4, d = 8$

Fig. 8. Examples of edge detection

In this example, the initial contour is placed on the background of the image. At the end of the deformation process, the main deformable curve corresponds to the image border, while the contours of the objects localized on the image background are detected by secondary generated closed curves. The example illustrates the influence of the parameters α and d on the ability of the model to separate very close objects. In such a situation, a large value of d does not allow the deformable contour to pass between the two objects, which are detected as a single object (fig8-b). With a larger value of d, the two objects can be separated (fig8-c). This example shows how the parameter d can be interpreted as a viscosity coefficient.

The model performances are illustrated in fig.9 in the case of a scanographic image of the heart. We note the good detection of the contours of the two ventricles.

4 Conclusion

We present a deformable contour model based on a geometrical deformation, which leads to a polygonal description of the contours. The initialization consists in pointing at the object to be segmented or on the background of the image. The model simulates a liquid expansion by an isotropic deformation of a closed curve.

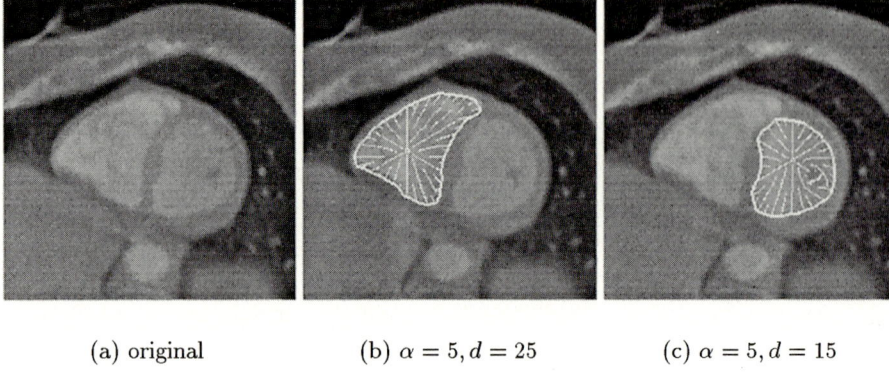

(a) original   (b) $\alpha = 5, d = 25$ (c) $\alpha = 5, d = 15$

Fig. 9. Detection of the endocardial contour

Two parameters are required to guide the dynamic deformation process : the radial deformation magnitude α, and the model resolution d. These parameters stand for the "liquid viscosity" that regulates the ability of the model to follow variations of small scale in the image features. The deformation constraints are inspired by region growing techniques, which leads to an algorithm that combines active contour models with region-based segmentation methods. The model is presented in a general framework, but it can be adapted to particular images. A method to quantify the interaction between the contour points is introduced to adapt the deformable contour curvature to the shape of the object contour. For processing a particular class of images, the model can be improved by adapting the similarity criterion to the characteristics of the image.

References

1. D. Terzopoulos, J. Platt, A. Barr, and K. Fleischer, *Elastically deformable models*, Comput. Graphics, vol. 21, no. 4, pp. 205-214, 1987.
2. M. Kass, A. Witkin, and D. Terzopoulos, *Snakes: Active contour models*, Int. J. Comput. Vision, pp. 321-331, 1988.
3. A. Witkin, K. Fleischer, and A. Barr, *Energy constraints on parameterized models*, Comput. Graphics, vol. 21, no. 4, pp. 225-232, 1987.
4. A. A. Amini, T. E. Weymouth, and R. Jain, *Using dynamic programming for solving variational problems in vision*, IEEE PAMI, vol. 12, no. 9, pp. 855-867, 1990.
5. J. V. Miller, D. E. Breen, W. E. W. E. Lorensen, R. M. O'Bara, and M. J. Wozny, *Geometrically deformed models: A method to extract closed geometric models from volume data*, Comput. Graphics, vol. 25, no. 4, pp. 217-226, 1991.
6. Y. F. Wang, and J. Wang, *Surface reconstruction using deformable models with interior boundary constraints*, IEEE PAMI, vol. 14, no. 5, pp. 572-579, 1992.
7. S. Lobregt, and M. A. Viergever, *A discrete dynamic contour model*, IEEE Trans. Med Imag., vol. 14, no. 1, pp. 12-24, March 1995.

Non-visible Deformations

Jean-Denis DUROU[1], Laurent MASCARILLA[1], and Didier PIAU[2]

[1] Université Paul Sabatier (Toulouse III) IRIT 118, route de Narbonne 31062 Toulouse Cedex, France tel: (+33)561.556.882; fax: (+33)561.556.258 E-mail: {durou,mascaril}@irit.fr
[2] Université Claude Bernard (Lyon I) Laboratoire de Probabilités 43, boulevard du 11 Novembre 1918 69622 Villeurbanne Cedex, France tel: (+33)472.431.260; fax: (+33)472.431.266 E-mail: piau@jonas.univ-lyon1.fr

Abstract. The number of non isomorphic solutions of the eikonal equation can be infinite. We show that there can exist a whole family of non isomorphic solutions, indexed by a continuous parameter. This implies, first, that the general problem of shape from shading can be ill-posed when no additional condition on the shape is imposed. This is in contradiction with what is sometimes stated in the literature of shape from shading. Furthermore, this implies that there can exist non visible deformations of a given surface, *i.e.*, continuous deformations of the surface which do not modify the image.

Keywords: *shape from shading, eikonal equation, ill-posed problem, shape reconstruction.*

1 Introduction

The determination of the shape of a surface from a single view of this surface can be achieved thanks to several criteria: texture, shadows, contours, shading. For a non textured surface, without edge nor hidden part, and if no silhouette is visible in the image, only the last of these criteria can function. This method is called *shape from shading*. B.K.P. Horn pioneered the work in this field in 1970 [1,2] and showed that its basis equation was:

$$R(f_x(x,y), f_y(x,y)) = E(x,y), \qquad (1)$$

in which the following notations have been chosen:

- Each point (x,y) in the image is characterized by its shading $E(x,y)$;
- The shape of the viewed surface is defined by the equation $z = f(x,y)$, where the zOz' axis points towards the observer, and where $f(x,y)$ designates the height at point (x,y);
- The behaviour of the surface, according to light re-emission, is described by the function $R(f_x, f_y)$, where f_x and f_y design the partial derivatives of f in relation to x and to y.

For a surface lit from the direction of the observer, that is from the direction of the zOz' axis, it can easily be proved that $R(f_x, f_y)$ is a function of form $r(f_x{}^2 + f_y{}^2)$, which means that the shading read in the image does not depend on anything else than the slope of the surface. In this case, equation (1) can be rewritten as:

$$f_x(x,y)^2 + f_y(x,y)^2 = g(x,y), \tag{2}$$

where $g(x,y) = r^{-1} \circ E(x,y)$ is known. This equation is called the *eikonal equation*. It is only a particular case of equation (1), but many authors have shown interest in its resolution [3–5]. It is a non linear partial derivatives equation of first order in $f(x,y)$, as well as (1). It is of major interest to state whether such an equation has a finite number of non isomorphic solutions (*i.e.*, solutions which describe really different shapes) or not, in the absence of boundary conditions, that is if one does not *a priori* know the height at any point in the image.

In this paper, we concentrate on the surface S_1, defined by the equation $z = 2x^2 + y^2$. This surface is represented in figure 1.

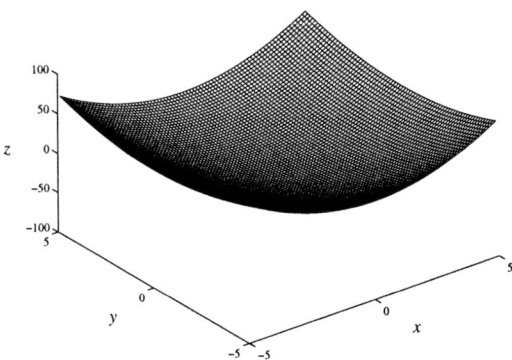

Fig. 1. Surface $z = 2x^2 + y^2$.

Its image, computed with the Matlab 4.2 software, under the hypotheses of front lighting and of diffuse reflection, is represented in figure 2.

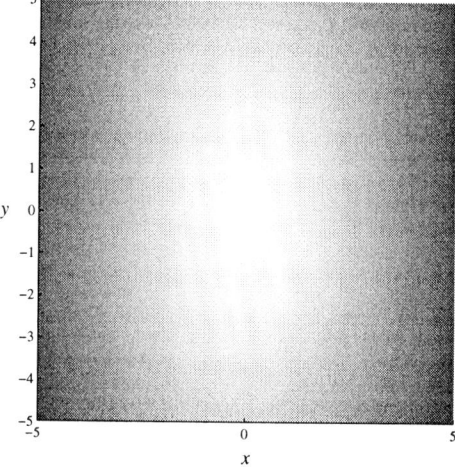

Fig. 2. Surface $z = 2x^2 + y^2$ lit from the direction of the observer.

We would like to answer the following question: how many surfaces correspond to the same image as that of S_1, with front lighting? This can be reformulated as the determination of the number of solutions of the eikonal equation corresponding to the surface S_1, which is:

$$f_x(x,y)^2 + f_y(x,y)^2 = 16x^2 + 4y^2. \tag{3}$$

It is obvious that if a surface S defined by $z = f(x,y)$ is a solution of (3), then the surfaces defined by $z = f(x,y) + c$ and $z = -f(x,y) + c$ are solutions too, where c designates any real value. These surfaces are what we call *isomorphic* to S solutions. So, S_2 defined by $z = -(2x^2 + y^2)$ is an isomorphic to S_1 solution of (3). Besides S_1 and S_2, it is obvious that the surface S_3 defined by $z = 2x^2 - y^2$ is a solution, as well as the isomorphic to S_3 surface S_4, defined by $z = -(2x^2 - y^2)$. We show that equation (3) has in fact many more analytical solutions than the four ones aforementioned. Actually, we prove the following theorem, which is the main result of this paper:

2 Theorem

There exists a family $(f_c)_{c \in \mathbb{R}_+}$ *of functions* $f_c : \mathbb{R}^2 \to \mathbb{R}$ *verifying the following properties:*

1. *For $c > 0$, f_c is analytical and real on the disc $D(0,c)$;*
2. *For $c > 0$, f_c is a solution of (3) in the neighbourhood $D(0,c)$ of $(0,0)$;*
3. *The function $(c,x,y) \mapsto f_c(x,y)$ is continuous on $\mathbb{R}_+ \times \mathbb{R}^2$;*

4. *If $c \neq c'$, then f_c and $f_{c'}$ are not isomorphic.*

Each function f_c, $c > 0$, corresponds thus to a surface, the image of which, when lit from the direction of the observer and restricted to the disc $D(0,c) = \{(x,y) \in \mathbb{R}^2; x^2 + y^2 < c^2\}$, coincides with the image of surface S_1 represented in figure 2. Therefore, it is possible to conceive, starting from one of the solutions f_c, $c > 0$, a continuous deformation for which the image does not change in a neighbourhood of the origin. This explains the somewhat surprising title of our paper: non visible deformations.

How does this theorem take place in the field of shape from shading? Among the papers dealing with the resolution of the eikonal equation, those of Brooks, Chojnacki and Kozera [5] and of Horn, Szelisky and Yuille [6] are particularly interesting. Actually, it is proved, in these papers, that certain eikonal equations may admit no continuously differentiable solution, which is yet the largest class of solution, since an eikonal equation cannot admit any non-differentiable solution. We provide a result corresponding to the opposite situation, when proving the existence of an eikonal equation for which there exists an infinity of analytical solutions in a neighbourhood of the origin, which are not only smooth, but also the most regular solutions that can exist. These two opposite extreme situations consequently show that the number of solutions of the eikonal equation (2) may vary from zero to infinity, with respect to its second member $g(x,y)$. A certain number of authors have searched for characteristics concerning this second member $g(x,y)$, so that equation (2) admits only one continuously differentiable solution [3,4,7]. Let us mention that all these theorems of unicity suppose the existence of a silhouette in the analysed image, *i.e.*, the existence of a closed curve, along which $g(x,y)$ is infinite (the image represented in figure 2 does not correspond to this situation). Such a situation is interesting, because it satisfies the initial wish of Horn [1], who wanted to compute, for any image, one surface exactly. Nevertheless, it is obvious that this situation can take place only in particular cases, since shape from shading is the inverse problem of image formation, and since the relation between all the possible surfaces and all the possible images is not a bijection. Here we prove that this inverse problem may be sometimes ill-posed, which means that there exists an infinity of solutions (this is in contradiction with what is claimed in [8]).

3 A one parameter family of solutions of (3)

In this section, we exhibit a family $(f_c)_{c \in \mathbb{R}_+}$ of functions $f : \mathbb{R}^2 \to \mathbb{R}$ satisfying properties 1-2-3-4 of the theorem. These functions verify also the two following properties:

5. *The function $(c,x,y) \mapsto f_c(x,y)$ is real and analytical on $(\mathbb{R}_+^* \times \mathbb{R}^2) \setminus \Delta$, with $\Delta = \{y = 0, x + c \leq 0\}$;*
6. — *If c tends towards $+\infty$, then $f_c(x,y)$ tends towards $f_\infty(x,y) = 2x^2 - y^2$;*
   — *If c tends towards 0^+, then $f_c(x,y)$ tends towards $f_0(x,y) = 2x|x| + y^2$.*

In fact, for $c \neq 0$, f_c is even analytical on $\mathbb{R}^2 \setminus \Delta_c$, with $\Delta_c =]-\infty, -c] \times \{0\}$. At $P_c = (-c, 0)$, f_c is differentiable, but not of class C^2. At any point on Δ_c different to P_c, f_c is continuous but not differentiable.

In this paper, we demonstrate only property 2 of the theorem, when proving that f_c is a solution of (3) on $\mathbb{R}^2 \setminus \Delta_c$, as well as at point P_c.

Let $c \geq 0$. We define f_c by the following:

$$f_c(x, y) = 2x^2 + y^2 + \frac{16}{3} g_c(x, y)(x + c - g_c(x, y)), \tag{4}$$

where the family of functions $(g_c)_{c \in \mathbb{R}_+}$ is going to be defined.

Remark: The definition of $f_c(x, y)$, as well as a certain number of other definitions which are going to be given, may appear to be very arbitrary. Nevertheless, the whole reasoning which allows us to find the functions $f_c(x, y)$ is tedious, and we have decided not to show it in full in this paper.

Let $(x, y) \in \mathbb{R}^2$. Let us note:

$$\begin{cases} a = x + c, & (5) \\ b = \dfrac{27}{32} cy^2, & (6) \\ r = a^3 + 4b, & (7) \\ s = 8b(a^3 + 2b). & (8) \end{cases}$$

Let us consider the following equation, where the unknown is g:

$$\Phi(g, a, b) = 0, \tag{9}$$

and where $\Phi(g, a, b)$ is defined by:

$$\Phi(g, a, b) = g^3 - \frac{3}{4} a^2 g - \frac{1}{4}(a^3 + 4b). \tag{10}$$

At any point (x, y), the function $g_c(x, y)$ equals, by definition, the greatest real solution of equation (9). It can be shown [9] that the analytical expression of $g_c(x, y)$ is:

$$g_c(x, y) = \frac{1}{2} \left(\sqrt[3]{r + \sqrt{s}} + \sqrt[3]{r - \sqrt{s}} \right), \tag{11}$$

where we designate by $\sqrt{s}$, when $s < 0$, the value $i\sqrt{-s}$, where $i^2 = -1$, and by $\sqrt[3]{\ }$, when the argument is complex, the complex cubic root having the greatest real part. It is easy to show that expression (11) is always equal to or greater than zero.

In figure 3, function $g_1(x, y)$ has been represented. In figure 4, function $f_1(x, y)$, computed from $g_1(x, y)$ thanks to (4), has been represented.

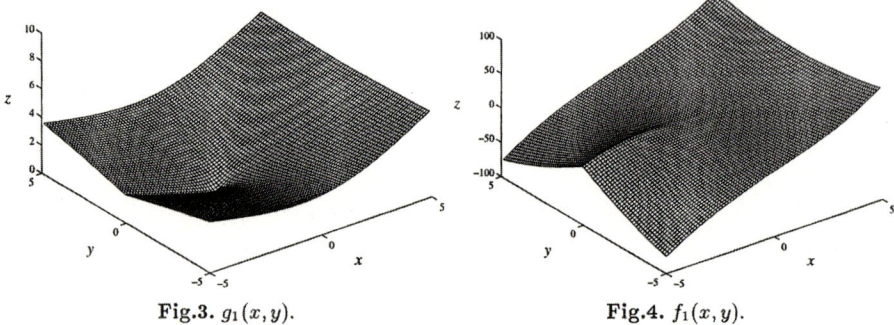

Fig.3. $g_1(x,y)$. **Fig.4.** $f_1(x,y)$.

4 $f_c(x,y)$ is a solution of equation (3)

Let $c > 0$. Let us compute the two partial derivatives g_x and g_y of $g_c(x,y)$ on $\mathbb{R}^2 \setminus \Delta_c$. When deriving equation (9) in relation to x and to y, and when using expression (10) of $\Phi(g,a,b)$, as well as expressions (5) and (6) of a and of b, we obtain:

$$\begin{cases} \left(g^2 - \dfrac{1}{4}a^2\right)g_x = \dfrac{1}{2}ag + \dfrac{1}{4}a^2, & (12) \\ \left(g^2 - \dfrac{1}{4}a^2\right)g_y = \dfrac{9}{16}cy. & (13) \end{cases}$$

It can be shown [9] that $2g + a$ never equals zero. Equations (12) and (13) can then be rewritten:

$$\begin{cases} (a - 2g)g_x = -a, & (14) \\ (a - 2g)g_y = -\dfrac{9}{4}\dfrac{cy}{2g+a}. & (15) \end{cases}$$

The partial derivatives of $f_c(x,y)$, defined by (4), in relation to x and to y, give on the other hand:

$$\begin{cases} f_x = 4x + \dfrac{16}{3}g + \dfrac{16}{3}(a-2g)g_x, & (16) \\ f_y = 2y + \dfrac{16}{3}(a-2g)g_y. & (17) \end{cases}$$

When using (14) and (15), these expressions become:

$$\begin{cases} f_x = 4x + \dfrac{16}{3}(g-a), & (18) \\ f_y = 2y - 12\dfrac{cy}{2g+a}. & (19) \end{cases}$$

The first member of equation (3) takes consequently the following form:

$$f_x{}^2 + f_y{}^2 = 16x^2 + 4y^2 + A, \tag{20}$$

with:

$$A = 8x\frac{16}{3}(g-a) + \left(\frac{16}{3}\right)^2 (g-a)^2 - 48\frac{cy^2}{2g+a} + 144\frac{c^2 y^2}{(2g+a)^2}. \tag{21}$$

Now, when using (6) and factorizing the expression of $\Phi(g, a, b)$ given in (10), equation (9) takes the form:

$$12\frac{cy^2}{(2g+a)^2} = \frac{32}{9}(g-a), \tag{22}$$

which gives, when reporting in (21):

$$A = \frac{64}{9}(g-a)(6x + 4(g-a) - 2(2g+a) + 6c). \tag{23}$$

The last factor in equation 23 equals zero, thus $f_c(x, y)$ is a solution of equation (3) on $\mathbb{R}^2 \setminus \Delta_c$. It is easy to verify that it also verifies (3) at point P_c, then finally $f_c(x, y)$ is a solution of equation (3) on $\mathbb{R}^2 \setminus]-\infty, -c[\times \{0\}$.

In figure 5, we represented the image of the surface $z = f_2(x, y)$, with front lighting.

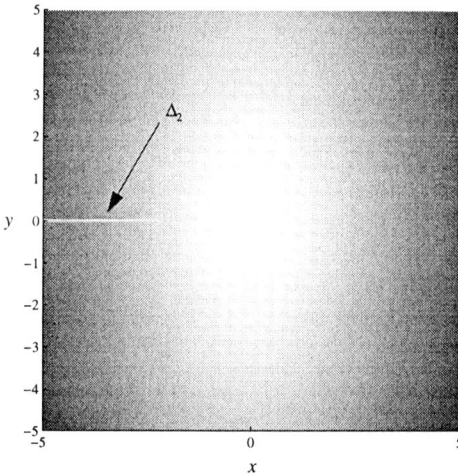

Fig. 5. Surface $z = f_2(x, y)$ lit from the direction of the observer.

This image is actually identical to that represented in figure 2, except on $\Delta_2 \setminus \{P_2\}$, where $f_2(x,y)$ is not derivable.

As for $f_0(x,y)$, it can be verified without any difficulty that it is a solution of (3) on $\mathbb{R}^2$, knowing that:

$$f_0(x,y) = 2x|x| + y^2. \tag{24}$$

5 Application

A surprising application of the theorem we just proved is the existence of non visible deformations. Such a deformation was simulated thanks to the Matlab 4.2 software. The image remains actually unchanged. On the other hand, if we modify, even slightly, the observation direction or the lighting direction, the deformation becomes visible. This means that such deformations are non visible within the context of monovision, but not within the context of stereovision. This situation shows, when necessary, the superiority of the stereovision methods over the monovision methods, with regard to shape reconstruction.

References

1. Horn B.K.P. "Shape from Shading: a Method for Obtaining the Shape of a Smooth Opaque Object from One View". PhD thesis, Department of Electrical Engineering, MIT, 1970.
2. Horn B.K.P. "Obtaining Shape from Shading Information". In *The Psychology of Computer Vision*, chapter 4, pages 115–155. P.H. Winston (ed.), New York, 1975.
3. Bruss A.R. "The Eikonal Equation: Some Results Applicable to Computer Vision". *Journal of Mathematical Physics*, 23(5):890–896, 1982.
4. Oliensis J. "Uniqueness in Shape from Shading". *International Journal of Computer Vision*, 6(2):75–104, 1991.
5. Brooks M.J., Chojnacki W., and Kozera R. "Shading without Shape". *Quarterly of Applied Mathematics*, L(1):27–38, 1992.
6. Horn B.K.P., Szeliski R.S., and Yuille A.L. "Impossible Shaded Images". *IEEE PAMI*, 15(2):166–169, 1993.
7. Blake A., Zisserman A., and Knowles G. "Surface Descriptions from Stereo and Shading". *Image and Vision Computing*, 3(4):183–191, 1985.
8. Durou J.D. "Reconnaissance du relief à partir de l'éclairement". PhD thesis, Université Paris XI-Orsay, 1993.
9. Durou J.D., Mascarilla L., and Piau D. "Non visible distortions". Submitted to *IEEE PAMI*.

Two-Step Parameter-Free Elastic Image Registration with Prescribed Point Displacements

Wladimir Peckar, Christoph Schnörr, Karl Rohr, and H. Siegfried Stiehl

Universität Hamburg, FB Informatik, AB Kognitive Systeme,
Vogt-Kölln-Str. 30, D-22527 Hamburg, Germany

Abstract. A two-step parameter-free approach for non-rigid medical image registration is presented. Displacements of boundary structures are computed in the first step and then incorporated as hard constraints for elastic image deformation in the second step. In comparison to traditional non-parametric methods, no driving forces have to be computed from image data. The approach guarantees the *exact* correspondence of certain structures in the images and does not depend on parameters of the deformation model such as elastic constants. Numerical examples with synthetic and real images are presented.

1 Introduction

Numerous applications in modern medical imaging deal with non-rigid image registration. Examples are image-atlas as well as multi-modality image registration in neurosurgery. There, a three-dimensional image (*deformable template*) has to be completely transformed onto another one (*study*).

One group of methods dealing with non-rigid image registration is the so-called *non-parametric* methods, where the degrees of freedom of admissible deformations are not defined by a fixed number of parameters [1], [3]. The non-parametric methods model the non-rigid transformations as deformations of physical bodies (solids, liquids) caused by applied forces. The traditional image registration scheme using non-parametric methods is the following: Applied forces are first derived from image data using some similarity measure and then used to deform the template image driving it to a correspondence with the study image.

In this paper, we propose a two-step registration approach based on elasticity theory. In the first step, we determine point correspondences of some boundary structures in both images by using an active contour model also known as snake model. In the second step, we elastically deform the template image by using the prescribed values of displacements of boundary structures obtained in the first step incorporated as hard constraints in addition to the conditions on the image boundary. This approach has several advantages compared to traditional methods. i) No driving forces have to be derived from image data. For multi-modality images using a local similarity measure, this is known to be a difficult problem. ii) As a consequence, the remaining parameters of the deformation model (elastic constants) drop out from the model and it becomes completely *parameter-free*. iii) It can always be guaranteed that the required deformation is obtained and that certain structures in the template are *exactly* matched with those of the study.

Relationship to Other Work Application of non-parameteric methods to medical image registration originated from the work of Broit [2], where images were represented as pieces of rubber and the cross-correlation coefficient between two images was used for the derivation of forces. This linear elastic model has been improved to increase the speed of computations and to avoid local minima [1], [15]. Two main drawbacks of this model are the assumption of small displacements and the usage of a local similarity measure. Gee *et al.* [11] proposed a probabilistic approach based on the finite element method which has been reported to have properties similar to those of [2], [1].

Another group of non-parametric methods which is based on the principles of fluid mechanics has been introduced by Christensen *et al.* [3], [4]. These methods use properties of fluids that do not carry memory about their initial state, thus allowing large deformations. However, a local similarity measure is still used, which considerably limits the applicability of the fluid model as a general model for registration problems.

The present paper describes a further development of the approach introduced in [14] and is closely related to the work of Davatzikos *et al.* [10], [9], where no local similarity measure is used and external forces are defined on the basis of correspondence of boundary structures such as the outer cortical boundary and the ventricles. The principal difference to our approach is, however, that we do not use any external forces. As a consequence, parameters of the deformation model, such as elastic constants, are not required for our approach. Incorporation of known displacements as hard constraints in the model allows the *exact* matching of the boundary structures.

In the following, we describe the two steps of our registration approach, its discrete representation, and present some numerical examples with synthetic and real images.

2 Two-Step Registration Model

In this section, we present the two steps of our registration approach. The snake model in the first step was used for demonstration purposes in order to define simple point correspondences between boundary structures in two dimensions. It will be replaced in the future by more efficient methods. The elastic model is presented for the three-dimensional case.

2.1 First Step: Snake Model

Snakes were first proposed by Kass *et al.* [13] as a general energy minimizing model which can be applied to numerous problems in computer vision (edge detection, tracking of moving objects, etc.). Since the time of introduction of snakes, several improvements have been made in the model (see, for example, [8], [7]). In this paper, we use the original model from [13].

Snakes are parametrically defined curves $\mathbf{v}(s) = (x(s), y(s))$. The snake model minimizes the following energy:

$$E_{snake} = \int_{\Omega} \{\alpha(s)\|\mathbf{v}'(s)\|^2 + \beta(s)\|\mathbf{v}''(s)\|^2 + P(\mathbf{v}(s))\} \, ds \qquad (1)$$

consisting of the internal energy of the curve and the potential P corresponding to the external forces derived from image data. Two parameters α and β control elastic properties of the snake. Since we are interested in finding edges, we use the following potential:

$$P(\mathbf{v}(s)) = -|\nabla I(\mathbf{v}(s))|, \qquad (2)$$

where $I(\mathbf{v}(s))$ denotes the image function. The result is a curve converging to the boundary of the object when placed close enough to it.

2.2 Second Step: Linear Elastic Model

Here, we present outlines of the variational formulation of the three-dimensional linearized elasticity problem.

Let Ω be an open, bounded, connected subset of $\mathbb{R}^3$ with a Lipschitz-continuous boundary Γ ($\Gamma = \Gamma_0 \cup \Gamma_1$, $\Gamma_0 \cap \Gamma_1 = \emptyset$). We define a normed vector space

$$\mathbf{V} := \{\mathbf{v} = (v_1, v_2, v_3)^t \in (H^1(\Omega))^3; \, v_i = 0 \text{ on } \Gamma_0 \subset \Gamma, \, i = 1, 2, 3\}.$$

The variational problem of the linearized elasticity which couples displacements in elastic materials with applied body and surface forces can be formulated as [5]: Find $\mathbf{u} \in \mathbf{V}$ such that

$$a(\mathbf{u}, \mathbf{v}) = f(\mathbf{v}), \quad \forall \mathbf{v} \in \mathbf{V}, \qquad (3)$$

where the symmetric bilinear form $a(\mathbf{u}, \mathbf{v})$ and the linear form $f(\mathbf{v})$ are defined as

$$a(\mathbf{u}, \mathbf{v}) = \int_{\Omega} \{\lambda(\nabla \cdot \mathbf{u})(\nabla \cdot \mathbf{v}) + 2\mu \sum_{i,j=1}^{3} e_{ij}(\mathbf{u}) e_{ij}(\mathbf{v})\} \, dx, \qquad (4)$$

$$f(\mathbf{v}) = \int_{\Omega} \mathbf{f} \cdot \mathbf{v} \, dx + \int_{\Gamma_1} \mathbf{g} \cdot \mathbf{v} \, d\gamma. \qquad (5)$$

Here, $\mathbf{f} = (f_1, f_2, f_3)^t \in (L^2(\Omega))^3$ and $\mathbf{g} = (g_1, g_2, g_3)^t \in (L^2(\Gamma_1))^3$ denote applied body and surface forces respectively,

$$e_{ij}(\mathbf{v}) = e_{ji}(\mathbf{v}) = \frac{1}{2}(\partial_j v_i + \partial_i v_j), \quad i, j = 1, 2, 3 \qquad (6)$$

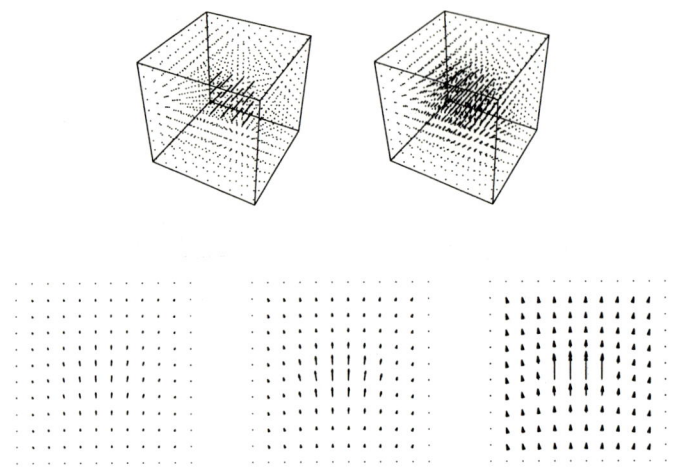

Fig. 1. Deformation with prescribed point displacements of a 3D synthetic image. Top/left: Prescribed displacements. Top/right: Computed 3D displacement field. Bottom: Horizontal layers 3, 4, 5 from the 3D computed displacement field above.

is the linearized strain tensor, $\mathbf{u}, \mathbf{v} \in \mathbf{V}$ denote the displacement field, λ and μ are Lamé elastic constants.

In the following, we use a simplified model by supposing $\Gamma = \Gamma_0$, and setting the parameter λ to zero. The last choice is quite common for non-rigid registration problems, because in that case objects in images are allowed to grow without being laterally shrunk [1], [4].

The elastic model is now: Find $\mathbf{u} \in \mathbf{V}$ such that

$$a(\mathbf{u}, \mathbf{v}) = f(\mathbf{v}), \quad \forall \mathbf{v} \in \mathbf{V}, \tag{7}$$

where

$$a(\mathbf{u}, \mathbf{v}) = \int_\Omega 2\mu\, (\mathbf{e}(\mathbf{u}), \mathbf{e}(\mathbf{v}))\, dx, \tag{8}$$

$$f(\mathbf{v}) = \int_\Omega \mathbf{f} \cdot \mathbf{v}\, dx. \tag{9}$$

Here $(\cdot, \cdot)$ denotes the usual matrix inner product.

From (8)-(9), one can see that the parameter μ can now be considered as a scaling coefficient for the applied forces. Since we use prescribed displacements instead of applied forces, the elastic model becomes completely parameter-free.

It can be shown that the problem (7) has a unique solution (see [14], [6] for details). In the next section, we will discuss the finite element discretization of the problem (7).

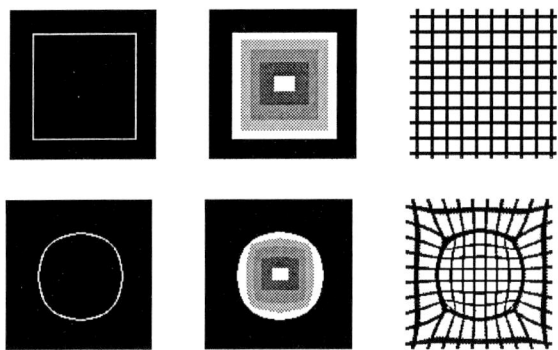

Fig. 2. Two-step deformation with prescribed displacements. Left: Two corresponding boundary structures. Middle: Deformation of a synthetic image. Right: The same deformation applied to a rectangular grid.

2.3 Discrete Representation

Following [13], we solved Euler equations corresponding to the snake model (1) iteratively by approximating the derivatives with finite differences and using constant parameters α and β.

For the discretization of the elasticity problem (7), we used the *Galerkin method* [5]: By replacing the space $\mathbf{V}$ with a finite-dimensional subspace $\mathbf{V}_N := \text{span}\{\phi_1, \ldots, \phi_N\}$, we seek a discrete solution to the problem: Find $\mathbf{u}_N \in \mathbf{V}_N$ such that

$$a(\mathbf{u}_N, \mathbf{v}_N) = f(\mathbf{v}_N), \quad \forall \mathbf{v}_N \in \mathbf{V}_N. \tag{10}$$

The solution vector $\mathbf{u}_N = \{u_i\}$ is obtained as a solution of the system of linear equations:

$$\sum_{i=1}^{N} u_i \int_{\Omega} \{\sum_{k,l=1}^{3} e_{kl}(\phi_i) e_{kl}(\phi_j)\} \, dx = \frac{1}{2\mu} \int_{\Omega} \mathbf{f}_j \cdot \phi_j \, dx, \quad j = 1, \ldots, N. \tag{11}$$

The system (11) can be written in matrix form as

$$\mathbf{A}\mathbf{u}_N = \mathbf{b} \tag{12}$$

with symmetric and positive definite stiffness matrix $\mathbf{A}$.

2.4 Prescribed Displacements

To incorporate prescribed values of u_i, we transform the matrix $\mathbf{A}$ in (12) by filling its i-th row and column with 0 and setting the element $\mathbf{A}_{ii}$ to 1. Since each row of $\mathbf{A}$ contains contributions of more than one finite element, the transformed matrix $\tilde{\mathbf{A}}$ will always be invertible. Then we set the initial vector $\mathbf{b}$ to zero, subtract the product $u_i \mathbf{A}^i$ from it ($\mathbf{A}^i$ denotes here the i-th column of $\mathbf{A}$), and put u_i in the i-th position in $\tilde{\mathbf{b}} = \mathbf{b} - u_i \mathbf{A}^i$ [16].

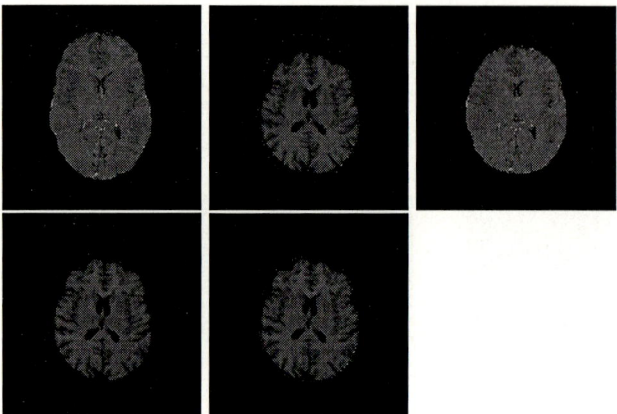

Fig. 3. Two-step medical image registration example. Left column: Two MR slices taken from different patients. Middle/top: Points from the outer brain contour of one image superimposed onto the second one. Midde/bottom: Result of the snake algorithm applied to the curve. Right: Deformed original image.

3 Numerical Examples

In this section, some numerical examples for our two-step registration approach will be presented. The deformations for all examples were based on prescribed point displacements, while no external forces have been used.

Figure 1 illustrates the usage of prescribed displacements for a 3D synthetic image which has the size of 12x12x12 voxels. At the top of the figure, the prescribed displacements (top/left) and the computed displacement field (top/right) are shown. In the bottom of the figure, there are three 2D horizontal layers from the displacement field above. For this example, layers 3, 4, and 5, starting from the top were chosen. This example illustrates also the usage of the homogeneous Dirichlet boundary condition ($\mathbf{v} = 0$ on Γ, where Γ is the image boundary).

Figure 2 (left) presents two corresponding boundary structures, where the image below is a result of some iterations of the snake algorithm applied to the curve above. In Figure 2 (middle), a synthetic image was deformed using the prescribed displacements obtained from the snake model. Figure 2 (right) shows the same deformation applied to a rectangular grid in order to illustrate topology preserving properties of elastic transformation. The size of images is 128x128 pixels. As a result of incorporating prescribed displacements as constraints, the outer boundary of the synthetic image after deformation (middle/bottom) *exactly* corresponds to the contour obtained from the snake algorithm (left/bottom).

Next, we illustrate the application of our two-step registration model to 256x256 MR slices of different patients.

In Figure 3 (left), two MR slices taken from different patients are shown. The outer contour from the upper image was next superimposed onto the lower image (middle/top) and then the snake algorithm was applied to the curve (middle/bottom). The right side of Figure 3 shows the deformation applied to the

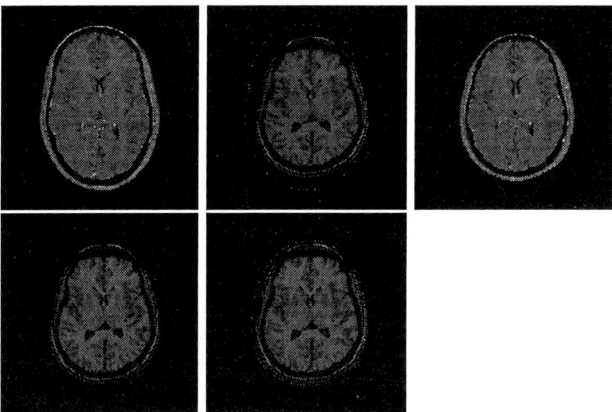

Fig. 4. Two-step medical image registration example. Left column: Two MR slices taken from different patients. Middle/top: Points from the outer skull contour of one image superimposed onto the second one. Midde/bottom: Result of the snake algorithm applied to the curve. Right: Deformed original image.

original image with prescribed displacements taken from the result of the snake algorithm.

In Figure 4, the same experiment with another pair of images is presented. Here, we took the outer skull contour instead of the outer brain contour as in the experiment above.

From the experiments in this section, one can see that a quite good global match can be obtained by using only outer contours. Local matching requires more fine structures (e.g. ventricle systems) to be brought into correspondence.

4 Summary and Further Work

We have presented a two-step parameter-free image registration approach, where images are elastically deformed with incorporated prescribed displacements. We assume that there exists a unique one-to-one mapping between two images, and constrain the global deformation by using local values of this mapping (known displacements of boundary structures). In contrast to traditional methods, our approach does not depend on parameters of the deformation model such as elastic constants and guarantees the *exact* matching of boundary structures.

Future research will address the development of a more efficient model that can be used instead of the snake model to provide point correspondences of boundary structures in the brain, since the usability of active contour models for practical purposes is quite limited because of high complexity of real medical data.

Another important point is the efficient numerical implementation of the model. Though the implemented conjugate gradient method with preconditioning [12] requires an acceptable amount of computation time for 2D images (several minutes on a SPARC 10 workstation), further development using explicit parallelization is required to make the application to 3D images practically feasible.

Acknowledgements

The support of the first author from the German Academic Exchange Service (DAAD) is kindly appreciated. Medical image data were provided by the UMDS Image Processing Group (London/UK), Philips Research Laboratories Hamburg, and Ramin Shahidi.

References

1. R. Bajcsy and S. Kovačič. Multiresolution elastic matching. *Computer Vision, Graphics, and Image Processing*, 46:1–21, 1989.
2. C. Broit. *Optimal Registration of Deformed Images*. Doctoral dissertation, University of Pennsylvania, August 1981.
3. G.E. Christensen. *Deformable Shape Models for Anatomy*. PhD thesis, Washington University, August 1994.
4. G.E. Christensen, R.D. Rabbitt, and M.I. Miller. Deformable templates using large deformation kinematics. *Submitted to the IEEE Transactions on Image Processing*, October 1994.
5. P.G. Ciarlet. *The Finite Element Method for Elliptic Problems*. North-Holland, Amsterdam, 1978.
6. P.G. Ciarlet. *Mathematical Elasticity. Volume I: Three-Dimensional Elasticity*. North-Holland, Amsterdam, 1988.
7. L.D. Cohen. Auxiliary variables and two-step iterative algorithms in computer vision problems. *Journal of Mathematical Imaging and Vision*, (6):59–83, 1996.
8. L.D. Cohen and I. Cohen. Finite-element-methods for active contour models and balloons for 2-d and 3-d images. *IEEE Transactions on Pattern Analysis and Machine Intelligence*, 15(11):1131–1147, 1993.
9. C. Davatzikos. Nonlinear registration of brain images using deformable models. In *Proc. of the IEEE Workshop on Math. Methods in Biomedical Image Analysis*, pages 94–103, San Francisco, June 1996.
10. C. Davatzikos, J.L. Prince, and R.N. Bryan. Image registration based on boundary mapping. *IEEE Transactions on Medical Imaging*, 15(1):112–115, 1996.
11. J.C. Gee, D.R. Haynor, M. Reivich, and R. Bajcsy. Finite element approach to warping brain images. In *Proc. SPIE Image Processing*, volume 2167, pages 327–337, 1994.
12. W. Hackbusch. *Iterative Solution of Large Sparse Systems of Equations*. Springer-Verlag, 1993.
13. M. Kass, A. Witkin, and D. Terzopoulos. Snakes: Active contour models. *International Journal of Computer Vision*, 1(4):321–331, 1988.
14. W. Peckar. FEM dicretization of the Navier equation with applications to medical imaging. Memo FBI-HH-M-266/96, Dept. of Computer Science, University of Hamburg, November 1996.
15. T. Schormann, S. Henn, and K. Zilles. A new approach to fast elastic alignment with applications to human brains. In *Visualization in Biomedical Computing (VBC'96)*, pages 337–342, Hamburg, Germany, September 1996. Springer-Verlag.
16. H.R. Schwarz. *Methode der finiten Elemente*. Teubner, Stuttgart, 1984.

Learning for Feature Selection and Shape Detection

Rita Cucchiara, Massimo Piccardi, Michele Bariani, Paola Mello

Dipartimento di Ingegneria University of Ferrara,
via Saragat 1 - I-44100 Ferrara, Italy
Tel +39-532-293800 Fax +39-532-768602
e-mail: rcucchiara@ing.unife.it

Abstract. The paper proposes a general framework for shape detection based on supervised symbolic learning. Differently from other visual systems exploiting machine learning, the proposed architecture does not follow the object segmentation - feature extraction and (learning based) classification approach. Instead, an initial data-driven processing selects points of interest in the scene by means of complex features which hypothesize the presence of the target shape; hypotheses are validated by a classifier defined by a machine learning algorithm. Learning is exploited not only for defining the model, i.e. the description of the target for the classifier, but also for defining the description language, i.e. the feature set useful in generating reliable object hypotheses. The proposed architecture of visual system has been implemented for an industrial application of unstructured shape detection: examples and results are reported in the paper

1. Introduction

Computer vision community is engaging in generalizing vision paradigms and devising complete and flexible approaches to visual tasks. Many proposals of visual systems include now a machine learning stage. Machine learning is adopted under different paradigms mainly to face uncertainty associated with noise, intending noise in a broad sense (including sensor noise, cluttering, target occlusions, signal-to-symbol distortion induced by view-centered observations) [1]. However, learning is often essential for handling the model definition and providing adequate classification parameters in the case of detection of unstructured objects or shapes. In fact, the model of object targets cannot always be a priori defined in terms of geometric, topologic or other metric features (as for instance it easily performed for many handmade objects), but in many contexts only a qualitative description of the target is available. Many application fields are covered by this last framework and may take advantage of learning for defining target models: examples are recognition of hand gesture, landscape inspection, medical images analysis, and appearance-based recognition[1,2,3,4]. Another interesting example is quality inspection, where humans are very skilful at processing visual stimuli and performing classification but they do not exhibit the same ability in inferring and formalizing the rules for classification [5]. According to these considerations, we propose an architecture for unstructured and qualitatively described shape recognition that makes a substantial use of learning with the goal of defining a reliable shape model used in the recognition process. Learning affects also the image analysis process since its results are applied for setting the description language used in the final recognition. Learning is used in conjunction with a hierarchical image analysis system based on a data-driven paradigm for performing layered parametric transformations. This proposal recalls other similar approaches, and in particular the one of Bolle and al. in [6], which describes a complete approach for object recognition based on parametric transformations on the feature spaces with the aim at providing a scene description in terms of complex or "generalized" features. In [6] parametric transformations are associated with a recognition network under a constraint satisfaction scheme for representing the object models. Instead in our proposal selection of visual features and related parametric transformations and classification are guided by the results of a symbolic learning process.

In the paper we describe the architecture of the visual system, by focusing on the layered and hierarchical image analysis block and the tree-based classifier. We present the object representation based on a hierarchical feature set together with an example of industrial application in the context of quality inspection.

2. The visual system architecture

According to the wide related literature, we aim at detecting objects in images on the basis of a set of visual features whose selection is goal-directed. Fig. 1 shows a scheme of the proposed architecture that is based on some fundamental issues: 1) the image analysis system for feature extraction providing object representation is hierarchical and layered; 2) the object representation is not performed by segmentation followed by feature measurements; on the contrary, significant features extracted from data hypothesize the presence of objects and then the feature tuples represent a possible object; 3) tuples are matched with the object model by means of a classifier which symbolically encodes the knowledge of the model; 4) the learning subsystem is based on a supervised symbolic learning algorithm which starting from a training set infers rules and the related classifier; moreover it selects the minimal and reliable set of features used in the hypothesis representation.

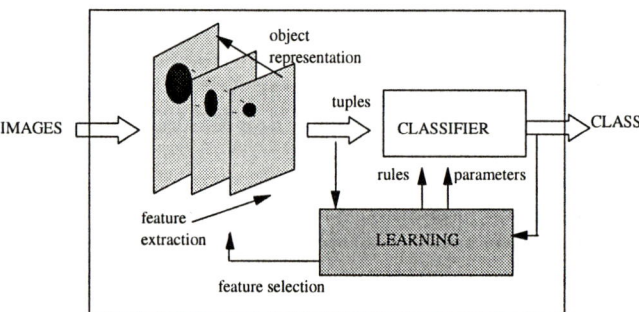

Fig. 1. The visual system architecture

The architecture defines a goal-directed overall process that is in essence data-driven at the lower levels while is strictly model-based at the highest level. The specific goal guides an initial selection of possible visual features. The result is a description of the scene in terms of features. Features are structured in hierarchical layers, with growing levels of abstraction in order to deal with the increasing complexity of scene aspects extracted by the visual process. This approach follows the proposal of "generalized features" of [6], so that the feature computation of a higher level is carried out by means of parametric transformations as a function of other features of the same or lower levels. Therefore, as indicated in Fig. 2, we define many layered multidimensional spaces: each Sj space contains points represented by a feature tuple $V^j = <v^j_1, \ldots v^j_{kj}>$. Therefore features of each level accumulate evidence of salient particulars of the scene, in order to generate goal-dependent hypotheses: in the case of shape detection, hypotheses are related to the possible presence of the target shape.

A peculiar aspect of our work is that not all the possible combinations of features of different levels will be used in the classification. Instead we select only those points of the highest level identified by "high" values of semantically significant features. The object hypothesis is generated with a top-down backward process, starting from top-level features and using only those features of the lower levels generating the ones of higher levels. In practice, the object hypothesis is computed by exploiting either the previously executed visual tasks (of the bottom-up approach), or suitable anti-transformations or, eventually, also by operating further processing in that zone of the image where the attention has been focused. Finally each computed multilevel tuple of features is a possible hypothesis on the presence of target, represented as
$O = \cup_{i=0\ldots j} V^i$.

Many works use different level features in the classification: but if all potential tuples are considered, a possible combinatorial explosion of hypotheses calls for the use of "ad-hoc" classification algorithms or special-purpose neural network based approaches [1]. Instead, the definition of a hierarchical and structured limited set of features allowed for adopting also a general-purpose symbolic approach.

Different representations of knowledge models have been used in visual systems, such as constraint networks [6], interpretation tree search [4], rules [3], or combination of them. Our visual architecture includes a classifier which validates the object hypotheses on the bases of a decision tree inferred after a learning phase.

The multi-level representation is often redundant and generally could cause a severe and undesirable cost of the overall visual process. This is the second reason for using machine learning: the exploitation of a general purpose symbolic learning tool enables for tuning some learning parameters, in order to specialize or generalize the classifier depending on application constraints or acceptable error-rates. The produced decision tree allows for selecting a very compact reduced set of features of various levels which is enough expressive of the problem and can be used for the final hypothesis validation.

3. Hierarchical feature spaces for an inspection application

The previously proposed architecture has been adopted for an application of defect inspection on industrial metallic products. In this problem, defects appear to human eyes as elongated, roughly straight and thin shapes, very bright with respect to the neighbor background. No quantitative information is however available for distinguishing them from other similar shapes (due to sensor noise, surface roughness or surface blobs). Inspection must be carried out under UV light, since the target shapes are visible only after a non destructive test, called MPI (Magnetic Particle Inspection) [7]. It allows for increasing the evidence of surface (and sub-surface) cracks, which can be recognized as thin long and roughly straight fluorescent strips.

The given qualitative description of the target suggests the adoption of some well assessed image analysis techniques, together with other special-purpose algorithms.

Gray level filtered images are used as input of the visual system. Features extracted by means of image processing are considered at the lowest level, so that the image space is the S0 space of the devised architecture. S1 is the space of simple primitives, such as curves or regions: in this application is the Hough space for lines obtained with the gradient-weighted Hough transform, which uses both information on the gradient magnitude and their orientation [8]. In this space some features can be extracted, such as the local peaks in the two semi-spaces H1 and H2 of the Hough space, between $[0,\pi]$ and $[\pi,2\pi]$ respectively which manifest the presence of straight edges with opposite gradient orientation.

The S2 space collects more structured primitives: in our system is the Correlated Hough Transform (CHT) space that is defined by a suitable correlation between meaningful points of the Hough space [5,9]. The CHT is a correlation of the H1 points with adequately filtered points of H2. It has been proven to be very robust to noise and non ideality for revealing the presence of thin, straight and elongated shapes, as the considered target shape is [9]. Therefore the S2 points, because of CH, accumulate the evidence of possible targets: the visual system selects points of interests in S2, representing detection hypotheses. An hypothesis is described by a tuple of features, comprehending CH, some features extracted previously of computing CH, and others obtained by an anti-transformation in the points of interest of the image. In the application, the computed tuple is $O = (CH^2, H1^1, H2^1, H2a^1, Tk^1, NP^0, Va^0, Ga^0)$, where the ceiling number indicates the level of the feature space. These features are briefly described as follows:

- CH is a "high" local peak in S2; it is the most salient feature computed according to the target model [9];
- H1 is the gradient weighted Hough value in $H1 \subset S1$ at the same coordinates of the point CH in S2; it indicates the "rectilinearity" of the first target edge, by supposing the target brighter than the background[9];
- H2 is the highest local peak in $H2 \subset S1$, computed in a neighbour of the point where the second peak (representing the second straight edge) should be found, in the case of an ideally straight defect;
- H2a $\in$ S1 represents the average value of the previous considered neighbour and accounts for possible not-ideality both in straight shape and in thickness.
- TK $\in$ S1 is the mutual Euclidean distance between the ideal edges of H1 and H2. It is the upper bound of the shape thickness. It is used also for delimiting a focus of attention in the image space;
- NP $\in$ S0 is the number of voting points of the image space which are transformed in the point of the Hough space corresponding to H1. It estimates the edge length.
- VA $\in$ S0 is the average vote of the voting points, i.e. the average Gradient of each point voting for the point corresponding to H1;
- GA$\in$ S0 is the average gradient of the image: differently from the others, it is a global feature of the whole image. It should be used as a corrective weight: images

containing cracks should have a "high" average gradient but not "too high", since for the MPI process the fluorescence is collected only in the zone of a possible crack and is not distributed on the surface.

It should be noted that the tuple representing the object has been computed starting by the most salient features, CH. This limits the number of possible hypotheses extracted from images, starting only from features of lower levels: for instance, the single presence of a high value of H1 is not enough for suggesting the presence of the target, since high values of H1 (without being supported from high values of other features) can be measured also in presence of spurious fluorescent blobs or other shapes not thin enough to be classified as a defect.

As a practical example, we report some measures on inspected images in Fig.2. They represent some particulars of workpieces with or without defect. Table 1 shows their correspondent feature values, computed on images with a gray level represented with a real number in the range [0,1].

Table 1. Tuples for images of Fig. 2

	CH	H1	H2	H2A	NP	VA	TK	GA x 10^3
I38	12.33	0.97	3.02	12.64	21	0.088	5	46.2
I34	13.63	1.00	3.02	12.51	21	0.084	5	44.9
I12	393.29	7.04	15.76	55.82	17	1.006	2	60.66
I17	8.44	0.59	2.72	14.10	17	0.085	3	45.44

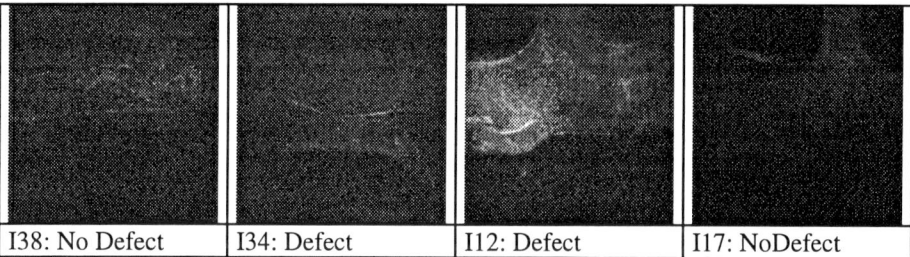

| I38: No Defect | I34: Defect | I12: Defect | I17: NoDefect |

Fig. 2. Example of images

4. The automatic learning subsystem

The highest level of the visual process addresses the decision making task with reference to the knowledge base on the target. Its implementation depends on the representation used for the target model: it can be encoded within the classifier, as in the case of connectionist systems, or can be symbolically defined by means of rules, decision trees, semantic nets and other symbolic techniques exploiting predicates on features.

Our visual system includes the Quinlan's C4.5 classifier based on learning by example algorithms which automatically generate both the decision tree and, after an adequate pruning operation, also a knowledge representation as a set of production rules [10]. It has been selected after a comparison with other approaches based on different paradigms, since it resulted very suitable for managing both symbolic and continuous values in accordance with the application requirements [10,11].

Since the goal is to validate hypotheses on the presence of the target shape (i.e. the defect) by means of the feature tuples, we address a classifier based on two classes of interest, namely *Defect* and *NoDefect*.

```
C4.5 [release 8] decision tree generator - Read 317 cases (8 attributes)
Decision Tree:
CH > 78.6283 : Defect (29.0)
CH <= 78.6283 :
|   CH <= 8.86457 : NoDefect (133.0)
|   CH > 8. 86457 :
|   |   GA > 47.0579 : NoDefect (98.0)
|   |   GA <= 47.0579 :
|   |   |   CH > 17.7267 : Defect (18.0)
|   |   |   CH <= 17.7267 :
|   |   |   |   GA > 45.6994 : NoDefect (13.0)
|   |   |   |   GA <= 45.6994 :
|   |   |   |   |   GA <= 44.6796 : NoDefect (3.0)
|   |   |   |   |   GA > 44.6796 :
|   |   |   |   |   |   H2a > 12.1436 : Defect (15.0)
|   |   |   |   |   |   H2a <= 12.1436 :
|   |   |   |   |   |   |   H2a <= 11.4278 : Defect (5.0)
|   |   |   |   |   |   |   H2a > 11.4278 : NoDefect (3.0)
```

Fig. 3. The decision tree generated by C4.5 on the training set

For each image, we considered the points of the S2 space having CH value higher than a given percentage (75%) of the maximum CH value, as possible positive hypotheses of the presence of the target shape. The used training set is composed of 317 tuples (67 Defect (21.1%) - 250 NoDefect (78.9%)). In order to train the classifier in the ambiguous region, the training set was composed so as to contain also NoDefect objects exhibiting high CH values, that are hard to be classified correctly.

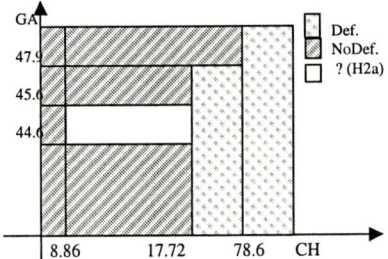

Fig. 4. Classification based on CH and GA features

Fig. 3 shows the decision tree produced by C4.5. The test set was generated by Cross-Validation, extracting one object at a time from the training set and submitting it to the classifier obtained from the remaining examples. Table 2 reports error results both in the training set and in the test set. In this last case also a partition between *false-identification* errors in the classes of Defect and NoDefect is provided. False-identification occurs when the decision tree assigns an element to a class i when it belongs to a class $j \neq i$ [12]. In the case of quality inspection the false identification in the NoDefect class is more critical, since it indicates the percentage of target which actually are not detected. Instead, the false identification rate in Defect class increases only the percentage of rejected workpieces. However it must be limited because it affects the final production efficiency.

Table 2. Results on the training and test sets

	Training Set	Test Set		
	Overall error(%)	Overall error(%)	Defect error(%)	NoDefect error(%)
C4.5 tree	0.0	2.8	4.5	2.3
C4.5 tree (no GA)	6.0	12.9	44.8	3.8

As a first important result, the use of a general-purpose machine learning tool provides a compact and efficient decision tree, by suggesting the use of specific numeric thresholds on the features and the related unary predicates. The decision tree of Fig. 5 is actually adopted by the classifier used for validating target hypotheses.

The results of learning validate the use of a hierarchical image analysis sub-system: obviously most of the predicates are carried on the highest level feature, CH, which is the most significant one; nevertheless the points of interest are represented by multi-level features. While the importance of the highest level feature, CH, (actually computed for the specific application) can be intuitively accepted, an important and less expected role is played by the average gradient of the image (GA). This feature takes into account both the image noise and the evidence of the defect with respect to its background; it turns out to be more valuable than a simple corrective weight factor since the relation between CH and GA variations and the Defect-NoDefect classification is not linear. This is described in Fig.6 by a graphical representation of the decision tree in the simplified feature space of CH and GA only.

The inner white rectangle indicat0es the ranges of CA and GA where the correct classification needs the information on other features (such as H2a).

Without the low-level GA feature the learning and the consequent classification obtain considerably worst results. In this case the decision tree turns out to be well more complex and specialized than the previous one. It takes into account six features of seven available, instead of only three of the tree of Fig.4. Moreover it exhibits a non negligible error-rate, as shown in Table 2.

Furthermore, a valuable contribution of this work is in the use of learning together with a layered image analysis subsystem: learning on tuples constructed starting from features of the highest level space handles for setting the minimal feature set useful for hypothesis description. In particular the decision tree suggests the use only of three features, one for each space level. The other five values in the tuple are not used by the classification tree. This is an important feedback on the vision process which enables for a priori avoiding measuring all those features, thus reducing computational load.

In this way the visual system used in a real on-line working environment is able for performing shape detection in a limited amount of time and the whole quality inspection process is provided in real-time. Several images of the workpieces are acquired so as to inspect different object poses, and a piece is classified as defective if at least an hypothesis on the defect presence is validated by the classifier. Moreover, the learning algorithm is available on-line for automatically updating the classifier in case of whichever changing of the working environment .

5. Conclusions

This work presented a visual architecture for detecting target shapes in images, based on a machine learning tool used with a hierarchical and layered image analysis system. The hierarchical structure of the image analysis block allows for maintaining the explicit relationships between features describing objects and it is exploited in two directions: a first bottom-up direction is used for focusing the attention; a second step in the opposite direction from the highest level down to the pixel level is used to compose a tuple representing one hypothesis. The hypothesis evaluation is carried out with decision trees automatically inferred from a training set.

The use of automatic learning is exploited for selecting the smallest subset of meaningful features and for easy updating the classifier when the external condition change. Learning showed that some of the objects features, even if they appeared to be salient, were not significant for the specific classification task, and therefore can be ignored. This feedback from the classification systems to the low-level image processing allowed for reducing the computational load and improving the reliability of the vision task that has been applied in a real industrial inspection process.

References

1. Cho, K., Dunn, S. M.: Learning shape classes. IEEE Trans. on PAMI **16** (1994) n. 9 882-887
2. Drapter, B. A., Brodley, C. E., Utgoff, P. E.: Goal directed classification using Linear Machine Decision tree. IEEE Trans. on PAMI **16** (1994) n. 9 888-893
3. Pellegretti, P., Roli, F., Serpico, S., Vernazza G.: Supervised learning of descriptions for image recognition purposes. IEEE Trans. on PAMI **16** (1994) n. 1 92-98
4. Murase, H., Nayar, S. K.: Learning by a generation approach to appearance based object recognition. Proc. of 13th ICPR, Vienna **1** (1996) 24-30
5. Cucchiara, R., Filicori, F., Andreetta, R.: Detecting micro-cracks in ferromagnetic material with automatic visual inspection. Proc. of QCAV95 Int. Conf. on Quality Control by Artificial Vision , Le Creusot France, (1995) 19-24
6. Bolle, R. M., Califano, A., Kjeldsen, R.: A complete and extendable approach to visual recognition. IEEE Trans. on PAMI **14** (1992) n. 5 534-548
7. Newman, T. S., Jain, A.K.: A Survey of automated visual inspection. Comp. Vision and Image understanding **61** (1995) n. 2 231-262
8. Illingworth, J., Kittler, J.: A Survey of the Hough transform. Comput. Vision Graphics, Image Process. **43** (1988) 221-238
9. Cucchiara, R., Piccardi, M.: Eliciting visual primitives for detecting elongated shapes, 3rd Int. Workshop on Visual Form Capri, Italy, (1997) (to appear)
10. Quinlan, J. R.: C4.5 Programs for machine learning. Morgan Kaufmann Publ. (1993)
11. Bariani, M., Cucchiara, R., Piccardi, M., Mello, P.: Data mining for automated visual inspection, Proc. of PADD97, London, UK, (1997) 51-64
12. Haralick, R., Shapiro, L.: Computer and Robot Vision vol. I, Addison Wesley (1991)

Experiments on the Decomposition of Arbitrarily Shaped Binary Morphological Structuring Elements *

Giovanni Anelli, Alberto Broggi**, and Giulio Destri

Dipartimento di Ingegneria dell'Informazione,
Università di Parma, I-43100 Parma, Italy

Abstract. The decomposition of binary structuring elements is a key problem in morphological image processing. So far only the decomposition of convex structuring elements and of specific subsets of non-convex ones have been proposed in the literature. This work presents the results of a new approach, based on a Genetic Algorithm, in which no constraints are imposed on the shape of the initial structuring element, nor assumptions are made on the elementary factors, which are chosen from a given set.

1 Introduction

Mathematical Morphology [8,12,6] concerns the study of shape using the tools of set theory. Mathematical morphology has been extensively used in low-level image processing and analysis applications, since it allows to filter and/or enhance only some characteristics of objects, depending on their morphological shape.

Within the mathematical morphology framework, a binary image A is defined as a subset of the two-dimensional Euclidean space E^2 ($Z \times Z$):

$$A = \{a = (a_i, a_j) \mid a_i, a_j \in Z\} \qquad (1)$$

In [6], monadic transforms acting on a generic image A (*complement, reflection,* and *translation*) and dyadic operators between sets (*dilation, erosion, opening,* and *closing*) are defined. This work will focus only on *dilations*,

$$A \oplus B \triangleq \{x \in E^2 \mid x = a + b, \text{ for some } a \in A, b \in B\} \qquad (2)$$

where A represents the image to be processed, and B is called *Structuring Element* (SE), i.e. another subset of E^2 whose shape parameterizes each operation.

A SE B is said to be *convex with respect to a given set of morphological operations (e.g. dilation) with a given set of SEs (factors)* $\{F_i, i = 1, ..., n\}$ if it can be expressed as a chain of dilations of the F_i elements:

$$B = F_{k_1} \oplus F_{k_2} \oplus F_{k_3} \oplus \ldots \oplus F_{k_m}, \text{ with } k_j \in [1, n], \text{ for } j = 1, ..., m \qquad (3)$$

---

* This work has been partially supported by the 'Progetto Finalizzato Trasporti 2' of the Italian CNR.
** E-Mail: broggi@CE.UniPR.IT

Otherwise B is said to be *non-convex with respect to the same set of SEs*, and thus it can only be expressed as a chain of boolean operations (e.g. unions and/or intersections) between convex elements (*partitions*):

$$B = B_1 \odot B_2 \odot B_3 \odot ... \odot B_z, \qquad (4)$$

where $\odot$ represents any boolean operation (such as unions $\cup$, intersections $\cap$,...) and B_i are convex elements that can be expressed as chains of dilations, as shown by equation (3).

The decomposition of a binary SE into a chain of operations involving only elementary factors is a key problem [1]. So far, only *deterministic* solutions have been analyzed and proposed in the literature [3,10,11,13], each relying on different assumptions (such as convex SEs, specific sets of elementary operators, etc.); on the other hand the optimal decomposition (with respect to a given set of optimality criteria) of non-convex generic SEs with a *deterministic* approach is still an open problem.

This paper addresses this problem utilizing a *stochastic* approach, based on Evolution Programs: starting from a population of potential solutions (individuals), an iterative process modifies the existing individuals and/or creates new ones in accordance to some given functions applied randomly. The best solutions in the population tend to replace the others, and, after a sufficient number of iterations, the algorithm tends to converge toward the optimal solution.

In particular, two are the purposes of this work:

- to apply the results to a real-world case in which the decomposition of SEs is a basic programming technique: the determination of the optimal decomposition with respect to the instruction set of the massively parallel architecture PAPRICA, in order to execute operations based on complex SEs;
- to determine a performance index for the instruction set of a generic massively parallel cellular system dedicated to morphological tasks.

Next section introduces the problem of SE decomposition; section 3 briefly summarizes the approach, and section 4 discusses the results of the stochastic decomposition. Section 5 concludes the paper with some remarks and future research directions.

2 Structuring Element Decomposition on SIMD Systems

This section addresses the problem of the optimal decomposition of a complex SE on *SIMD cellular systems*, whose operations are based on a neighborhood smaller than the size of the SE. In the following examples, a *dilation* between a generic image A and a complex SE B is considered; due to the properties of unions and intersections discussed in [6], namely

$$A \oplus (B_1 \cup B_2) = (A \oplus B_1) \cup (A \oplus B_2) \qquad (5)$$

$$A \oplus (B_3 \cap B_4) \subseteq (A \oplus B_3) \cap (A \oplus B_4), \qquad (6)$$

in the following we prefer to express a non-convex SE as a chain of *unions* of convex SEs, as in equation (5), instead of using *intersections* or other boolean operations.

In cellular systems the set of all possible operations (known as *Instruction Set*, IS) is generally based on 3×3 SEs. Thus the main constraint that must be considered in the decomposition of complex SEs is that each elementary operation must belong to the Instruction Set.

Assuming a system capable of performing horizontal and vertical dilations, and translations in the 8 main directions, the SE B of the following dilation $R = A \oplus B$, where

$$A = \begin{array}{c}\text{[grid]}\end{array} \quad \text{and} \quad B = \begin{array}{c}\text{[grid]}\end{array}, \tag{7}$$

is non-convex with respect to the IS of the system. It can be expressed as a union of convex sets, for example:

$$R = A \oplus B = A \oplus (C_1 \cup C_2) = (A \oplus C_1) \cup (A \oplus C_2) =$$

$$= \left(A \oplus \boxed{\cdot} \oplus \boxed{\cdot} \oplus \boxed{\cdot} \oplus \boxed{\cdot} \right) \cup \left(A \oplus \boxed{\cdot} \oplus \boxed{\cdot} \right) \tag{8}$$

Eq. (8) contains 6 elementary dilations and 1 logical union. Using the chain rule property, $R = A \oplus B$ can be expressed with a two-level solution as:

$$R = \left[\left(A \oplus \boxed{\cdot} \oplus \boxed{\cdot} \oplus \boxed{\cdot} \right) \cup \left(A \oplus \boxed{\cdot} \right) \right] \oplus \boxed{\cdot} \tag{9}$$

This solution requires only 5 dilations and 1 logical union.

2.1 Optimality criteria

The decomposition of a SE can be aimed to many different goals, such as:

- the minimization of the number of decomposing sets (to reduce the number of dilations);
- the minimization of the total number of elements in the decomposing sets (to reduce the size of the data structures and thus also the memory requirements in serial systems);
- the minimization of the total number of computations (for speed-up reasons);
- the possibility to implement complex morphological operations on cellular systems whose IS is based on simple, elementary operations (to overcome the problem caused by the simple interconnection topology that limits the size of possible SEs);
- or even the determination of factors with a given shape (to ease the recognition of 2D objects).

The optimality criterion addressed in this work is the reduction of the computational complexity of the processing, namely the minimization of the number of elementary operations required to perform morphological processings based on large and complex SEs.

2.2 A case study: PAPRICA system

PAPRICA [4] is a special-purpose SIMD massively parallel coprocessor designed to be installed on a moving vehicle for vision-based obstacle detection and lane keeping tasks [2]. Special care has been devoted to the design of the IS of the machine since it affects directly the system performance and effectiveness: the minimization of the number of elementary instructions that must be combined together to synthesize an operation based on a complex SE is a task that extends far beyond the optimization of a specific algorithm, involving also the definition of a sufficiently general IS.

Beside logical operations, PAPRICA IS is composed of the following morphological operations:

$$\text{IS}_\text{P} = \left\{ \oplus \boxed{}, \oplus \boxed{}, \oplus \boxed{}, \oplus \boxed{}, \oplus \boxed{}, \oplus \boxed{}, \oplus \boxed{}, \oplus \boxed{} \right\} \tag{10}$$

3 Implementation of the Genetic Approach

Genetic Algorithms (GAs) are optimization algorithms based on a stochastic search [7], widely used in various fields [5]. They use ideas taken from the biology mechanism to drive the search toward an optimal, or nearly optimal solution: the terminology used in GAs has thus been imported from biology. GAs operate on a *population* of potential solutions for the considered problem (*individuals*) by means of *genetic operators*. Each individual contains a *Genome* or *Chromosome*, that is composed by a set of *Genes*, representing the function parameters and by a *Fitness value*, the result of the *evaluation function*, measuring the "goodness" of the solution encoded in this individual. The genetic search is driven by the *fitness* values of the individuals: each individual must be evaluated to give some quantitative measure of its *fitness*, that is the "goodness" of the solution it represents. At each iteration (*generation*) the fitness evaluation is performed on all individuals. Then, at the following iteration, a new population of potential solutions (*Offspring*) is generated, starting from the individuals with the highest fitness, and replacing, completely or partially, the previous generation.

The genetic operators used to generate new individuals are subdivided into two main categories: *unary* operators, creating new individuals and replacing the existing ones with a modified version of them (e.g. *mutation*,introduction of random changes of genes), and *binary* operators, creating new individuals through the combination of data coming from two individuals (e.g. *crossover*, exchange of genetic material between two individuals). Each iteration step is called *generation*.

The study of GAs led to the more general Generalized GAs or *Evolution Programs* (EPs) [9]. In "standard" GAs an individual is represented by a fixed-length binary string, encoding the parameter set, which corresponds to the solution it represents; the genetic operators act on these binary codes. In EPs, individuals are represented as generalized data structures without the fixed-length constraint. The programmer can choose the most appropriate data structure with respect to the specific problem, for example operating in the same parameters space of the application. In addition, ad-hoc operators are defined to act on these data structures.

When the genetic approach is applied to the SE decomposition problem, its intrinsic nature, that is the *varying number* of elementary items forming a solution, does not allow to know a priori the size of a generic solution, that is the length of the coding of a generic individual. The data structure representing the individual must explicitly encode both the number and the shape of each single elementary operation composing the solution. Moreover, the coding must allow also a quick and easy evaluation phase. An ad-hoc EP has thus been developed, exploiting a method similar to the solution of the bin-packing problem.

4 Analysis of the Results

Let us now consider the decomposition of the following non-convex SE B, whose optimal decomposition is definitely non-trivial:

$$B = \quad (11)$$

After 300 generations on a population of 2000 individuals, taking less than one hour of processing time on a Sun Sparc station 20, considering PAPRICA IS shown in (10), the stochastic decomposition led to 50 elementary dilations and 8 logical unions. The corresponding two-level solution comprises 22 elementary dilations and 8 logical unions, as shown in the following:

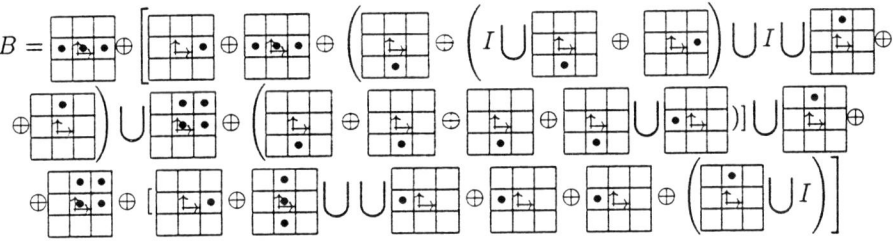

where I is the identity image. This decomposition allows the direct translation of the solution into PAPRICA Assembly code (more details can be found in [4]):

```
L2  = HEXP(L1)      L10 = WMOV(L9)      L18 = SMOV(L17)     L26 = (L15) | L25
L3  = NMOV(L1)      L11 = EMOV(L4)      L19 = SMOV(L18)     L27 = (L19) | L26
L4  = NEEXP(L3)     L12 = VEXP(L11)     L20 = WMOV(L7)      L28 = (L20) | L27
L5  = HEXP(L2)      L13 = SMOV(L6)      L21 = NMOV(L10)     L29 = (L21) | L28
L6  = EMOV(L5)      L14 = NMOV(L6)      L22 = SMOV(L13)     L30 = (L10) | L29
L7  = NEEXP(L2)     L15 = NMOV(L14)     L23 = EMOV(L22)     L31 = (L12) | L30
L8  = WMOV(L4)      L16 = SMOV(L7)      L24 = (L13) | L23
L9  = WMOV(L8)      L17 = SMOV(L16)     L25 = (L6)  | L24
```

4.1 Instruction set evaluation

Different ISs have been evaluated for the decomposition of the same set of SEs, starting from the simplest IS_1, extending it to a RISC-oriented non-symmetric set IS_2, including single direction dilations (as implemented on PAPRICA system) IS_3, replacing the first 4 elements with symmetrical ones, according to a CISC-oriented implementation IS_4, up to a completely unusual set IS_5.

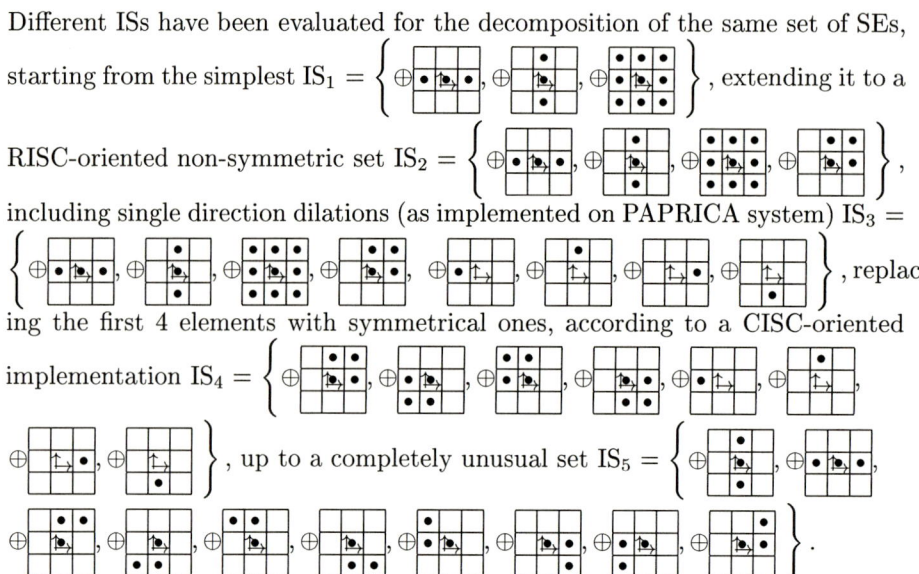

The average number of operations N required to synthesize a sufficiently large set of different SEs (normalized with respect to the number of operations $N(IS_1)$ required by IS_1) is given in table 1, showing that the solution adopted for PAPRICA system represents a good trade-off between the complexity of the IS (in terms of number of factors) and its potentialities.

In the following results are compared in two different cases: with different SE (a) size and (b) shape.

a) The size of the 6 following SEs

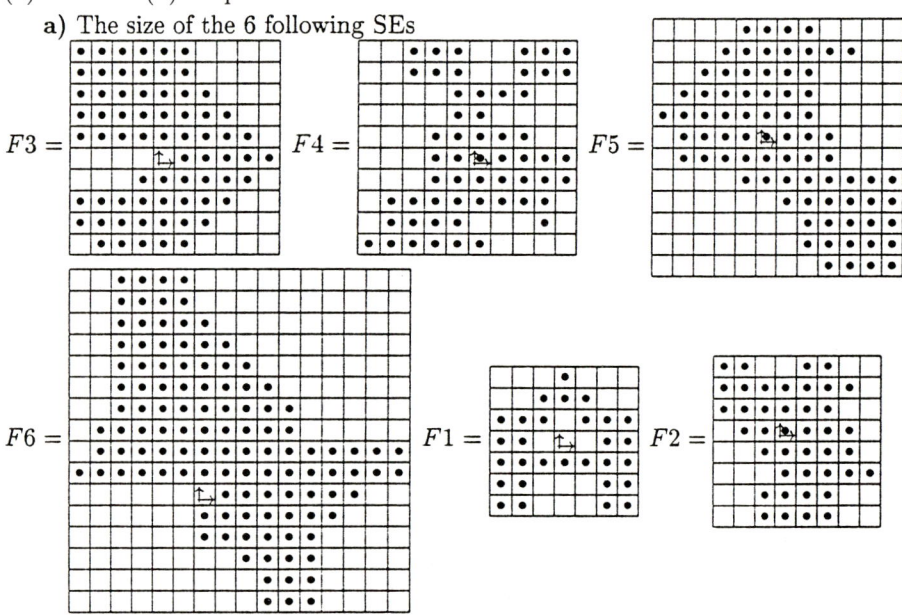

varies from 7×7 up to 16×16; the single-level decomposition has been performed with respect to the 5 different instruction sets presented above; the results are shown in table 2.

Instruction Set	IS_1	IS_2	IS_3	IS_4	IS_5
$N/N(IS_1)$	1	0.93	0.82	0.84	0.71

Table 1. Average IS performance

SE	IS_1	IS_2	IS_3	IS_4	IS_5
F1	24(16/8)	24(16/8)	20(15/5)	21(16/5)	23(16/7)
F2	35(24/11)	30(22/8)	26(20/6)	26(20/6)	19(15/4)
F3	32(23/9)	32(24/8)	29(22/7)	30(23/7)	23(19/4)
F4	52(38/14)	47(35/12)	39(31/8)	41(33/8)	43(34/9)
F5	42(31/11)	38(30/8)	36(29/7)	38(31/7)	29(24/5)
F6	62(47/15)	55(42/13)	–	–	–

Table 2. Results of the decomposition of SEs $F1, F2, ..., F6$. Only two of the decompositions of $F6$ could be performed due to the extremely large memory requirements. The first number represents the total number of instructions; between brackets the number of morphological and boolean operations, respectively.

b) Table 3 shows the results of the decomposition of the following three 7×7 SEs (used in the search for planes in aerial images):

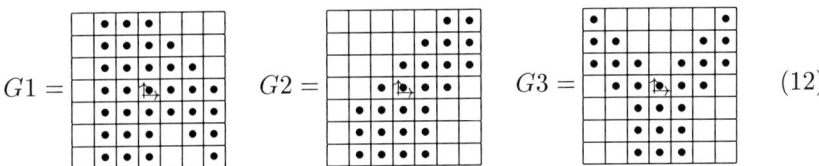

(12)

is shown in table 3.

SE	IS_1	IS_2	IS_3	IS_4	IS_5
G1	16(11/5)	15(10/5)	13(10/3)	14(11/3)	11(9/2)
G2	17(10/7)	16(9/7)	13(9/4)	13(9/4)	7(6/1)
G3	21(14/7)	21(14/7)	20(14/6)	20(14/6)	15(11/4)

Table 3. Results of the decomposition of SEs $G1, G2, G3$.

The example shows that, as a general rule, the larger the number of operations in the IS, the shorter the decomposition; but it also shows that established a fixed number of factors in the IS, a risc-oriented solution provides better results.

5 Conclusion

This paper presented the results of the use of a stochastic approach to the decomposition of arbitrarily shaped binary morphological structuring elements into chains of elementary factors. The application of this technique to *convex* SEs leads to the optimal decomposition discussed in the literature [1]; in addition, this paper addressed the decomposition of generic *non-convex* SEs.

In particular, two are the main results reported in this work:

- first, it is possible to derive automatically the PAPRICA assembly program from the result of the decomposition; when the optimality criterion used in the decomposition is the minimization of the number of elementary morphological operations, this corresponds to the determination of the PAPRICA program requiring the minimum computational time.
- second, the decomposition of a large number of SEs was used to derive a performance index for different ISs; in particular the one implemented on the PAPRICA system demonstrated to be a good trade-off between performance and complexity (intended as the number of elements in the IS).

Due to the extremely high computational load and to the large memory requirements needed by the iterative approach, the genetic engine is now being ported to the MPI parallel environment in order to speed-up the processing and to allow the decomposition of very large SEs.

A graphical interface is also under development to ease the definition of both the initial SE and the IS, as well as the introduction of parameters. The interface, based on Java, will allow remote users to run their own decompositions using the new Web technology. The first release of the complete tool running on Unix systems under the MPI environment will be shortly available as Public Domain software.

References

1. G. Anelli, A. Broggi, and G. Destri. Decomposition of Arbitrarily Shaped Binary Morphological Structuring Elements using Genetic Algorithms. *IEEE Trans PAMI*, 1997. In press.
2. M. Bertozzi and A. Broggi. GOLD: a Parallel Real-Time Stereo Vision System for Generic Obstacle and Lane Detection. *IEEE Trans Image Processing*, 1997. In press.
3. A. Broggi. Speeding-up Mathematical Morphology Computations with Special-Purpose Array Processors. In *Procs of the 27th HICSS*, vol I, pages 321–330, 1994.
4. A. Broggi, G. Conte, F. Gregoretti, C. Sansoè, R. Passerone, and L. M. Reyneri. Design and Implementation of the PAPRICA Parallel Architecture. *The Journal of VLSI Signal Processing*, 1997. In press.
5. D. Goldberg. *Genetic Algorithms in Search, Optimization and Machine Learning*. Addison Wesley, Readings, MA, 1989.
6. R. M. Haralick, S. R. Sternberg, and X. Zhuang. Image Analysis Using Mathematical Morphology. *IEEE Trans PAMI*, 9(4):532–550, 1987.
7. J. Holland. *Adaption in Natural and Artificial Systems*. University of Michigan Press, Ann Arbor, MI, 1975.
8. G. Matheron. *Random Sets and Integral Geometry*. John Wiley, New York, 1975.
9. Z. Michalewicz. *Genetic Algorithms + Data Structures = Evolution Programs*. Springer-Verlag, Berlin, 1992.
10. H. Park and R. T. Chin. Optimal Decomposition of Convex Structuring Elements for a 4-Connected Parallel Array Processor. *IEEE Trans PAMI*, 16(3), March 1994.
11. H. Park and R. T. Chin. Decomposition of Arbitrarily Shaped Morphological Structuring Elements. *IEEE Trans PAMI*, 17(1), January 1995.
12. J. Serra. *Image Analysis and Mathematical Morphology*. Academic Press, 1982.
13. X. Zhuang and R. M. Haralick. Morphological structuring element decomposition. *Computer Vision, Graphics, and Image Processing*, 35:370–382, September 1986.

Bézier Modelling of Cracks

Andrew Varley and Peter Rayner

Department of Engineering, University of Cambridge, Trumpington Street, Cambridge, CB2 1PZ, UK

Abstract. In this paper we show how arbitrary patterns of cracks can be fitted by Bézier curves of unknown order. The Reversible Jump MCMC technique is used to estimate both the number of curves in the image, and the positions of the knots and control points for each curve. The technique described in this paper is suited to a variety of line fitting applications.

1 Introduction

There has been a lot of interest recently in statistical image analysis using global image models [1–4], as opposed to pixel-based methods. Here we present an application that is ideally suited to a global parameterisation.

Art historians are keen to be able to establish the period and region of old paintings, as this can help identify fakes. One way of doing this is by examination of the patterns of cracks on the surface of the paint layer [5]. These cracks evolve over many years and are characteristic of the material and paint composition. Research in this area is still in its early stages. Some examples of some actual crack patterns are shown in Fig. 1.

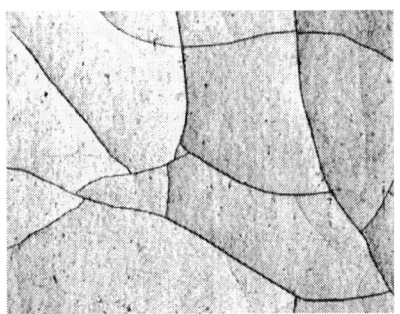
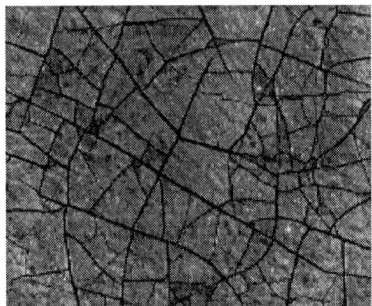

Fig. 1. Examples of crack patterns

2 The Model

The crack images (**Y**) are modelled as an unknown number of Bézier curves, each curve having an unknown number of segments. A Bézier segment is shown in

Fig. 2 and is completely parameterised by the positions of the 2 endpoints, plus the magnitude, μ, and gradient, θ, of the tangent vectors at the endpoints. A Bézier curve is a number of these segments joined end to end, with the gradients having the same direction (but not necessarily the same magnitude) at the joins (Fig. 3). By varying the number and position of the knots, Bézier curves can be fitted to almost all lines in images.

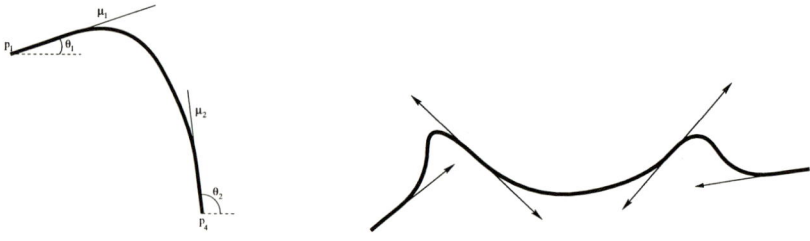

Fig. 2. A Bézier segment   **Fig. 3.** A Bézier curve

The ith curve is completely parameterised by its width, w, greyscale intensity, c, the location of the segment endpoints, P (known as "knots" from now on), and the magnitude, μ, and gradient, θ, of the tangent vectors at each knot. These parameters are collected into a single, variable length, parameter vector ψ_i.

The parameters of the complete model are therefore:

k — the number of curves in the image
s_i — the number of knots in the ith curve
ψ_i — the curve parameters for the ith curve.

2.1 The Likelihood

The likelihood term, $p(\mathbf{Y}|\psi, k)$, is a model of the observational noise present in the original image. We assume that each pixel has independent Gaussian random noise added, the likelihood is therefore

$$p(\mathbf{Y}|\psi, k) = \frac{1}{\sigma\sqrt{2\pi}} \exp\left(\frac{1}{2\sigma^2} \sum_{pq}(Y_{pq} - S_{pq})^2\right) \qquad (1)$$

where the sum is over all the pixels in the image, and S is the model realisation at this iteration. We like to think of the likelihood as a measure of *goodness of fit* of the model to the image data — if the current model configuration is similar to the original image then the likelihood has a high value, whereas a model configuaration that doesn't match the image very well leads to a low value of the likelihood.

2.2 The Priors

The priors are an indication of the range of the parameters that occur in real images. In this application especially, these are very important in determining the sort of images that get simulated when the algorithm is run. There are two parts to the prior — the first is a prior on the number of curves in the image ($p(k)$), and the second a prior on the curve parameters, given the number of curves ($p(\psi|k)$).

Model Order Prior($p(k)$). Prior knowledge of the number of curves in the image is incorporated in the prior $p(k)$, where we use a Poisson distribution with mean λ_k,

$$p(k) = \frac{\exp(-\lambda_k)\lambda_k^k}{k!} \ . \tag{2}$$

Curve Prior ($p(\psi_i)$). The number of knots per curve is also modelled by a Poisson distribution,

$$p(s_i) = \frac{\exp(-\lambda_s)\lambda_s^{s_i}}{s_i!} \ . \tag{3}$$

The greyscale intensity of a curve is modelled with a ramp pdf,

$$p(c_i) = \mathrm{ramp}(0, 255) \ . \tag{4}$$

This biases towards dark curves, and prevents a large buildup of small bright curves, a similar problem to that noticed by Ripley [6]. The width of a curve is modelled by a Poisson distribution with mean 8 pixels. The knot positions are given a uniform pdf. over the image plane. We have found that reasonable results are obtained by giving the knot positions a non-informative prior like this. We could of course include some sort of repulsion term to prevent there being too many knots too close together, but we haven't really found this to be a problem.

The curve prior for the ith curve, $p(\psi_i)$, is simply the product of the priors for all the individual curve parameters, and the overall image prior is

$$p(\psi|k)p(k) = p(k) \prod_{i=1}^{k} p(\psi_i) \ . \tag{5}$$

3 The sampler

Now the model is specified, it remains for it to be fitted to actual images. Since the number of curves present in the image is variable, we have adopted a reversible-jump MCMC sampling scheme to give estimates of both the number of curves present, and their parameters.

The algorithm begins by "guessing" a random number of curves, in random positions on the image (their parameters are chosen from the respective priors). At each iteration from then on a change is proposed in one or more of the parameters. The change may or may not be accepted, depending on the value of the Metropolis-Hastings-Green acceptance ratio for that particular change,

$$p(\text{acceptance}) = \min(1, \text{likelihood ratio} \times \text{prior ratio} \times \text{proposal ratio} \times |J|) \tag{6}$$

where the likelihood ratio is the likelihood value for the current model configuartion divided by the likelihood of the previous model configuration, the prior ratio is the current prior value divided by the previous prior value, the proposal ratio is the probability of choosing the reverse step divided by the probability of choosing the forward step, and $|J|$ is the determinant of the Jacobian of the transformation between the old and new parameters.

A detailed derivation of the acceptance probability formula can be found in [7], but all that need concern us here is the fact that if the proposed change leads to a "better" fit between model and image we accept the change. A "worse" fit may or may not be accepted (the fact that we may accept a worse fit means that the algorithm can escape from local minima).

3.1 The moves in detail

We have defined fourteen different "moves" that we can propose to make at each iteration. The moves have been designed so that the sampler can easily jump out of local minima. For example, the split type 1 move below is really just a combination of a death of a sequence of knots on one curve, followed by a birth of a new curve. The huge reduction in likelihood resulting from a death of a portion of an almost correct curve means it is highly unlikely to be accepted, hence the curve would never be able to be split.

Curve birth/death (Fig. 4). For a birth, a new curve is added to the model. The width and intensity of the curve are chosen from the respective prior distributions, and the knot and control point positions are chosen randomly (subject to maintaining continuity over the segment joins). A death is the opposite, whereby the curve to be deleted is chosen at random and removed from the model. The proposal ratio for a birth is found to be $\frac{d_{k+1}}{b_k(k+1)}$, and for a death $\frac{k b_{k-1}}{d_k}$, where b_k (d_k) is the probability that a birth (death) move is chosen, given there are k curves currently present in the model. The new parameters introduced in the birth step are accounted for by drawing new random variables from the prior distributions, so the Jacobian is 1.

Interior knot birth/death (Fig. 5). For a knot birth, a curve is chosen at random, and a position on that curve is chosen at random. A new knot and associated control points are added in that position, so that the shape of the curve is unchanged (this is a standard procedure, details of which can be found in [8]). The new knot is then displaced a random distance from its original position and new values for the gradient and magnitude of the tangent vector at the new knot are chosen. A death is the opposite, where we randomly choose an interior knot to be deleted.

Width change. A curve is chosen at random and a new width for that curve is proposed from a uniform distribution centred at the original width.

Intensity change. A curve is chosen at random and a new intensity for that curve is proposed from a uniform distribution centred at the original intensity.

Knot move (Fig. 6). A curve is chosen at random, and a knot on that curve is chosen. A new position for that knot is proposed from a uniform distribution centred on the original position of the knot. The tangent vectors at the moved knot remain the same as before.

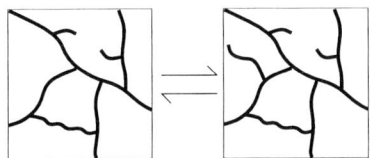

Fig. 4. A curve birth/death

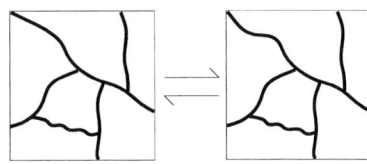

Fig. 5. An interior knot birth/death

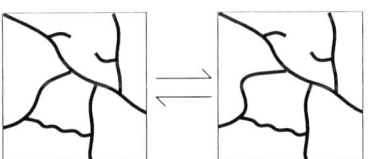

Fig. 6. A knot position change

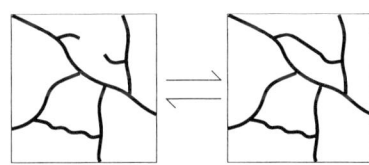

Fig. 7. A split-join type 1 change

Tangent gradient change (Fig. 8). A curve is chosen at random, and a knot on that curve is chosen at random. The tangent gradient at that knot is changed by an angle α.

Tangent magnitude change (Fig. 9). A curve is chosen at random, and a knot on that curve is chosen at random. The tangent magnitude of one of the curve segments at that knot is changed.

Fig. 8. A tangent gradient change   **Fig. 9.** A tangent magnitude change

Move/gradient combined. This is simply a knot move, followed by a gradient change. We found that much better mixing was achieved by including this combined move.

Split/join type 1 (Fig. 7). Two curves are chosen at random. We propose to join one of the ends of the first curve to one of the ends of the second curve with an additional Bézier segment. The split move is the reverse of this — we propose that a segment of an existing curve gets deleted. The join move introduces 2 new variables into the model (the new segment tangent magnitudes), which are chosen based on the existing tangent magnitudes at the new segment endpoints, plus a random perturbation.

Split/join type 2 (Fig. 10). A curve is chosen at random. If there is another curve endpoint within a circle of radius 50 pixels centred at one of the original curve's endpoints, then we propose to move one of the endpoints to the other. A split move (type 2) is the opposite, so we split a curve in a random place, then propose to move one of the two new endpoints away from the other.

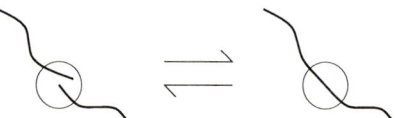

Fig. 10. A split-join type 2 change

Our experience with algorithms of this type has shown that we need to propose births relatively often, as they have a high rejection rate. Deaths should be proposed infrequently, as once a Bézier segment is in the right place, it is not likely to get removed. Various authors have suggested basing the probability of choosing a birth or death on the current value of k [1,2,7,9], however we have found that it makes little difference to have a constant birth and death probability throughout the simulation. We therefore propose a birth with probability 0.2, a death with probability 0.02, and all the other moves with equal probability. Due to the huge size of the parameter space, the sampler needs to be run for a large number of iterations, typically a few million.

4 Results and Discussion

We have tested our algorithm on a selection of real data sets, that is digitised close-up photographs of the paint layer of various paintings. Figure 11 shows the original painting, followed by the (random) starting point for the sampler and samples from the posterior at various times afterwards. Examination of the output of the sampler shows that once a curve appears in the model in the correct position on the image, it tends to stay there and grow to adapt to the original curve. The example shown here is typical of the results obtained on other similar images.

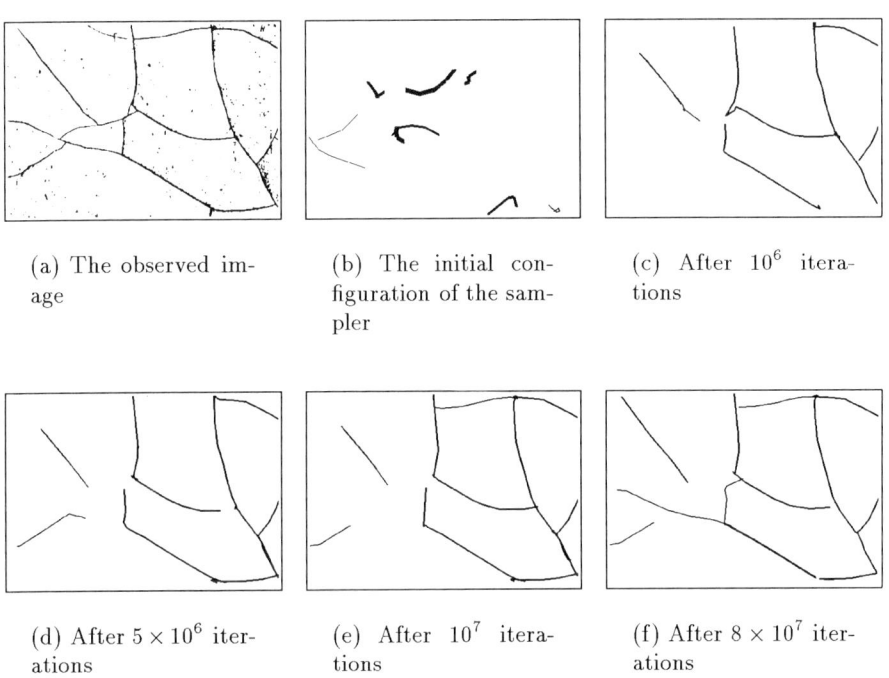

(a) The observed image

(b) The initial configuration of the sampler

(c) After 10^6 iterations

(d) After 5×10^6 iterations

(e) After 10^7 iterations

(f) After 8×10^7 iterations

Fig. 11. Samples from the posterior during the simulation

The design of plausible moves is extremely important in algorithms of this kind. In preliminary tests, we had not included the split/join type 2 move. Figure 12 shows the model configuration after 10^6 iterations in this case. It can be seen that there are a number of curves whose ends have not joined where it seems likely that they should have done. The join type 1 move is not very likely in these cases, as it would introduce a large loop which would immediately get rejected, and no other combination of moves is able to overcome the decrease in likelihood arising from deleting one of the existing curves.

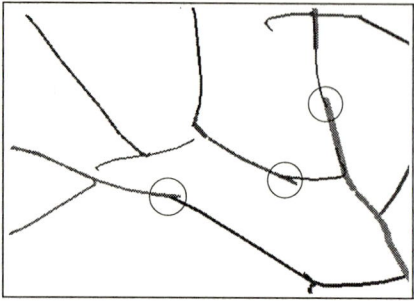

Fig. 12. Sampler output after 10^6 iterations without the split/join type 2 move. The circles show the curve endpoints which would be able to join given the existence of the type 2 join move.

Although a couple of million iterations are needed before most of the lines have been detected, it has to be remembered that each iteration requires very little CPU time. The likelihood (which is a pixel by pixel comparison between the model and the observed image) only has to be recalculated at the pixel positions that have actually changed, in this case along the particular line that is being altered. On a SUN Sparc-20, the algorithm runs at approximately 500 iterations per second.

References

1. Morris, R.: Image sequence restoration using Gibbs distributions. PhD. thesis, Cambridge University, (1995)
2. Rue, H., Syversveen, A. R.: Bayesian object recognition using Baddeley's delta loss. Department of Mathematical Sciences, NTH, Norway, (1995)
3. Clifford, P., Nicholls, G.: A Metropolis sampler for polygonal image reconstruction. Department of Statistics, Oxford University, (1994)
4. Nicholls, G.: Bayesian image analysis with Markov chain Monte Carlo and colored continuum triangulation models. Department of Mathematics, Aukland University, (1996)
5. Bucklow, S.: Formal connoisseurship and the characterisation of craquelure. PhD. thesis, Cambridge University, (1997)
6. Ripley, B., Sutherland, A.: Finding spiral structures in images of galaxies. Philosophical Transactions of the Royal Society of London **332** (1990) 477–485
7. Green, P.: Reversible jump Markov chain Monte Carlo computation and Bayesian model determination. Biometrika **82** (1996) 711–732
8. Foley, J., van Dam, A., Feiner, S., Hughes, J.: Computer graphics: principles and practice. Addison-Wesley (1991)
9. Denison, D., Mallick, B., Smith, A.: Automatic Bayesian curve fitting. Department of Mathematics, Imperial College, (1996)

An Adaptive Deformable Template for Mouth Boundary Modeling

Ali Reza Mirhosseini[1], Kin-Man Lam[2], and Hong Yan[1]

[1] Department of Electrical Engineering, University of Sydney, Australia
[2] Electrical Engineering Department, Hong Kong Polytechnic University, Hong Kong

Abstract. The authors propose an algorithm to automatically extract mouth boundary model in human face images using deformable templates. Our vision algorithm is based on a hierarchical model adaptation scheme. In this paper it will be shown that the role of priori knowledge of the domain is essential for perceptual organization in our algorithm. The knowledge about the shape of the object is used to define its initial deformable template. Each mouth boundary curve is initially formed based on three control points whose locations are found through an optimization process using a suitable cost functional. The cost functional captures the essential knowledge about the shape to perceptually organize image information. Two of the control points are the mouth corners that are primarily located using the priori knowledge of the properties of edge map of the mouth image at its corners. They are used as the initial location of the mouth after an approximate mouth window is found based on locating the head boundary. The model is hierarchically improved in the second stage of the algorithm. Each boundary curve is finely tuned using more control points. An old model is adaptively replaced by a new model only if a secondary cost is further reduced. The results show that model adaptation technique satisfactorily enhances the mouth boundary model in an automated fashion.

1 Introduction

Modeling natural and flexible objects is a complicated task that has not been thoroughly investigated. For instance automatic detection and description of salient features is crucially important in building a human face image processing and analysis system. This difficult task was initiated by [3] and continued by many others [2, 1, 7]. Conventional edge detectors are not able to find the boundary of the natural objects especially facial features such as mouth and eyes partially because the edges are seldom sharp as the idealized detectable edges by edge detectors. Furthermore the contrast of gray level around the boundaries is very low. However even the local edges exist they still can not be organized into a global percept [7]. Recent works by [1] using active contours and deformable models pioneered modeling the shape of boundary for natural objects. Deformable models make use of global information, and hence increase the reliability of locating a contour. A deformable model is specified by a set of parameters, which allows inclusion of a priori knowledge of the expected shape of an object.

2 Mouth Deformable Model

A mouth can have large degrees of variations due to its physical flexibility. Existence of various facial expressions together with its shape variations among individuals make modeling the shape of mouth contours difficult by rigid templates. Therefore it seems appropriate to deal with these problems using deformable templates [7]. A mouth model can be represented by three boundary curves passing over inner and outer mouth outlines. Each curve passes through a number of control points that can be found through optimizing a suitably defined cost function that includes the required knowledge and constraints.

2.1 Adaptive Modeling Procedure

An automated procedure to produce a mouth model can be described as:

1. Locate approximate mouth corners as the initial location of the mouth,
2. Initialize mouth model using three parametric curves with three control points for each boundary curve,
3. Optimize the model for the best control points using a primary cost function. The search for a better control point takes place within a neighborhood window. Replace the old model with new model if cost further decreases,
4. Add more control points and go to step 3 if the model was renewed earlier, otherwise end the procedure.

The order of the mouth outline curves and the position of their control points are the major parameters that are to be optimized. We propose an order adaptation algorithm for fine tuning the model using higher order models for mouth outline curves. Finding the model parameters is cast into an optimization problem.

3 Mouth Corner Detection

The approximate location of a mouth can be worked out with anthropometric standards, using extracted head boundaries [6, 10]. The mouth boundaries are located between two mouth corners, denoted by X_0 and X_1 as illustrated in Fig.2. These two points can form the initial location of the mouth. A corner detection method is used based on the scheme proposed by [4] and applied by [9] to facial corners. It provides information about corner orientations and location. A corner is formed by intersection of two straight lines. A detected corner points is classified as a mouth corner candidate if it satisfies certain properties that are based on image information, as tabulated in Table.1.The curvature β is defined as the acute angle between the two lines. The orientation α represents the orientation of the line which bisects the acute angle between the two lines. The region dissimilarity, D, is a measure of the difference between the gray level averages in the two regions R_1 and R_2 bounded by the edges [9]. The first step for corner detection is to detect the edges of an image. Each edge pixel is then considered as a candidate for a corner. In order to detect the two lines of a

	β	α	D
C_0	$52.5° - 90°$	$-45° - 45°$	< -10
C_1	$52.5° - 90°$	$135° - 225°$	< -10

Table 1. *Properties of the corners of the mouth.*

corner accurately we use the matching scheme used in [9]. This method has the advantage of being insensitive to noisy edge pixels. The existence of noisy edge pixels will not affect the identification of the lines and the corner features.

3.1 Selection and Detection of Corner Positions

A procedure is introduced to simplify the selection of the best corner among corner candidates [9]. The candidates for a corner type are clustered into groups. Two points, (x_i, y_i) and (x_j, y_j), of the same corner type belong to the same cluster if

$$|y_i - y_j| + |x_i - x_j| < d \qquad (1)$$

where d is a threshold. The representative of a cluster is the one with the largest $|D|$. The next procedure is to choose one of the clusters to represent the corner. In this procedure, different cost functions are defined for the different corner types.

By assuming that the face is rotated to a limited degree, $dx = x_1 - x_0$ represents the approximate length of the mouth. Pairs of corners are thus formed if

$$0.3 \times face\_width < dx < 0.5 \times face\_width \qquad (2)$$

where $face\_width$ is the width of the face which is obtained from the face boundary. A pair is then selected to represent the two corners if the value of the following cost function is a minimum.

$$cost = K_1 \frac{|dy|}{|dy|_{max}} + K_2 \frac{|D|_{max}}{|D|} \qquad (3)$$

where K_1 and K_2 are the weighting factors for the two normalized cost terms, $|dy|_{max}$ is the maximum value of $|dy| = |y_1 - y_0|$; and $|D|_{max}$ is the maximum value of $|D| = |D(C_0) + D(C_1)|$ among all the corner pairs.

4 Initial Mouth Model

The initial mouth model is composed of three mouth outline curves, denoted by P_1, P_2 and P_3 and illustrated in Fig.2. Each curve is initially a parabola with three control points including $\mathbf{X_0}$ and $\mathbf{X_1}$ and a third point, $\mathbf{X_2}$, which is to be located through optimizing a cost function.

4.1 Cost function for Mouth Shape

An optimization process is employed to precisely locate mouth corners and find mouth boundaries for the initial mouth model. We use a primary cost function based on a set of potential energy functions, composed of a set of conditions that suitably constraints each mouth outline curve. Each potential energy captures the features of each mouth boundary and its relation with the other parts of the mouth model. These features include distance between mouth outlines, corner point potential, and edge and valley potentials.

The total primary potential energy $E1_{P_i}$ for each boundary curve P_i formed by three control points $\mathbf{X_i}$, $\mathbf{X_j}$ and $\mathbf{X_k}$ is defined as:

$$E1_{P_i}(\mathbf{X_i}, \mathbf{X_j}, \mathbf{X_k}) = E_c(P_i, P_j) + E_{d1} + E_e(P_i, \mathbf{X_i}, \mathbf{X_j}) + E_v(P_i, \mathbf{X_i}, \mathbf{X_j}) \quad (4)$$

where

1. The corner potential term, E_c, is a measure of the fitness of a point to be a mouth corner. The potential is defined as being equal to the region dissimilarity of the corner at the point:

$$E_c(P_i, P_j) = -|D(R_1, R_2)| \quad (5)$$

where R_1 and R_2 are the two regions in the window separated by the two boundary curves P_i and P_j.

2. The valley energy term, E_v, is given as the integral of valley forces Φ_v along boundary curve P_i between two mouth corners $\mathbf{X_i}$ and $\mathbf{X_j}$:

$$E_v(P_i, \mathbf{X_i}, \mathbf{X_j}) = -\frac{1}{L_{P_i}} \int_{P_i} \Phi_e(\mathbf{x}) ds \quad (6)$$

3. The edge potential term, E_e, is a measure of the edge intensities Φ_e along boundary curve P_i between two mouth corners $\mathbf{X_i}$ and $\mathbf{X_j}$:

$$E_e(P_i, \mathbf{X_i}, \mathbf{X_j}) = -\frac{1}{L_{P_i}} \int_{P_i} \Phi_e(\mathbf{x}) ds \quad (7)$$

4. Distance potential term, E_{d1}, is defined in order to control the shape of the mouth models:

$$E_{d1} = |h_{tb} - d(\mathbf{x}_i, \mathbf{x}_j)| + K_3|b - b_1| + K_4|b - b_2| \quad (8)$$

where $h_{tb} = \lambda|\mathbf{X_1} - \mathbf{X_0}|$. The definitions of these two potentials are illustrated in Fig.1 (a). It measures the absolute difference between the expected thickness h_{tb} of the upper or lower lip and the corresponding thickness $d(P_i, P_j)$ as represented by the model. K_3 and K_4 are the weighting factors for their corresponding terms.

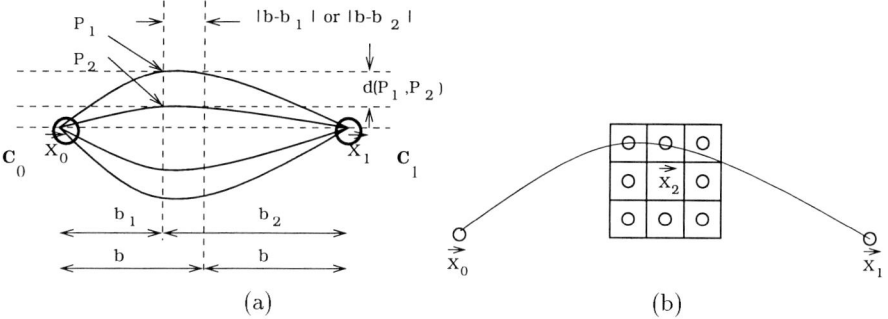

Fig. 1. (a) Dimensions of the mouth for measuring E_1 and E_2 and two mouth corners are C_0 and C_1, (b) Neighbors searched in the greedy algorithm.

where L_{P_i} denotes the length of the parametric curve P_i in pixels and the edge intensities and valley forces, Φ_e and Φ_v, are extracted by morphological edge and valley detectors [5].

Distance potential, $E_d(P_i, P_j)$, is a measure of the shape of the mouth and controlling the distance between two mouth boundaries and is defined with the minimum distance between P_i and P_j.

4.2 Search for Cost Minimization

The minimum value of E_{P_i} is found by moving the control points of each outline curve. The total potential energy for each mouth outline curve is optimized by searching for the control points within a neighborhood region around each curve. The search process is based on a greedy algorithm that is applied in active contour model optimization technique as shown in Fig.1 (b) [10]. Each control point can move to an eight connected neighborhood window if the total potential energy decreases. After the initial model is optimized in this step, it is passed to step 4 of procedure in Section 2.1., for fine tuning.

5 Adapting Mouth Model

The old model extracted in step 3 of procedure in Section 2.1., can be improved by increasing the order of each boundary curve, through adding more control points to each mouth outline curve. Therefore for instance a cubic spline curve, as a new model, can be formed using four control points for each outline curve, by adding an extra control point to an old parabolic boundary curve model.

The total secondary potential energy $E2_{P_i}$ for a generalized new model with a boundary curve P_i formed using m control points $\{\mathbf{X_1}, \cdots, \mathbf{X_{m-1}}\}$ is defined as:

$$E2_{P_i}(\mathbf{X_1}, \cdots, \mathbf{X_{m-1}}) = E_e(P_i, \mathbf{X_1}, \cdots, \mathbf{X_{m-1}}) + E_v(P_i, \mathbf{X_1}, \cdots, \mathbf{X_{m-1}}) + E_{d2} \quad (9)$$

where

1. The valley energy term, E_v, is given by Equation (7),
2. The edge potential term, E_e, is given by Equation (8),
3. The distance potential, E_{d2} is a measure of the distance and shape of the mouth for a higher order boundary curve that can is defined similar to Equation (9):

$$E_{d2} = |h_{tb} - d(P_i, P_j)| \qquad (10)$$

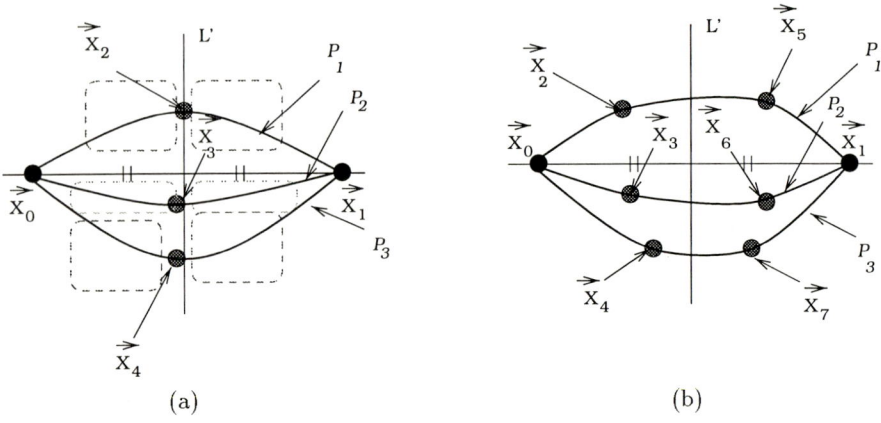

Fig. 2. A sketch of a deformable mouth template modeled with three boundary curves P_1, P_2 and P_3 for (a) initial model with 3 control points $X_1, \cdots, X_3$, and (b) final model with 4 control points $X_1, \cdots, X_4$.

For instance in the new model of the upper lip outline curve, the corner points $\mathbf{X}_0$ and $\mathbf{X}_1$ are retained, but $\mathbf{X}_2$ in the old model is replaced by two middle points $\mathbf{X}_2$ and $\mathbf{X}_3$ in the new model, as illustrated in Fig.2(b). However both new points $\mathbf{X}_2$ and $\mathbf{X}_3$ are initially located on $\mathbf{X}_2$ that of the old model. The dashed rectangular areas in Fig.2(a) shows an example of search area of control points of a new model. This methodology can be easily extended to further increase the number of control points. The secondary cost function is partially based on the potential energy functional employed in the old model. However in the secondary the corner potential terms are canceled since we assume that the corner points are optimally found in the primary optimization process. The secondary optimization process is mainly applied to locate the best control points for the new model. The new model captures more information about mouth outlines. If the cost function does not decrease in optimizing the new model, the old model is retained as the final model. It is always possible to increase the number of control points to achieve a control points leads to higher computational cost. However a higher order model is more prune to image noise and it is also computationally expensive. Therefore there is a trade-off between the order of the model, the amount of image noise and the speed of the algorithm.

6 Experiments

We have tested our algorithm on a set of human face images obtained from Olivetti face image database, Cambridge. Our algorithm was implemented using C and run on *SUN Sparc 2* workstation. To remove image background and also to locate an initial mouth region the boundary of a head is located using active contours. The mouth is modeled by three adaptive deformable templates. The model is initialized by parabolic curves using three control points and each control point is located using Equation (5). Then it is finely tuned by increasing the order of the model with adding more control points for each curve and optimizing Equation (10). We have found empirically that parabolic and cubic curve are usually sufficient to model various forms of a mouth image. Examples of mouth models superimposed on the mouth region are illustrated in Fig.3.

7 Conclusion

We propose a new mouth model based on a hierarchical adaptive deformable template. This modeling scheme employed a priori knowledge of the shape of mouth contour to define a proper energy function. The energy function captures the properties of a mouth contour and its relation with other part of the mouth. Each contour line is modeled using a set of control points. The location of each control point is found by a greedy search algorithm. The model is tuned by increasing the number of control points to achieve a better model. The experimental showed satisfactory results on a database of human face images.

Fig. 3. The examples of mouth model using hierarchical adaptive deformable template.

References

1. A.Blake and A.L.Yuille, *Deformable Templates in Active Vision*, M.I.T. Press, Cambridge, (1992).
2. A.L.Yuille, P.W.Hallinan and D.S. Cohen, *Feature Extraction from Faces using Deformable Templates*, International Journal of Computer Vision, 8(2), 99–111, (1992).

3. T. Kanade, *Computer Recognition of Human Faces*, Bikhauser Verlag, Basel and Stuttgart, (1977).
4. X.Xie, R.Sudhakar, and H.Zhuang, *Corner Detection by a Cost Minimization Approach*, Pattern Recognition, **26** (8): 1235–1243, (1993).
5. J.Serra, *Image Analysis and Mathematical Morphology*, Academic Press, New York, (1982).
6. M.Vezjak and M.Stefancic, *An Anthropological Model for Automatic Recognition of the Male Face*, Annals of Human Biology, **21** 363, (1994).
7. A.L.Yuille, *Deformable Templates for Face Recognition*, Journal of Cognitive Neuroscience, **3** (1), 59-70, (1991).
8. M.Kass, A.Witkin and D.Terzopoulos, *Snakes: Active Contour Models*, in Proc. First Int. Conf. on Computer Vision, 259–269, London, (1995).
9. Kin-Man Lam, *Computerized Human Face Recognition*, PhD thesis, Department of Electrical Engineering, University of Sydney, Sydney, Australia, (1996).
10. A.R.Mirhosseini and H.Yan, *An optimaly Fast Greedy Algorithm for Active Contours*, IEEE International Symposium on Circuits and Systems, Hong Kong, (1997).

A Two-Stage Framework for Polygon Retrieval Using Minimum Circular Error Bound

Lun Hsing Tung and Irwin King

{lhtung, king}@cse.cuhk.edu.hk
Department of Computer Science and Engineering,
The Chinese University of Hong Kong,
Shatin, New Territories, Hong Kong

Abstract. We have proposed a two-stage framework for polygon retrieval [12, 11] which incorporates both qualitative and quantitative measures of polygons in the first and second stage respectively. In this paper, we introduce an extension to our two-stage framework. We propose a new polygon matching technique using Circular Error Bound and describe how this technique works under translation and scaling of polygons. Base on this technique, we propose a new translation invariant similarity measure for polygons named Minimum Circular Error Bound, which can be used in the second stage of the two-stage framework. We compare the Minimum Circular Error Bound method with the Hausdorff Distance method and demonstrate the advantages of our method.

1 Introduction

Shape matching and measuring similarity between shapes are important issues in pattern recognition. They can be applied in many applications, such as providing query-by-shape facility in image database systems and constructing hand writing recognition systems. In this paper, we focus on the matching of polygonal shapes instead of arbitrary shapes, since shapes are often represented by polygons and polygon approximation of shapes is acceptable in many applications.

In [12, 11], we proposed a two-stage framework for polygon retrieval in image databases. The first stage of the framework uses the Binary Shape Descriptor (BSD) [2] technique to perform polygon classification and prune the search space in order to speed up query processing. The second stage of the framework uses any available polygon similarity measuring technique for quantitative measurement of the similarity between polygons. We proposed the Multi-Resolution Area Matching (MRAM) technique in [12, 11] as the technique to be incorporated at the second stage. In this paper, we propose an extension to the two-stage framework which allows systematic control on the degree of search space pruning.

Considerable works have been carried out on the polygon matching problem. Most of these researches extract features from polygons and use these features as similarity measure [9, 6, 4, 10]. However, the similarity ranking of polygons produced by these methods may not coincide with human perception. In this paper, we propose a polygon matching technique using Circular Error Bound

(CEB) which is based on an intuitive human concept of polygon resemblance. Using the same idea, we propose a polygon similarity measure named Minimum Circular Error Bound (MCEB) which produces polygon rankings resembling human rankings. The MCEB method can be used at the second stage of the two-stage framework for shape matching.

The two-stage framework is implemented in the *Montage* image database system [8] currently under development at the Chinese University of Hong Kong. The *Montage* system is an image database system designed for the fashion, textile, and clothing industry in Hong Kong. It supports feature based retrieval by color histogram, color sketch, shape, and texture.

This paper is organized as follows. We present the extension to the two-stage framework in Section 2. We propose the polygon matching technique using Circular Error Bound in Section 3. In Section 4, we present the Minimum Circular Error Bound similarity measure for polygons. The experimental results of our work will be discussed in Section 5. Conclusion is made in Section 6.

2 Extension to the Two-Stage Framework

The two-stage framework approach we proposed in [12, 11] may fail to produce good matching results because of the first stage filtering mechanism. In Figure 1, there are three polygons named P, Q, and R. Using polygon P as the target polygon, the proposed method will not be able to produce the result that polygon Q is more similar to polygon P than polygon R does. It is because polygon Q is not in the equivalent class as polygon P so it will not be selected at the first stage. On the other hand, polygon R is in the same equivalent class as polygon P so it is selected for the second stage matching. Yet, polygon Q appears to be more similar to polygon P than polygon R to polygon P.

To tackle this problem, we propose the following extension to the first stage of the two-stage framework. When a query is initiated, the SBSD of the target polygon is computed. All polygons inside the database having a SBSD within a user specified Hamming Distance to the SBSD of the target polygon are selected for the second stage processing. For example, if user specifies a Hamming Distance of 1, both polygon Q and polygon R in Figure 1 will be selected for second stage matching so polygon Q will have the chance to be compared with polygon P. Whether polygon Q is said to be more similar to polygon P or not will still depends on the polygon similarity measuring method used at the second stage.

This extension provides a systematic way for controlling the degree of pruning database entries. Small Hamming Distance value has larger pruning effect but with higher risk of producing worse matching results. On the other hand, large Hamming Distance value may produce better matching result but will result in inefficiency since a lot of polygons will be selected for second stage matching.

3 Polygon Matching using Circular Error Bound

We propose a polygon matching technique based on an intuitive human definition of polygon resemblance. The intuitive definition of similar polygons is as follows. If two polygons are similar (or matched), then each vertex of one polygon is close to its corresponding vertex of another polygon when the two polygons are overlapped. The correspondence between vertices is an one-to-one mapping. Therefore, the definition and the technique we proposed only work on polygons that have the same number of vertices. Before the two polygons are overlapped, translation, scaling and rotation are allowed to be performed on the polygons.

Definition 1. A polygon P is represented by an ordered list of vertex coordinates, $P = (V_1, \ldots, V_n)$ where $V_i \in \mathbb{R}^2$ and n is the number of vertices of P.

Definition 2. A transformation T is a vector, i.e. $T = (t_x, t_y, s_x, s_y, \theta)$ where t_x is translation in X direction, t_y is translation in Y direction, s_x is the scaling in X direction, s_y is the scaling in Y direction, and θ is the rotation about the origin. $T(Q)$ denotes the polygon (or vertex) obtained by applying T to Q.

Definition 3. Given a tolerance vector $E = (\epsilon_1, \ldots, \epsilon_n)$, Q is said to be matched with P if there exists a transformation T such that $Q' = T(Q) = (U'_i, \ldots, U'_n)$ and $\forall_{1 \leq i \leq n} \|V_i - U'_i\| \leq \epsilon_i$, where $\|\cdot\|$ denotes the Euclidean norm.

Definition 4. Given V_i, ϵ_i and U_i, the ith Circular Error Bound, C_i, is a circle with ϵ_i as its radius and $(V_i - U_i)$ as its center.

Note that Definition 3 assumes we already know the pairing of vertices between the two polygons, i.e. V_i should match U_i.

The polygon matching task is formulated as follows. Given two polygons P and Q with a tolerance vector E, the task is to determine whether a transformation T exists such that Q is said to be matched with P under Definition 3. By Definition 3, the transformation T is an arbitrary tuple $(t_x, t_y, s_x, s_y, \theta)$. However, in nowadays applications, the transformations in polygon matching task are often restricted to some special cases, for example, translation and (or) scaling only. With restricted transformations, we have efficient solutions for the polygon matching task. In the following sections, we will present the solution for the polygon matching task when (1) only translations are allowed, (2) only translations and uniform scaling in X, Y direction are allowed, and (3) only translations and independent scaling in X, Y direction are allowed.

3.1 Translation

Assume that transformation T in Definition 3 is restricted to $T = (t_x, t_y, 1, 1, 0)$.

Proposition 1. Given $P = (V_1, \ldots, V_n)$, $Q = (U_1, \ldots, U_n)$, $E = (\epsilon_1, \ldots, \epsilon_n)$, if the n Circular Error Bounds $C_1, \ldots, C_n$ of P and Q have common intersection, then Q is matched with P.

Proof. Assuming $V_i = (a_i, b_i)$ and $U_i = (c_i, d_i)$, by Definition 4, Circular Error Bound C_i is a circle with ϵ_i as its radius and $(a_i - c_i, b_i - d_i)$ as its center. If $C_1, \ldots, C_n$ have common intersection, then for any point (t_x, t_y) in the common intersection, the distance between this point and the center of any C_i is larger or equal to the radius of C_i. Figure 2(a) illustrates this idea when both P and Q are triangles. Thus, $\forall_{1 \leq i \leq n}, [(a_i - c_i) - t_x]^2 + [(b_i - d_i) - t_y]^2 \leq \epsilon_i^2$. Re-arranging this equation, we have $[a_i - (c_i + t_x)]^2 + [b_i - (d_i + t_y)]^2 \leq \epsilon_i^2$ which is equivalent to $\|V_i - U'_i\| \leq \epsilon_i$ where $U'_i = T(U_i)$ and $T = (t_x, t_y, 1, 1, 0)$. By Definition 3, Q is matched with P.

3.2 Translation and uniform scaling in X, Y direction

Assume that transformation T in Definition 3 is restricted to $T = \langle t_x, t_y, s, s, 0 \rangle$. Let $U_i = (c_i, d_i)$ and apply the scaling transformation $S = (0, 0, s, s, 0)$ to Q, we have $U_i' = S(U_i) = (sc_i, sd_i)$. Thus, Circular Error Bound C_i of P and Q', where $V_i = (a_i, b_i)$, is a circle with ϵ_i as its radius and $(a_i - sc_i, b_i - sd_i)$ as its center. Two Circular Error Bounds C_i and C_j intersect each other if and only if

$$[(a_i - sc_i) - (a_j - sc_j)]^2 + [(b_i - sd_i) - (b_j - sd_j)]^2 \leq (\epsilon_i + \epsilon_j)^2 \quad (1)$$

Re-arranging Equation (1), we have

$$[(c_i - c_j)^2 + (d_i - d_j)^2]s^2 - 2[(a_i - a_j)(c_i - c_j) + (b_i - b_j)(d_i - d_j)]s \\ + [(c_i - c_j)^2 + (d_i - d_j)^2 - (\epsilon_i + \epsilon_j)^2] \leq 0 \quad (2)$$

Solving Equation (2), we get a range, $\mathbb{S}_{ij}$, for s that the inequality holds (Figure 2(b)).

Proposition 2. *If $\bigcap_{1 \leq i,j \leq n} \mathbb{S}_{ij} \neq \emptyset$, then Q is matched with P.*

Proof. If $\bigcap_{1 \leq i,j \leq n} \mathbb{S}_{ij} \neq \emptyset$, then $\exists S = (0, 0, s, s, 0) \in \bigcap_{1 \leq i,j \leq n} \mathbb{S}_{ij}$ such that Circular Error Bounds $C_1, \ldots, C_n$ of P and Q' have common intersection, where $Q' = S(Q)$. By Proposition 1, Q' is matched with P. Thus, $\exists T = (t_x, t_y, 1, 1, 0)$ such that $\forall_{1 \leq i \leq n} \|V_i - U_i''\| \leq \epsilon_i$ where $U_i'' = T(U_i')$. Therefore, $\exists T' = T \circ S = (t_x, t_y, s, s, 0)$ such that $\forall_{1 \leq i \leq n} \|V_i - U_i''\| \leq \epsilon_i$ where $U_i'' = T'(U_i)$. By Definition 3, Q is matched with P.

3.3 Translation and independent scaling in X, Y direction

Assume that transformation T in Definition 3 is restricted to $T = \langle t_x, t_y, s_x, s_y, 0 \rangle$. Let $U_i = (c_i, d_i)$ and apply the scaling transformation $S = (0, 0, s_x, s_y, 0)$ to Q, we have $U_i' = S(U_i) = (s_x c_i, s_y d_i)$. Thus, Circular Error Bound C_i of P and Q', where $V_i = (a_i, b_i)$, is a circle with ϵ_i as its radius and $(a_i - s_x c_i, b_i - s_y d_i)$ as its center. Two Circular Error Bounds C_i and C_j intersect each other if and only if

$$[(a_i - s_x c_i) - (a_j - s_x c_j)]^2 + [(b_i - s_y d_i) - (b_j - s_y d_j)]^2 \leq (\epsilon_i + \epsilon_j)^2 \quad (3)$$

Re-arranging Equation (3), we have

$$[(a_i - a_j) - (c_i - c_j)s_x]^2 + [(b_i - b_j) - (d_i - d_j)s_y]^2 \leq (\epsilon_i + \epsilon_j)^2 \quad (4)$$

Equation (4) defines an ellipse, $\mathbb{E}_{ij}$, on the s_x-s_y plane. A point (s_x, s_y) in $\mathbb{E}_{ij}$ defines a transformation $S = (0, 0, s_x, s_y, 0)$ such that when S is applied to Q, the Circular Error Bounds C_i and C_j, of $S(Q)$ and P, intersect each other (Figure 2(c)).

Proposition 3. *If $\forall_{1 \leq i,j \leq n} \mathbb{E}_{ij}$ have common intersection, then Q is matched with P.*

Proof. If $\forall_{1 \leq i,j \leq n} \mathbb{E}_{ij}$ have common intersection, then for any point (s_x, s_y) in the intersection, Circular Error Bounds $C_1, \ldots, C_n$ of P and Q', intersect each other where $Q' = S(Q)$ and $S = (0, 0, s_x, s_y, 0)$. By Proposition 1, Q' is matched with P. Thus, $\exists T = (t_x, t_y, 1, 1, 0)$ such that $\forall_{1 \leq i \leq n}, \|V_i - U_i''\| \leq \epsilon_i$ where $U_i'' = T(U_i')$. Therefore, $\exists T' = T \circ S = (t_x, t_y, s_x, s_y, 0)$ such that $\forall_{1 \leq i \leq n}, \|V_i - U_i''\| \leq \epsilon_i$ where $U_i'' = T'(U_i)$. By Definition 3, Q is matched with P.

4 Minimum Circular Error Bound

The results presented in Section 3.1, 3.2, and 3.3 only deal with queries of whether a polygon Q is matched with another polygon P subject to some tolerances and under certain transformation restrictions. It is also useful to find out how similar a polygon Q is comparing to another polygon P. For example, we may want to rank a list of polygons according to the similarity between these polygons and a target polygon. We propose a translation invariant similarity measure of polygons named Minimum Circular Error Bound (MCEB) based on the Circular Error Bound technique we described above.

Definition 5. The Minimum Circular Error Bound of a polygon $Q = (U_1, \ldots, U_n)$ comparing to another polygon $P = (V_1, \ldots, V_n)$ is defined as

$$\xi = \min_{\forall t_x, t_y T = (t_x, t_y)} \max_{1 \leq i \leq n} \|V_i - T(U_i)\|$$

ξ can be calculated as follows. Let $V_i = (a_i, b_i)$ and $U_i = (c_i, d_i)$. Further assume that the tolerance vector $E = (\epsilon_1, \ldots, \epsilon_n)$ where $\epsilon_1 = \cdots = \epsilon_n$. The Circular Error Bound C_i is a circle with ϵ_i as its radius and $(a_i - c_i, b_i - d_i)$ as its center. If two Circular Error Bounds C_i and C_j intersect each other, we have

$$[(a_i - c_i) - (a_j - c_j)]^2 + [(b_i - d_i) - (b_j - d_j)]^2 \leq (\epsilon_i + \epsilon_j)^2 \qquad (5)$$

Since $\epsilon_i = \epsilon_j$, we denote the value of ϵ_i and ϵ_j as ϵ_{ij}. The minimal value of ϵ_{ij} that Equation (5) holds is $\epsilon_{ij} = \frac{1}{2}\sqrt{[(a_i - c_i) - (a_j - c_j)]^2 + [(b_i - d_i) - (b_j - d_j)]^2}$. The MCEB of the two polygons Q and P is $\xi = \max_{1 \leq i,j \leq n} \epsilon_{ij}$, such that for $\epsilon_1 = \cdots = \epsilon_n \geq \xi$, $\forall_{1 \leq i,j \leq n} C_i$ and C_j intersect each other. That is, for $\epsilon_1 = \cdots = \epsilon_n \geq \xi$, Circular Error Bounds $C_1, \ldots, C_n$ of Q and P have common intersection and Q is matched with P under Proposition 1.

5 Experimental Results

We compare the MCEB method with the Hausdorff Distance method [3, 5].

Definition 6. Given two finite point sets $A = \{a_1, \ldots, a_n\}$ and $B = \{b_1, \ldots, b_m\}$, the Hausdorff Distance is defined as $H(A, B) = \max(h(A, B), h(B, A))$, where $h(A, B) = \max_{a \in A} \min_{b \in B} \|a - b\|$ and $\|\cdot\|$ is some underlying norm on the points of A and B.

We choose Hausdorff Distance method for comparison since both methods measure polygon similarity based on the distance between polygon vertices. We will compare their similarity ranking results as well as the running time complexity.

5.1 Similarity Ranking

We compare the polygon rankings produced by the MCEB and the Hausdorff Distance method. The experiments are conducted as follows. In each experiment, two polygons are selected for generating the input polygons. For example, Figure 3(a) and Figure 3(b) show the two polygons used in one of experiments where

Figure 3(a) is used as the first polygon and Figure 3(b) is used as the last one. By interpolating these two polygons, 48 intermediate polygons are generated, which gives us 50 polygons in total.

Using this method, we obtain a list of polygons ranked by their relative similarity to the first polygon, which resembles human ranking. We then rank these 50 polygons using the MCEB method and the Hausdorff Distance method accordingly using the first polygon as the target polygon. We use the number of polygons having different relative ranking from the original list as the quality measure of the rankings produced. A small number indicates a good ranking which means that the ranking produced is similar to the original list as well as human visual ranking.

Figure 3(e) and Figure 3(f) show the rankings produced by the MCEB method and the Hausdorff Distance method using Figure 3(a) and Figure 3(b) as the input data. The quality measure of the two rankings are 0 and 9 respectively. Figure 3(g) and Figure 3(h) show the rankings produced by the MCEB method and the Hausdorff Distance method using Figure 3(c) and Figure 3(d) as the input data. The quality measure of the two rankings are 0 and 13 respectively. In these two experiments, the MCEB method produces better rankings than the Hausdorff Distance method.

5.2 Running Time Complexity

The computational complexity of the Hausdorff Distance is $O(n^2)$ and it is $O(n^2)$ for the MCEB method, if the correspondence of vertices is known, or $O(n^3)$ if the correspondence of vertices is unknown since we have to exhaust the n possible correspondences in order to find out the overall MCEB for a n-gon. However, note that the MCEB method gives similarity measure between two polygons under the optimal translation, but the Hausdorff Distance method does not. We may want to use the Hausdorff Distance under optimal translation for ranking polygons instead of the original Hausdorff Distance in order to produce better ranking of polygons. The computational complexity for optimal Hausdorff Distance under translation is $O(n^4 \log^3(n^2))$ [1], which is much larger than that of the MCEB method.

Table 1 shows the average query processing time of the MCEB method, the Hausdorff Distance method and the MRAM method we that proposed in [12, 11]. The experiments are conducted using the simple system we described in [12]. There are 9000 polygons in each testing database and each database is consisted of polygons with a specific number of sides, from 3 to 8. The table shows that MCEB has an average query processing time less than the Hausdorff Distance method, while the MRAM has the smallest average query processing time among the three methods.

6 Conclusion

In this paper, we introduce an extension to the two-stage framework for polygon retrieval we proposed to augment the framework with systematic control on search space pruning. We also propose a new polygon matching technique using Circular Error Bound (CEB) method which is based on an intuitive human idea of polygon resemblance. Based on CEB, we propose a translation invariant similarity measure of polygons named Minimum Circular Error Bound (MCEB). We find that the MCEB method gives rankings of polygons similar to human rankings and it is more efficient than the Hausdorff Distance method.

Right now, the CEB method only handles polygon matching under translation and scaling. A natural extension to the CEB method is to incorporate rotation. Another possible extension to our work is to enhance the translation invariant MCEB method such that it is scaling invariant as well.

References

1. P. K. Agarwa, M. Sharir, and S. Toledo. Applications of Parametric Searching in Geometric Optimization. *Proc. of the 3rd Annual ACM-SIAM Symposium on Discrete Algorithms*, pages pp. 72–82, 1992.
2. B. Bhavnagri. A Method for Representing Shape Based on an Equivalent Relat ion on Polygons. *Pattern Recognition*, Vol. 27(No. 2):pp. 247–260, 1994.
3. Daniel P. Huttenlocher and William J. Rucklidge. A Multi-Resolution Technique for Comparing Images Using the Hausdorff Distance. *Department of Computer Science, Cornell University, Technical Report TR92-1328*, 1992.
4. H. V. Jagadish. A Retrieval Technique for Similar Shapes. *Proc. of the ACM SIGMOD Int. Conf. on the Management of Data*, pages pp. 208–217, May 1991.
5. Helmut Alt, Bernd Behrebds, and Johnnes Blömer. Approximate Matching of Polygonal Shapes. *Annals of Mathematics and Artificial Intelligence*, Vol. 13:pp. 251–265, 1995.
6. Jia Guu Leu. Computing a Shape's Moments from its Boundary. *Pattern Recognition*, Vol. 24(No. 10): 949–957, 1991.
7. Jonathan Ashley, Ron Barber, Myron Flickner, James Hafner, Denis Lee, Wayne Niblack, and Dragutin Petkovic. Automatic and Semi-Automatic Methods for Image Annotation and Retrieval in QBIC. *SPIE*, Vol. 2420:pp. 24–35, 1995.
8. I. King and T. K. Lau. A Feature-Based Image Retrieval Database for the Fashion, Textile, and Clothing Industry in Hong Kong. *Proc. of International Symposium Multi-Technology Information Processing '96*, pages pp. 233–240, 1996.
9. Pedro Cox, Henri Maitre, Michel Minoux, and Celso Ribeiro. Optimal Matching of Convex Polygons. *Pattern Recognition Letters*, Vol. 9:pp. 327–334, 1989.
10. Rajiv Mehrotra and James E. Gary. Feature-Based Retrieval of Similar Shape. *Proc. 9th Int. Conf. on Data Engineering*, pages pp. 108–115, 1993.
11. L. H. Tung, I. King, P. F. Fung, and W. S. Lee. A Two-Stage Framework for Efficient Simple Polygon Retrieval in Image Databases, journal = Proc. of International Symposium Multi-Technology Information Processing '96. 1996.
12. L. H. Tung, I. King, P. F. Fung, and W. S. Lee. Two-Stage Polygon Representation for Efficient Shape Retrieval in Image Databases. *Proc. of the 1st International Workshop on Image Databases and Multimedia Search*, 1996.
13. W. Niblack, R. Barber, W. Equitz, M. Flickner, E. Glasman, D. Petkovic, and P. Yanker. The QBIC Project: Querying Images By Content Using Color, Texture, and Shape. *SPIE*, Vol. 1908:pp. 173–187, 1993.

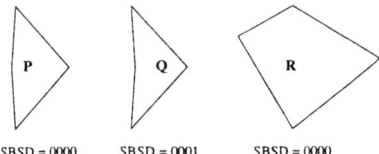

Fig. 1. The problem of the two-stage framework

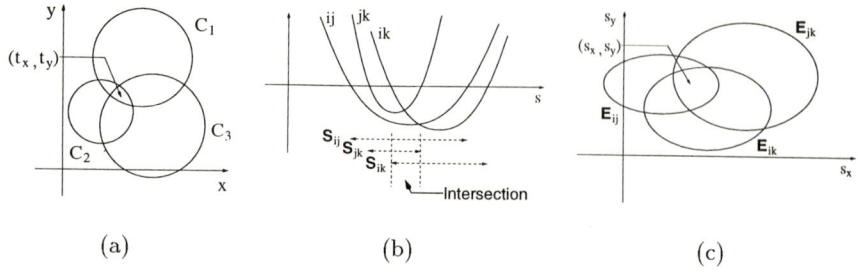

Fig. 2. (a) Intersection of Circular Error Bounds (b) $\mathbb{S}_{ij}$ and their intersection (c) $\mathbb{E}_{ij}$ and their intersection

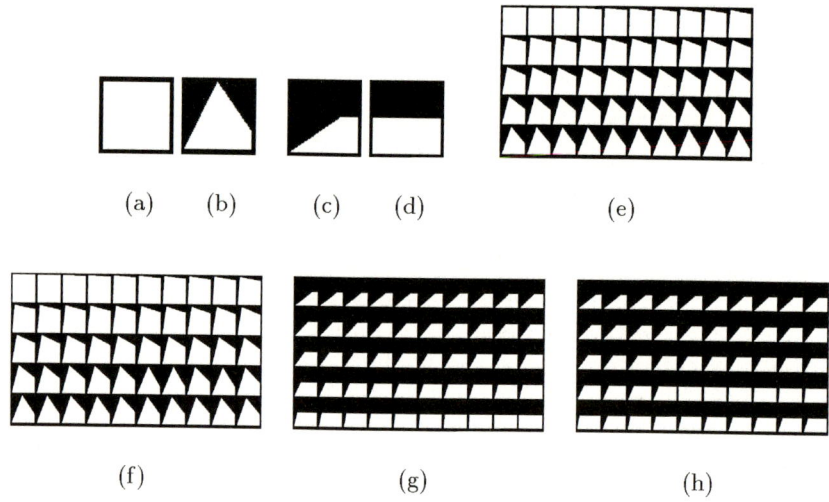

Fig. 3. Similar ranking experiments

Table 1. Experiment results

N-gon	Number of polygons	MRAM	MCEB	Hausdorff Distance
3	9000	2.04 sec	3.24 sec	3.40 sec
4	9000	1.97 sec	2.40 sec	2.62 sec
5	9000	1.74 sec	2.00 sec	2.21 sec
6	9000	1.64 sec	1.78 sec	1.87 sec
7	9000	1.62 sec	1.68 sec	1.72 sec
8	9000	1.63 sec	1.67 sec	1.68 sec

Topology and Shape Preserving Parallel Thinning for 3D Digital Images — A New Approach

P.K. Saha and D. Dutta Majumder

Electronics and Communication Sciences Unit
Indian Statistical Institute, 203 Barrackpur Trunk Road
Calcutta 700035   INDIA

Abstract. This paper is concerned with a new parallel thinning approach for three dimensional (3D) digital images that preserves the topology and maintains their shape. We introduce a new approach of selecting shape points and outer-layer used for erosion during each iteration. The approach produces good skeleton for different types of corners. The concept of using two image versions in thinning is introduced and its necessity in parallel thinning is justified. The robustness of the algorithm under pseudo random noise with respect to shape properties is studied and the results are found to be satisfactory.

1 Introduction

The processing of 3D digital images has become of increasing interest with the rapid growth of 3D image analysis and computer vision based applications in different fields such as astrophysics, geophysics, industrial inspection and medical imaging. Of all these 3D medical imaging is the most popular one such as Computed Tomography (CT), Magnetic Resonance Imaging (MRI), Ultrasound Echography (UE), Positron Emission Tomography (PET), Digital Subtraction Angiography (DSA) etc. This paper is concerned with a new parallel thinning approach for three dimensional (3D) digital images. Many image processing methodologies such as smoothing, filtering, thinning and segmentation are of interest in various applications to 3D image processing. The objective of 3D image thinning is to produce a medial surface representation that preserves the topology and maintains the shape of an object as much as possible. Thinning makes a compact representation of an object and hence is computationally attractive for future analysis.

One of the important uses of thinning is to decompose an object into meaningful segments [1]. Although 3D thinning has a lot of interest in 3D image processing [2], only a few publications [3,4,5,6,7,8,2] are found in 3D thinning. Unfortunately none of the previous researchers studied the behavior of their algorithms around different types of corners. Also they did not study the robustness of their algorithms under noise. In this paper we consider the aspects of thinning algorithm stressing its behavior and robustness under pseudo random noise.

Theoretical aspect of the proposed thinning approach that produces a medial surface representation of 3D object is described in Section 2. The parallel thinning algorithm is described and the experimental results are presented in Sections 3 and 4 respectively.

2 The Thinning Approach

We consider thinning as an approach of producing a medial surface representation of a 3D digital object that preserves the topology and maintains the shape of the object to the maximum extent. The proposed thinning approach is an iterative erosion process that consists of two steps namely primary-thinning and final-thinning. The results of these steps are called primary-skeleton and final-skeleton respectively.

Our thinning approach exploits the information from two versions of an image implicitly stored throughout the thinning procedure. One image version denotes the black/white configuration before the current iteration while the other denotes the current stage of the processed image. Here, it is worthy to mention that simple points[9,10] are always detected on the current version of the image while the shape preserving constraints are mostly defined on the image version before each current iteration. This idea is quite different from other works [8,4] where only one version of image is used for thinning. At this point it should be made clear that in this work, iteration and scan are two completely different concepts. A scan is a (point by point) traversal of the entire image when subjected to the thinning process. On the other hand, an iteration is completed after considering the entire outer-layer of an object through proper topology and shape constraints. An iteration may consist of one or more scans in which case the operation in each scan is generally different. The set of points considered for erosion during an iteration defines the outer-layer for that iteration. Before we describe the thinning procedure let us present some definitions and conditions in this context. In this paper we follow conventional definitions [10,1] of adjacency, path, connectivity etc.

Definition 1. $\mathcal{N}(p)$ denotes the set of 27 points in the $3 \times 3 \times 3$ neighborhood of p. An *s-point* of $\mathcal{N}(p)$ is 6-adjacent to p. An *e-point* of $\mathcal{N}(p)$ is 18-adjacent but not 6-adjacent to p. A *v-point* of $\mathcal{N}(p)$ is 26-adjacent but not 18-adjacent to p. Two s-points $a, b \in \mathcal{N}(p)$ are called *opposite* if they are not 26-adjacent. Otherwise, they are called *non-opposite s-points*. Let a, b, c denote three non-opposite s-points of $\mathcal{N}(p)$. Five points are defined with respect to a, b, c, p as follows: (1) $e(a, b, p)$ is the point q such that $q \in \mathcal{N}^*(p)$ and 6-adjacent to a, b, (2) $v(a, b, c, p)$ is the point q such that $q \in \mathcal{N}^*(p)$ and 6-adjacent to $e(a, b, p), e(b, c, p), e(c, a, p)$, (3) $f_1(a, p)$ is the point q such that $q \notin \mathcal{N}(p)$ and 6-adjacent to a, (4) $f_2(a, b, p)$ is the point q such that $q \notin \mathcal{N}(p)$ and 6-adjacent to $f_1(a, p), e(a, b, p)$ and (4) $f_3(a, b, p)$ is the point q such that $q \notin \mathcal{N}(p)$ and 6-adjacent to $f_2(a, b, p), f_2(b, a, p)$. Let $p = (i, j, k)$ be a point. The set $\{(i-1, j, k), (i, j-1, k), (l, j, k-1)\}$ of three non-opposite s-point of p is denoted as $\mathcal{T}(p)$. Thus $\mathcal{T}(p)$ contains the s-points of $\mathcal{N}(p)$ which are at the west, south and bottom sides of p.

In the following definitions and conditions $(a,d), (b,e)$ and (c,f) denote three distinct unordered pairs of opposite s-points of $\mathcal{N}(p)$ unless stated otherwise.

Definition 2. Let condition C_1 be '$x \in \{b,e,c,f\}$'; condition C_2 be '$x,y \in \{b,e,c,f\}$ and x,y are non-opposite'; condition C_3 be '$x \in \{b,e,c,f\}$ and $x \in \mathcal{T}(p)$'; condition C_4 be '$x,y \in \{b,e,c,f\}$; x,y are non-opposite and $x \in \mathcal{T}(p)$' and condition C_5 be '$x,y \in \{b,e,c,f\}$; x,y are non-opposite and $x,y \in \mathcal{T}(p)$'. We define a middle plane and an extended middle plane of $\mathcal{N}(p)$ as follows:
$\mathcal{M}(a,d,p) = \{x \mid C_1\} \cup \{e(x,y,p) \mid C_2\}$,
$\mathcal{E}\mathcal{M}(a,d,p) = \mathcal{M}(a,d,p) \cup \{f_1(x,p) \mid C_3\} \cup \{f_2(x,y,p) \mid C_4\} \cup \{f_3(x,y,p) \mid C_5\}$.

Definition 3. During an iteration a black point p is an *s-open point* if at least one s-point of $\mathcal{N}(p)$ is white before the iteration. During an iteration a black point p is an *e-open point* if p is not an s-open point and an e-point $e(a,b,p)$ is white while the points $f_1(a,p)$, $f_1(b,p)$ are black before the iteration. During an iteration a black point p is a *v-open point* if p is neither an s-open point nor an e-open point and a v-point $v(a,b,c,p)$ is white while the points $f_1(a,p)$, $f_1(b,p)$, $f_1(c,p)$ are black before the iteration. The set of s-open, e-open and v-open points defines the outer-layer in an iteration. It is understood from the above definitions that the labeling of points as s-open, e-open and v-open points is made once before each iteration.

Condition 4. During an iteration a point p satisfies Condition 4 if there exist two opposite s-points $a,d \in \mathcal{N}(p)$ such that $\mathcal{E}\mathcal{M}(a,d,p)$ contains a 6-closed path of white points encircling p and each of $surface(a,p)$ and $surface(d,p)$ contains at least one black point before the iteration.

Condition 5. During an iteration a point p satisfies Condition 5 if there exists a pair of opposite s-points (a,d) such that $d \in \mathcal{T}(p)$, a is white, d or $f_1(d,p)$ is white and each of the sets $\{e(a,b,p), b, e(b,d,p)\}$, $\{e(a,c,p), c, e(c,d,p)\}$, $\{e(a,e,p), e, e(d,e,p)\}$, $\{e(a,f,p), f, e(d,f,p)\}$, $\{v(a,b,c,p), e(b,c,p), v(b,c,d,p)\}$, $\{v(a,b,f,p), e(b,f,p), v(b,d,f,p)\}$, $\{v(a,c,e,p), e(c,e,p), v(c,d,e,p)\}$, $\{v(a,e,f,p), e(e,f,p), v(d,e,f,p)\}$ contains at least one black point before the iteration.

Condition 6. During an iteration a point p satisfies Condition 6 if for each middle plane $\mathcal{M}(a,d,p)$ of $\mathcal{N}(p)$ — either all e-points in $\mathcal{M}(a,d,p)$ are black before the iteration or the current black points of $\mathcal{M}(a,d,p)$ generate single 26-component without any tunnel [10,1].

Definition 7. During an iteration a function is defined on the black/white configuration before the iteration as follows:

$$thick(a,d,p) = \begin{cases} true & \text{if } a \text{ and } f_1(d,p) \text{ are white while } d \text{ is black,} \\ false & \text{otherwise.} \end{cases}$$

Condition 8. A point p satisfies Condition 8 if $thick(a,d,p)$, where $d \in \mathcal{T}(p)$, is true and the current black points of each of $\mathcal{M}(b,e,p)$ and $\mathcal{M}(c,f,p)$ generate single 26-component without any tunnel.

Condition 9. A point p satisfies Condition 9 if $thick(a,d,p)$ and $thick(b,e,p)$, where $d,e \in \mathcal{T}(p)$, are true and the current black points of $\mathcal{M}(c,f,p)$ generate single 26-component without any tunnel.

Condition 10. A point p satisfies Condition 10 if $thick(a,d,p)$, $thick(b,e,p)$ and $thick(c,f,p)$, where $d,e,f \in \mathcal{T}(p)$, are true.

Definition 11. During an iteration a black point is an *erodable point* if it is a simple point and satisfies any of the Conditions 8, 9 or 10.

2.1 Primary-Thinning

As mentioned earlier primary-thinning is an iterative procedure and iterations are continued as long as any point is deleted in the last iteration. Each iteration is completed in three successive scans. During the first scan an unmarked *s*-open point is marked if it is a shape point. When it is not a shape point, it is deleted if it is a simple point, otherwise it is left unmarked. During the second scan an unmarked *e*-open point is deleted if it is a simple point and satisfies Condition 6. During the third scan an unmarked *v*-open point is deleted if it is a simple point.

2.2 Final-Thinning

From the definition of shape point it may be understood that a two-point thick slanted surface may occur in primary-skeleton. Final-thinning is necessary to get a proper skeleton for such cases. This is a single iteration procedure and the iteration consists of single scan. During this scan a black point p (irrespective of whether p is marked or unmarked) is deleted if it is an erodable point.

3 The Parallel Thinning Approach

In this section we describe a parallel thinning algorithm based on the approach discussed in Section 2. For parallelization we use the concept of sub-fields [3,4]. An image is partitioned into eight disjoint subsets such that no two members p, q in the same subset are 26-adjacent. The members of each subset may be used for parallel erosion. Eight subsets $O_0, O_1, \cdots O_7$ are defined as follows:

$$O_l = \{(2 \times i + f, 2 \times j + g, 2 \times k + h) \mid i,j,k = 0, \pm 1, \pm 2, \cdots ; \\ f,g,h \in \{0,1\} \text{ and } 2^2 \times f + 2^1 \times g + 2^0 \times h = l\} \quad (1)$$

such that two points $p, q \in O_l$ are never 26-adjacent. Each scan of the thinning algorithm may be completed in eight cycles and at lth cycle the image subset O_l is subjected to parallel erosion. The parallel algorithm requires $m^3/8$ processors for an image of size $m \times m \times m$ and it needs 8 cycles to complete each scan. The image size means the size of the smallest rectangular parallelepiped that encloses the set of black points.

4 Results and Discussion

To test the effectiveness of the proposed algorithm it is applied on several 3D objects. We present three of these results in Figures 1-3. In all these figures backgrounds are made dark to render a better visual effect. Both noiseless and noisy versions (generated in a pseudo random manner) are shown. The surface representation of both noiseless and noisy versions of Figures 1-3 are visually satisfactory.

Fig. 1. Results of thinning. Top row (from left to right): original object and skeleton. Bottom row (from left to right): original object with noise and skeleton.

A new parallel thinning algorithm of 3D digital images with topology and shape preserving properties has been developed in this chapter. To preserve topology we have applied the concept of simple points [9,10]. On the other hand the concept of sub-fields [3] has been used for parallel implementation. The concepts of open points and shape points have been introduced and applied to 3D thinning. The concept of open points produces proper skeleton around different types of corners has been justified. Also, the shape points are found to be robust under noise. We have used two versions of the image — one before the current iteration while the other being the currently processed image. This concept has

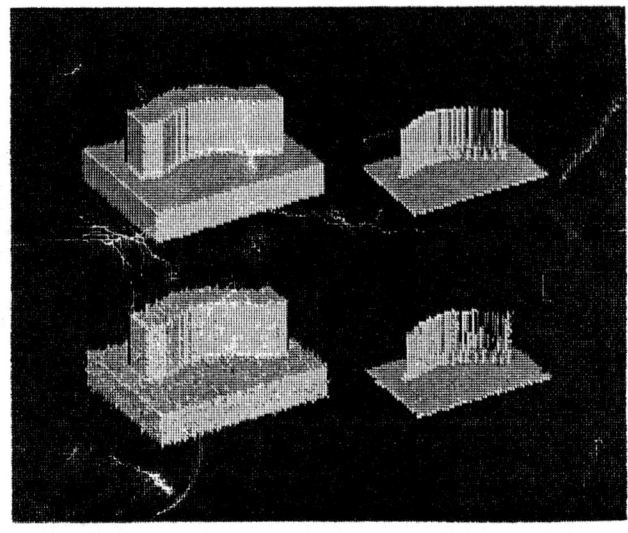

Fig. 2. Results of thinning. Top row (from left to right): original object and skeleton. Bottom row (from left to right): original object with noise and skeleton.

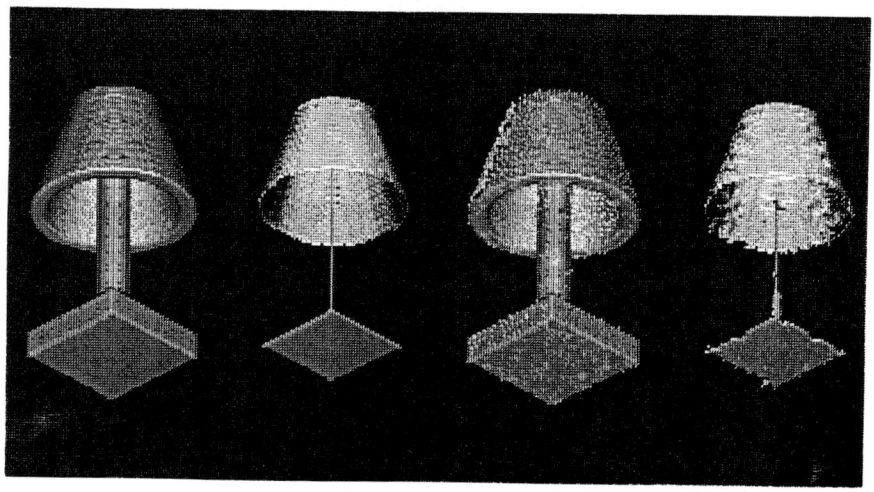

Fig. 3. Results of thinning. From left to right: original object, skeleton, original object with noise and skeleton.

made a major improvement in the quality of thinned image. The results of application of the parallel thinning algorithm on several synthetically generated 3D objects and their noisy versions have been presented.

References

1. Saha, P. K., Chaudhuri, B. B.: 3D Digital topology under binary transformation with applications. CVGIP: Image Understanding **63** (1996) 418-429
2. Ma, C. M.: On topology preservation in 3D thinning. CVGIP: Image Understanding **59** (1994) 328-339
3. Golay, M. J.: Hexagonal Parallel Pattern Transformations. IEEE Trans. Comput. **C-18** (1969) 733-740
4. Hafford, K. J., Preston Jr, K.: Three-dimensional Skeletonization of Elongated Solids. Comput. Vision Graphics Image Process. **27** (1984) 78-91
5. Lobregt, S., Verbeek, P. W., Groen, F. C. A.: Three-dimensional skeletonization: principle and algorithm. IEEE Trans. Pattern Anal. Mach. Intell. **PAMI-2** (1980) 75-77
6. Mukherjee, J, Das, P. P., Chatterjee, B. N.: Thinning of 3-D images using the Safe Point Thinning Algorithm (SPTA). Pattern Recognition Letters **10** (1989) 167-173
7. Srihari, S. N.: Representation of three-dimensional digital images. ACM Comput. Surveys **13** (1981) 400-424
8. Tsao, Y. F., Fu, K. S.: A parallel thinning algorithm for 3D pictures. Comput. Graphics Image Process. **17** (1981) 315-331
9. Saha, P. K., Chaudhuri, B. B., Chanda, B., DuttaMajumder, D.: Topology preservation in 3D digital space. Pattern Recognition **27** (1994) 295-300
10. Saha, P. K., Chaudhuri, B. B.: Detection of 3-D simple points for topology preserving transformations with application to thinning. IEEE Trans. on Pattern Anal. Mach. Intell. **16** (1994) 1028-1032

Convergence of Model Based Shape From Shading

Michael S. Lew   Michel Chaudron   Nies Huijsmans
Computer Science Dept.
Leiden University
Postbus 9512
2300 RA Leiden
Netherlands
email: mlew@cs.leidenuniv.nl

Alfred She   Thomas S. Huang
Electrical Engineering Dept.
University of Illinois
Urbana, IL 61801
USA

Abstract. Three dimensional models are used by a wide variety of applications such as face recognition and model based coding for teleconferencing. In face recognition and model based coding the generic 3D model is used to generate images of the individual from any viewpoint. Recent work has been focussed on using shape from shading for refinement of the generic model toward the 3D shape of the individual. However, the previous methods have been based on iteration schemes in which convergence could not be proven. The main goal of this paper is to introduce a provably convergent algorithm for solving the problem of model based shape from shading. Bounds on the number of iterations for attaining error of order h are also given.

1   Introduction

In a previous paper, we integrated shape from shading techniques with a generic model [Lew, She, and Huang 1995]. However, in the previous work, we had not considered the problem of convergence of the iterative algorithm, nor the time complexity. Thus, the goal of this paper is to introduce a provably convergent algorithm for adjusting the generic model to the individual, and to give bounds on the number of iterations for attaining error of order h.

In Section 2 we integrate the generic model with shape from shading and discuss the convergence of the solution. In Section 3, we discuss time complexity and integration of knowledge of discontinuities. In Section 4, conclusions and directions for future work are discussed.

2   Shape from Shading

Humans have the ability to estimate shape from one image of a scene. Although it can be shown that the surface shape recovered from the intensity distribution of one image is not unique in general, the ability would still be useful in many applications such as video compression in telecommunications, terrain mapping, robotic navigation, and face recognition. In the past, most work on shape from shading has not proven robust in the presence of noise which occurs in real images.

In this section, we review the adaptation of standard shape from shading methods to face recognition by the inclusion of relevant constraints of the generic model in the derivation. Furthermore, we prove convergence and uniqueness using a matrix reformulation of the solution.

2.1 Global Shape from Shading

First, we consider the case for diffuse cosine shading. These surfaces reflect light as the cosine of the angle between the surface normal and the vector toward the light source as depicted in Figure 1.

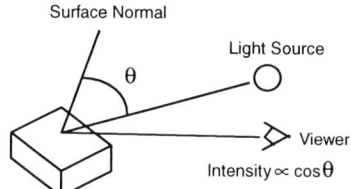

Figure 1. The imaging setup for diffuse cosine shading.

In the literature (i.e. Ballard and Brown [1982]) the diffuse cosine shading is referred to as Lambertian shading. Thus, $R = r_0 \cos\theta$. Let $p = \partial f/\partial x$ and $q = \partial f/\partial y$ for a surface defined by $z = f(x,y)$, then the Lambertian reflectance, R_L, can be written as

$$R_L(p,q) = \frac{ap + bq - c}{\sqrt{p^2 + q^2 + 1}}$$

where [a,b,c] is the unit vector toward the light source. Horn[1975] solved for the depth map using a characteristic strip method. To reduce the noise sensitivity [Ikeuchi and Horn 1981], a term was added which constrains p and q to be smooth functions. Using variational calculus Euler-Lagrange equations [Courant and Hilbert 1953], we minimize

$$\varepsilon^2 = \sum_x \sum_y [I(x,y) - R(p,q)]^2 \lambda + [p_x^2 + p_y^2 + q_x^2 + q_y^2] \qquad (1)$$

with respect to p and q, where $p_x = \partial p/\partial x$, $p_y = \partial p/\partial y$, $q_x = \partial q/\partial x$, $q_y = \partial q/\partial y$, and λ is a user set constant, which weighs the contributions from the shading term and the smoothing term.

In order to solve for the unknown surface from the gradient vectors p and q, boundary conditions which are usually unknown must be assumed. One commonly applied assumption is that the occluded contours have the same gradient vectors. However, the gradient vectors p and q approach infinity at the occluded contours. This problem can be overcome by using a coordinate system based on the Gaussian sphere [Ikeuchi 1980]. Another solution to the problem of boundary conditions is to impose an additional constraint such as the quadratic surface assumption in local shape from shading.

2.2 Method: GLOBAL

In this section we briefly review our previous method found in Lew, She, and Huang [1995]. We derive a shape from shading method based on an arbitrary reflectance map $R(p,q)$ which can be measured from real images. Specifically, the surface partial derivatives, p and q, can be measured from the 3D model using a laser range finder, and the reflected intensity from the real image is proportional to $R(p,q)$. The reflectance map found from the paired 3D model and real image is denoted the *generic reflectance map*.

In equation (1), ε can be zero when $I(x,y)$ equals $R(p,q)$ and the gradient vectors are zero, which will be the case when the surface is a plane. Thus, the smoothing term in (1) biases the surface toward a plane. Instead of enforcing gradient smoothness to a *plane*, we enforce gradient smoothness to the *generic model* f^M. Let p^M and q^M refer to the first partial derivatives of f^M with respect to x and y. Consequently, the generalized criterion function is

$$\varepsilon^2 = \sum_x \sum_y [I(x,y) - R(p(x,y), q(x,y))]^2 \lambda + \left[(p_x - p_x^M)^2 + (p_y - p_y^M)^2 + (q_x - q_x^M)^2 + (q_y - q_y^M)^2 \right]$$

(2)

The traditional method to determine the minimum of equation (2) is to apply variational calculus to derive the necessary conditions for a minimum. Following such an approach and creating an iterative scheme using the Gauss-Seidel method gives

$$p_{t+1}(i,j) = \tilde{p}_t(i,j) + p^M(i,j) - \tilde{p}^M(i,j) + \frac{\lambda}{4}\left(I(i,j) - R[p_t(i,j), q_t(i,j)]\right)\frac{\partial R}{\partial p}[p_t(i,j), q_t(i,j)]$$

(3)

$$q_{t+1}(i,j) = \tilde{q}_t(i,j) + q^M(i,j) - \tilde{q}^M(i,j) + \frac{\lambda}{4}\left(I(i,j) - R[p_t(i,j), q_t(i,j)]\right)\frac{\partial R}{\partial q}[p_t(i,j), q_t(i,j)]$$

(4)

where

$$\tilde{p}_t(i,j) = \frac{p_t(i-1,j) + p_t(i+1,j) + p_t(i,j-1) + p_t(i,j+1)}{4}$$

$$\tilde{p}^M(i,j) = \frac{p^M(i-1,j) + p^M(i+1,j) + p^M(i,j-1) + p^M(i,j+1)}{4}$$

and

$$\tilde{q}_t(i,j) = \frac{q_t(i-1,j) + q_t(i+1,j) + q_t(i,j-1) + q_t(i,j+1)}{4}$$

$$\tilde{q}^M(i,j) = \frac{q^M(i-1,j) + q^M(i+1,j) + q^M(i,j-1) + q^M(i,j+1)}{4}$$

2.3 Existence and Convergence

Two interesting aspects of an iterative solution are existence and convergence. If $R(p,q)$ is continuous then there exists an **x** which satisfies equation (2) [Lee 1989]. One method to prove convergence is to show that

$$\frac{\|x_t - x^*\|}{\|x_{t-1} - x^*\|} < \alpha < 1 \tag{5}$$

where x_t is the estimate of the solution at iteration t, and x^* is the correct solution, and α is a fixed positive constant. We can rewrite (5) as $\|x_t - x^*\| < \alpha^t \|x_0 - x^*\|$ or as $\|x_t - x^*\| < \varepsilon$, and for $\alpha < 1$, $\lim_{t \to \infty} \alpha^t = 0$ and $\lim_{t \to \infty} x_t = x^*$

First, we formulate the solution to the criterion function in a manner similar to Lee [1989]. Let S denote the region of interest. Rewriting equations (3) and (4) in matrix form and accounting for boundary conditions yields

$$M(x_t - x^M) = -\lambda h^2 b(x_t) + r \tag{6}$$

where

λ is the weighting constant between the shading term and the smoothing term.

h is the mesh size

$M^{-1} = \text{diag}(A^{-1}, A^{-1})$

A is the Laplacian Matrix [Isaacson & Keller 1966]

$$A = \begin{bmatrix} B & -I & . & . \\ -I & B & \ddots & . \\ . & \ddots & \ddots & -I \\ . & . & -I & B \end{bmatrix} \quad B = \begin{bmatrix} 4 & -1 & . & . \\ -1 & 4 & \ddots & . \\ . & \ddots & \ddots & -1 \\ . & . & -1 & 4 \end{bmatrix}$$

Let $N = nm$ = total number of grid points.
A is N x N, B is n x n tridiagonal matrix
and I is an n x n identity matrix

$x_t = (..., p_t(i,j), ..., ..., q_t(i,j), ...)^T$
$x^M = (..., p^M(i,j), ..., ..., q^M(i,j), ...)^T$
$r = (..., r(i,j), ...)^T$ boundary point constants which satisfy (6) at the noninterior points.
$b(x_t) = (..., (R(p_t,q_t) - I(i,j))(\partial R(p_t,q_t)/\partial p), ..., , ..., (R(p_t,q_t) - I(i,j))(\partial R(p_t,q_t)/\partial q), ...)^T$

Since **M** is positive definite and symmetric, it has an inverse. Multiplying by M^{-1} and solving for **x** yields

$$x_t = -\lambda h^2 M^{-1} b(x_{t-1}) + M^{-1} r + x^M \tag{7}$$

We denote the solution to equation (7) as x^* where

$$x^* = -\lambda h^2 M^{-1} b(x^*) + M^{-1} r + x^M \tag{8}$$

We assume that $(R(p_t,q_t) - I(i,j))(\partial R(p_t,q_t)/\partial p)$ and $(R(p_t,q_t) - I(i,j))(\partial R(p_t,q_t)/\partial q)$ are Lipschitz functions for all (i,j). This results in
$$\|b(x) - b(x')\| \leq v\|x - x'\| \tag{9}$$
where v is the maximum Lipschitz constant of $(R(p_t,q_t) - I(i,j))(\partial R(p_t,q_t)/\partial p)$ and $(R(p_t,q_t) - I(i,j))(\partial R(p_t,q_t)/\partial q)$ for all (i,j).
From Isaacson and Keller[1966] and Lee [1989],
$$\|M^{-1}\| = \|A^{-1}\| \leq \frac{1}{4s(n,m)} \tag{10}$$
where $s(n,m) = \sin^2\frac{\pi}{2(m+1)} + \sin^2\frac{\pi}{2(n+1)}$

m = number of rows in S
n = maximum number of grid points in a single row in S

Then subtracting (8) from (7) gives
$$x_t - x^* = -\lambda h^2 M^{-1}(b(x_{t-1}) - b(x^*)) \tag{11}$$
Using the Cauchy-Schwarz Inequality gives
$$\frac{\|x_t - x^*\|}{\|b(x_t) - b(x^*)\|} \leq \lambda h^2 \|M^{-1}\| \tag{12}$$
Substituting equations (9) and (10) into (12) gives
$$\frac{\|x_t - x^*\|}{\|x_{t-1} - x^*\|} \leq \frac{\lambda h^2 v}{4s(n,m)} \tag{13}$$

If $\lambda \in \left[0, \frac{4s(n.m)}{h^2 v}\right)$ then $\frac{\|x_t - x^*\|}{\|x_{t-1} - x^*\|} < \alpha < 1$, \hfill (14)

and x_t converges to x^*.

3 Issues

In the previous section, we proved that for a certain range of lambda, the algorithm would converge. In this section we discuss the time complexity to attain error of order g, and we comment on inclusion of discontinuity information into the objective criterion function in equation (2).

3.1 Accuracy and Computational Complexity

Let $\alpha = \frac{\lambda h^2 v}{4s(n,m)} < 1$, then from equation (13), the error of x_{t+1} is of order α^t.

Thus, an approximation to the solution can be constructed of order g in $O(\log(1/g))$ iterations. Furthermore for a rectangular region of size N, the total cost of constructing an estimate with error of order g is $O(N \log N \log(1/g))$ using a Fast Fourier Transform approach [Lee 1989].

3.2 Discontinuities

In many applications, apriori knowledge about discontinuities in the depth map is given. One method of modeling discontinuities in the criterion function is to alter λ from a constant to a function such as $\lambda(x,y)$. When λ is large then the contribution of the smoothing term will be negligible which allows for a closer modeling of discontinuities. It is important to note that the condition on λ in (14) becomes a condition on the maximum of $\lambda(x,y)$ over all (x,y). Thus, convergence is not guaranteed for a discontinuous depth map. Convergence is guaranteed when the minimum and maximum $\lambda(x,y)$ over all (x,y) satisfies (14).

4 Conclusions

This paper presented theoretical sufficient conditions for convergence of the shape from shading algorithm of Lew, She, and Huang[1995]. The criterion function integrating the generic model was solved using the traditional variational calculus method and using the matrix formulation. The advantage to the matrix formulation is that sufficient conditions for convergence can be proved analytically. Furthermore, accuracy, computational complexity and incorporation of depth discontinuites were discussed. Regarding future work, it would be interesting to pursue the incorporation of discontinuities into the algorithm

References

Ballard, D. and C. M. Brown, **Computer Vision**, Prentice-Hall, Englewood Cliffs, New Jersey, 1982.

Courant, R. and D. Hilbert, **Methods of Mathematical Physics**, vol. 1, Interscience Publishers, New York, 1953.

Horn, B. K. P., "Obtaining Shape from Shading Information," *Psychology of Machine Vision*, McGraw-Hill, New York, pp. 115-155, 1975.

Ikeuchi, K., "Shape from Regular Patterns - An Example of Constraint Propagation in Vision," *Proceedings of the Fifth International Joint Conference on Pattern Recognition*, Miami Beach, Florida, 1980.

Ikeuchi, K., and B. K. P. Horn, "Numerical Shape from Shading and Occluding Boundaries," *Artificial Intelligence*, vol. 17, pp. 141-184, 1981.

Isaacson, E., and H. B. Keller, **Analysis of Numerical Methods**, John Wiley & Sons, New York, 1966.

Lee, David, "A Provable Convergent Algorithm for Shape From Shading," **Shape From Shading** edited by Horn and Brooks, MIT Press, Cambridge, pp. 349-373.

Lew, M., A. C. She and T. S. Huang, "Intermediate Views For Face Recognition," Lecture Notes in Computer Science: Proceedings of the 6th International Conference on Computer Analysis of Images and Patterns, pp. 138-145, 1995.

Quantitative Assessment of Two Skeletonization Algorithms Adapted to Rectangular Grids

M. CIUC, D. COQUIN, Ph. BOLON

Laboratoire d'Automatique et de Micro-Informatique Industrielle
LAMII/CESALP - Université de Savoie, BP 806, F. 74016 Annecy Cedex (France)
e-mail: coquin@esia.univ-savoie.fr

Abstract
In this paper, an adaptation to rectangular grids of two skeletonization algorithms is presented. Skeletonizations are quantitatively compared by using Baddeley's distance between the original pattern and the one reconstructed from the skeleton.

Keywords: skeletonization, rectangular grids, Baddeley distance

1. Introduction

Among the different approaches to skeletonize objects, there are the methods based on a distance transformation. A Distance Map (DM) of the object is computed before extracting the skeleton, each pixel in the object being assigned to a label which represents the nearest distance of the pixel to the complement of the object (i.e. background). This property is useful, since it allows reconstruction of the pattern starting from its skeleton.

Several skeletonization algorithms based on a distance transformation have been developed. Most of them treat the case of images digitized on a square grid. However, most industrial vision systems digitize images with sampling steps which are different, depending on whether they are in rows or columns. Hence, images are digitized on a rectangular grid. To avoid the resampling of such images, algorithms which compute the distance map of a binary object from the background with respect to this particular shape of the pixel have been introduced. Skeletonization methods proposed for a square grid need some modifications.

In this paper, we propose an adaptation of the cases of images digitized on a rectangular grid, to two methods. The first one, proposed by Arcelli and Sanniti di Baja in [2] is based on the extraction of ridge points. The second, proposed by Thiel in [1] is based on the extraction of the centres of maximal discs by means of look-up table.

Results and comparisons between the two methods are provided in section 6.

2. Distance map on rectangular grids

Many methods to extract the distance map of a binary object in an image have been developed: the city-block distance (d_4) [9], the chessboard distance (d_8) [3], the chamfer distance [1]. They are characterized by a local mask and are computed within a two raster-scan of the original binary image, regardless of the size of the object. The best approximation of the Euclidean distance is obtained by chamfer distances. This makes the result less dependent on rotation and scaling of the object than the other ones. Moreover, chamfer distances can be adapted to rectangular grids. The coefficients a,b,c,d,e (fig. 1) are optimized according to the method proposed in [4].

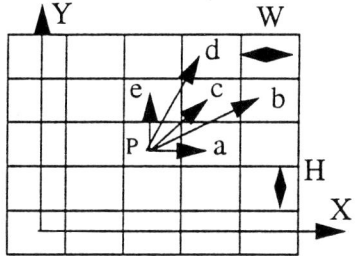
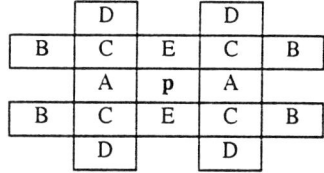

Fig. 1. real coefficients on rectangular grid, W=pixel width, H=pixel height

Fig. 2. 5x5 chamfer mask used for computation of the **DM** on rectangular grid

Subsequently, we use a 5x5 chamfer mask for the computation of the **Distance Map**. DM can be obtained with either a sequential or a parallel algorithm [5].
For reasons of computation time and memory availability, integer coefficients A,B,C,D,E, obtained by means of scaling and rounding the real ones are used instead.

$$\{A,B,C,D,E\} = \text{round}(\{a,b,c,d,e\}.N)$$

where N is chosen with regard to the errors that appear by rounding and to the limits of memory required to store the distance map.

3. Centre of Maximal Discs extraction

In this section, the method of Thiel [1] is extended to the case of a rectangular grid. Notations and definitions are kept. We use the same notation for pixels as well as for their labels. Any pixel p belonging to the object can be regarded as a centre of a disc D_{Rp} which includes each pixel q such that: $dc(p,q) < Rp$
$dc(p,q)$ denotes the chamfer distance between pixels p and q.

Pixels in the DM for which the associated disc is not completely covered by any other disc is declared Centre of a Maximal Disc (CMD) and is marked as belonging to the Median Axis (MA). The set of CMD constructed in this way is sufficient to allow a perfect reconstruction of the original pattern. Hence, it is not minimal, since it contains pixels from the associated disc included not in another disc, but in the union of other discs. Our main goal is to extract the skeleton. Hence, these pixels must be kept as skeletal in order to achieve connectivity.

3.1 Look-Up Table computation

It was shown that, generally, it is sufficient to verify for each pixel the following condition:
$p \in MA \Leftrightarrow \forall i \quad n_i(p) \leq LUT_i(p)$ with $n_i(p)$ is the mask,
$LUT_i(p)$ is the look-up table associated to the chamfer distance dc.

For a very fast and easy computation, a look-up table is precalculated. For each possible label p, the LUT stores the minimal value q that the pixel located in one of the five directions A,B,C,D,E should have, such that the disc D_{Rq} completely overlaps D_{Rp}.

For the computation of the LUT we used the method of Thiel adapted to the rectangular grid. The only difference due to this distinctive feature of the CDM is that the mask is no more 8-symmetrical, but 4-symmetrical. Instead of building up 1/8 of the distance image to the origin (gray part), we are obliged to construct a quarter of it in order to obtain correct results.

0	42	84	126	168
20	47	86	128	170
40	57	94	133	172
60	77	104	141	180
80	97	114	151	188

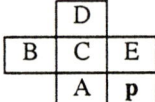
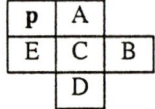

Fig. 3. quarter of the distance image to the origine with H=1, W=2 (A=42, B=86, C=47, D=57, E=20)

Fig. 4. Two quater of mask

The steps to be applied in order to compute the LUT (Fig. 5) are:

- **step 1:** We start from an image having all pixels set at a very high value, except for the origin, which is set at zero. The quarter of the image distance to the origin is computed by means of a single raster-scan, column by column from top-left to down-right with the quarter of the mask in Fig. 4 following the same principle as in the computation of the CDM (fig. 3) see section 3.2.
- **step 2:** The LUT is filled in a single raster-scan of the previous image too. For each pixel, labels of pixels located in each of the five directions as showed in Fig. 4 are inspected. These values, added with 1, will be stored at the LUT entry corresponding to the label next to the current pixel's one.
- **step 3:** For each of the five directions A,B,C,D,E the increasing order is established in the LUT.

p	a	b	c	d	e
20	43	87	48	58	21
40	48	95	58	78	41
42	58	105	78	98	61
47	85	129	87	98	61
57	87	134	95	105	61
60	95	142	105	115	78
77	95	142	105	118	81

Fig. 5. the beginning of LUT for A=42, B=86, C=47, D=57, E=20, (H=1, W=2)

3.2 CMD extraction

The set of CMD is extracted within a single raster-scan of the image containing the CDM by means of a single inspection of the neighbourhood of the pixels. For each pixel we inspect the labels of pixels located in one of the five directions A,B,C,D,E. If none of these pixels has a label which is equal to or greater than the LUT entry corresponding to the current pixel label in the same direction, then the current pixel is marked as a CMD (Fig. 6). The set of CMD is not sufficient to permit connectivity. Also *saddle points*, i.e. pixels which connect two parts of the DM with greater labels must be detected. The conditions a pixel must satisfy to be declared a saddle point are kept with no modifications from [1].The set of CMD and saddle points represents the median axis (MA) of the object. Starting from it, skeletons will be obtained after the steps in section 5 are applied.

0	20	20	40	57	77
20	40	40	57	77	97
40	57	60	77	97	114
57	60	77	80	97	100
40	40	57	60	77	86
20	20	40	40	57	60

Fig. 6. portion of the CDM with A=20,B=86,C=47,D=57,E=20.
gray pixels are detected as CMD

All median axis pixels, when detected, are introduced in a separate list, from which they will be directly accessed in order to increase the speed of the algorithm.

4. Ridge points extraction

In this section we will extend the method proposed by Arcelli and Sanniti di Baja in [2], to the case of rectangular grids. Notations and definitions are kept. As in the previous section, we will denote by p both the pixel and its label.

If we consider the DM as a landscape in three dimensions, the label of the pixel indicating its height, skeletal pixels are those located on the ridges, i.e. pixels having two neighbours with smaller labels at least in one direction. This is the principle of the method. Ridge points are detected in a one-pass raster-scan of the DM by means of a simple inspection of their neighbourhood [2]. Operators were applied only for pixels labelled more than one, i.e. for pixels not belonging to the 8-connected border of the pattern. For border pixels, two other sets of operands were especially designed. These operators can no more be applied in the case of rectangular grid DM (Fig. 7), since in this case there are two types of border pixels: the ones which are labelled A, the others labelled E (Fig. 2).

57	40	0	40	57
40	20	20	20	40
20	0	40	0	20
0	0	42	0	0

Fig. 7. neither of the gray pixels is detected as a ridge point
with the operators designed for square grids

We propose the following modifications: we apply the operators described in [2] to the border pixels too. In addition, for pixels labelled E, we check if *crossing number* ($X_4(p)>1$) from the detection of saddle points in section 3.2 is satisfied, and we mark them too in this case. As we did in the CMD extraction case, pixels detected as ridge points are introduced in a separate list.

5. Skeleton extraction

Starting from the two sets of points determined in section 3 and 4, called intrinsic skeletal pixels in what follows, the same steps are applied in order to obtain the skeleton.

5.1 The connection step

The set of intrinsic skeletal pixels generally does not preserve the pattern connectivity order. Hence a connection step must be applied. It consists of growing paths following the steepest gradient direction from any intrinsic skeletal pixel.

Let p be the current pixel. The gradient: $\mathbf{grad_p(p_i) = k \cdot (p_i - p)}$
is computed for all neighbours p_i of pixel p, with:

$k = \frac{1}{A}$ if p_i is an A-neighbour $\qquad$ $k = \frac{1}{D}$ if p_i is a D-neighbour

$k = \frac{1}{B}$ if p_i is a B-neighbour $\qquad$ $k = \frac{1}{E}$ if p_i is an E-neighbour

$k = \frac{1}{C}$ if p_i is a C-neighbour

The pixel p_i which maximizes the gradient is marked as a skeletal pixel too (it is called an induced skeletal pixel) and we continue to grow the path starting from it. Path growing stops when either the pixel which maximizes the gradient is already marked, or all the gradients computed are negative, i.e. a local maximum in the CDM is reached.

If a B-neighbour or a D-neighbour is chosen as the next pixel in the path, one of the two pixels that connect this neighbour with the current pixel has to be marked too in order to preserve 8-connectivity. We choose the one with the maximum gradient. Also, if two or three pixels are maximized the gradient, only one of them is choose to continue the path in order to avoid useless thickening.

Connectivity is achieved in the same raster scan as the intrinsic skeletal pixels extraction. If such a pixel is detected, we try to grow a path from it and, when completed, we continue the raster-scan. By using this technique, checking marked pixels can occur, since paths are grown in directions not yet inspected by the raster-scan too. In the case of *CMD extraction* (section 3), the pixel already marked are skipped by the test, whereas in the *ridge point detection* case (section 4) the pixels must be inspected anyway in order to determine whether one of its neighbours is a ridge point.

After connectivity is achieved, the skeleton still doesn't preserve the topology of the pattern, since it contains spurious holes. To guarantee correct results, a **hole filling step** is applied. At this point of the algorithm, the set of skeletal pixels allows a perfect reconstruction of the original pattern. But the skeleton is not unit-wide. Hence, it must be thinned. **Thinning** is accomplished in two steps by directly accessing the pixels through the list, which considerably reduces the time required for this operation [3].

5.2 Pruning

The skeleton obtained after thinning generally has many branches. Only some of them are significant for the description of the pattern, the others existing due only to the border roughness. Hence, a pruning step must be applied in order to get rid of the useless branches. Moreover, pruning is important because it considerably reduces the sensitivity of the skeleton with respect to rotation and scaling of the pattern.

Pruning is started from end points, i.e. pixels having only one neighbour. For each pixel p in the branch that ends by q we compute the quantity:

$$r(p) = \frac{1}{Q}[q - p + dc(p, q)]$$

where: $Q = A$ on square grid, and $Q = E$ on rectangular grid, and $dc(p,q)$ represents the chamfer distance between pixels p and q, which represents the loss of information we get if we prune the branch from q to p. If this quantity is smaller than a given **threshold Th**, we decide to unmark pixel p and inspect the pixel next to p, and so on. Pruning is stopped for each branch when either **r(p)** becomes greater than the threshold, or the other end of the branch is reached and continuing to prune would disconnect the skeleton.

As for the chamfer distance $dc(p,q)$, it must be adapted to the rectangular grid. Let the coordinates of pixel p be i_p and j_p and the coordinates of pixel q i_q and j_q. We introduce the quantities K and G such that:

$$G = \min(|i_p - i_q|, |j_p - j_q|) \qquad K = \max(|i_p - i_q|, |j_p - j_q|) - G$$

G represent the number of diagonal displacements, and **K** the number of horizontal or vertical displacements between p and q. With these notations, the chamfer distance adapted to the rectangular grid between pixels p and q can be written:

if $|i_p - i_q| > |j_p - j_q|$ $\qquad dc = AxK + CxG + (B-C-A) \times min(K,G)$

if $|j_p - j_q| > |i_p - i_q|$ $\qquad dc = ExK + CxG + (D-C-E) \times min(K,G)$

Pruning must not be restricted only to the peripheral branches. Each time a connection point is reached, i.e. an intersection between two or more branches, the value of the end point p as well as its coordinates are stored. Hence, when all branches intersecting in this point are pruned the connection point will become an end point. To avoid additional loss of information in this case, we compute the quantity **r(q)** with respect to all former end-points and we stop when one of these quantities reaches the threshold.

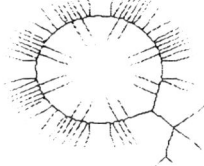
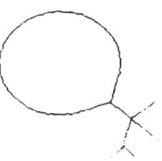

original pattern   skeleton before pruning   skeleton after pruning, **Th = 2**

Fig. 8. Pruning step

5.3 Beautifying

The aim of this step is to improve the aesthetics of the skeleton by straightening the zig-zags that appear after thinning and eliminating one-pixel branches and superfluous pixels that may remain after pruning. Details are given in [1].

5.4 Reconstruction

We use for the reconstruction the inverse chamfer distance transformation, which is mentioned in [10] and is achieved, as the direct transformation, in a two-pass raster scan of the image containing the skeleton. Using a 5x5 mask yields, a skeleton is more accurate and robust against rotation and scaling of the pattern.

6. Results and comparison

There are no great differences between the results obtained following the two methods. They appear only for the intrinsic skeletal pixels detection, since the same steps are applied afterwards in order to determine the skeleton.

The CMD set is thicker than the ridge points one, but it contains fewer peripheral branches. As for the time required, the CMD extraction is quicker, since detecting a pixels as CMD requires fewer operations in its neighbourhood and a pixels marked in the path growing step is no longer inspected.

The dissimilarity between the original pattern and the reconstructed one can be assessed by means of the Baddeley distance [8], which is introduced in section 6.1

6.1 Principle of the Baddeley distance

Let $R \subset Z^2$ be the domain on which images are defined. Binary image A (respectively B) can be regarded as a function $f_A : R \rightarrow \{0, 1\}$ (resp. f_B).

$f_A(x,y) = 1$ if pixel (x,y) is in the pattern
$f_A(x,y) = 0$ else.

Let A and B be the pattern to be compared. The Baddeley distance between A and B is defined by:

$$D(A, B) = \left[\frac{1}{N} \sum_{p \in R} |d_A(p) - d_B(p)|^m \right]^{\frac{1}{m}}$$

where N is the number of pixels in R and $d_A(p)$ (respectively $d_B(p)$) is the distance between pixel p and the binary set characterizing pattern A (resp.B) (Fig. 9). Index m is equal to 2 in our application.

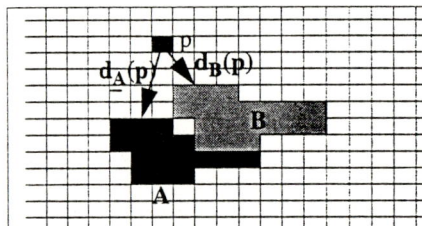

Fig. 9. Distance between pixels and patterns A and B.

Distances $d_A(p)$ and $d_B(p)$ are computed using the chamfer distance transformation introduced in section 2, considering the patterns A (respectively B) as background in the image.

It should be noticed that, Baddeley distance D(A,B) is a global criterion. It depends not only on the characteristics of the two pattern A and B, but also on the number of pixel in domain R. From this point of view, the skeleton obtained using Arcelli and Sanniti di Baja's method allows a reconstruction which is more accurate, i.e. generally, the Baddeley distance of the reconstructed pattern to the original pattern is smaller for the ridge points skeleton for most of the patterns

6.2 Experimental results

An example is given in fig. 10. The original pattern is shown in Fig. 10a.
Fig 10b and 10d are obtained after applying the different steps just mentioned, and the same threshold (**Th=2**)is applied for the pruning step. The reconstructed patterns are shown in fig. 10c and 10e. **D(A,B)** is computed over the same domain R.

The best reconstruction is achieved by using the set of ridge points. This result was obtained with other patterns.

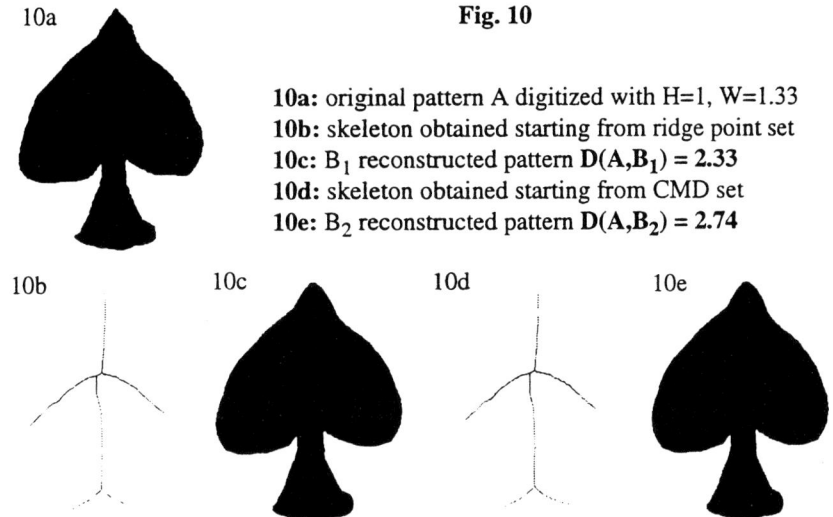

Fig. 10

10a: original pattern A digitized with H=1, W=1.33
10b: skeleton obtained starting from ridge point set
10c: B_1 reconstructed pattern $D(A,B_1) = 2.33$
10d: skeleton obtained starting from CMD set
10e: B_2 reconstructed pattern $D(A,B_2) = 2.74$

ridge point set, Th = 2, $D(A,B_1) = 2.33$ **CMD set, Th = 2, $D(A,B_2) = 2.74$**

7. Conclusion

In this paper, we have adapted the skeletonization methods described in [1] and [2] to the case of rectangular grids. By comparing them according to a quantitative criterion based on the Baddeley distance, it can be shown that the skeleton based on the ridge points extraction is better than the one based on the CMD extraction. However, the latter requires less computation time.

8. References

[1] E. Thiel, *"Unification de la squelettisation menée en distance"* 9th congrès RFIA, vol 1, pp 349-358, Paris, Janvier 1994. AFCET..

[2] C.Arcelli, G. Sanniti di Baja, *"Ridge points in Euclidean distance maps"*. Pattern Recognition Letters, vol. 13, 1992.

[3] C.Arcelli, G. Sanniti di Baja, *"Euclidean skeleton via centre-of-maximal disc extraction"*. Image and Vision Computing, vol. 11, 3 April 1993.

[4] D. Coquin, Ph. Bolon, *"Discrete distance operators on rectangular grids"*, Pattern Recognition Letters, vol. 16, pp 911-923, 1995.

[5] A. Rosenfeld, J. Pflatz, *"Sequential operators in digital picture processing"*, J. ACM, 13, pp 471-494, 1966

[6] D. Coquin, Ph. Bolon, *"Comparaison d'opérateurs locaux de distance"*, Proc. 3e Colloque de Geometrie Discrete: Fondements et Applications, pp 182-191, 1993

[7] G. Borgefors, *"Digital transformations in digital images"*, Computer Vision, Graphics and Image Processing, vol. 34, pp 344-371, 1986.

[8] A.J.Baddeley, *"An error metric for binary images"*, Robust Computer Vision, Wichmann, Karlsruhe, pp 59-78, 1992

[9] C. Arcelli, G. Sanniti, *"A one-pass two operation process to detect the skeletal pixels on the 4-distance transform"*, IEEE Transaction Pattern Analysis, Machine Intelligence, vol.11, no. 4, pp 411-414, 1989.

[10] J.M.Chassery, A. Montanvert, *"Géométrie discrète en analyse d'images"*. Hermes 1991.

An Algorithm for the Global Solution of the Shape-from-Shading Model

M. Falcone and M. Sagona

Dipartimento di Matematica, Università di Roma "La Sapienza",
P. Aldo Moro 2, 00185 Roma, e-mail: `falcone@caspur.it`

Abstract. A numerical scheme to solve the Dirichlet type problem for the first order Hamilton–Jacobi equation related to the shape–from–shading model is proposed. The algorithm computes the maximal solution of the problem provided a compatibility condition on the discretization steps is satisfied. This global formulation allows to include in the model the informations brought by the shadows in a rather natural way avoiding cumbersome boundary conditions on the interfaces between light and shadows and the use of additional informations on the surface.

1 Introduction

Let S be a Lambertian surface given as a graph $z = u(x)$, $x \in \mathbb{R}^2$ and let us assume that there is a unique light source at infinity whose direction is indicated by the unit vector $\omega = (\omega_1, \omega_2, \omega_3) \in \mathbb{R}^3$. We will assume in the sequel that ω is known. As it is well known, (see f.e. [6]) the partial differential equation related to the Shape-from-Shading (SFS) model can be derived by the Image Radiation Equation

$$R(\hat{n}(x)) = I(x) \tag{1}$$

where $I(x)$ is the brightness function measured at all points x in the domain of u, $\hat{n}(x)$ is the unit normal to the surface at the point $(x, u(x))$ and $R(\hat{n}(x))$ is the reflection map giving the value of the light reflection on the surface as a function of its orientation (i.e. of the normal) at each point. The brightness function I is known in the model since it is measured on each pixel of the "picture" for example in terms of a grey level (from 0 to 255). To construct a continuous model we will assume that I takes (real) values in the interval $[0, 1]$.
Assuming that u has a compact support Ω and recalling that for a Lambertian surface $R(\hat{n}(x)) = \hat{n}(x) \cdot \hat{\omega}$ equation (1) can be written in the form

$$I(x)\sqrt{1+ \mid \nabla u(x) \mid^2} + (\omega_1, \omega_2) \cdot \nabla u(x) - \omega_3 = 0, \text{ for } x \in \Omega \tag{2}$$

which is a first order nonlinear partial differential equation of Hamilton-Jacobi type. Moreover, we complement the equation with the natural Dirichlet boundary condition

$$u(x) = 0 \qquad \text{for } x \in \partial\Omega \tag{3}$$

---

[*] This work was supported by the EEC Program Human Capital "Mathematical Modelling of Image Processing" (Contract ERB CHRX–CT93–0095)

which corresponds to the assumption that the surface is standing on a background. The solution of the above Dirichlet problem (2), (3) will give the surface corresponding to the brightness $I(x)$ measured in the "picture" representing Ω.

There are several technical difficulties related to this problem. The first is that the surface can be non differentiable at some points so that we have to consider solutions in a "weak sense". The theory of viscosity solutions for Hamilton–Jacobi type equations provides the right framework for the analysis of the problem (see [4] for an up-to-date presentation of that theory) although to get uniqueness, the problem is usually solved adding some informations such as the height at the points where the brightness has a maximum or the complete knowledge of a level curve (see e.g. [10]). Recent results of the theory of viscosity solutions allow to characterize the maximal solution without extra informations besides the equation and to construct an algorithm which converges to that solution. From the numerical point of view one would like to have an algorithm able to compute non smooth solutions.

The addition of "shadows" makes the problem even more complicated since in principle the equation is defined only where $I(x) > 0$ and the interfaces between light and shadows could be non regular curves. The approach briefly described here allows to compute a global solution with just one single computation and does not require any preliminary reconstruction of the light/shadow interfaces (for a more detailed analysis of the algorithm we refer to [7]).

It should be noted that other global algorithms have been proposed in [9]. They use different numerical schemes to solve the associated Hamilton–Jacobi equation (f.e. the finite difference scheme introduced in [11]) and they do not include shadows as we are.

2 Approximation of the maximal solution of the eikonal equation

Let us start examining the theoretical results available for the case of a vertical light source. That case corresponds to $(\omega_1, \omega_2, \omega_3) = (0, 0, 1)$ so the general equation (2) becomes

$$I(x)\sqrt{1+ \mid \nabla u(x) \mid^2} - 1 = 0, \qquad x \in \Omega \qquad (4)$$

Writing (4) in explicit and adding the homogeneous Dirichlet boundary condition we obtain

$$\begin{cases} \mid \nabla u(x) \mid = f(x), & x \in \Omega \\ u(x) = 0 & x \in \partial\Omega \end{cases} \qquad (5)$$

where

$$f(x) = \sqrt{\frac{1}{I(x)^2} - 1} \qquad 0 < I(x) \leq 1. \qquad (6)$$

Note that with a vertical light we will never have shadows since our surface is a graph. This implies that $I(x)$ can only vanish at some single points. Moreover,

$f(x) = 0$ where $I(x) = 1$, i.e. the right–hand side vanishes at all points of maximum brightness and this causes the lack of uniqueness of "classical" viscosity solutions. However, a result by Ishii-Ramaswamy [8] gives a characterization of the maximal solution of (2), (3) for the eikonal equation corresponding to the case of a vertical light source.

In order to construct a numerical approximation, let us observe first that the eikonal equation (4) can also be written as

$$\begin{cases} \max_{a \in B_2(0,1)} \{-a \cdot \nabla u(x)\} = f(x), & x \in \Omega \\ u(x) = 0 & x \in \partial\Omega \end{cases} \quad (7)$$

since the maximum will be achieved for $a^* \equiv -\nabla u(x)/|\nabla u(x)|$.

Here and in the sequel we will assume for simplicity that $u \geq 0$. This is not restrictive, since the equation just depends on ∇u and we can always add to u the constant $u_0 \equiv \min_{x \in \Omega}\{u(x)\}$ to satisfy that requirement. In order to obtain an approximation scheme in the form of a fixed point problem it is useful to introduce the new variable

$$v(x) = 1 - e^{-u(x)} \quad (8)$$

Note that by definition $0 \leq v \leq 1$. The problem for the new variable v is

$$\begin{cases} v(x) + \max_{a \in B_2(0,1)} \left\{ -\frac{a}{f(x)} \cdot \nabla v(x) - 1 \right\} = 0, & x \in \Omega \\ v(x) = 0 & x \in \partial\Omega \end{cases} \quad (9)$$

where f is given by (6).

It is known (see e.g. [1]) that (9) has a unique continuous viscosity solution provided f is bounded and never vanishes in Ω.

Let us introduce the fully discrete scheme. To simplify, we will assume that the "picture" is a rectangle $\Omega \subset \mathbf{R}^2$. Let us consider a mesh of the set $\Omega_\delta = \Omega + \delta B_2(0, 1)$. We will denote by $\mathcal{I}_{in}$ the set of indices of the nodes x_i belonging to Ω, by $\mathcal{I}_{out}$ the set of indices of the nodes belonging to $\Omega_\delta \setminus \Omega$ (where we will impose the boundary condition) and by $\mathcal{I}$ their union. We will also denote by N_{in}, N_{out} and N respectively the number of nodes belonging to Ω_{in}, Ω_{out} and to their union.

Let k be the size of the mesh and let W^k denote the space of piecewise affine functions which are linear on the cells (P^1 finite element approximation). We look for a solution $w \in W^k$ of

$$w(x_i) = \min_{a \in B_2(0,1)} \left\{ e^{-h} w \left(x_i + h \frac{a}{f(x_i)} \right) \right\} + (1 - e^{-h}), \text{ for } i \in \mathcal{I}_{in} \quad (10)$$

$$w(x_i) = 0, \text{ for } i \in \mathcal{I}_{out}. \quad (11)$$

In [5] it has been proved that the numerical solution of (10) exists and it is unique. The numerical solution of our problem can be computed by a fixed point iteration on the operator $T : \mathbb{R}^N \to \mathbb{R}^N$

$$(T(V))_i \equiv \begin{cases} \min_{a \in B_2(0,1)} \{e^{-h} \Lambda(a) V\}_i + 1 - e^{-h}, & i \in \mathcal{I}_{in} \\ 0 & i \in \mathcal{I}_{out} \end{cases} \quad (12)$$

where V is the N-dimensional vector containing the values at the nodes of the mesh, i.e. $V_i = v(x_i)$, and $\Lambda(a)$ is the matrix of the local coordinates of the points $x_i + h \frac{a}{f(x_i)}$.

It is easy to prove that the operator T is a contraction mapping and a monotone operator from $[0,1]^N$ to $[0,1]^N$ so that there exists a unique fixed point $V^\star$, $V^\star = T(V^\star)$. By the monotonicity property starting from a subsolution $(V_0 \leq T(V_0))$ the sequence will monotically converge to the fixed point. This property is crucial to speed–up convergence and it also helps to compute the maximal solution.

3 Approximation of the maximal solution with shadows

Let us go back to our general equation corresponding to a light source in the direction of ω. If the light is oblique we will have shadows so that we can divide the support of the surface (the domain of u) into two regions,

$$\Omega_l \equiv \{x : I(x) > 0\}, \qquad \Omega_s \equiv \{x : I(x) = 0\} \quad (13)$$

which represent respectively the "light" and the "shadow" regions. Naturally, $\Omega = \Omega_l \cup \Omega_s$ and we will assume that also the projection of the shadows on the background will fall in Ω.

In Ω_l the equation is always the same, whereas in the "shadow" region the surface can have any shape since the model is not able to describe the real surface there. This is why other authors have included boundary conditions (f.e. Neumann boundary conditions) on $\partial \Omega_l$ to treat the problem in Ω_l just ignoring the region Ω_s. This can in turn create some difficulties in the construction of the numerical algorithm since the boundary of Ω_l can be non smooth and it will not belong to the mesh (unless a special mesh is constructed starting from that boundary).

Following [3] we include the region Ω_s in the computation just defining there a conventional surface to replace the (unknown) surface. We will substitute to the surface the "separation ray" (or "shadow ray") i.e. the ray separating light from shadows. That ray has the same direction of ω. This means that in Ω_s we have to solve the equation

$$(\omega_1, \omega_2) \cdot \nabla u(x) - \omega_3 = 0, \quad x \in \Omega_s \quad (14)$$

Note that (2) coincides with (14) since $I = 0$ in Ω_s. Then, we can use the same equation everywhere in Ω and we will not need to introduce any boundary condition on Ω_l, i.e. we can write the global problem as

$$\begin{cases} I(x)\sqrt{1+|\nabla u(x)|^2}+(\omega_1,\omega_2)\cdot\nabla u(x)-\omega_3=0, & x\in\Omega \\ u(x)=0 & x\in\partial\Omega \end{cases} \quad (15)$$

Following the same lines of Section 2, we obtain the fully discrete scheme

$$w(x_i) = \min_{a\in B_3(0,1)}\left\{\beta w(x_i+hb(x_i,a))-(1-\beta)\frac{I(x_i)\,a_3}{\omega_3}(1-w(x_i))\right\}+1-\beta \quad (16)$$

where $\beta = e^{-h}$ and $b(x,a) = \frac{1}{\omega_3}(I(x)a_1-\omega_1, I(x)a_2-\omega_2)$.

The new operator T_s corresponding to the oblique light source is

$$(T_s u)(x) \equiv \min_{a\in B_3(0,1)}\left\{\beta u(x+hb(x,a))-(1-\beta)\frac{I(x)\,a_3}{\omega_3}(1-u)\right\}+1-\beta$$

Under appropriate conditions, the operator T_s has the same properties of T and the algorithm will converge monotonically to the maximal solution.

On the boundary we just impose the homogeneous Dirichlet boundary condition. This condition implies that the shadows must not cross the boundary of Ω, so the choice $\omega_3 = 0$ corresponding to an infinite shadow behind our surface is not admissible. For the analysis of the algorithm and for some hints on its implementation we refer to [7].

4 Numerical experiments

The experiments has been made using curves and surfaces with different types of regularity (just continuous or differentiable) and with different numbers of maximum brightness points. In the vertical and in the oblique case we have made a cut-off on I at the level $\bar{I}$, defining the new brightness function

$$\widehat{I}(x) = \begin{cases} I(x) & \text{if } 0 \le I(x) < \bar{I} \\ \bar{I} & \text{if } \bar{I} \le I(x) \le 1 \end{cases} \quad (17)$$

The cut-off is necessary in the vertical light case since when $I(x) = 1$ the vectorifield "blows-up" at the points of maximum brightness. As far as the choice of the discretization steps is concerned, we observe that the convergence to the maximal solution takes place when k and h are such that $\frac{k}{h} \ge 1$.

Let us consider some 2-dimensional surface reconstructions with vertical light source. In the first example we have reconstructed a pyramid. In that case the brightness image (Fig. 1b) has a unique grey level. The reconstruction is represented in Figure 1a, it has been obtained with $\Delta x = \Delta y = h = 0.05, \bar{I} = 0.99$ and gives an L^1 error of about 0.0195. As a second example we considered a smooth surface. Figures 1c, 1d and 1e respectively show the real solution surface, the brightness image and the maximal solution. In this case the image has five internal maximum points where we have to make a cut-off (here $\bar{I} = 0.96$). The numerical solution has been computed for $\Delta x = \Delta y = 0.02$ and $h = 0.01$. Figure 1f is the numerical reconstruction of the same surface obtained fixing the height at the maximum brightness points as in [12]. We conclude with the reconstruction of a real grey-level image representing a vase (Figure 2a). Figure 2b shows the computed surface for $\Delta x = \Delta y = h = 0.01$ and $\bar{I} = 0.99$. It is interesting to compare the original image with the images obtained computing I from the computed surface. Infact, once we have the surface we can compute $I(x_{ij})$ at each node x_{ij} simply making the scalar product $\omega \cdot \eta(x_{ij})$. However, note that the numerical solution has been computed over a grid so that the normal is not uniquely defined at each node. In principle, there are four admissible normals corresponding to the triangles having x_{ij} among their vertices and several (non equivalent) choices are possible. Figure 2c shows the intensity obtained taking the normal of the upper rightmost triangle. In that case one can note that the numerical surface is non smooth along three curves. One can also compare the original image with the one in Figure 2c obtaining an L^1 error is about 0.0082. Taking all the possible normals (four as we said) one can compute four different values of the intensity at every internal node x_{ij}. Figure 2d shows the image corresponding to the minimum of those values at each node, the L^1 error is about 0.0076. In that case the picture looks smoother and closer to the original.

References

1. M. Bardi and M. Falcone, *An approximation scheme for the minimum time function*, SIAM Journal of Control and Optimization, **28** 1990, 950-965.
2. F. Camilli, *Qualche applicazione delle soluzioni viscosità a problemi di evoluzione di fronti*, Ph. D. Thesis, Dipartimento di Matematica dell'Università di Roma "La Sapienza", October 1995.
3. F. Camilli and M. Falcone, *An approximation scheme for the maximal solution of the shape-from-shading model*, Proceedings ICIP 96 International Conference on Image Processing, IEEE Inc., 1996, 49-52.
4. M. G. Crandall, H. Ishii and P. L. Lions, *User's guide to viscosity solutions of second order partial differential equations*, Bull. Amer. Math. Soc., **27** (1992), 1-67.
5. M. Falcone, *The minimum time problem and its applications to front propagation*, in A. Visintin e G. Buttazzo (eds), "Motion by mean curvature and related topics", De Gruyter Verlag, Berlino, 1994
6. B.K.P. Horn and M.J. Brooks, *Shape from Shading*, The MIT Press, 1989.
7. M. Falcone and M.Sagona, forthcoming.
8. H. Ishii and M. Ramaswamy, *Uniqueness results for a class of Hamilton-Jacobi equations with singular coefficients*, preprint, 1995.
9. R. Kimmel and A.M. Bruckstein, *Global Shape fron Shading*, Center for Intelligent System Report CIS #9327,Technion-Israel Inst. of Tech., Israel, November 1993.
10. P.L. Lions, E. Rouy and A. Tourin, *Shape from shading, viscosity solution and edges*, Numerische Mathematik, **64**, (1993), 323-353.
11. S.J. Osher and J.A. Sethian, *Fronts Propagating with curvature dependent speed: Algorithms based on Hamilton-Jacobi formulations*, J. of Comput. Phys., **79** (1988), 12-49.
12. E. Rouy and A. Tourin, *A viscosity solutions approach to Shape from Shading*, SIAM J. Numer. Anal., **29** (1992), 867-884.

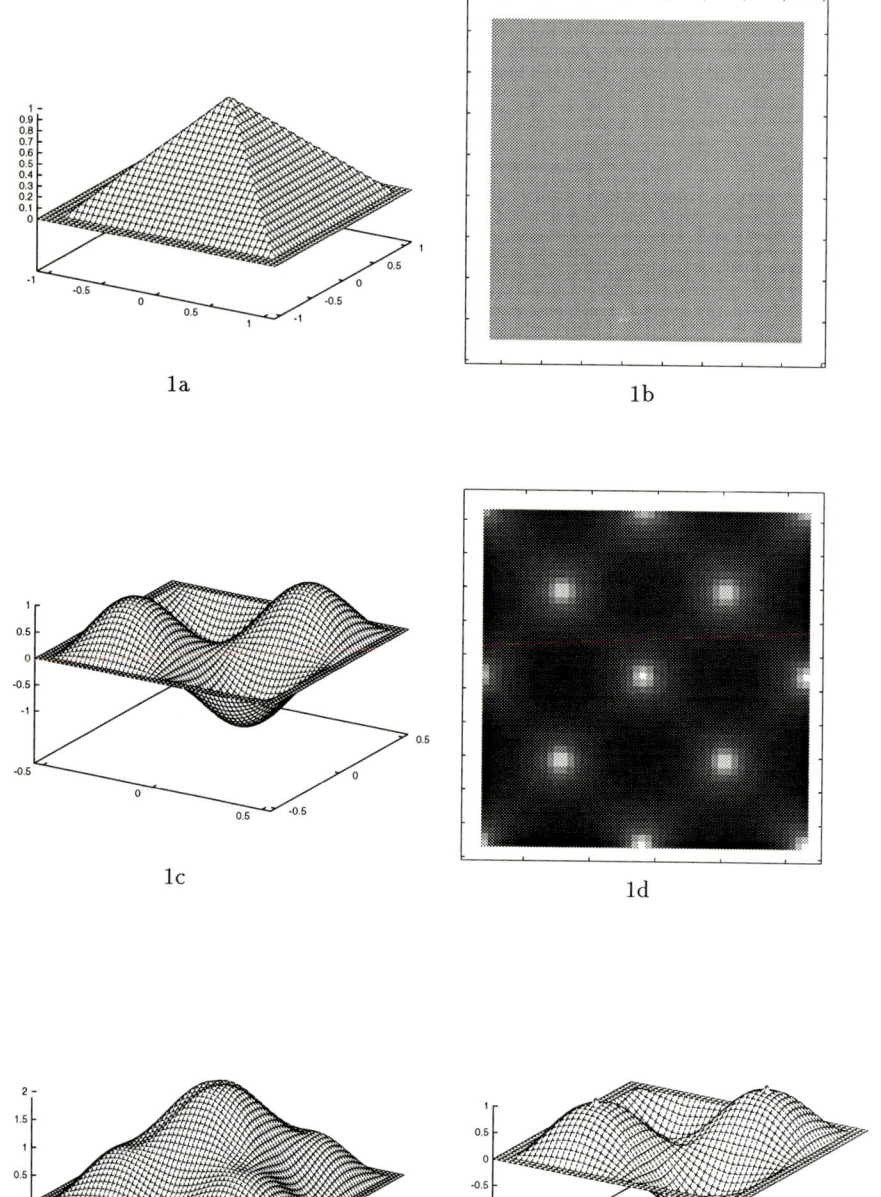

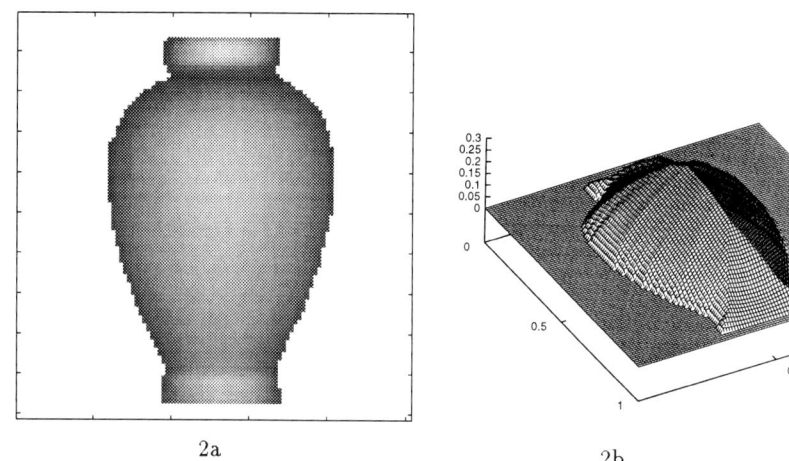

2a 2b

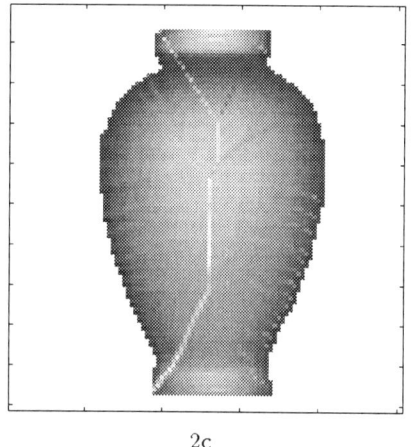 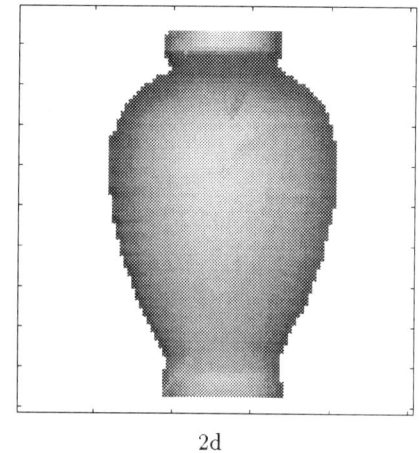

2c 2d

A Statistical Classification Method for Hierarchical Irregular Objects

Markus Peura

Helsinki University of Technology
Laboratory of Computer and Information Science
P.O. Box 2200, FIN-02015 HUT, Finland**

Abstract. This paper introduces a method for classifying structured visual objects that appear frequently in meteorological, medical and biological imagery. The focused objects are taken to be highly irregular and composed of subobjects in a hierarchical manner. The approach consists three principal steps. At first, hierarchical objects are detected in an segmented image. Secondly, shape descriptors are used to extract information of the contours of the objects. Finally, a global description for an object is obtained by applying statistical moments. As the goal is to classify natural objects, the most challenging task is to tolerate irregularity present in both spatial and hierarchical levels. Experiments with artificial images show that the method combines succesfully shape descriptors and object hierarchy.

1 Introduction

Contour irregularity is a pronounced feature of many real-world natural objects. Object hierarchy, as far as it exists, might also be irregular and explicable only by statistical means. Some artificial objects are shown in Fig. 1. In each column, visual similarity between the two objects is obvious. Moreover, one easily gets an intuition of how new instances of each class might look like. The problem is to find a possibly general method for classifying objects of this type.

Sometimes there are no clear, separate "objects" in the imagery – or they remain difficult to extract due to their continous nature, noise or effects of illumination. Depending on the application, there are several other approaches to choose from: histograms, color information or texture.

On the other hand, if objects appear clearly it is possible to utilize image geometry. In this paper, we use *shape descriptors* as the features of an object as they complete the topological information in a natural fashion. The shape descriptors measure an object's overall geometrical properties such as elongation, curvature and rectangularity[6, Ch. 6]. Efficiency of some descriptors for irregular forms is discussed in [5].

** This article was written during a visit to **Université Louis Pasteur**, CNRS, LSIIT (URA 1871), 7, rue de René Descartes, 67084 Strasbourg, France.

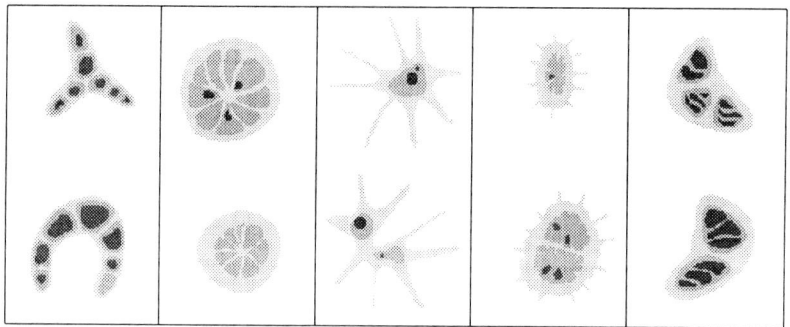

Fig. 1. Examples of irregular hierachical objects.

Hierarchical properties have been recently studied to exploit semantics, the highest level of image understanding. A retrieval system of medical images presented by [3] involved *spatial relations* of the type "nearby","slightly touching" and "invading". Hierarchy was also applied in calculating distances between the relations.

The emphasis in this paper is on extracting information from both contours and hierarchical structure in a combined fashion. The statistics used to provide an overall descriprion of an object belong to the mathematical family of *moments*, which was early applied in visual pattern recognition [4]. Related methods have also been adapted directly as shape descriptors [1] and in texture description [2]. Input images are taken to have been preprocessed with some appropriate segmentation algorithm. Nevertheless, the suggested method does not require any additional segment attributes - only forms and spatial organization are considered. If further information exists, it can be flexibly added to the method as proposed in discussion.

This paper is organized as follows. The suggested method is presented in Sec. 2 and consists of definitions of applied object model (2.1), an algorithm for detecting object hierarchy (2.2), the applied object descriptors (2.3) and statistical measures (2.4).

2 Suggested method

2.1 Object hierarchy

A simple visual object is depicted in Fig. 2a. The object consists of segments labeled as 1,2,3 and 4. The background, marked with 0, can is seen as a special segment – its definition remains application-specific. The simplest interpretation of the object is to regard it as an equal composition of segments 1, 2, 3 and 4.

As far as hierarchy is considered there exists several alternatives for modeling. Clearly, the essential properties are connectivity and nesting. Three different interpretations of hierarchy are illustrated in Fig. 2. The one applied in this

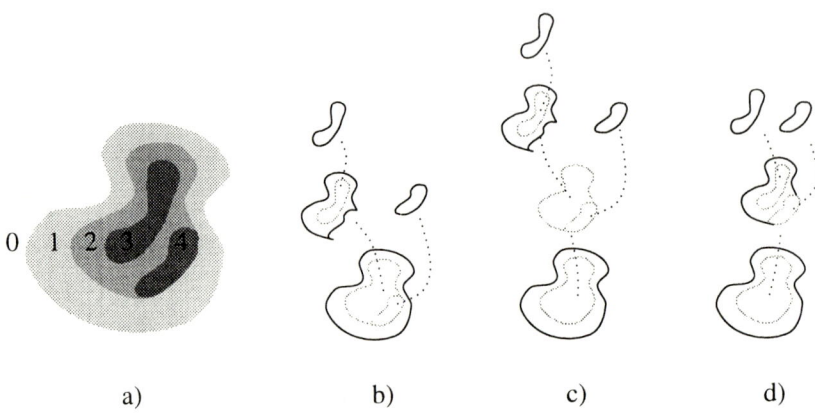

Fig. 2. Alternative interpretations of object hierarchy.

paper is presented in Fig. 2d and applies an additional rule: among connected segments, the segment that has the maximal portion of the common contour *is not* an object. This concept of ignoring the "dominating" segment is motivated by the intuitive concept of altitude: an object is treated like a set curves on a topographical map. Notice that this rule is not based on topology but on a computational measure. Also the gray tones in Fig. 2a only illustrate the preferred interpretation: as metioned above, colors (or any labels) do not affect the hierarchy.

2.2 TRIM algorithm

In this section, an algorithm is presented for constructing a data structure that corresponds to the tree hierarchy described in 2.1. The algorithm, from this on called TRIM (TRee IMage), scans an image in a recursive manner. Each time a *nested* segment is encountered, a definite border, *HardBorder*, is drawn around the aggregate of all the connected segments to force traversing the complete object before quitting. This constraint facilitates the bookkeeping: the hierarchy is built one branch at the time. In the hierarchy tree, let us call the nested objects within an object as *descendants*.

On the other hand, each time a *connected* segment is found, an one-directional border, *SoftBorder* is drawn around it: the traversal is allowed to exit the segment but *not to return*. This ensures that each segment is entered exactly once, thus no multiple instances of the same object will occur. In the hierarchy tree, let us call the connected objects as *parallel* objects for obvious reasons.

Practically, the algorithm requires an additional image for bookkeeping. Fortunately, *HardBorders*, *SoftBorders* and visited pixels can be marked in the same bookkeeping image - eight bits per pixel are sufficient for storing all information involved. The high-level syntax of TRIM is listed in Table 1. An application of the algorithm for the object shown in Fig. 2 is illustrated in Fig. 3.

Table 1. Top-level description of TRIM algorithm

```
TRIM
set previous pixel to Visited
if current pixel is Visited then
    return;
if HardBorder was crossed then
    return;
if SoftBorder was crossed then
    trace and mark contour around current segment as SoftBorder
    add object as parallel node to Tree ;
    call TRIM in neighboring pixels;
    return;
if new segment was entered without crossing Soft/HardBorder then
    trace and mark contour around all connected segments as HardBorder;
    add object as descendant node to Tree;
    trace and mark contour around current segment as SoftBorder;
    add object as descendant node to Tree ;
call TRIM in neighboring pixels;
detect and remove dominant descendant;
return;
```

Inside an object (a) the traversal proceeds recursively in all directions. As far as a nested object is encountered (b), *HardBorder* is drawn around (c) the set of connected segments. Then, *SoftBorder* is drawn (d) around the encountered segment. Recursion continues inside the segment but is allowed to pass trough the *SoftBorder*. For any subsequent connected segments (e) this procedure will be repeated (f). For new nested segments (g), the procedure of (c-d) will be repeated (h). Recursion completes (i) inside the nested segment and continues (j) within the parent object. Before quitting a nested segment, the possible dominant segment is supprimed from the tree structure.

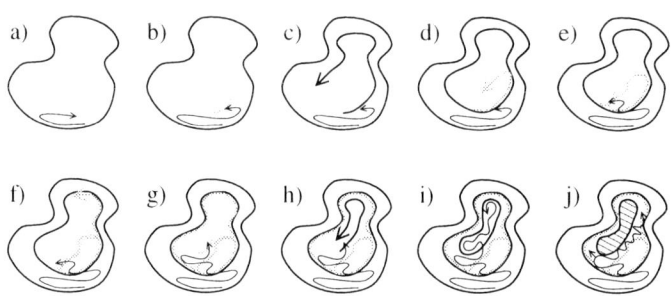

Fig. 3. TRIM algorithm applied to the object shown in Fig. 2a.

2.3 Object descriptors

In order to describe an object one has to carry out measurements. The *descendant count* (N_{desc}), also called the *branching factor*, is a straightforward topological feature. For example, the object in the upper left corner of Fig. 1 has seven descendants, each of which have one descendant. *Proportional area* (A_{pr}) is defined as the area of the object divided by the area of its parent object. Clearly, A_{pr} is not a topological feature but shows practically some inverse correlation with N_{desc}.

In addition, we wish to extract information of object geometry. This is carried out by means of four *shape descriptors* illustrated in Fig. 4. All these shape descriptors can be calculted in a time linearly proportional to the contour length. Detailed definitions can be found in [5], for example.

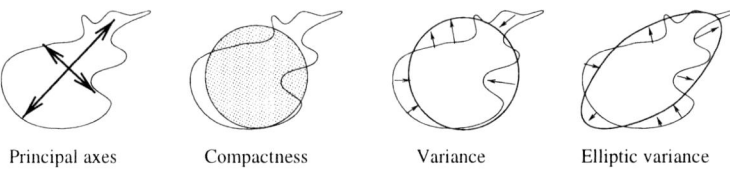

Principal axes   Compactness   Variance   Elliptic variance

Fig. 4. Four simple shape descriptors.

2.4 Statistical modelling

As explained in 2.2, TRIM transforms an image to a tree structure. The structure of a tree corresponds to the object topology in the image. All that actually has to be stored in a node is a feature vector consisting of N_{desc}, A_{pr} and the four shape descriptors - there is no need to store the contour itself. An example of a resulting tree object is depicted in the left hand side of Fig. 5. It is clear that a tree will not preserve all the information present in the original image. However, memory requirements are remarkably decreased, unless an object consists of a large number of small specks.

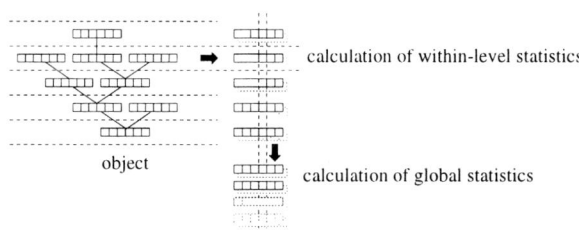

Fig. 5. An object modeled as a tree. Calculation of global statistics.

For purposes of classification, one has to be able to compare trees. Practically, one applies a codebook, a collection of samples labelled according to correct classes. In our case, a codebook could consist of trees presented by some appropriate syntax. However, the comparison operation between a given and stored samples might be computationally elaborate. In addition, we desire to make our approach invariant of the hierarchical depth of objects. Thus, instead of matching the tree graphs directly we suggest global statistical measures.

The basic idea is depicted in Fig. 5. The global measures are obtained by applying repeatedly four basic statistics listed in Table 2.

Table 2. Simple statistical moments for variable x.

$\frac{1}{N}\sum_{i=1}^{N} x_i$	$\frac{1}{N\mu^2}\sum_{i=1}^{N}(x_i - \mu_x)^2$	$\frac{1}{N\Sigma_x}\sum_{i=1}^{N} x_i i$	$\frac{1}{N^2\Sigma_x}\sum_{i=1}^{N} x_i i^2$
mean $[\mu]$	variance $[\sigma^2]$	centroid $[c]$	inertia $[\mathcal{I}]$

Given a tree presentation of an object, μ and σ^2 are calculated at each layer for each element of the feature vector. As the feature vectors consist of six elements, twelve statistics result at each layer. Then, all the statistics (μ, σ^2, c and $\mathcal{I}$) are calculated for the obtained set, resulting in 48 statistics. When calculating the global statistics, the index i equals the height of a layer. Thus, c and $\mathcal{I}$ provide valuable information on how the features are distributed *vertically*. For example, $\mathcal{I}$ will tell that the compactness of the objects in the leftmost column of Fig. 1 *tends to increase* along the levels of the tree hierarchy. On the other hand, the compactness of the objects in the rightmost column seems relatively constant troughout the levels.

There is no motivation for calculating the layer-wise c or $\mathcal{I}$, because the horizontal order of branches is insignificant.

3 Experiments

The discriminatory power of the proposed method was tested with a set of artificial images consisting of 72 samples: 12 classes, 6 samples in each. The objects shown Fig. 1 belong to this test set and give some impression of intra- and interclass diversity. To keep the evaluation of results simple, the *nearest neighbour* (NN) and *3–nearest neighbours algorithm* (3NN) were used in the experiments. Implementation of cross-validation was straightforward: every sample, in turn, was picked out from the set and the codebook consisted of the remaining 71 samples. The results for some combinations of features are listed in Table 3. There are a couple of interesting issues. At first, the within-level variance (σ^2) seems to be redundant; without the mean (μ) reasonable results would not be obtained at all. Secondly, classification using no shape descriptors but hierarchical information only (A_{pr} and N_{desc}) shows suprisingly poor performance. In general, using either shape descriptors or hierarchical information leaves performance around 70% but their cooperation enhances classification remarkably.

shape descriptors	×	×	×	×	×		×	×	×	×		×	×
A_{pr}	×	×		×		×	×		×		×	×	×
N_{desc}	×	×	×			×	×	×			×	×	×
within-level μ	×	×	×	×	×	×	×	×	×	×	×		
within-level σ^2	×							×	×	×	×	×	×
global μ,σ^2	×	×	×	×	×	×	×	×	×	×	×	×	×
global $c,\mathcal{I}$	×	×	×	×	×	×		×	×	×	×		×
NN	85	94	94	82	76	67	72	81	82	75	60	71	47
3NN	78	93	85	79	71	60	76	76	81	72	63	67	49

Table 3. Classification percentages for some combinations of features and operators.

4 Discussion

A flexible method was presented for recognizing and classifying natural objects. The statistical aproach provided means for treating irregularity and hierarchy in a scale-invariant manner. The experiments showed that the proposed global descriptors are able to extract central information of the structrural nature of an object. The test set was relatively small, 72 samples, but clearly suggested general applicability of the method.

As the studied objects have tree-like hierarchy, graph matching might be another method for classification. The choice between these two rather different approaches is dependent on the application.

As far as tree hierarchy is considered, visual objects have some special practical restrictions. Inside the contour of an object there is limited space available for subobjects. In natural objects, sizes of subsequent contours tend to change exponentially – for a complicated phenomenon, the resolution runs out in "both ends": some details are smaller than a pixel and some properties are beyond the scope of the whole image. Practically, this means that captured visual objects have relatively simple hierarchy. Forcing more levels to tree hierarchy will naturally not bring any new information: instead of more refined hierarchy (more branching) the tree will only get "streched" as the countours of subsequent objects resemble each other.

Paradoxally, simplicity of visual objects might be a problem when applying statistical approach. For example, the variance is mathematically well-defined for a single sample but its statistical interpretation and significance becomes vague. This problem is visible in the results of the experiments (Table 3); the variance seems to produce noise rather than information. This suggests that in classifying a sample, the variance should be used only if the decision, based on the mean only, remains ambiguous.

In the lowest levels of a tree, relatively few nodes contribute to in-level statistics. On the other hand, intuition suggests that the most significant information is located near the root: apparently, the set of artificial test objects could have been recognized well by applying specially designed shape descriptors to the outer contours only. But as mentioned in Sec. 1, the goal was to develop a possibly general approach – which finally showed promising behaviour for the test set containing a diversity of forms.

Practically, success of high-level image understanding requires reliable segmentation, techniques of which were beyond the scope of this paper. For example, due to some unstable segmentation algorithm separate objects might become conneted (or vice versa) producing severe problems in shape description, especially when detecting elongation. Moreover, topology may be distorted if faulty connected objects create closed chains and thus new "irreal" subobjects.

The background and the motivation for this study was in interpretation of meteorological radar images but the general applicability of the proposed method is obvious. Moreover, the model is directly adjustable to utilize any application-specific segment information: existing feature vectors are easily augmented with further elements. This information can be obtained during segmentation or by means of more complex shape descriptors.

Acknowledgement

The author wishes to thank the Finnish Foundation of Technology Development (Tekniikan Edistämissäätiö) for funding and professor J. Korczak at Louis Pasteur University, Strasbourg, for providing research facilities. The weather radar reseach team at the Finnish Meteorological Institute is acknowledged for cooperation.

References

1. L. Gupta and M.D. Srinath. Contour sequence moments for the classification of closed planar shapes. *Pattern Recognition*, 20(3):262–272, 1987.
2. Robert M. Haralick, K. Shanmugam, and Its'hak Dinstein. Textural features for image classification. *IEEE Transactions on Systems, Man, and Cybernetics*, SMC-6(6):610–621, November 1973.
3. Chih-Cheng Hsu, Wesley W. Chu, and Ricky K. Taira. A knowledge-based approach for retrieving images by content. *IEEE Transactions on Knowledge and Data Engineering*, 8(4):522–532, August 1996.
4. Ming-Kei Hu. Visual pattern recognition by moment invariants. *IRE Transactions on Information Theory*, 8(2):179–187, 1962.
5. Markus Peura and Jukka Iivarinen. Efficiency of simple shape descriptors. In *IWVF3, 3th International Workshop on Visual Forms*, Capri, Italy, May 1997.
6. Milan Sonka, Vaclav Hlavac, and Roger Boyle. *Image Processing, Analysis and Machine Vision*. Chapman-Hall Computing, London, 1995.

Multi-level Dynamic Programming for Axial Motion Stereo Line Matching

Raymond K K Yip
Department of Electronic Engineering
City University of Hong Kong
83, Tat Chee Avenue, Kowloon, Hong Kong.
email: eeryyip@cityu.edu.hk

ABSTRACT

In this paper, a multi-level dynamic programming approach is used to solve the line segment based correspondence problem in axial motion stereo. In this method, a *Local Similarity Measure* is calculated for each line segment pair between the *Front* and *Back* images. In level 1, the *matching probability* between line segments is represented by their *Local Similarity Measure (LSM)*. Line segment pair that have a *matching probability* larger than a threshold T_1 is selected as potential matching pair. T_1 is set to a relative high value so that the probability of correct match of level 1 is very high. Dynamic programming is then used to search for their best match. Based on the geometric properties between the matched and the unmatched line segments, a *Global Similarity Measure (GSM)* is calculated for each unmatched line segment pair. An overall *Similarity Measure (matching probability)* is then obtained by the *LSM* and the *GSM*. Then, the algorithm begin the second match but with a slightly lower threshold T_2. The new matched results are then used to modify the *GSM* and the overall *Similarity Measure*. These processes are repeated until a predefined level n_{stop} (or a predefined condition) is reached. By using the *GSM* and multi-level searching technique, the proposed technique increases the matching accuracy and reduce the number of unmatched line segment due to misordering when dynamic programming is used for axial motion stereo matching.

1. INTRODUCTION

One of the major area in computer vision is the recovery of 3D information (depth map) using stereo vision analysis. In stereopsis, one of the most important and difficult stage is stereo matching. It is the process of identifying the corresponding 2D features between images that are belongs to the same physical point in the 3D scene. This problem is also known as the *correspondence problem*.

The most popular cameras arrangement for multiple images is the lateral stereo model [2, 4-6]. Recently, researchers have been moved towards studying images of a moving camera system. Axial motion stereo [1, 3] is one of the technique in which the camera system is moved along its optical axis.

Many lateral stereo matching techniques have been proposed in the past decades [2, 4-6]. A review [2] is provided by U. R. Dhond and J. K. Aggarwal. One of the important approach is the use of dynamic programming [2, 4, 5]. It has been proved to be one of the simple and efficiency technique for stereo matching. In this paper, a line segment based multi-level dynamic programming method is proposed to solve the corresponding problem of axial motion stereo images from a single camera. The proposed technique increases the matching accuracy and reduce the number of unmatched line segment when dynamic programming is used for axial motion stereo matching.

2. MULTI-LEVEL LINE MATCHING PROCESS

In this paper, the corresponding problem of axial motion stereo is formulate as:
1) Assign unique labels to each line segment of the front F (f_i) and back B (b_j) images.
2) Calculate a *Local Similarity Measure*, $LS(f_i,b_j)$, for each line segment pair (f_i,b_j).
3) Use $LS(f_i,b_j)$ as the overall *Similarity Measure*, $SM(f_i,b_j)$, of (f_i,b_j) and select a threshold T.
4) Select those line segment pairs whose $SM(f_i,b_j)$ is greater than T.
5) Use dynamic programming to find the best match for those selected from step 3.
6) Use the matched results to obtain a *Global (Structure) Similarity Measure*, $GS(f_i,b_j)$, for each unmatched line segment pair (f_i,b_j).
7) Calculate the $SM(f_i,b_j)$ by $LS(f_i,b_j)$ and $GS(f_i,b_j)$ for each unmatched line pair.
8) Lower the threshold T and repeat step 4 to 8 for all unmatched line segments until a predefined condition is meet.

The proposed method is a multi-level line matching process with the first level matching those line segments that have a very high similarity value ($SM(f_i, b_j) > T$). Thus, the probability of correct match in this level is very high which provides a strong basis (accurate references) to assist the matching process of the next level for weaker matches. The matching criterion is gradually decreases (by lowering T) as the level increases so stronger matches are matched first then followed by weaker matches. Furthermore, matched line segments in all previous levels are used to assist the matching process of the next level so as to improve the chances of correct match of weaker matches.

2.1 Ordering, Constraints and Similarity Measure

In each level (level n), unmatched line segments in the front image F and back image B are sorted in ascending order w.r.t. x-coordinates of their mid-points. For those points that have the same x-coordinates, their y-coordinates are used to sorted the order between them.

If any line segment b_j in B satisfied all the following constraints, b_j is a potential matching line of f_i and a similarity measure will be calculated to represent the similarity between these two lines.

1. Focus of Expansion (FOE) Constraint and FOE Similarity Measure

As the camera only moves along the optical axis. The displacement vector between correspondence features will intersect at a point known as the focus of expansion (FOE) [1, 3], see figure 1.

In this paper, f_i and b_j are considered as satisfying the FOE constraint

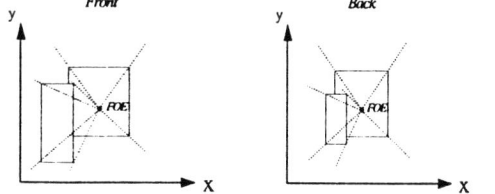

Figure 1. Example illustrates features of an image moves away form the focus of expansion (FOE) when the camera moves along its optical axis.

if $\quad S_{FOE}(f_i, b_j) > T_n \quad$ (1)

with $\quad S_{FOE}(f_i, b_j) = e^{\dfrac{-|n\_dist| \cdot S_1}{Theoretical\_len(f_i, b_j)}} \quad$ (2)

and $S_1 = 4$, *n_dist* is the normal distance between the mid-point of b_j to the line formed by the mid-point of f_i and FOE (Figure 2), *Theoretical_len(f_i, b_j)* is the length of the equation formed by b_j overlapped with the triangle formed by the end points of f_i and FOE (figure 3), and T_n is the threshold of level *n*.

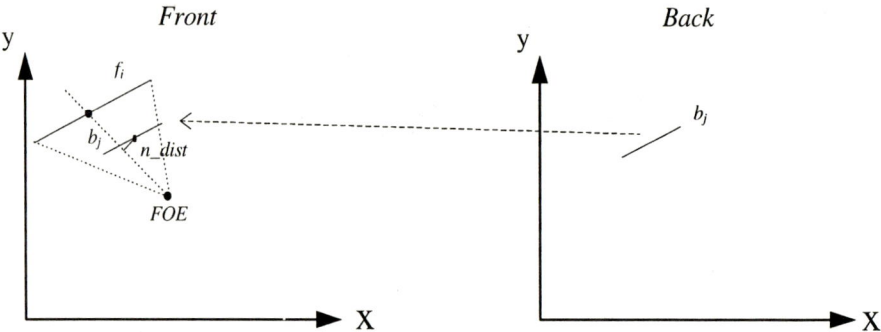

Figure 2. *n_dist* is the normal distance between the mid-point of b_j to the line formed by the mid-point of f_i and FOE.

2. *Overlap Similarity Constraint*

f_i and b_j are considered as overlapping

if $\quad S_{OLP}(f_i, b_j) > T_n \quad$ (3)

with $\quad S_{OLP}(f_i, b_j) = \dfrac{\min[\, Real\_len(f_i, b_j)\,,\, Theoretical\_len(f_i, b_j)\,]}{\max[\, Real\_len(f_i, b_j)\,,\, Theoretical\_len(f_i, b_j)\,]} \quad$ (4)

and *Real_len(f_i, b_j)* is the length of b_j overlapped with the triangle formed by f_i and FOE, *Theoretical_len(f_i, b_j)* is the length of the equation formed by b_j overlapped with the triangle formed by the end points of f_i and FOE (figure 3), and T_n is the threshold of level *n*.

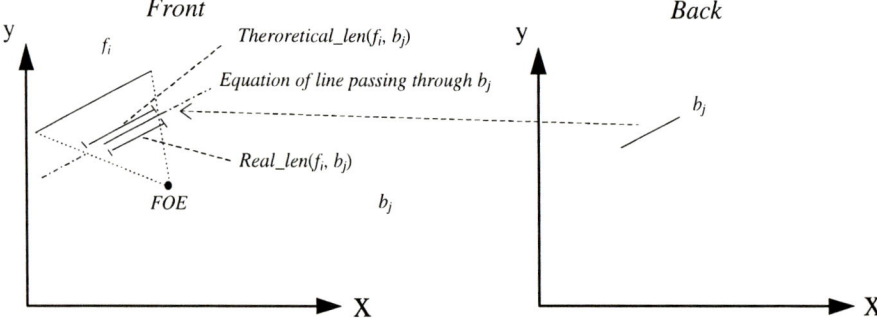

Figure 3. Example showing the *Real_len(f_i, b_j)* and *Theoretical_len(f_i, b_j)* of the overlap similarity constraint.

3. Orientation Constraint and Orientation Similarity Measure

The orientations of f_i and b_j are considered as similar

if $\quad OS(f_i, b_j) > C_{OS}$ \hfill (5)

with $\quad OS(f_i, b_j) = \left[\cos\left(\theta_{f_i} - \theta_{b_j}\right)\right]^{S_2}$ \hfill (6)

and $C_{OS} = 0.9$, $S_2 = 4$.
The definition of orientation of a line segment is shown in Figure 4.

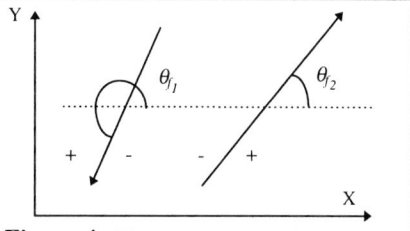

Figure 4. The orientation of two line segments of an image. The gray value of the + side is greater than that of the - side.

4. Global (Structure) Similarity Constraint and Measure

Once when matching pair(s) is/are obtained, a factor $GS(f_i, b_j)$ is added in the next level to measure the *Global (Structure) Similarity* between f_i and b_j. In addition, a factor $SE(f_i, b_j)$ is used to calculated the *Global (Structure) error* between f_i and b_j. If $SE(f_i, b_j)$ is larger than a threshold, then $GS(f_i, b_j)$ is set to 0. The structure of f_i and b_j are considered as similar

if $\quad GS(f_i, b_j) > T_n$ \hfill (7)
and $\quad SE(f_i, b_j) < 0.2k$ \hfill (8)

with $\quad GS(f_i, b_j) = \dfrac{1}{2k} \sum_{p=1}^{k} (e^{-|\Delta \cos m\theta_p| \cdot S_3} + e^{-|\Delta \cos m\phi_p| \cdot S_3})$ \hfill (9)

$$SE(f_i, b_j) = \sum_{p=1}^{k} \delta_p \quad (10)$$

$$\delta_p = \begin{cases} 1 & \text{if } \Delta \cos m\theta_p > 0.2 \text{ or } \Delta \cos m\phi_p > 0.2 \\ 0 & \text{otherwise} \end{cases} \quad (11)$$

$\Delta \cos m\theta_p = \cos m\theta_{f(p,i)} - \cos m\theta_{b(p,j)}$ \hfill (12a)
$\Delta \cos m\phi_p = \cos m\phi_{f(p,i)} - \cos m\phi_{b(p,j)}$ \hfill (12b)

and $S_3 = 10$; T_n is the threshold of level n; k is the number of all matched pairs; $\cos m\theta_{f(p,i)}$ is defined as the horizontal angle between the mid-points of f_p and f_i; $\cos m\theta_{b(p,j)}$ is defined as the horizontal angle between the mid-points of b_p and b_j; $\cos m\phi_{f(p,i)}$ is defined as the vertical angle between the mid-points of f_p and f_i; $\cos m\phi_{b(p,j)}$ is defined as the vertical angle between the mid-points of b_p and b_j; $\Delta \cos m\theta_p$ is defined as the horizontal angles difference between (f_p, f_i) and (b_p, b_j); $\Delta \cos m\phi_p$ is defined as the vertical angles difference between (f_p, f_i) and (b_p, b_j); f_p and b_p are matched pair obtained from all previous level(s). Figure 5 illustrates this idea.

From equation (9) and figure 5, it can be seen that the $GS(f_i, b_j)$ measures the angles difference between the potential matching pair (f_i, b_j) with all matched (f_p, b_p) line segments. Equation (11) defines the maximum allowed angle deviation (0.2) for each

(f_p, f_i) and (b_p, b_j). If either the horizontal or vertical angle difference between (f_p, f_i) and (b_p, b_j) are larger than the maximum allowed angle deviation, (f_p, f_i) and (b_p, b_j) is considered as a mismatched case (i.e. $\delta_p = 1$). If the total number of mismatched cases of (f_i, b_j) (equation (10)) is larger than *20%* of the number of matched line segments (equation (8)), (f_i, b_j) is considered as mismatched and $GS(f_i, b_j)$ will be set to 0.

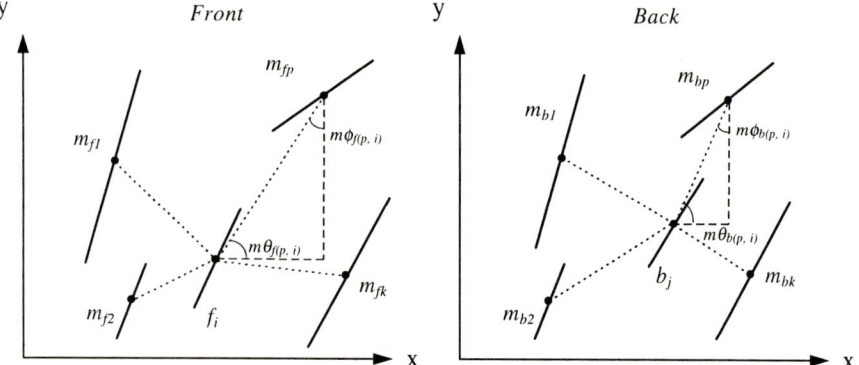

Figure 5. The relationship of line segments f_i and b_j with all previous matched results $m_{f1}, m_{f2}, ..., m_{fp}, ..., m_{fk}$ and $m_{b1}, m_{b2}, ..., m_{bp}, ..., m_{bk}$ respectively.

2.2 Matching with Dynamic Programming

A multi-stage weighted search graph [3][5] G_n (level n) is then constructed after the above process. A primitive G_1 (practically, the graph is more complicated) is shown in Figure 5. A null vertex $v_{i,0}$ is added to each stage V_i to represent the condition of no matching occurs between the front image and the back image. All possible path must start from stage s and terminate in t. And each vertex $v_{i,j}$ of the weighted graph G_n has an associated weight $SM(f_i, b_j)$ to indicate the *Similarity Measure* between f_i and b_j,

$$SM(f_i, b_j) = \begin{cases} \dfrac{w_1 S_{FOE}(f_i, b_j) + w_2 S_{OLP}(f_i, b_j) + w_3 OS(f_i, b_r) + w_4 GS(f_i, b_j)}{\sum_{1}^{4} w_a} & \text{if } (f_i, b_j) \text{ satisfies all constraints} \\ 0 & \text{otherwise} \end{cases} \quad (13)$$

with $w_1 = 1$, $w_2 = 1$, $w_3 = 1$ and $w_4 = 6$.

An accumulated weight $ASM(v_{i,j})$ is used in dynamic programming to search the maximal weighted path of G_n. It is defined as

$$ASM(v_{i,j}) = \max_{\forall v_{i-1,p} \in V_{i-1}} \left[ASM(v_{i-1,p}) \right] + SM(l_i, r_j) \quad (14)$$

and the maximal path is back traced from stage V_t to V_s. The matched result is stored in M which contains the index pairs $[(m_{l1}, m_{r1}), (m_{l2}, m_{r2}), ..., (m_{lk}, m_{rk})]$ of line segments.

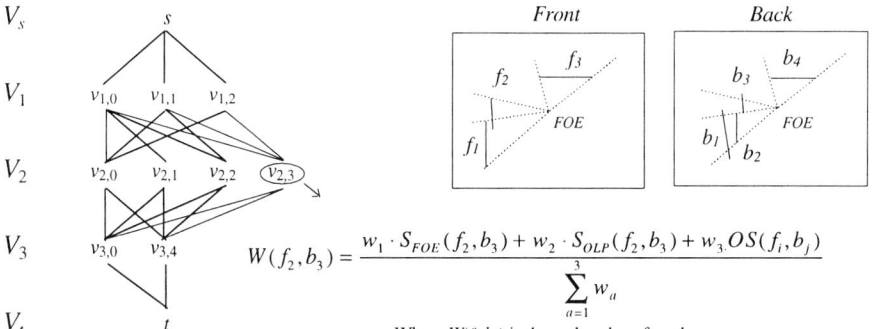

Figure 5. A primitive weighted graph G_1 is shown where all possible paths must start from s and terminate at t.

$$W(f_2,b_3) = \frac{w_1 \cdot S_{FOE}(f_2,b_3) + w_2 \cdot S_{OLP}(f_2,b_3) + w_3 \cdot OS(f_i,b_j)}{\sum_{a=1}^{3} w_a}$$

Where $W(f_2,b_3)$ is the node value of node $v_{2,3}$.

2.3 Update the Similarity Measure and Threshold Decreasing

The matched results of each level is used to update the $GS(f_i, b_j)$ of all the unmatched potential matching pairs. A lower threshold T_n is then used for the next level. The process of matching, updating and lowering threshold is repeated until a predefined level n_{stop} (or a predefined condition) is reached.

By using this multi-level (repeating) searching technique, errors due to broken lines, partially occlusion and misordering are greatly reduced. Figure 6 shows the problem of misordering between the front and back images. Figure 6 indicates the first two line segments of the front and back images are labelled in reverse order. A one pass dynamic programming method will result in either (f_1, b_2), (f_3, b_3), (f_4, b_4) or (f_2, b_1), (f_3, b_3), (f_4, b_4). But for the proposed multi-level searching technique, the remaining pair will be matched in the next level or some level afterwards.

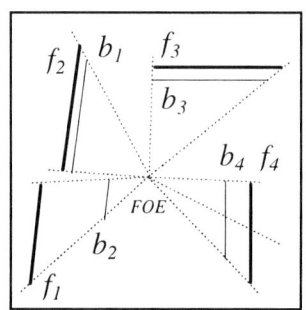

Figure 6. Example showing the misordering problem of axial motion stereo using dynamic programming. The first two line segments of the front and back images are in reverse order.

3. EXPERIMENTAL RESULTS AND DISCUSSION

In this paper, an image pairs are used to demonstrate the capacity of the proposed method. An imperfect line segment detector is used to extractor line segments form the images, line segments shorter than 4 pixels are eliminated from the results. The value of T_n in each level and experimental results are summarized in Table 1. For comparison, results of applying a one pass dynamic programming method are shown in Table 2. Experiments are done with the threshold T being set to 0.5 and 0.0.

Figure 7 shows the results of matching a block world. Figure 7 (a) and (b) show the origin front and back images. Figure 7 (c) and (d) show the final results with the total number of matched pairs = 113 and 101 are correctly matched; successful rate = 89.38%.

Results of the experiment show that the proposed multi-level dynamic programming method achieves a good successful rate under complex environments. Compare its results with the one pass method, it can be seen that the proposed method improved both the number of correctly matched line segments and the successful rate of the one pass dynamic programming.

From Table 2, it can be observed that although lowering the threshold value T (from 0.50 to 0.00) improves the number of correct matches of the one pass approach, it also increases the number of mismatched and lower the accuracy (successful rate) of the overall result. On the other hand, the multi-level method increases the number of correctly matched line segments but with a fairly constant successful rate. Furthermore, as the threshold of the one pass method is already set to 0 and the multi-level method can obtained a higher number of correctly matched lines, it can be concluded that some of the unmatched line segments of the one pass method are due to misordering; and the multi-level approach does reduce the number of unmatched line segments due to misordering when dynamic programming is used.

In this paper, we presented a multi-level dynamic programming method for solving the correspondence problem of axial motion stereo. Experimental results show the feasibility of the proposed method in handling these situations. By using the *Global Similarity Measure* and multi-level searching technique, the proposed technique increases the matching accuracy and reduce the number of unmatched line segment due to misordering when dynamic programming is used for axial motion stereo matching.

Table 1. Experimental results, Figure 7, Blocks - $F = 175$ lines and $B = 192$ lines.

		Matched lines						
		In each level			Accumulated			
n	T_n	Correct	Wrong	Total	Correct	Wrong	Total	Failure rate	Successful rate
1	0.90	4	0	4	4	0	4	0.00%	100.0%
2	0.80	16	2	18	20	2	22	9.09%	90.91%
3	0.70	25	3	28	45	5	50	10.00%	90.00%
4	0.60	15	1	16	60	6	66	9.09%	90.91%
5	0.50	15	1	16	75	7	82	8.54%	91.46%
6	0.40	11	1	12	86	8	94	8.51%	91.49%
7	0.30	6	1	7	92	9	101	8.91%	91.09%
8	0.20	4	3	7	96	12	108	11.11%	88.89%
9	0.10	2	0	2	98	12	110	10.91%	89.09%
10	0.00	3	0	3	101	12	113	10.62%	89.38%

Table 2. Experimental results of the one pass dynamic programming approach.

		Matched lines				
	T	Correct	Wrong	Total	Failure rate	Successful rate
Blocks	0.50	57	7	62	11.29%	88.71%
	0.00	91	13	104	12.50%	87.50%

REFERENCES

[1] Nicolas Alvertos, Dragana Brzakovic and Rafael C. Gonzalez, "Camera Geometries for Image Matching in 3-D Machine Vision", *IEEE Trans. on PAMI*, Vol. 11, No. 9, pp. 897-915, Sept. 1989.

[2] U.R. Dhond and J.K. Aggarwal, "Structure from stereo - A Review", *IEEE Trans. Syst. Man Cybern.*, Vol. 19, No., 6, pp. 1489-1510, Nov./Dec. 1989.
[3] X.Y. Jiang and H. Bunke, "Line Segment Based Axial Motion Stereo", *Pattern Recognition*, Vol. 28, no. 4, pp. 553-562, 1995.
[4] S. H. Lee and J. J. Leou, "A dynamic programming approach to line segment matching in stereo vision", *Pattern Recognition*, vol. 27, no. 8, pp. 961-986, 1994.
[5] Z. N. Li, "Stereo correspondence based on line matching in Hough space using dynamic programming", *IEEE Trans. Syst. Man Cybern.*, vol. 24, no. 1, pp. 144-152, 1994.
[6] R.K.K. Yip and W.P. Ho, "Multi-level Based Stereo Line Matching with Structure Information Using Dynamic Programming", *Proceeding of the 1996 International Conference on Image Processing*, Lausanne, Switzerland, September 16-19, 1996.

(a) Front image of a block world.

(b) Back image of a block world.

(c) Final matched results.

(d) Final matched results.

Figure 7. Matched results of a block world. Total number of match = 113, correct match = 101, successful rate = 89.38%.

Analysis of Grey-Level Features for Line Segment Stereo Matching

Oliver Schreer, Irmfried Hartmann and Roger Adams

Institute for Measurement and Automation Technology
Department of Electrical Engineering, Technical University Berlin
E-mail: schreer@rtws18.ee.TU-Berlin.DE

Abstract - In order to recover 3D-information from stereo image pairs, a number of stereo matching methods are available. These methods mainly differ in various tokens which are applied to solve the correspondence problem between left and right image. In most of the line segment based stereo matching algorithms, geometrical and structural information is used to estimate the correspondence. Unfortunately these features are dependent on the accuracy of segmentation and the angle of view. This might cause an ambiguity regarding the solution. Grey-level features (GLF) are introduced as robust features independent of segmentation errors and the angle of view. Concerning different line segmentation algorithms, e.g. sequential-oriented or global-oriented algorithms, two methods are proposed for a proper estimation of grey-level information. Finally this additional information will be applied in a line segment based stereo matching algorithm. Regarding these different grey-level features the uniqueness of the solution and the computational effort will be compared with geometrical features. Experimental results are presented.

1 Introduction

Many vision based applications, e.g. navigation of autonomous robots, involve recovering 3D-information from 2D-images. In order to solve this problem, two different images of the same scene are needed, and the correspondence between different tokens in two images [2] is to establish. There is a wide range of tokens, e.g. simple points, straight line segments and more complex image structures. The matching strategy is definitely depending on the type of token respectively their combination. Straight line segments are suitable tokens for the description of industrial or man-made environment. The most important constraint for stereo matching is the epipolar constraint, but for an unambigious match additional constraints like geometrical, structural and contrast constraints are needed due to parallel line segments or segmentation errors [6] [7] [8]. The next chapter defines grey-level features, based on the original image, which extend the contrast constraint. Different strategies are proposed to estimate them depending on the segmentation algorithm. In chapter 3, this additional features are applied in a line segment based stereo matching algorithm [1], and their reliability and complexity will be discussed.

2 Estimation of Grey-Level Features (GLF)

Geometrical and structural features are unfortunately sensitive to segmentation errors and depending on the angle of view, which results in an ambiguous solution. Considering the grey-level image and the intensity change regarding line segments, the following three features become obvious:

- the intensity difference at the edge, the gradient
- the direction of intensity change, the sign of the gradient as a binary feature
- the mean value across the edge

The advantages of these features are their simplicity, robustness against segmentation errors and view-point independency.

First of all, the edge image has to be calculated with an operator, which should show the following characteristics: a one-pixel-width edge image, correct placement and robustness against noise. In the experiments a simplified edge operator based on Haralick's facet model [5] is used, which meets these requirements. Furthermore, line segments will be extracted from edge images, to which sequential [9] or more global segmentation algorithms [4] like Hough transform can be applied.

2.1 Direct Feature Estimation in Sequential-Oriented Segmentation Algorithms

Concerning sequential-oriented algorithms, the grey-level features could be estimated during the segmentation process. The gradient *(Grad)* of a segment is defined as the mean value of the gradient of all edge points which are part of the segment.

$grad_e(i)$: Gradient value of an edge point

$$Grad_{seg} = \frac{1}{n}\sum_{i=1}^{n} grad_e(i), \quad n: \text{number of edge points} \qquad (2.1)$$

This gradient definition depends on the single gradient values of each edge point determined by the the type of edge detector. Regarding this fact, the accuracy of estimation depends on the used method. The direction of intensity change *(DIC)* is the sign of the gradient. It is also determined by the applied algorithm.

$$DIC_{seg} = sign(Grad_{seg}), \qquad (2.2)$$

The mean grey value *(MGV)* has to be estimated as the mean sum of intensity values in a defined neighbourhood around the edge points of the segment.

$I_e(i,j)$: Intensity value of an edge point in a defined neighbourhood

$$MGV_{seg} = \frac{1}{nm}\sum_{i=1}^{n}\sum_{j=1}^{m} I_e(i,j) \quad \begin{array}{l} n: \text{number of edge points} \\ m: \text{number of neighbours} \end{array} \qquad (2.3)$$

In the figure below an example is given of edge points corresponding to an edge. The neighbourhood is determined by a mask of size=5. The distance between the centers of masks has to be the size of the mask, guaranteeing no overlap of the masks, i.e. no double summation of pixels.

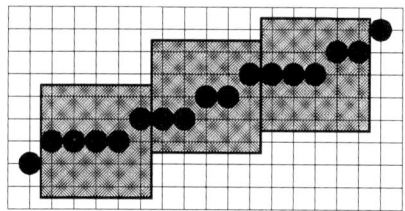

Fig.1. Example of neighbourhood definition

2.2 Feature Estimation in Global-Oriented Segmentation Algorithms

Regarding the feature estimation from a more global point of view, it is obvious to define the gradient at an edge as the difference between the intensity of each side of the edge. The definition of each side is determined by the orientation and slope *m* of the segment, given by the line segment parameters, e.g. x/y-coordinates of the end points or start point, angle and length of the segment.

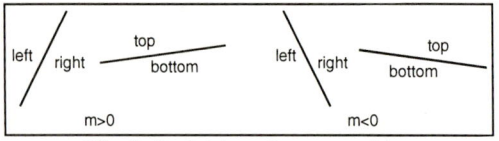

Fig.2. Definition of segment sides

$I_{l/r/t/b}$: Mean Intensity with l: left, r: right, t: top, b: bottom

$$Grad_{seg} = I_{l/t} - I_{r/b}, \quad (2.4)$$

The direction of intensity change is the sign of the gradient again, and the mean grey value is the mean sum of mean intensities.

$$MGV_{seg} = (I_{l/t} + I_{r/b})/2, \quad (2.5)$$

What do we understand by mean intensity? Considering a real edge, approached with a line segment, there are two regions on the left/right or top/bottom side, each of them with approximately the same intensity values. These regions are limited in straight line direction by a cut-off parameter, called *border*. A second parameter, the *neighbourhood* is limiting the region parallel to the segment. Fig.3 shows an example of two segments with different parameters. The choice of each parameter will be investigated in the next chapter.

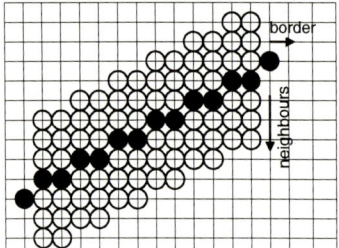 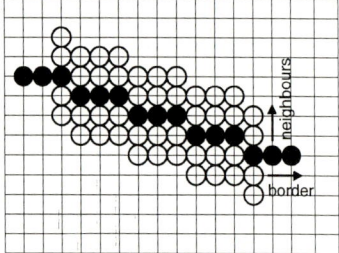

Fig.3. Region parameters: (left) neighbours = 3, border = 1, (right) neighbours = 2, border = 2

When a fixed set of region parameters is given, the Bresenham-algorithm is very helpful to determine the mean intensity [3]. This algorithm calculates pixel positions only with integer arithmetic and is often used in computer graphics.

2.3 Experimental Results

The following analysis is based on five indoor scenes. In order to rate the accuracy of feature estimation subject to different parameters, three criteria are used.
- The standard deviation is calculated taking all pixels in each observed region on each side of the segment into account. The lower the deviation, the more homogenous is the region and the more accurate the mean intensity. The analysis was carried out for a defined percentage of segments with a maximum standard deviation,

basing on the assumption that in a region with less standard deviation there will be a greater accuracy of the mean intensity estimation.
- The mean gradient over all segments is calculated. The higher the gradient, the more accurate is the estimation, as error pixels shift each mean intensity towards the mean and reduce the distance between both of them, i.e. the gradient.
- Finally a visual measure: the neighbourhood regions of each segment are marked in the original image with the estimated mean intensities.

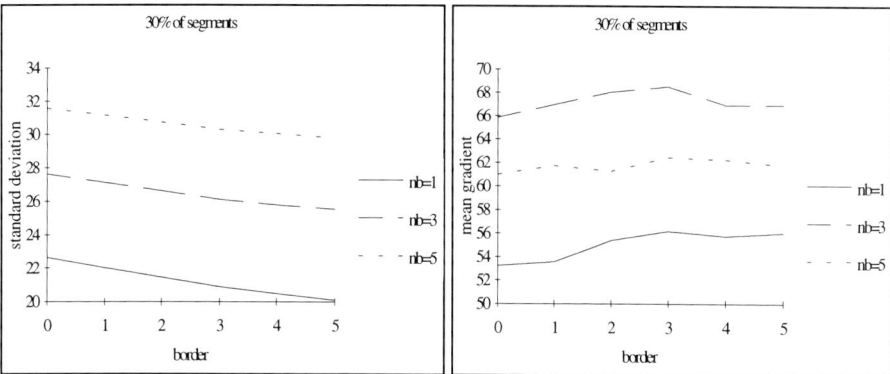

Fig.4. Standard deviation (left) and mean gradient (right) depending on border and neighbourhood (nb)

Regarding Fig.4, the standard deviation decreases with greater border, while the mean gradient increases. This fact is motivated from the segmentation faults at the end of the segment. However, the small change of the mean gradient points to a small influence of the border value.

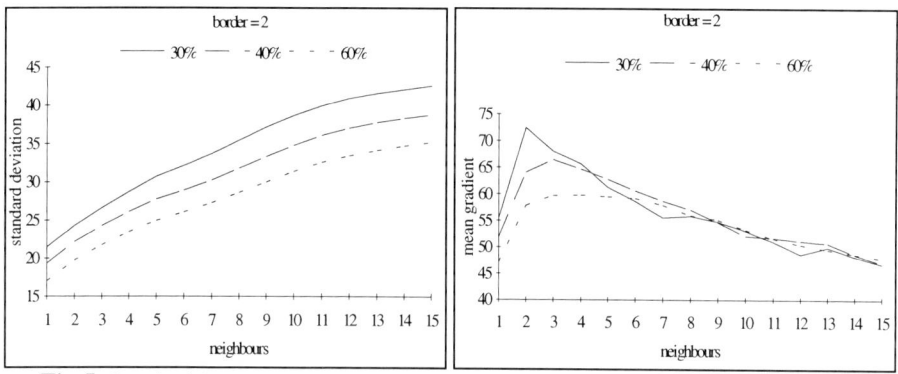

Fig.5. Standard deviation (left) and mean gradient (right) depending on neighbourhood

Fig.5 shows that, when a fixed border value is given, the standard deviation increases with greater neighbourhood, because the region of true intensity of the edges has been left. It is evident, that the optimal region is where the mean gradient has its maximum, i.e. where the two regions are not influenced by other grey values or neighbouring edges. The analysis, concerning different percentages of considered segments, shows, that segments with high standard deviation need a smaller region than those with lower standard deviation.

Fig.6. Stereo image pair „scene2"

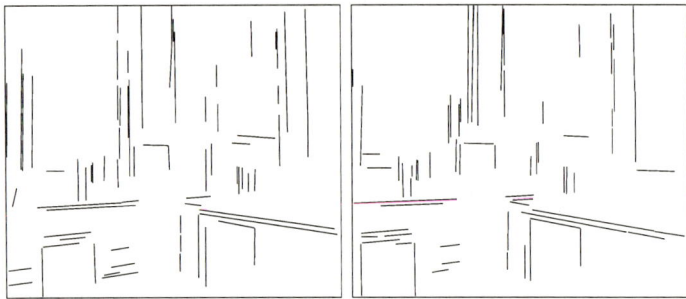

Fig.7. Segment image of „scene2"

The segment image is generated with a sequential segmentation algorithm [9]. The figure below shows a part of the office scene in Fig.6., where the line segments and the region with the estimated mean intensity at each side of the segment are marked.

Fig.8. Sequential (left) and global (right) oriented method

In the left image, the sequential-oriented method is applied leading to some errors of the gradient, e.g. at the edges of the drawer. Regarding the global-oriented method, the mean intensities have been estimated with higher accuracy. The computational costs are in the same order.

3 Stereo Matching

The stereo matching algorithm applied in this experiment is based on line segments that are matched according to a three step strategy containing prediction, propagation and validation [1]. This strategy is based on the epipolar and geometrical constraints. In a recursive algorithm different hypotheses will be analysed with respect to the continuity constraint of the disparity for neighbouring segments. This allows a decision whether two segments are possibly homologue, i.e. both segments in the left and right image correspond to the same 3D-edge.

Concerning multiple stereo matching methods in general, it is useful to add further features provided that they meet the following requirements:
- The number of wrong matches tends to decrease, while the number of correct matches increases, which leads to a greater reliability of the matching results.
- The larger computational effort in feature estimation has to be compensated by the subsequent matching algorithm at least.

3.1 The GLF Constraints

Defining suitable GLF-constraints, the thresholds have to be determined. A number of stereo image pairs were analysed, and the following characteristics were established:
- The intensity change of homologue segments is always the same. If there is a difference in the sign of gradient, the segments are not homologue and can be excluded, the constraint is binary.
- The mean grey value differs only by 30 grey values between homologue segment pairs. Therefore, it is worthwhile taking into account the different mean values of both images. The constraint is of integer-type.
- The greatest deviation is established at the gradient and differs in some cases by 40 grey level differences. Likewise, this constraint is of integer-type.

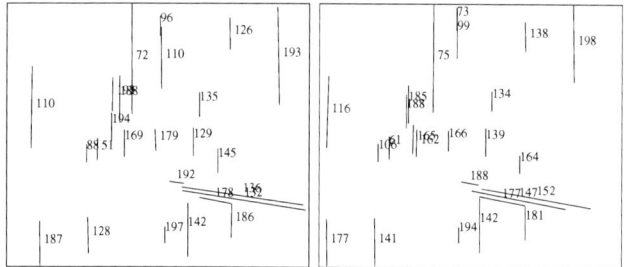

Fig.9. The left and right homologue segments of „scene2" with registered MGV

3.2 Matching with GLF

The segment images, which are to be matched, can be seen in Fig.7. The main characteristics of this image pair are the large number of parallel segments, while only about 65 percent of all segments are part of the same 3D-edge. Fig.10 shows the corresponding line segments, obtained without using GLF. There are some false matches, marked with corresponding numbers. These segments fulfil the geometrical and epipolar constraint. However, the paralell case could not be resolved.

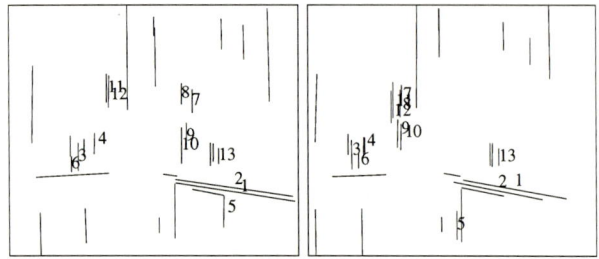

Fig.10. The left and right homologue segments without GLF

The investigation of reliability is based on five indoor scenes with different complexity. The table below shows the results of various features applied. In the first line, the percentage of all matched segments in relation to the total amount of possible matches is registered.

method	without	DIC	DIC+GRAD	DIC+MGV	all GLF's
% of possible matches	54.3	45.57	46.42	43.93	42.34
% of correct matches	71.11	87.31	87.53	90.31	92.43

Tab. 1. Matching results with different combination of GLF's

If more features are included, the total number of matched segments is lowered, i.e. some of the segments are eliminated. Regarding the second line, it shows that false matches were eliminated, while the percentage of correct matches increases. Considering the factor of computational time, the effort of stereo matching is reduced to about 50 percent, regarding Fig.11. However, there is an additional effort of estimating the GLF, which leads to a total amount which is even lower than without GLF. The effort of estimating the GLF increases in a linear way with the number of segments, while the effort for matching increases exponentially. The use of Direction of Intensity Change (DIC)-feature improves the matching results, but the additional features Gradient (GRAD) and the Mean Grey Value (MGV) are again reducing the computational costs, and the matching is more reliable.

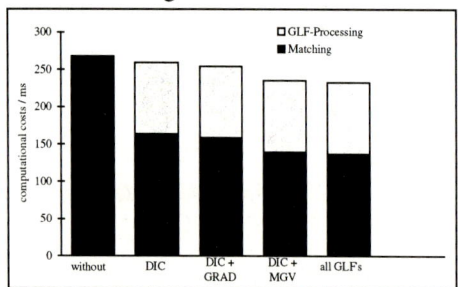

Fig.11. Computational effort with and without GLF

In Fig.12. all three GLF's are applied, observing the following results. The majority of false matches are eliminated and correct match candidates are added. Only three segments have been mismatched in this scene. To conclude, the disambiguity as well as the computational costs have been reduced.

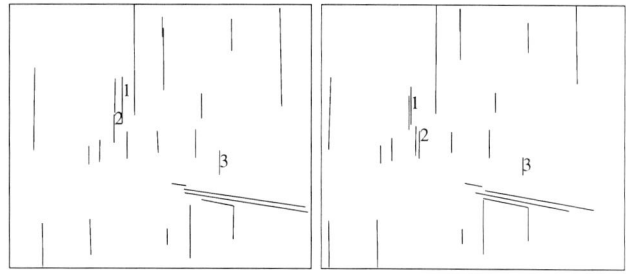

Fig.12. The matching results of „scene2" to which GLF are applied

4 Conclusion

Concerning the estimation of grey-level-features of straight line segments, we described two methods, a sequential-oriented and a global-oriented one. The sequential-oriented method is depending on the edge operator, which estimates the local gradient at each edge point. The global-oriented method can be regarded as the more robust method, using the line segment parameters and providing a fast estimation of the GLF with the Bresenham-algorithm. A statistical analysis and a visual comparison of estimated mean intensity at each side of the edges showed the reliability of the estimated values. In order to investigate the effect of these additional features, they were applied in a line segment based stereo matching algorithm. In general, the additional grey-level features lead to better matching results combined with reduced computational costs. The advantage of this features is the reduction of disambiguity, which primarily occurs in parallel line segments. If the image contains repetitive line structures, these additional features might fail as well. This problem could be solved by a detailed structural analysis of neighbouring line segments.

5 References

[1] N. Ayache: „Artificial Vision for Mobile Robots", MIT Press, Cambridge, Massachusetts, 1991.
[2] O. D. Faugeras: „Three-Dimensional Computer Vision", MIT Press, Cambridge, Massachusetts, London, England, 1993.
[3] W. D. Fellner: Computer Grafik, Reihe Informatik, Band 58, BI Wissenschaftsverlag, 1988, pp. 95-98.
[4] N. Guil, J. Villalba, E. Zapata: "A Fast Hough Transform for Segment Detection", IEEE Transactions on Image Processing, Vol.4, No.11, Nov. 1995, pp.1541-1548.
[5] R. M. Haralick: "A Facet Model for Image Data, Computer Graphics and Image Processing", Nr. 15, pp. 113-129, from R. Chellappa, A. Sawchuk: Digital Image Processing and Analysis: Vol. 1, IEEE Computer Society Press, Silver Spring 1981.
[6] D. Torkar, N. Pavesic: „Feature Extraction from Aerial Images and Structural Stereo Matching", Int. Conf. on Pattern Recognition, Aug. 1996, Austria.
[7] A. Ude, T. E. Ekre: „Stereo Grouping for Model-Based Recognition", Int. Conf. on Pattern Recognition, Aug. 1996, Austria.
[8] V. Venkateswar and R. Chellappa: „Hierarchical Stereo Matching Using Feature Groupings", Proc. of Image Understanding Workshop, January 1992.
[9] K. Wall and P. E. Danielson:"A Fast Sequential Method for Polygonal Approximation of Digitized Curves", Pattern Recognition Vol.12, Pergamon Press Ltd., England 1980, pp. 327-331.

3-D Object Positioning from Monocular Image Brightnesses

Tadayoshi SHIOYAMA, Hai Yuan WU,
Wen Biao JIANG and Susumu TERAUCHI

Department of Mechanical and System Engineering
Kyoto Institute of Technology
Matsugasaki, Sakyo-ku, Kyoto 606, JAPAN
FAX: 075-724-7300, Email: shioyama@ipc.kit.ac.jp

Abstract. This paper proposes an estimation scheme for three-dimensional (3-D) object positioning from monocular image brightnesses without point correspondences while previous 3-D positioning methods need point correspondences. In the present scheme, at the first step, 3-D shapes of object are recovered from image brightnesses before and after rotation motion. At the second step, two vectors which are transformed according to rotation matrix under 3-D rotation, are constructed from central moments of the recovered 3-D shapes of object before and after motion, respectively. At the third step, rotation matrix is computed from these vectors by singular value decomposition. At the fourth step, from the rotation matrix the rotation axis and the rotation angle around the axis are estimated.

1 Introduction

Object positioning is an important area of research in computer vision. Estimation schemes for three-dimensional(3-D) object positioning from two image frames with a single camera have been developed in literatures [1]-[5]. However, in the previous estimation schemes it is assumed that image point (or line) correspondences are known. It is time-consuming and difficult to solve point correspondence problem for the case of many points. Therefore, it is desired to develope an estimation method for 3-D object positioning without point correspondences.

This paper proposes an estimation scheme for 3-D object positioning from two monocular image brightnesses. The present scheme has an advantageous property that it does not need point correspondences. In the present scheme, at the first step, 3-D shapes of object are recovered from image brightnesses before and after motion, respectively. At the second step, two vectors which are transformed according to rotation matrix under 3-D rotation, are constructed from central moments of the recovered 3-D shapes of object before and after motion, respectively. At the third step, rotation matrix is computed from these vectors by singular value decomposition(SVD). At the fourth step, from the rotation matrix the rotation axis and the rotation angle around the axis are estimated. The present scheme is evaluated quantitatively.

2 3-D Shape Recovery

We assume orthographic image projection in observing objects and take the viewing direction to be parallel to the z-axis. Then, the shape of the object can be described by its height z at coordinate (x,y) in the image plane. The coordinate (x,y,z) is called the camera coordinate. We denote by n_i a unit vector with the direction of light, and by n a unit vector normal to the surface of the object. The vector n is called the surface normal. The unit vectors n and n_i are described by points on the sphere with unit radius called Gaussian sphere. The stereographic projection is given as the projection of points on the Gaussian sphere through rays from the south pole onto a tangent plane (which is called the stereographic plane) at the north pole. The coordinates (f,g) in the stereographic plane are given as follows:

$$f = 2p[\sqrt{1 + p^2 + q^2} - 1]/(p^2 + q^2), \qquad (1)$$
$$g = 2q[\sqrt{1 + p^2 + q^2} - 1]/(p^2 + q^2), \qquad (2)$$

where

$$p \equiv \partial z/\partial x, \qquad q \equiv \partial z/\partial y. \qquad (3)$$

Unit vectors n and n_i are described in terms of f and g as follows:

$$\mathbf{n} = (-4f, -4g, 4 - f^2 - g^2)/(4 + f^2 + g^2), \qquad (4)$$
$$\mathbf{n}_i = (-4f_i, -4g_i, 4 - f_i^2 - g_i^2)/(4 + f_i^2 + g_i^2). \qquad (5)$$

We assume that the viewing direction coincides with the north pole of the Gaussian sphere and that points on the northern hemisphere of the Gaussian sphere are considered. Therefore, the considered points (f,g) and (f_i, g_i) in the stereographic plane are constrained to the following regions:

$$f^2 + g^2 \leq 4, \qquad f_i^2 + g_i^2 \leq 4. \qquad (6)$$

We denote by $F(f, g; f_i, g_i)$ or F(f,g) the reflectance map. Let E_{ij} denote the observed brightness at a point with coordinate (i,j) in the image plane, where E_{ij} is normalized so that the maximum observed brightness is equal to 1. Then, it holds that

$$F(f_{ij}, g_{ij}; f_i, g_i) = E_{ij}, \qquad (7)$$

where (f_{ij}, g_{ij}) represents the stereographic coordinate of the surface normal at the point with coordinate (i,j) in the image. Denote by $(f_{ij}^{(k)}, g_{ij}^{(k)})$ the estimated (f_{ij}, g_{ij}) at k-th iteration and by (f_{ij}^*, g_{ij}^*) the true (f_{ij}, g_{ij}). We assume the following relation:

$$f_{ij}^* = f_{ij}^{(k)} + \Delta f_{ij}^{(k)}, \qquad (8)$$
$$g_{ij}^* = g_{ij}^{(k)} + \Delta g_{ij}^{(k)}. \qquad (9)$$

Here, $\Delta f_{ij}^{(k)}$ and $\Delta g_{ij}^{(k)}$ are the improvement at k-th iteration. We obtain $\Delta f_{ij}^{(k)}$ and $\Delta g_{ij}^{(k)}$ by Marquardt method with the constraint (7) as follows.

$$\Delta f_{ij}^{(k)} = \frac{\lambda[E_{ij} - F(f_{ij}^{(k)}, g_{ij}^{(k)})] F_f(f_{ij}^{(k)}, g_{ij}^{(k)})}{1 + \lambda[F_f(f_{ij}^{(k)}, g_{ij}^{(k)})^2 + F_g(f_{ij}^{(k)}, g_{ij}^{(k)})^2]}, \qquad (10)$$

$$\Delta g_{ij}^{(k)} = \frac{\lambda[E_{ij} - F(f_{ij}^{(k)}, g_{ij}^{(k)})] F_g(f_{ij}^{(k)}, g_{ij}^{(k)})}{1 + \lambda[F_f(f_{ij}^{(k)}, g_{ij}^{(k)})^2 + F_g(f_{ij}^{(k)}, g_{ij}^{(k)})^2]}, \qquad (11)$$

where
$$F_f \equiv \frac{\partial F}{\partial f_{ij}}, \qquad F_g \equiv \frac{\partial F}{\partial g_{ij}}.$$

Hence, the iterative algorithm for inferring (f_{ij}, g_{ij}) is given by

$$f_{ij}^{(k+1)} = f_{ij}^{(k)} + \Delta f_{ij}^{(k)}, \tag{12}$$

$$g_{ij}^{(k+1)} = g_{ij}^{(k)} + \Delta g_{ij}^{(k)}, \tag{13}$$

here $\Delta f_{ij}^{(k)}$ and $\Delta g_{ij}^{(k)}$ are given by (10) and (11).

The line of sight lies in the tangent plane at a point on the occluding boundary. Since we assume an orthographic image projection, the line of sight is perpendicular to the image plane. Hence, the tangent plane at a point on the occluding boundary is perpendicular to the image plane and is projected as a line in the image plane. This line is tangent to the silhouette of the occluding boundary in the image plane. Thus, a normal to the silhouette in the image plane is parallel to the surface normal at the corresponding point on the occluding boundary. That is, (f_{ij}, g_{ij}) is known at the point (i,j) corresponding to the occluding boundary. In the 3-D shape inference algorithm, at the initial iteration (k=0) only (f_{ij}, g_{ij}) corresponding to the occluding boundary are known, and (f_{ij}, g_{ij}) in the region other than the occluding boundary are set as $f_{ij} = g_{ij} = 0$. The region where $f_{ij} = g_{ij} = 0$ and (f_{ij}, g_{ij}) is not yet improved, is called "unknown region". In the unknown region, for $k \neq 0$, we approximate $f_{ij}^{(k)}$ and $g_{ij}^{(k)}$ by the following steps (1) and (2).

(1) In the case where at least one of 8-neighbors of point (i,j) belong to the known region,

$$f_{ij}^{(k)} = a\bar{f}_{ij}^{(k)} + b\hat{f}_{ij}^{(k)}, \qquad g_{ij}^{(k)} = a\bar{g}_{ij}^{(k)} + b\hat{g}_{ij}^{(k)},$$

where
$$\begin{aligned}
\bar{f}_{ij} &= f_{i+1,j} + f_{i,j+1} + f_{i-1,j} + f_{i,j-1}, \\
\bar{g}_{ij} &= g_{i+1,j} + g_{i,j+1} + g_{i-1,j} + g_{i,j-1}, \\
\hat{f}_{ij} &= f_{i-1,j-1} + f_{i+1,j-1} + f_{i-1,j+1} + f_{i+1,j+1}, \\
\hat{g}_{ij} &= g_{i-1,j-1} + g_{i+1,j-1} + g_{i-1,j+1} + g_{i+1,j+1},
\end{aligned}$$

$(a,b) = (1/c_1, 0)$ when at least one of 4-neighbors belong to the known region or $(a,b) = (0, 1/c_2)$ otherwise. c_1 (or c_2) denotes the number of 4-(or 8-)neighbors belonging to the known region.

(2) In the case other than the case of (1),

$$f_{ij}^{(k)} = g_{ij}^{(k)} = 0.$$

In this case, we set as $\Delta f_{ij}^{(k)} = \Delta g_{ij}^{(k)} = 0$.

As the iterative algorithm proceeds, the unknown region decreases. Hence, the steps (1) and (2) are only the initial transient process. After obtaining (f,g) by the above algorithm, we can compute (p,q) from equations (1) and (2), and we can obtain the height z by integrating p and q. Thus we can recover 3-D shape.

3 Vectors Derived from Moments

We define a moment of order $(\ell+m+n)$ as follows

$$M_{\ell m n} \equiv \int_{-\infty}^{\infty} \int_{-\infty}^{\infty} \int_{-\infty}^{\infty} x^\ell y^m z^n \rho(x,y,z) dx dy dz, \tag{14}$$

where $\rho(x,y,z)$ denotes the density function of a 3-D object, and it is assumed that the origin of the coordinate system coincides with the centroid of the density function.

We consider a unit vector $\vec{u} \equiv (u_1, u_2, u_3)^T$, where T denotes transpose operation. Using the polar coordinate $(r = 1, \xi, \eta)$, the vector $\vec{u}$ is represented as follows

$$\vec{u} = (sin\xi cos\eta, sin\xi sin\eta, cos\xi)^T, \quad 0 \leq \xi \leq \pi, \quad 0 \leq \eta \leq 2\pi. \tag{15}$$

We define u as a vector whose components are monomials of u_1, u_2 and u_3 with order p:

$$u \equiv (u_1^p, u_2^p, u_3^p, ..., u_1^i u_2^j u_3^k, ...)^T, \quad p = i + j + k. \tag{16}$$

Using moments corresponding to components of the vector u, we define a vector m as

$$m \equiv (M_{p00}, M_{0p0}, M_{00p}, ..., \frac{p!}{i!j!k!}M_{ijk}, ...)^T. \tag{17}$$

Furthermore, we define y as a vector whose components are spherical harmonics Y_ℓ^m, $m = -\ell, -\ell+1, ..., \ell-1, \ell$:

$$y = (Y_p^p, ..., Y_p^{-p}, Y_{p-2}^{p-2}, ..., Y_{p-2}^{-p+2}, ..., Y_\ell^m, ...)^T, \tag{18}$$

where the minimal value of ℓ is 0 for even p and is 1 for odd p. The monomials $u_1^i u_2^j u_3^k$ in (16) are functions defined on the unit sphere and are represented by linear combinations of spherical harmonics Y_ℓ^m in (18), that is, there is a nonsingular complex matrix A such that

$$u = Ay, \tag{19}$$

$A = (A_{ijk,\ell m})$, $i+j+k = p$, $\ell = p, p-2, ..., \ell_0$, $\ell_0 = 0$ for even p, $= 1$ for odd p, $m = -\ell, -\ell+1, ..., \ell-1, \ell,$

$$A_{ijk,\ell m} = \int_0^{2\pi} \int_0^\pi [Y_\ell^m(\xi, \eta)]^* u_1^i u_2^j u_3^k sin\xi d\xi d\eta, \tag{20}$$

where * denotes a complex conjugate. Then, we define a complex moment vector ν as [6]

$$\nu \equiv A^\dagger m, \tag{21}$$

$$\nu = (\nu_p^p, ..., \nu_p^{-p}, \nu_{p-2}^{p-2}, ..., \nu_{p-2}^{-p+2}, ..., \nu_\ell^m, ...)^T, \tag{22}$$

where $A^\dagger$ is the hermitian conjugate of A. It is known [6] that under rotation the vector ν is transformed according to the same block-diagonal matrix as the vector y. Since spherical harmonics Y_1^1, Y_1^0 and Y_1^{-1} are represented as

$$Y_1^{\pm 1} = \mp \frac{1}{2}\sqrt{\frac{3}{2\pi}} (x \pm jy), \quad Y_1^0 = \frac{1}{2}\sqrt{\frac{3}{2\pi}} z, \quad j \equiv \sqrt{-1}, \quad x^2 + y^2 + z^2 = 1, \tag{23}$$

we obtain the following relations

$$x = \sqrt{2\pi/3} (Y_1^{-1} - Y_1^1), \quad y = j\sqrt{2\pi/3} (Y_1^{-1} + Y_1^1), \quad z = 2\sqrt{\pi/3} Y_1^0. \tag{24}$$

Since the vector $(x, y, z)^T$ is transformed by rotation matrix R under rotation, and the complex moment ν_l^m is transformed in the same fashion as the spherical harmonics Y_l^m under rotation, the vector $\mathbf{b} = (b_1, b_2, b_3)^T$ defined as

$$b_1 = \sqrt{2\pi/3} \, (\nu_1^{-1} - \nu_1^1), \quad b_2 = j\sqrt{2\pi/3} \, (\nu_1^{-1} + \nu_1^1), \quad b_3 = 2\sqrt{\pi/3} \, \nu_1^0, \qquad (25)$$

is transformed by rotation matrix R. Let vectors $\mathbf{b}$ and $\mathbf{b}'$ be vectors computed by (25) from density functions of 3-D object before and after rotation, respectively. Then, it holds that

$$\mathbf{b}' = R\mathbf{b}. \qquad (26)$$

4 Rotation Inference by Singular Value Decomposition

Let $\mathbf{b}_1$ and $\mathbf{b}_2$ be two column vectors defined as (25) for the 3-D object before rotation, and $\mathbf{a}_1$ and $\mathbf{a}_2$ be two similar vectors corresponding to $\mathbf{b}_1$ and $\mathbf{b}_2$ for the object after rotation. We define 3×2 dimensional matrices A and B as

$$A \equiv [\mathbf{a}_1 \; \mathbf{a}_2], \qquad B \equiv [\mathbf{b}_1 \; \mathbf{b}_2]. \qquad (27)$$

Let AB^T be represented by singular value decomposition as

$$AB^T = UDV^T, \qquad (28)$$

where

$$UU^T = VV^T = I, \quad D = diag(d_1, d_2, d_3), \quad d_1 \geq d_2 \geq d_3 \geq 0, \quad I = identity \; matrix.$$

For the case where $rank(AB^T) \geq 2$, the rotation matrix R minimizing $\| A - RB \|^2$ under the constraints $\det(R)=1$ and $RR^T = I$ is given by [7]

$$R = USV^T, \qquad (29)$$

where

$$S = diag(1, 1, det(U)det(V)). \qquad (30)$$

5 3-D Object Positioning Algorithm

The proposed estimation algorithm for 3-D object positioning from two monocular image brightnesses without point correspondences is summarized as follows. In this section, we use the camera coordinate system (x,y,z).

(Step 1) Estimate (f_{ij}, g_{ij}) from monocular image brightnesses of object before and after rotation by iterative algorithm (12) and (13). Infer the 3-D shapes z_{ij} of object before and after rotation by integrating (p_{ij}, q_{ij}) computed from (f_{ij}, g_{ij}), where p_{ij} and q_{ij} are p and q defined as (3) at the point (i,j) in the image plane, and z_{ij} is the height at an image coordinate (x,y)=(i,j). The density function $\rho(x', y', z')$ in (14) where $x' = x - \bar{x}$, $y' = y - \bar{y}$, $z' = z - \bar{z}$ and $(\bar{x}, \bar{y}, \bar{z})$ is the centroid, is given by

$$\begin{aligned} \rho(x', y', z') &= 1 \quad for \; x' = i - \bar{x}, \; y' = j - \bar{y}, \; z' = z_{ij} - \bar{z}, \\ &= 0 \quad otherwise. \end{aligned} \qquad (31)$$

(Step 2) For the density function $\rho(x', y', z')$ in step 1 of object before rotation, compute moments $M_{\ell mn}$, $\ell+m+n=3$, and $M_{\ell'm'n'}$, $\ell' + m' + n' = 5$ by (14), and construct vectors $\mathbf{b}_1$ and $\mathbf{b}_2$ corresponding to $M_{\ell mn}$ and $M_{\ell'm'n'}$ by (25), respectively. For $\rho(x', y', z')$ of object after rotation, compute vectors $\mathbf{a}_1$ and $\mathbf{a}_2$ similar to $\mathbf{b}_1$ and $\mathbf{b}_2$ by (25).

(Step 3) Obtain the rotation matrix R by (27) through (29).
(Step 4) Estimate rotation axis $e = (e_1, e_2, e_3)$, $\| e \| = 1$, and rotation angle θ around the axis by the following relations.

$$R \equiv \begin{pmatrix} r_1 & r_2 & r_3 \\ r_4 & r_5 & r_6 \\ r_7 & r_8 & r_9 \end{pmatrix}, \quad d^2 \equiv (r_8 - r_6)^2 + (r_3 - r_7)^2 + (r_4 - r_2)^2,$$

$$\sin\theta = \pm d/2, \quad \cos\theta = \frac{d^2 r_1 - (r_8 - r_6)^2}{d^2 - (r_8 - r_6)^2},$$

$$e_1 = \pm(r_8 - r_6)/d, \quad e_2 = \pm(r_3 - r_7)/d, \quad e_3 = \pm(r_4 - r_2)/d,$$

where $\pm$ denotes the sign of same upper or lower side.

6 Experimental Results

In order to evaluate the performance of the present 3-D positioning algorithm quantitatively, we show the results of numerical experiments. We use an ellipsoidal object which is represented in the world coordinate system (X,Y,Z) as follows

$$\frac{X^2}{a^2} + \frac{Y^2}{b^2} + \frac{Z^2}{c^2} = 1.$$

We assume that the rotation axis coincides with Z-axis, that the direction of light coincides with the viewing direction perpendicular to the image plane and that the surface material exhibits Lambertian reflection. Examples of the image brightnesses of the ellipsoid of a=25, b=40 and c=25 before and after the rotation whose angle θ is 40°, are illustrated in Fig.1 (a) and (b), respectively, in case of the light direction of $\theta_i = 90°$ and $\phi_i = 10°$, where θ_i and ϕ_i represent the azimuth and zenith angles of the light direction n_i in the world coordinate system (X,Y,Z). Figure 2 shows the corresponding 3-D shapes recovered from the image brightnesses shown in Fig.1. The estimation errors of rotation angle θ are shown against view angle ϕ_i in Fig.3 as the average over $\theta = 10°, 20°, ..., 90°$, where $\theta_i = 90°$ and we denote by model 1 the ellipsoid of a=c=25 and b=40, and by model 2 the ellipsoid of a=20, b=30 and c=25. It is found that the rotation angle is estimated with the accuracy of error less than few degrees in case of small view angle. In the experiments, at the step(2) in algorithm in the preceding section, the principal axis of object is used as the vector b_2 or a_2 instead of the vector constructed from the moment $M_{\ell'm'n'}$, $\ell' + m' + n' = 5$, because the moment of order 5 is too sensitive to noise.

7 Conclusion

We have proposed the algorithm for 3-D object positioning from two monocular image brightnesses. The algorithm has the advantageous property that it does not need point correspondence while previous methods need the point correspondence. The numerical experimental results show that by the proposed algorithm, rotation angle is estimated with the accuracy of error less than few degrees in case of small view angle.

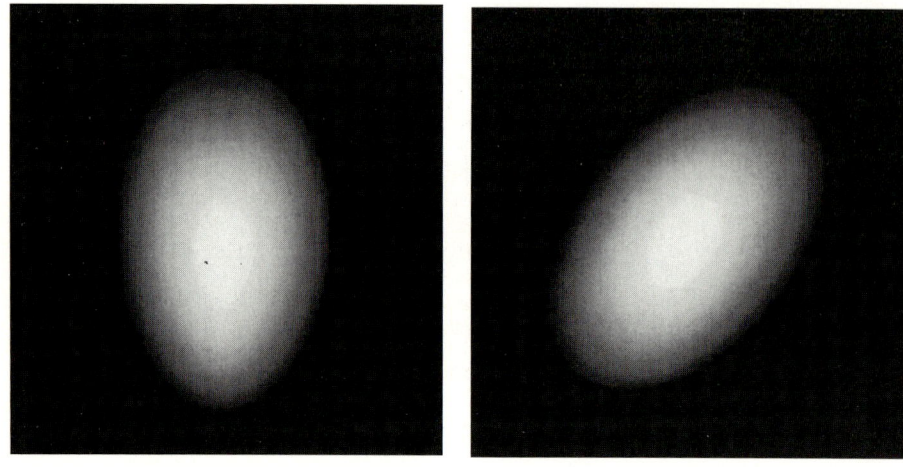

(a) (b)

Fig.1 Image brightnesses of an ellipsoidal object of a=c=25 and b=40 (a)before and (b)after rotation of angle $\theta = 40°$, where $\theta_i = 90°$, $\phi_i = 10°$.

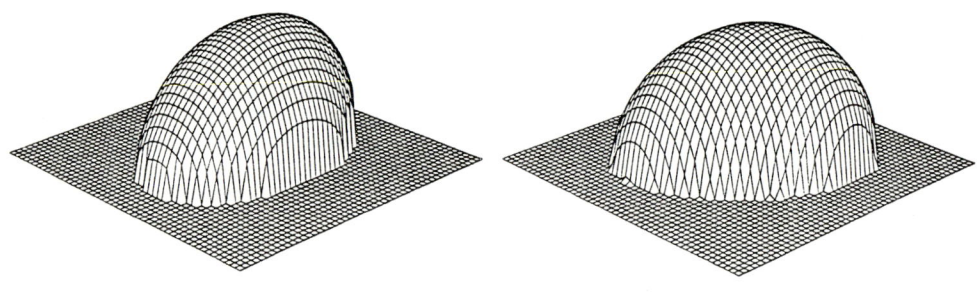

(a) (b)

Fig.2 The corresponding 3-D shapes recovered from the image brightnesses shown in Fig.1.

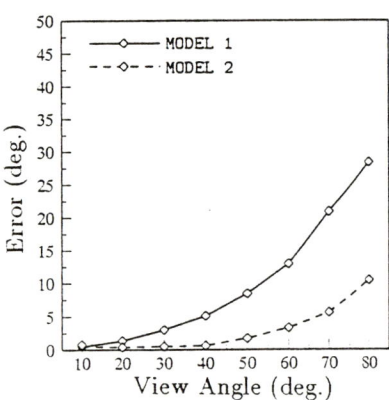

Fig.3 The estimation errors of rotation angle against view angle ϕ_i; we denote by model 1 the ellipsoid of a=c=25 and b=40, and by model 2 that of a=20, b=30 and c=25.

References

[1] R.Y.Tsai and T.S.Huang,"Uniqueness and Estimation of Three-Dimensional Motion Parameters of Rigid Objects with Curved Surfaces," IEEE Trans. Vol.PAMI-6, No.1 (1984) pp.13-27.

[2] O.D.Faugeras and M.Hebert,"A 3-D Recognition and Positioning Algorithm Using Geometrical Matching between Primitive Surfaces," Proc.8th Int.Joint Conf. Artificial Intell.(1983)pp.996-1002.

[3] J.Weng, T.S.Huang and N.Ahuja,"Error Analysis of Motion Parameter Estimation from Image Sequences,"Proc.1st Int.Conf.Comput.Vision (1987)pp.703-707.

[4] J.Philip,"Estimation of Three-Dimensional Motion of Rigid Objects from Noisy Observations,"IEEE Trans. Vol.PAMI-13,No.1(1991)pp.61-66.

[5] K.Kanatani,"Unbiased Estimation and Statistical Analysis of 3-D rigid Motion from Two Views,"IEEE Trans. Vol.PAMI-15,No.1(1993) pp.37-50.

[6] C.H.Lo and H.S.Don,"3-D Moment Forms:Their Construction and Application to Object Identification and Positioning," IEEE Trans. Vol.PAMI-11,No.10(1989)pp.1053-1064.

[7] S.Umeyama,"Least-Squares Estimation of Transformation Parameters Between Two Point Patterns,"IEEE Trans. Vol.PAMI-13,No.4(1991)pp.376-380.

Camera Calibration Based on 3D-Point-Grid *

X.-F. Zhang[1], A. Luo[2], W. Tao[1] and H. Burkhardt[3]†

[1] Tech. Informatik I, Hamburg-Harburg University of Technology, 21071 Hamburg
[2] Mikroelektronik Anwendungszentrum (MAZ) Hamburg GmbH, 21079 Hamburg
[3] Institut für Informatik, University of Freiburg, 79085 Freiburg
Germany

Abstract. In this paper an effective technique of camera calibration based on a 3D-point-grid is presented. The properties of structured space points, defined as 3D-point-grid, and their relations under perspective transformation are analyzed. Upon the correspondence between these 3D-points and their image points a calibration is made not only to compute the parameters of camera model by a linear method simply from the independent points in 3D-point-grid but also to verify, modify and regulate extrinsic parameters of the orientation and position simultaneously from the structure constraint implied in these points. On the basis of the verification, modification and regulation the intrinsic parameters of camera can be computed with more objective criteria. Experimental results show the necessity and advantage of the verification and modification of the orientation and position parameters, and prove that our calibration technique can improve the accuracy of both extrinsic and intrinsic parameters greatly.

1 Introduction

Camera calibration is the process of determining the internal camera geometric and optical characteristics (represented by intrinsic parameters such as focus f, center C_x, C_y and scale factor s_x etc.) and/or the 3D position and orientation of the camera relative to a certain world coordinate system (called extrinsic parameters like rotation matrix R and translation vector T). For a real camera this process includes also a determination of distortion parameters. A lot of papers dealing with the computation of these parameters have been published, e.g. [Tsai86], [LT88], [BM91], [LC93], [WM94], [Faug93], etc.

In most cases the extrinsic and intrinsic parameters are computed by neglecting the distortion. The intrinsic parameters can be verified by some physical methods. But the orientation and position are difficult to verify unless there are exact mechanical settings such as CMM in [BM91]. The methods in above listed

---

* This work was supported by DAAD for the first author and by the BMBF-project MOVIS partly
† email: x.zhang or tao@tu-harburg.d400.de, luo@maz-hh.de, or burkhardt@ informatik. uni-freiburg.de

papers are based on a sequence of correspondent 3D- and 2D-points, but no relationship exists among them, or has been considered and used. The verification of the model parameters can only be made until after all the parameters are computed.

Upon these considerations, a more effective technique, which uses structure-related space points during calibration with the process of verification, modification and regulation, is proposed in this paper. It is proceeded as follows: (1) analysis of the structure of 3D-point-grid, its relation to the transformation parameters, its construction and location of its image counterpart; (2) improved calibration of camera parameters and results; (3) conclusions.

2 3D-Point-Grid

2.1 Definition

Suppose a sequential triple set $\mathcal{L} = \{(i,j,k),\ i = 1,...,M,\ j = 1,...,N,\ k = 1,...,L\}$, $\Delta x, \Delta y$ and Δz are the unit lengthes along the three orthogonal directions in space $\mathcal{R}^3$, then the set $\mathcal{Q} = \{(i\Delta x, j\Delta y, k\Delta z), (i,j,k) \in \mathcal{L}\}$ is defined as a 3D-point-grid, $\mathcal{L}$ is its label set. When $(I_0, J_0, K_0) \in \mathcal{L}$ is chosen as the origin of a coordinate system, $\{((i-I_0)\Delta x, (j-J_0)\Delta y, (k-K_0)\Delta z), (i,j,k) \in \mathcal{L}\}$ is the coordinates set of $\mathcal{Q}$.

2.2 Characteristics in structure

Coplanar and parallel relation Let $i = I_1$, a fixed number, then the elements in subset $q_{I_1} = \{(I_1\Delta x, j\Delta y, k\Delta z),\ j = 1,...,N,\ k = 1,...,L\}$ are coplanar. The M planes π_{I_1}, formed by q_{I_1} ($I_1 = 1,...,M$), are parallel, and $\mathcal{Q} = \cup_{I_1=1}^{M}(q_{I_1})$.

Coplanar and perpendicular relation Let $i = I_1$ and $j = J_1$, then we get two subsets $q_{I_1} = \{(I_1\Delta x, j\Delta y, k\Delta z),\ j = 1,...,N,\ k = 1,...,L\}$ and $q_{J_1} = \{(i\Delta x, J_1\Delta y, k\Delta z),\ i = 1,...,M,\ k = 1,...,L\}$ respectively. They form two planes π_{I_1} and π_{J_1} with $\pi_{I_1} \perp \pi_{J_1}$. In the same way we can get q_{K_1} and its π_{K_1}, and $\pi_{J_1} \perp \pi_{K_1}$ or $\pi_{K_1} \perp \pi_{I_1}$.

2.3 The relations among parameters of orientation and position from 3D-point-grid

Between subsets of the coplanar points that form parallel planes Generally, if any coordinate system takes a point in 3D-point-grid, that is Δx_w, Δy_w and Δz_w away from its previous origin in x, y and z direction respectively, as its current origin and is acquired by translation in three directions, we have

$$R_1 = R_0, \quad T_1 = T_0 + R_0(\Delta x_w, \Delta y_w, \Delta z_w)^T \tag{1}$$

in which $\Delta y_w = \Delta z_w = 0$ corresponds to the situation of parallel plane in x direction, while $\Delta x_w = \Delta z_w = 0$, $\Delta x_w = \Delta y_w = 0$ to y and z direction respectively.

Between subsets of coplanar points that form perpendicular planes
Suppose that an origin is selected with a label (I_0, J_0, K_0), then (see Fig. 1(a))
$q_{I_0} = \{(I_0 \Delta x, j \Delta y, k \Delta z), j = 1, ..., N, k = 1, ..., L\}$ forms π_{I_0} i.e. y_0-O-z_0 plane
$q_{J_0} = \{(i \Delta x, J_0 \Delta y, k \Delta z), i = 1, ..., M, k = 1, ..., L\}$ forms π_{J_0} i.e. x_0-O-z_0 plane
$q_{K_0} = \{(i \Delta x, j \Delta y, K_0 \Delta z), i = 1, ..., M, j = 1, ..., N\}$ forms π_{K_0} i.e. x_0-O-y_0 plane in x_0-y_0-z_0 coordinate system. Suppose that different coordinate systems

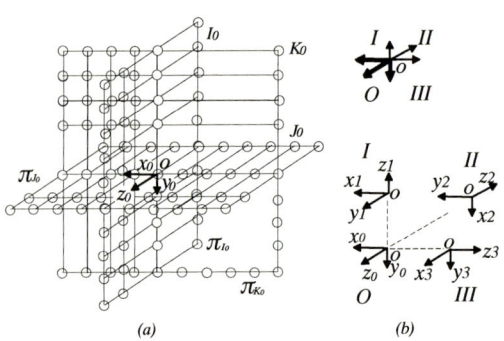

Fig. 1. (a) Points on perpendicular planes and (b) Selection of coordinate systems

are selected on these three planes by rules: (1) the origins are the same as that of x_0-y_0-z_0; (2) all are right-handed systems; (3) all x-o-y planes are on these three planes; (4) all z axes are in counter direction of the axes of x_0-y_0-z_0, as shown in Fig. 1(b). The rotation matrices of these coordinate systems, which transform points in coordinate systems O, I, II, III into camera coordinate system, are supposed to be R_0, R_1, R_2, R_3, then

(i) if R_{ij} stands for a matrix that transforms a point (x_i, y_i, z_i) from ith coordinate system into jth coordinate system, then the matrices for the coordinate transformations among O, I, II and III are listed in Table 1.

(ii) if a point in coordinate system I is transformed into coordinate system O, and then to camera coordinate system, then the rotation matrices have relations $R_1 = R_0 R_{10}$. Similarly we have $R_1 = R_2 R_{12}$, $R_1 = R_3 R_{13}$. Same relations hold for other coordinate systems and can be concluded in Table 2(a).

2.4 Construction of 3D-Point-Grid and extraction of image points

In experiment a 3D-point-grid $Q = \cup_{I_1=1}^{M}(q_{I_1})$ is constructed by the horizontal movement of a square pattern, or specially made cube with surfaces of square patterns, as shown in Fig.2. By steps: (1) subpixel edge detection; (2) edge tracing; (3) corner detection; (4) computation of corner-coordinates (see [Zhang96]), the image points can be extracted.

Table 1. Transformations of the coordinate systems on perpendicular planes

$i \diagdown j$	O	I	II	III
O	$R_{00}: \begin{bmatrix} 1 & 0 & 0 \\ 0 & 1 & 0 \\ 0 & 0 & 1 \end{bmatrix}$	$R_{01}: \begin{bmatrix} 1 & 0 & 0 \\ 0 & 0 & 1 \\ 0 & -1 & 0 \end{bmatrix}$	$R_{02}: \begin{bmatrix} 0 & 1 & 0 \\ 1 & 0 & 0 \\ 0 & 0 & -1 \end{bmatrix}$	$R_{03}: \begin{bmatrix} 0 & 0 & 1 \\ 0 & 1 & 0 \\ -1 & 0 & 0 \end{bmatrix}$
I	$R_{10}: \begin{bmatrix} 1 & 0 & 0 \\ 0 & 0 & -1 \\ 0 & 1 & 0 \end{bmatrix}$	$R_{11}: \begin{bmatrix} 1 & 0 & 0 \\ 0 & 1 & 0 \\ 0 & 0 & 1 \end{bmatrix}$	$R_{12}: \begin{bmatrix} 0 & 0 & -1 \\ 1 & 0 & 0 \\ 0 & -1 & 0 \end{bmatrix}$	$R_{13}: \begin{bmatrix} 0 & 1 & 0 \\ 0 & 0 & -1 \\ -1 & 0 & 0 \end{bmatrix}$
II	$R_{20}: \begin{bmatrix} 0 & 1 & 0 \\ 1 & 0 & 0 \\ 0 & 0 & -1 \end{bmatrix}$	$R_{21}: \begin{bmatrix} 0 & 1 & 0 \\ 0 & 0 & -1 \\ -1 & 0 & 0 \end{bmatrix}$	$R_{22}: \begin{bmatrix} 1 & 0 & 0 \\ 0 & 1 & 0 \\ 0 & 0 & 1 \end{bmatrix}$	$R_{23}: \begin{bmatrix} 0 & 0 & -1 \\ 1 & 0 & 0 \\ 0 & -1 & 0 \end{bmatrix}$
III	$R_{30}: \begin{bmatrix} 0 & 0 & -1 \\ 0 & 1 & 0 \\ 1 & 0 & 0 \end{bmatrix}$	$R_{31}: \begin{bmatrix} 0 & 0 & -1 \\ 1 & 0 & 0 \\ 0 & -1 & 0 \end{bmatrix}$	$R_{32}: \begin{bmatrix} 0 & 1 & 0 \\ 0 & 0 & -1 \\ -1 & 0 & 0 \end{bmatrix}$	$R_{33}: \begin{bmatrix} 1 & 0 & 0 \\ 0 & 1 & 0 \\ 0 & 0 & 1 \end{bmatrix}$

Table 2. (a) Relations of coordinates' transformations on three perpendicular planes and (b) illustration of modification of R_0, R_1, R_2, R_3

R_{ij}	R_0	R_1	R_2	R_3
R_0	$R_0 I$	$R_0 R_{10}$	$R_0 R_{20}$	$R_0 R_{30}$
R_1	$R_1 R_{01}$	$R_1 I$	$R_1 R_{21}$	$R_1 R_{31}$
R_2	$R_2 R_{02}$	$R_2 R_{12}$	$R_2 I$	$R_2 R_{32}$
R_3	$R_3 R_{03}$	$R_3 R_{13}$	$R_3 R_{23}$	$R_3 I$

(a)

$$R_0 \Rightarrow R_0 I = R_0^0 \searrow \qquad \nearrow \bar{R}_0 I \Rightarrow \bar{R}_0$$
$$R_1 \Rightarrow R_1 R_{10} = R_0^1 \searrow \qquad \nearrow \bar{R}_0 R_{01} \Rightarrow \bar{R}_1$$
$$R_2 \Rightarrow R_2 R_{20} = R_0^2 \nearrow \bar{R}_0 \searrow \bar{R}_0 R_{02} \Rightarrow \bar{R}_2$$
$$R_3 \Rightarrow R_3 R_{30} = R_0^3 \nearrow \qquad \searrow \bar{R}_0 R_{03} \Rightarrow \bar{R}_3$$

(b)

3 Improved Camera Calibration: Verification, Modification and Regulation, and their Results

3.1 R's Verification and Modification

Case of parallel planes According to (1) there should be $R_0 = R_1 = ... = R_{M-1}$ for R_j $(j = 0, ..., M-1)$ that are computed initially by Tsai's coplanar methods. In the light of this relation the validation can be verified by checking R_js' elements or the angles from them. A common rotation matrix can be modified by averaging the orientation angles from them.

Case of perpendicular planes The verification can be made by checking the relations listed in Table 1. And furthermore based upon these relations, R_{ij}s can be modified by the process listed in Table 2(b). The verifying results in cube-suface experiment are given in Table 3.

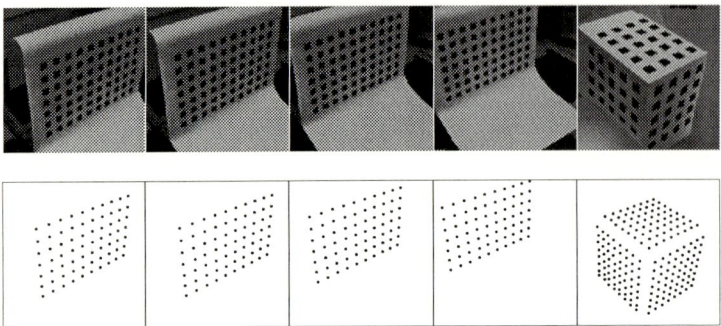

Fig. 2. Original images and the extracted corners for the construction of a 3D-point-grid with parallel planes or surfaces of a cube.

3.2 T's Verification and Regulation

Case of parallel planes Considering the orthogonality of R_j and the real geometrical relation among these coordinate systems, we have $|T_{(j+1)} - T_j| = D_{(j+1)j}$, $j = 0, ..., M - 2$, where $D_{(j+1)j}$ are the distances between the origins of the coordinate system on these planes. These relations can serve as a verification of T_j. They can be changed into one objective function

$$F = [\sum_{j=0}^{M-2} (|T_{(j+1)} - T_j| - D_{(j+1)j})^2]^{1/2} = min \quad (2)$$

and its minimization acts as a goal to the regulation of T_j.

In our experiment the origins of the world coordinate systems are selected at the top-right corner of the top-right square (see Fig.3(a)), $D_{(j+1)j} = Z_{(j+1)j}$. The results show that $|T_2 - T_1|$, $|T_3 - T_2|$ in Table 4 are much nearer to the real distances. Figure 4 gives the shape and contour of the objective function at final iteration, in which the minimum corresponds to optimal values of T_i, f and C_y.

Case of perpendicular planes In this case, when the coordinate systems are chosen on each plane, as shown in Fig. 3(b), if D_{12}, D_{23} and D_{31} stand for the distances among the origins of the selected coordinate systems, the objective function can be constructed as

$$F = [(|T_{II} - T_I| - D_{12})^2 + (|T_{III} - T_{II}| - D_{23})^2 + (|T_I - T_{III}| - D_{31})^2]^{1/2} = min \quad (3)$$

With this function the regulation is similar to that of the parallel plane. In experiment we choose $D_{12} = D_{23} = D_{31} = 0$, the results are shown in Table 5.

Finally improvements are made by optimalization of the total residiual error of all correspondent 2D- and 3D-points for the image center (C_x, C_y). Distortion parameters k_{1x}, k_{2x}, k_{1y} and k_{2y} are computed based on Tsai's Radial Alignment Constraint model. A set of calibrated parameters are given in Table 6.

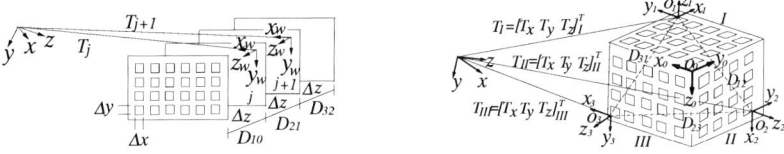

(a) Case of parallel planes   (b) Case of perpendicular planes

Fig. 3. Verification and modification of R, T.

4 Conclusions

A method of camera calibration that computes the parameters from 3D-point-grid, a group of structure related 3D points, is proposed in this paper. According to objective criteria deduced from the 3D-point-grid the camera extrinsic parameters are verified, modified and regulated. Based on the accurate estimation, the intrinsic parameters of the camera continue to be optimized. It is proved experimentally that our method can improve the accuracy of both extrinsic and intrinsic parameters greatly. The method can be applied not only in a well-defined lab situation but also in a nature scene with a precisely made structure-pattern.

References

[Tsai86] R. Y. Tsai, An Efficient and Accurate Camera Calibration Technique for 3D Machine Vision, Proc. of IEEE Int. Conf. on Computer Vision and Pattern Recognition, Miami Beach,FL, 1986, pp 364-374

[LT88] R. K. Lenz and R. Y. Tsai, Techniques for Calibration of the Scale Factor and Image Center for high Accuracy 3D Machine Metrology, IEEE Trans. on Pattern Analysis and Machine Intelligence, Vol. 10, May 1988, pp 713-720

[BM91] E.Bruzzone and F.Magli, Calibration of a CCD Camera on a Hybrid Coordinate Measuring Machine for Industrial Meterology, SPIE, Vol. 1526, Industrial Vision Metrology (1991), pp96-112

[LC93] Jim Z. C. Lai, M. Chao, et al., The Effects of a Camera's Intrinsic Parameters on the Determination of a Coordinate Frame, Jounal of the Chinese Institute of Engineerings, Band 16 (1993), Vol. 5, pp621-630.

[WM94] G.-Q. Wei and S. -D. Ma, Implicit and Explicit Camera Calibration, Theory and Experiments, IEEE, PAMI, Vol. 16, No. 5, May 1994, pp 469-488

[Faug93] O.Faugeras, Three-Dimensional Computer Vision — A Geometric Viewpoint, MIT Press, 1993

[Zhang96] X.-F. Zhang, Untersuchung, Implementierung und Verbesserung der Kamerakalibrierung, Technical Report, TI-1, TUHH, Nov. 1996

Table 3. Verification of R_i's Relations of the Coordinate System on the Cube

$i \diagdown j$		O		I
O	R_{00}:	$\begin{bmatrix} 1 & 0 & 0 \\ 0 & 1 & 0 \\ 0 & 0 & 1 \end{bmatrix}$	R_{01}:	$\begin{bmatrix} 1.00 & 0.03 & 0.00 \\ 0.00 & 0.06 & 1.00 \\ 0.03 & -1.00 & 0.06 \end{bmatrix}$
I	R_{10}:	$\begin{bmatrix} 1.00 & 0.00 & 0.03 \\ 0.03 & 0.06 & -1.00 \\ 0.00 & 1.00 & 0.06 \end{bmatrix}$	R_{11}:	$\begin{bmatrix} 1 & 0 & 0 \\ 0 & 1 & 0 \\ 0 & 0 & 1 \end{bmatrix}$
II	R_{20}:	$\begin{bmatrix} 0.02 & 1.00 & -0.06 \\ 1.00 & -0.02 & 0.05 \\ 0.05 & -0.06 & -1.00 \end{bmatrix}$	R_{21}:	$\begin{bmatrix} 0.05 & 1.00 & -0.05 \\ 0.11 & -0.06 & -1.00 \\ -1.00 & 0.05 & -0.11 \end{bmatrix}$
III	R_{30}:	$\begin{bmatrix} -0.08 & 0.00 & -1.00 \\ -0.05 & 1.00 & 0.01 \\ 1.00 & 0.05 & -0.08 \end{bmatrix}$	R_{31}:	$\begin{bmatrix} -0.09 & 0.03 & -1.00 \\ 0.99 & 0.11 & -0.08 \\ 0.10 & -0.99 & -0.04 \end{bmatrix}$
$i \diagdown j$		II		III
O	R_{02}:	$\begin{bmatrix} 0.02 & 1.00 & 0.05 \\ 1.00 & -0.02 & -0.06 \\ -0.06 & 0.05 & -1.00 \end{bmatrix}$	R_{03}:	$\begin{bmatrix} -0.08 & -0.05 & 1.00 \\ 0.00 & 1.00 & 0.05 \\ -1.00 & 0.01 & -0.08 \end{bmatrix}$
I	R_{12}:	$\begin{bmatrix} 0.05 & 0.11 & -0.99 \\ 1.00 & -0.06 & 0.05 \\ -0.05 & -0.99 & -0.11 \end{bmatrix}$	R_{13}:	$\begin{bmatrix} -0.09 & 0.99 & 0.10 \\ 0.03 & 0.11 & -0.99 \\ -1.00 & -0.08 & -0.04 \end{bmatrix}$
II	R_{22}:	$\begin{bmatrix} 1 & 0 & 0 \\ 0 & 1 & 0 \\ 0 & 0 & 1 \end{bmatrix}$	R_{23}:	$\begin{bmatrix} 0.00 & -0.14 & -0.99 \\ 1.00 & -0.02 & 0.00 \\ -0.02 & -0.99 & 0.14 \end{bmatrix}$
III	R_{32}:	$\begin{bmatrix} 0.00 & 1.00 & -0.02 \\ -0.14 & -0.02 & -0.99 \\ -0.99 & 0.00 & 0.14 \end{bmatrix}$	R_{33}:	$\begin{bmatrix} 1 & 0 & 0 \\ 0 & 1 & 0 \\ 0 & 0 & 1 \end{bmatrix}$

Table 4. Verification of the Distances among Origins

	Real Distance	before regulation		after regulation	
		Computed Dis.	Differences	Computed Dis.	Differences
$\lvert T_1 - T_0 \rvert$	32.35	30.78	-1.57	29.97	-2.38
$\lvert T_2 - T_1 \rvert$	49.95	55.23	5.28	50.59	0.64
$\lvert T_3 - T_2 \rvert$	49.25	89.85	40.60	49.63	0.38

Note: $(\sum_{i=0}^{2} (\lvert T_{(j+1)} - T_j \rvert - Z_{(j+1)j})^2)^{1/2}$ = 40.97 (before reg.), 2.49 (after reg.)

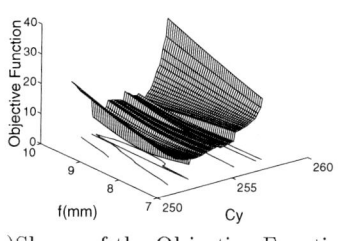
(a) Shape of the Objective Function

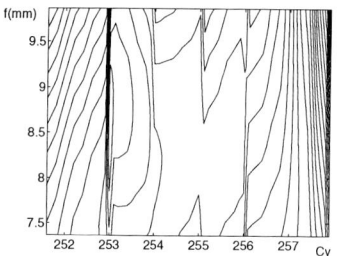
(b) Contour of the Objective Function

Fig. 4. Objective Function at the Final Iteration

Table 5. Verification of T_i'of the Coordinate Systems on the Cube

unit(mm)	T_I	T_{II}	T_{III}	$|T_I - T_{II}|$	$|T_{II} - T_{III}|$	$|T_{III} - T_I|$
T_x	-0.93	-0.11	-0.60	0.82	0.49	0.32
T_y	6.57	6.56	8.61	0.02	2.05	2.04
T_z	272.18	321.87	249.44	49.69	72.43	22.74
distance				49.69	72.46	22.83

Before Regulation, $\left[(|T_{II} - T_I|)^2 + (|T_{III} - T_{II}|)^2 + (|T_I - T_{III}|)^2\right]^{1/2} = 90.78$

unit(mm)	T_I	T_{II}	T_{III}	$|T_I - T_{II}|$	$|T_{II} - T_{III}|$	$|T_{III} - T_I|$
T_x	-0.93	-0.11	-0.60	0.82	0.49	0.32
T_y	6.57	6.56	8.61	0.02	2.05	2.04
T_z	226.26	226.7	223.12	0.53	3.67	3.14
distance				0.97	4.24	3.76

After Regulation, $\left[(|T_{II} - T_I|)^2 + (|T_{III} - T_{II}|)^2 + (|T_I - T_{III}|)^2\right]^{1/2} = 5.75$

Table 6. Calibrated Parameters from the Square Patterns on four Parallel Planes

Transformation Parameters (the unit for elements of T is mm)		Physical Parameters	Distortional Parameters
R $\begin{matrix} -0.8488 & 0.0144 & 0.5285 \\ 0.3335 & 0.7903 & 0.5140 \\ -0.4102 & 0.6125 & -0.6756 \end{matrix}$ $\begin{matrix} \theta \\ \phi \\ \psi \end{matrix}$ $\begin{matrix} -31.90° \\ 142.74° \\ -0.98° \end{matrix}$		f=8.559 C_x=260.0 C_y=253.0 s_x=1.0214	k_{1x}=0.002535 k_{2x}=0.002532 k_{1y}=-0.001040 k_{2y}=0.000791
$T_0 = (142.61, -102.64, 546.97)^T$ $T_1 = (121.62, -116.12, 563.82)^T$ $T_2 = (90.19, -137.14, 597.45)^T$ $T_3 = (59.12, -158.02, 630.20)^T$			

*NOTE: the selection of Vector T_i depends on the selected world coordinate system

A Geometric Modeling Tool for Stereo-Matching and Reconstruction of a Model of 3D-Scene

L. Sommellier, E. Tosan, and D. Vandorpe

Laboratoire d'Informatique Graphique Image et Modélisation (L.I.G.I.M.)
Université Claude Bernard Lyon I,
bâtiment 710, 43 Bld du 11 Novembre 1918,
69622 Villeurbanne CEDEX, FRANCE

Abstract. In this paper, we present how a geometrical modeling tool enables to accelerate the matching of stereo-images and to obtain a 3D-model of the reconstructed scene. In the first time, we represent the topology of images of segments using combinatorial maps. This representation enables to efficiently match the images and to construct, during the matching process, the topological model of the matched scene. A boundary representation of the 3D-scene is obtained by embedding this topological model in $\mathbb{R}^3$.

1 Introduction

Computer vision does not consist any more to only find the three-dimensional (3D) information. Many applications need to reconstruct a model of the scene, for example to use it in a CAD modeler or in a virtual world. The analysis and the synthesis domain must therefore cooperate [6,16]. In this article, we propose to introduce a geometric modeling tool within a stereo-vision process. This will enable to accelerate the matching of the stereoscopic views and to easily obtain a B-rep model of the 3D-scene.

The matching of the stereoscopic views is an important stage of the stereo-vision process : it enables to reconstruct the scene by triangulation. The matching consists to find couples of image primitives which are the projections of the same 3D-entity. It remains difficult due to its combinatorial aspect : given a geometric primitive in an image, what is its equivalent in the another image? Two broad classes of methods are used to reduce the extent of the search for matches :

- methods based on *geometric constraints* [1,4,10] (epipolar, disparity, order...). These constraints being local, they are used by combinatorial methods which verify the coherence of the local matches (dynamic programming [14], neural networks [15], integration of matching and surface interpolation [9]...).
- methods based on *topological information* [5] (graphs [1], grouping [11], neighbouring information [7]). These methods use this information to find the matches (graph isomorphism, clique search...). These methods are also combinatorial because the topological information is local and is used like a primitive attribute or an additional constraint.

After the matching process, the matches are reconstructed by triangulation. Then the set of 3D-points can be modelize. This step is difficult too.
We have approached this problem of matching in a context of modeling for the CAD. Our goal is, at last, to automatically construct the CAD model of the viewed object. Therefore it seems natural to use modeling tools during the matching process. We propose to construct geometrical models of the images using topological informations with the double-aim of reducing the combinatorial and of obtaining the 3D-model of the scene.

2 Principle of our approach

The principle of our approach [18,19] lies in the fact that the topological structure (faces, edges, vertices) of a polyhedral object can partly be found in the topological structure of the images of this object (regions, segments, junctions). For example (figure 1), the faces *face1* and *face2* are neighbors. These two faces

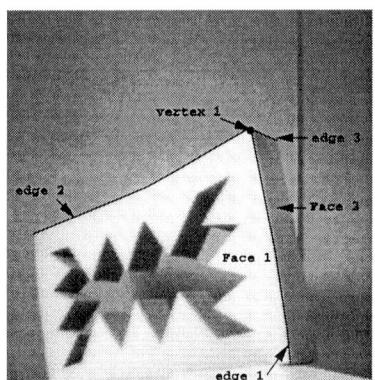
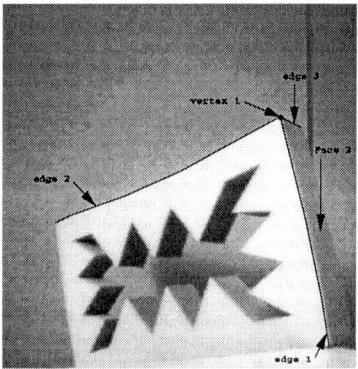

Fig. 1: Topological information about the 3D-object in the images

are the images of two 3D-faces of the viewed object, which are neighbors too. We propose to code this topological information concerning the 3D-object to reconstruct. They will be very useful :

- to structure the matched primitives : for example, we know that the 3D-faces reconstructed using *face1* and *face2* are 3D-neighbors.
- during the matching process :
  - to compare the neighborhoods of two primitives to match.
  - to move coherently in the two images : to go over a face, to follow a contour, etc. Thus, the research of matches becomes a traversal of the topological structures. For example, if the edges *edge1* match, then, by turning around *vertex1*, it is possible to verify that *edge2* and *edge3* match in the two images.

The principle of our approach can be resumed as follows : the perspective projection projects a 3D-structure faces/edges/vertices on two images. Matching the images structures enables, after triangulation, to embed again this structure in $\mathbb{R}^3$ and to find again a partial object model.

We choose to code this topological information with a well known tool in geometric modeling and in CAD : the *boundary representation* (B-rep model) [17].

3 Topological structures

The boundary representation has the particularity of representing surfaces according to their face, edge and vertex subdivision. This surface subdivision is represented by distinguishing the topological structure and the geometrical data. The topological structure [2][20] corresponds to the way in which the faces, edges and vertices are joined in the 3D-model, and the regions, segments and junctions are joined in the image model. The geometrical data correspond to the coordinates or equations of the vertices, edges and faces in the 3D-model, and to the attributes of the primitives (segments and junctions[1]) in the images.

In this paper, we use *combinatorial maps* [13] to code the topological aspect of the boundary representation. When applied to the primitives the map will give informations on their relative positions. It is therefore a tool for primitives grouping.

3.1 Combinatorial maps

Definition 1 - Combinatorial map [12] [3]
A *combinatorial map* $\mathcal{C} = (B, \alpha, \sigma)$ is given by :

- B, a finite set of darts
- α, an involution without fixed points acting on B : $\forall b \in B$, $\alpha(b) \neq b$ and $\alpha^2(b) = b$. α allows the passing from one dart to its opposite in an edge.
- σ, a permutation acting on B : $\sigma(b_{out}) = b_{next}$ with b_{out} the outgoing dart at a vertex and b_{next} the next outgoing dart at this vertex. σ allows the passing from one outgoing dart to the next at a vertex.

A permutation acting on B, ϕ, can also be used and is defined by : $\phi : \begin{vmatrix} B \to B \\ b \mapsto \sigma(\alpha(b)) \end{vmatrix}$

ϕ allows the passing from one dart to the next in a face.
The data of ϕ or σ are equivalent since $\phi = \sigma \circ \alpha \iff \sigma = \phi \circ \alpha$.

Example : in figure 2, edges are $\alpha(b_i^+) = b_i^-$, $\alpha(b_i^-) = b_i^+$; one vertex is $\sigma(b_1^+) = b_2^-$, $\sigma(b_2^-) = b_7^-$, $\sigma(b_7^-) = b_1^+$; one face is $\phi(b_1^-) = b_2^-$, $\phi(b_2^-) = b_3^-$, $\phi(b_3^-) = b_4^+$, $\phi(b_4^+) = b_1^-$.

---

[1] An *n*-branch junction [8] is the intersection of *n* segments.

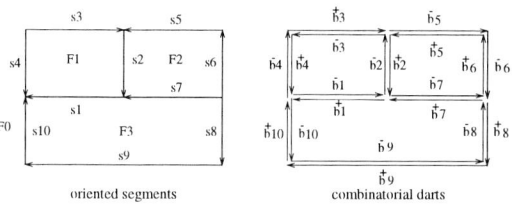

Fig. 2: Embedding of a map in an image of segments

Embedded combinatorial map The combinatorial map $\mathcal{C}$ previously defined is a set of linked darts : it is a purely combinatorial structure. To use it to represent a surface subdivision, it is necessary to link the darts of $\mathcal{C}$ to the geometrical components of the subdivision which they represent. The association of a geometrical element to each dart of the map defines the *embedding* of the combinatorial map.

In our approach, we use combinatorial maps to code the topology of contours images. This contours are approximated by segments. So, we will embed the maps in images of segments.

The embedding of the darts of $\mathcal{C}$ consists to create bijective links between the set of darts and the set of segments. First we link each edge to a segment. As each edge splits up into two darts, we *orient the segments* to associate to each dart a segment and a direction (direct or inverse). This orientation is an *arbitrary* orientation.

Definition 2 - Embedded combinatorial map [19]
The embedding of the map $\mathcal{C}$ in an image of oriented segments is defined by a bijection Q between B and $S \times \{-1, +1\}$ given by :

$$Q : \begin{vmatrix} B \to S \times \{-1, +1\} \\ b \mapsto (s, \pm 1) \end{vmatrix} \qquad Q^{-1} : \begin{vmatrix} S \times \{-1, +1\} \to B \\ (s, +1) \mapsto b \\ (s, -1) \mapsto \alpha(b) \end{vmatrix}$$

Example : in figure 2, the links between the darts and the segments are :
$Q(b_i^+) = (s_i, +1)$, $Q(b_i^-) = (s_i, -1)$, $Q^{-1}(s_i, +1) = b_i^+$, $Q^{-1}(s_i, -1) = b_i^-$.

This notion of embedded combinatorial map is used to create :

- *topological image models* which represent the topological structure of the primitives.
- a *topological matching model* which represent the structure of the matched scene.
- a *topological matched scene model* which is the matched scene map embedded in $\mathbb{R}^3$.

4 Topological modeling of an image of segments

The stereo images are segmented to obtain images of segments. Then, we construct a topological image model by embedding a combinatorial map in each image. This is done in three stages (more details can be found in [19]).

- **Creation of the embedded edges** : each segment is linked to an edge $\{b, \alpha(b)\}$ with Q and Q^{-1}.
- **Detection of the junctions** : after the previous stage, we obtain a set of independent edges. To create vertices, we must find topological relations between the segments. So, we search junctions of segments. We only search 2- and 3-branch junctions because most junctions in the images are such junctions. Junctions with more than 3 branches will be treated in the topological matching model (§5.2).
- **Creation of the junctions in the map** : each segment junction enables to create an edge junction in the map. We link the edges with the permutation ϕ using the Möbius rule to obtain coherently oriented faces.

5 Construction of the topological scene model

In the first paragraph, we only give the *principle* of the algorithm for junction matching. A detailed description of this algorithm can be found in [19,18]. Then we show how to construct a *topological matching model* during the junction matching.

5.1 Principle of matching

Let $\mathcal{C}$ and $\mathcal{C}'$ be two embedded combinatorial maps corresponding to the left and right image. Let's assume that a couple of 3-branch junctions (j_1, j_1') is matched by a geometrical method. The principle of matching is to move, stemming from two matched darts, to the neighbouring couple of 3-branch junctions (j_2, j_2'). Then we verify that (j_2, j_2') match geometrically.

5.2 Construction of the topological matching model

The topological matching model is constructed using the topological image models and the traversal moves of the matching process in these models. This is a new map $\mathcal{C}_m$ created to code the topology of the matched scene. A basic topological matching model is constructed during the 3-branch junction matching. Then we improve it during the matching of 2-branch junctions and we close some of its faces.

During the 3-branch junctions matching We suppose that each 3-branch junction match corresponds to a 3D-junction in the scene. So, for each new match, we create a junction in the map $\mathcal{C}_m$. This new junction is linked to the previously created junction by using the moves in the maps (cf. figure 3). This corresponds to the principle given in §2.

Each vertex of $\mathcal{C}_m$ corresponds to a couple of matched junctions. The edges are embedded in the contours between the matched junctions.

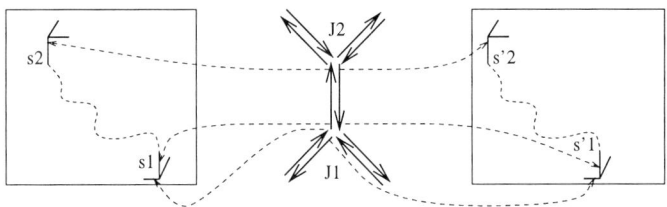

Fig. 3: Embedding of the scene model

Updating during the 2-branch junction matching We update the topological matching model during the 2-branch junction matching. The most robust matches are undertaken first : the matches of contours (broken lines) between two matched junctions. There is an other type of broken lines which match *a*

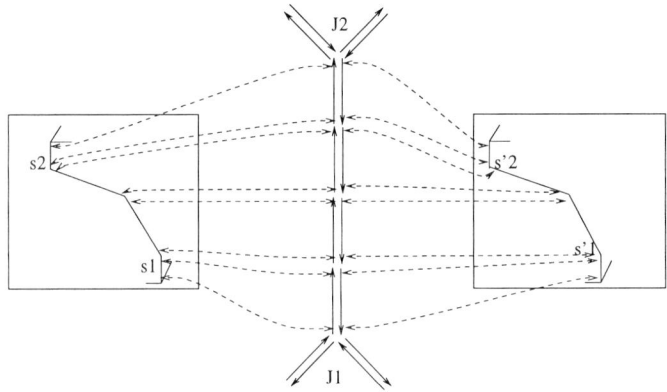

Fig. 4: Embedding of a broken line

priori : the edges branching out from a vertex in the scene model. To match them, we proceed as to match the edges situated between two matched vertices. The edges which are not connected to a matched 3-branch junction are matched as the 3-branch junctions (cf. §5.1).

Closing faces in the topological matching model By closing opened faces of this model, we create new vertices in $\mathcal{C}_m$ which correspond to new matches of junctions in the images. A face of the model will be closed if the corresponding faces in the images are closed at this junction and if the geometrical constraints on the junctions (epipolarity, position, type) are validated.
The closure of the faces in the map $\mathcal{C}_m$ enables :

- the 2-branch/3-branch matching : this case is important when the third branch of a junction was not detected or when a branch is in fact a noise segment.
- the creation of n-branch ($n > 3$) junctions in the model $\mathcal{C}_m$: therefore the detection of only 2- and 3-branch junctions has not impoverished the model of the scene to be reconstructed.

The evolution of the scene model during its construction is shown on a simple example[2] (cf. §"Some experimental results").

5.3 The topological matched scene model

After the matching stage, we have obtained a topological model of the matched scene. This model is embedded on couples of segments. This enables to calculate, by triangulation, the 3D-coordinates of each vertex of the model. After the triangulation, the model is embedded in $\mathbb{R}^3$ (figure 5). It is a boundary repre-

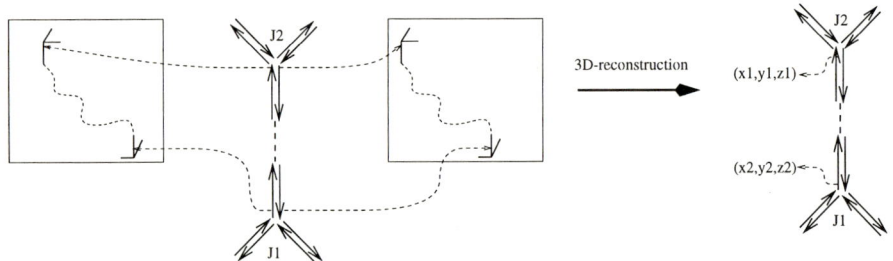

Fig. 5: Embedding of the scene model in $\mathbb{R}^3$

sentation which topological structure is represented by the combinatorial map $\mathcal{C}_m$ and the geometrical data by the 3D-coordinates of the vertices.

5.4 Results and conclusions

We have presented an algorithm for the stereo-matching and the 3D-reconstruction based on the use of geometrical models of the stereoscopic images. These models enable to code the topological informations contained in the stereo-images using combinatorial maps. This approach enables :

---
[2] Examples of stereo matching can be found in [19].

- a substantial reduction of the combinatorial of the matching :
    - The traversal of the models enables to find potential matches and guides the geometrical matching method.
    - The use of the topological matching model enables to eliminate all combinatorial during the matching of the 2-branch junctions linked to previously matched 3-branch junctions. It is just a going over of an edge.

  To assess the reduction in number of matching tests obtained with this cooperative approach, we have compared it to the geometrical matching method used alone. The tests, carried out on real images, lead to a reduction in the number of matching tests varying between 20% and 50% [19,18].
- the construction of the B-rep model of the 3D matched scene during the matching of the images. We have shown how to use the two geometric image models and the traversal moves of the matching process in these models. This enables us to easily obtain the B-rep model of the scene.

Today, we only have a model of the visible part of the object. A possible direction for future researches is to compute the complete CAD model of an object from multiple stereo views by turning around the object.

References

1. N. Ayache. *Artificial vision for mobile robots: stereo vision and multisensory perception.* MIT Press, 1991.
2. Baumgart. A polyhedron representation for computer vision. In *National Computer Conference*, pages 589–596, Arlintgton, 1975. AFIPS Press.
3. R. Cori. Un code pour les graphes planaires et ses applications. *Astérisque*, 27, 1975.
4. O. Faugeras. *Three-dimensional computer vision.* Artificial Intelligence Series. MIT Press, 1993.
5. M.M. Fleck. A topological stereo matcher. *International journal of computer vision*, 6(3):197–226, 1991.
6. A. Gagalowicz. Towards a vision system for a domestic robot. In *Journée analyse/synthèse d'images*, pages 63–82, 1994.
7. A. Gagalowicz and L. Vinet. Regions matching for stereo pairs. In *Sixth scandinavian conference on image analysis*, pages 63–70, Juin 1989.
8. M. Herman and T. Kanade. Incremental reconstruction of 3D scenes from multiple complex images. *Artificial intelligence*, 30:289–341, 1986.
9. W. Hoff and N. Ahuja. Surfaces from stereo: integrating feature matching, disparity estimation, and contour detection. *IEEE Transactions on pattern analysis and machine intelligence*, 11(2), Février 1989.
10. R. Horaud and O. Monga. *Vision par ordinateur.* série informatique. Traité des Nouvelles technologies, 1993.
11. R. Horaud and T. Skordas. Stereo correspondence through feature grouping and maximal cliques. *IEEE Transactions on pattern analysis and machine intelligence*, 11(11):1168–1180, Novembre 1989.
12. A. Jacques. Constellations et graphes topologiques. In *Combinatorial theory and applications*, pages 657–673, 1970.

14. P. Lienhardt. Topological models for boundary representation: a comparison with n-dimensional generalised maps. *Computer-aided design*, 23(1):59–82, Février 1991.
15. G. Medioni and R. Nevatia. Segment-based stereo matching. *Computer vision, graphics, and image processing*, 31:2–18, 1985.
16. R. Mohan. Constraint satisfaction networks for vision. Technical report, Exploratory computer vision group IBM Thomas J. Watson research center Yorktown, 1989.
17. Pun and Blake. Relationship between image synthesis and analysis : towards unification? *Computer Graphics Forum*, 9(2):149–163, 1990.
18. A. Requicha. Representations for rigid solids : theory, methods and systems. *Computing surveys*, 12(4):437–464, 1980.
19. L. Sommellier, E. Tosan, S. Bouakaz, and D. Vandorpe. Using combinatorial maps for stereo correspondence and reconstruction. In A. Behrooz, editor, *Proc. Int'l Conf. on Knowledge Transfer Visualization & Graphics'96*, pages 540–546, London, England, Juillet 1996.
20. K. Weiler. Edge-based data structure for solid modeling in curved-surface environments. *IEEE computer graphics and applications*, 5(1), 1985.

Some experimental results

A reconstructed object

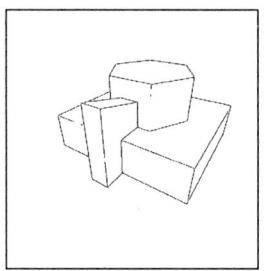
Left image of contours

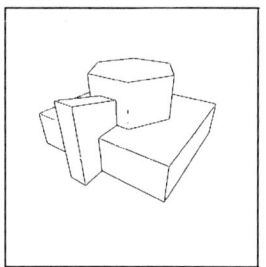
Right image of contours

Construction of the 3D matched scene :

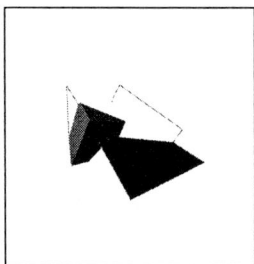
After the 3-branch junction matching

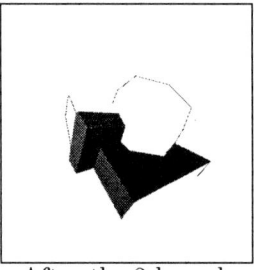
After the 2-branch junction matching

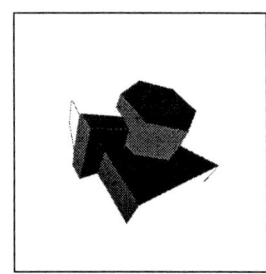
After closing some faces

Estimating Translation/Deformation Motion through Phase Correlation *

Filiberto Pla[†] and Miroslaw Bober[‡]

[†]Dept. of Computer Science.
University Jaume I, 12071 Castelló, Spain.
[‡]Dept. of Electrical and Electronic Engineering.
University of Surrey, Guilford GU2 5XH, UK.
pla@uji.es   m.bober@ee.surrey.ac.uk

Abstract. Phase correlation techniques have been used in image registration to estimate image displacements. These techniques have been also used to estimate optical flow by applying it locally. In this work a different phase correlation-based method is proposed to deal with a deformation/translation motion model, instead of the pure translations that the basic phase correlation technique can estimate. Some experimentals results are also presented to show the accuracy of the motion paramenters estimated and the use of the phase correlation to estimate optical flow.

Key Words : Motion, Optical Flow, Image Registration, Phase Correlation.

1 Introduction

Estimating visual motion is a valuable information for many machine vision applications, like traffic monitoring, surveillance, image coding, etc. The work presented here has been aimed at developing some techniques for accurate enough visual motion estimation and optical flow. Optical flow estimation methods measure the velocity and displacement vectors perceived from the time-varying image intensity pattern, that is, they measure the apparent 2D image plane motion which is the projected 3D motion of the objects of the scene in the image plane. Most of existing techniques for optical flow estimation can be classified into *Gradient-based techniques*, that is, methods based on the so-called *optical flow equation* [7][12]; *Block matching-based techniques* try to overcome the aperture problem by assuming that all pixels in a block undergo the same motion [6][11]; *Spatiotemporal Energy-based techniques* which exploits the equivalent optical flow equation in the frequency domain [9]; *Bayesian techniques* utilize probabilistic smoothness constraints, usually in the form of a Gibbs ramdom field [10][8].

According to Tekalp's classfication [13], the above mentioned methods correspond to the *non parametric models*. *Parametric models* try to work out the

---

* This work has been partially supported by the *JP/ij-estades/96* grant from the *Generalitat Valenciana*, and ESPRIT project EP21007 IOTA.

motion model of the ortographic or perspective projection of the 3D motion in the image plane. Motion models are also used in block matching techniques, thus, in order to look for the best matching block in the other image, some motion model has to be assumed. There are generalized 2D image motion models. One of the most used is the affine coordinate transformation,

$$x' = a_1 x + b_1 y + c1 \; ; \qquad y' = a_2 x + b_2 y + c2 \qquad (1)$$

with (x', y') being the image transformed coordinates, and $a_1, b_1, c_1, a_2, b_2, c_2$ the motion parameters of the affine model. The affine tranformation corresponds to the ortographic projection of a 3D rigid motion of a planar surface. A particular instance of this model is a translation/deformation model, which takes into account tranlations and scale changes in the image plane,

$$x' = a_1 x + c1 \; ; \qquad y' = b_2 y + c2 \qquad (2)$$

In the work we are presenting here, a block matching technique is used to estimate motion and optical flow. This block matching technique will be analyzed and extended from the translational model to the deformation/translation motion model (equation 2). Next section will introduce the basics of the phase correlation technique, which will be further analyzed and extended in a subsequent section. Some experimental results are also presented to shown the accuracy of the extended method.

2 Phase correlation techniques

Another type of techniques which could be included in the block matching techniques are the *phase correlation* techniques. Phase corelation techniques have been used in image registration [1][5]. Although in image registration phase correlation techniques are applied to the entire image, the phase correlation method has also been used in optical flow estimate applying it locally, using a small window around the point of interest where the image flow velocity is being estimated [13].

The basic phase correlation method estimates the relative shift between two image blocks by means of a normalized cross-correlation function computed in the 2D spatial Fourier domain. It is also based on the principle that a relative shift in the spatial domain results in a linear phase term in the Fourier domain. Therefore, let $f_k(x, y)$ be the image intensity function at time k, and let $f_{k+1}(x, y)$ be the image intensity function at time $k + 1$.

If we assume that $f_k(x, y)$ has undergone a translation (x_0, y_0), then $f_{k+1}(x, y) = f_k(x - x_0, y - y_0)$. Taking the Fourier tranformation of both functions, the image displacement (x_0, y_0) can be calculated by means of the normalized cross-power spectrum funtion, as

$$\overline{C}_{k,k+1}(u, v) = \frac{F_{k+1}(u, v) F_k^*(u, v)}{|F_{k+1}(u, v) F_k^*(u, v)|} \qquad (3)$$

Since $F_k(u,v)$ and $F_{k+1}(u,v)$ are related as $F_{k+1}(u,v) = e^{-j(ux_0+vy_0)}F_k(u,v)$, the inverse of the above equation results in $\overline{c}_{k,k+1}(x,y) = \delta(x-x_0, y-y_0)$ which is the cross correlation function consisting of an impluse whose location gives the displacement vector. Ideally, we would expect to find a single impulse in the phase correlation function, but in practice several factors contribute to the degeneration of the phase correlation function.

On the other hand, the phase correlation method has some advantages. One important feature is that the phase correlation method is relatively insensitive to changes in illumination, because variations in the mean value or multiplication for a constant do not affect the Fourier phase. Since the phase correlation function is normalized with the Fourier magnitude, the method is also insensitive to any other Fourier-magnitude-only degradation.

The use of the DFT to compute the phase correlation function has some undesirable efects:

Boundary effects. To obtain a perfect impulse, the shift in the spatial domain has to be ciclic. Since things appearing at one end of the block (window) generally do no appear at the other end, the impulse degenerate into a peak. Further, since the 2D DFT assumes periodicity in both directions, discontinuities from left to right boundaries, and from top to bottom, may introduce spurious peaks.

It is well known that the boundary effects due to the finiteness of the image (block) frame become less relevant if the image function has small values near the frame boundaries. Therefore, the rectangular window representing the framing process, may be substituted by a weighting window $w(x,y)$ that produces the decay of the image function values near the boundaries. In this case, we have addopted a Gaussian-like windowing function to our approach [1],

Spectral leakage. In order to observe a perfect impulse, the components of the displacement vector must correspond to an integer multiple of the fundamental frequency. Otherwise, the impulse degenerates into a peak due to the well known spectral leakage phenomenom. Thus, if we assume that the peak values are normally distributed around its maximum, then the actual maximum would be the mean of this distribution.

Range of displacement estimates. Since the 2D DFT is periodic with the block size (N,M), only displacements (x_0, y_0) can be detected if they satisfy that $-N/2 \leq x_0 \leq N/2$ and $-M/2 \leq y_0 \leq M/2$ due to the wrapping effect of the Fourier transform. Therefore, in order to estimate displacements in the range $(-d,d)$ along a spatial direction, the block size has to be theoretically at least of $2d$ size in this spatial direction.

3 Extending the phase correlation technique to estimate translation/deformation motion

So far, the phase correlation function has been used to estimate displacements assuming a pure translation model within the image or block. Some work has been also done in image registration to estimate pure rotation of images by means of an iterative-search procedure [5]. In order to extend the phase correlation

techniques to a different motion model and taking into account some properties of
the Fourier transform, let us analyze what would happen if the motion undergone
for the pixels in the block is not a pure translation.

Let us consider that an image (block or window) $f_k(x,y)$ at instant k undergoes a translation/deformation motion as shown in equation 2. Therefore,
the corresponding gray level values $f_{k+1}(x,y)$ of that block at instant $k+1$ are
related with $f_k(x,y)$ as $f_{k+1}(x,y) = f_k(a_1 x + c1, b_2 y + c_2)$. Hence, taking the
definition of the Fourier Transform of the above expression, and making the
transformation $x' = a_1 x + c1$ and $y' = b_2 y + c_2$, the expression for $F_{k+1}(u,v)$
becomes

$$F_{k+1}(u,v) = \frac{1}{|a_1 b_2|} e^{j(\frac{uc_1}{a_1} + \frac{vc_2}{b_2})} \int_{-\infty}^{+\infty} \int_{-\infty}^{+\infty} f_k(x',y') \, e^{-j(\frac{ux'}{a_1} + \frac{vy'}{b_2})} dx' dy'$$

which is a combination of the *shift* and *similarity* theorems of the Fourier transform. In the above expression, only positive values of a_1 and b_2 will be considered,
since negative values would mean a deformation plus a symmetry transformation in the image, situation that cannot occur in a real moving scene. Hence, and
rewriting the above equation as a function of the Fourier transform of $f_k(x,y)$,

$$F_{k+1}(u,v) = \frac{1}{a_1 b_2} e^{j(\frac{uc_1}{a_1} + \frac{vc_2}{b_2})} F_k(\frac{u}{a_1}, \frac{v}{b_2}) \tag{4}$$

In order to calculate the parameters of the motion undergone by $f_k(x,y)$,
note that from the above equation, the magnitude of $F_{k+1}(u,v)$ is $|F_{k+1}(u,v)| = \frac{1}{a_1 b_2} |F_k(\frac{u}{a_1}, \frac{v}{b_2})|$. Thus, from the above equation we can obtain the values of the
deformation parameters a_1 and b_2 as follows. Let us calculate the weighted energy
of the spectrum for F_{k+1} as

$$\int_{-\infty}^{\infty} \int_{-\infty}^{\infty} |u\, F_{k+1}(u,v)| \, du dv = \frac{1}{a_1 b_2} \int_{-\infty}^{\infty} \int_{-\infty}^{\infty} |u\, F_k(\frac{u}{a_1}, \frac{v}{b_2})| \, du dv$$

making the transformation $u' = u/a_1$ and $v' = v/b_2$ yields

$$\int_{-\infty}^{\infty} \int_{-\infty}^{\infty} |u\, F_{k+1}(u,v)| \, du dv = a_1 \int_{-\infty}^{\infty} \int_{-\infty}^{\infty} |u'\, F_k(u',v')| \, du' dv'$$

Solving for a_1, and analogously for b_2, and exchanging integrals by summations for the discrete case, we reach to the expression

$$a_1 = \frac{\sum_{u=-N/2}^{N/2} \sum_{v=-M/2}^{M/2} |u\, F_{k+1}(u,v)|}{\sum_{u=-N/2}^{N/2} \sum_{v=-M/2}^{M/2} |u\, F_k(u,v)|} \;;\quad b_2 = \frac{\sum_{u=-N/2}^{N/2} \sum_{v=-M/2}^{M/2} |v\, F_{k+1}(u,v)|}{\sum_{u=-N/2}^{N/2} \sum_{v=-M/2}^{M/2} |v\, F_k(u,v)|} \tag{5}$$

Given $F_k(u,v)$ and $F_{k+1}(u,v)$, equations 5 provide the values for a_1 and b_2.
On the other hand, the translation parameters, c_1 and c_2, can be obtained as
follows. Given $F_{k+1}(u,v)$, let us transform it into $G_{k+1}(u,v)$ using the estimated
a_1 and b_2 as $G_{k+1}(u,v) = F_{k+1}(a_1 u, b_2 v)$. Applying now equation 4, yields

$$G_{k+1}(u,v) = \frac{1}{a_1 b_2} e^{j(uc_1+vc_2)} F_k(u,v)$$

Note that if we now calculate the normalized cross power spectrum of $G_{k+1}(u,v)$ and $F_k(u,v)$ we obtain $\overline{C}_{k,k+1}(u,v) = e^{j(uc_1+vc_2)}$ whose inverse Fourier transform will give an impulse function at $(-c_1,-c_2)$, corresponding to the translation parameters.

Therefore, given the estimated motion parameters for a pixel, to estimate the optical flow field we can compute the velocity vectors (v_x, v_y) at every pixel as $(v_x, v_y) = (x - (a_1 x + c_1), y - (b_2 y + c_2))$.

4 Experimental Results

Series	$\sigma(a_1)$	$\sigma(b_2)$
1	0.00197139	0.00750315
2	0.0128401	0.00464964
3	0.015433	0.00788849

Table 1. Standard deviations of estimated deformation parameters.

In order to test the proposed approach, two type of experiments are shown in this section. The first experiments are directed to show the accuracy of the estimated parameters. A second set of experiments will compare the optical flow estimation made by the proposed method with respect to other methods.

To compute the DFT we use the Fast Fourier Transform (FFT) algorithm. Therefore, for image block sizes not multiple of 2, the zero padding technique is used to complete a power 2 block size. Moreover, we utilized the nearest neighbour technique as interpolation function, but calculating $F_{k+1}(u,v)$ in a mesh k times the size of the block, usually $k = 2$.

Figure 1.a shows the *Tree* image used in these experiments, the same as usued in [2][3] and it will be used later for comparison purposes. This image was transformed using three series of deformation parameters. The transformed images $g(x,y)$ were calculted from the original image $f(x,y)$ using the ideal interpolator function. The series of transformed images were calcuated as follows:

1. Fixing $a_1 = 1$, $c_1 = 0$ and $c_2 = 0$ and varying b_2 from 0.9 to 1.1 incrementing it in steps of 0.02.
2. The same as in the above serie, but fixing $b_2 = 1$ and varying a_1 in the same way.

3. The same as in the above serie, but now varying a_1 and b_2 in the same way as in the series 2 and 1, respectively.

The motion parameters were estimated using the whole 150 × 150 image as a block. In table 1 we can see the standard deviations of the estimated values for a_1 and b_2. Note that the maximum standard deviation is 0.015433, so the expected accuracy estimate is around this order.

	Translating			Diverging		
Technique	Av. Error	Std	Density	Av. Error	Std	Density
Lucas [12]	1.75	1.43	40.8%	3.05	2.53	49.4%
Barron [2]	0.36	0.41	76%	1.08	0.52	64.3%
Bober [3]	0.33	0.25	100%	3.69	4.39	100%
defor. model	1.79	1.10	100%	9.00	3.65	100%

Table 2. Errors for the optical flow of the *Translating* and *Diverging Tree* (Ref. Bober et al., 1994).

Another experiment to test the accuracy of the method is related to the estimation of dense optical flow. Figure 1.b represents the ground optical flow field of the *Translating Tree*, and Figure 1.c represents the optical flow field of the *Diverging Tree* of figure 1.a. Both examples were used in [2][3]. The error measure utilized is the one defined in [2], and the same as the one used in the above mentioned works.

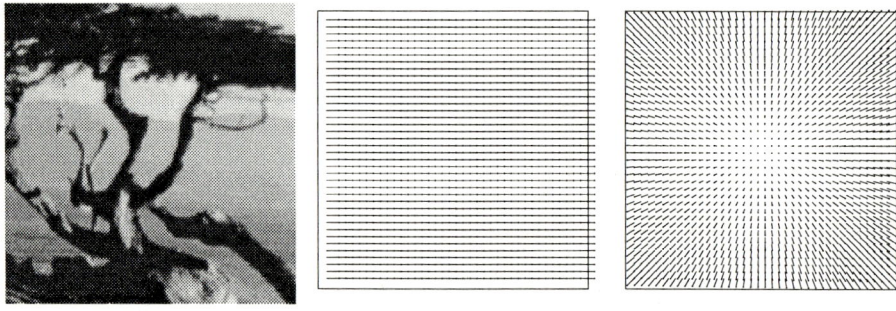

(a) Original image (b) Ground translating flow (c) Ground diverging flow

Fig. 1. The ground flow fields of the *Tree* image.

Table 2 shows the results presented in [3] comparing serveral optical flow methods on the same *Translating* and *Diverging Tree* images, where only the 3 best results for each example are shown, together with the results obtained by the proposed method. The flow estimated by all these variations in table 2 have been computed using a 32x32 window centered at every pixel.

Although errors from the proposed method are bigger with respect to the errors shown in table 2, note that errors provided by Lucas [12] and Barron [2] are estimated over 40% and 76% of the flow field for the *Translating Tree*, and 49% and 64% for the *Diverging Tree* respectively. Bober's method [3] has got the advantage that it is a combined flow estimation and segmentation method, so it also uses global information to estimate the velocity at each point, however, the approach presented here estimates the flow using only local information. In addition, neither smoothing or any type of filtering have been applied as a postprocessing to the raw flow field.

Figure 2 shows an example of the flow field computed where the background (road) is approximately stationary and the cars move in both directions. Looking at the figures we can notice that the optical flow discontinuities are tackled quite satisfactorily. The same applies to figure ??, where we can notice even the different motions undergone from different parts of the car, for example, the wheel. In this example, the car is almost stationary with respect to the camera and the background moves, however the rotation of the wheels can be appreciated with respect to the camera.

(a) Original image (b) Estimated flow field

Fig. 2. The *n234* sequence.

5 Conclusions

In this paper, the phase correlation method has been combined with a deformation/translation motion model, and some techniques have been developed to estimate the motion parameters assuming this motion model. Experimental results have shown the accuracy of the estimated deformation paramenters by the proposed method.

The developed phase correlation techniques have also been applied to estimate optical flow in image sequences providing satisfactory and encouraging results. A more accuarte subpixel estimation method would be necessary to obtain better optical flow estimates.

Futur extensions of the present work are directed to several issues. First of all, work is being carried out to try to extend the proposed deformation/translation motion model using phase correlation techniques to an affine motion model, which can deal with rotation, deformation and translation.

References

1. Alliney, S. and Morandi, C.; "Digital image registration using projections", *IEEE Trans. Patt. Anal. Mach. Intel.*, vol. PAMI-8, no. 2, pp. 222-233, 1986.
2. Barron, J.; Fleet, D. and Beauchemin, S.; "Performance of optical flow techniques", *Int. J. Comp. Vision*, vol. 12, pp. 43-77, 1994.
3. Bober, M. and Kittler, J.; "Robust motion analysis", Proc. of IEEE Conf. on Computer Vision and Pattern Recognition, Seatle, pp. 947-952, 1994.
4. Bracewell, R. N.; *The Fourier transform and its applications*, McGraw Hill, 1986.
5. De Castro, E. and Morandi, C.; "Registration of translated and rotated images using finite Fourier transforms", *IEEE Trans. Patt. Anal. Mach. Intel.*, vol. PAMI-9, no. 5, pp. 700-703, 1987.
6. Ghraravi, H. and Mills, M.;"Block-matching motion estimation algorithms: New results", *IEEE Trans. Circ. and Syst.*, vol. 37, pp. 649-651, 1990.
7. Horn, B.K.P. and Schunk, B. G.; "Determining optical flow", *Artificial Intelligence*, vol. 5, no. 3,pp. 276-287, 1993.
8. Iu, S. L.; "Robust estimation of motion vector fields with discontinuity and occlusion using local outliers rejection", *SPIE*, vol. 2094, pp. 588-599, 1993.
9. Jacobson, L. and Wechsler, H.; "Derivation of optical flow using a spatiotemporal-frequency approach", *Computer Vision, Graphics and Image Processing*, vol. 38, pp. 29-65, 1987.
10. Konrad, J. and Dubois, E.; "Comparison of stochastic and deterministic solution methods in Bayesian estimation of 2D motion", *Image and Vision Computing*, vol. 9, pp. 215-228, 1991.
11. Liu, B. and Zaccarin, A.;"New fast algorithms for the estimation of block motion vectors", *IEEE Trans. Circ. and Syst. Video Tech.*, vol. 3, no. 2, pp. 148-157, 1993.
12. Lucas, B. D. and Kanade, T.; "An iterative image registration technique with an application to stereo vision", *Proc. DARPA Image Understanding Workshop*, pp. 121-130, 1981.
13. Tekalp, A. M.; *Digital video processing*, Prentice Hall, 1995.

Robust Fitting of 3D CAD Models to Video Streams

Christophe Meilhac and Chahab Nastar

INRIA B.P. 105, 78153 Le Chesnay Cedex, France
Contact E-mail: Christophe.Meilhac@inria.fr
http://www-syntim.inria.fr/syntim/recherche/meilhac/

Abstract. We present a robust and accurate semi-automatic algorithm for registering and tracking a 3D geometric model in a 2D video stream. The algorithm is a generalization of the "Iterative Closest Point" technique. Each iteration is composed of two steps: computation of camera parameters, and 3D/2D vertex matching. This last step is performed by polygon fitting in an edge image. To account for false matches, we use a robust M-estimation both for camera parameter estimation and 2D feature extraction. Experimental results show that accurate registration can be obtained even with very noisy outdoor images and incomplete data. Error analysis proves that the accuracy is obtained at the pixel level.

1 Introduction

Model-based vision leads to a vast improvement in performance, at the cost of *a priori* knowledge of the 3D geometry of one (or several) objects of the scene, and maybe of some user interactivity. In this framework, we show interest in fitting a geometrical CAD-model of an object to a 2D image of the object in a complex scene (figure 4 and 5). In other words, the projection of the model into the image should match image features. In mathematical terms, the goal is to determine a 3D translation/rotation aligning (registering) the model with the image data. Our goal is to perform a model-based tracking of the object in a sequence of 2D images, thus partially reconstructing significant 3D information from the video stream. Although this is a general and important vision problem *per se*, our application is augmented reality for entertainment, i.e. mixing virtual and real-world objects while ensuring visual and physical interactions [3]. For our application domain, *accuracy* is a major issue in the 3D information recovery, in order to avoid visual aberration in the augmented sequence. On the other hand, as we deal with real-world images, the object of interest is not restricted to lie on a uniform background, yielding essentially in a noisy pre-processing stage (e.g. edge extraction). Therefore, we need a *robust* algorithm for fitting and tracking. Indeed, our method needs to be accurate and robust for performing 3D/2D registration and tracking.

Model-based registration and tracking is a rather recent issue in computer vision. Lowe [8] performs the registration of a 3D model in a segmented (edge) image, and develops a robust tracking method based on Bayesian decision theory. Wunsch and Hirzinger [10] propose a matching algorithm between an image and a polyhedral model based on the inverse perspective in order to give a 3D-3D

match. Lavallée and Szeliski [6] compute a 2D-3D match of the occluding contour
and use 3D distance maps and octrees to speed up the matching process. These
methods are interesting but they assume a clean segmentation of the object
of interest and lack the registration accuracy that we need in our applicative
framework. Other methods of model-based pose estimation have been developed
with emphasis on vehicle tracking [9,4].

Our method is mostly related to Kumar and Hanson [5] who use lines as 2D
primitive and present a robust estimation of pose. We use the Iterative Closest
Point (ICP) algorithm [1,11] combined with robust statistics [2,12].

2 A robust 3D/2D registration technique

The main idea of our iterative algorithm is to improve a prediction of camera parameters by (i) computing the best match between a vertex point of a 3D-polyhedral CAD model and a corner point of a 2D-image data, then (ii) running a calibration algorithm based on the previous match. For outlier rejection and stabilization of the iterative process, we use robust statistics both for line estimation and for calibration.

2.1 Statement of the problem

The goal is to find the camera parameters that make the projection of the 3D-model consistent with the 2D-image, i.e. to minimize the following objective function: $\sum_{\mathbf{x} \in V} \|\sigma(\mathbf{P}, \mathbf{x}) - \Pi(\mathbf{P}, \mathbf{x})\|$ (1). w.r.t. camera parameters $\mathbf{P}$. $\sigma(\mathbf{P}, .)$ is the matching operator, $\Pi(\mathbf{P}, .)$ the projection operator and V the set of vertices of the 3D polyhedral model. Thus the problems are : (1) estimating the optimal $\mathbf{P}$, (2) computing operators $\sigma(\mathbf{P}, .)$ and $\Pi(\mathbf{P}, .)$, and (3) being robust to noise and spatial uncertainty. Formally, the operator of projection is defined as a perspective transformation:

$$\Pi(\mathbf{P},.) : \mathbb{R}^3 \to \mathbb{R}^2 \quad \mathbf{x} = (x\ y\ z)^t \mapsto \begin{pmatrix} u \\ v \end{pmatrix} = \begin{pmatrix} u_0 + \alpha_u \frac{x'}{z'} \\ v_0 + \alpha_v \frac{y'}{z'} \end{pmatrix} \quad (2)$$

with $(x'\ y'\ z')^t = \mathbf{R}(x\ y\ z)^t + \mathbf{T}$ where $\mathbf{R}$ is the rotation and $\mathbf{T}$ the translation. $\alpha_u, \alpha_v, (u_0, v_0)$ are intrinsic camera parameters.

Let us now introduce a robust version of the objective function and one that can be iteratively minimized (see section 2.2):

$$f(\mathbf{P}, \mathbf{Q}) = \sum_{\mathbf{x} \in V} \rho(\sigma(\mathbf{P}, \mathbf{x}) - \Pi(\mathbf{Q}, \mathbf{x})) \quad (3)$$

where ρ is a M-estimator[1] (the classical least squares problem is obtained for $\rho(x) = \frac{x^2}{2}$). We have implemented many M-estimators (Welsh, Cauchy,...); for simplicity, we use the robust Geman-McClure M-estimator: $\rho(x) = \frac{x^2}{1+x^2}$.

---
[1] ρ is a generalization of classical M-estimators to $\mathbb{R}^2$: $\rho(x, y) = \rho(x) + \rho(y)$.

2.2 Description of the algorithm

Our purpose is to minimize the objective function $\mathbf{P} \mapsto f(\mathbf{P},\mathbf{P})$ w.r.t. $\mathbf{P}$. The main idea is to adapt the Iterative Closest Point (ICP) algorithm [1] as follows:

Initialization: interactive initialization (see section 2.3) of $\mathbf{P}^0$.
Repeat
Matching: computation of $\sigma(\mathbf{P}^{(m)}, \mathbf{x})$ for each 3D vertex $\mathbf{x}$.
  Sampling: each projected line $s \in$ Model is sampled, giving $M_s = \{(u_{i,s}, v_{i,s})\}$.
  Searching: $(x_{i,s}, y_{i,s}) := \texttt{ClosestPoint}(u_{i,s}, v_{i,s})^2$ in the contour image.
  Regression: estimation of the best polygon fitting $(x_{i,s}, y_{i,s})$, (see section 2.4)
  Computation of corners: for all 3D vertex x, $\sigma(\mathbf{P}^{(m)}, \mathbf{x}) =$ polygon corner.
Calibration: computation of $\mathbf{P}^{(m+1)}$ that minimizes $\mathbf{P} \mapsto f(\mathbf{P}, \mathbf{P}^{(m)})$. This is a classical non-linear calibration problem. To obtain a fasterconvergence we use a quasi-Newton technique. This is a robust version of [7].
Until stabilization

2.3 Interactive Initialization

For initializing the process, we have to compute a camera estimation that provides approximate projection of the model onto the image. The user can intuitively and easily match a few vertices of the 3D model with corresponding image features (figure 1). The problem is then a classical calibration problem. This step provides the visible part of the 3D model in the 2D image.

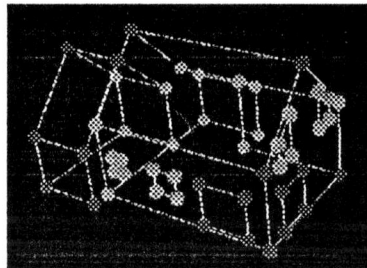

3D Model   interactive image matching

Fig. 1. Initialization is computed using interactive matching followed by calibration algorithm.

2.4 Robust polygon estimation

The 2D polygon is defined by the graph (vertices and edges) of the visible part of the 3D model. It is parameterized by the line equation of each edge. A polygon corner is defined as a line intersection. To find the polygon in edge image, we minimize the following objective funtion w.r.t. $(a_s, b_s, c_s)_s$: $\sum_{i,s} \rho(r_{i,s})$ (4). where $r_{i,s} = a_s x_{i,s} + b_s y_{i,s} + c_s$ is the residual of line. With no other information, finding this minimum is equivalent to finding each line minimum independently. There are some constraints that could be added to estimate a coherent polygon,

---
[2] Note that this operator is not differentiable.

e.g. the polygon should not be null and the intersection point is unique when more than two lines intersect (this is a graph constraint). We want to find a coherent polygon with respect to graph knowledge. A necessary condition that expresses that lines l_i, l_j, l_k intersect on a same point is $\det(l_i, l_j, l_k) = 0$ with $l = (a, b, c)$. This means that if a point is defined as intersection of $n > 2$ lines, $(n-2$ independent constraints).

The main interest of this global minimization (instead of independent estimation of each line) is that we take into account "good" lines versus "bad" lines (partially occluded lines or "short" lines) with respect to constraints (see fig. 2).

Note that the error distribution over the polygon has to be normalized for efficient M-estimation.

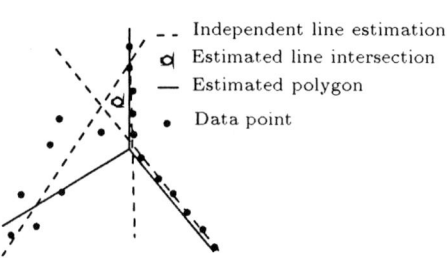

Fig. 2. Stability of intersection computation

2.5 Error analysis

There are many sources of error in the general registration problem. The model-based approach allows to compute the errors w.r.t. the model. There are many errors involving the errors of computation of the transformation: the matching error, the regression error, the projection error, the camera parameters error and the vertex projection error (see figure 3 for a geometric interpretation).

The **matching error** is the error between the image data and the 3D projection. Let $M = \bigcup_{s \in Seg} M_s$ the set of all the sample points computed from section 2.2, and where Seg is the set of all edges of the 3D-model. The error is defined as: $m_x = \|\text{ClosestPoint}(\mathbf{x}) - \mathbf{x}\|, \forall \mathbf{x} \in M$ (5). The matching error gives an idea of the presence of data. It is very sensitive to occlusion and noise.

The **regression error** is the error between image data and polygon estimation. It is defined as: $r_x = d(\text{ClosestPoint}(\mathbf{x}), s), \forall \mathbf{x} \in \bigcup_{s \in Seg} M_s$ (6). where $d(\mathbf{x}, s)$ is the distance of point $\mathbf{x}$ to segment s. This is a pure 2D error. The regression error describes the corelation between the data and the estimated polygon.

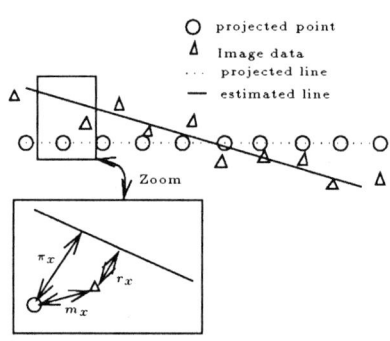

Fig. 3. Matching, regression and projection errors

The **projection error** is the error between the estimated 2D-polygon of the 3D projection of the model: $\pi_x = d(\mathbf{x}, s), \forall \mathbf{x} \in M$ (7). The projection error gives an idea of the quality of the reprojection. It is not sensitive to occlusion or noise since they have been dealt with by polygon regression.

Computing the vertex projection error v_x is more complex and has to integrate the following covariance matrix of camera parameters, polygon corners and line parameters. See [13] for a full explanation.

3 Results and experiments

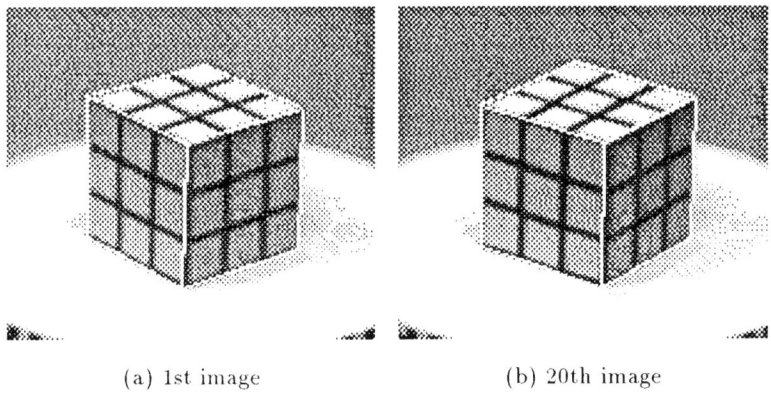

(a) 1st image (b) 20th image

Fig. 4. Tracking of 3D CAD model "Rubik cube" (in wireframe).

First, we test our algorithm on an easy case. The Rubik cube sequence contains twenty images of a Rubik cube in pure rotation with a fixed camera. Rotation is sampled very regularly. For all these reasons, the input data (edge image) is very clean and produces a large stability of view point.

Our algorithm gives very good results even with non robust estimation. Only the first and the last frames of our tracking are displayed on figure 4.

Our second data set is a video stream of 88 images of a complex aerial view of the famous "Arche de la Défense" monument in Paris. This is a real-world application where the object is still and the camera has an unstable trajectory due to the helicopter. This is a complex example of tracking because the input data is very noisy (low resolution, noise on video, missing data) and because the camera motion is not smooth.

Figure 5 displays the tracking result. Our algorithm with full robust estimation is able to track the arch through the whole sequence. The algorithm provides a good reprojection of the 3D model into the video stream. Note the wide variations of point of view (rotation and translation).

We now detail the performance of the algorithm on the complex arch sequence. Table 1 shows typical numerical results of errors in the 3D/2D registration process. We sample each line at one point per pixel. Estimated segments have an average length of 40 pixels (maximum length is 70 pixels and minimum length is 8 pixels). In this sequence, each corner moves over 40 pixels.

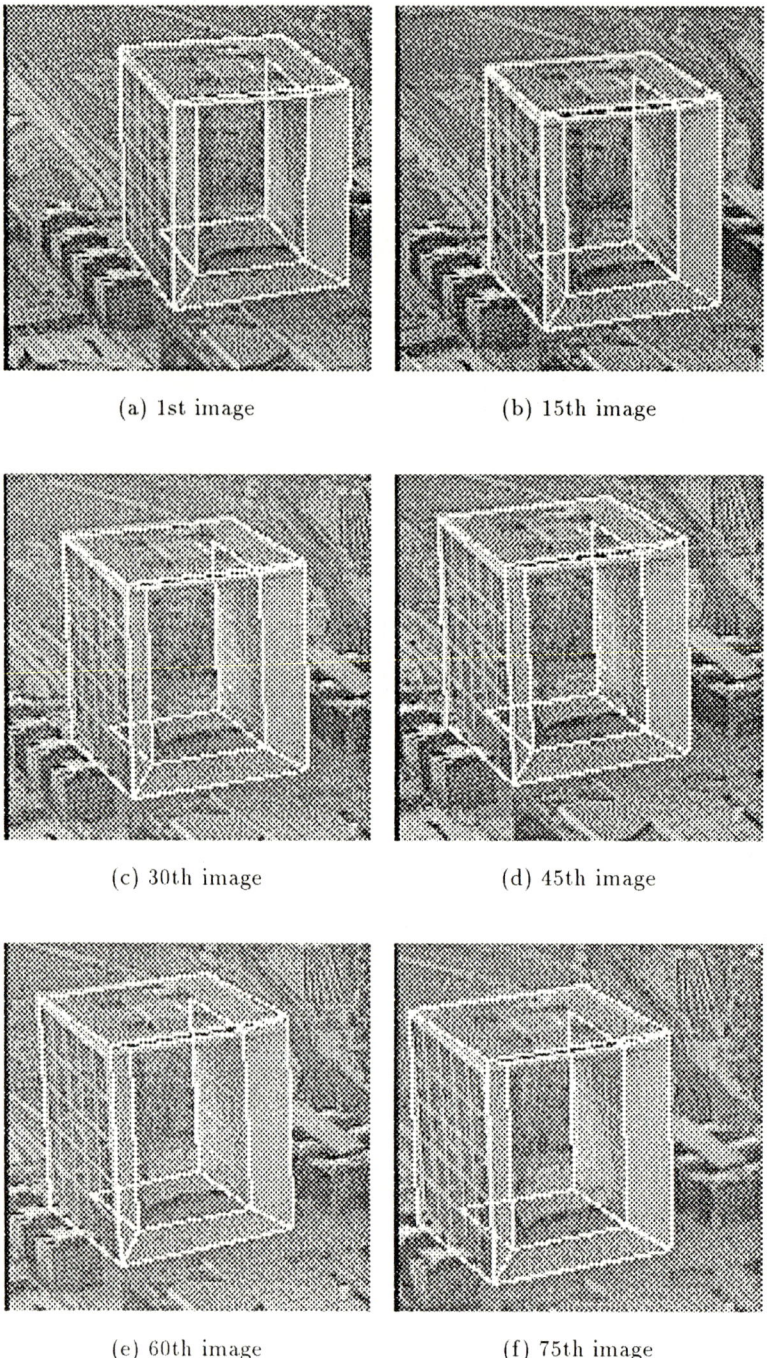

Fig. 5. Tracking of the arch. In wireframe, reprojection of the 3D model. Note the wide pose variations, and the correct estimation of occluded edges of the arch.

error type	v_x	m_x (equation (2.5))	r_x (equation (2.5))	π_x (equation (2.5))
mean error	0.5	1.23	0.63	0.16
deviation	0.2	1.23	0.75	0.14
max error	0.9	8.64	6.66	0.5
min error	0.2	0.05	7.6e-3	2e-2
# of points	14	892	892	892

camera parameter	a_x	a_y	a_z	t_x	t_y	t_z	α_u	α_v	u_0	v_0
value v	153	140	-0.71	-0.18	-0.21	-179	-90	909	536	594
deviation σ	1.2e-3	5.1e-4	3.5e-4	1.6	2.1	4.8	2.8	3.3	0.9	1.3

Table 1. Error analysis of projection and of parameters of camera. $(a_x\ a_y\ a_z) = \mathbf{r} \tan \frac{\theta}{2}$ where $(\mathbf{r}, \theta)$ is the rotation, $(t_x\ t_y\ t_z)$ is the translation, $\alpha_u, \alpha_v, u_0, v_0$ the intrinsic camera parameters (see equation (2)).

The vertex location is obtained with an accuracy of 0.5 pixel (table 1, v_x). This is the main error we wish to minimize, and the obtained precision is satisfactory. The regression error is about 0.6 pixel (table 1, regression error r_x). It gives an idea of a presence of polygon in edge image. It is sensitive to corrupted data. The matching error is about 1.2 pixels (table 1, matching error m_x), and basically reflects missing image data (occlusion etc.). Finally the projection error is 0.16 pixels and reflect the 3D/2D coherence of the polygon. Our robust approach allows to deal with this rather large error. The estimation of camera parameters is very good for all parameters.

We observe that through the video stream, each corner moves more than 40 pixels, and still, our estimation is accurate. On Figure 6 we display the distribution of vertex projection error. Note that it is less than 1 pixel.

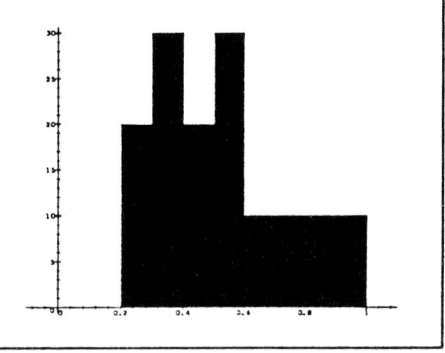

Fig. 6. Distribution of the vertex projection error. Note that the error is always less than 1 pixel.

4 Conclusion and future work

We presented an algorithm for robust and accurate registration and tracking of a 3D-model in video streams. The algorithm uses information about edge location. It is based on an ICP minimization technique. All estimations (2D-extraction, camera parameter computation) are robust. We prove experimentally the robustness of the approach to very noisy data and important occlusion.

The experimental results are very encouraging. We are currently improving the method by performing temporal stabilization, e.g. Kalman filtering, post-computation regularization. The 2D-feature extraction technique can also be extended to more generic 2D-models like ellipses, or any parametric description of 2D-curves.

Finally, the subpixel error obtained by our method should allow a straightforward application to mixing virtual objects into video streams while ensuring 3D coherence.

References

1. P. J. Besl and N. D. McKay. A method for registration of 3-D shapes. *IEEE Transactions on Pattern Analysis and Machine Intelligence*, 14(2):239–256, February 1992.
2. P.J. Huber. *Robust Statistics*. Wiley series in probability and mathematical statistics, 1981.
3. P. Jancène, C. Meilhac, F. Neyret, X. Provot, J.-P. Tarel, J.-M. Vezien, and A. Verroust. RES: computing the interactions between real and virtual objects in video sequences. In *the second IEEE workshop on networked realities*, Boston Mass. (USA), October 26-28 1995.
4. H. Kollnig and H.H. Nagel. 3D pose estimation by fitting image gradients directly to polyhedral models. In *International Conference on Computer Vision*, pages 569–574, 1995.
5. R. Kumar and A. R. Hanson. Robust methods for estimating pose and sensitivity analysis. *Computer Vision, Graphics, and Image Processing: Image Understanding*, 60(3):313–342, November 1994.
6. S. Lavallée and R. Szelisky. Recovering the position and orientation of free-form objects from image contours using 3D distance maps. *IEEE Transactions on Pattern Analysis and Machine Intelligence*, 17(4):378–390, April 1995.
7. R.K. Lenz and R.Y. Tsai. Techniques for calibration of the scale factor and image center for high accuracy 3-D machine vision metrology. *IEEE Transactions on Pattern Analysis and Machine Intelligence*, 10(5):713–720, 1988.
8. D.G. Lowe. Fitting parameterized three-dimensional models to images. *IEEE Transactions on Pattern Analysis and Machine Intelligence*, 13(5):441–450, May 1991.
9. A.D. Worrall, G.D. Sullivan, and K.D. Baker. Pose Refinement of Active Models Using Forces in 3D. In *European Conference on Computer Vision*, pages A:341–350, 1994.
10. P. Wunsh and G. Hirzinger. Registration of CAD-Models to Images by Iterative Inverse Perspective Matching. In *International Conference on Pattern Recognition*, 1996.
11. Z. Zhang. Iterative point matching for registration of free-form curves and surfaces. *International Journal of Computer Vision*, 13(2):119–152, 1994.
12. Z. Zhang. Parameter estimation techniques: A tutorial with application to conic fitting. Technical Report 2676, INRIA, October 1995.
13. Z. Zhang. Determining the epipolar geometry and its uncertainty: A review. *International Journal of Computer Vision*, 1997.

Experiments with a New Area-Based Stereo Algorithm

A. Fusiello[1], V. Roberto[1], and E. Trucco[2]

[1] Machine Vision Laboratory, Dept. of Informatics
University of Udine, Italy
[2] Dept. of Computing and Electrical Engineering
Heriot-Watt University, UK

Abstract. We present a new, efficient stereo algorithm addressing robust disparity estimation in the presence of occlusions. The algorithm uses multiple windows and left-right consistency to compute disparity and its associated uncertainty. We demonstrate and discuss performances with both synthetic and real stereo pairs, and show how our results improve on those of closely related techniques for both robustness and efficiency.

1 Introduction

The aim of computational stereopsis is to reconstruct the 3-D geometry of a scene from two (or more) views, which we call *left* and *right*, taken by pinhole cameras (for a comprehensive review on computational stereo, see [8]). A well-known problem is *correspondence*, i.e., finding which points in the left and right images are projections of the same scene point *(a conjugate pair)*. This is approached as search: finding the element in the right image which is most similar, according to a similarity metric, to a given element in the left image (a point, region, or generic feature).

Area-based (or *correlation-based*) algorithms [1,3] match small image windows centered at a given pixel, assuming that the gray levels are similar. They yield dense depth maps, but fail within occluded areas and poorly textured regions. *Feature-based* algorithms [6,12] match local cues (e.g., edges, lines, corners) and provide robust, but sparse, disparity maps requiring interpolation. These algorithms rely on feature extraction.

Several factors make the correspondence problem difficult: (i) its inherent *ambiguity* requires the introduction of physical and geometrical constraints; (ii) *occlusions*, i.e., points in one image with no corresponding point in the other; (iii) *photometric distortions* [2] arising when conjugate pair pixels have significantly different intensities; and (iv) *figural distortion* [9] that makes the projected shapes different in the two images.

This paper presents a new robust area-based algorithm, addressing all problems (i)-(iv) listed above by exploiting symmetry in matching and multiple windows. For this reason it will be called *Symmetric Multi-Window* (SMW) algorithm in the following. Preliminaries needed to meet some assumptions on image pairs is first illustrated (Sect. 2). The SSD correlation method is presented (Sect. 3), followed by our adaptive, multi-window scheme (Sect. 4), which contrasts distortions and yields accurate disparities. Robust disparity estimates in the presence of occlusions are achieved thanks to the *left-right consistency constraint* (Sect. 5); the associate uncertainty is estimated too (Sect. 6). SMW algorithm implementation is sketched in Sect. 7. Experimental results are presented in Sect.s 8 and 9.

2 Assumptions

Our algorithm for disparity computation assumes that conjugate pairs lie along raster lines. In general this is not true, therefore stereo pairs need to be *rectified* – after appropriate camera calibration – to achieve epipolar lines parallel and horizontal in each image [5].

The SMW algorithm also assumes that the image intensity of a 3D point is the same on the two images. If this is not true, the images must be *normalised*. This is done by a simple algorithm [2] which computes the parameters of the gray-level transformation

$$I_l(x,y) = \alpha I_r(x,y) + \beta \quad \forall (x,y)$$

by fitting a straight line to the plot of the left cumulative histogram versus the right cumulative histogram.

3 Solving Correspondence

Similarity scores are computed, for each pixel in the left image, by comparing a fixed small window centered on the pixel with a window in the right image, shifting along the raster line. As a similarity measure we adopt the Euclidean distance, which is also called SSD (*Sum of Squared Differences*) error:

$$C(x,y,d) = \frac{\sum_{(\xi,\eta)} [I_l(x+\xi, y+\eta) - I_r(x+\xi+d, y+\eta)]^2}{\sqrt{\sum_{(\xi,\eta)} I_l(x+\xi, y+\eta)^2 \sum_{(\xi,\eta)} I_r(x+\xi+d, y+\eta)^2}} \quad (1)$$

where $\xi \in [-n, n]$, $\eta \in [-m, m]$. The computed disparity is the one that minimises the SSD error. *Subpixel precision* is achieved by fitting a curve to the errors in the neighbourhood of the minimum [1].

If one computes SSD by a straightforward implementation of (1), the asymptotic complexity of the resulting algorithm is $O(N^2 nm)$, with N the image size. However, one should observe that squared differences need to be computed only once for each disparity and that the sum over the window should not be recomputed from scratch when it is moved by one pixel. The optimised implementation that follows from this observation [3] has a computational complexity of $O(4N^2)$, which is independent of the window size.

4 Window Shaping

As observed by Kanade and Okutomi [9], when the correlation window covers a region with non-constant disparity, area-based matching is likely to fail, and the error in the depth estimates grows with the window size. Reducing the latter, on the other hand, makes the computed disparity more noise-sensitive. To overcome such difficulties, Kanade and Okutomi proposed a statistically sound, adaptive technique which selects at each pixel the window size that minimises the uncertainty in the disparity estimates.

In this work we take the multiple window approach in the simplified version proposed by [7]. For each pixel we perform the correlation with nine 7×7 different windows (shown in Fig. 1), and retain the disparity with the smallest SSD error value. The idea is that a window yielding a smaller SSD error is more likely to cover a constant depth region; in this way, *the disparity profile itself drives the selection of an appropriate window.*

5 Left-Right Consistency

Occlusions create points that do not belong to any conjugate pairs. In many cases, occlusions occur at depth discontinuities: indeed, one may observe that occlusions on one image correspond to disparity jumps on the other. Although occlusions help the human visual system in detecting object boundaries, in computational stereo they are a major source of errors.

A key observation to address the occlusion problem is that *matching is not a symmetric process*: when searching for conjugate pairs, only the visible points in one image are matched. If the role of left and right images is reversed, new conjugate pairs are found. The so-called *left-right consistency constraint* [4] states that feasible conjugate pairs are those found with both direct and reverse matchings. Consider for instance an occluded point, e.g., B in the left image of Fig. 2: although it has no corresponding point in the right image, the SSD minimisation matches it to some point (C') anyhow. One can see that the latter point, in turn, corresponds to a different point in the left image, but this information is available only by searching from right to left.

In our approach, occlusions are detected by checking the left-right consistency, and suppressing unfeasible matches accordingly. For each point (x, y) on the left image the disparity $d_l(x, y)$ is computed as described in Sect. 2. The process is repeated after reversing the two images, in order to compute $d_r(.,.)$. If $d_l(x, y) = -d_r(x + d_l(x, y), y)$ the point keeps its computed left disparity, otherwise it is marked as occluded and a disparity is assigned heuristically: following [10], we assume that occluded areas, occurring between two planes at different depth, take the disparity of the deeper plane.

6 Uncertainty Estimates

Area-based algorithms are likely to fail not only in occluded regions, but also in poorly textured regions, which make disparity estimates more uncertain. It is

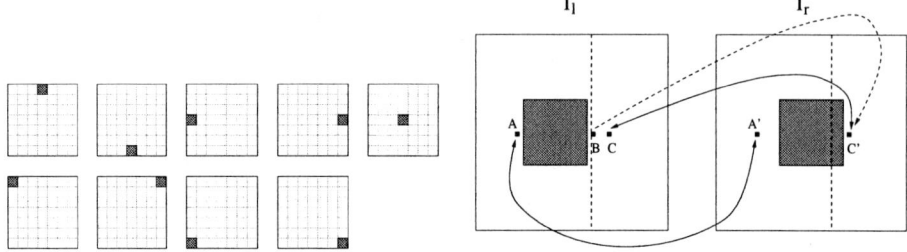

Fig. 1. The nine asymmetric correlation windows. The pixel for which disparity is computed is highlighted.

Fig. 2. Left-right consistency. Point B is given C' as a match, but C' matches $C \neq B$. The pair (B, C') can be suppressed.

therefore essential to compute confidence measures for disparities, which enables one to fill in gaps of the depth maps by fusing multiple views. Several techniques are available to estimate uncertainty, most of them based on the shape of the SSD error function [1,11,13].

In our approach we take advantage of the fact that disparity values computed with different windows are sensitive to the signal-to-noise ratio (SNR): as the latter decreases, the variance of the disparity values increases (see Fig. 8). Hence, we take it as an uncertainty measure for the computed disparity; occluded points are assigned infinite uncertainty.

7 The SMW Algorithm

The SMW algorithm can be implemented in the following steps (disparity is assumed to be positive, that is the right view is right-shifted with respect to the left view):

1. Compute disparity values with SSD correlation from left to right, using the asymmetric windows, and retain the lowest SSD disparity.
2. Compute uncertainty as the variance of disparity values.
3. Do Step 1 by reversing left and right images.
4. Check the left-right consistency and suppress matches accordingly. Unmatched pixels are marked as occluded.
5. Compute subpixel refinement of disparity values.
6. Set to infinite the uncertainty of occluded pixels and fill occluded regions with disparity values from right to left.

To facilitate reimplementations and experiments with the SMW, the C code of the algorithm is available via anonymous ftp at
taras.dimi.uniud.it/pub/code/smw.tar.gz

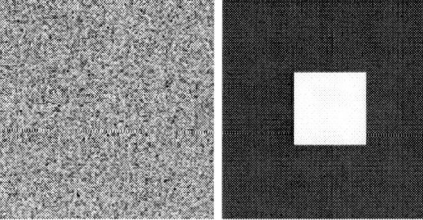  

Fig. 3. Square random-dot stereogram. The left image of the stereogram is shown (left). The right one is computed by warping the latter with a given disparity pattern (right). The square has disparity 10 pixel, the background 3 pixel.

Fig. 4. Computed disparity map by SSD correlation for the square random-dot stereogram in Fig. 5 with 3 × 3 window (left) and 7 × 7 window (right); MAE is 0.240 and 0.144, respectively.

8 Experiments with Synthetic Data

We first performed experiments on uncorrupted random-dot stereograms (Fig. 3), in order to assess the algorithm in a simple, albeit not trivial, case. Disparity maps are gray-level encoded (the brighter the closer). Images have been equalised to improve readability, subpixel-accuracy values have been computed and rounded to integers. The estimated Mean Absolute Error (MAE), that is the mean of the absolute value of differences between computed disparity and ground true disparity, was computed.

Simple SSD correlation applied to random-dot stereograms shows how most of the problems outlined in the previous sections affect the disparity computation. Fig. 4 shows the disparity maps computed with the SSD correlation algorithm, with fixed 3 × 3 and 7 × 7 windows. In both pictures it is visible the effect of disparity jumps (near the left and horizontal borders of the square patch) and occlusions (near the right border of the square patch).

The SMW algorithm was applied to the square random-dot stereograms of Fig. 3 and to a circular random-dot stereogram, not shown here. Fig. 5 and Fig. 6 show the disparity maps computed by SMW and the estimated uncertainty maps (the darker the lower) in both cases. The estimated MAE is negligible and may be ascribed to the subpixel accuracy only. The occluded points, shown in white in the uncertainty maps are recovered with 100% accuracy, in both cases. The circle random-dot stereogram shows that the algorithm is not biased toward square disparity patterns, as it may seem due to the shape of the windows. The reader may compare the present results to those reported in [2].

As for efficiency, running on a SUN SparcStation 4 - 110MHz under SunOS 5.5, the SMW algorithm takes 8 seconds, on the average, to compute the depth maps on 128×128 input images. Although accuracy results are comparable to those of closely related techniques, such as [9], the efficiency of SMW is clearly superior.

Further experiments with noisy random-dot stereograms show a graceful degradation when noise increases. Gaussian noise with zero mean and increasing variance was added independently to both images of the square random-dot stereogram. Fig. 7 shows the MAE vs noise standard deviation for SMW and SSD correlation. Each point depicts the average result of 20 independent trials.

In order to assess the uncertainty map produced by SMW, the average uncertainty computed over a square patch of uniform disparity was plotted against the SNR (Fig. 8). The plot shows that the computed uncertainty consistently increases as the SNR decreases.

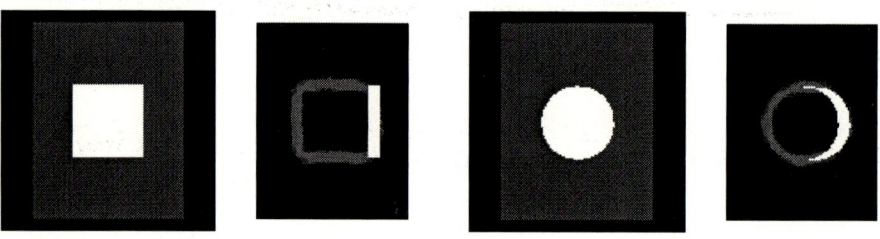

Fig. 5. Computed disparity map (left) by SMW for the square random-dot stereogram and its uncertainty (right). MAE is 0.019.

Fig. 6. Computed disparity map (left) by SMW for the circle random-dot stereogram and its uncertainty (right). MAE is 0.026.

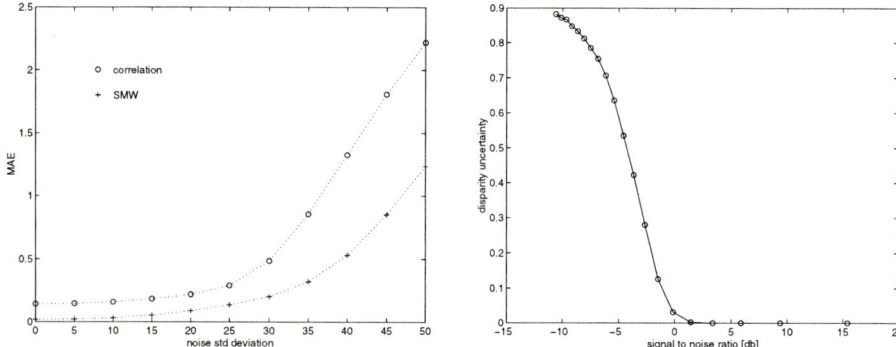

Fig. 7. MAE vs noise standard deviation for the square random-dot stereogram. Window size is 7×7.

Fig. 8. Mean uncertainty vs SNR for a constant disparity region of the square random-dot stereogram.

9 Experiments with Real Data

We performed experiments on standard image pairs from the JISCT (JPL-INRIA-SRI-CMU-TELEOS) stereo test set. Only the "Parking meter"(Fig 9) is reported here for reason of space.

Small values cannot be appreciated in spite of histogram equalisation, due to the large difference between high-uncertainty occlusion points and the rest of the image. Although a quantitative comparison with published results was not possible with real images, the quality of SMW results seems perfectly comparable to that of results reported, for example, in [14,2].

Running on the same hw/sw platform, our current implementation takes 50 seconds, on the average, to compute depth maps from 256×256 pairs, and a disparity range of 10 pixels.

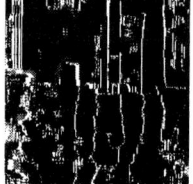

Fig. 9. The "Parking meter" stereo pair; the disparity (left) and uncertainty maps (right).

10 Conclusions

We have introduced SMW, a new, efficient algorithm for stereo reconstruction, based on a multi-window approach, and taking advantage of left-right consistency. Our tests have shown the advantages offered by SMW. The adaptive, multi-window scheme yields robust disparity estimates in the presence of occlusions, and clearly outperforms fixed-window schemes.

The left-right consistency check proves very effective in eliminating false matches and identifying occluded regions (notice that this can be regarded as a segmentation method in itself). In addition, disparity is assigned to occluded points heuristically, thereby achieving reasonable depth maps even in occluded areas. Uncertainty maps are also computed, allowing the use of SMW as a module within more complex data fusion frameworks [13]. Areas of lower SNR are consistently marked with higher uncertainty.

The main disadvantage is that the window size remains a free parameter; we are considering a multi-resolution extension to the SMW algorithm, where correlation is performed with a 3 × 3 window at different resolution levels.

Work is in progress also to embed the SMW module in a dynamic stereo fusion system.

Acknowledgements

This work was partially supported by a British Council-MURST/CRUI grant.

References

1. P. Anandan. A computational framework and an algorithm for the measurement of visual motion. *International Journal of Computer Vision*, 2:283–310, 1989.
2. I. J. Cox, S. Hingorani, B. M. Maggs, and S. B. Rao. A maximum likelihood stereo algorithm. *Computer Vision and Image Understanding*, 63(3):542–567, May 1996.
3. O. Faugeras, B. Hotz, H. Mathieu, T. Viéville, Z. Zhang, P. Fua, E. Théron, L. Moll, G. Berry, J. Vuillemin, P. Bertin, and C. Proy. Real-time correlation-based stereo: algorithm, implementation and applications. Technical Report 2013, Unité de recherche INRIA Sophia-Antipolis, Août 1993.
4. P. Fua. Combining stereo and monocular information to compute dense depth maps that preserve depth discontinuities. In *Proceedings of the International Joint Conference on Artificial Intelligence*, Sydney, Australia, August 1991.
5. A Fusiello, E. Trucco, and A. Verri. Rectification with unconstrained stereo geometry. Submitted to ICCV'98.
6. W.E.L. Grimson. Computational experiments with a feature based stereo algorithm. *IEEE Transactions on Pattern Analysis and Machine Intelligence*, 7(1):17–34, 1985.
7. S. S. Intille and A. F. Bobick. Disparity-space images and large occlusion stereo. In Jan-Olof Eklundh, editor, *European Conference on Computer Vision*, pages 179–186, Stockholm, Sweden, May 1994. Springer-Verlag.
8. M. R. M. Jenkin, A. D. Jepson, and J. K. Tsotsos. Techniques for disparity measurements. *CVGIP: Image Understanding*, 53(1):14–30, 1991.
9. T. Kanade and M. Okutomi. A stereo matching algorithm with an adaptive window: Theory and experiments. *IEEE Transactions on Pattern Analysis and Machine Intelligence*, 16(9):920–932, September 1994.
10. J. J. Little and W. E. Gillett. Direct evidence for occlusions in stereo and motion. *Image and Vision Computing*, 8(4):328–340, 1990.
11. L. Matthies, T. Kanade, and R. Szelisky. Kalman filter based algorithms for estimating depth from image sequences. *International Journal of Computer Vision*, 3:209–236, 1989.
12. Y. Ohta and T. Kanade. Stereo by intra- and inter-scanline search using dynamic programming. *IEEE Transactions on Pattern Analysis and Machine Intelligence*, 7(2):139–154, 1985.
13. E. Trucco, V. Roberto, S. Tinonin, and M. Corbatto. SSD disparity estimation for dynamic stereo. In R. B. Fisher and E. Trucco, editors, *Proceedings of the British Machine Vision Conference*, pages 342–352. BMVA Press, 1996.
14. Y. Yang and A. L. Yuille. Multilevel enhancement and detection of stereo disparity surfaces. *Artificial Intelligence*, 78(1-2):121–145, October 1995.

Adaptive Stereo Matching in Correlation Scale-Space*

Christian Menard and Walter G. Kropatsch

Pattern Recognition and Image Processing Group,
Vienna University of Technology, Treitlstraße 3/183/2,
A-1040 Vienna, Austria, *e-mail: men@prip.tuwien.ac.at*

Abstract. Stereo computes the distance of objects, "their depth", from two images of two cameras using the triangulation principle. Points of imaged objects are mapped in different locations in the two stereo images. A central problem in stereo matching using correlation techniques lies in selecting the size of the search window. Small windows contain only a small number of data points, and thus are very sensitive to noise and therefore result in false matches. Whereas large search windows contain data from two or more different objects or surfaces, thus the estimated disparity is not accurate due to different projective distortions in the left and the right image. The new method introduces a continuous scale parameter for the matching process. It allows the adaption of the scale for every individual region and overcomes the drawbacks of fixed window sizes which is impressively demonstrated by the experimental results.

1 Introduction

In order to reconstruct the *three-dimensional information* of a scene, computer vision uses two different views of the same scene. This technique is known under the term *stereo-vision* or *binocular vision*. It is important for a vision system that interesting structures for certain scales are found by the stereo-vision process. According to Barnard and Fischler [1] a stereo-vision system performs the following steps: *image acquisition, camera calibration, extraction of points of interest, depth determination, and depth interpolation.*

All of these steps play important roles in the design of a stereo system and it is of interest to know how the different tasks depend on each other. Even small mistakes that are made during this processes can result in wrong depth information for the object currently under consideration. Starting from the image acquisition, noise can be introduced for certain reasons into the images, which complicates the modeling of the used cameras and, of course, the search for the corresponding points in a stereo pair. But the success of the approach mostly depends on its ability to solve the problem of stereo matching, a topic upon which this work concentrates.

The paper is organized as follows: In the next section we review basic notation of area-based stereo matching. Section 3 discusses the problem of the window size. A new adaptive matching method is proposed using a correlation scale-space. Experimental results are described in section 4. Finally we give a brief summary.

---

* This work was supported by the Austrian Science Foundation (FWF) under the grant Nr. P09954-SPR and Nr. S7002-MAT.

2 Standard Area-Based Stereo Matching

The epipolar geometry is the starting point for this method. For a given pair of stereo images, the corresponding points are supposed to lie on the epipolar lines [10]. Since a parallel camera alignment is used in this paper, the epipolar lines are the scanlines in both images. In Fig. 1 a synthetic stereo pair is depicted

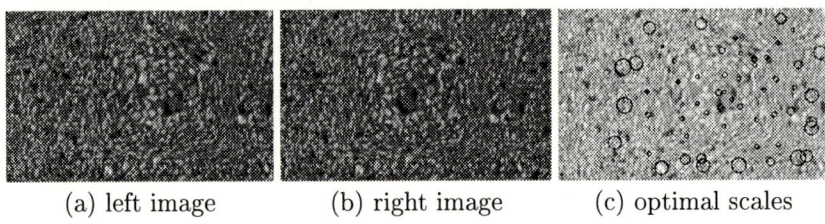

(a) left image   (b) right image   (c) optimal scales

Fig. 1. Synthetic stereo pair: *Pyramid* on a flat ground with natural texture added on the surface. (c) optimal scales for points of interest.

which consists of a pyramid on a plane ground with natural texture added on the surface. If for a given point (x_L, y_L) in the left image, a corresponding point (x_R, y_R) in the right image can be found, the three dimensional position of the object point can be computed with the additional information about the camera parameters. As a similarity measure the correlation of gray-level intensities is used. 1D-correlation $C(x_L, x_R, w)$ between two intervals of size w can be written as convolution with a rectangular function $\delta_{1/w}$ [8, 6].

$$C(x_L, x_R, w) = \frac{[I_L(x_L) I_R(x_R)] * \delta_{1/w} - \mu_L(x_L, w) \mu_R(x_R, w)}{\sqrt{[I_L^2(x_L) * \delta_{1/w} - \mu_L^2(x_L, w)][I_R^2(x_R) * \delta_{1/w} - \mu_R^2(x_R, w)]}}, \quad (1)$$

$$\text{with } \mu(x, w) = I(x) * \delta_{1/w}(x) \text{ , where } \delta_{1/w}(x) = \begin{cases} \frac{1}{w} & |x| \leq \frac{w}{2} \\ 0 & elsewhere \end{cases}. \quad (2)$$

The 2D-correlation $C(x_L, y_L, x_R, y_R, w)$ for two-dimensional regions can easily be extended from equation (1). For each point in the left stereo image the disparity information $D(x_L, y_L)$ is computed using the correlation function C as

$$D(x_L, y_L) = \begin{cases} x_L - x_R & for \; x_R = argmax\{C(x_L, y_L, x_R, y_R, w)\} > T \\ -1 & otherwise \end{cases}, \quad (3)$$

where T defines the minimal threshold for accepting a corresponding point. The fact that the maximum of C is below this threshold T may be caused by occlusion, highlights or depth discontinuities. The maximum of the correlation function is accepted if it is above the threshold T. In this case the position of the distinct maximum defines the corresponding point in I_R.

A central problem in finding correspondences lies in selecting the size of the search window w. Small windows contain only a small number of data points, and thus are very sensitive to noise and therefore result in false matches. Whereas large search windows contain data from two or more different objects or surfaces, thus the estimated disparity is not accurate due to different projective distortions in the left and the right image. The strategy is to find a corresponding point with the smallest window size w. There exist various works dealing with this problem; Levine et al. use an adaptive correlation window, the size of which varies inversely with the variance of the region which is currently considered [4]. In another adaptive approach Kanade and Okutomi proposed that the size and shape of the matching window is chosen adaptively on the basis of local evaluation of the variation in both the intensity and the disparity [3]. In all these works the search window is rectangularly shaped and the size is changed in a discrete way.

In the next section a new method is proposed by changing the size of the window in a continuous way, thus making it possible to determine an optimal size for a given region.

3 Adaptive Stereo Matching

The problem of changing the size of the search window during the matching process depends on the objects which are considered. If a window contains data from two or more objects or surfaces the correlation for that region does not show a clear maximum, unless the window is decreased and contains only data points from one single object. Another problem occurs when a search window contains occluded regions. In this case the computed disparity value is not correct. In order to find an optimal size of the search window the function C has to be modified in such way that the scale can be changed in a continuous way.

3.1 Gaussian Weighted Correlation

The products of intensity values in equation (1) are constantly weighted with the normalizing function $\delta_{1/w}$ (2). The formulation as convolution allows us to substitute $\delta_{1/w}$ by the Gaussian weights. In the following, the scale-space notation[2] is used, where the scale is defined by t. For a 1D-image $I : \mathbb{R} \mapsto [0,1]$ we define the Gaussian weighted local mean $\mu : \mathbb{R} \times \mathbb{R} \mapsto [0,1]$

$$\mu(x;t) = \int_{-\infty}^{\infty} I(x-\xi)g(x+\xi,t)\,d\xi = I(x)*g(x;t)\,, \text{ with } g(x;t) = \frac{1}{t\sqrt{2\pi}}e^{\frac{-x^2}{2t^2}}. \tag{4}$$

The Gaussian weighted standard deviations are defined by

$$\sigma_k^2(x;t) = I_k^2(x) * g(\bullet;t) - \mu_k^2(x;t) \quad k = L, R\,, \tag{5}$$

and the covariance can be written as

$$\sigma_{LR}^2(x_L, x_R; t) = [I_L(x_L)I_R(x_R)] * g(\bullet;t) - \mu_L(x_L;t)\mu_R(x_R;t) \quad k = L, R\,. \tag{6}$$

The correlation can be written as convolution with the Gaussian kernel in the one-dimensional case (4) as follows:

$$C_\Gamma(\bullet;t) = \frac{[I_L(x_L)I_R(x_R)] * g(\bullet;t) - \mu_L(x_L;t)\mu_R(x_R;t)}{\sqrt{[I_L^2(x_L) * g(\bullet;t) - \mu_L^2(x_L;t)][I_R^2(x_R) * g(\bullet;t) - \mu_R^2(x_R;t)]}}. \tag{7}$$

---

[2] The notation $C_\Gamma(\bullet;t)$ stands for $C_\Gamma(x_L, x_R; t)$ and $C_\Gamma(\bullet,\bullet;t)$ for $C_\Gamma(x_L, y_L, x_R, y_R; t)$.

For the two-dimensional case the two-dimensional Gaussian kernel is used and the 2D-weighted correlation function $C_\Gamma(\cdot,\cdot\,;t)$ can be extended from equation (7). Instead of using all the data points in the correlation window equivalently weighted, the window is weighted with the Gaussian kernel. The size of the search window can be controlled by the scale parameter t. The influence of pixels far from the center of the window diminish at a rate controlled by t. Furthermore the shape of the search window has changed from rectangular to circular. The function C_Γ defines a *Correlation Scale-Space (CSS)* for one point $I_L(x,y)$ in the left stereo image. In the *CSS* the similarity value is available at different scales driven by the parameter t of the scale-space kernel. The main advantages of the *CSS* compared to standard methods, such as the hierarchical approach, is that the scale can be changed in a continuous way. Furthermore in this representation all levels of scale are immediately accessible.

3.2 Optimal Scale Selection

In general situations it is not possible to know in advance at what scales interesting structures can be expected to appear. Size variations of image structures in a stereo pair can occur for several reasons:

- objects in the scene have different physical size;
- surface textures contain structures at different scales; and
- scale variations appear due to perspective distortions.

There are many ways to select the best scale for a given problem. A very interesting work in this field was presented by Lindeberg in which he describes the "scale-space primal sketch" [5]. An operator gives maximal output if its size is best tuned to the object. Other approaches study the variation of the information content over scale [2]. For the correspondence establishment it is possible so far to change the scale parameter t in a continuous way using the correlation function C_Γ. But the problem to be solved is to find the "best scale(s)" t_{opt} for certain regions in a stereo pair. Basically, the scale at which a maximum over scales is attained will be assumed to give information about the window size for that region. The maximum over scale for a region defines the optimal scale. In the next step the *CSS* for different placements of a point is analyzed: on a *plane* parallel to the image plane; on a *roofed surface*; near a *depth edge*; and in an *occluded area*. For these situations the correlation values at the corresponding

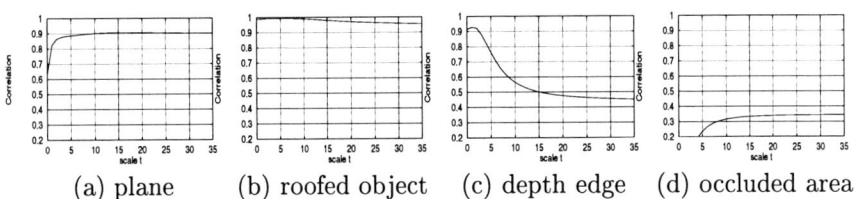

(a) plane (b) roofed object (c) depth edge (d) occluded area

Fig. 2. Optimal scale values for different situations: (a) $C_\Gamma=0.91$ at $t=15.0$, (b) $C_\Gamma=0.99$ at $t=3.0$, (c) $C_\Gamma=0.92$ at $t=1.5$ and (d) $C_\Gamma=0.35$ at $t=35$.

position in the right stereo image are tested by tracing along the local maxima in the direction of the scanline from high to low scale. In Fig. 2 the correlation function at the corresponding point along scale t is visualized for these different situations.

Plane: For the plane lying parallel to the image plane there are only small variations along the correlation function. The correlation value decreases if no or not enough gray-level information is available in the correlation window. The scale-space maximum for this example lies at $t=15$, with $C_\Gamma(\bullet,\bullet\,;t) \in [0.6..0.9]$, which means that for this scale the correlation value can be maximized.

Roofed surface: In the case of a roofed surface, by tracing from large to smaller scale the value of the correlation increases successively until it decreases because of too little information in the correlation window. For this example a small scale obtains the highest correlation value at scale $t=3$, with $C_\Gamma(\bullet,\bullet\,;t) \in [0.95..0.99]$.

Depth Edge: Near a depth edge it is similar. At larger scales the correlation value is low because the window contains data from two objects which have different disparity values. By decreasing the scale parameter t the correlation value is maximized at a very low scale, since the window contains only data from one single object. The obtained scale for this situation lies at $t = 1.5$, with $C_\Gamma(\bullet,\bullet\,;t) \in [0.45..0.93]$.

Occluded area: In the last test the correlation value over scale for an occluded region is determined. It can be seen in Fig. 2 (d) that the maximum of the correlation function along scale is at $t = 35$ with $C_\Gamma(\bullet,\bullet\,;t) \in [-0.1..0.35]$. This represents the highest scale and the correlation value decreases successively to lower scale. But the correlation value is very low anyway, thus the position of this maximum represents an average depth information of the surrounding regions, which contain the occluded area.

In order to establish correspondences, the function C_Γ is tested on the pyramid (Fig. 1). In the left image some points of interest are chosen for which the corresponding points are determined in the right stereo image. Every scale-space maximum is graphically illustrated by a circle centered at the point in the left stereo image for which the correspondence is established. The size of the circle corresponds to the scale of maximum correlation. The circles are superimposed on a bright copy of the left image. The result is visualized in Fig. 1 (c). The optimal scale of points on the surface of the pyramid is small, since the planes form an oblique angle with the image plane, thus not all points in the search window have the same disparity value, whereas for regions on the flat ground the selected scale value is large. One way to determine the optimal scale t_{opt} for finding correspondences is to determine the scale-space maximum by tracing from high to low scale along the correlation maxima. By following the correlation maximum from high to low scale a change in the direction x_R defines a variation in the disparity value. The algorithm can be outlined:

1. Compute the initial disparity value D_0 using the correlation function C_Γ starting at a large scale $t_0=t_{max}$. If no unique maximum can be determined there is no corresponding point.
2. Decrease the scale with $t_{n+1} = t_n/\Delta t$
3. Compute the new disparity value D_{n+1} by using the previously estimated disparity D_n.
4. Iterate steps 2. and 3. until a global maximum along the scale is found or a maximum number of iterations is reached.

4 Experimental Results on Synthetic and Real Images

The adaptive matching algorithm using function C_Γ is applied on two synthetic and real images and is compared to the standard stereo approach.

As synthetic stereo pairs a *pyramid* and a *sphere* on a flat ground are used in each case with texture added onto the surface. The disparity values for these synthetic sets of stereo pairs are known to compare the absolute accuracy of the matching method. For each test the accuracy is compared to the results

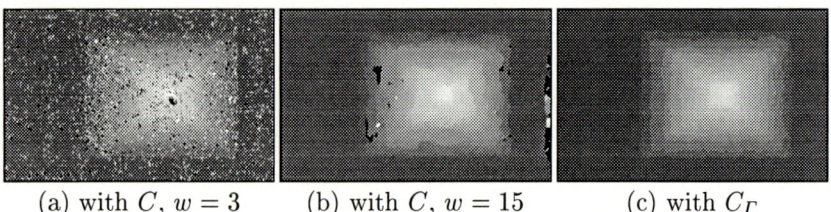

(a) with C, $w = 3$ (b) with C, $w = 15$ (c) with C_Γ

Fig. 3. Pyramid on a plane ground with a natural texture added onto the surface: (a)(b) disparity maps computed with the standard stereo method using fixed window sizes $w = 3$ and $w = 15$ and (c) with the adaptive approach.

computed with the standard matching method using the MSE^3. Fig. 3 shows the results for the pyramid. The true disparity range for the pyramid is 15 to 28 and for the sphere 15 to 22 pixels. These stereo pairs are tested first with the standard stereo method using fixed window sizes $w = 3$, $w = 7$ and $w = 15$ and then with the adaptive matching method using the function C_Γ (Fig. 3). For flat surfaces a large search window obtains good results, whereas on depth edges the disparity values are blurred. The disparity values along depth edges are more accurate using small search windows, but the disparity over the complete image is very noisy. The adaptive approach obtains good results both on depth discontinuities and on flat surfaces. Table 1 illustrates the $MSE(C(w \in \{3, 7, 15\}) - ideal)$ and the $MSE(C_\Gamma - ideal)$. The adaptive matching method has the smallest MSE.

MSE	$MSE(C(w = 3) - ideal)$	$MSE(C(w = 7) - ideal)$	$MSE(C(w = 15) - ideal)$	$MSE(C_\Gamma - ideal)$
Pyramid	280	203	266	50
Sphere	320	280	303	65

Table 1. Difference between true disparity and the computed disparity.

This approach reduces two types of errors,
- large random errors all over the image caused by a small search window and
- systematic errors along depth discontinuities which occur when using large search windows.

Archaeological fragments are used as real objects [9, 7]. The experimental setup consists of two 8Bit-CCD cameras with a resolution of 768×568 pixels and

---

[3] The notation $MSE(< method > (< w >) - ideal)$ is used for the comparison between the ideal and the computed disparity maps. The MSE is only computed for the region which is visible in both stereo images.

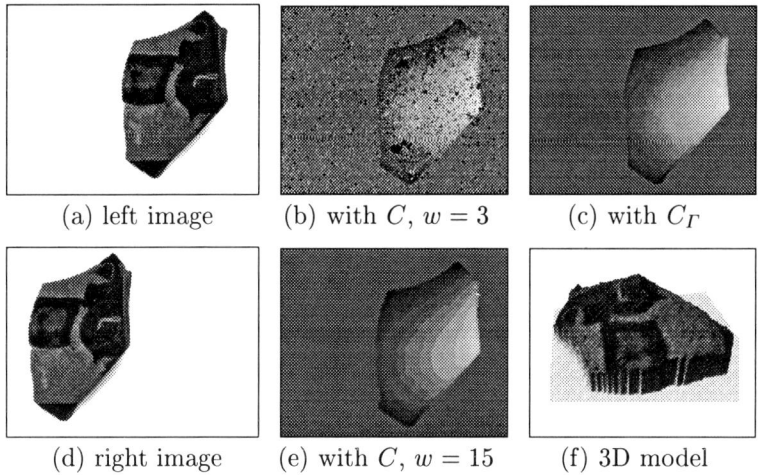

Fig. 4. Archaeological fragment 1: (a) and (d) stereo pair, (b)(e) disparity maps computed with the standard method using fixed window sizes $w = 3$ and $w = 15$ and (c) with the adaptive approach and (f) 3D model of the object.

a 133-Pentium running OS Linux. We tested about twenty different fragments, each of them with different orientation parameters and under varying lighting conditions. The stereo pairs of two different fragments are depicted in Figs. 4 and Figs. 5 (a) and (d).

The curved surface of both fragments is recovered. The objects have a smooth surface without any false matches and the edge of the fragments is also recovered. For comparison the disparity values computed with the standard method using fixed window sizes are shown in Figs. 4 (b),(e) and Figs. 5 (b),(e). A small window size produces several mismatches on the surface of both fragments whereas large search windows smooth the disparity values. Together with both gray-level and disparity information three-dimensional models of the objects with the texture added on the surface can be created(Figs. 4 (f) and 5 (f)).

5 Summary

In this paper a new method detects the optimal scale to determine the corresponding region for each location in a given stereo pair. A correlation scale-space defines the scale in a continuous way. For each region in the stereo pair, depending on the gray-level and disparity information, the size of the search window can be changed adaptively in a continuous way by changing the scale parameter t of the correlation scale-space. Furthermore the shape of the search window has changed from rectangular to circular. Experience demonstrates that tilted regions with varying disparity values need a small scale, whereas flat regions with low gray-level variations favor a larger scale which produces a distinct maximum. The global scale-space maximum for a certain region, which maximizes the correlation value is defined as the optimal size of the search window. The adaptive matching strategy combines the benefits of small and large scales

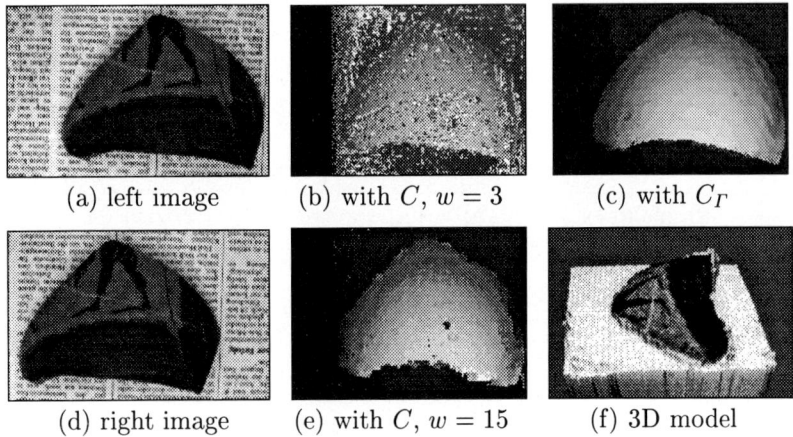

Fig. 5. Archaeological fragment 2: (a) and (d) stereo pair (b)(e) disparity maps computed with the standard method using fixed window sizes $w = 3$ and $w = 15$ and (c) with the adaptive approach and (f) 3D model of the object.

which is demonstrated by experiments with synthetic data and archaeological fragments.

References

1. S.T. Barnard and M.A. Fischler. Computational stereo. In *ACM Computing Surveys*, volume 14, pages 553–572, Dec. 1982.
2. M. Jaegersand. Saliency maps and attention selection in scale and spatial coordinates: An information theoretic approach. In *Proceedings Fifth Intern. Conf. on Computer Vision*, pages 195–202, Cambridge, MA, June 20-23 1995. MIT, IEEE. Catalogue no 95CB35744.
3. T. Kanade and M. Okutomi. A stereo matching algorithm with an adaptive window: Theory and experiment. *PAMI*, 16(9):920–932, September 1994.
4. M.D. Levine, D.A. O'Handley, and G.M. Yagi. Computer determination of depth maps. *Comput. Graphics Image Processing*, 2:131–150, September 1973.
5. L. Lindeberg. *Scale-Space Theory in Computer Vision*. Kluwer Academic Publishers, 1994.
6. C. Menard. *Robust Stereo and Adaptive Matching in Correlation Scale-Space*. PhD thesis, TU Wien, Institut für Automation, PRIP, Wien, 1996.
7. C. Menard and R. Sablatnig. Computer based Acquisition of Archaeological Finds: The First Step towards Automatic Classification. In Hans Kamermans and Kelly Fennema, editors, *Interfacing the Past. Computer Applications and Quantitative Methods in Archaeology. CAA95*, number 28, pages 413–424, Leiden, March 1996. Analecta Praehistorica Leidensia.
8. Azriel Rosenfeld and Avinash C. Kak. *Digital Picture Processing Volume 2*. Academic Press, Inc., 1982.
9. R. Sablatnig and C. Menard. Stereo and Structured Light as Acquisition Methods in the Field of Archaeology. In New York Springer Verlag Berlin, Heidelberg, editor, *Mustererkennung 92 14. DAGM Symposium Dresden*, pages 398–404. Fuchs S., 1992.
10. J. Weng. Camera calibration with distortion models and accuracy evaluation. *IEEE Trans. Patt. Anal. Machine Intell.*, 14:965–980, 1992.

Hierarchical Depth Mapping from Multiple Cameras

Jong-Il Park and Seiki Inoue

ATR Media Integration & Communications Research Labs.
2-2, Hikaridai, Seika-cho, Soraku-gun, Kyoto 619-02, Japan

Abstract. We present a method to estimate a dense and sharp depth map using multiple cameras. A key issue in obtaining sharp depth map is how to overcome the harmful influence of occlusion. Thus, we first propose an occlusion-overcoming strategy which selectively use the depth information from multiple cameras. With a simple sort and discard technique, we resolve the occlusion problem considerably at a slight sacrifice of noise tolerance. Another key issue in area-based stereo matching is the size of matching window. We propose a hierarchical scheme that attempts to acquire a sharp depth map such that edges of the depth map coincide with object boundaries on the one hand, reduce noisy estimates due to insufficient size of matching window on the other hand. We show the hierarchical method can produce a sharp and correct depth map.

1 Introduction

Recently, computer vision technology is widely used for achieving high degree of freedom and efficiency in video content creation. Thus, it is even said to be a kind of media technology [11].

We are developing a video component database in order to realize a flexible and versatile framework for video content creation [5]. It is based on the layered representation of video where a video sequence is regarded as a spatio-temporally ordered set of video components [17]. Video components are stored with various property information such as camera work, key words, depth, and the like in the database. We can freely select some video components from the database and enjoy arranging them in a spatio-temporal domain to make a new video and/or creating new video expressions by exploiting the given property information.

Among the information, one of the most important one would be depth. It is Z value of the camera-centered coordinate of the corresponding object point for each pixel, where Z axis is set to optical axis. Depth information corresponds to the spatial part in the spatio-temporal description of a scene. Thus, it takes a crucial role in making natural-looking videos and/or creating various video expressions with high degree of freedom using the video component database. Virtualized Reality [7], Z-keying for video composition [8], 3D special video effects [13], arbitrary view generation [16], and automatic scene description [14] are typical application using the depth information. Moreover, we can automatically generate multi-layer description of a scene using depth information.

In such application, dense and sharp depth map is strongly required. Here, "sharp" means that object boundary and/or depth discontinuity should be correct. Correctness of depth map in shape is sometimes more important than precision of depth value depending on application. In this paper, how to get such a dense and sharp depth map is the main theme. The proposed method is a hierarchical scheme combined with an occlusion-overcoming disparity estimator using stereo images from 5 cameras. Considering hardware feasibility, we confine the method to a signal-level processing.

2 Related Work

Stereo matching is a useful method in obtaining depth map from image. However, stereo matching faces several problems such as lack of texture, occlusion, photometric change, repetitive pattern, and so on [1][2].

A considerable amount of effort has been exerted to cope with such problems in computer vision [2][10]. Almost methods to obtain dense depth map are computationally expensive or iterative. Among some exception is multiple-baseline stereo matching [12][18]. It demands more cameras but alleviates the problems of lack of texture and repetitive texture without much increase of computational complexity. Recently, a real-time depth mapper has been developed on the basis of multiple-baseline method [8].

However, little attention has been paid to clearing the occlusion problem in stereo matching. In fact, occlusion is one of the main culprits to prevent from obtaining correct depth map in shape. In two-view stereo, occlusion problem is unavoidable and it is impossible to get correct match. Only some appropriate interpolation can fill such area based on some assumption and knowledge [1]. From the standpoint of correct match, multiple view(more than two) can give a clue to resolve the occlusion problem. When an area is occluded in an image from a camera, another camera located at a different position can see the area and give a correct match. Kanade et al.'s depth mapper does not seem to explicitly exploit this property although they mentioned the occlusion problem a little [8]. Recently, Nakamura et al. extensively studied the occlusion problem [9]. Using eye array camera, they analyze occlusion patterns quantitatively and propose a disparity estimation scheme. However, it demands at least 9 cameras and furthermore, it does not provide a strategy for controlling the effect of matching-window size.

A very important issue underlying the area-based matching is the size of matching window [6][10]. It should be large enough to include enough intensity variation for characterizing an area. But it should be small enough to avoid projective distortion. Toward resolving such a dilemma, two approaches have been proposed. One is to use a locally adaptive window [6]. The other is to use hierarchical coarse-to-fine scheme [2][3][4][10].

In this paper, we propose a hierarchical scheme focusing on getting correct depth map around discontinuities. Two kinds of implementation are presented. One is fine-to-fine approach which requires large amount of computation [15].

The other is a coarse-to-fine method which alleviates computational burden considerably. We will explain the details in the following section.

3 Depth Estimation

3.1 Configuration of 5 Camera System

We put a camera at the center and a total of 4 cameras of the same specification to each direction of upper, lower, right, and left, separated by the same distance L. Optical axes of the 5 cameras are parallel and the cameras are synchronized. What we are to acquire is the depth map of the image from the center camera. Other cameras work as sensors in this sense.

In the configuration, the disparity d_t of an object point is given by $d_t = \frac{FL}{Z}$. By estimating the disparity, we can obtain the depth of an object point.

The camera configuration is based on the assumption that when a camera cannot give a correct match for a pixel because of occlusion, another camera located at the other side can give a good one. This holds good for almost occluding cases.

3.2 Occlusion-Overcoming Stereo Match

We use the sum of squared-difference as a matching measure. At a point $\mathbf{x}$ on the image plane of the base camera, the matching measure is calculated at each displacement d for each camera C_i by

$$e_i(\mathbf{x}, d) = \sum_{\mathbf{b} \in W} [I_0(\mathbf{x} + \mathbf{b}) - I_i(\mathbf{x} + \mathbf{b} + \mathbf{d}_i)]^2 \tag{1}$$

where I_i is the intensity and W is a matching window.

A straightforward implementation of multiple-baseline stereo [8] would be $\hat{d}(\mathbf{x}) = arg\min_d \sum_{i=1}^{4} e_i(\mathbf{x}, d)$. It gives a good result if there is no discontinuity of depth. However, when there is a discontinuity of depth near the matching window for a pixel, we cannot expect all of the matching data from the 4 directions gives us useful information. On the contrary, some data, especially from the direction of occluded area, affect the estimation harmfully, which should be eliminated for a good estimation.

If we can eliminate such bad observations during the matching, a considerable improvement can be expected. Thus, we devise a simple sort and discard method based on the observation from e_i curves.

We assume that at least two data sets among the four are not corrupted by occlusion. At each displacement, we sort the difference $e_i(\mathbf{x}, d)$ and discard the largest 2 data among the given 4 data. We sum the two data that are considered as useful for the estimation of disparity. By repeating the collection and summation of data along the epipolar line, we get a 1-D curve for the estimation and consider the displacement which gives the minimum of the curve as the disparity of the pixel. In short, the estimation scheme can be described by

$$\hat{d}(\mathbf{x}) = arg\min_d \sum_{i=1}^{2} \tilde{e}_i(\mathbf{x}, d), \qquad (2)$$

where $\tilde{e}_i$ is the sorted one of e_i such that $\tilde{e}_i \leq \tilde{e}_j$ for all $i < j$.

As we see in the Fig. 1, a considerable improvement is achieved around depth discontinuity by the above scheme at the slight sacrifice of noise tolerance as is expected [12]. The loss of noise tolerance will be compensated for by a hierarchical scheme in the following subsection. In the depth maps, we can observe two kinds of distortion. One is from occlusion. We see many noisy estimates around object boundaries in the left depth map while no such estimates in the right depth map. The other is *boundary overreach*. As Cochran *et al.* pointed out, the more strongly textured surface tends to leak into the less textured region. We see the disparity of higher texture tends to reach over the true edge of disparity and out to lower texture area in both of the depth map. The difference is that the depth map by the proposed occlusion-overcoming strategy shows clear and abrupt change around discontinuity. Moreover, the amount of the boundary overreach is roughly half of the size of the matching window [15]. This observation gives us an idea to implement a novel hierarchical scheme.

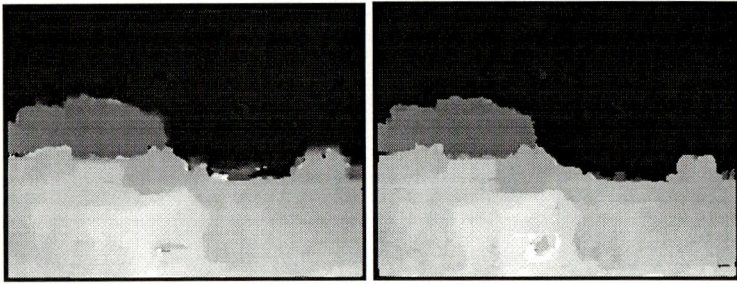

Fig. 1. Depth map by multi-camera matching. Matching window size is 15x15.

Smaller matching window is favorable in the aspect of reducing boundary overreach. But, smaller matching window gives us less reliable results of estimation. There is a trade-off between reliability and geometric correctness of estimation with respect to the size of matching window. Thus, we propose to use hierarchical schemes in order to cope with the trade-off problem as follows.

3.3 Hierarchical Estimation Scheme

The proposed hierarchical method is based on the observation that, when we use the occlusion-overcoming strategy, the correct disparity for boundary overreach area exists near(within half of the size of the matching window) the point in the disparity map in most cases. We first briefly explain the fine-to-fine implementation [15]. Then we present an efficient implementation using a resolution pyramid.

Fine-To-Fine Hierarchical Method Figure 2 illustrates the concept of the fine-to-fine hierarchical method.

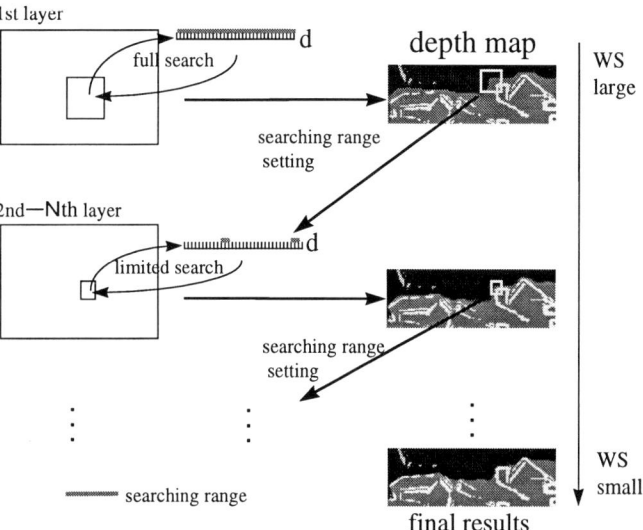

Fig. 2. Illustration of the fine-to-fine hierarchical scheme. Searching range is restricted to the estimates in the matching window of previous(=large matching window) layer except for the 1st layer.

We start by obtaining a disparity map with a large matching window using the occlusion-overcoming strategy. We assume the true disparity value of a pixel exists within the matching window of the pixel in the disparity map of the 1st layer. Now, we reduce the size of the matching window by half. Then, the disparity is estimated similarly except that the searching range is restricted to a set consisting of the disparity values of the upper layer within the window of the upper layer at the position. This restriction of searching range is based on the above observation about boundary overreach. The procedure is repeated until the last layer where the size of matching window is 3x3.

When there is no disparity discontinuity around a point(within the matching window of the point), we don't need to estimate the disparity of the point again in the successive layers. Instead, we just enhance the resolution of the disparity value to sub-pixel accuracy by quadratic fitting and no more update in the successive layers.

Coarse-to-Fine Hierarchical Method Figure 3 shows the concept of the proposed coarse-to-fine hierarchical method.

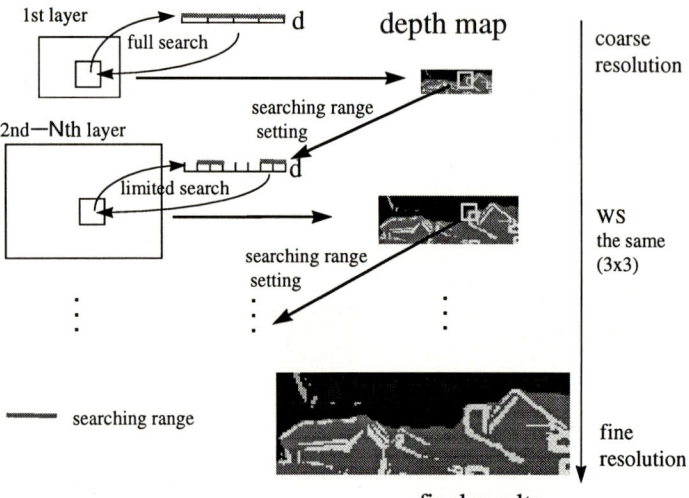

Fig. 3. Illustration of the coarse-to-fine hierarchical scheme. Searching range is restricted to the estimates in the matching window of the previous layer except for the 1st layer.

A resolution pyramid is first constructed by low-pass filtering and decimation. We use a 4-layer pyramid. The size of matching window is set to 3x3[pixels] through the layers. Searching is performed to a pixel accuracy at each layer.

We start from the most coarse layer. Using the occlusion-overcoming strategy, we acquire the disparity map of the first layer. One pixel distance at this layer corresponds to 8 pixels at the finest layer. Thus, the disparity map is coarse in both spatial resolution and accuracy. The amount of computation is thus reduced substantially. Now we move to the 2nd layer. The spatial resolution and the accuracy are twice those of the previous layer. We repeat stereo matching by the occlusion-overcoming strategy. But the searching range is restricted to a set consisting of the estimates in the matching window at the corresponding position of the previous layer and their neighboring disparities(See Fig.3). We repeat the same processing as of the 2nd layer for the successive layers.

In Fig.4, we show the disparity map obtained by the hierarchical method. The fine-to-fine result is obtained by using 3-layer(15x15 to 7x7 to 3x3) hierarchy. We see very clear boundaries in both of the depth maps. The coarse-to-fine result is slightly inferior to the fine-to-fine one, but not substantially.

4 Concluding Remarks

In this paper, we have investigated on obtaining a sharp and dense depth map from multiple cameras. The method is based on a simple discard and summation of disparity information from 5 cameras implemented on a hierarchical structure. It is confirmed that the method achieves shape-correctness of depth map.

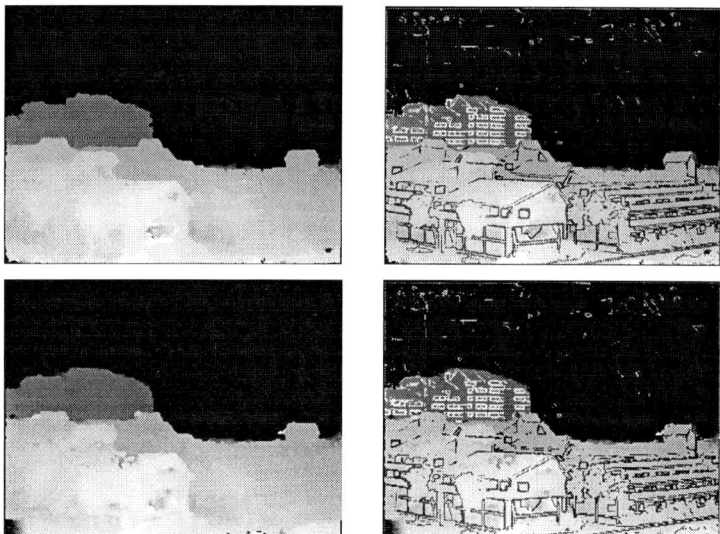

Fig. 4. Depth map obtained by the fine-to-fine [upper] and the coarse-to-fine [lower] hierarchical methods. Edge map is overwritten in the right maps.

We have tested the method over a wide variety of real images under various camera setup and confirmed the validity of the method. It is observed that the performance depends on noise level, scene complexity, and accuracy of camera calibration. The method sometimes fail to give a correct match for strongly concave areas, small holes, and narrow valley as can be reasonably predicted from the configuration of the camera system. The limitation does not seem to be resolved within the scope of non-iterative signal-level processing. Some high-level recognition and/or interactive scheme would be necessary for overcoming this limitation.

Currently, we are developing application techniques using the depth map obtained, which include arbitrary view generation for image-based rendering, Z-key method for 3D video composition, automatic multi-layer description of a scene.

References

1. S.D.Cochran and G.Medioni, "3-D surface description from binocular stereo," *IEEE Trans. PAMI*, vol.14, no.10, pp.981-994, Oct. 1992.
2. U.Dhond and J.Aggarwal, "Structure from stereo: A review," *IEEE Trans. System, Man, and Cybernetics*, vol.19, no.6, pp.1489-1510, Nov./Dec. 1989.
3. W.E.L.Grimson, "A computer implementation of a theory of human stereo vision," *Phil. Trans. Royal Soc. London*, vol.B292, pp.217-253, 1981.

4. M.J.Hannah, "Bootstrap stereo," *Proc. ARPA Image Understanding Workshop*, pp.201-208, College Park, MD, Apr. 1980.
5. S.Inoue, "Mental image expression by media integration - COMICS," *Proc. of 1st International Workshop on New Video Media Technology*, pp.47-52, Seoul, Korea, March 1996.
6. T.Kanade and M.Okutomi, "A stereo matching algorithm with an adaptive window: Theory and experiment," *IEEE Trans. PAMI*, vlo.16, no.9, pp.920-932, Sept. 1994.
7. T.Kanade et al., "Virtualized Reality: Concepts and early results," *Proc. IEEE Workshop on Representation of Visual Scenes*, pp.69-76, June 1995.
8. T.Kanade et al., "A stereo machine for video-rate dense depth mapping and its new applications," *Proc. IEEE CVPR'96*, pp.196-202, San Francisco, June 1996.
9. Y.Nakamura et al., "Occlusion detectable stereo - Occlusion patterns in camera matrix," *Proc. IEEE CVPR'96*, pp.371-378, San Francisco, June 1996.
10. V.S.Nalwa, *A Guided Tour of Computer Vision*, Addison-Wesley, 1993.
11. Y.Ohta, "Computer vision as media technology," *Proc. Image Sensing Symposium*, pp.265-270, 1996 (in Japanese).
12. M.Okutomi and T.Kanade, "A multiple-baseline stereo," *IEEE Trans. PAMI*, vol.15, no.4, pp.353-363, April 1993.
13. J.Park et al., "Extraction of depth information for scene description and its application," *ITE'96*, pp.112-113, Nagoya, Japan, July 1996 (in Japanese).
14. J.Park and S.Inoue, "Image expression based on disparity estimation from multiple cameras," *Proc. 3rd Joint Workshop on Multimedia Communications*, 7-1, Taegu, Korea, Oct. 1996.
15. J.Park and S.Inoue, "Toward occlusion-free depth estimation for video production," *Proc. International Workshop on New Video Media Technology '97*, 6-2, pp.131-136, Tokyo, Japan, Jan. 1996.
16. J.Park and S.Inoue, "New view generation from multi-view image sequence," *Technical Report of IEICE*, Sapporo, Japan, Feb. 1997 (in Japanese).
17. M.Shibata et al., "Scene describing method for video production," *ITEJ Tech. Report*, vol.16, no.10, pp.19-24, Jan. 1992 (in Japanese).
18. R.Tsai, "Multiframe image point matching and 3-D surface reconstruction," *IEEE Trans. PAMI*, vol.5, no.2, pp.159-174, March 1983.

Fast Error-Correcting Graph Isomorphism Based on Model Precompilation

B.T. Messmer and H. Bunke

Institut für Informatik und angewandte Mathematik, Universität Bern,
Neubrückstr. 10, Bern, Switzerland, bunke@iam.unibe.ch

Abstract. In this paper we present a fast algorithm for the computation of error-correcting graph isomorphisms. The new algorithm is an extension of a method for exact subgraph isomorphism detection from an input graph to a set of a priori known model graphs, which was previously developed by the authors.

1 Introduction

In pattern recognition and image analysis, graphs are often used for the representation of structured objects. If the problem is to recognize instances of known objects in an image, then often models of the known objects are represented by means of model graphs and stored in a database. The unknown objects in the input image are extracted by means of suitable preprocessing and segmentation algorithms, and represented by input graphs. Thus the problem of object recognition can be solved by searching for graph or subgraph isomorphisms between the models and the input graph. In a real world application, however, there is usually a certain amount of noise and distortion present in an input graph. Therefore, perfect correspondences between the models and the input do usually not exist. Hence, it is necessary to provide means for error-tolerant matching. In the past, different methods for finding exact and error-tolerant graph and subgraph isomorphisms have been proposed, such as heuristic search[BA83, SH81, Ull76], probabilistic relaxation[KCP92] or simulated annealing[HHVN90]. One of the major problems of error-correcting graph or subgraph isomorphism detection is its exponential time complexity, which is due to the fact that the problem is NP-complete. Combinatorial search methods such as A^* are guaranteed to always find the optimal solution. However, they require exponential time in the worst case. Stochastic optimization methods such as relaxation or simulated annealing, on the other hand, have only polynomial time complexity, but they are not guaranteed to always yield the correct solution. Another problem with graph matching arises if the number of models in the database is large. In case of many models, it may become impossible to sequentially match each model against the input graph.

In this paper, we propose a method which is capable of finding all error-correcting graph isomorphisms between an input and a set of model graphs in time that is only polynomial in the number of vertices of the input graph. In particular, the time complexity of the new method is completely independent of the number of model graphs in the database. The new algorithm is an extension of the method for exact subgraph isomorphism detection that was previously presented by the authors in [MB95].

2 Definitions and Notations

Definition 1: A *labeled graph* G is a 4-tuple, $G = (V, E, \mu, \nu)$, where (1) V is the set of vertices, (2) $E \subseteq V \times V$ is the set of edges, (3) $\mu : V \to L_V$ is a function assigning labels to the vertices, (4) $\nu : E \to L_E$ is a function assigning labels to the edges. □

In this definition, L_V and L_E are finite sets of symbolic labels. Let $G = (V, E, \mu, \nu)$ be a graph with $V = \{v_1, v_2, \ldots, v_n\}$. Then G can also be represented by its adjacency matrix $M = (m_{ij}), i, j = 1, \ldots, n$, where $m_{ii} = \mu(v_i)$ and $m_{ij} = \nu((v_i, v_j))$ for $i \neq j$. Clearly, the matrix M is not unique for a graph G. If M represents G, then any permutation of M is also a valid representation of G.

Definition 2: A binary $n \times n$-matrix $P = (p_{ij})$ is called a *permutation matrix* if the sum of the elements of each row and the sum of the elements of each column is equal to one. □

If a graph G is represented by an $n \times n$ adjacency matrix M and P is an $n \times n$ permutation matrix, then the $n \times n$ matrix

$$M' = PMP^T \qquad (1)$$

where P^T denotes the transpose of P, is also an adjacency matrix of G. If $p_{ij} = 1$ then the j-th vertex in M becomes the i-th vertex in M'.

Definition 3: Let G_1 and G_2 be two graphs and M_1 and M_2 their corresponding adjacency matrices. G_1 and G_2 are *isomorphic* if there exists a permutation matrix P such that

$$M_2 = PM_1P^T \qquad (2)$$

□

Notice that the matrix P can be understood as a bijective function f that maps the vertices of G_1 to G_2, and vice versa. That is, $f(v_j) = v_i$ iff $p_{ij} = 1$. We will call both P and f a *graph isomorphism* between G_1 and G_2. Thus, the problem of finding a graph isomorphism between two graphs G_1 and G_2 is equivalent to finding a permutation matrix P for which Eq.(2) holds true.

In order to integrate the concept of error correction into graph matching, we define a distance measure for graphs which is based on the idea of correcting distortions in an input graph by means of edit operations [WF74, BA83]. The graph edit operations are used to modify either the model or the input graph until there exists a graph isomorphism between the model and the input. In order to model the fact that certain distortions, i. e. edit operations, are more likely than others, each graph edit operations δ is assigned a cost $C(\delta) \geq 0$. The graph distance from a model to an input graph is then defined to be the minimum cost taken over all sequences of edit operations that are necessary for the correction of the distortions in the input graph. In this paper, we consider the following distortions in a graph: vertex and edge label substitution, and missing and extraneous edges. For each type of distortion, a corresponding graph edit operation is defined.

Definition 4: Given a graph $G = (V, E, \mu, \nu)$, a *graph edit operation* δ on G is any of the following:

- $\mu(v) \to l$, $v \in V$, $l \in L_V$: substituting the label $\mu(v)$ of vertex v by l (for the correction of vertex label distortions).
- $\nu(e) \to l'$, $e \in E$, $l' \in L_E$: substituting the label $\nu(e)$ of edge e by l' (for the correction of edge label distortions).
- $e \to \$$, $e \in E$: deleting the edge e from G (for the correction of missing edges).
- $\$ \to e = (v_1, v_2)$, $v_1, v_2 \in V$, $(v_1, v_2) \notin E$: inserting an edge between two existing vertices v_1, v_2 of G (for the correction of extraneous edges). □

Definition 5: Given a graph $G = (V, E, \mu, \nu)$ and an edit operation δ, the *edited graph*, $\delta(G)$, is a graph $\delta(G) = (V_\delta, E_\delta, \mu_\delta, \nu_\delta)$ with

1. $V_\delta = V$
2. $E_\delta = \begin{cases} E \cup \{e\} & \text{if } \delta = (\$ \to e) \\ E - \{e\} & \text{if } \delta = (e \to \$) \\ E & \text{otherwise} \end{cases}$
3. $\mu_\delta(v) = \begin{cases} l & \text{if } \delta = (\mu(v) \to l) \\ \mu(v) & \text{otherwise} \end{cases}$
4. $\nu_\delta(e) = \begin{cases} l' & \text{if } \delta = (\nu(e) \to l') \\ \nu(e) & \text{otherwise} \end{cases}$ □

Definition 6: Given a graph $G = (V, E, \mu, \nu)$ and a sequence of edit operations $\Delta = (\delta_1, \delta_2, \ldots, \delta_k)$, $k \geq 1$, the edited graph, $\Delta(G)$, is a graph $\Delta(G) = \delta_k(\ldots \delta_2(\delta_1(G))) \ldots)$. The total cost of the transformation of G into $\Delta(G)$ is given by $C(\Delta) = \sum_{i=1}^{k} C(\delta_i)$. □

Definition 7: Given two graphs G and G', an *error-correcting (ec) graph isomorphism* from G to G' is a 2-tuple (Δ, P) where Δ is a sequence of edit operations and P is a graph isomorphism from $\Delta(G)$ to G'. The cost of an ec graph isomorphism (Δ, P) is the cost $C(\Delta)$. □

It follows from this definition that, if there is an *ec* graph isomorphism (Δ, P) from a graph G to a graph G', then $G' = PM_{\Delta(G)}P'$ where $M_{\Delta(G)}$ is the adjacency matrix of the graph $\Delta(G)$. It is also easy to see that the permutation matrix P is implied by G, G' and Δ.

Usually, there is more than one sequence Δ of edit operations such that a graph isomorphism from $\Delta(G)$ to G' exists. Consequently, there is usually more than one ec graph isomorphism from G to G'. For our graph distance measure, we are particularly interested in the ec graph isomorphism with minimum cost.

Definition 8: Let G and G' be two graphs. The *graph distance* from G to G', $d(G, G')$, is given by the minimum cost taken over all ec graph isomorphisms from G to G':

$$d(G, G') = MIN_\Delta \{C(\Delta) \mid (\Delta, P) \text{ is an } ec \text{ graph isomorphism from } G \text{ to } G'\}$$

□

The ec subgraph isomorphism (Δ, P) associated with $d(G, G')$ is called the *optimal* error-correcting (*oec*) graph isomorphism from G to G'. In the rest of this paper, we assume that the costs for substituting labels and for inserting and deleting edges are symmetric, i.e., $C(l_1 \to l_2) = C(l_2 \to l_1)$ for all labels l_1, l_2 and $C(e \to \$) = C(\$ \to e)$ for all edges e^1. It is easy to see that this assumption guarantees that $d(G, G') = d(G', G)$ for any pair of graphs, G and G'.

3 Error-correcting Graph Isomorphism by Decision Tree

Given a set of model graphs $G_1, \ldots, G_L$ and an input graph G_I we want to find the *oec* graph isomorphism (Δ^i, P^i) between G_i and G_I such that the cost $C(\Delta^i)$ is minimal over all model graphs, i.e. $C(\Delta^i) = \min\{C(\Delta^j); j = 1, \ldots, L\}$. Traditionally, this problem is solved by applying an A^*-based algorithm to each model-input graph pair [SH81, SF83, Won90, CYS+96]. Any such algorithm has an exponential time complexity in the worst case. Moreover, the method must be applied individually to each model-input pair. Consequently, the time complexity is also linearly dependent on the number of model graphs. In the case of exact graph isomorphism detection, both of these disadvantages can be avoided by using the decision tree approach described in [MB95]. Due to extensive preprocessing of the model graphs, the time complexity of this method is only polynomial in the number of nodes in the input graph. In particular, the time complexity is completely independent of the number and size of models in the database. For all further details, the reader is refered to [MB95].

The basic idea of applying the decision tree approach to error-correcting graph isomorphism detection is to separate the graph isomorphism search from the error-correction process. That is, given two graphs G and G', we propose to generate all distorted copies of G such that the graph distance from each copy to G is not larger than a certain threshold ϑ. Each of the distorted copies of G is then separately matched with the graph G'. Clearly, if the graph distance of G and G' is not larger than ϑ, there exists a distorted copy of G in the generated

---

[1] This assumption is not essential, but it simplifies the description of the proposed method.

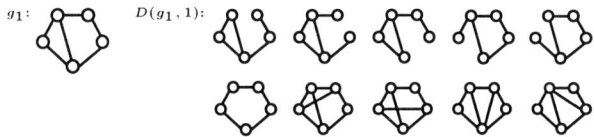

Fig. 1. The graph g_1 and the set $D(g_1, \vartheta)$ for $\vartheta = 1$.

set that is isomorphic to G'. Hence, this copy will be detected by the exact graph isomorphism process. Formally, let

$$D(G, \vartheta) = \{\Delta(G) | \Delta \text{ is a sequence of edit operations with } C(\Delta) \leq \vartheta\} \quad (3)$$

denote the set of all edited copies of G with cost less than or equal to ϑ. Clearly, for any graph G', if $d(G, G') \leq \vartheta$, then $G' \in D(G, \vartheta)$. An example is shown in Fig. 1. A graph g_1 and the set $D(g_1, \vartheta)$ for $\vartheta = 1$ are displayed. As there are no labels the only possible edit operations are the insertion and the deletion of an edge. Each such operation is assigned a cost equal to 1. Hence, $D(g_1, 1)$ consists of exactly ten edited copies of g_1.

After $D(G, \vartheta)$ has been computed, the optimal error-correcting graph isomorphism (if it exists) can be determined by testing each graph $G'' \in D(G, \vartheta)$ with G' for graph isomorphism. There are two possible implementations of this idea. Given a model graph G and an input graph G_I, we can either compute the set $D(G, \vartheta)$ or the set $D(G_I, \vartheta)$. In the first case, $D(G, \vartheta)$ can be computed off-line and at run time it is tested if $G_I \in D(G, \vartheta)$. In the second case, due to the fact that G_I becomes available at run time only, $D(G_I, \vartheta)$ must be computed on-line before the condition $G \in D(G_I, \vartheta)$ can be tested. The first case is described in [MB96] in greater detail, while the second case will be discussed in this paper. In Fig. 2 the on-line error-correcting graph isomorphism algorithm based on a decision tree is outlined.

It can be shown that for L model graphs with n vertices each the size of the decision tree is $O(L3^n)$ [MB96]. For a constant cost of 1 of each edit operation, the size of $D(G_I, \vartheta)$ for a graph G_I with n vertices is bounded by $O(\vartheta n^{2\vartheta})$. Therefore, the time complexity of the on-line ec graph isomorphism algorithm based on a decision tree is bounded by $O(\vartheta n^{2(\vartheta+1)})$. It is important to note that this time complexity is completely independent of the number of model graphs in the database. We conclude that the on-line error-correcting algorithm based on a decision tree is especially efficient if the database of model graphs is large and the maximal degree of distortion to be considered is rather small.

4 Experimental Results

In order to examine the efficiency of the new algorithm in practice, we have performed a number of experiments with randomly generated graphs. Both the new decision tree algorithm and a conventional, A^*-based algorithm were implemented in C++ and run on a SUN Sparc10 Workstation. For each experiment,

ERROR-CORRECTING_GRAPH_ISOMORPHISM(GRAPH G_I, ϑ)

1. generate $D(G_I, \vartheta)$ by applying all edit operations and combinations of edit operations to G_I
2. for each $G'_I \in D(G_I, \vartheta)$
   (a) classify the adjacency matrix of G'_I with the decision tree representing the model graphs $G_1, \ldots, G_L$
   (b) if G'_I is successfully classified by the decision tree as being isomorphic to the graph G_i, then add the error-correcting graph isomorphism between G_I and G_i to the list F.
3. output the error-correcting graph isomorphism with the least cost in F.

Fig. 2. Algorithm *error-correcting_graph_isomorphism*.

we generated one or more model graphs and used these model graphs to create input graphs that were distorted copies of the model graphs. All of the graphs generated for the experiments in this section were undirected and unlabeled. Each experiment was repeated 20 times and the average computation time was recorded. The size of the decision trees in terms of disk space is also given for each experiment. In the experiment described in the following, the performance of the on-line error-correcting algorithm was tested for varying model and database sizes and varying degrees of distortion.

In the 1st experiment, the size of the model graph was increased from 6 vertices and 12 edges to 16 vertices and 32 edges. The error threshold was kept at $\vartheta = 1$. The computation time required by the new and by the conventional algorithm are displayed in Fig. 5. Notice that the decision tree method is much faster than the conventional algorithm. Graphs with up to 16 vertices can be easily handled. In the 2nd experiment, the error threshold was set to $\vartheta = 2$. The size of the model graphs was again increased from 6 vertices and 12 edges to 16 vertices and 32edges. The results of this experiment are displayed in Fig. 6. Note that for graphs with less than 14 vertices, the conventional algorithm is faster than the decision tree approach. For larger graphs, however, the decision tree approach outperforms the conventional algorithm.

In the 3rd and the 4th experiment, we tested the performance of the on-line decision tree approach for a growing number of model graphs. In both experiments, the number of models in the database was gradually increased starting at one and ending at five graphs. Each model graphs consisted of 11 vertices and 22 edges. In the 3rd experiment documented in Fig. 5, the error threshold was kept at $\vartheta = 1$ while in the 4th experiment documented in Fig. 6, the error threshold was set to $\vartheta = 2$. Notice that in both experiments, the time required by the decision tree method was independent of the number of model graphs while the conventional method's performance was linearly dependent on the size of the

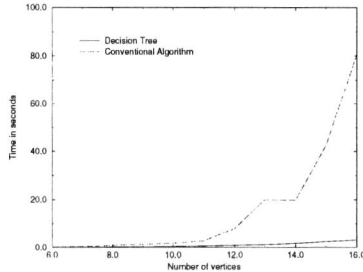

Fig. 3. Computation time in seconds for $\vartheta = 1$ and a growing number of vertices (1st experiment).

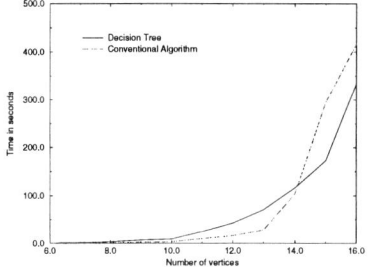

Fig. 4. Computation time in seconds for $\vartheta = 2$ and a growing number of vertices (2nd experiment).

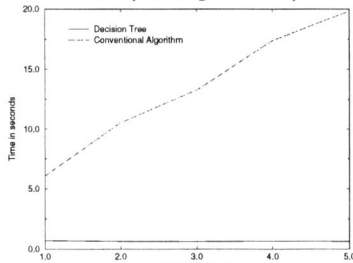

Fig. 5. Computation time in seconds for $\vartheta = 1$ and a growing number of models (3rd experiment).

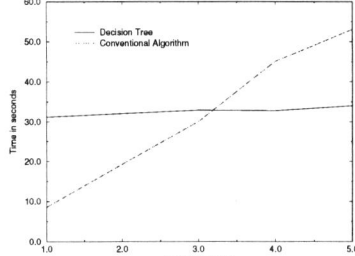

Fig. 6. Computation time in seconds for $\vartheta = 2$ and a growing number of models (4th experiment).

database. We can observe that for $\vartheta = 1$ the decision tree approach is always faster than the conventional algorithm. For $\vartheta = 2$, the decision tree approach is slower than the conventional algorithm when the database is small. But for more than 3 models in the database, the decision tree method becomes superior.

5 Conclusions

We have presented a new algorithm for the problem of error-correcting graph isomorphism detection based on the decision tree paradigm. The new algorithm is an extension of an algorithm for exact graph and subgraph isomorphism detection that was developed by the authors before [MB95]. Its time complexity is quadratic in the size of the model graphs and exponential in the error threshold that is to be considered. Very important is the fact that the method is completely independent of the number of model graphs that are represented by the decision tree. On the other hand, the size of the decision tree grows exponentially with the size of the model graphs for both new algorithms.

The results of the theoretical complexity analysis have been confirmed in a number of practical experiments with randomly generated graphs. The advantages of the new algorithm in terms of computational performance were demonstrated in these experiments for graphs with up to 16 vertices.

Acknowledgment

This work has been part of a project of the Priority Program SPP IF, No: 5003-34285, funded by the Swiss National Science Foundation.

References

[BA83] H. Bunke and G. Allerman. Inexact graph matching for structural pattern recognition. *Pattern Recognition Letters 1*, 4:245–253, 1983.

[CYS+96] L. Cinque, D. Yasuda, L.G. Shapiro, S. Tanimoto, and B. Allen. An improved algorithm for relational distance graph matching. *Pattern Recognition*, 29(2):349–359, 1996.

[HHVN90] L. Herault, R. Horaud, F. Veillon, and J.J. Niez. Symbolic image matching by simulated annealing. In *Proc. British Machine Vision Conference*, pages 319–324. Oxford, 1990.

[KCP92] J. Kittler, W. J. Christmas, and M. Petrou. Probabilistic relaxation for matching of symbolic structures. In H. Bunke, editor, *Advances in Structural and Syntactic Pattern Recognition*, pages 471–480. World Scientific, 1992.

[MB95] B.T. Messmer and H. Bunke. Subgraph isomorphism detection in polynomial time on preprocessed model graphs. In *Proceedings of the Asian Conference on Computer Vision ACCV*, pages 151–155, 1995.

[MB96] B.T. Messmer and H. Bunke. Fast error-correcting graph isomorphism based on model precompilation. Technical Report IAM-96-012, University of Bern, 1996.

[SF83] A. Sanfeliu and K.S. Fu. A distance measure between attributed relational graphs for pattern recognition. *IEEE Transactions on Systems, Man, and Cybernetics*, 13:353–363, 1983.

[SH81] L.G. Shapiro and R.M. Haralick. Structural descriptions and inexact matching. *IEEE Transactions on Pattern Analysis and Machine Intelligence PAMI*, 3:504–519, 1981.

[Ull76] J.R. Ullman. An algorithm for subgraph isomorphism. *Journal of the Association for Computing Machinery*, 23(1):31–42, 1976.

[WF74] R.A. Wagner and M.J. Fischer. The string-to-string correction problem. *Journal of the Association for Computing Machinery*, 21(1):168–173, 1974.

[Won90] E. K. Wong. Three-dimensional object recognition by attributed graphs. In H. Bunke and A. Sanfeliu, editors, *Syntactic and Structural Pattern Recognition- Theory and Applications*, pages 381–414. World Scientific, 1990.

Function-Described Graphs
Applied to 3D Object Representation

Francesc Serratosa
Departament d'Enginyeria Informàtica
Universitat Rovira i Virgili
email fserrato@etse.urv.es

Alberto Sanfeliu
Institut de Robòtica i Informàtica Industrial
Universitat Politècnica de Catalunya
email asanfeliu@iri.upc.es

Abstract

The aim of this work is the characterization of a new structure called Function-Described Graphs (FDG) which can be used to represent objects in computer vision. The FDGs are useful in synthesizing structural information from a set of objects described through their structure. The FDG nodes and arcs are characterized by the probability distribution of the attributes of the ARGs nodes and arcs from where they have been synthesized. The FDG incorporates information of the family of the synthesized ARG and of the antagonistic node and arcs. In this work we apply this new structure to 3D object labeling.

1 Introduction

High level computer vision is used to analyze a scene and carry out subsequent reasoning tasks, such as identifying the objects in a scene. Attributed Relational Graphs (ARGs) [7] have been used in scene analysis to represent complex objects. Examples of ARGs and their use can be seen in the identification of English letters [2], hand written symbols [3], aerial road images [4,9] or 3-D objects [5,6]. When the number of ARGs is high, the number of matching becomes an important issue from the complexity point of view. Random Graphs [2] have been suggested as a solution for this problem, however they can not represent the complete structural information due to their representation weakness.

In this paper, we present a new representation graph scheme called a Function-Described Graph (FDG) which extends the capabilities of Random Graphs by means of adding *control functions* into the nodes and arcs of a classical ARG . Moreover we apply a labeling method to identify nodes of the FDG with nodes of an ARG. The same methodology used for the labeling process can be used for matching and synthesis. In order to reduce the time complexity we use a gradient ascendant labeling technique to find local optimal solutions, different from the probabilistic relaxation [1],[8],[9].

2 The New Structure

The new Function-Described Graphs structure arose from the idea of representing objects for computer vision. The FDG is a general structure, for this reason is applicable in any type of pattern. For instance, voice recognition, 3D objects, hand written characters, etc. An FDG is composed by two main parts. The first one is a relational structure that has the local and structural information of the object. The second part is composed by the *control functions* witch control and generate the structure information. These functions are applied during the labeling, synthesis and matching processes.

2.1 FDG Relational Structure

The relational structure is composed by a set of nodes and a set of arc relations between the nodes. The attributes of an FDG node or arc consists in local properties that come from the node or arc attributes of the synthesized ARG nodes and arcs. Each FDG node attribute or FDG arc attribute value is a probability distribution. We have characterized these probability distribution by means of the average, the standard deviation and the number of elements.

2.2 FDG Control Functions

Control Functions are used in the labeling, synthesis and matching processes of the FDG. We have defined six Control Functions which are the following ones:

Node Compatibility Function (R_ω) and **Arc Compatibility Function** (R_ε): R_ω and R_ε are the compatibility functions that are used in the diverse processes (labeling, ...) to support or no support mappings between node (or arc) of an ARG and an FDG. Take into account that always we compare an ARG attribute with the probability distribution of an FDG attribute. See section "mathematical representation" for additional details. These functions are applied to the attribute values of the ARG nodes and arcs. See section 5 for an application example.

Node Population Function (F_ω) and **Arc Population Function** (F_ε): The Population Functions are used to incorporate a new element in an FDG probability distribution or create new ones. This element is the attribute of an ARG node or arc. Before this process, the compatibility between the elements have been tested with the compatibility functions. Take into account that a probability distribution can be empty.

Node Antagonistic Function (A_ω) and **Arc Antagonistic Function** (A_ε): These functions are used to describe the antagonism between nodes and arcs in an FDG. For example, if two vertices of an object can not be seen in a single perspective view, they are antagonistic. If the vertices are the nodes of an FDG, then both nodes will be antagonistic.

2.3 Mathematical Representation

As it has been said, two types of relational structures are used in this system. The first one is the Attributed Relational Graphs (ARG) that have the information of the features of the input object. The mathematical representation of an Attributed Relational Graphs is as follows:

An ARG is (Σ_v, Σ_e, Δ_v, Δ_e, γ_v, γ_e) where $\Sigma_v = \{v^i\}_{i=1..n}$ is a set of nodes. Assume there are n nodes. $\Sigma_e = \{e^{ij}\}_{i,j=1..n}$ is a set of arcs. $\Delta_v = \{a^i\}_{i=1..n}$ and $\Delta_e = \{b^{ij}\}_{i,j=1..n}$ are the set of attributes of the nodes and the arcs. The attributes of the nodes and of the arcs are $a^i = (a^i_1, a^i_2, ... a^i_t)$ and $b^i = (b^i_1, b^i_2, ... b^i_s)$. Assume t attributes in the nodes and s attributes in the arcs. The domains are $a_k=\{D_{vk}\}_{k=1..t}$ and $b_k=\{D_{ek}\}_{k=1..s}$. Finally, γ_v: $\Sigma_v \rightarrow \Delta_v$ and γ_e: $\Sigma_e \rightarrow \Delta_e$ are two applications that assign attributes to the nodes and to the arcs. $\forall v^i$: $\gamma_v(v^i) = (a^i_1, a^i_2, ... a^i_t)$ and $\forall e^{ij}$: $\gamma_e(e^{ij}) = (b^{ij}_1, b^{ij}_2, ... b^{ij}_s)$

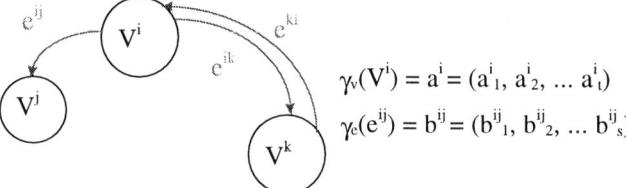

$$\gamma_v(V^i) = a^i = (a^i_1, a^i_2, ... a^i_t)$$
$$\gamma_e(e^{ij}) = b^{ij} = (b^{ij}_1, b^{ij}_2, ... b^{ij}_s)$$

Figure 1. An ARG.

The second relational structure is the Function-Described Graphs (FDG), that contains the synthesis of a set of prototype ARGs. A FDG is composed by $(\Sigma_\omega, \Sigma_\varepsilon, \Delta_\omega, \Delta_\varepsilon, \gamma_\omega, \gamma_\varepsilon, F_\omega, F_\varepsilon, A_\omega, A_\varepsilon, R_\omega, R_\varepsilon)$ where $\Sigma_\omega = \{\omega^\lambda\}_{\lambda=1..m}$ is a set of nodes. Assume there are m nodes. $\Sigma_\varepsilon = \{\varepsilon^{\lambda\lambda'}\}_{\lambda,\lambda'=1..m}$ is a set of arcs. $\Delta_\omega = \{\alpha^\lambda\}_{\lambda=1..m}$ and $\Delta_\varepsilon = \{\beta^{\lambda\lambda'}\}_{\lambda,\lambda'=1..m}$ are the set of attributes of the nodes and the arcs. Attributes in the nodes and in the arcs are composed by $\alpha = (\alpha_1, \alpha_2, ... \alpha_t)$ and $\beta = (\beta_1, \beta_2, ... \beta_s)$. Assume τ attributes in the nodes and σ attributes in the arcs. The domains are $\alpha_k = \{D_{\omega k}\}_{k=1..\tau}$ and $\beta_k=\{D_{\varepsilon k}\}_{k=1..\sigma}$. Finally, γ_ω: $\Sigma_\omega \rightarrow \Delta_\omega$ and γ_ε: $\Sigma_\varepsilon \rightarrow \Delta_\varepsilon$ are two applications that assign attributes to the nodes and to the arcs. $\forall \omega^\lambda$: $\gamma_\omega(\omega^\lambda) = (\alpha^\lambda_1, \alpha^\lambda_2, ... \alpha^\lambda_\tau)$ and $\forall \varepsilon^{\lambda\lambda'}$: $\gamma_\varepsilon(\varepsilon^{\lambda\lambda'}) = (\beta^{\lambda\lambda'}_1, \beta^{\lambda\lambda'}_2, ... \beta^{\lambda\lambda'}_\sigma)$

The second part is composed by six sets of structure control functions. (F_ω, F_ε, A_ω, A_ε, R_ω, R_ε)

Node Antagonistic Function
$A_\omega(\omega^\lambda, \omega^{\lambda'}) = 1$ (Antagonistic) 0 (Non antagonistic)

Arc Antagonistic Function
$A_\varepsilon(\varepsilon^{\lambda\lambda'}, \varepsilon^{\lambda''\lambda'''}) = 1$ (Antagonistic) 0 (Non antagonistic)

Node Population Function:
$F_\omega = \{F_{\omega k}\}_{k=1..\tau}$;
$F_{\omega k}(\{a_k^{i\{p\}}\}_{p=1..|ARG|}, \{x_{\omega k}, \Gamma_{\omega k}, n_{\omega k}\}^\lambda) = \{x_{\omega k}, \Gamma_{\omega k}, n_{\omega k}\}^\lambda_{k=1..\tau, \lambda=1..m; i=1..n}$
$F_{\omega k}: D_{vk}^{\{1\}} \times D_{vk}^{\{2\}} \times ... D_{vk}^{\{p\}} \times D_{\omega k} \rightarrow D_{\omega k \, p=1..|ARG|, k=1..\tau}$

Arc Population Function:
$F_\varepsilon = \{F_{\varepsilon k}\}_{k=1..\sigma}$;
$F_{\varepsilon k}(\{b_k^{ij\{p\}}\}_{p=1..|ARG|}, \{x_{\varepsilon k}, \Gamma_{\varepsilon k}, n_{\varepsilon k}\}^\lambda) = \{x_{\varepsilon k}, \Gamma_{\varepsilon k}, n_{\varepsilon k}\}^\lambda_{k=1..\sigma; \lambda,\lambda'=1..m; i,j=1..n}$
$F_{\varepsilon k}: D_{ek}^{\{1\}} \times D_{ek}^{\{2\}} \times ... D_{ek}^{\{p\}} \times D_{\varepsilon k} \rightarrow D_{\varepsilon k \, p=1..|ARG|, k=1..\sigma}$

Node Compatibility Function:
$R_\omega = \{R_{\omega k}\}_{k=1..\tau}$; $R_{\omega k}(v^i, \omega^\lambda) \in \{0,1\}_{k=1..\tau, \lambda=1..m, i=1..n}$
$R_{\omega k}: \Sigma_v \times \Sigma_\omega \rightarrow \{0,1\}_{k=1..\tau}$

Arc Compatibility Function:
$R_\varepsilon = \{R_{\varepsilon k}\}_{k=1..\tau}$; $R_{\varepsilon k}(e^{ij}, \varepsilon^{\lambda\lambda'}) \in \{0,1\}_{k=1..\tau; \lambda,\lambda'=1..m; i,j=1..n}$
$R_{\varepsilon k}: \Sigma_e \times \Sigma_\varepsilon \rightarrow \{0,1\}_{k=1..\tau}$

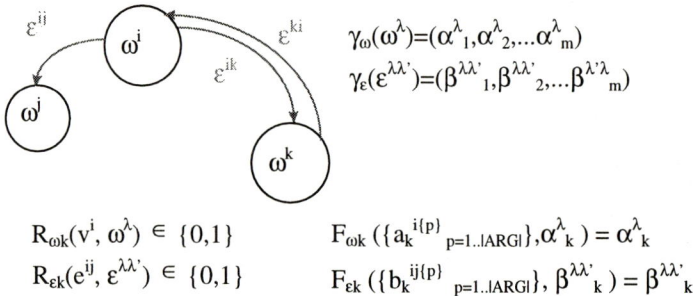

$R_{\omega k}(v^i, \omega^\lambda) \in \{0,1\}$ $F_{\omega k}(\{a_k^{i\{p\}}\}_{p=1..|ARG|}, \alpha^\lambda_k) = \alpha^\lambda_k$

$R_{\epsilon k}(e^{ij}, \epsilon^{\lambda\lambda'}) \in \{0,1\}$ $F_{\epsilon k}(\{b_k^{ij\{p\}}\}_{p=1..|ARG|}, \beta^{\lambda\lambda'}_k) = \beta^{\lambda\lambda'}_k$

Figure 2. An FDG.

3 Application to 3D Object Recognition through perspective views

We have tested our relational structure with 3D objects with plane faces. All the object vertices have three faces. The aim of the application is to recognize them. As the synthesis process has not been studied yet, a data base with some FDGs has been created manually [Figure 3]. We have scenes with some objects and we assume parallel perspective. We have extracted their local features and studied the labeling process behavior. We have also studied the system behavior with different Control Functions and their influence on the representation power of the FDG. Integration of this application in the new structure is explained in the following sections: 1) Extracted features from the objects, 2) Integration of these features in the Attributed Relational Graph, 3) Description of the Function-Described Graph and finally 4) Results.

3.1 Extracted features from the objects

Three features of the object have been extracted of the scenes. As the objects can be partially occluded and it is not possible to have the whole object view on the scene, local features are used. The first feature is the edge type. There are three kinds of edges in a scene: convex, concave and occluded[10]. An occluded edge appears when a face is not visible from the camera point of view. The second feature is the vertex type. There are twelve types of vertices[10] depending on the type vertex junction. And finally, the third feature is the orthographic invariant.

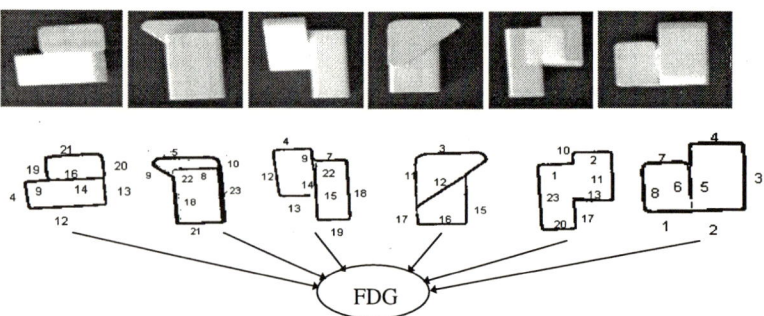

Figure 3. The six pattern views used to create an FDG.

3.2 Attributed Relational Graph

The three aforementioned features have been structured on an Attributed Relational Graph [Figure 4]. The nodes of the Graph represent straight edges whereas the arcs represent relations between them. The nodes have one attribute and the arcs have two attributes. These attributes are:

The attribute of the node v^i is a^i_1 = Edge type. If the feature of the straight segment is concave or convex, then the attribute of the node is the same, whereas if the feature is occluded the node attribute will be convex.

The first attribute of the arcs is b^{ij}_1 = orthographic invariant. Orthographic invariant is applicable when the straight segments are parallels. This attribute will be "not parallels (Ω)" if the straight segments are not parallels.

The second attribute of the arcs is b^{ij}_2 = Type of vertex. Arcs that represent vertices well extracted, their property will be the vertex type. Whereas vertices partially extracted (they have at most two edges), their property will be "not classified (θ)". Also when there is not a vertex, their property will be "not a vertex (Ω)".

Domain of first node attribute: D_{v1} = {Concave, Convex}
Domain of first arc attribute: D_{e1} = R U { Ω }
Domain of second arc attribute: D_{e2} = {{Vertex type} U {Ω, θ}}

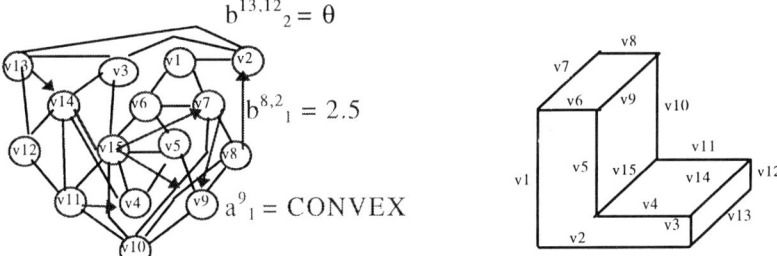

Figure 4. An Attributed Relational Graph and the represented object.

3.3 Function-Described Graph

Each Function-Described Graph has the synthesized information of an object, therefore the information of the nodes or of the arcs between them have to include properties of the object, if possible invariant to the point of view. Since an object can be partially occluded, these properties will usually be local properties. No global features are considered. The relational structure and the control functions of the FDG are as follows. Figure 5 shows 3 objects and their synthesized FDG.

3.3.1 Relational Structure

The nodes of the structure are represented by the probability distributions of the attributes of the ARG nodes and arcs. The domain of the first arc attribute is $D_{\epsilon 1}$ = {$x_{\epsilon 1}$, $\Gamma_{\epsilon 1}$, $n_{\epsilon 1}$} and the second domain is $D_{\epsilon 2}$ = { $x_{\epsilon 2}$, $\Gamma_{\epsilon 2}$, $n_{\epsilon 2}$}. The domain of the node attribute is $D_{\omega 1}$ = {$x_{\omega 1}$, $\Gamma_{\omega 1}$, $n_{\omega 1}$}

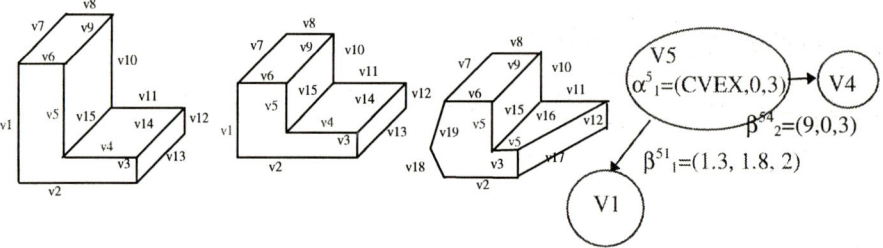

Figure 5. Three objects that belong to a class. Three nodes of the Function-Described Graph that represents their class.

3.3.2 Population Functions
As we have mentioned earlier, we have not studied the synthesis process yet therefore the probability distributions domains are known but their Population Functions have not been studied.

3.3.3 Compatibility Functions
Depending on the probability distributions and the attributes of the ARG nodes and ARG arcs, there are three Compatibility Functions. The Edge Type Compatibility Function ($R_{\omega 1}$) in the nodes, the Orthographic Invariant Compatibility Function ($R_{\epsilon 1}$) on the arcs and the Vertex Type Compatibility Function ($R_{\epsilon 2}$) on the arcs:

$R_{\omega 1}(v^i, \omega^\lambda)$ = 1 if $a^i_1 = x^\lambda_{\omega 1}$ and $n^\lambda_{\omega 1} \neq 0$
= 0 Otherwise

$R_{\epsilon 1}(e^{ij}, \epsilon^{\lambda\lambda'})$ = 1 if Mahalanobis-Dist.(b^{ij}_1, $x^{\lambda\lambda'}_{\omega 1}$, $\Gamma^{\lambda\lambda'}_{\omega 1}$) < Threshold
  & $n^{\lambda\lambda'}_{\omega 1} \neq 0$
= 0 Otherwise

$R_{\epsilon 2}(e^{ij}_2, \epsilon^{\lambda\lambda'}_2)$ = 1 if $b^{ij}_2 = x^{\lambda\lambda'}_{\omega 2}$ and $n^{\lambda\lambda'}_{\omega 2} \neq 0$
= 0 Otherwise

3.4 Results
The process consists on combining different Compatibility Functions. We have used a Goodness Function $A(R_i)$ that gives information about the labeling process. $A(R_i) \in [0,1]$. If $A(R_i)=1$ all the nodes have been correctly labeled and if $A(R_i)=0$ there is no labeled node. We have used an iterative process based on a gradient descendent method to update the probabilities in the labeling process. A class represented by an FDG has been created taking as samples six perpendiculars views of the object of figure 3. The application of Consistent Labeling to the 8 views of figure 6 are shown in table 1 where the columns represents: Column 1: $A(R_i)$ when the Compatibility Function Type of edge is applied. Column 2: $A(R_i)$ when the Compatibility Function Orthographic Invariant is applied. Column 3: $A(R_i)$ when the Compatibility Function Vertex Type is applied. Column 4: The $A(R_i)$ when the three Compatibility Functions are applied. Column 5: The number of edges totally labeled and column 6: The number of nodes of each graph.

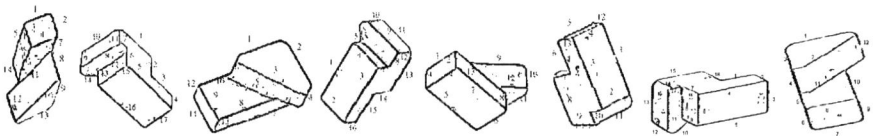

Figure 6. Eight points of view ARG1..ARG8 used to test the FDG.

A summary of the results is the following. When only one Compatibility Function is used in the labeling process, the results are poor. Whereas, when some Compatibility Functions are combined, the results are quite good. However, there are cases that the results are low (ARG 3, ARG5, ARG8) since the individual Compatibility Functions have not been capable to obtain good results. We have not studied the Validation process yet that will improve these results.

In the ARG5 case, although there is not any labeled node, the goodness function is not 0. This is due to the fact that any ARG node is correctly labeled but some labeling hypothesis have been discarded.

	$A(h,R_{\omega 1})$	$A(h,R_{e1})$	$A(h,R_{e2})$	$A(h,R_{\omega 1},R_{e1},R_{e2})$	$L(h,R_{\omega 1},R_{e1},R_{e2})$	$N(R_{\omega 1},R_{e1},R_{e2})$
ARG1	0.01	0.22	0.09	0.75	3	17
ARG2	0.07	0.27	0.52	1.00	17	17
ARG3	0.01	0.07	0.09	0.50	2	13
ARG4	0.08	0.16	0.54	0.89	12	16
ARG5	0.01	0.04	0.06	0.34	0	13
ARG6	0.01	0.37	0.07	1.00	14	14
ARG7	0.07	0.35	0.51	0.98	16	17
ARG8	0.01	0.05	0.09	0.41	1	12

Table 1. Results from eight points of view.

We have considered, also, a view partially occluded [Figure 7]. This means that there are some edges (therefore nodes of the graph) that have disappeared and some edges that their attributes have changed (therefore the node still remains in the graphs but the attribute values have changed). We have studied the results using the three Compatibility Functions. We have to read the results [Table 2], having in mind that the type of vertex and the edge type is not possible to be changed by the occlusion but the length of the edge can be modified.

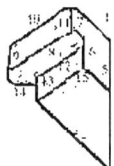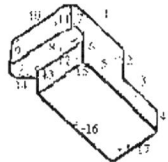

Figure 7. On the left, view partially occluded and on the right the same view without occluding. These views have been structured in the graphs ARG2.1 and ARG2. Five edges have been totally occluded and three edges have been partially occluded

	$A(h,R_{\omega 1})$	$A(h,R_{e1})$	$A(h,R_{e2})$	$A(h,R_{\omega 1},R_{e1},R_{e2})$	$L(h,R_{\omega 1},R_{e1},R_{e2})$	$N(R_{\omega 1},R_{e1},R_{e2})$
ARG2	0.07	0.27	0.52	1.00	17	17
ARG2.1	0.09	0.19	0.52	0.68	5	14

Table 2. Results from the partially occluded view.

4 Conclusions

We have defined a new representation graph scheme called Function-Described Graph which allows to integrate diverse ARG into one unified structure. This new structure incorporates all the information of the ARG including the antagonistic information. We have used the FDG to synthesize projective views of 3D polyhedral objects. Scene objects represented by ARGs have been labeled with objects represented by FDGs assuming parallel perspective. Examples of them can be seen in table 1 and 2.

References

[1] J.T.L.WANG, K.ZHANG, G-W. CHIRN, "The approximate Graph Matching Problem", IEEE pp. 284-288 , 1994.
[2] A.K.C.WONG, J.CONSTANT and M.L.YOU, "Random graphs", (H. BUNKE and A. SANFELIU eds.). "Syntactic and structural pattern recognition: Theory and applications", World Scientific Publishing Co. Pte. Ltd pp 179-195, 1990.
[3] A.SANFELIU and K.FU," A Distance Measure Between Attributed Relational Graphs for Pattern Recognition", IEEE Trans. on Sys. man and cybern. Vol. smc 13 No 3 May/June, 1983.
[4] R.C.WILSON, A.N.EVANS and E.R.HANCOCK, "Relational matching by discrete relaxation", Image and vision computing, Vol. 13, pp 411-421, 1995.
[5] W.Y.KIM and A.C.KAK, "3D object recognition using bipartite matching embedded in discrete relaxation", IEEE Trans. Pattern anal. mach. intell. 13 pp 224-251, 1991.
[6] L.SHAPIRO and R.M.HARALICK, "Matching relational structures using discrete relaxation", (H. BUNKE and A. SANFELIU eds.). "Syntactic and structural pattern recognition: Theory and applications", World Scientific Publishing Co. Pte. Ltd pp 179-195, 1990.
[7] A. SANFELIU, "Matching Methods", 8th Scandinavian conference on image analysis, 1993.
[8] E.R.HANCOCK and J.KITTLER, "Discrete relaxation", Pattern recognition Vol. 23 pp.711-733, 1990.
[9] W.J.CHRISTMAS, J.KITTLER and M.PETROU, "Structural matching in computer vision using probabilistic relaxation", PAMI, Vol. 17, No 8 pp 749-764, 1995.
[10] D.H.BALLARD and C.M.BROWN, "Computer Vision", Prentice Hall 1982.

Cooperative Vision in a Multi-Agent Architecture

Norbert Oswald and Paul Levi

University of Stuttgart,
Institute of Parallel and Distributed High-Performance Systems,
Applied Computer Science - Image Understanding, 70565 Stuttgart, Germany

Abstract. We present the concept of cooperative vision and its application to a multi-agent system with special attention to the integration of vision. Cooperative vision can be described as a type of distributed vision, where several agents working in a shared environment are involved. The object recognition task was distributed to several agents in order to demonstrate the concept of cooperative vision. This enables, on the one hand, a verification of objects by several agents and, on the other hand, a localization of spatial positions of other agents. A Bayesian approach is used for the combination of conclusions of several agents. Experiments done so far show significant results with regard to both tasks.

1 Introduction

Cooperative vision can be described as a type of distributed vision, where several agents working in a shared environment are involved. It can be sensible in a series of vision tasks but it requires arrangements for a local and temporal coordination between agents as well as strategies for the combination of individual conclusions. To achieve cooperative vision in a multi-agent system the design of an agent architecture has to meet certain criteria. One criterion is a modular concept for a multi-agent system that facilitates communication among agents, another is the design of a single component in form of a general frame to enable the integration of sensor perception and action. When vision is integrated into such an architecture, each kind of tight coupling of hardware components with control processes, as usually required in active vision applications, has to be bursted in order to avoid exclusive occupations of shared resources.

There are mainly two sorts of approach with regard to the design of agent architectures: The functional ones as suggested in [1] that support the planning aspects; and the behavior-based ones as in [2] that support reactivity. Attempts to combine both concepts were made by e.g. [6] and [4]. Crowley [4] developed a framework of how to build an active vision system that fulfils both low-level and higher-level tasks. The system aspect as a central role is also emphasized in [5]. There, use and integration of multiple cues and attention is described with an example of figure-ground segmentation. Dynamic attention and selective processing by static and dynamic belief nets is used for control in the VIEWS project [3]. To achieve computational efficiency in performing multiple visual tasks a tight coupling scheme is proposed. An early approach to an integration of multiple sources of information was proposed in the VISIONS project [7]. Previous research in the field of artificial intelligence has been related to bringing together uncertain data of several independent sources with Bayesian belief nets [10]. In [12], Bayesian nets were suggested to provide a general framework in visual tasks as control or decision making.

2 A cooperative agent architecture

In *CoMRoS* (Cooperative Mobile Robots Stuttgart) we use a matrix shaped agent architecture (fig. 1 (a)) as presented in detail in [11] [8]. Its modular design enables planning components, reactive behavior and cooperative task solving in a multi-agent environment.

The architecture is divided into three levels of abstraction, each containing a set of concurrent processes with equal rights. The *strategical* level is responsible for mission planning of agents in a wider range. The *tactical* level exercises contextual planning and execution control. It receives tasks from the strategical level, it plans solution strategies and supervises their execution. The *reflexive* level is composed of processes with a basically reactive behavior, some of them directly coupled to hardware components. Each of these processes has a sensor-actuator coupling at one's disposal. This coupling represents an abstract control circuit that enables reactive behavior. We call these abstract control circuits *autonomy cycles (AC)*. Each *AC* consists of units to *decide*, *plan*, *learn*, and several, parallel operating units to *do* and to *monitor*. Cooperation in the ar-

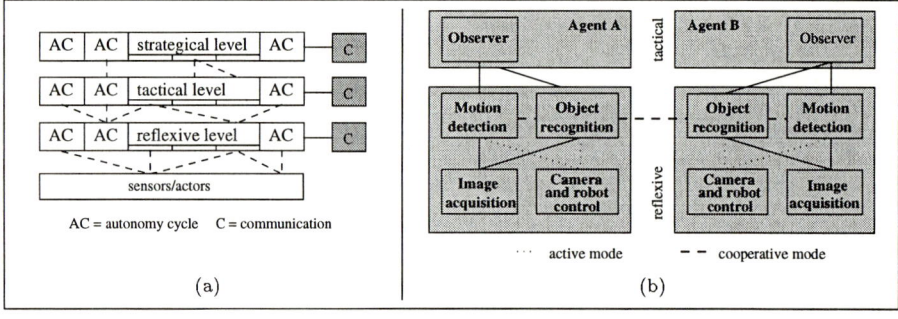

Fig. 1. Agent architecture (a) and integrated vision system with possible connection structures (b)

chitecture occurs, on the one hand, *inside* an agent between processes of the same level that can either be competitive, if accessing the same resources, or cooperative, if connected for data exchange, and thus requires strategies for conflict removal. On the other hand, cooperation occurs *between* multiple agents.

Therefore, corresponding levels of the architecture are linked via dedicated communication modules to facilitate communication and arrangements.

3 Integration of vision

In our architecture each sensor type can be represented through a *system* that consists of several autonomy cycles on the reflexive and tactical level. A minimal vision system is currently embedded into the architecture as shown in fig. 1 (b). It is composed of an autonomy cycle on the tactical level, the *Observer* (AC_{Obs}) that builds a behavior pattern to fulfil a given task, and four autonomy cycles on the reflexive level. The latter independently carry out elementary vision tasks such as motion detection, image recognition, respectively image acquisition, as well as camera and robot control.

Modes of the vision system. The actual connection structure within the behavior pattern between autonomy cycles on the reflexive level is given by the chosen plan of AC_{Obs}. This plan establishes whether autonomy cycles operate *passive* or *active*. Connection structures for autonomy cycles in active mode are shown by dotted lines in fig. 1 (b). While a camera control is not needed in passive mode, it is required in active mode to do well-directed camera movements. Only autonomy cycles that are not directly linked to any physical hardware components are able to operate in passive or active mode. In the vision system, this applies to the *Motion detection AC_{Emo}* and to the *Object recognition AC_{Sor}*.

AC_{Emo} pursues regions of motion in passive mode on the image plane without any camera movements. In active mode, AC_{Emo} supplies a particular moving region to the camera cycle AC_{Cam} for object tracking. In contrary to common active vision applications, the camera is not occupied exclusively. This facilitates a simultaneous usage of the camera or robot by any other autonomy cycle (e.g. AC_{Sor}). AC_{Sor} recognizes objects according to [9] and operates in passive mode like a consumer. It works up all input but does not manipulate; it follows that the recognition method has to cope with arbitrary views. In active mode, AC_{Sor} is able to influence the input data by choosing a particular view position. This action is supported by an appearance-based model database that additionally contains a description of how to move from one aspect to another. To guarantee an image from the new view position, the taking requires a synchronization between camera control and image acquisition.

Types of cooperation. Irrespective of the operation mode of autonomy cycles, the architecture enables two types of cooperation: *internal* and *external*. In the vision system, *internal cooperation* occurs in tasks where autonomy cycles AC_{Emo} and AC_{Sor} are linked as shown in fig. 1 (b) by broken lines. AC_{Emo} is, on the one hand, used to simplify AC_{Sor} by separating moving objects from background, on the other hand, to guarantee a continuous object recognition in image sequences by region identification. To select such a region of interest for

AC_{Sor}, we currently use a selection algorithm as output filter in AC_{Emo}. This attention mechanism always tries to supply corresponding regions except the object recognition has found out that a particular region does not contain the object sought after. Though, as a consequence of internal cooperation, AC_{Sor} can apply strategies to handle hypotheses that result from analysis of succeeding regions of motion in order to obtain more robust conclusions.

External cooperation in the vision system occurs in tasks where AC_{Sor} of several agents are connected. The aim of external cooperation is to solve a task with several agents working in a shared environment. In this way, an agent can exploit potentially better view positions of other agents for processsing. For cooperation, at first communication links between agents have to be built up to establish a work group at first. Therefore, we use a modified contract net protocol [13] between involved agents. The protocol is instantiated by any agent with request of support. Based on the established work group, external cooperation requires mechanisms to combine single evidences of involved agents to receive a common conclusion. A possible mechanism is presented subsequently.

4 Aspects of cooperation

In order to demonstrate the concept of cooperative vision we have distributed the recognition task to several agents. Following, we introduce feasible recognition tasks that appear in external cooperation. Then, a method for combining hypotheses of single runs is presented which can be used in internal or in external cooperation. Finally, we show how a distributed reasoning is done either by single or by continuous image analysis.

Cooperative recognition tasks. In general, we distinguish between two types of cooperative recognition tasks: *verification* and *localization*. The aim of the first task is to verify the object recognition result of a single agent by combining it with recognition results of all agents. In the second task, a suspected object is used to localize the relative spatial positions of other agents involved in the recognition. The basic pre-condition for cooperative identification is that agents observe the same environment. Geometrically this means, that the aperture angles of all cameras build a *spatial section*. To construct such a spatial section, the relative positions of all agents as well as their orientations of camera and robot have to be known respectively suspected. Fig. 2 (a) shows a spatial section between two cameras with a circle within and a square and triangle outside. In the verification task, known positions are assumed. Nevertheless, even in a spatial section agents have to make sure that they observe the same object. Objects like the circle not only have to be in the section but also must have been chosen from the attention mechanism. In the localization task, agents try to locate relative spatial positions of other agents according to a set of hypotheses. All hypotheses from two or more suspected viewpoints are combined by transformation to a *reference viewpoint*. The results are hypotheses about the *orientation angle* between two cameras.

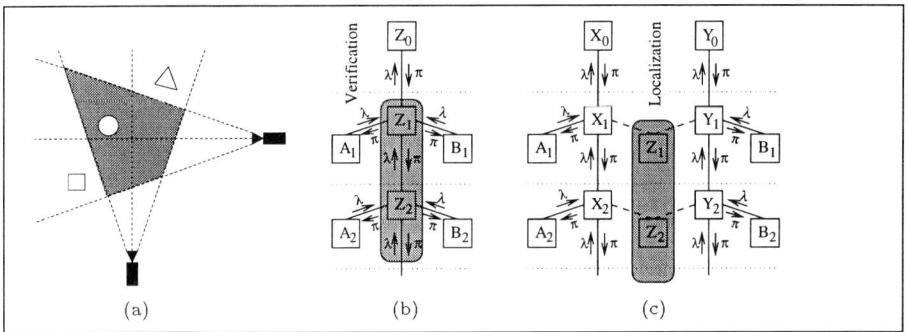

Fig. 2. Objects in 2 view fields (a), belief nets for verification (b) and localization (c)

Continuous recognition. With a continouos temporal analysis we avoid a strong rating of outliers from a single run and obtain assessments about objects and their orientation for each moment. Assume AC_{Sor} operates in active mode and observes a stationary unknown object. If the current maximum hypothesis about an object indicates the real aspect, then this hypothesis can be verified by a defined movement in the model space. Through such a movement that corresponds to a specific camera movement in the world a particular new aspect is expected to be rated strong in the next recognition run. The corresponding former aspect respectively hypothesis is verified, if the recognition algorithm will calculate a high energy value.

In general, a change from one aspect of an object to another occurs with a particular transition probability. Assumed, we have a prior knowledge about the transition probabilities as in the active mode at one's disposal, we can calculate predictions for the next observed aspects. Aspects a_i and transition probabilities $r_{i,j}$ build an *aspect transition matrix* T_{t_{k-1},t_k} for each model at moments t_{k-1} and t_k of the following type:

$t_{k-1} \backslash t_k$	a_1	a_2	a_3	...	a_n
a_1	$r_{1,1}$	$r_{1,2}$	$r_{1,3}$	...	$r_{1,n}$
a_2	$r_{2,1}$	$r_{2,2}$	$r_{2,3}$	...	$r_{2,n}$
...	...	...	...	...	...
a_n	$r_{n,1}$	$r_{n,2}$	$r_{n,3}$	...	$r_{n,n}$

Each row indicates the probability $r_{i,j}$, that aspect a_i at time t_{k-1} will change to a_j at t_k. The sum of all $r_{i,j}$ for each row is supposed to be 1. Like this, at time t_k we expect a new set of hypotheses $h_{t'_k}$ resulting from combination of the sets $h_{t_{k-1}}$ and h_{t_k} with ratings $r_{i,j}$. In the continuous analysis, a different aspect transition matrix can be used at each time step. If we are absolutely certain about an aspect transition like in active mode, we set the corresponding $r_{i,j}$ to 1 and the remaining $r_{i,j}$ to 0. In passive mode, a moving object has to be observed, otherwise there are no aspect transitions. Assuming a homogeneous motion of an unknown object, we know in advance that the new aspect is situated in an interval around the current aspect. Though, we set aspect ratings according to a distribution function, e.g. uniform or gaussian.

Combining several conclusions To combine resulting hypotheses, we use a Bayesian approach of information integration as proposed in [10]. In a cooperation protocol subsequent to the contract net, agents agree to a reference viewpoint to which all hypotheses are transformed and to a minimal parameter setting. In the verification task the known orientation angle between two cameras is used for the transformation of hypotheses. As shown in fig. 2 (b), the combination of results follows, with new evidences entering as diagnostic supports from camera A in A_i and from camera B in B_i. Z_i calculates new belief values for each diagnostic support together with the aspect transition matrices $T^{A_i}_{t_{i-1},t_i}$ or $T^{B_i}_{t_{i-1},t_i}$. The calculated belief values are simultaneously the new causal support π to Z_{i+1}.

In the localization task, the orientation angle between two observers has to be determined. To be more robust, only values from a continuous analysis are considered for combination (fig. 2 (c)). From that we get two belief nets X and Y for camera A and B. One camera is chosen to be the reference viewpoint with the *reference set* of hypotheses. Hypotheses of the other camera are transformed by any orientation angle. Then the belief net combines both sets as in the verification task and calculates belief values for Z_i which denote aspects. The combination with the reference set is done for varying orientation angles. The number of possible orientation angles depends on the current model set of aspects. As a result from that, the maximum belief value represents an estimation about the suspected orientation angle.

Experimental results To present (the first) experimental results for both cooperative recognition tasks, we observed an unknown moving object with two agents operating in passive mode. Fig. 3 (a) shows a part of that image sequence where the object was mutually observed by the two agents. The orientation angle between the cameras of agent A and B was approximately 120°. The used model set contained 24 aspects of our robot in 15° steps from 0° to 360° taken from a fixed distance. All aspect transition matrices were filled according to a uniform distribution in the interval $[i-1, i+1]$ around an aspect a_i.

In the *verification task*, object analysis and the combination of resulting hypotheses was performed in a continuous belief net according to fig. 2 (b). The camera position of agent B was chosen as a reference viewpoint. In fig. 3 (b), the horizontal axis shows time steps t_i of the image sequence, the vertical axis describes the suspected aspect in degree. We see a comparison between continuous and cooperative analysis in relation to the approximated real object aspect over a time interval. Each graph shows aspects corresponding to the calculated maximum hypothesis for each time step. Of course, all hypotheses depend on results of single runs which, in turn, depend on the quality of the segmentation. Observer B has a more homogeneous background than observer A and distortion that influences the object recognition is mostly compensated.

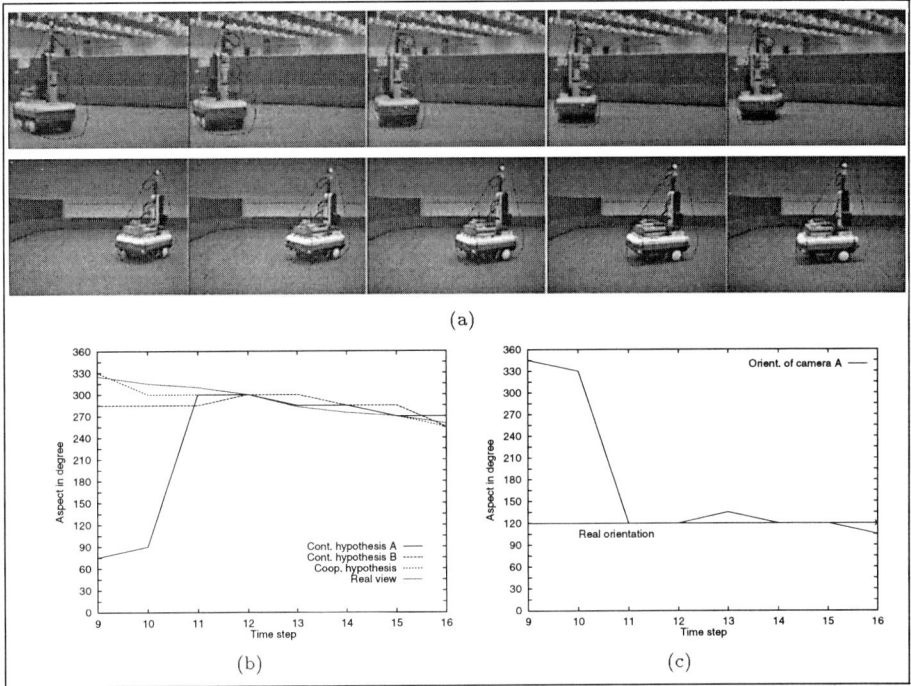

Fig. 3. Image sequence of a scene observed from two view points (a), calculated aspects of an object observed by two agents and projected to the reference viewpoint of camera B for each time step Δt (b), estimated orientation of camera A as a result of object analysis from the reference viewpoint of camera B (c).

Thus, its propositions from single runs about the suspected object aspect is approximately correct. In contrast to that, observer A has to cope with changes in illumination because the extracted region of interest from AC_{Emo} contains neon strips. Thus predictions vary, although the convex hull has correctly been cut. From that, camera A needs three time steps to adapt the real object aspect. A combination of recognition results from camera A and B leads to a quick and good approximation of the real object aspect. That means, that agent A gets fairly accurate hypotheses at each time step.

In the *localization task*, agents combine both sets of hypotheses continuously to calculate for each time step an estimation about the current orientation angle (fig. 2 (c)). The angle is determined by calculating the maximum hypothesis for each possible orientation related to viewpoint B. The combination (fig. 3 (c)) supplies a fairly good evidence for the actual orientation angle. The accuracy of any result depends on the aspect difference in the model set. If segmentation supplies poor input, the recognition results will not be very accurate. But as it was demonstrated above with agent A, if at least one agent is able to calculate correct recognition results, other agents that deal with poor input will profit.

5 Conclusion

We presented the concept of cooperative vision applied to a multi-agent system with an example of a distributed recognition task. In the proposed architecture, a vision system is integrated as a system of loosely coupled autonomy cycles which facilitates interlaces with autonomy cycles that belong to other sensor types. The architecture makes possible two types of cooperation, on the one hand, internal cooperation that appears inside an agent, on the other hand, external cooperation that appears between several agents. To combine calculated conclusions of several agents to single evidences we used belief nets. This mechanism is applied to the verification and localization task, where objects are verified and combined by several agents in order to get more robust recognition result. Results show that agents can profit through a common recognition, especially when the quality of its own input is poor. The localization task facilitates the determination of local spatial positions of other agents based on individual recognition results. Experiments show a fairly good approximation of the actual relative orientation.

References

1. Albus, J.S., McCain, H.G., Lumia, R.: NASA/ NBS Standard Reference Model for Telerobot Control System Architecture. Technical Note **1235** (1987)
2. Brooks, R.A.: A Layered Control System for a Mobile Robot. 3rd Symposium. MIT Press (1986) 367–372
3. Buxton, H., Gong, S.: Advanced Visual Surveillancce using Bayesian Networks. AI-Journal (1997) (to appear)
4. Crowley, J. et al.: Integration and Control of Reactive Visual Processes. Lecture Notes in Computer Science **801** (1994) 47–58
5. Eklundh, J.-O., Nordlund, P., Uhlin, T.: Issues in active vision: attention and cue integration/ selection. BMVC (1996)
6. Fleury S., Herrb M., Chatila R.: Design of a Modular Architecture for Autonomous Robot. IEEE Robotics and Automation (1994) 3508–3513
7. Hanson, A.R., Riseman, E.M.: The VISIONS Image Understanding System – 1986. COINS Technical Report 86–92 (1986)
8. Levi, P.: Architectures of individual and distributed autonomous agents. IAS-2 (1989) 315–324
9. Oswald, N., Gerl, S., Biedert, R.: Konfigurationsbasiertes Verfahren zur schnellen Identifikation komplexer Objekte. DAGM (1996) 187–195
10. Pearl, J.: Distributed Revision of Composite Beliefs. Artificial Intelligence **33** (1987) 173–215
11. Rausch, A., Oswald, N.: Cooperative Crossing of traffic intersections in a distributed robot system. SPIE **2589** (1995) 218–229
12. Rimey, R.D.: Control of Selective Perception Using Bayes Nets and Decision Theory. University of Rochester. Technical Report **468** (1993)
13. Smith, R.G.: The Contract Net Protocol: High-Level Communication and Control in a Distributed Problem Solver. IEEE Transactions on Computers **C-29** No. 12 (1980) 1104–1113

Author Index

Abbasi S. II-140
Abbate M.G. II-584
Abbott P. I-239,II-672
Achten E. II-688
Adams R. I-620
Aggarwal J.K. I-343
Aiazzi B. I-87
Albanesi M.G. II-276
Albert J. I-454
Aldon M.-J. I-103
Almeida L.B. II-254
Alparone L. I-87
Amin A. II-616
Andrade T. II-117
Andrews D. II-445
Androutsos D. I-119
Androutsos P. I-119
Anelli G. I-543
Anh V. I-287
Arica N. II-608
Armstrong W.W. II-332
Arrebola F. I-422,II-477
Artigas J.M. I-454
Attolico G. II-584
Aubert D. II-757
Austin J. II-765
Bae S.C. I-271
Badenas J. I-502
Baglietto P. II-46
Ballou B.T. II-663
Barcucci E. I-166
Bariani M. I-535
Barni M. I-446
Baronti S. I-14,I-87
Bartolini F. I-446
Basman A. I-223
Batlle Grabulosa J. I-79
Beil W. II-380
Bello F. II-428
Bellon O.R.P. I-279
Berman L.H. II-412
Biancardi A. I-142,II-109
Bigun J. II-727
Bloch I. I-30
Blosseville J.-M. II-493
Bober M. I-502,I-653
Boland F. I-377
Boldrin E. I-335
Bolon P. I-588
Bonifazzi C. II-436
Borgefors G. I-369
Bottoni P. I-430
Bouchafa S. II-757
Bozzano R. II-576
Branca A. II-584
Braun M. I-239,II-672
Broggi A. I-543
Brown M.K. II-647
Bruckstein A.M. II-30
Bruyelle J.L. II-757
Bunke H. I-693
Burge M. II-316
Burger W. II-316
Burkhardt H. I-636
Buzug T.M. II-380
Cabello D. II-364
Cabestaing F. II-757
Cafforio C. I-134
Camacho P. I-422,II-477
Cappellini V. I-87,I-446
Carbonaro A. I-247
Çarkacioğlu A. I-127
Carlà R. I-87
Carpentieri B. II-54
Carstensen J.M. II-532
Casini A. I-14
Castagno R. II-735
Cepornyuk S.V. II-246
Chao J. II-284
Chaudhuri S. II-356
Chaudron M. I-582
Chen L.-H. II-469
Cheng C.-M. II-164
Chetverikov D. I-95,II-781
Chianese A. II-600
Chmielewski L. II-781
Cho S.I. II-262
Choi S. II-196

Cinque L.	I-430,II-188
Cipolla R.	I-223,I-414
Ciuc M.	I-588
Cohen M.	I-385
Colchester A.C.F.	II-428
Colombo C.	II-204
Coquin D.	I-588
Cordella L.P.	I-46
Cross A.D.J.	I-406
Crowley J.L.	II-340
Cucchiara R.	I-535
Cucurachi G.	II-228
Cudny W.	II-781
Cuisenaire O.	I-263
Dalton K.J.	II-412
Dambra C.	II-781
Dance C.R.	II-412
Daoudi M.	II-220
De Floriani L.	II-308
De Santo M.	II-600
Del Bimbo A.	II-180
Del Lungo A.	I-166
Del Ninno E.	II-632
Dellepiane S.G.	I-255,II-560
Denteneer D.	II-22,II-148
Deparis J.P.	II-757
Desachy J.	I-174,I-198
Destri G.	I-543
Di Lecce V.	II-404
Di Ruberto C.	I-214
Di Sciascio E.	I-134
Di Stefano L.	I-377
Dimauro G.	II-592
Distante A.	II-93,II-584
Distasi R.	II-101
Djeziri S.	I-510
Dobashi T.	II-697
Dornaika F.	I-478
Du C.	II-663
Duchamp T.	I-385
Duffy L.	II-639
Durou J.-D.	I-519
Duta N.	I-398
Eguchi K.	II-420
Emptoz H.	II-639
Fairhurst M.C.	II-624
Falcone M.	I-596
Fan K.-C.	I-22
Farkas D.L.	II-663
Ferraro M.	I-361
Ferretti M.	II-77
Ferri F.J.	I-206
Ferri F.	I-454
Fisher G.W.	II-663
Foggia P.	I-46
Foresti G.L.	II-749
Fusiello A.	I-669
Gadiou J.	I-198
Garcia C.	I-478
García-Silvente M.	II-372
Garibotto G.	II-705
Garrido A.	II-372
Gertsiy A.A.	II-246
Genovesi I.	II-204
Gevers T.	I-319
Ghassemian Yazdi M.H.	II-544
Ghorbel F.	II-220
Giacinto G.	I-38,II-765
Godsill S.J.	II-719
Gökmen M.	I-303
Gonzalez C.	II-445
Gorodnichy D.O.	II-238,II-332
Gosling J.P.M.	II-412
Grau A.	I-70
Gregori M.	II-509
Grosso E.	II-727
Grudin M.A.	II-246
Guaragnella C.	I-134
Guerriero A.	II-404
Guillaume S.	II-14
Hahn F.	II-517
Haindl M.	I-295
Han C.-C.	II-469
Hancock E.R.	I-150,I-406,II-172
Hartmann I.	I-620
Harvey D.M.	II-246
He W.X.	II-132
Hedley M.	I-470
Heng T.C.H.	II-568
Heras J.	II-364

Hermida X.F.	II-552	Kropatsch W.G.	I-677
Herodotou N.	I-494	Kunishi T.	II-524
Holt R.J.	II-30	Kuno Y.	II-568
Hong K.-S.	II-196	Kuramoto K.	II-524
Hoppe H.	I-385	Kurt B.	I-303
Hsieh J.-W.	I-22	Kutaev Y.F.	II-246
Hu J.	II-647	Kweon I.-S.	I-54,I-271
Huang T.S.	I-582	Lalonde M.	I-111
Huijsmans N.	I-582	Lam K.-M.	I-559
Huijsmans D.P.	II-22,II-148	Lambardi F.	I-446
Hung H.-L.	I-22	Lasenby J.	I-223
Iglesias M.P.	II-552	Lau C.	II-663
Iisaku S.-i.	II-396	Lees K.	II-765
Imade M.	II-524	Lebourgeois F.	II-639
Impedovo S.	II-592	Lemahieu I.	II-85,II-688
Iñesta J.M.	I-231	Lemoine J.	I-510
Inokuchi S.	II-697	Leonardi R.	II-124
Inoue S.	I-685	Levenson R.M.	II-663
Ishikawa T.	II-270	Levi P.	I-709
Ishizaka T.	I-287	Levialdi S.	I-430,II-188
Iwanaga T.	II-445	Lew M.S.	I-582,II-22,II-148
Jain A.K.	I-303	Li M.	I-438
Jain R.	II-1,II-38	Li X.	II-332
Jiang H.	II-396	Li Y.	I-111
Jiang X.	I-182	Liao H.-Y.M.	I-22,II-469
Jiang W.B.	I-628	Lijò J.L.F.	II-552
Jozwik A.	II-781	Livens S.	I-327
Kanazawa K.	II-420	Lombardi L.	I-142,II-276,II-509
Kanellopoulos I.	II-765	Lončarić S.	II-388
Karasudani A.	II-284	Lorenz C.	II-380,II-680
Kavianifar M.	II-616	Lotti F.	I-14
Kawata Y.	II-420	Lovergine F.P.	II-584
Khoudour L.	II-757	Lugg M.	II-781
Kim C.-Y.	I-54	Luo A.	I-636
Kim H.	II-196	MacKie R.M.	II-453
Kimura T.D.	II-713	Mack J.	II-445
King I.	I-567	Maeda J.	I-287
Kingsbury N.	I-486	Magarey J.	I-486
Kittler J.	II-140	Magillo P.	II-308
Knobler R.	II-445	Maino G.	II-436
Koga K.	II-655	Maître H.	I-30
Kokaram A.C.	I-486,II-719,II-773	Majumder D.D.	I-575
Kontinen J.	II-453	Malizia A.	II-188
Kosugi M.	II-501	Malo J.	I-454
Kovačević D.	II-388	Manabe Y.	II-697

Mancini R. II-188	Ohmatsu H. II-420
Maresca M. II-46	Ojala T. I-311
Mari M. II-781	Okada S. II-524
Marini D. I-62	Olsen O.F. I-6
Marino F. II-93	Oswald N. I-709
Martin J. II-340	Otsuki M. II-270
Martinez J. II-14	Ottaviani E. II-632
Mascarilla L. I-519	Ozawa S. II-262
Matsakis P. I-198	Özdil M.A. II-608
McDonald J. I-385	Pacaccio V. I-142
Meghini C. II-156	Pahor V. II-735
Meilhac C. I-661	Pala P. II-180,II-212
Mello P. I-535	Pardo J.M. II-364
Menard C. I-677	Paries A. I-174
Mérigot A. II-109	Park J.-I. I-685
Messmer B.T. I-693	Peckar W. I-527
Migliardi M. II-46	Pei S.-C. II-164
Minowa K. II-284	Pérez De La Blanca N. II-372
Mirhosseini A.R. I-559	Petit E. I-510
Miyauchi H. II-524	Petkovic D. I-1
Mizukami Y. II-655	Peura M. I-604
Mokhtarian F. II-140	Philips W. II-85
Mori S. II-461	Piau D. I-519
Moriyama N. II-420	Piazza F. II-228
Mortelli L. I-87	Picariello A. II-600
Moss S. II-172	Piccardi M. I-535
Mota R. II-517	Pieroni G.G. II-749
Motamed C. II-493	Pietikäinen M. I-311
Mugnaini M. II-180	Pinho A.J. II-254
Murino V. II-749	Pinzani R. I-166
Mussio P. I-430	Pirlo G. II-592
Nagata N. II-697	Piscitelli G. I-134
Nakamura S. II-461	Pla F. I-206,I-502,I-653
Nappi M. I-214,II-101	Plataniotis K.N. I-119
Nastar C. I-661	del Pobil Á.P. I-231
Nebbia B. I-430	Pontil M. II-300
Netravali A.N. II-30	Postaire J.-G. II-493
Neumann A. II-680	Porcinai S. I-14
Nicchiotti G. II-632	Prager R.W. II-412
Nielsen M. I-6	Pujas P. I-103
Nieniewski M. II-781	Puliti P. I-247
Niki N. II-420,II-461	Pulli K. I-385
Nishitani H. II-461	Puppo E. II-308
Nivat M. I-166	Rachid S. II-461
Nomura A. I-462	Rahman A.F.R. II-624

Raji A. I-510
Ramella G. I-369
Ramponi G. II-735
Rayner P. I-551
Regazzoni C.S. II-485
Regincós Isern J. I-79
Reznik A.M. II-238
Rizzi A. I-62,II-204
Rizzo D. II-77
Roberto V. I-669
Rodríguez F.M. II-552
Rohr K. I-527
Roli F. I-38,II-765
Röll S.A. II-428
Röning J. II-453
Rosenthal A.S. II-647
Sacerdoti C. II-727
Sagona M. I-596
Saha P.K. I-575
Saito H. II-262
Salden A.H. I-158
Saludes J. I-70
Salzo A. II-592
Sánchez J.S. I-206
Sande F.P. II-552
Sandoval F. I-422,II-477
Sanfeliu A. I-701
Sanniti di Baja G. I-369
Sansone C. I-46
Santini S. II-38,II-212
Santos-Victor J. II-727
Sanz P.J. I-231
Saraceno C. II-124
Sarkar S. II-356
Sato J. I-414
Sato Y. II-270
Satoh H. II-420
Savini M. II-509
Scagliola C. II-705
Schettini R. I-335
Scheunders P. I-327
Schnörr C. I-527
Schreer O. I-620
Schultz N. II-532
Scianna A. II-509

Sebastiani F. II-156
Seo Y.-S. I-54
Seo Y. II-196
Serpico S.B. II-743
Serra J.R. II-324
Serratosa F. I-701
Shah S. I-343
Shapiro L. I-385
She A. I-582
Shioyama T. I-628
Shirai Y. II-568
Siccardi A. II-576
Sirakov N.M. II-292
Sklodowski M. II-781
Sloboda F. I-190
Smeulders A.W.M. I-319
Smits P.C. I-255,II-743
Sommellier L. I-644
Sonka M. I-398
Spiess S. I-438
Stella E. II-93
Stiehl H.S. I-527
Straccia U. II-156
Stråhlén K. II-348
Stringa E. II-485
Stuetzle W. I-385
Subirana J.B. II-324
Sumimoto T. II-524
Suzuki Y. I-287
Syn M.H. II-412
Sze C.-J. I-22
Tao W. I-636
Tartari A. II-436
Tascini G. II-228
Teixeira L. II-62,II-117
Terauchi S. I-628
Teschioni A. II-485
Tieng Q. I-287
Timchenko L.I. II-246
Tistarelli M. II-727
Tiu W. II-396
Tortorella F. I-46
Tosan E. I-644
Toshioka S. II-420
Tozzi C.L. I-279

Trajković M. I-470
Trucco E. I-669
Tung L.H. I-567
Turco F. II-180
Usami T. II-697
Vaccaro R. II-560
Van Assche S. II-85
Van Dyck D. I-327
Van de Walle R. II-688
Van de Wouwer G. I-327
Vandorpe D. I-644
Vannoorenberghe P. II-493
Varley A. I-551
Vass G.G. II-220
Velastin S.A. II-757
Venetsanopoulos A.N. ... I-119,I-494
Veneziani N. II-93
Vento M. I-46
Verestoy J. II-781
Vernazza G. I-255,II-765
Vernon D. II-727
Verri A. II-300
Verzucoli L. II-180
Vicario E. II-132
Vincencio-Silva M.A. II-757
Vinitski S. II-445
Vitulano S. I-214,II-101
Weese J. II-380
Wendling L. I-174
Wherett M. II-757
Wiatr K. II-69
Wilkinson G. II-765
Wilson R.C. I-150
Wu H.Y. I-628
Yamamoto H. II-524
Yamamoto S. II-396
Yamashita K. II-501
Yan H. I-559
Yarman-Vural F.T. I-127,II-608
Yip R.K.K. I-612
Yu G.-J. II-469
Yuille A.L. I-361
Zat'ko B. I-190
Zhang T. I-361
Zhang X.-F. I-636
Zhang Z. I-239
Žid P. I-295
Zingaretti P. I-247
Zingirian N. II-46
Zurli A. I-166

Springer and the environment

At Springer we firmly believe that an international science publisher has a special obligation to the environment, and our corporate policies consistently reflect this conviction.

We also expect our business partners – paper mills, printers, packaging manufacturers, etc. – to commit themselves to using materials and production processes that do not harm the environment. The paper in this book is made from low- or no-chlorine pulp and is acid free, in conformance with international standards for paper permanency.

Lecture Notes in Computer Science

For information about Vols. 1–1238

please contact your bookseller or Springer-Verlag

Vol. 1239: D. Sehr, U. Banerjee, D. Gelernter, A. Nicolau, D. Padua (Eds.), Languages and Compilers for Parallel Computing. Proceedings, 1996. XIII, 612 pages. 1997.

Vol. 1240: J. Mira, R. Moreno-Díaz, J. Cabestany (Eds.), Biological and Artificial Computation: From Neuroscience to Technology. Proceedings, 1997. XXI, 1401 pages. 1997.

Vol. 1241: M. Akşit, S. Matsuoka (Eds.), ECOOP'97 – Object-Oriented Programming. Proceedings, 1997. XI, 531 pages. 1997.

Vol. 1242: S. Fdida, M. Morganti (Eds.), Multimedia Applications, Services and Techniques – ECMAST '97. Proceedings, 1997. XIV, 772 pages. 1997.

Vol. 1243: A. Mazurkiewicz, J. Winkowski (Eds.), CONCUR'97: Concurrency Theory. Proceedings, 1997. VIII, 421 pages. 1997.

Vol. 1244: D. M. Gabbay, R. Kruse, A. Nonnengart, H.J. Ohlbach (Eds.), Qualitative and Quantitative Practical Reasoning. Proceedings, 1997. X, 621 pages. 1997. (Subseries LNAI).

Vol. 1245: M. Calzarossa, R. Marie, B. Plateau, G. Rubino (Eds.), Computer Performance Evaluation. Proceedings, 1997. VIII, 231 pages. 1997.

Vol. 1246: S. Tucker Taft, R. A. Duff (Eds.), Ada 95 Reference Manual. XXII, 526 pages. 1997.

Vol. 1247: J. Barnes (Ed.), Ada 95 Rationale. XVI, 458 pages. 1997.

Vol. 1248: P. Azéma, G. Balbo (Eds.), Application and Theory of Petri Nets 1997. Proceedings, 1997. VIII, 467 pages. 1997.

Vol. 1249: W. McCune (Ed.), Automated Deduction – CADE-14. Proceedings, 1997. XIV, 462 pages. 1997. (Subseries LNAI).

Vol. 1250: A. Olivé, J.A. Pastor (Eds.), Advanced Information Systems Engineering. Proceedings, 1997. XI, 451 pages. 1997.

Vol. 1251: K. Hardy, J. Briggs (Eds.), Reliable Software Technologies – Ada-Europe '97. Proceedings, 1997. VIII, 293 pages. 1997.

Vol. 1252: B. ter Haar Romeny, L. Florack, J. Koenderink, M. Viergever (Eds.), Scale-Space Theory in Computer Vision. Proceedings, 1997. IX, 365 pages. 1997.

Vol. 1253: G. Bilardi, A. Ferreira, R. Lüling, J. Rolim (Eds.), Solving Irregularly Structured Problems in Parallel. Proceedings, 1997. X, 287 pages. 1997.

Vol. 1254: O. Grumberg (Ed.), Computer Aided Verification. Proceedings, 1997. XI, 486 pages. 1997.

Vol. 1255: T. Mora, H. Mattson (Eds.), Applied Algebra, Algebraic Algorithms and Error-Correcting Codes. Proceedings, 1997. X, 353 pages. 1997.

Vol. 1256: P. Degano, R. Gorrieri, A. Marchetti-Spaccamela (Eds.), Automata, Languages and Programming. Proceedings, 1997. XVI, 862 pages. 1997.

Vol. 1258: D. van Dalen, M. Bezem (Eds.), Computer Science Logic. Proceedings, 1996. VIII, 473 pages. 1997.

Vol. 1259: T. Higuchi, M. Iwata, W. Liu (Eds.), Evolvable Systems: From Biology to Hardware. Proceedings, 1996. XI, 484 pages. 1997.

Vol. 1260: D. Raymond, D. Wood, S. Yu (Eds.), Automata Implementation. Proceedings, 1996. VIII, 189 pages. 1997.

Vol. 1261: J. Mycielski, G. Rozenberg, A. Salomaa (Eds.), Structures in Logic and Computer Science. X, 371 pages. 1997.

Vol. 1262: M. Scholl, A. Voisard (Eds.), Advances in Spatial Databases. Proceedings, 1997. XI, 379 pages. 1997.

Vol. 1263: J. Komorowski, J. Zytkow (Eds.), Principles of Data Mining and Knowledge Discovery. Proceedings, 1997. IX, 397 pages. 1997. (Subseries LNAI).

Vol. 1264: A. Apostolico, J. Hein (Eds.), Combinatorial Pattern Matching. Proceedings, 1997. VIII, 277 pages. 1997.

Vol. 1265: J. Dix, U. Furbach, A. Nerode (Eds.), Logic Programming and Nonmonotonic Reasoning. Proceedings, 1997. X, 453 pages. 1997. (Subseries LNAI).

Vol. 1266: D.B. Leake, E. Plaza (Eds.), Case-Based Reasoning Research and Development. Proceedings, 1997. XIII, 648 pages. 1997 (Subseries LNAI).

Vol. 1267: E. Biham (Ed.), Fast Software Encryption. Proceedings, 1997. VIII, 289 pages. 1997.

Vol. 1268: W. Kluge (Ed.), Implementation of Functional Languages. Proceedings, 1996. XI, 284 pages. 1997.

Vol. 1269: J. Rolim (Ed.), Randomization and Approximation Techniques in Computer Science. Proceedings, 1997. VIII, 227 pages. 1997.

Vol. 1270: V. Varadharajan, J. Pieprzyk, Y. Mu (Eds.), Information Security and Privacy. Proceedings, 1997. XI, 337 pages. 1997.

Vol. 1271: C. Small, P. Douglas, R. Johnson, P. King, N. Martin (Eds.), Advances in Databases. Proceedings, 1997. XI, 233 pages. 1997.

Vol. 1272: F. Dehne, A. Rau-Chaplin, J.-R. Sack, R. Tamassia (Eds.), Algorithms and Data Structures. Proceedings, 1997. X, 476 pages. 1997.

Vol. 1273: P. Antsaklis, W. Kohn, A. Nerode, S. Sastry (Eds.), Hybrid Systems IV. X, 405 pages. 1997.

Vol. 1274: T. Masuda, Y. Masunaga, M. Tsukamoto (Eds.), Worldwide Computing and Its Applications. Proceedings, 1997. XVI, 443 pages. 1997.

Vol. 1275: E.L. Gunter, A. Felty (Eds.), Theorem Proving in Higher Order Logics. Proceedings, 1997. VIII, 339 pages. 1997.

Vol. 1276: T. Jiang, D.T. Lee (Eds.), Computing and Combinatorics. Proceedings, 1997. XI, 522 pages. 1997.

Vol. 1277: V. Malyshkin (Ed.), Parallel Computing Technologies. Proceedings, 1997. XII, 455 pages. 1997.

Vol. 1278: R. Hofestädt, T. Lengauer, M. Löffler, D. Schomburg (Eds.), Bioinformatics. Proceedings, 1996. XI, 222 pages. 1997.

Vol. 1279: B. S. Chlebus, L. Czaja (Eds.), Fundamentals of Computation Theory. Proceedings, 1997. XI, 475 pages. 1997.

Vol. 1280: X. Liu, P. Cohen, M. Berthold (Eds.), Advances in Intelligent Data Analysis. Proceedings, 1997. XII, 621 pages. 1997.

Vol. 1281: M. Abadi, T. Ito (Eds.), Theoretical Aspects of Computer Software. Proceedings, 1997. XI, 639 pages. 1997.

Vol. 1282: D. Garlan, D. Le Métayer (Eds.), Coordination Languages and Models. Proceedings, 1997. X, 435 pages. 1997.

Vol. 1283: M. Müller-Olm, Modular Compiler Verification. XV, 250 pages. 1997.

Vol. 1284: R. Burkard, G. Woeginger (Eds.), Algorithms — ESA '97. Proceedings, 1997. XI, 515 pages. 1997.

Vol. 1285: X. Jao, J.-H. Kim, T. Furuhashi (Eds.), Simulated Evolution and Learning. Proceedings, 1996. VIII, 231 pages. 1997. (Subseries LNAI).

Vol. 1286: C. Zhang, D. Lukose (Eds.), Multi-Agent Systems. Proceedings, 1996. VII, 195 pages. 1997. (Subseries LNAI).

Vol. 1287: T. Kropf (Ed.), Formal Hardware Verification. XII, 367 pages. 1997.

Vol. 1288: M. Schneider, Spatial Data Types for Database Systems. XIII, 275 pages. 1997.

Vol. 1289: G. Gottlob, A. Leitsch, D. Mundici (Eds.), Computational Logic and Proof Theory. Proceedings, 1997. VIII, 348 pages. 1997.

Vol. 1290: E. Moggi, G. Rosolini (Eds.), Category Theory and Computer Science. Proceedings, 1997. VII, 313 pages. 1997.

Vol. 1291: D.G. Feitelson, L. Rudolph (Eds.), Job Scheduling Strategies for Parallel Processing. Proceedings, 1997. VII, 299 pages. 1997.

Vol. 1292: H. Glaser, P. Hartel, H. Kuchen (Eds.), Programming Languages: Implementations, Logigs, and Programs. Proceedings, 1997. XI, 425 pages. 1997.

Vol. 1294: B.S. Kaliski Jr. (Ed.), Advances in Cryptology — CRYPTO '97. Proceedings, 1997. XII, 539 pages. 1997.

Vol. 1295: I. Prívara, P. Ružička (Eds.), Mathematical Foundations of Computer Science 1997. Proceedings, 1997. X, 519 pages. 1997.

Vol. 1296: G. Sommer, K. Daniilidis, J. Pauli (Eds.), Computer Analysis of Images and Patterns. Proceedings, 1997. XIII, 737 pages. 1997.

Vol. 1297: N. Lavrač, S. Džeroski (Eds.), Inductive Logic Programming. Proceedings, 1997. VIII, 309 pages. 1997. (Subseries LNAI).

Vol. 1298: M. Hanus, J. Heering, K. Meinke (Eds.), Algebraic and Logic Programming. Proceedings, 1997. X, 286 pages. 1997.

Vol. 1299: M.T. Pazienza (Ed.), Information Extraction. Proceedings, 1997. IX, 213 pages. 1997. (Subseries LNAI).

Vol. 1300: C. Lengauer, M. Griebl, S. Gorlatch (Eds.), Euro-Par'97 Parallel Processing. Proceedings, 1997. XXX, 1379 pages. 1997.

Vol. 1301: M. Jazayeri (Ed.), Software Engineering - ESEC/FSE'97. Proceedings, 1997. XIII, 532 pages. 1997.

Vol. 1302: P. Van Hentenryck (Ed.), Static Analysis. Proceedings, 1997. X, 413 pages. 1997.

Vol. 1303: G. Brewka, C. Habel, B. Nebel (Eds.), KI-97: Advances in Artificial Intelligence. Proceedings, 1997. XI, 413 pages. 1997. (Subseries LNAI).

Vol. 1304: W. Luk, P.Y.K. Cheung, M. Glesner (Eds.), Field-Programmable Logic and Applications. Proceedings, 1997. XI, 503 pages. 1997.

Vol. 1305: D. Corne, J.L. Shapiro (Eds.), Evolutionary Computing. Proceedings, 1997. X, 313 pages. 1997.

Vol. 1307: R. Kompe, Prosody in Speech Understanding Systems. XIX, 357 pages. 1997. (Subseries LNAI).

Vol. 1308: A. Hameurlain, A M. Tjoa (Eds.), Database and Expert Systems Applications. Proceedings, 1997. XVII, 688 pages. 1997.

Vol. 1309: R. Steinmetz, L.C. Wolf (Eds.), Interactive Distributed Multimedia Systems and Telecommunication Services. Proceedings, 1997. XIII, 466 pages. 1997.

Vol. 1310: A. Del Bimbo (Ed.), Image Analysis and Processing. Proceedings, 1997. Volume I. XXI, 722 pages. 1997.

Vol. 1311: A. Del Bimbo (Ed.), Image Analysis and Processing. Proceedings, 1997. Volume II. XXII, 794 pages. 1997.

Vol. 1312: A. Geppert, M. Berndtsson (Eds.), Rules in Database Systems. Proceedings, 1997. VII, 214 pages. 1997.

Vol. 1313: J. Fitzgerald, C.B. Jones, P. Lucas (Eds.), FME '97: Industrial Applications and Strengthened Foundations of Formal Methods. Proceedings, 1997. XIII, 685 pages. 1997.

Vol. 1314: S. Muggleton (Ed.), Inductive Logic Programming. Proceedings, 1996. VIII, 397 pages. 1997. (Subseries LNAI).

Vol. 1315: G. Sommer, J.J. Koenderink (Eds.), Algebraic Frames for the Perception-Action Cycle. Proceedings, 1997. VIII, 395 pages. 1997.

Vol. 1317: M. Leman (Ed.), Music, Gestalt, and Computing. IX, 524 pages. 1997. (Subseries LNAI).

Vol. 1320: M. Mavronicolas, P. Tsigas (Eds.), Distributed Systems. Proceedings, 1997. X, 333 pages. 1997.

Vol. 1321: M. Lenzerini (Ed.), AI*IA 97: Advances in Artificial Intelligence. Proceedings, 1997. XII, 459 pages. 1997. (Subseries LNAI).

Vol. 1324: C. Peters, C. Thanos (Ed.), Research and Advanced Technology for Digital Libraries. Proceedings, 1997. X, 423 pages. 1997.